Standing between Life and Extinction

Standing between Life and Extinction

Ethics and Ecology of Conserving Aquatic Species in North American Deserts, with a Foreword by Senator Tom Udall

Edited by
David L. Propst, Jack E. Williams,
Kevin R. Bestgen,
and Christopher W. Hoagstrom

THE UNIVERSITY OF CHICAGO PRESS
Chicago & London

The University of Chicago Press, Chicago 60637
The University of Chicago Press, Ltd., London

Published 2020
Printed in China

29 28 27 26 25 24 23 22 21 20   1 2 3 4 5

ISBN-13: 978-0-226-69433-7 (cloth)
ISBN-13: 978-0-226-69447-4 (paper)
ISBN-13: 978-0-226-69450-4 (e-book)
DOI: https://doi.org/10.7208/chicago/978-0-226-69450-4.001.0001

Funding to support publication of this volume was provided by the Desert Fishes Council, US Bureau of Land Management, US Forest Service, Texas Parks and Wildlife Department, and Dixon Water Foundation.

Library of Congress Cataloging-in-Publication Data

Names: Propst, David L., editor. | Williams, Jack Edward, editor. | Bestgen, Kevin R., editor. | Hoagstrom, Christopher W. (Christopher William), editor.
Title: Standing between life and extinction : ethics and ecology of conserving aquatic species in North American deserts / [edited by] David L. Propst, Jack E. Williams, Kevin R. Bestgen, and Christopher W. Hoagstrom; with a Foreword by Senator Tom Udall.
Other titles: Battle against extinction
Description: Chicago ; London : The University of Chicago Press, 2020. | Includes bibliographical references and index.
Identifiers: LCCN 2019052169 | ISBN 9780226694337 (cloth) | ISBN 9780226694474 (paperback) | ISBN 9780226694504 (ebook)
Subjects: LCSH: Fishes—Conservation—Southwest, New. | Rare fishes—Southwest, New. | Aquatic ecology—Southwest, New. | Desert biology—Southwest, New.
Classification: LCC QL617.73.U6 S73 2020 | DDC 333.95/680979—dc23
LC record available at https://lccn.loc.gov/2019052169

This paper meets the requirements of ANSI/NISO Z39.48-1992 (Permanence of Paper)

Dedicated to the next generation of citizens, scientists, students, and artists who will fall in love with, strive to understand, and fight to protect the incredible aquatic fauna and precious places of the desert

Contents

Foreword *xi*
Senator Tom Udall

Preface *xiii*
Edwin P. (Phil) Pister

SECTION 1 ENGAGING THE BATTLE

1 **The Battle to Conserve Aquatic Species in Lands of Water Scarcity Continues** *3*
Jack E. Williams and David L. Propst

2 **The Protagonists** *13*

2a Carl Leavitt Hubbs and Robert Rush Miller *13*
Robert J. Edwards

2b W. L. Minckley *17*
Chuck O. Minckley

2c Salvador Contreras-Balderas *20*
María de Lourdes Lozano-Vilano and Armando Jesús Contreras-Balderas

2d James E. Deacon *22*
Cindy Deacon Williams

2e Clark Hubbs *26*
Gary P. Garrett

2f Robert J. Behnke *30*
Kevin R. Bestgen and Kurt D. Fausch

2g Edwin P. (Phil) Pister *33*
Kathryn Boyer

3 **Biodiversity, Biogeography, and Conservation of North American Desert Fishes** *36*
Christopher W. Hoagstrom, Derek D. Houston, and Norman Mercado-Silva

4 **Living with Aliens: Nonnative Fishes in the American Southwest** *69*
Peter B. Moyle

5 **Current Conservation Status of Some Freshwater Species and Their Habitats in México** *79*
María de Lourdes Lozano-Vilano, Armando J. Contreras-Balderas, Gorgonio Ruiz-Campos, and María Elena García-Ramírez

6 **Ghosts of Our Making: Extinct Aquatic Species of the North American Desert Region** *89*
Jack E. Williams and Donald W. Sada

SECTION 2 RACING TO COLLAPSE

7 **Running on Empty: Southwestern Water Supplies and Climate Change** *109*
Bradley H. Udall

8 **Mining Hidden Waters: Groundwater Depletion, Aquatic Habitat Degradation, and Loss of Fish Diversity in the Chihuahuan Desert Ecoregion of Texas** *124*
Gary P. Garrett, Megan G. Bean, Robert J. Edwards, and Dean A. Hendrickson

9 **Southwestern Fish and Aquatic Systems: The Climate Challenge** *137*
Jonathan T. Overpeck and Scott A. Bonar

10 **Novel Drought Regimes Restructure Aquatic Invertebrate Communities in Arid-Land Streams** *153*
Kate S. Boersma and David A. Lytle

11 **The Exotic Dilemma: Lessons Learned from Efforts to Recover Native Colorado River Basin Fishes** *167*
Brandon Albrecht, Ron Kegerries, Ron Rogers, and Paul Holden

SECTION 3 IMPROVING THE ODDS

12 **Applying Endangered Species Act Protections to Desert Fishes: Assessment and Opportunities** *183*
Matthew E. Andersen and James E. Brooks

13 The Value of Specimen Collections for Conserving Biodiversity *199*
Adam E. Cohen, Dean A. Hendrickson, and Gary P. Garrett

14 Conservation Genetics of Desert Fishes in the Genomics Age *207*
Thomas F. Turner, Thomas E. Dowling, Trevor J. Krabbenhoft, Megan J. Osborne, and Tyler J. Pilger

15 Long-Term Monitoring of a Desert Fish Assemblage in Aravaipa Creek, Arizona *225*
Peter N. Reinthal, Heidi Blasius, and Mark Haberstich

16 Human Impacts on the Hydrology, Geomorphology, and Restoration Potential of Southwestern Rivers *239*
Mark C. Stone and Ryan R. Morrison

17 Conservation and Ecological Rehabilitation of North American Desert Spring Ecosystems *255*
Donald W. Sada and Lawrence E. Stevens

SECTION 4 SEARCHING FOR RECOVERY

18 Oases: Finding Hidden Biodiversity Gems in the Southern Sonoran Desert *272*
Michael T. Bogan, Carlos Alonso Ballesteros-Córdova, Scott E. K. Bennett, Michael H. Darin, Lloyd T. Findley, and Alejandro Varela-Romero

19 Recent Discoveries and Conservation of Catfishes, Genus *Ictalurus*, in México *285*
Alejandro Varela-Romero, Carlos Alonso Ballesteros-Córdova, Gorgonio Ruiz-Campos, Sergio Sánchez-Gonzalez, and James E. Brooks

20 Ecology, Politics, and Conservation of Gila Trout *295*
David L. Propst, Thomas F. Turner, Jerry A. Monzingo, James E. Brooks, and Dustin J. Myers

21 Large-River Fish Conservation in the Colorado River Basin: Progress and Challenges with Razorback Sucker *316*
Kevin R. Bestgen, Thomas E. Dowling, Brandon Albrecht, and Koreen A. Zelasko

22 Assisting Recovery: Intensive Interventions to Conserve Native Fishes of Desert Springs and Wetlands *335*
Sean C. Lema, Jennifer M. Gumm, Olin G. Feuerbacher, and Michael R. Schwemm

23 **Restoration of Aquatic Habitats and Native Fishes in the Desert: Some Successes in Western North America** *353*
Anthony A. Echelle and Alice F. Echelle

SECTION 5 EXPLORING OUR FUTURE

24 **The Devils Hole Pupfish: Science in a Time of Crises** *378*
Kevin P. Wilson, Mark B. Hausner, and Kevin C. Brown

25 **Politics, Imagination, Ideology, and the Realms of Our Possible Futures** *391*
Christopher Norment

26 **Searching for Common Ground between Life and Extinction** *407*
Christopher W. Hoagstrom, Kevin R. Bestgen, David L. Propst, and Jack E. Williams

Acknowledgments *419*

List of Contributors *421*

Index *425*

Foreword

You would think being a fish in the desert would be hard. Fish need water. Deserts don't have much of that. So, on the surface, it seems like a rough go.

Yet it turns out that fish have survived quite well in the deserts of North America for thousands of years. The Gila trout *Oncorhynchus gilae gilae,* for example, is thought to have entered the Gila River basin in New Mexico 500,000 to 1 million years ago. Gila trout take their colors from the New Mexican sunset—gold sides blend to copper gill covers, small black spots dot the deeply golden upper half. This ancient fish is part of the culture and heritage of my home state.

Over the millennia, the 300-plus fish lineages endemic to North American deserts have adapted to high temperatures, high salinity, and high turbidity of their aquatic habitat, and to dramatic population fluctuations.

But desert fishes' adaptations to nature are ill-fitted for the challenges we humans have given them. Many of these impressive creatures have survived longer than we have existed. In just decades we have managed to jeopardize too many. Loss and destruction of habitat, degradation of water quality, introduction of nonnative species—and now climate change—all threaten the existence of many species. Our Gila trout, for example, was listed as endangered in 1967, under the precursor to the Endangered Species Act. Through overfishing, habitat deterioration, and the introduction of nonnative trout, their habitat had been reduced to just four streams in New Mexico.

As a major contributor to the threats to their existence, do humans bear responsibility to help them survive? I emphatically believe we do.

And so do the remarkable scientists and others who have contributed to *Standing between Life and Extinction.*

We know that biodiversity is a positive. The greater diversity in nature, the more resilient the ecosystem, the better chance animals and plants have to withstand threats.

We know that diversity in our environment is useful to humanity. We get food, shelter, clothing, medicine, and so much more from animals and plants—necessities that ensure our own survival.

But the value of biodiversity is more than utilitarian to humankind. Each species has its own intrinsic value.

Our laws recognize that value.

The Endangered Species Act of 1973 recognizes that threatened and endangered "species of fish, wildlife, and plants are of esthetic, ecological, educational, historical, recreational, and scientific value to the Nation and its people."

The Wilderness Act of 1964 protects areas "where the earth and its community of life are untrammeled by man, where man himself is a visitor who does not remain," that have "ecological, geological, or other features of scientific, educational, scenic, or historical value."

The 1968 Wild and Scenic Rivers Act declares that "selected rivers of the Nation which, with their immediate environments, possess outstandingly remarkable scenic, recreational, geologic, fish and wildlife, historic, cultural or other similar values, shall be preserved in free-flowing condition."

That nature has its own innate worth is a bedrock principle in American culture. This principle is embedded in our important environmental laws.

I'm proud that my father, Stewart Udall, helped shepherd many of these early laws through Congress. At that time, there was broad bipartisan support for carving out certain species, lands, and waters for full protection from development.

We now have deep cleavages in our society between protecting species "whatever the cost," as the Endangered Species Act directs, and pushing development forward. How do we bridge these gaps? Can we hike to common ground?

The scientists in this book are part of that hike—through study, good science, and educating the public.

Sometimes, in Washington, it seems that scientists are on the endangered species list (or at least the threatened list). Our "post-fact" world is turning science upside down. It is hard to believe—with the overwhelming consensus among scientists and the overwhelming scientific evidence—that anyone would deny that the climate is warming and that humans are the major cause. Yet there are those in the highest levels of our government who deny the science of climate change.

We all must do our part to protect scientists and the integrity of their work. And those of us who are policy makers must accept the science as is, and we must base our decisions on what is supported by the scientific record, not on made-up facts or pseudoscience.

Twenty-nine years ago, some of the foremost ichthyologists produced the seminal *Battle against Extinction: Native Fish Management in the American West*. In the foreword, my father said, "As citizens of the planet, we must educate our children to do what we have not done well—to act as stewards of the Earth and all its inhabitants and to pass the planet off to their offspring in a better condition than they found it. The more we know, the better we can apply ourselves to such aims."

That is true. We do know more now than we did then. And the news is sobering.

We know that climate change is here and now, and that if we don't act faster than we are, it will be devastating. And we know now that we are in the middle of a "sixth mass extinction"—also caused by us.

Desert fishes and their habitats are under siege. We cannot act too quickly.

We have had our successes. The Gila trout was reclassified to threatened in 2006. And though two major fires in the Gila Wilderness destroyed 8 of 17 Gila trout populations, extraordinary evacuation efforts and a robust interagency recovery program have brought the Gila trout back to about where they were before the fires.

It is humans that "stand between life and extinction" of desert fishes. We have created much of the threat to their existence, and we hold the keys to their continued survival. We owe it to these exceptional creatures to leave them in better condition than we found them.

US Senator Tom Udall
Santa Fe, New Mexico

Preface

Life is full of fateful twists and turns. As I write this at age 87, I reflect upon a long career and the times when my direction in life changed dramatically—first as a student at the University of California, Berkeley, and later as a young fisheries biologist working out of Bishop, California. My career turned from feeding the insatiable appetite of opening-day anglers for the limit of hatchery-produced trout to the farthest corners of my district, where tiny pupfish lived in small desert springs and fragile wetlands. More than once, I would be standing between life and extinction for these desert fishes.

I was born in Stockton, California, in 1929, at the start of the Great Depression. My parents were both schoolteachers, and we had a small farm with a few dairy cattle. I attended the University of California, Berkeley, and initially enrolled in the pre-med program. I would often drive the 70 miles from school on weekends to milk the cows for my dad.

In 1947, my brother Karl, with whom I had spent much of my childhood in the mountains, had been reading the new general catalog at UC Berkeley. Karl phoned and suggested that I contact Professor A. Starker Leopold in the Life Science Building. Starker was a marvelous guy and, as Aldo's eldest son, brought with him the thinking upon which *A Sand County Almanac* was built. He quickly turned my educational pursuits toward conservation.

That was my first big career change that would stay with me over time. I was with Starker for about six years through my graduate program. I drew heavily upon Aldo's Land Ethic, especially the truth of his observation that "a thing is right when it tends to preserve the integrity, stability, and beauty of the biotic community. It is wrong when it tends otherwise."

During summers, I was fortunate to obtain a job as a fisheries research biologist at the US Fish and Wildlife Service's Convict Creek Experiment Station near Bishop, California. We studied in great detail the lakes of the upper Convict Basin, ranging in elevation from 7,500 to 11,000 feet. During this time, I was fortunate to meet Carl and Laura Hubbs, who were collecting fishes from nearby waters. This turned out to be a very fortuitous acquaintance.

In 1952, the Interior Department suffered major budget cuts, which caused me to seek employment with the California Department of Fish and Game (now Fish and Wildlife) in nearby Bishop. The focus of the department's efforts was to provide good angling, often to the exclusion of habitat preservation or protecting native species. Such ethical concerns were excluded from my job description. One of my first assignments was Crowley Lake, the largest reservoir in the Los Angeles aqueduct system. Most of the anglers at Crowley were from Southern California.

On the opening day of trout season in 1961, 17,000 anglers caught more than 40 tons of rainbow trout at Crowley, requiring the use of several trucks to haul the guts to a nearby dump. Even at this early stage of my career, I began to feel uneasy about the direction in which my department was taking me. With the glamour and publicity the department was receiving from the Crowley Lake openers, nothing was being done for the native fishes. Most department employees didn't even know what they were, let alone where they lived.

Robert Rush (Bob) Miller studied for his PhD under Carl Hubbs at the University of Michigan during the 1940s. His dissertation covered the cyprinodont fishes of the Death Valley hydrographic system. One of his study areas was here in the Owens Valley at Fish Slough, not far from my home.

Sometime in early 1964, Bob phoned me from Ann Arbor. He stated that he and Carl Hubbs were planning a trip to Owens Valley and wondered if I could accompany them to Fish Slough, a short distance north of Bishop and the type locality (the area from which the species was taken and initially described scientifically) of the Owens pupfish *Cyprinodon radiosus*. When Bob described the fish for his dissertation using previously collected material, *C. radiosus* was thought to be extinct. He and Carl wanted to see if they might find a remnant population somewhere in Fish Slough. My boss in Los Angeles was a bit suspicious of academics, but gave me permission to accompany Hubbs and Miller for one day, adding that often "Hubbs and Miller" types expected us to "drop everything" when they showed up.

We entered the marsh on a very hot day in July carrying dip nets. I can hear Carl's voice even now, more than 50 years later, as he exultantly called out, "Bob! They are still here." At that point, I not only "dropped everything," but I never picked it up again. We were trying to save an entire species, not to provide trout for people to catch and eat. My change in values was profound and irreversible as we began to structure a recovery plan for *C. radiosus*. That program is ongoing, headed up by Steve Parmenter of California Department of Fish and Wildlife since my retirement in 1990.

During the late sixties and early seventies, Devils Hole (now in Death Valley National Park) was threatened by pumping from the all-important underground aquifers supplying water not only to Devils Hole, but to all of Ash Meadows. Spurred on by concerns for the Devils Hole pupfish *Cyprinodon diabolis* and the Owens pupfish, representatives of several state and federal agencies met in Death Valley National Park in April 1969 and discussed what we might do to preserve these marvelous resources in the eastern Sierra and desert regions of California and western Nevada. This meeting constituted the start of the Desert Fishes Council.

The first meeting of the council drew 44 very concerned university and government agency scientists to Death Valley in November 1969 to assess and define the problems that we faced. Two individuals, W. L. Minckley (Minck) of Arizona State University and Jim Deacon of the University of Nevada, Las Vegas, were key to our eventual success through their enthusiasm, graduate students, and research. Jim would lead the council's efforts all the way to the US Supreme Court, where in 1976 a unanimous court ruling guaranteed water for the Devils Hole pupfish, whose existence was threatened by a nearby ill-conceived development scheme. Working together, Minckley and Deacon (1991), with the assistance of other council members, brought about and edited *Battle against Extinction*, the book upon which this volume is based, and which it follows. I contributed a chapter on those early days of the Desert Fishes Council. Although Minck and Jim have now passed on, their dedication to their students, to native fish conservation, and to the early successes of the council will endure.

I was asked some years ago to write an essay for *Natural History* (Pister 1993) describing an incident that resulted in my holding the entire population of Owens pupfish in two buckets while standing in a marsh area in Fish Slough. This incident encompasses the problems we currently face throughout the American West in terms of water use and native fish management.

> As I walked back to my truck following the final transplant within Fish Slough, the sun had long

ago set. In my dip net remained a few dead pupfish. I glanced up at the darkening sky and thought of Pierre Teilhard de Chardin's concept of the infinitely large, the infinitely small, and the infinitely complex, represented here (in order) by the Milky Way, the pupfish, and the difficulty in pointing out the paramount value of such things to an increasingly materialistic society.

The day had been long. We had won an early round in a fight that will inevitably continue as long as we have a habitable planet. As a realist, I could not help but ponder the ultimate fate not only of the Owens pupfish, but all of the southwestern fishes and species in general. I wondered about our own future. Can the values driving the industrial nations be modified sufficiently to allow for the perpetuation of all species, including humans? Will we ever realize the potential in *Homo sapiens*, the wise species? I hope the day will come when public policy will be guided by the wisdom of Aldo Leopold: "A thing is right when it tends to preserve the integrity, stability, and beauty of the biotic community. It is wrong when it tends otherwise" (Leopold 1949). Such recognition could constitute perhaps the first major step toward creating the society upon which our long-term survival obviously depends.

That August day twenty-three [now forty-seven] years ago had been a very humbling experience for me. The principles of biogeography and evolution I had learned many years ago at Berkeley had taught me why the pupfish was here; but it took the events of those few hours in the desert to teach me why I was. Such are the reflections of a biologist who, for a few frightening moments long ago, held an entire species in two buckets, with only himself standing between life and extinction.

It seems appropriate that this preface should end with a cautionary word. We need to be honest with ourselves and temper the optimism shown implicitly in this volume. The populations of the US Southwest and northern México (barring some unforeseen catastrophe) will continue to grow, as will the accompanying demands for water. Global warming enters into this equation. Competent research hydrologists predict that if the current climate patterns continue, within 20 years Lake Mead will be nothing but a mud flat with the Colorado River flowing through it. The same thing will occur upstream at Lake Powell and at other water storage facilities. Referring back to the delight in Carl Hubbs's voice 50 years ago when the Owens pupfish was rediscovered in Fish Slough, I fear that we will hear dialog to the effect that many currently extant southwestern populations and species are gone. This situation should not discourage us, but cause us to go forward and do our best with conservation and preservation plans for the continued existence of this marvelous fauna. But we need to do this with consideration of the realities involved in our work.

In this new volume, the next generation of conservation scientists reflects on how far we have come and what we have learned since *Battle against Extinction*. *Standing between Life and Extinction: Ethics and Ecology of Conserving Aquatic Species in North American Deserts* tells the story of new discoveries and insights, as well as our successes and failures in efforts to save desert fishes, other aquatic species, and their fragile habitats. So, this was what happened when I was exposed to the words of Aldo and Starker Leopold and the sight of pupfish. My life changed, and so might yours. Read on . . .

Edwin P. (Phil) Pister
from the World Headquarters of the Desert Fishes Council, Bishop, California
May 2016

REFERENCES

Leopold, A. 1949. *A Sand County Almanac, with Essays on Conservation*. New York: Oxford University Press.

Minckley, W. L., and J. E. Deacon, eds. 1991. *Battle against Extinction: Native Fish Management in the American West*. Tucson: University of Arizona Press.

Pister, E. P. 1993. Species in a bucket: For a few frightening moments, there was only myself standing between life and extinction. *Natural History* 102 (January), 14–17.

∴Indigenous Faces

SECTION 1

Engaging the Battle

1 The Battle to Conserve Aquatic Species in Lands of Water Scarcity Continues

Jack E. Williams and David L. Propst

Possibly the most compelling reason for preserving species is the value such a program has in demonstrating the importance of restraint. An "endangered species" program is imperative, not only for the sake of the species being studied but also because of what it can teach us about the possibilities for continued survival of other species, including man.
—Minckley and Deacon (1968)

DAWNING OF DESERT FISH CONSERVATION

The 1950s and 1960s were a time of rapid change in North America. Technological advances and the availability of heavy equipment following World War II led to large-scale road building, forest clearing, and home construction. Widespread ecosystem disruption followed the surge in development as baby boomers swelled human population growth. New water development projects were proposed and constructed to meet the growing demand as cities bloomed in the arid West. It may be hard to imagine a time before the era of strong environmental protection laws like the Endangered Species Act and Clean Water Act, but this was that time.

In 1960, Clark County, Nevada, the home of Las Vegas, had a total population of 127,016. Death Valley was a national monument, not a national park, and the idea of an Ash Meadows National Wildlife Refuge would not materialize for decades, as desert fishes had few protectors at that time. In fact, few knew of their existence outside a scientific community that mostly consisted of Carl Hubbs, his son Clark, his son-in-law Bob Miller, and a scattering of ichthyologists in western universities.

In 1961, Bob Miller authored a seminal review of the emerging dilemma for desert fishes in a paper entitled "Man and the Changing Fish Fauna of the American Southwest." Miller described the depauperate fish fauna of the desert region as "only about 100 species of strictly freshwater fishes . . . characterized by relicts, monotypic genera, and much regional endemism" (Miller 1961, 365–66). Miller went on to chronicle how the building boom and human population growth were impacting these fishes by (1) destruction of vegetation, (2) dam construction and irrigation, (3) mining operations,

(4) depletion of groundwater, and (5) introduction of nonnative species. Basin by basin, he summarized extinctions, extirpations, and general endangerment of the fishes in the Southwest. Miller's paper stands as an early warning of what was to come and how difficult the battle would be to save these species.

During the 1960s, if a fish was not a trout, bass, catfish, or food for one of these "game fish," it was seldom valued by management agencies. This clearly was the justification for treating the San Juan River with rotenone to reduce "trash" fish abundance in September 1961 (Olson 1962). Although the focus of the treatment was the 56 km river reach that would be inundated by Navajo Reservoir, the fish kill extended another 64 km downstream of Navajo Dam. No effort was made to inventory fishes killed, but several Colorado pikeminnow *Ptychocheilus lucius* specimens were deposited in the University of New Mexico Museum of Southwestern Biology, and roundtail chub *Gila robusta* was one of the most common species killed. The project was deemed successful. But perhaps the most remarkable manifestation of this philosophy was the poisoning of a huge section of the Green River in Wyoming and Utah. In anticipation of the construction of Flaming Gorge Dam on the Green River, the river was to be purged of nongame fish in preparation for a new reservoir and a legion of nonnative game fishes. The operation began on September 4, 1962, on the Upper Green and New Forks Rivers in Wyoming. All told, 715 km of the Green and its tributaries were treated for three days using 81,350 L of rotenone (Holden 1991). Dead and dying fish were found as far downstream as the mouth of the Yampa River in Colorado. Large numbers of razorback sucker *Xyrauchen texanus*, humpback chub *Gila cypha*, bonytail *Gila elegans*, and Colorado pikeminnow, all of which would be listed in the future as endangered species, were killed.

But some challenged such destruction and began to seek change in management of native species. A rising league of agency biologists and other scientists were becoming keenly aware of the pace of habitat destruction in the Southwest and how their job priorities needed to shift to address these threats. Aldo Leopold (1949) wrote that "conservationists were notorious for their dissensions. . . . In each field a group (A) regards the land as soil, and its function as commodity-production; another group (B) regards the land as a biota, and its function as something broader." During this era, a number of agency biologists were changing their priorities and making the transition from Group A to Group B. Occasionally this transition was abrupt and striking, turning on one event that overshadowed the daily routine and pointed the scientist in a new philosophical direction.

Such an event occurred on August 19, 1969, for Phil Pister, at the time a fisheries biologist for the California Department of Fish and Game (CDFG) in Bishop. On that date, during a long, hot summer, Phil learned that the marsh containing the last population of Owens pupfish *Cyprinodon radiosus* was in danger of drying. That night, Phil faced the critical task of carrying the surviving members of that species in two buckets across Fish Slough to better water, thereby literally saving the Owens pupfish from extinction (Pister 1993). Phil recalls being scared and praying that he would not stumble as he carried the heavy buckets and their precious cargo across the marshy terrain. It was a pivotal moment, not only for the pupfish, but for Phil's career as well.

[Box 1.1]

For a few frightening moments, there was only myself standing between life and extinction.

—PHIL PISTER (1993, 14)

On November 19 of that year, Phil organized a group of 44 agency and university scientists that met at Furnace Creek in Death Valley to discuss the plight of the region's desert fishes and their dwindling habitats. The group would ultimately comprise the founding members of the Desert Fishes Council (Pister 1991). Suddenly there was a convergence of federal and state agency biologists with scientists from the University of Michigan; University of Nevada, Las Vegas; Arizona State University; and University of California, Los Angeles, all of whom were in agreement that something must be done to save these species.

Battles to save the environment never really end. Often "big victories" to save one area or another are followed in subsequent years by a new threat that was not even envisioned earlier. The fight to save the Devils Hole pupfish *Cyprinodon diabolis* was one of the earliest challenges to face the Desert Fishes Council, and despite some huge conservation victories for the species,

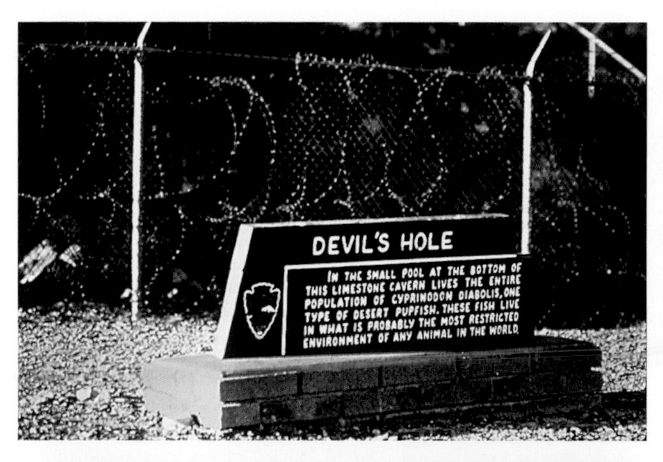

Fig. 1.1 Devils Hole, as the historic Park Service sign notes, is perhaps the most restricted environment of any animal in the world. Lower right image shows 1981 Desert Fishes Council members peering down into Devils Hole. (Devils Hole pupfish image by Olin Feuerbacher, US Fish and Wildlife Service; others by Jack Williams.)

it continues to be one of the most endangered species fully 50 years after conservation efforts began (fig. 1.1).

The Devils Hole pupfish is restricted in distribution to a spring-fed pool within a collapsed limestone cavern, which has been called "the most restricted environment of any animal in the world." Devils Hole, as it is appropriately named, was first set aside for protection by a January 17, 1952, presidential proclamation declaring Devils Hole and the surrounding 16 ha as a disjunct part of what was then Death Valley National Monument. By 1969, nearby farming operations were pumping enough groundwater from the area that the water level within Devils Hole began to sink, and along with it, the population of pupfish (Deacon and Deacon Williams 1991). Spring outflows were ditched and wetlands drained to facilitate farming (fig. 1.2).

Jim Deacon and other members of the newly formed Desert Fishes Council immediately knew that the force of public attention needed to be brought to bear on this crisis. They gained public support for efforts to save the pupfish through a documentary on water overuse entitled *Timetable for Disaster*, which won an Emmy Award for best television documentary of 1970, and through an episode of *Bill Burrud's Animal World* that aired about the same time. The issue caught the attention of major western newspapers and found a sympathetic ear in Walter Hickel, then secretary of the interior. On August 17, 1971, the US government filed a complaint in US district court seeking to enjoin Spring Meadows Inc. from pumping at the three wells that were the closest to Devils Hole. Ultimately, the case, *Cappaert v. United States*, would land in the US Supreme Court, which ruled for the pupfish on June 7, 1976, by upholding a permanent injunction issued by the lower court and directing the district court to establish a minimum water level that would tend to ensure survival of the pupfish. In *Cappaert*, the Supreme Court ruled that when President Truman set aside the 16 ha tract of land surrounding Devils Hole, the federal government also implicitly reserved the necessary groundwater to support the land. It seemed at last that the Devils Hole pupfish was saved.

But the battle raged on. The conflict between the need to supply water to ever-growing farming opera-

Fig. 1.2 Point of Rocks Spring, Ash Meadows, as it appeared on September 15, 1981, following land clearing and spring creek channelization. (Photo by Donald W. Sada.)

tions and the need to support the springs at Devils Hole and nearby Ash Meadows continued. The Nye County commissioners and the Nevada legislature strongly supported farming interests and rejected attempts by Senator Alan Cranston to establish a broader Pupfish National Monument. It was not until February 7, 1984, that the land was purchased by The Nature Conservancy for eventual transfer to the US Fish and Wildlife Service (USFWS) for establishment of the Ash Meadows National Wildlife Refuge, which occurred in June of that year, once more seeming to save the Devils Hole pupfish (Deacon and Deacon Williams 1991). Despite the national monument proclamation, court restrictions on groundwater withdrawals, and establishment of a national wildlife refuge on the surrounding lands, the future of the Devils Hole pupfish remains uncertain, as the population has repeatedly crashed and new larger proposals for regional water withdrawal loom (Deacon et al. 2007).

EXPANDING THE BATTLE AGAINST EXTINCTION

As public awareness of environmental problems increased during the 1960s, Congress passed the Endangered Species Preservation Act in 1966, which resulted in the first list of endangered and threatened species one year later. That 1967 list included 22 fishes. Sixteen of those 22 fishes occurred in the Southwest within the purview of the Desert Fishes Council (fig. 1.3), perhaps signaling the breadth of conflict between water development and fish conservation in this region. With passage of the Endangered Species Act (ESA) in 1973, even more desert fishes would be added to the list.

Passage of the ESA has saved numerous desert fishes from extinction. The Ash Meadows pupfish *Cyprinodon nevadensis mionectes* and Ash Meadows speckled dace *Rhinichthys osculus nevadensis* were listed under emergency provisions of the ESA in 1982, which ultimately encouraged establishment of the Ash Meadows National Wildlife Refuge. Unfortunately, by the time the ESA was passed, the Ash Meadows poolfish *Empetrichthys merriami* was extinct.

Conservation concerns in the desert Southwest quickly spread from spring-dwelling pupfishes to native Gila trout *Oncorhynchus gilae*, Apache trout *O. apache*, and Paiute cutthroat trout *O. clarkii seleniris*, and on to native large-river fishes such as the humpback chub and razorback sucker (fig. 1.4). It followed that scientists from México soon joined the ranks of their colleagues from across the border, and shortly thereafter it became standard procedure for the Desert Fishes Council to meet every third year in México.

By the late 1980s, W. L. Minckley, from Arizona State University, and James E. Deacon, from the University of Nevada, Las Vegas, who had known each other since their graduate school days at the University of Kansas, decided it was time to critically examine the effectiveness of conservation efforts for desert fishes,

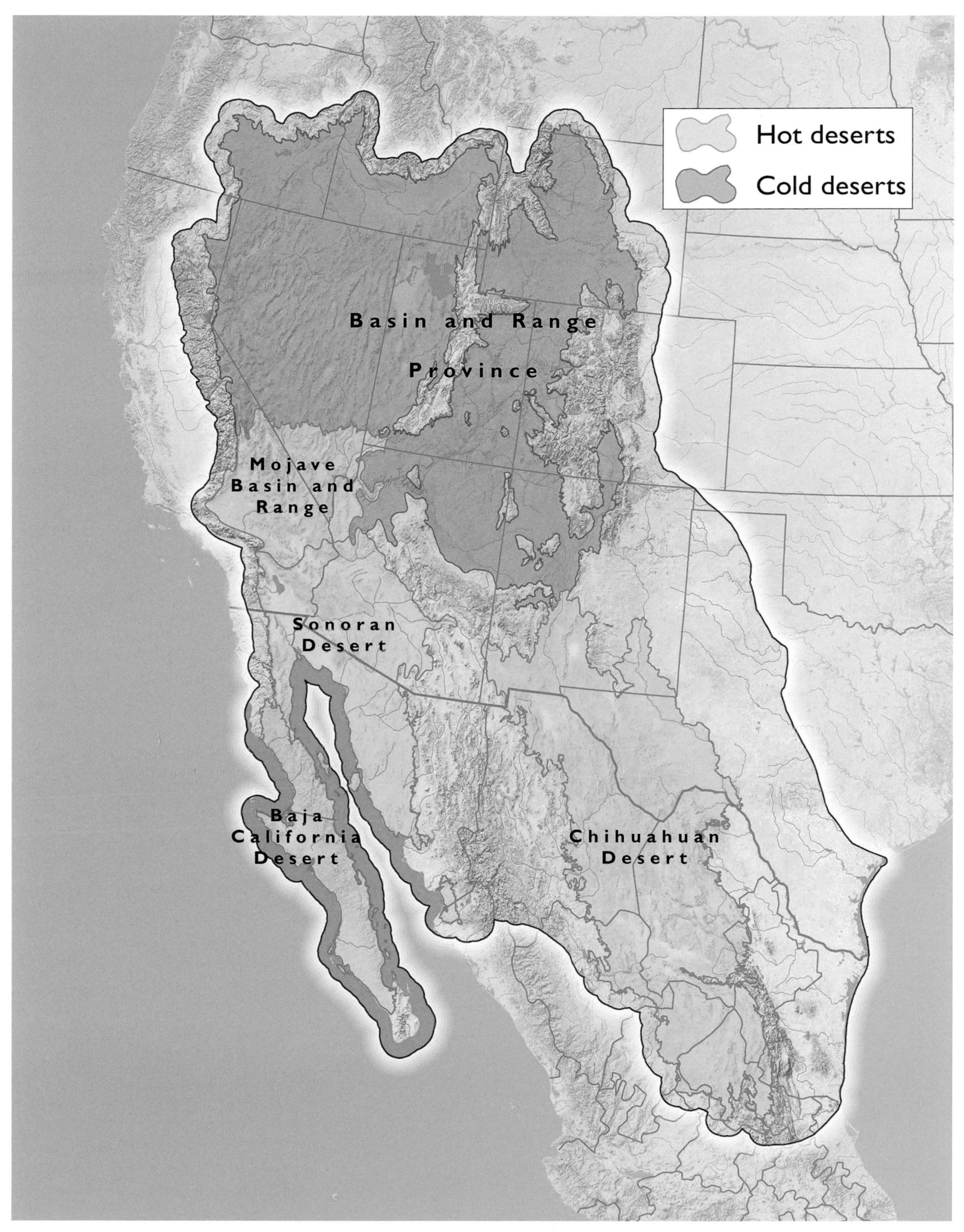

Fig. 1.3 Deserts of western North America. The red line broadly demarks the area within the interest of the Desert Fishes Council.

Fig. 1.4 The native big-river fishes of the Colorado River basin. *Top to bottom:* bonytail, humpback chub, Colorado pikeminnow, and razorback sucker. Note the distinct morphology for life in the fast, turbid conditions that characterized much of the Colorado River prior to dams and impoundments: prominent nuchal hump (in chubs and razorback); large, falcate caudal fins; and narrow caudal peduncle. (Illustrations by Joseph Tomelleri.)

review successes and challenges in the region, and compile what had been learned to date into a single volume of conservation literature. In 1991, with Minckley and Deacon acting as editors, and with other members of the Desert Fishes Council authoring many of the chapters, *Battle against Extinction: Native Fish Management in the American West* was published (Minckley and Deacon 1991). It would quickly become a conservation classic and the bible for those struggling to save desert fishes and their habitats. In addition to chapters describing the founding of the Desert Fishes Council (Pister 1991) and the struggle to save Ash Meadows and Devils Hole (Deacon and Deacon Williams 1991), *Battle against Extinction* included the challenges of native fish conservation in the Colorado River and in México, whose desert fishes and their dire condition had recently been revealed by Salvador Contreras-Balderas (1991) and his colleagues; and in large western lakes, where several long-lived yet endangered catostomid fishes were present (Scoppettone and Vinyard 1991). Southwestern rivers, such as the Colorado and Rio Grande, continue to pose difficult challenges because of the one-two punch of impoundments containing large populations of nonnative fishes and the growing demands on diminishing water supplies from rapidly expanding southwestern cities (Minckley 1983; Minckley et al. 1991).

Battle against Extinction also focused on what were then emerging technologies and novel approaches to conservation. Captive propagation of southwestern fishes began as early as the 1920s and 1930s with efforts to maintain stocks of Gila trout and Apache trout, but the use of hatcheries to conserve desert fishes did not become a mainstay of southwestern conservation until the Dexter National Fish Hatchery (now Southwestern Native Aquatic Resources and Recovery Center) began working with threatened and endangered fishes in 1974 (Johnson and Jensen 1991). Similarly, Echelle (1991) reported on the emerging field of conservation genetics and how new insights into genetic variation in western fishes should inform population management.

Although most conservation efforts remain focused on the species, with the Endangered Species Act often driving this emphasis, there was a recognition that conservation should occur at broader community levels. High-quality habitats that support multiple native fish species are rare in the Southwest, but provide a logical emphasis for protective measures where they occur (Moyle and Sato 1991; Williams 1991). Conservation at this level not only has the potential to conserve multiple species, including those not federally protected, but also typically has a large enough geographic extent to promote the ecological functions needed to maintain habitats and thereby minimize the need for human intervention to maintain or restore habitat values (Poff et al. 1997).

As the momentum for conservation efforts increased during the 1970s and 1980s, it became clear that without strong intervention, the pressures of development and human population growth were poised to cause a new wave of extinctions among desert species that were dependent on water resources. Newly minted laws such as the Clean Water Act and Endangered Species Act would have a profound impact on protecting native species and their habitats. In *Battle against Extinction*'s epilogue, Deacon and Minckley (1991) described the success of efforts to protect key desert fish habitats, but also noted that because of their restricted distributions and small population sizes, many desert fishes were likely to remain vulnerable into the foreseeable future and would probably need permanent protection afforded by the Endangered Species Act.

Despite conservation victories, especially through efforts to save pupfishes from extinction, there were dark clouds on the horizon that gave pause to Minckley, Deacon, and others within the Desert Fishes Council. Human population growth in places like Phoenix, El Paso–Juarez, Las Vegas, and the Los Angeles metropolitan area continued unfettered as water and other resources poured in from surrounding regions to support their growth. Nonnative species continued to plague conservation efforts as they competed with, preyed upon, or hybridized with native desert fishes. Unfortunately, their likelihood of introduction spread with the sprawling cities and modified habitats. While newly created desert impoundments were stocked with nonnative sport fish predators, desert springs and streams proved to be favorable habitats for a suite of tropical fishes that were released from fish farms and home aquaria.

If conservation efforts in the Southwest were to succeed in the long run, it was becoming clear that human values must shift from exploitation to stewardship. The lack of scientific data no longer was the factor limiting conservation success. It was the will of the people. How could strong environmental protection laws survive in

[Box 1.2]

During the past two decades it has become evident that knowledge no longer limits our ability to protect native fishes. Most endangered species can be recovered, if we choose.

—DEACON AND MINCKLEY (1991, 410)

the face of growing conflicts between human population growth and increasingly rare species?

By the last decade of the twentieth century, it seemed that progress was being made in the conservation of arid-land aquatic resources. No longer was protecting dwindling aquatic fauna an afterthought for agencies, both state and federal; it was now a substantial component of their overall missions. In 1990, most state agencies had only one or two biologists dedicated to native fishes, but by 2000, most had several biologists, if not entire sections, devoted to research and management of native aquatic fauna. Federal agencies also increased their personnel responsible for native fauna. Increased academic attention was reflected in the addition of courses with an emphasis on conservation and the amending of department names to include "conservation biology." Certainly, the reach, power, and authority of the Endangered Species Act was a critical motivator, but there was also increasing public support for conservation of rare species and protection of the ecosystems they depended upon (Czech and Krausman 1999). Legacy nongovernmental organizations (NGOs), such as the Audubon Society, the Sierra Club, and The Nature Conservancy, continued to exert influence and pressure on agencies to include all native species in their management prospectuses, but new NGOs, such as Center for Biological Diversity and WildEarth Guardians, also arose that would challenge state and federal agencies in court to adhere to regulations in support of native fauna and flora.

Although support for native aquatic species conservation was comparatively strong through the first decade of the twenty-first century, there remained the nagging problem of balancing increasing demands for finite natural resources, especially water, between natural systems and human communities. Allocation of water among southwestern states, and between the United States and México, is governed by compacts and international treaties. A complex web of dams, diversions, canals, and ditches has been constructed to dispense water among municipal and agricultural users holding water rights (Reisner 1986). When these compacts and treaties were negotiated in the first half of the twentieth century, no provision was made for the welfare of aquatic species. But with passage of the ESA, avoiding or minimizing impacts to federally protected species by federally funded, authorized, or implemented water projects was mandated. Often the mere threat of the issuance of a "jeopardy" biological opinion by USFWS on a proposed water development project was sufficient to bring all project participants and beneficiaries to the table. The Upper Colorado River Endangered Fish Recovery Program was born from such a gathering (Wydoski and Hamill 1991), and it became a template, with modifications, for subsequent collaborative programs. One of its early progeny was the San Juan River Basin Recovery Implementation Program that arose from the biological opinion issued for the proposed Animas–La Plata Project (USFWS 1991). In addition to requiring a collaborative program, the opinion required that Navajo Dam be operated to mimic a natural hydrograph to benefit native fishes, including federally protected Colorado pikeminnow and razorback sucker. Natural flow regime mimicry was central to avoiding a "jeopardy" biological opinion (Gosnell 2001). Other multiagency recovery programs, such as Middle Rio Grande Endangered Species Collaborative Program and Lower Colorado River Multi-species Conservation Program, followed. While all programs focused on protected species, their scopes typically extended beyond the target species. Success of these programs has been variable, with some yielding measurable progress in species recovery (e.g., Franssen et al. 2014) while others struggled to maintain their target species in the wild (Archdeacon 2016).

As conservation of aquatic species became more multidisciplinary in scope, significant advances were made in an array of fields that greatly influenced evolving approaches and strategies. Advances in computer technology made possible or facilitated numerous advances in fields more directly related to conservation biology. The simple task of organizing scientific data became considerably more efficient and robust, which in turn contributed to development of increasingly sophisticated statistical methods that could rapidly analyze huge datasets, yielding insights that otherwise might be missed. Numerous technical advances improved upon existing methods. For example, passive integrated transponder (PIT) tags have largely replaced external tags in movement studies and demographic parameter estimation, and the global positioning system (GPS) has made spatial data collection precise. Among biological disciplines, perhaps the greatest advances have occurred in genetics, especially conservation genetics. When Echelle penned his chapter on genetics for *Battle against Extinction*, the state-of-the-art genetic technique was protein electrophoresis to examine allozymic variation within and among populations. Since then, considerable strides have been made, such that now it is possible to sequence the entire genome of an organism for about the same cost as a well-designed allozyme study. These advances have led to a more comprehensive understanding of the genetic profiles of populations as well as those of higher-level units, insights into evolutionary relationships among populations and species, documentation of exchanges of genetic material among populations, understanding of the role of local adaptations, revelations of hybridization, and detection of species in aquatic habitats using environmental DNA. More sophisticated genetic analyses have led to a greater understanding of the diversity and richness of aquatic life in deserts. With increasingly powerful computers in the hands of agency biologists, academics, and NGOs, it is possible to model the effects of global warming on the quality and extent of aquatic environments and the extent to which climate change will impose demographic and genetic consequences on aquatic species (Ruhí et al. 2016; Whitney et al. 2017).

WHAT THE FUTURE HOLDS

As the second decade of the twenty-first century closes, the reality of climate change and human responsibility is broadly recognized, at least in the scientific community. There is little doubt that many aquatic organisms of arid lands will be adversely affected by climate change (Whitney et al. 2017). This fact, coupled with an ever-expanding human population and an inexorable increase in demand on a finite and diminishing water supply (Dettinger et al. 2015), will sorely test the will and means to conserve desert fishes and the myriad of other species dependent upon water in the desert. Whether sufficient measures are taken to conserve and protect dwindling aquatic habitats and their native species will depend less on our knowledge of the species and their habitats and more on public awareness, appreciation, support, and engagement. Opposition to many regulations that provide environmental protections, especially for imperiled aquatic species rarely seen or appreciated by many, has not diminished in the years since publication of *Battle against Extinction* and remains a potent rallying cry in many quarters. It is best countered by pragmatic, science-based, and comprehensive conservation strategies. But, as Deacon and Pister realized in the early battles to save Devils Hole, public appreciation of these rare desert species and their likely fate is vital as well.

When they conceived *Battle against Extinction* in the late 1980s, Minckley and Deacon meant the volume to be a synthesis of existing knowledge, a demonstration of the need to approach conservation from an ecosystem perspective, and a vehicle to help resist ongoing efforts to erode legislated protections of imperiled organisms. Our intent in this volume is to accomplish similar goals, but with an expanded breadth of topics, including chapters on homogenization of surface waters, groundwater mining, desert oases, and aquatic invertebrates. The uncertainty of what climate change means for aquatic fauna and how it will be accommodated imbues this volume. For many species, a warming and drying environment bodes a bleak future. Critical evaluation of past and current practices and strategies, modification or change in methods as dictated by new information, and new innovative technologies are essential to tip the scales toward the persistence of aquatic organisms. All this is made more difficult by wild swings in political support and increased polarization of conservation efforts.

The challenges of protecting aquatic species in desert ecosystems remain formidable and will become greater with global warming. We hope the chapters that follow serve to inform the conservation challenges, point to effective approaches in this endeavor, and inspire a greater appreciation of the values and fates of aquatic resources in the arid regions of North America.

REFERENCES

Archdeacon, T. P. 2016. Reduction in spring flow threatens Rio Grande silvery minnow: Trends in abundance during river intermittency. *Transactions of the American Fisheries Society* 145: 754–65.

Contreras-Balderas, S. 1991. Conservation of Mexican freshwater fishes: Some protected sites and species, and recent federal legislation. In W. L. Minckley and J. E. Deacon, eds., *Battle against Extinction: Native Fish Management in the American West*, 191–97. Tucson: University of Arizona Press.

Czech, B., and P. R. Krausman. 1999. Public opinion on endangered species conservation and policy. *Society and Natural Resources* 12: 469–79.

Deacon, J. E., and C. Deacon Williams. 1991. Ash Meadows and the legacy of the Devils Hole pupfish. In W. L. Minckley and J. E. Deacon, eds., *Battle against Extinction: Native Fish Management in the American West*, 69–87. Tucson: University of Arizona Press.

Deacon, J. E., and W. L. Minckley. 1991. Western fishes and the real world: The enigma of "endangered species" revisited. In W. L. Minckley and J. E. Deacon, eds., *Battle against Extinction: Native Fish Management in the American West*, 405–13. Tucson: University of Arizona Press.

Deacon, J. E., A. E. Williams, C. Deacon Williams, and J. E. Williams. 2007. Fueling population growth in Las Vegas: How large-scale groundwater withdrawal could burn regional biodiversity. *BioScience* 57: 688–98.

Dettinger, M., B. Udall, and A. Georgakakos. 2015. Western water and climate change. *Ecological Applications* 25: 2069–93.

Echelle, A. A. 1991. Conservation genetics and genic diversity in freshwater fishes of western North America. In W. L. Minckley and J. E. Deacon, eds., *Battle against Extinction: Native Fish Management in the American West*, 141–53. Tucson: University of Arizona Press.

Franssen, N. R., S. L. Durst, K. B. Gido, D. W. Ryden, V. Lamarra, and D. L. Propst. 2014. Long-term dynamics of large-bodied fishes assessed from spatially intensive monitoring of a managed desert river. *River Research and Applications* 32: 348–61.

Gosnell, H. 2001. Section 7 of the Endangered Species Act and the art of compromise: The evolution of a Reasonable and Prudent Alternative for the Animas-La Plata Project. *Natural Resources Journal* 41: 561–626.

Holden, P. B. 1991. Ghosts of the Green River: Impacts of Green River poisoning on management of native fishes. In W. L. Minckley and J. E. Deacon, eds., *Battle against Extinction: Native Fish Management in the American West*, 43–54. Tucson: University of Arizona Press.

Johnson, J. E., and B. L. Jensen. 1991. Hatcheries for endangered freshwater fishes. In W. L. Minckley and J. E. Deacon, eds., *Battle against Extinction: Native Fish Management in the American West*, 199–217. Tucson: University of Arizona Press.

Leopold, A. 1949. *A Sand County Almanac: With Sketches Here and There*. New York: Oxford University Press.

Miller, R. R. 1961. Man and the changing fish fauna of the American Southwest. *Papers of the Michigan Academy of Science, Arts, and Letters* 46: 365–404.

Minckley, W. L. 1983. Status of the razorback sucker, *Xyrauchen texanus* (Abbott), in the lower Colorado River basin. *Southwestern Naturalist* 28: 165–87.

Minckley, W. L., and J. E. Deacon. 1968. Southwestern fishes and the enigma of "endangered species." *Science* 159: 1424–32.

Minckley, W. L., and J. E. Deacon, eds. 1991. *Battle against Extinction: Native Fish Management in the American West*. Tucson: University of Arizona Press.

Minckley, W. L., P. C. Marsh, J. E. Brooks, J. E. Johnson, and B. L. Jensen. 1991. Management toward recovery of the razorback sucker. In W. L. Minckley and J. E. Deacon, eds., *Battle against Extinction: Native Fish Management in the American West*, 303–57. Tucson: University of Arizona Press.

Moyle, P. B., and G. M. Sato. 1991. On the design of preserves to protect native fishes. In W. L. Minckley and J. E. Deacon, eds., *Battle against Extinction: Native Fish Management in the American West*, 155–69. Tucson: University of Arizona Press.

Olson, H. F. 1962. *Rehabilitation of the San Juan River*. Job Completion Report, Federal Aid Project F-19-D-4. Santa Fe: New Mexico Department of Game and Fish.

Pister, E. P. 1991. The Desert Fishes Council: Catalyst for change. In W. L. Minckley and J. E. Deacon, eds., *Battle against Extinction: Native Fish Management in the American West*, 55–68. Tucson: University of Arizona Press.

Pister, E. P. 1993. Species in a bucket: For a few frightening moments, there was only myself standing between life and extinction. *Natural History* 102 (January), 14–17.

Poff, L. N., J. D. Allan, M. B. Bain, J. R. Karr, K. L. Prestegaard, B. D. Richter, R. E. Sparks, and J. C. Stromberg. 1997. The natural flow regime. *BioScience* 47: 769–84.

Reisner, M. 1986. *Cadillac Desert*. New York: Penguin Viking.

Ruhí, A., J. D. Olden, and J. L. Sabo. 2016. Declining streamflow induces collapse and replacement of native fish in the American Southwest. *Frontiers in Ecology and the Environment* 14: 465–72.

Scoppettone, G. G., and G. Vinyard. 1991. Life history and management of four endangered lacustrine suckers. In W. L. Minckley and J. E. Deacon, eds., *Battle against Extinction: Native Fish Management in the American West*, 359–77. Tucson: University of Arizona Press.

Secretary of the Interior. 1967. Native Fish and Wildlife: Endangered Species. *Federal Register* 32 (48): 4001.

USFWS (US Fish and Wildlife Service). 1991. *Final Biological Opinion for the Animas-LaPlata Project, Colorado and New Mexico*. Denver: United States Department of the Interior Fish and Wildlife Service Region 6. https://www.fws.gov/southwest/sjrip/pdf/DOC_Animas-LaPlata_BO_FINAL_1991-10-25.pdf.

Whitney, J. E., J. B. Whitter, C. P. Paukert, J. D. Olden, and A. L. Strecker. 2017. Forecasted range shifts of arid-land fishes in response to climate change. *Reviews in Fish Biology and Fisheries* 27: 463–79.

Williams, J. E. 1991. Preserves and refuges for native western fishes: History and management. In W. L. Minckley and J. E. Deacon, eds., *Battle against Extinction: Native Fish Management in the American West*, 171–89. Tucson: University of Arizona Press.

Wydoski, R. S., and J. Hamill. 1991. Evolution of a cooperative recovery program for endangered fishes in the upper Colorado River basin. In W. L. Minckley and J. E. Deacon, eds., *Battle against Extinction: Native Fish Management in the American West*, 123–35. Tucson: University of Arizona Press.

2 The Protagonists

Each field has its pioneers, and the field of desert fish conservation is no exception. Efforts to conserve aquatic life in the desert have been fueled equally by the ecological understandings developed by science and conservation actions directed by firmly held ethical beliefs. A few individuals are capable of combining scientific knowledge and conservation ethics into an undeniable passion that becomes contagious. These pioneering giants marveled at the ability of certain species to survive in seemingly harsh environments and, when necessary, mounted a vigorous defense when outside forces threatened their survival. This chapter traces the careers of a small handful of highly dedicated and influential scientists who recognized the value of desert fishes and came to their aid at a time when few cared for, or even knew of, such species. They saved species and habitats while influencing so many of us to follow in their footsteps. In this chapter, we honor the legacy of eight giants in our field, and we hope that through these biographies they will continue to inspire new professional, ethical, committed warriors to join the battle against extinction. Each biography was written by a close friend, colleague, or family member.

2A CARL LEAVITT HUBBS AND ROBERT RUSH MILLER

Robert J. Edwards

Carl Leavitt Hubbs was born in Williams, Arizona, on October 19, 1894. He moved with his family to San Diego when he was two years old and grew up in the then sparsely populated area of Southern California around Los Angeles and the Central Valley. He began college at Los Angeles Junior College, where he was mentored by George Bliss Culver, who had worked with David

Starr Jordan, president of Stanford University. Culver convinced Carl to give up studying birds and, instead, study fishes in the Los Angeles area. Consequently, Carl transferred to Stanford University, where he studied under Charles Henry Gilbert, another close associate and collaborator of Jordan's (Miller and Shor 1997). While still an undergraduate, he made his first research trip to the Great Basin with John Otterbein Snyder, another of Jordan's students. Hubbs later characterized this 70-day collecting trip, which took them first through the Bonneville Basin in Utah and Idaho, as "a trip from Heaven to Hell" (Miller and Shor 1997; Brittan and Jennings 2008). While this may have been his first trip to the desert, it would certainly not be his last.

While at Stanford University, Carl began his tremendous publishing effort, which continued throughout his life. After receiving his bachelor's degree (1916) and his master's degree (1917), he took a position as assistant curator of fishes, amphibians, and reptiles at Chicago's Field Museum of Natural History. While at Stanford, he collected salamanders for a biological supply company with fellow student Frances Clark, who introduced him to her older sister, Laura Clark, a mathematics major at Stanford (Miller and Shor 1997). Frances received her PhD at the University of Michigan, and as a fisheries biologist, later became a researcher and director of the California State Fisheries Laboratory at Terminal Island, California (Brown 1994). Carl married Laura in 1918. In 1920, he moved again and became the curator of the Fish Division at the University of Michigan's Museum of Zoology. Shortly thereafter, he returned to the Great Basin to study its endemic fishes.

Carl and Laura Hubbs had three children, Frances, Clark, and Earl. The Hubbs family built the University of Michigan fish collection into one of the largest in North America, and the entire family would spend summers in the field, collecting a wide variety of organisms. The parents developed an allowance system with incentives for finding species, especially new ones. This system was quite lucrative for the children and served as a means to keep them happy (Miller and Shor 1997).

It was during his time at the University of Michigan that Carl was awarded his PhD, in a most unusual manner. As a part of a faculty upgrading at the university, Carl was asked by the university president to submit a dissertation. Carl is rumored to have pointed to a number of his publications and asked, "Which one?" (Shor et al. 1987; Miller and Shor 1997). The paper eventually selected had been submitted for publication to *American Naturalist* and dealt with modifications of developmental rates in fishes as they relate to evolution (Hubbs 1926). This contribution was an extension of his earlier studies of phenotypic variation in fishes due to temperature and salinity, and he cited no fewer than nine of his previous papers on the subject in this submission.

Robert Rush (Bob) Miller was born on April 23, 1916, in Colorado Springs, Colorado. His family had a ranch in Wyoming, but they moved to Los Angeles when Bob was two years old. He took his first trip to Death Valley with his family in December 1934, and a highlight was visiting Badwater and Zabriskie Point, where the highest and lowest points in the contiguous United States are visible (Cashner et al. 2003). In 1934, he entered Pomona College and stayed until 1936, when he transferred to the University of California, Berkeley, where he finished his undergraduate degree. He started out studying geology; it was on a geologic field trip to Afton Canyon that he noticed some small fishes in the Mojave River in the middle of the desert. His professor did not know about these fishes, so Bob collected them to study. Thus, his career as an ichthyologist was launched; he just didn't know it yet (Cashner et al. 2003). He revisited Death Valley a number of times after that first trip, as part of college field trips and camping trips with classmates (Hendrickson et al. 2002; Cashner et al. 2003), and tried to track down the rumors that fishes could be found there. He succeeded. Ray Coles, a herpetologist at the University of California, Los Angeles, wrote to Carl Hubbs and noted Bob's collection of fishes in Death Valley and his interest in fishes. Carl wrote to Bob, then a college junior, and after some correspondence, invited him to join his upcoming 1938 field trip to the Great Basin in July and August, after the Hubbs family's trip to the Big Bend region of Texas in June (Hubbs 1940; Miller and Shor 1997; Cashner et al. 2003). Bob and a friend arrived at the scheduled meeting point at Lehman Cave, Nevada, a day early and waited for the Hubbs family to join them on July 5, 1938. In anticipation of their arrival, Bob cleaned up, put on fresh clothes and combed his hair, intending to make a good impression when he met Carl for the first time. Imagine his surprise when the Hubbs family arrived and Carl looked much like a bum (Cashner et al. 2003). Frances, then 19 years old, had been less than thrilled about this whole trip to the desert with her family and would rather have stayed in Ann Arbor, but this changed when she saw Bob for the first time, and upon meeting, both she and Bob fell madly in love! (Cashner et al. 2003). Thus, the Hubbs and Miller connection was initiated. Carl convinced Bob to come to Michigan with him and study fishes instead of rocks. Bob and Frances married in 1940 and had five children (Gifford, Frances, Roger, Ben, and Lawrence). Bob received his MS degree in 1943 and his PhD in 1944 under the direction of Carl. It must have been very interesting having Carl as both his major advisor and father-in-law (Hendrickson et al. 2002).

Carl, Laura, and their children had made significant collections of fishes in the summers of 1934, 1938, and 1942. These collections were extensively documented in a chapter by Miller et al. (1991) in *Battle against Extinction*. In addition to the fish they collected during these expeditions, when aquatic habitats were few and far between, the Hubbs family would collect non-fish samples, and more than 10% of their 1938 collections included lizards, snakes, frogs, toads, snails, and insects (Miller and Shor 1997). After graduating from the University of Michigan, Bob took a position from 1944 to 1948 as an assistant curator of fishes at the US National Museum, but returned to the University of Michigan as a faculty member of the Fish Division in the university's Museum of Zoology (Hendrickson et al. 2002; Cashner et al. 2003). A number of monumental works came from the early work of the Hubbs-Miller collaboration, including "The Zoological Evidence: Correlation between Fish Distribution and Hydrographic History in the Desert Basins of Western United States" in 1948 and "Hydrographic History and Relict Fishes of the North-Central Great Basin" in 1974. Hubbs and Miller were both prolific authors. Carl published 712 titles in his lifetime, and Bob had 349 publications, abstracts, and reviews, both remarkable numbers (Shor et al. 1987; Cashner et al. 2003).

Both Carl and Bob had strong feelings about the conservation of nature and protecting natural environments. In one of his earlier papers, which appeared in *Science* in 1946, Bob wrote passionately about the need for study of the fishes of the western rivers because he saw the damaging consequences of the anthropogenic changes, including dams, water diversions, pollution, adverse effects of livestock grazing, and the haphazard introduction of nonnative species, that were occurring (Miller 1946). He predicted in this paper that these changes were likely to continue at an accelerated rate in the future. A later paper extensively documented the changes that had occurred due to the hand of man throughout the desert Southwest (Miller 1961). The 1960s were a very productive time for Bob's conservation efforts. He co-authored a book on the fishes of Utah (Sigler and Miller 1963); became the chairman of the Freshwater Fish Group of the International Union for Conservation of Nature (IUCN), and developed the first "Red Data List" of endangered fishes worldwide (Miller 1977). He also became the chairman of the Endangered Species Committee of the American Fisheries Society in 1969–1971 (Cashner et al. 2003).

Carl also shared an extensive history of conservation. He and Bob worked to protect the Devils Hole pupfish *Cyprinodon diabolis* beginning in the 1940s. In 1952, Carl wrote to President Harry Truman and convinced him to make Devils Hole a separate part of Death Valley National Monument to protect both its unique geologic features and its pupfish. Because there was no Endangered Species Act at the time, its unusual travertine limestone formation became the basis for the protection of Devils Hole (Riggs and Deacon 2004).

The Desert Fishes Council was formed and first met in Death Valley in November 1969. Both Carl and Bob, along with their wives, were there. The group was composed of academic scientists and agency biologists and managers who were concerned that a great number of threats to and lack of protection for desert fishes were causing harm to this unique fauna. Their interest in the Desert Fishes Council and its work lasted for the rest of their lives, although each branched out significantly in his later years (fig. 2a.1). Carl had moved to Scripps Institute of Oceanography in La Jolla, California, in 1944 and expanded his earlier studies of marine fishes as well as marine mammals. Bob continued his interest in desert fishes of southwestern North America, taking new trips to the Chihuahuan Desert in Texas (Hubbs et al. 1977; Miller 1978), and continued his lifelong interest in the freshwater fishes of México, which culminated in his final tome, published posthumously, entitled *Freshwater Fishes of México* (Miller et al. 2006), with the Spanish spelling of México as a sign of respect. Throughout both of their long careers, their wives played co-starring roles, often behind the scenes, but always as coequals to those who knew them well (Cashner et al. 2011).

Fig. 2a.1 Robert R. Miller, Carl L. Hubbs, and Salvador Contreras-Balderas at the 1973 Desert Fishes Council meeting in Tempe, AZ. Photo courtesy of Desert Fishes Council Archives.

Carl died in La Jolla, California, on June 30, 1979. He was 85 years old. Laura Hubbs died in La Jolla on June 24, 1988, at the age of 95. Frances Hubbs Miller died in Ann Arbor, Michigan, on October 17, 1987, at the age of 68, and Bob Miller died in Ann Arbor on February 10, 2003, two months shy of his 87th birthday (Croker 1987; Chernoff 1988; Cashner et al. 2011).

REFERENCES

Brittan, M. R., and M. R. Jennings. 2008. Stanford University's John Otterbein Snyder: Student, collaborator, and colleague of David Starr Jordan and Charles Henry Gilbert. *Marine Fisheries Review* 70: 24–29.

Brown, P. S. 1994. Early women ichthyologists. *Environmental Biology of Fishes* 41: 9–30.

Cashner, F. M., G. R. Smith, R. C. Cashner, R. R. Miller, and F. H. Miller. 2003. Robert R. Miller and Frances H. Miller. *Copeia* 2003: 910–16.

Cashner, R. C., G. R. Smith, G. H. Miller, F. M. Cashner, and B. Chernoff. 2011. Robert Rush Miller (1916–2003). *Copeia* 2011: 342–47.

Chernoff, B. 1988. Frances Voorhees Hubbs Miller, 1919–1987. *Copeia* 1988: 520–23.

Croker, R. 1987. In Memoriam: Frances N. Clark. *California Cooperative Oceanic Fisheries Investigations Report* 28: 5.

Hendrickson, D. A., S. M. Norris, and J. J. Schmitter-Soto. 2002. Dedication to Dr. Robert "Bob" Rush Miller. *Reviews in Fish Biology and Fisheries* 12: 113–18.

Hubbs, C., R. R. Miller, R. J. Edwards, K. W. Thompson, E. Marsh, G. P. Garrett, G. L. Powell, D. J. Morris, and R. W. Zerr. 1977. Fishes inhabiting the Rio Grande between New Mexico and the Pecos confluence. In R. R. Johnson and D. Jones, eds., *Importance, Preservation and Management of Riparian Habitat, A Symposium*, 91–97. USDA Forest Service General Technical Report RM-43.

Hubbs, C. L. 1926. The structural consequences of modifications of the developmental rate in fishes, considered in reference to certain problems of evolution. *American Naturalist* 60: 57–81.

Hubbs, C. L. 1940. Fishes from the Big Bend Region of Texas. *Transactions of the Texas Academy of Science* 23 (1938–1939): 3–12.

Hubbs, C. L., and R. R. Miller. 1948. The zoological evidence: Correlation between fish distribution and hydrographic history in the desert basins of western United States. In E. Blackwelder, *The Great Basin, with Emphasis on Glacial and Postglacial Times*, University of Utah Biological Series vol. 10, no. 7, 17–166. Salt Lake City: University of Utah.

Hubbs, C. L., R. R. Miller, and L. C. Hubbs. 1974. Hydrographic history and relict fishes of north-central Great Basin. *Memoirs of the California Academy of Sciences* 7: 1–259.

Miller, R. R. 1946. The need for ichthyological surveys of the major rivers of western North America. *Science* 104: 517–19.

Miller, R. R. 1961. Man and the changing fish fauna of the American Southwest. *Papers of the Michigan Academy of Science, Arts, and Letters* 46: 365–404.

Miller, R. R. 1977. *Red Data Book*. Vol. 4., *Pisces: Freshwater Fishes*. Morges, Switzerland: International Union for the Conservation of Nature and Natural Resources.

Miller, R. R. 1978. Composition and derivation of the native fauna of the Chihuahuan Desert region. In R. H. Wauer and D. H. Riskind, eds., *Transactions of the Symposium on Biological Resources of the Chihuahuan Desert Region, U.S. and Mexico*, 365–81. US National Park Service, Transactions and Proceedings, no. 3.

Miller, R. R., C. Hubbs, and F. H. Miller. 1991. Ichthyological exploration of the American west: The Hubbs-Miller era, 1915–1950. In W. L. Minckley and J. E. Deacon, eds., *Battle against Extinction: Native fish management in the American West*, 19–40. Tucson: University of Arizona Press.

Miller, R. R., W. L. Minckley, and S. M. Norris. *Freshwater Fishes of México*. Chicago: University of Chicago Press.

Miller, R. R., and E. N. Shor. 1997. Carl L. Hubbs (1894–1979): Collection builder extraordinaire. *Collection Building in Ichthyology and Herpetology* 19: 367–76.

Riggs, A. C., and J. E. Deacon. 2004. Connectivity in aquatic ecosystems: The Devils Hole story. In D. W. Sada and S. E. Sharpe, eds., *Conference Proceedings. Spring-Fed Wetlands;*

Important Scientific and Cultural Resources of the Intermountain Region, May 7–9, 2002. Paradise, NV: Desert Research Institute. http://www.dri.edu/images/stories/conferences_and_workshops/spring-fed-wetlands/spring-fed-wetlands-riggs-deacon.pdf.

Shor, E. N., R. H. Rosenblatt, and J. D. Isaacs. 1987. Carl Leavitt Hubbs 1894–1979: A biographical memoir. *Biographical Memoirs of the National Academy of Sciences* 56:215–49.

Sigler, W. F., and R. R. Miller. 1963. *Fishes of Utah*. Salt Lake City: Utah State Department of Fish and Game.

2B W. L. MINCKLEY

Chuck O. Minckley

W. L. Minckley was born during November 1935 in Ottawa, Kansas. "Minck"'s childhood was uneventful until the summer of 1945, when he contracted polio. This experience included a diagnosis that he would never walk again, time spent tied to a board to straighten his spine, and time in an iron lung. His recovery included foot surgery and days of rehabilitation. This had to be one of the most significant events in his life and may be where he developed his tenacity for life and work.

Over time, his health improved, and despite problems with walking, he went through the school system in Ottawa, where he was the equipment manager for athletic teams, as he could not participate. As a teenager, he was a hell raiser, like the Fonz from the TV show *Happy Days*. He tried to join the Navy, but was unsuccessful due to his polio-related disability. At some point during high school, he decided to attend the University of Kansas. The day after high school graduation, he broke a jar of gasoline, soaking himself. He changed clothes in a small bathroom, where the pilot light on the water heater ignited the gas fumes, causing a flash fire that burned his legs severely. His mother put the flames out, and he was rushed to the hospital. Surprisingly, he started at the university that fall (1953).

Minck attended the University of Kansas (KU) for one semester, but walking the "seven hills of KU" so soon after the fire was too difficult, so he transferred to Kansas State University. While there, he was a member of the gymnastics team, working the flying rings and pommel horse. He excelled, and in his senior year competed in the Olympic trials, but did not qualify. In the fall of 1957, he started a master's program, returning to the University of Kansas, where Dr. Frank Cross became his major professor. There, he also met Jim Deacon, a PhD candidate with whom he formed a professional and personal friendship that lasted through their lives. His master's thesis was "Fishes of the Big Blue River in Kansas" (Minckley 1959). While at KU, he gained his first exposure to México on a field zoology trip.

Following graduation from KU, he entered the PhD program at University of Louisville (1958). Dr. Louis A. Krumholtz was his major professor. There, he worked on Ohio River fishes, but his primary research was on Doe Run, a small spring-fed stream. His dissertation, "The Ecology of a Spring Stream: Doe Run, Meade County, Kentucky" (Minckley 1963), is still an important contribution to aquatic ecology.

He graduated from the University of Louisville and accepted a job as an assistant professor of biology at Western Michigan University. However, on April 1, 1963, he received a telegram from the president of Arizona State University, stating, "Offer position Assistant Professor Zoology at salary $7,600 for Academic Year, beginning September 1, 1963. Wire reply." Thereafter, the future of native fishes in the North American desert changed forever.

In 1963, Arizona State University was small and beginning its transition from the Arizona Territorial Normal School into a major university. The zoology department was new, just beginning to train graduate students. Minck was there, developing programs and working with other faculty in growing the department. During this time, he and his students intensively surveyed the fishes of Arizona and northern México. Collections and observations made at that time formed the basis for his ideas on the management of native fishes and their habitats. They also provided the training ground for his students, many of whom found employment within federal and state agencies, where they carried his philosophy forward.

These early years were also when he developed relationships with Carl L. Hubbs and Robert R. Miller. Initially competitors, they came to realize the need to work together to save a vanishing resource: native desert fishes. Perhaps one of the earliest benefits of this collaboration involved litigation to save Devils Hole, home to the endemic Devils Hole pupfish *Cyprinodon diabolis*. Through their efforts, along with those of Jim Deacon, Phil Pister, and other members of the fledgling Desert Fishes Council, the US Supreme Court ruled that the Devils Hole pupfish had the "right to life" and that groundwater pumping affecting the water level in Devils Hole must be stopped. As part of this general effort, Minck became one of the founding members of the Desert Fishes Council in 1969.

Minck ultimately authored and co-authored 222 scientific papers. Some of his more significant works include "Southwestern Fishes and the Enigma of 'Endangered Species'" (Minckley and Deacon 1968); "Environments of the Bolsón of Cuatro Ciénegas, Coahuila, México" (Minckley 1969); "Fishes of the Río Yaqui Basin, México and United States" (Hendrickson et al. 1980); "Ciénegas: Vanishing Climax Communities of the American Southwest" (Hendrickson and Minckley 1984); "Geography of Western North American Freshwater Fishes: Description and Relationships to Intracontinental Tectonism" (Minckley et al. 1986); and "A Conservation Plan for Native Fishes of the Lower Colorado River" (Minckley et al. 2003). He also authored, edited, and/or contributed to four books: *Fishes of Arizona* (Minckley 1973), *Battle against Extinction: Native Fish Management in the American West* (Minckley and Deacon 1991), *Freshwater Fishes of México* (Miller et al. 2005), and *Inland Fishes of the Greater Southwest: Chronicle of a Vanishing Biota* (Minckley and Marsh 2009). Additionally, he wrote innumerable reports, chaired or served on many committees (e.g., Desert Fishes Recovery Team), and was a well-known expert on fishes and aquatic systems in the North American desert.

During his 43-year career (1958–2001), he taught a variety of graduate and undergraduate courses, retiring from active teaching in 2000, when he was appointed professor emeritus (fig. 2b.1). By then, he had instructed hundreds of undergraduates and produced 39 master's and 22 PhD students whose research focused on aquatic ecosystems and desert fishes. To support his research and departmental programs, he wrote proposals to over 30 agencies and organizations during his career, receiving major funding (Collins et al. 2002). Ultimately, his efforts helped establish an internationally recognized research program in aquatic ecology, systematic ichthyology, and conservation biology.

Fig. 2b.1 W. L. Minckley, 1990. Photograph taken for Arizona State University Outstanding Mentor award.

As important as his research and teaching, and perhaps one of his most important achievements, was the positive working relationships he developed over the years with private individuals, organizations, and agency biologists. He worked well with many of them, participating on committees, attending meetings, co-authoring reports, and publishing articles with them. Minck was not an ivory-tower academic, but a practitioner of conservation biology, as shown by his legacy.

He also began several research projects that are ongoing, including research on the Lake Mohave, Arizona, population of razorback suckers *Xyrauchen texanus* and studies on the fishes of Aravaipa Creek, Arizona. The razorback work began with a trip by his ichthyology class to Lake Mohave, where they camped at Carp Cove and surveyed the lake. Razorbacks were abundant then, and that trip spawned research resulting in the continued survival of a genetically diverse population of razorbacks in Lake Mohave. The Aravaipa Creek project, which also began as a class trip, resulted in the protection of Aravaipa Canyon and its native fish populations by The Nature Conservancy and the US Bureau of Land Management, who manage it as a designated wilderness area.

A third project, first envisioned in 1958 while Minck was a student at KU, resulted in the investigations of the endorheic Cuatro Ciénegas Basin, Coahuila, México. The basin lies along the eastern edge of the Mesa del Norte, about 270 km by air south-southeast of the Big Bend of the Río Bravo–Rio Grande (Minckley 1969). The small intermountain valley was first explored biologically by E. G. Marsh, Jr., in 1939, but its unique biota was not described until 1965 (Hubbs and

Miller 1965). Minck first visited the valley in 1958, but a random event at KU raised his curiosity before that trip, when he found a series of preserved Coahuilan box turtles *Terrapene coahuila* while working in the KU vertebrate collection. What caught his attention was the presence of an obligate aquatic alga on their shells and the locality: Cuatro Ciénegas, Coahuila.

Given that box turtles are normally terrestrial, why were the algae present? This question was answered during the 1958 trip, when it was found that the turtles were semiaquatic and occurred only in aquatic habitats of Cuatro Ciénegas. This first trip to the basin resulted in Minck beginning research there while at the University of Louisville, where, with two other graduate students, he cobbled together a few hundred dollars, purchased a 1955 Dodge panel truck they named "Nellie," and began making research trips to Cuatro Ciénegas. There, they discovered a unique aquatic ecosystem theretofore basically unnoticed by the scientific world. Such trips consumed his time for the next 40 years. A 1969 paper by Minck, detailing what had been discovered by that time, represented just the start of a stream of publications describing the aquatic and terrestrial flora and fauna, limnology, geology, paleoecological record, and prehistoric peoples of the basin.

The importance of the Cuatro Ciénegas Basin in terms of biodiversity was soon recognized, and Minck began working to have it designated a protected area, making the scientific community aware and enlisting others to join the fight. He also worked with the local community through Mayor Don Francisco Manrique Dávila and the city council to gain their support (Gutierrez 2001). Eventually, the town, the state, and the Mexican government were all made aware of the basin's importance, and it was designated and officially protected as a biological reserve by México in 1994 (Grall 1995). Today, the basin is known to the world and represents a unique ecosystem with some 150 different plants and animals endemic to the valley and its surrounding mountains. They include some 30 aquatic species, including 8 species of fish. Four endemic organisms are named for W. L. Minckley: the Cuatro Ciénegas cichlid (Cichlidae: *Herichthys minckleyi*), the flower-loving fly (Apioceridae: *Apiocera minckleyi*), a scorpion (Vaejovidae: *Vaejovis minckleyi*), and an aquatic snail (Hydrobiidae: *Nymphophilus minckleyi*). The water-penny beetle (Psephenidae: *Psephenus minckleyi*) of central Arizona also bears his name.

Minck entered Desert Samaritan Hospital in Mesa, Arizona, in mid-June 2001 for cancer treatment and died June 22, 2001, from a MRSA infection. His ashes were scattered in the spring Poso Azul, in Cuatro Ciénegas. He was mourned by friends, family, and the scientific community. The people of Cuatro Ciénegas devoted a page in their newspaper detailing his work in the basin, what they thought of him, and how he had benefited the valley (Gutierrez 2001).

Minck left behind an eclectic group of adversaries, friends, former students, and family members who remember him in their own ways. In my case, looking back, I cannot help but remember being hunkered down around a windy, dying campfire at Two-Cave Canyon, drinking Sangre de Cristo with limón, listening to the exploits of the day and plans for tomorrow. The hardships were forgotten, at least for the night, and new adventures awaited the dawn.

ACKNOWLEDGMENTS

Much of this material was taken from Minck's obituary, written by his colleagues (Collins et al. 2002). Mr. Humberto Rodriquez, San Bernardino National Wildlife Refuge, translated the newspaper article from Cuatro Ciénegas. Their participation and those of the reviewers are gratefully acknowledged. Minck's publications can be accessed at http://www.nativefishlab.net.

REFERENCES

Collins, J. P., J. Deacon, T. Dowling, and P. Marsh. 2002. Obituary, W. L. Minckley. *Copeia* 2002 (1): 258–62.

Grall, G. 1995. Cuatro Ciénegas, Mexico's desert aquarium. *National Geographic* 186 (4): 85–97.

Gutierrez, C. 2001. Wendell Lee Minckley, pioneer discoverer of the Cuatro Ciénegas Valley. *Zocala* (Monclova, Coahuila), June 19, 2001, 3D, in the Minckley Library (website).

Hendrickson, D. A., and W. L. Minckley. 1984. Ciénegas: Vanishing climax communities of the American Southwest. *Desert Plants* 6: 131–75.

Hendrickson, D. A., W. L. Minckley, R. R. Miller, D. J. Siebert, and P. H. Minckley. 1980. Fishes of the Río Yaqui basin, México and United States. *Journal of the Arizona-Nevada Academy of Science* 15: 65–106.

Hubbs, C. L., and R. R. Miller. 1965. Studies of the cyprinodont fishes. XXII. Variation in *Lucania parva*, its establishment in Western United States, and description of a new species from the interior basin in Coahuila, Mexico. *Miscellaneous Publications of the Museum of Zoology, University of Michigan* 127: 1–113.

Miller, R. R., W. L. Minckley, and S. M. Norris. 2005. *Freshwater Fishes of México*. Chicago: University of Chicago Press.
Minckley, W. L. 1959. Fishes of the Big Blue River basin, Kansas. *University of Kansas Publications, Museum of Natural History* 11: 401–42.
Minckley, W. L. 1963. The ecology of a spring stream: Doe Run, Meade County, Kentucky. *Wildlife Monographs* 11: 3–124.
Minckley, W. L. 1969. Environments of the Bolsón of Cuatro Ciénegas, Coahuila, Mexico with special reference to the aquatic biota. *University of Texas, El Paso, Science Series* 2: 1–65.
Minckley, W. L. 1973. *Fishes of Arizona*. Phoenix: Arizona Game and Fish Department.
Minckley, W. L., and J. E. Deacon. 1968. Southwestern fishes and the enigma of "endangered species." *Science* 159: 1424–32.
Minckley, W. L., and J. E. Deacon, eds. 1991. *Battle against Extinction: Native Fish Management in the American West*. Tucson: University of Arizona Press.
Minckley, W. L., D. A. Hendrickson, and C. L. Bond. 1986. Geography of western North American freshwater fishes: Description and relationships to intracontinental tectonism. In C. H. Hocutt and E. O. Wiley, eds., *The Zoogeography of North American Freshwater Fishes*, 519–613. New York: John Wiley & Sons.
Minckley, W. L., and P. C. Marsh. 2009. *Inland Fishes of the Greater Southwest: Chronicle of a Vanishing Biota*. Tucson: University of Arizona Press.
Minckley, W. L., P. C. Marsh, J. E. Deacon, T. E. Dowling, P. W. Hedrick, W. J. Matthews, and G. Mueller. 2003. A conservation plan for native fishes of the lower Colorado River. *BioScience* 53: 219–34.

2C SALVADOR CONTRERAS-BALDERAS

María de Lourdes Lozano-Vilano and Armando Jesús Contreras-Balderas

Salvador Contreras Balderas was born in México City, México, on February 19, 1936. When he was a teenager, he and his family moved to the City of Monterrey, Nuevo León, where he completed his high school studies. Because of his strong ties to both cities, he claimed to be "chilango" (of México City) by birth and "regiomontano" (of Monterrey City) by conviction. Salvador completed his bachelor's degree in biological sciences in High School No. 1 (1952–1953) and No. 3 (1955–1957), and because of his love of nature, he immediately began his academic career as a biologist at the School of Biological Sciences, University of Nuevo León (1957–1961). He defended his thesis, entitled "Contributions to the Knowledge of the Ichthyofauna of the Río San Juan, Province of Bravo, México," in September 1962. Soon after, he searched for financial support to attend graduate school for his MS and PhD degrees in the United States and was awarded an assistantship at Tulane University in New Orleans, Louisiana, working under the direction of Dr. Royal D. Suttkus. His MS studies were conducted from 1962 through 1966, and his PhD dissertation, entitled "Zoogeography and Evolution of *Notropis lutrensis* and *'Notropis' ornatus* in the Rio Grande Basin, México and United States (Pisces: Cyprinidae)" was completed in 1975.

Early in his career, Salvador built and maintained strong personal and collegial relationships with other prominent ichthyologists, including Carl L. Hubbs, Clark Hubbs, Robert R. Miller, and W. L. Minckley in the United States, and in México with José Álvarez del Villar (Edwards, chap. 2a, this volume; Garrett, chap. 2e, this volume). Salvador's love of fishes extended to interacting with aquarists as a means to conserve organisms, an effort that has been recognized internationally.

For more than 35 years, he focused on teaching and research at the Universidad Autónoma de Nuevo León (UANL), where he instructed many generations of students who were also seeking careers as biologists. Although he received offers to work in other institutions, he refused to leave UANL because of his interest in teaching as well as his devotion to research in his primary area of interest, the native fishes of México and their habitats. He worked hard to maintain facilities and his standing at the university and became a highly regarded ichthyologist in México and internationally. One of the most important achievements built through his professional life was his collection of fish specimens. Beginning in 1958, and using personal funds, Salvador began to archive specimens collected during field work, housing them in a closet in his home. Over time, the collection was moved to the university and has grown to more than 15,000 lots, containing about 1,000 species and nearly 1,000,000 specimens. The collection features species from México, the United States, Europe, and Africa and now supports the teaching and research activities of UANL personnel and others.

Salvador was also the founder of the graduate degree program in the biology department of UANL in 1973 (now the Division of Postgraduate Studies of the Faculty of Biological Sciences [FCB]), which grants both MS and PhD degrees. He directed that program from 1973 to 1984. Those responsibilities were in addition to serving as an instructor in basic biology courses as well as in aquatic ecology, his area of expertise, for more than 30 years.

As a professional, he always had great interest in and knowledge of natural resources and was committed to their conservation. For example, Salvador was one of the first Mexican scientists to raise concerns about environmental pollution, overuse of natural resources, and the abuse and waste of water. He was equally committed to sharing his knowledge and experience with and mentoring new professionals and organizations capable of continuing the struggle for conservation of natural resources, particularly desert fish.

After his retirement in 1994, he continued to be active, both independently and in close collaboration with the Ichthyology Laboratory of the FCB, UANL, in the production of various scientific contributions and papers for national and international congresses, as well as in mentoring and collaborating with graduate students. Salvador was also associated with research and teaching programs in several other universities, including those in Baja California, Jalisco, Michoacán, Quintana Roo, and Chiapas. For his outstanding contributions, he was recognized by the Universidad Autónoma de Nuevo León as professor emeritus.

In his professional career, he produced more than 100 publications in national and international journals (Contreras-Balderas 1974, 1975, 1977, 1984, 1987, 1991; Williams et al. 1989; García-Ramírez et al. 2006), in addition to 60 contributions regarding the profession of teaching, and authored or co-authored nearly 160 presentations given at national and international professional meetings. He also supervised to completion 38 bachelor's degree theses, 8 MS theses, and 6 PhD dissertations. For these and other contributions, he gained the respect, recognition, and encouragement of the National Biodiversity Commission. Salvador was also a main driver in efforts to found the Mexican Society of Zoology in 1977 and the Mexican Ichthyology Society in 1985. He served as president of both societies and was part of the foundation of Bioconservación, A. C., which supports various species conservation projects and environmental education programs.

Salvador received many awards and recognitions from professional societies as well, including Honorary Member of the Mexican Society of Zoology; the George M. Sutton Award in Conservation (1996, Southwestern Association of Naturalists); the acknowledgment of research productivity (Consejo Estatal de Flora y Fauna Silvestre de Nuevo León); the Donald W. Tinkle Research Excellence Award (1999, Southwestern Association of Naturalists); the "W. L. Minckley" award for Conservation of the Valle de Cuatro Ciénegas, Coahuila (2001, H. Ayuntamiento de Cuatro Ciénegas); and the President's Award in Conservation of Fish and Fisheries (2002, American Fisheries Soci-

Fig. 2c.1 Salvador Contreras-Balderas in 2007 (*right*) with Gorgonio Ruiz-Campos (*left*), Oasis San Ignacio, Baja California Sur, México. (Image courtesy of G. Ruiz-Campos.)

ety). Other awards included the Civil Merit in Research (Nuevo León, 1993); Distinguished Researcher and Distinguished Professor (1993), and the Dr. Eduardo Aguirre Pequeño award (1993), the latter three being from UANL. Salvador was also honored by having several aquatic animal species named after him.

A symposium celebrating the career of Dr. Salvador Contreras-Balderas was organized and held at the Universidad Autónoma de Nuevo León in 2002, with many collaborators, colleagues, and close friends in attendance. It included 21 papers describing different aspects of ichthyology and aquatic ecology. Those papers included descriptions of a pupfish *Cyprinodon salvadori* by María de Lourdes Lozano-Vilano and of a Neotropical silverside *Chirostoma contrerasi* by Clyde D. Barbour, both named in Salvador's honor. The resulting Libro Jubilar volume (Lozano-Vilano 2002) also chronicled Salvador's scientific contributions.

Throughout his career, and even after retirement, Salvador and his students were a strong presence at annual meetings of the Desert Fishes Council, whether they were held in the United States or México. As a natural outgrowth of his research program and mentoring at UANL, they reported findings on the conservation status and ecology of desert fishes and their habitats. In addition to reporting his own research, Salvador was a conduit of information on desert fish conservation for other colleagues in México who could not always attend the meetings. His contributions fostered stronger international relationships in the Desert Fishes Council and promoted broader participation by Mexican colleagues and others in supporting conservation of desert aquatic systems. His legacy in making the Desert Fishes Council a truly international society in scope and membership is evident and will be celebrated well into the future.

REFERENCES

Contreras-Balderas, S. 1974. *Cambios de composicion de especies en comunidades de peces en zonas semiaridas de México*. Contribution no. 15, Laboratorio de Vertebrados, Universidad Autónoma de Nuevo León.

Contreras-Balderas, S. 1975. Zoogeography and evolution of *Notropis lutrensis* and *"Notropis" ornatus* in the Rio Grande basin, México and United States (Pisces: Cyprinidae). PhD diss., Tulane University.

Contreras-Balderas, S. 1977. Biota endemica de Cuatro Ciénegas, Coahuila, México. *Memoires Primer Congreso Nacional Zoologia* 1: 106–13.

Contreras-Balderas, S. 1984. Environmental impacts in Cuatro Ciénegas, Coahuila, México. *Journal of the Arizona-Nevada Academy of Sciences* 19: 58–65.

Contreras-Balderas, S. 1987. Lista anotada de especies de peces mexicanos en peligro o amenazados de extincion. *Proceedings of the Desert Fishes Council XVI* (1984): 58–65.

Contreras-Balderas, S. 1991. Conservation of Mexican freshwater fishes: Some protected sites and species, and recent federal legislation. In W. L. Minckley and J. E. Deacon, eds., *Battle against Extinction: Native Fish Management in the American West*, 191–97. Tucson: University of Arizona Press.

García-Ramírez, M. E., S. Contreras-Balderas, and M. L. Lozano-Vilano. 2006. *Fundulus philpisteri* sp. nov. (Teleostei: Fundulidae) from the Río San Fernando basin, Nuevo León, México. In M. L. Lozano-Vilano and A. J. Contreras-Balderas, eds., *Studies of North American Desert Fishes in Honor of E. P. (Phil) Pister, Conservationist*, 13–19. Monterrey: Universidad Autónoma de Nuevo León.

Lozano-Vilano, M. L., ed. 2002. *Libro jubilar en honor al Dr. Salvador Contreras Balderas*. Monterrey: Universidad Autónoma de Nuevo León.

Williams, J. E., J. E. Johnson, D. A. Hendrickson, S. Contreras-Balderas, J. D. Williams, M. Navarro-Mendoza, D. E. McAllister, and J. E. Deacon. 1989. Fishes of North America endangered, threatened, or of special concern: 1989. *Fisheries* 14: 2–20.

2D JAMES E. DEACON

Cindy Deacon Williams

The *Las Vegas Review Journal* (Brean 2015) announced Dad's death by noting, "James Deacon was a biologist, environmentalist and staunch defender of desert fish who launched UNLV's environmental studies program and helped spawn new generations of scientists in both the classroom and his own family." Dad probably wouldn't have quibbled with that summation, though I would argue that he also was a born philosopher, scientific warrior, and despite it all, a politely stubborn quixotic optimist.

He was born in small-town White, South Dakota (population around 500), during the Great Depression and grew up tramping around the outdoors, spending his summers at Pickerel Lake. After moving to the big town of Aberdeen, Dad picked up the clarinet, a talent that garnered a musical scholarship that allowed him and his high school sweetheart and future wife, Maxine, to attend Midwestern State University in Texas. Upon completion of his bachelor's degree in 1956 and following the birth of his first child, Jim was accepted into the PhD program at the University of Kansas (KU), where he worked under Frank Cross. His PhD dissertation, "Fish Populations of the Neosho and Marais Des Cygnes Rivers, Kansas, Following Drought," was completed in 1960. During the course of his study he not only examined fish community responses to environmental stressors, but also learned how to "noodle" for catfish. More importantly, at KU he met W. L. Minckley, who then was one of Cross's master's students. "Minck" and Dad forged a professional and personal friendship that would grow and endure through both their lives.

In the summer of 1960, the Deacon family (Jim, Maxine, and Cindy) moved to Las Vegas, Nevada, where Dad prepared to take up his new position as the second biology professor at the Southern Regional Extension Division of the University of Nevada, Reno, joining W. G. Bradley in the fledgling department. The first campus-wide commencement graduated 29 students in 1964. The following year, the campus would become Nevada Southern University, and in 1969 it became the University of Nevada, Las Vegas (UNLV). Dad was excited to start his career; Mom couldn't quite believe that a fish biologist would be able to find anything to do in the desert . . . after all, fish need water! Furthermore, the campus was rather unprepossessing when they arrived, consisting of only two small buildings at the end of a long dirt road well beyond the outskirts of town.

It didn't take long for Dad to become enamored with desert fishes. Within a year of his arrival, his family had grown by one with the addition of his son David, and he had taken his first scuba dive in Devils Hole, beginning a career-long study of the pupfish *Cyprinodon diabolis* that called the water-filled cavern home (Deacon and Bunnell 1970; Deacon and Deacon 1979). His research interests during the first decade of his career also encompassed other isolated desert fishes, including other pupfish in Ash Meadows and Death Valley as well as the Pahrump poolfish *Empetrichthys latos latos* and springfish *Crenichthys* spp. found in various southern Nevada spring systems, and the effects of introduced fishes on the region's native fish fauna (Deacon et al. 1964; Minckley and Deacon 1968).

Very early on, Dad was confronted with the reality that much of his effort to gain an understanding of the ecology of these desert fishes was being undertaken in an environment where the fish populations were either in decline or facing imminent threat. Not surprisingly, his work increasingly branched out to include a focus on understanding the causes of endangerment and prospects for conservation of these uniquely adapted creatures. By the late 1960s, Dad was grappling with this problem in a serious way, and more than that, he was struggling to create a framework for scientific activism. By that time, his old friend from KU, W. L. Minckley, had assumed a professorship at Arizona State University and was confronting many of the same issues in Arizona that Dad was in Nevada. Together, they wrote a seminal paper exploring the "enigma of endangered species" in the desert Southwest, which was published in the March 29, 1968, issue of *Science*. Their joint interest in scientific activism—that is, in bringing solid scientific knowledge and understanding to bear on the conservation effort—is documented by the many papers Dad and Minck co-authored on the topic over the years, an endeavor culminating in their co-editorship of the book *Battle against Extinction* (1991).

Coincidentally, the same month the "enigma of endangered species" paper was published, major habitat disturbances resulting from agricultural development appeared in Ash Meadows, raising concerns over the long-term survival of the Devils Hole pupfish and other endemic species there. For a full year, habitat modifications continued while state and federal agencies proved unable to grapple with the destruction. Growing concerns led Dad to join Phil Pister (CDFG) and others from the BLM, USFWS, USNPS, and NVDFG in an informal April 1969 field meeting to discuss a preservation plan for Ash Meadows and schedule a November 1969 symposium. Out of that November symposium, attended by 44 individuals, the Desert Fishes Council was born and became a test-bed for scientific activism. Dad embraced this approach wholeheartedly. In the 1970s, he testified in defense of the Devils Hole pupfish during a legal fight over groundwater pumping in Ash Meadows that landed in the US Supreme Court and resulted in a landmark 1977 decision favoring endangered species protection. Unfortunately, as Dad found out, that wasn't the end of this particular battle against extinction. He and others were not able to secure protection for the habitat until 1984, when the area was acquired for public ownership and the Ash Meadows National Wildlife Refuge was established (fig. 2d.1).

Fig. 2d.1 James Deacon points out pupfish at Devils Hole in 1984, following the establishment of the Ash Meadows National Wildlife Refuge. (Photo by Gary Thompson, Las Vegas Review Journal.)

Fig. 2d.2 Jim Deacon and Robert Miller making sure the water pump is maintaining good circulation in the holding tank containing the "rescued" Pahrump poolfish during the great goldfish eradication project at Manse Spring, 1967. (Photo by Maxine Deacon.)

The work to restore the habitat and recover the endemic species remained an interest throughout his life, and that work continues today.

At the same time Dad was engaged in the battle to ensure that the Devils Hole pupfish and other fishes of Ash Meadows persisted, he was faced with the threatened demise of the Pahrump poolfish in Manse Spring. He and one of his graduate students were studying various aspects of the life history of this only extant member of the genus; the Ash Meadows poolfish *E. merriami* became extinct in the 1940s, and the other two subspecies of *E. latos*, the Raycraft Ranch poolfish *E. latos concavus* and Pahrump Ranch poolfish *E. latos pahrump*, were lost to extinction in the 1950s due to destruction and dewatering of their spring habitats. By 1967, things were becoming critical for the Pahrump poolfish, too. Among other aspects of poolfish ecology, Dad and his graduate student were examining the response of Pahrump poolfish to the unauthorized introduction of goldfish to the spring. Goldfish numbers had exploded, and poolfish numbers were crashing. In an effort to forestall what seemed to be the poolfish's imminent demise, a group of university scientists (including Dad and R. R. Miller), agency personnel, and graduate students undertook the massive effort to capture all fish in Manse Spring using seines and minnow traps, placing the few remaining poolfish in temporary holding tanks and killing all the captured goldfish (fig. 2d.2). The effort to rid the system of goldfish even included the use of dynamite when a single goldfish was seen swimming in the spring pool after the water had cleared following the end of collection efforts. Unfortunately, that goldfish survived the experience of being blown out of the water; it managed to flop back into the spring before any of the harried workers could grab it. The effort did have a short-term beneficial effect. Poolfish numbers soared, at least until goldfish numbers also recovered. Despite their best efforts, the "dynamite fish" was not the only goldfish to remain in Manse Spring following the effort to eradicate them.

The battle to save Pahrump poolfish in Manse Spring ultimately was lost in 1975, when nearby groundwater pumping dried the spring, but not before some individuals were removed by officially unnamed "scientists and officials" and transplanted into three pools in different locations in Nevada. Dad always hoped that Manse Spring could be rewatered and restored and the Pahrump poolfish returned to its native habitat. To help inform that objective, Dad and his youngest grandson, Josh Williams, dug out the data that were originally collected back in the 1960s but never published due to the intervention of the Vietnam War into the aspirations of Dad's graduate student. In 2010, the results of their analysis were published (Deacon and Williams 2010), but the aspiration of restoring Manse Spring remains unfulfilled.

Dad's research interests continued to expand throughout the 1960s, 1970s, and 1980s, ultimately including not only significant work with Death Valley and Ash Meadows pupfishes, but also a focus on fishes of the Moapa and Virgin River systems and Pahrump Valley, Great Basin springsnails, Lake Mead limnology, and impacts of increasing water demands on aquatic systems throughout the region. And everywhere he worked, Dad found a need to bring the lessons regarding scientific activism to bear. He was fond of telling his students that as a professional biologist, he was obligated to share not only his scientific knowledge, but also his wisdom.

Dad loved working in the field, and he loved pulling results of his research together to tell coherent "stories" about the world. This undoubtedly helped him live through the 1983 tragedy of the death of his son, in a boating accident on Lake Mead, and the dissolution of his marriage. Unfortunately, further personal difficulties arose to challenge him. By the late 1980s, he no longer was physically able to work independently in the field; a creeping numbness was impairing his ability to walk and posing a serious health risk, as he was unable to feel any injury he received. In fact, the problem originally was identified when he suffered a severe burn on his arm while changing the oil in his truck . . . an injury he didn't notice until well after completing the oil change. Initially, Dad maintained a field presence by hiring UNLV football players to lug him to a creekside chair, where he took field notes for his graduate students.

Ultimately, despite his greatest exercise of stubbornness and optimism, Dad was unable to overcome the debilitating effects of a spinal tumor that eventually resulted in him dealing with the realities of being a partial quadriplegic. This did not quench his optimism or result in the end of his career . . . it simply led to its second chapter. Dad resigned as chair of UNLV's Biology Department and became the founding director and later chair of the university's Department of Environmental Studies, a program he designed to take an interdisciplinary approach to environmental studies, combining policy analysis with hard science to solve today's complex ecological problems (fig. 2d.3). With the support of his second wife, MaryDale, he mentored an entire new crop of students that tackled issues like campus recycling, water conservation at Las Vegas hotels, effects of endocrine disruptors on human health, and yes, fish conservation issues.

Fig. 2d.3 Jim Deacon presiding over the Department of Environmental Studies thesis presentation, April 29, 2009. (Photo courtesy of UNLV.)

During this period, he continued his work on Devils Hole pupfish, helped secure water rights for Zion National Park, and assessed streamflow and temperature requirements for Virgin River fishes. He also added a major new area of interest: assessing the potential impacts of proposals by the Southern Nevada Water Authority (SNWA) to mine the state's groundwater aquifers. In fact, helping the public understand the consequences of the hyper-growth of Las Vegas and its consequences for Nevada's water supplies and water-dependent ecosystems occupied much of his later career. He actively engaged in the public processes associated with the SNWA's proposed "water grab," writing opinion columns in Las Vegas newspapers, attending and testifying at public meetings, and helping to organize public resistance to the various proposals. During this battle, he was quite proud to co-author a major article, published in *BioScience*, with his older grandson Austin Williams, son-in-law Jack Williams, and daughter Cindy Deacon Williams on the consequences of fueling the runaway growth of Las Vegas with statewide groundwater withdrawals (Deacon et al. 2007).

Dad was instrumental in establishing UNLV's first MS and PhD programs in both biology and environmental science. He mentored more than 25 master's and doctoral students and taught "across the (undergraduate and graduate) curriculum" in both biological sciences and environmental studies during his 42-year UNLV career. In 1988, Dad was honored with the title of Distinguished Professor by UNLV. He was a Fellow

Fig. 2d.4 Jim Deacon was the first recipient, in 2012, of the annual E. O. Wilson Award for Outstanding Science in Biodiversity Conservation from the Center for Biological Diversity. The award included an original sculpture of an ant, Wilson's favorite species. (Photo courtesy Center for Biological Diversity.)

of the American Association for the Advancement of Science and received many major awards, including those from the American Fisheries Society, the National Wildlife Federation, and The Nature Conservancy. He was perhaps most proud to receive the first annual E. O. Wilson Award for Outstanding Science in Biodiversity Conservation from the Center for Biological Diversity in 2012 (fig. 2d.4).

Upon his death, Kierán Suckling, executive director of the Center for Biological Diversity, summarized Dad's contribution as follows:

> Dr. Deacon and a few colleagues arrived in the Southwest as young professors in the early 1960s, at a time when the western universities had no fish departments, no ecology departments, no institutionalized conservation at all. The field of "fisheries" also lacked a conservation focus. Jim and his allies worked to build the needed foundations for desert fish conservation from the ground up, with no models to guide them. And because of their intense activist spirit, they injected aggressive conservation values and action right into the institutions, which has inspired the generations of biologists that have followed in their footsteps.

REFERENCES

Brean, H. 2015. Noted UNLV biologist James Deacon has died. *Las Vegas Review Journal.* Posted February 27, 2015, 11: 05 a.m.

Deacon, J. E. 1960. Fish populations of the Neosho and Marais Des Cygnes Rivers, Kansas, following drought. PhD diss., University of Kansas.

Deacon, J. E., and S. Bunnell. 1970. Man and pupfish: A process of destruction. In *Save the Pupfish! A Task Force Report,* 14–21. San Francisco: California Tomorrow.

Deacon, J. E., and M. S. Deacon. 1979. Research on endangered fishes in the National Parks with special emphasis on the Devils Hole pupfish. *Proceedings of the First Conference on Scientific Research in the National Parks,* vol. 1. US National Park Service Transactions and Proceedings, no. 5.

Deacon, J. E., C. Hubbs, and B. J. Zahuranec. 1964. Some effects of introduced fishes on the native fish fauna of southern Nevada. *Copeia* 1964: 384–88.

Deacon, J. E., A. E. Williams, C. Deacon Williams, and J. E. Williams. 2007. Fueling population growth in Las Vegas: How large-scale groundwater withdrawal could burn regional biodiversity. *BioScience* 57: 688–98.

Deacon, J. E., and J. E. Williams. 2010. Retrospective evaluation of the effects of human disturbance and goldfish introduction on endangered Pahrump poolfish. *Western North American Naturalist* 70(4): 425–36.

Minckley, W. L., and J. E. Deacon. 1968. Southwestern fishes and the enigma of "endangered species." *Science,* new series, 159: 1424–32.

Minckley, W. L., and J. E. Deacon, eds. 1991. *Battle against Extinction: Native Fish Management in the American West.* Tucson: University of Arizona Press.

2E CLARK HUBBS

Gary P. Garrett

Clark Hubbs was an inspiration to many, but some of us were also fortunate to have him as a mentor. As the son of Carl and Laura Hubbs, he carried on the family tradition of ichthyology. However, his early encounters with fishes were more influenced by financial reward than by scientific curiosity. Accompanying his parents on field trips in the 1930s, Clark and his siblings were paid five cents per species collected, one dollar for each new species, and five dollars for a new genus. It wasn't long before he developed a deep appreciation for the fishes themselves and the environments they depended upon. He ultimately devoted his life to cataloging and protecting the fish species of Texas. During his 60 years at the University of Texas, Clark sampled more streams and springs in the state, and placed more fish

Fig. 2e.1 1979—Clark Hubbs and graduate students work to protect habitat of the federally endangered Clear Creek gambusia *Gambusia heterochir*. *Left to right, back row:* Dr. Gary Garrett (University of Texas), Dr. Dave Marsh (Angelo State University; deceased), Dr. Bob Edwards (UT-Pan American; retired), Dr. Mike Dean (orthopedic surgeon), Dr. Clark Hubbs (University of Texas); *front row:* Dr. Edie Marsh-Mathews (University of Oklahoma). (Photo by Deb Edwards.)

in the museum collections, than anyone else in Texas ichthyological history. Clark founded the University of Texas fish collection, which evolved into the Texas Natural History Collections and is now part of the University of Texas Biodiversity Collections.

Despite a lengthy battle with cancer, Clark continued collecting fish and associated data until a few weeks before he died in 2008, at the age of 87. In his later years, it became hard for him to get around, but he still insisted on it, so we always found ways to take him on field trips to the Devils River and other favorite spots.

As one of his graduate students, I was most impressed by his single-minded devotion to understanding the biology of fishes of the American Southwest and his deep passion for conservation of natural resources. That impression never left me. His pure joy and excitement for being in the field typically meant that we graduate students would have to trot behind him just to keep up. It was also instructive to watch him talk to private landowners and try to explain why he was so interested in fishes and why one small fish might have some importance. He understood that it was hard for a rancher, who was trying to scrape a living out of the land, to understand the value of a minnow. He would patiently explain to the landowners that when fish that have been there for millions of years start dwindling, something is wrong with the larger system and is ignored at one's peril. Sometimes his efforts bore fruit, and the landowner came to understand, and even became quite concerned. Clark often told us that if we could achieve good conservation in the harsh environments of West Texas and figure out ways to work with reluctant landowners, we could do it anywhere.

Clark was a tireless supporter of Texas natural resources and defender of endangered species and habitats. He was largely responsible for the first fish, and several subsequent lists of fishes, on the Texas Threatened and Endangered Species list. The Texas Parks and

Fig. 2e.2 1991—Clark Hubbs proving he truly could walk on water. (Photo by Dean Hendrickson.)

Fig. 2e.3 2001—Clark Hubbs at Phantom Lake Springs working on conservation of the federally endangered Pecos gambusia *Gambusia nobilis* and Comanche Springs pupfish *Cyprinodon elegans*. (Photo by Gary Garrett.)

Wildlife Department recognized that his knowledge of the fishes and their status was far superior to anyone else's. The standard statement, when anyone asked why a certain species was listed, was "Because Clark said so." He also served on the boards of several organizations, including The Nature Conservancy, Rio Grande Fishes Recovery Team, Hubbs-SeaWorld Research Institute, the University of Texas Marine Sciences Institute, and the southwestern division of the Environmental Defense Fund.

Clark was a key player in getting a small refuge canal constructed at Balmorhea State Park in 1974. The purpose of the canal was to provide needed habitat for the federally endangered Pecos gambusia *Gambusia nobilis* and Comanche Springs pupfish *Cyprinodon elegans*. The natural habitat for these species had been destroyed decades before, and the canal was at least better than the swimming pool and irrigation ditches to which they had been relegated. In the mid-1990s, Clark provided the insight and guidance that was instrumental in developing a naturally functioning ciénega (San Solomon Ciénega) at the park. This improved habitat resulted in a pupfish population increase of at least an order of magnitude. In 2009, park staff created more habitat by building an additional ciénega. This one, the Clark Hubbs Ciénega, was named in his honor.

However, Clark Hubbs's interests were not limited to desert fishes. He published more than 300 articles between 1941 and 2008 on a wide array of fish topics, and he supervised 22 graduate students earning MA degrees, 22 graduate students earning PhD degrees, and 2 postdoctoral fellows. He served as an expert witness in the landmark Edwards Aquifer lawsuit, which not only resulted in the protection of the San Marcos gambusia *Gambusia georgei* and fountain darter *Etheostoma fonticola*, but ensured adequate spring flows for fishes and people. He was a leader in many professional organizations, including the Southwestern Association of Naturalists (president, 1966–1967), Texas Academy of Science (president, 1972–1973; editor, 1957–1961), Desert Fishes Council (chairman—Chihuahuan Desert 1974–1976), Texas Organization for Endangered

Species (president, 1978–1979), American Society of Ichthyologists and Herpetologists (president, 1987; managing editor, 1972–1984), and American Institute of Fisheries Research Biologists (president, 1995–1997). He also edited the *Southwestern Naturalist* and *Transactions of the American Fisheries Society*. He was chair emeritus of the Research Committee at Hubbs-SeaWorld Research Institute (1989 until his death). One example of the admiration and respect his students had for him was the Clark Hubbs Symposium, held in 1993 at the American Society of Ichthyologists and Herpetologists' annual meeting—a first for the society. A T-shirt designed especially for the symposium held a special place for Clark, who had it signed not only by his former students, but also by over 1,000 of his colleagues and professional friends. Yet another example of how he was revered was the establishment of the annual Clark Hubbs Student Research Scholarship by the Texas chapter of the American Fisheries Society in 2010.

Although Clark published on a wide variety of fish-related subjects, some of the contributions that stand out in particular include those that address reproductive strategies (e.g., Hubbs 1959; Hubbs 1964; Hubbs 1985; Hubbs and Schlupp 2008) and the importance of habitat integrity (e.g., Hubbs et al. 1997; Hubbs 2001). However, I believe he felt his most important work dealt with problems (and some solutions) concerning rare and endangered fishes (e.g., Hubbs 1959; Hubbs and Echelle 1973; Hubbs et al. 1977; Hubbs 1980; Hubbs 2003). These and many other discoveries and insights by Clark provided a foundation for many of the conservation efforts we see today.

One of Clark's most enduring contributions—detailing the distribution and status of Texas freshwater fishes—began in the 1950s (Jurgens and Hubbs 1953; Hubbs 1957a; Hubbs 1957b), and was periodically updated through the 1980s. In 1991, Bob Edwards and I joined Clark in evolving the product into "An Annotated Checklist of the Freshwater Fishes of Texas, with Keys to Identification of Species" (Hubbs et al. 1991). There were further improvements when Dean Hendrickson joined the effort in 2006 and we began to focus on compiling, standardizing, and georeferencing all museum specimen–based data on the state's fish fauna (dating back to 1851) and creating a web interface for it. This effort became the Fishes of Texas Project (http://www.fishesoftexas.org) and is a lasting legacy of the father of conservation and biology of Texas fishes.

REFERENCES

Hubbs, C. 1957a. *A Checklist of Texas Fresh-Water Fishes*. Texas Game and Fish Commission, IF Series, no. 3: 1–11.

Hubbs, C. 1957b. Distributional patterns of Texas fresh-water fishes. *Southwestern Naturalist* 2: 89–104.

Hubbs, C. 1959. Population analysis of a hybrid swarm between *Gambusia affinis* and *G. heterochir*. *Evolution* 13: 236–46.

Hubbs, C. 1964. Interactions between a bisexual fish species and its gynogenetic sexual parasite. *Bulletin Texas Memorial Museum* 8: 1–72.

Hubbs, C. 1980. The solution to the *Cyprinodon bovinus* problem: Eradication of a pupfish genome. *Proceedings of Desert Fishes Council* 10: 9–18.

Hubbs, C. 1985. Darter reproductive seasons. *Copeia* 1985: 56–68.

Hubbs, C. 2001. Environmental correlates to the abundance of spring-adapted versus stream-adapted fishes. *Texas Journal of Science* 53: 299–326.

Hubbs, C. 2003. Spring-endemic *Gambusia* of the Chihuahuan Desert. In G. P. Garrett and N. L. Allan, eds., *Aquatic Fauna of the Northern Chihuahuan Desert*, 127–33. Museum of Texas Tech University, Special Publication no. 46.

Hubbs, C., and A. A. Echelle. 1973. Endangered non-game fishes of the upper Rio Grande basin. In W. C. Huey, ed., *Symposium on Rare and Endangered Wildlife of the Southwestern United States*, 147–67. Santa Fe: New Mexico Department of Game and Fish.

Hubbs, C., R. J. Edwards, and G. P. Garrett. 1991. An annotated checklist of the freshwater fishes of Texas, with keys to identification of species. *Texas Journal of Science* 43: Supplement.

Hubbs, C., E. Marsh-Matthews, W. J. Matthews, and A. A. Anderson. 1997. Changes in fish assemblages in East Texas streams from 1953 to 1986. *Texas Journal of Science* 49: 67–84.

Hubbs, C., R. R. Miller, R. J. Edwards, K. W. Thompson, E. Marsh, G. P. Garrett, G. L. Powell, D. J. Morris, and R. W. Zerr. 1977. Fishes inhabiting the Rio Grande, Texas-Mexico, between El Paso and the Pecos confluence. In R. R. Johnson and D. Jones, eds., *Importance, Preservation and Management of Riparian Habitat, A Symposium*, 91–97. USDA Forest Service General Technical Report RM-43.

Hubbs, C., and I. Schlupp. 2008. Juvenile survival in a unisexual/sexual complex of mollies. *Environmental Biology of Fishes* 83: 327–30.

Jurgens, K. C., and C. Hubbs. 1953. A checklist of Texas freshwater fishes. *Texas Game and Fish* 11: 12–15.

2F ROBERT J. BEHNKE

Kevin R. Bestgen and Kurt D. Fausch

The life story of Robert J. Behnke is one marked with improbable events and unlikely circumstances, which ultimately produced a deep and long-lasting legacy of trout and coldwater habitat conservation. A young lad from a working-class family in a company town, Bob Behnke did not envision that one could actually earn a living studying the fish he loved to catch. In one of his early experiences, he described his first angled salmonid, a stocked brook trout *Salvelinus fontinalis* from a stream near his home, as one of the most beautiful things he had ever seen. Bob was born in Stamford, Connecticut, on December 30, 1929, into a family of modest means, who sometimes dined on young Robert's catch, according to postings in his personal diary, "Piscatorial" (fig. 2f.1).

Bob was a gifted student in high school, earning awards in subjects such as history, which foretold the keen mind and encyclopedic memory that was retained until the end of his life, on September 13, 2013. Indeed, all of his students and colleagues marveled at his ability to survey large stacks of literature in his office and reach into one to remove, on the first try, a timeless article that he handed us to read or assigned to his class. In addition, even though he might not have read it for some years, he would recall its contents better than anyone else would.

After graduating from high school, Behnke worked at the Yale and Towne lock company, but was drafted into the US Army shortly thereafter. During his service in the Korean War as a communications specialist, a remarkable photograph was taken that summed up the depth of his interest in fish and fishing (fig. 2f.2). In pursuit of trout in northern Japan during leave, he recalled catching native trout (whitespotted charr *Salvelinus leucomaenis*) that had a strong morphological similarity to North American brook trout, and he probably wondered what the relationships between the two species were, given the distance separating them. After completing his service, his Veterans Administration counselor informed a surprised Bob Behnke that one could study fish for a living, so he enrolled immediately at the University of Connecticut. His wife, Sally, relates that being drafted was a major turning point in his life, making the improbable a reality, because without that service and the associated GI Bill, Bob Behnke never would have sought higher education.

Nearing the end of his education at the university

Fig. 2f.1 Robert J. Behnke (*seated, center*) with high school fishing buddies. Above, his personal diary, a three-ring binder of notes and writings called "Piscatorial," where details of many fishing adventures and observations on the natural history of fish and aquatic environments were recorded.

(Behnke and Wetzel 1960), Bob corresponded with Dr. Paul Needham at the University of California, Berkley, who invited him to spend the summer catching, with a fly rod, western North American trouts for systematic studies. He accepted, and with classmate Phil Pister close at hand, the California connection for great early ichthyologists, described in other sections of this chapter, was strengthened. Behnke's trout research began with his master's thesis (Behnke 1960), which was later expanded to studies of the family Salmonidae for his PhD dissertation (Behnke 1965). That work was summarized in one of his seminal publications (Behnke 1972), which features no tables or figures. It was an extraordinary contribution because it explored in detail the systematics of the diverse and widely distributed salmonid subfamilies Salmoninae and Coregoninae of the Northern Hemisphere. It was also remarkable because Behnke made an eloquent and early appeal for the recognition and conservation of intraspecific genetic diversity (Echelle 1991) and its use in fish management programs, even before the taxon-based Endangered Species Act was passed in 1973.

After completing his PhD, Behnke spent 10 months as an American Academy of Sciences exchange scholar in the Soviet Union to continue his systematic work on salmonids, and he maintained connections with Russian colleagues throughout his life. In 1966, he was a temporary instructor of ichthyology at the University of California, Berkeley, after which he landed a position as the assistant unit leader of the US Fish and Wildlife Service, Colorado Cooperative Fishery Research Unit, and a faculty member at Colorado State University (CSU). He continued to teach courses at CSU in ichthyology, ecological zoogeography, and conservation biology (the first of its kind at CSU) until 1999 and attained the rank of professor, even though he was not granted tenure (fig. 2f.3). During his career, he traveled extensively to study salmonids and other fishes in western North America as well as in distant lands, including Iran, Japan, Mongolia, Siberia, and several European countries.

Throughout his career, Bob Behnke was never one to seek large grants to fund his research. Instead, he maintained his scholarly pursuits with lesser support from other sources: service work involving salmonid identification for agencies across the Southwest, consulting, and English translation services for the Russian *Journal of Ichthyology*. Sally, Bob's lifelong companion and a nurse by training, would attest to that as an unlikely means to make a living, and stated, "He pretty much lived the life he wanted." Perhaps it is because of limited connections to grants and agencies that Bob Behnke always felt he could, and often did, speak his mind on issues important to conservation.

Bob Behnke was passionate about salmonid systematics and conservation, to which he made major contributions in two broad areas (Behnke et al. 1962;

Fig. 2f.2 Robert J. Behnke in northern Japan, during his service in the Korean War, toting rifle and fly rod.

Fig. 2f.3 Robert J. Behnke, later in life (ca. 75 years old), with Pancho the pet donkey.

Behnke 1968, 1972, 1986a, b, 1988, 1989, 2007). The first was sorting out the messy and confused taxonomy and systematics of western North American trouts, especially cutthroat trout *Oncorhynchus clarkii* (Behnke 1992, 2002). Because of their popularity as sport fish, salmonids were some of the first fishes given the attention of early ichthyologists such as David Starr Jordan (Jordan 1891; Jordan and Evermann 1896). After those early efforts, however, they received little scholarly attention until Bob Behnke began his work. He discovered a profusion of named "taxa" that were erroneously based on slight morphological differences, errors in localities and specimen descriptions made during early expeditions, and a paucity of early specimens available for study. The situation was confounded further by early, repeated, and often unrecorded stockings of hatchery-reared cutthroat trout of various lineages, often into waters with existing native populations, which rendered understanding the historical distribution of native cutthroat trout lineages difficult.

It is not an overstatement to say that Bob Behnke defined the taxonomy and phylogeny of cutthroat trout (Behnke 1988; 1992). He was the first to use a drainage basin and associated lineage approach, while often admitting broad morphological overlap of taxa such as Colorado River and greenback cutthroat trout subspecies (*Oncorhynchus clarkii pleuriticus* and *O. c. stomias*, respectively), and his classification remains valid today. He also made important contributions by rediscovering and helping to save presumably extinct subspecies such as the greenback cutthroat trout and Great Basin Pyramid Lake Lahontan cutthroat trout (*O. c. henshawi*), which still persist on the landscape, present taxonomic and systematic issues notwithstanding (Hickman and Behnke 1979; Metcalf et al. 2012; Bestgen et al. 2013; Bestgen et al. 2019). One of us (KDF) observed that without Bob Behnke's insightful work, we probably would have lost much of the diversity of trout throughout the West because nobody would have known why it was important. So, in his unassuming manner, Bob Behnke raised the profile of trouts in arid-land streams of western North America, and many ongoing and future conservation efforts can be traced back to his work, through which agencies and individuals knew what to save (Penaluna et al. 2016; Bestgen et al. 2019).

A second important contribution of Bob Behnke was to convey knowledge and awareness of trout conservation not only to scientists and fisheries managers, but to the public, which served to unite these powerful groups in conserving lineages of native trout and salmon (Li et al. 2014). In his regular series, "About Trout," which appeared for over 25 years in *Trout* magazine, Behnke described the unique life histories and ecology of various salmonids throughout the world. In each essay, his last thoughts were typically about how the life history diversity of salmonids could be used to great advantage by fisheries managers to achieve their end goal of providing recreational opportunities to the public in diverse habitats, a strategy he advocated for over 35 years (Behnke 1972). Even though a magazine aimed at anglers was an unlikely forum for work by an academic scientist, it was highly effective for furthering a conservation message that was broad and powerful. Bob once wryly reminded us that it was unlikely that 35,000 people, then the readership of *Trout*, would ever read any scientific journal article yet written.

As it was for many academics and agency professionals, the annual meeting of the Desert Fishes Council was one of Bob's favorite events. This was not only because of the many friends and colleagues that gathered there, but also because of the unified focus on the conservation of unique aquatic animals and their habitats in the unlikely region where he often worked, the desert Southwest of the United States and México. We would often travel to DFC meetings in vans and listen to Bob's descriptions of fish assemblages and their zoogeographic history as we traversed arid-land drainages on our way from Fort Collins, Colorado, to Death Valley, California. We would classify Dr. Behnke's responses to our questions in miles, where a typical answer would last 30 miles, but one that involved trout, or a particularly colorful personality, would often engender a "50-miler," which made the trip faster and enormously fun. Once we reached the meeting, Behnke would be there in the front row, with Carl, Robert Rush, Deacon, Pister, and Salvador, and Clark a few rows back, with students and professionals alike often secretly hoping the inevitable hand would not rise after their talk. We owe much of the historical success of the Desert Fishes Council to these pillars of wisdom, whose work, along with the on-the-ground efforts of many agency biologists, stabilized the imperiled faunas that live on today. We should be grateful to have Bob Behnke and the others to look back on and look up to as we continue to draw wisdom and inspiration from the writings they left us. We honor their contributions and legacy in our continuing battle to conserve biota and aquatic habitat in an unlikely setting, the desert Southwest.

REFERENCES

Behnke, R. J. 1960. Taxonomy on the cutthroat trout of the Great Basin with notes on the rainbow series. MS thesis, University of California, Berkeley.

Behnke, R. J. 1965. A systematic study of the family Salmonidae with special reference to the genus *Salmo*. PhD diss., University of California, Berkeley.

Behnke, R. J. 1968. A new subgenus and species of trout, *Salmo (Platysalmo) platycephalus*, from south-central Turkey, with comments on the classification of the subfamily Salmoninae. *Mitteilungen aus dem Hamburgischen Zoologischen Museum und Institut* 66: 1–15.

Behnke, R. J. 1972. The systematics of salmonid fishes of recently glaciated lakes. *Journal of the Fisheries Research Board of Canada* 29: 639–71.

Behnke, R. J. 1986a. Pyramid Lake and its cutthroat trout. *American Fly Fisher* 13: 18–22.

Behnke, R. J. 1986b. Redband trout. *Trout* 27: 34–39.

Behnke, R. J. 1988. Catch-and-release: The last word. In R. A. Barnhart and T. D. Roelofs, eds., *Catch-and-Release Fishing—A Decade of Experience*, 291–99. California Cooperative Fishery Research Unit, Humboldt State University.

Behnke, R. J. 1989. We're putting them back alive. *Trout* 29: 34–39.

Behnke, R. J. 1992. *Native Trout of Western North America*. Monograph 6. Bethesda, MD: American Fisheries Society.

Behnke, R. J. 2002. *Trout and Salmon of North America*. New York: Free Press.

Behnke, R. J. 2007. *About Trout: The Best of Robert J. Behnke from Trout Magazine*. Guilford, CT: Lyons Press.

Behnke, R. J., T. P. Koh, and P. R. Needham. 1962. Status of the landlocked salmonid fishes of Formosa with a review of *Oncorhynchus masou* (Brevoort). *Copeia* 1962: 400–407.

Behnke, R. J., and R. M. Wetzel. 1960. A preliminary list of the fishes found in the fresh waters of Connecticut. *Copeia* 1960: 141–43.

Bestgen, K. R., K. B. Rogers, and R. Granger. 2013. *Phenotype Predicts Genotype for Lineages of Native Cutthroat Trout in the Southern Rocky Mountains*. Colorado State University Larval Fish Laboratory Contribution 177. Denver: US Fish and Wildlife Service.

Bestgen, K. R., K. B. Rogers, and R. Granger. 2019. Distinct phenotypes of native cutthroat trout emerge under a molecular model of lineage distributions. *Transactions of the American Fisheries Society* 148: 442–63.

Echelle, A. A. 1991. Conservation genetics and genetic diversity in freshwater fishes of western North America. In W. L. Minckley and J. E. Deacon, eds., *Battle against Extinction: Native Fish Management in the American Southwest*, 141–53. Tucson: University of Arizona Press.

Hickman, T. J., and R. J. Behnke. 1979. Probable discovery of the original Pyramid Lake cutthroat trout. *Progressive Fish-Culturist* 41: 135–37.

Jordan, D. S. 1891. Report of explorations in Colorado and Utah during the summer of 1889, with an account of the fishes found in each of the river basins examined. *Bulletin of the US Fish Commission* 9: 1–40.

Jordan, D. S., and B. W. Evermann. 1896. *The Fishes of North and Middle America*. US National Museum Bulletin 47, part 1.

Li, H., C. B. Schreck, K. D. Fausch, and K. R. Bestgen. 2014. In Memoriam, Robert J. Behnke, Ph.D., 1929–2013. *Fisheries* 39:128–29.

Metcalf, J. L., S. L. Stowell, C. M. Kennedy, K. B. Rogers, D. McDonald, J. Epp, K. Keepers, A. Cooper, J. J. Austin, and A. P. Martin. 2012. Historical stocking data and 19th century DNA reveal human-induced changes to native diversity and distribution of cutthroat trout. *Molecular Ecology* 21: 5194–5207.

Penaluna, B. E., A. Abadía-Cardoso, J. B. Dunham, F. J. García de León, R. E. Gresswell, A. Ruiz Luna, E. B. Taylor, B. B. Shepard, R. Al-Chokhachy, C. C. Muhlfeld, K. R. Bestgen, K. Rogers, M. A. Escalante, E. R. Keeley, G. Temple, J. E. Williams, K. Matthews, R. Pierce, R. L. Mayden, R. P. Kovach, J. C. Garza, and K. D. Fausch. 2016. Conservation of native Pacific trout diversity in western North America. *Fisheries* 41: 286–300.

2G EDWIN P. (PHIL) PISTER

Kathryn Boyer

This sequel to *Battle against Extinction*, entitled *Standing between Life and Extinction*, would be incomplete without formal acknowledgment of the accomplishments, contributions, dedication, and fortitude of Edwin Philip Pister, or Phil, as most friends and colleagues know him. Phil's focus on the conservation of freshwater ecosystems and their native biota spans over 60 years, first as a student of A. Starker Leopold at the University of California, Berkeley, and then as a professional fish biologist for the California Department of Fish and Wildlife (CDFW; then called the California Department of Fish and Game, CDFG). Phil began his career with CDFW in 1953 and continued Leopold's efforts to educate our profession on the conservation ethic throughout his career. Early in his tenure at CDFG, the conventional fish and wildlife management philosophy of "put and take" fisheries and the dearth of initiatives to protect native faunas challenged Phil. Inspired by Leopold, Phil was undaunted in pursuing an aquatic management ethic that focused on conservation of biodiversity. In his career, Phil conducted research on the diverse aquatic ecosystems of the Sierra Nevada and Owens Valley of California and executed pivotal restoration measures that literally prevented the extinction of Owens pupfish *Cyprinodon radiosus* and conserved California golden trout *Oncorhynchus mykiss aguabonita* (Pister 2010; Bonham and Pister 2014). Through relationships with Robert Rush Miller, Carl L. Hubbs, and Starker Leopold, Phil developed a profound conservation ethic that influenced the federal protection of Devils Hole pupfish *Cyprinodon diabolis* under the Endangered Species Act. Later in his career, Phil became a respected supervisor for the CDFG, a position that provided opportunities to mentor and inspire a sustained succession of young scientists and managers that filled leadership positions in many branches of the federal government, state wildlife agencies, public and private universities, and conservation organizations. Throughout his career, Phil has worked to bridge the institutional gap between academic biologists and fish and wildlife agency personnel, encouraging their close coordination and collaboration on policies and actions to protect imperiled aquatic species in California and the arid West. His central role in founding and sustaining the Desert Fishes Council (DFC) ensured a venue for the reporting of rigorous and transparent research and conservation actions by academics, agency biologists, fish managers, and private citizen conservationists (Pister 1991). Phil mentored members of all ages in the art of conservation ethics and results-oriented actions. His keen attention to detail and his resolve have made him a sought-after speaker and instructor on aquatic conservation science for three decades. Phil's most valuable gift as a mentor may be his candid and intrepid nature, which inspires conservation biologists and concerned citizens alike. He has been honored by myriad professional and nongovernmental organizations, including the American Society of Ichthyologists and Herpetologists, the American Fisheries Society, and the Society for Conservation Biology for his tenacity and forthright efforts to protect and celebrate threatened aquatic species and their waning habitats.

Fig. 2g.1 Phil Pister at Fish Slough, near Bishop, California. (Photo credit K. Milliron.)

Before and after retirement, Phil also taught conservation science at the university and graduate level and authored numerous scientific and popular publications about imperiled desert fishes and their deteriorating habitats. His greatest accomplishments and conservation legacies have been shared with an illustrious list of desert fish scientists, including W. L. Minckley, James Deacon, Carl L. Hubbs, and Robert Rush Miller—all of who participated in the founding, development, and growth of the Desert Fishes Council.

Phil's status as a key figure in the conservation history of California and the western United States and México is well known to those of us who worked with him through the Desert Fishes Council, and he will continue to be recognized by future generations. Phil is included among luminaries such as Ansel Adams, David Brower, and Wallace Stegner as the subject of a full biographical oral history by the Bancroft Library (Pister 2009). A total of 52 individuals and 8 institutions made donations to support the project, which provides an invaluable archive of Phil's wisdom, knowledge, and countless contributions to the conservation of desert aquatic ecosystems and the native species that are dependent on them. This important archive includes 510 pages of transcripts of the extensive and meticulous interviews of Phil done in 2007–2008, as well as 13 videos totaling over 1,000 minutes, and is available online (http://bancroft.berkeley.edu/ROHO/collections/subjectarea/natres/parks_envir.html). It encapsulates not only Phil's lifelong work in conservation, but also the history and accomplishments of the Desert Fishes Council, which he has served as executive secretary for 50 years. On behalf of the DFC, let this chapter pay tribute to our friend, confidante, warrior, and gifted spokesperson. Phil's accomplishments and role as ambassador for conservation in general, and native aquatic fauna of the North American West in particular, continues to inspire his colleagues and future generations of scientists.

REFERENCES

Bonham, C. H., and E. P. Pister. 2014. Introduction to volume 100: The special fisheries issue. *California Fish and Game* 100: 587.

Pister, E. P. 1991. The Desert Fishes Council: Catalyst for change. In W. L. Minckley and J. E. Deacon, eds., *Battle against Extinction: Native Fish Management in the American West*, 55–68. Tucson: University of Arizona Press.

Pister, E. P. 2007–2008. Preserving native fishes and their ecosystems: A pilgrim's progress, 1950s–present. An oral history conducted by Ann Lage. Regional Oral History Office, Bancroft Library, University of California, Berkeley. http://bancroft.berkeley.edu/ROHO/collections/subjectarea/natres/parks_envir.html.

Pister, E. P. 2010. Just a few more yards. In K. D. Moore and M. P. Nelson, eds., *Moral Ground: Ethical Action for a Planet in Peril*, 220–24. San Antonio: Trinity University Press.

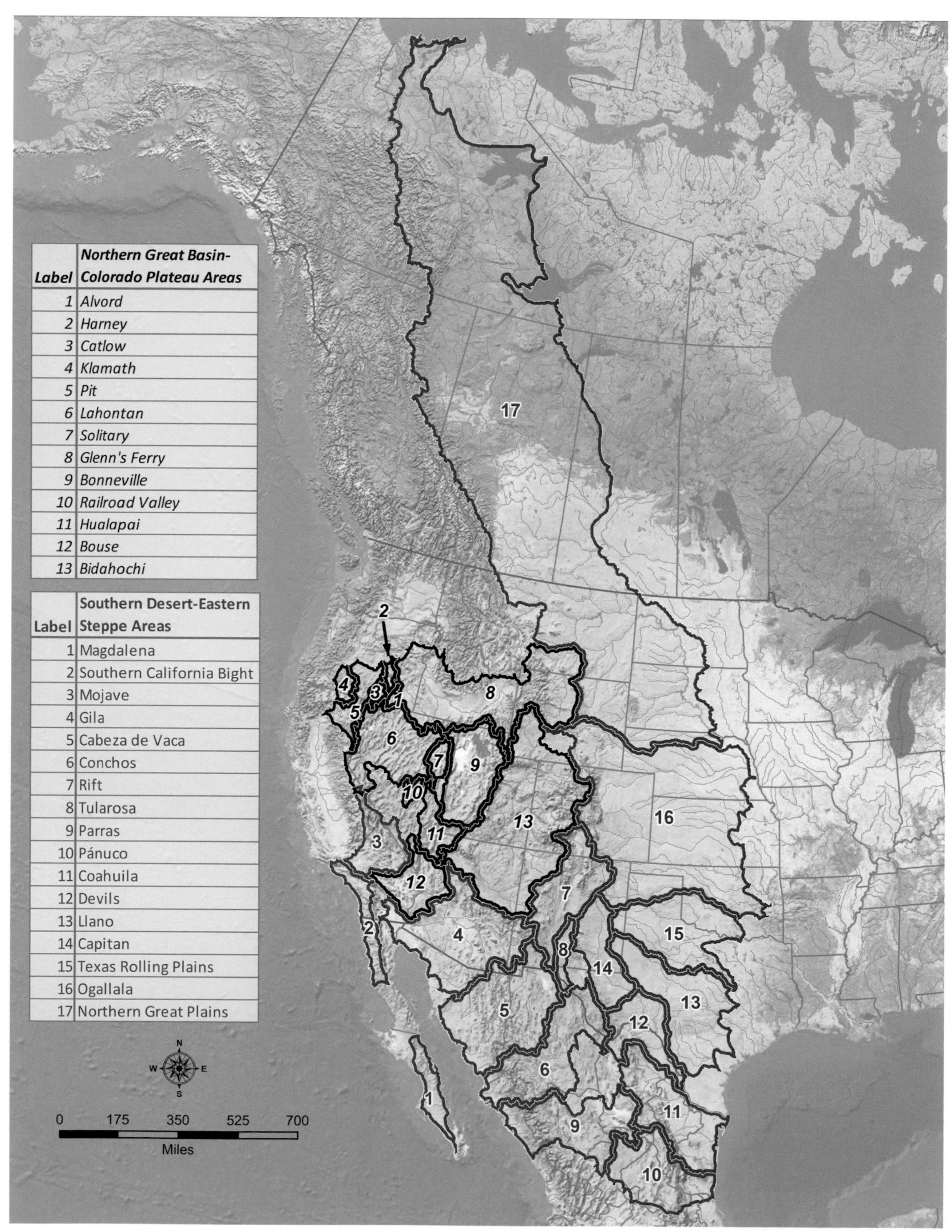

Fig. 3.1 Shaded relief map of western North America with areas of endemism for desert fishes outlined (see table 3.1). Fishless basins are excluded from these areas.

3

Christopher W. Hoagstrom,
Derek D. Houston,
and Norman Mercado-Silva

Biodiversity, Biogeography, and Conservation of North American Desert Fishes

In North America, the concept of a "desert fish" arose in the twentieth century with the recognition of distinct arid-land taxa (Hubbs 1940) and the emergence of desert fish biology (Deacon and Minckley 1974). Prominent ichthyologists began to promote awareness of imperiled desert fishes (Miller 1946, 1961) and advance ecological understanding (Minckley and Deacon 1968; Contreras-Balderas 1969). The Desert Fishes Council became a clearinghouse for information, strengthening conservation efforts (Pister 1990, 1991).

However, although increasing taxonomic precision has greatly improved documentation of desert fish biodiversity, there is no comprehensive list of desert fishes. Further, no study has assessed their distributional patterns throughout North America. Thus, this chapter is a comprehensive inventory of North American desert fishes. It establishes a benchmark of biodiversity within a biogeographic context that putatively explains the biodiversity of desert fishes, with implications for their conservation.

NORTH AMERICAN DESERT

By definition, a *desert* has an aridity index of less than 0.65 (aridity index = precipitation/potential evapotranspiration). This includes arid, semiarid, and dry-subhumid regions (Thomas 2011). Diversity among deserts precludes a detailed, universal definition, but all deserts share (1) little and sporadic precipitation, (2) low biomass and little vegetation structure, and (3) high abundance of bare ground (Parsons and Abrahams 2009). Such areas occupy a vast, contiguous expanse of western North America (Thomas 2011) (fig. 3.1).

ENDEMISM

Endemic lineages represent distinct populations restricted to a "specific place" (Parenti and Ebach 2009). Thus, a desert-endemic lineage is a distinct population restricted to the desert region. High regional aridity tends to increase the potential for geographic isolation among aquatic habitats, promoting geographic diversification (Smith et al. 2010). Upon isolation, populations confined to deserts must adapt to cope with the environmental harshness (Deacon and Minckley 1974).

In North America, desert-endemic fishes are widespread and occur at a range of spatial scales from individual springs to major river basins, making a simple definition of an endemic lineage impractical. Criteria used here are (1) geographic distribution centered within some portion of the desert region and (2) an association with desertlike habitats. In exceptional cases, where desertlike habitats occur beyond the desert boundary, the distribution of a desert endemic may extend outside the desert region sensu stricto. For example, the distribution of *Platygobio gracilis* centers on the semiarid Great Plains (Page and Burr 2011), but fringes of its distribution extend into sediment-rich and turbid desertlike habitats of the lower Mackenzie River (McPhail and Lindsey 1970; Culp et al. 2005), the Assiniboine River and Lake Winnipeg (Stewart and Watkinson 2004), and the lower Mississippi River (Robison and Buchanan 1988). In contrast, lineages with distributions centered outside the desert region are not desert endemics, even if their native range extends into the desert. For example, of three lineages within *Richardsonius balteatus*, one (Bonneville-Snake) is endemic to the desert region, whereas the others have distributions extending to the fringe of the desert region, but centered elsewhere (Houston et al. 2014).

Endemic lineages do not necessarily represent a particular taxonomic level (i.e., species or subspecies), but include all lineages with evidence of distinct evolutionary histories in the desert. Although some readers might disagree with the recognition of certain lineages, and it is acknowledged that systematic hypotheses are dynamic, meaning that some lineages may never gain formal recognition, it is also probable many will, and that even more unrecognized lineages are yet to be discovered. Fish names follow Page et al. (2013) or literature providing more detailed information on lineage diversity (see references in tables 3.2–3.4). Nomenclature is also debatable and a work in progress. To ensure clarity, names presented in the cited references, those used to document lineages, are followed here.

AREAS OF ENDEMISM

Delimiting "areas of endemism"—discrete regions with unique faunas—is a basic step in comparative biogeography (Parenti and Ebach 2009). In the desert, modern drainage boundaries often separate distinct fish assemblages (Hubbs and Miller 1948; Smith et al. 2002). However, this is not always true, because some drainages that are presently disconnected have a history of interconnection, which allowed episodic, trans-divide dispersal in the past (e.g., Schönhuth et al. 2011). In other cases, modern drainages have a history of prehistoric subdivision that remains evident among many, but not necessarily all, endemics. In these cases, such as in the Colorado River (Spencer et al. 2008) and Río Bravo–Rio Grande (Smith and Miller 1986), subdrainages separated for millions of years in the Miocene or Pliocene have since become integrated, allowing some endemic lineages to expand their distributions (see below) while others remain confined near their apparent areas of origin.

Thus, to define areas of endemism, phylogenetic and geologic evidence was synthesized to determine faunal boundaries throughout the desert region. Although the standard definition of an area of endemism requires the presence of multiple endemic lineages with similar range limitations (Parenti and Ebach 2009), this criterion fails for desert fishes because some distinct biogeographic areas harbor just one fish lineage (e.g., Miller and Echelle 1975; Houston et al. 2015). Despite this technicality, these one-fish-species areas represent separate areas of endemism with independent phylogenetic and hydrographic histories.

Sørensen's coefficient of similarity (S) was used to assess faunal similarity among putative areas of endemism. This asymmetric, binary coefficient computes faunal similarity between a pair of areas based on how many of the total endemic lineages found between both areas are either shared (double presence) or present in just one or the other area (Legendre and Legendre 1998). Lower similarity values indicate higher assemblage distinctiveness (S ranges from 0 to 1, no similarity to total similarity). For each area of endemism, mean similarity was computed for pairwise comparisons with all other areas. The statistical distribution of these pairwise similarities was then calculated for each area of endemism as an indication of overall faunal distinctiveness among areas.

AREA-OF-ENDEMISM BOUNDARIES VERSUS ENDEMIC LINEAGE DISTRIBUTIONS

Although assemblages of desert fishes define areas of endemism, distributions of individual lineages commonly differ somewhat. A "specific place" wherein a lineage is restricted can have a volume of less than 30 cubic meters, as for *Cyprinodon diabolis* in Devils Hole, Nevada (Andersen and Deacon 2001), or can extend over about 24° latitude in rivers, as for *Platygobio gracilis* of the Great Plains (Page and Burr 2011). These distributions differ because the habitat of *C. diabolis* is isolated (Echelle 2008), whereas the habitat of *P. gracilis* is not (McPhail and Lindsey 1970; Nelson and Paetz 1992). Hence, an area of endemism can include multiple water bodies, each with distinct populations of endemics. Although such areas could, theoretically, be subdivided into smaller sub-areas, endemics of different water bodies within an area are often members of the same regional clade that has a history of intra-area diversification, as in *Cyprinodon*, Mojave Area (Echelle 2008) and *Xiphophorus*, Pánuco Area (Kang et al. 2013). These patterns of intra-area relatedness indicate evolutionary unity among sub-areas within areas of endemism.

At the other extreme, not all endemic lineages are restricted to a single area of endemism because area boundaries are not necessarily impassable to all fishes at all times. In the field of comparative biogeography, lineages occupying multiple areas of endemism are termed "widespread" (Parenti and Ebach 2009). Widespread lineages are less useful for defining area boundaries. However, if restricted to the desert region or its characteristic habitats (as defined above), they still represent desert endemics and provide insight into inter-area connectivity.

Widespread lineages of desert-endemic fishes most commonly inhabit river corridors where they have gained access to multiple areas of endemism during development of modern river basins, such as when the upper Colorado River began to flow through Grand Canyon, allowing endemic fishes of the Bidahochi Area (e.g., *Ptychocheilus lucius* and *Xyrauchen texanus*) to disperse downstream into the Hualapai, Bouse, and Gila Areas (Spencer et al. 2008). Other widespread lineages have exploited headwater connectivity (e.g., *Campostoma ornatum*; Schönhuth et al. 2011) or have been transferred via river captures (e.g., *Catostomus ardens*; Mock et al. 2006). Thus, although areas of endemism are not always 100% faunally distinct, geomorphic evidence for hydrographic connectivity typically explains cases of inter-area lineage sharing (Hubbs and Miller 1948; Smith et al. 2002).

CENSUS RESULTS

In the North American desert, there is evidence for 338 endemic fish lineages distributed among 30 areas of endemism (table 3.1). The number of lineages per area ranges from 43 (Coahuila) to 1 (Tularosa, Solitary, Alvord; mean = 14 ± 11.0 SD). Faunal composition is unique in all areas (mean $S < 0.05$) (fig. 3.2*A*) and the range of pairwise similarities is very low for every area (fig. 3.2*B*). Across the region, there is very little lineage sharing among areas of endemism.

Overall, only 62 of the endemic lineages (18%) inhabit more than one area (fig. 3.3). Only four areas

Table 3.1 Areas of fish endemism in the North American desert. Area abbreviations correspond to tables 3.3 and 3.4 and figure 3.2.

Area of endemism	Description
Southern Desert–Eastern Steppe subdivision	Tributaries to Gulf of Mexico, Gulf of California, Southern California Bight, Hudson Bay, and Arctic Ocean
Magdalena (Mag)	Sierra de La Giganta and Magdalena ecoregions, Baja California peninsula (Ruiz-Campos et al. 2002)
Southern California Bight (SCB)	Tributaries to Southern California Bight: Ríos Agua Escondida, El Rosario, Huatamote, Santa Clara, San Gabriel, Santa Ana, Santa Margarita, San Isidro, San Rafael, Santo Domingo, San Luis Rey, San Diego, and Tijuana (Moyle 2002; Abadía-Cardoso et al. 2016)

Area of endemism	Description
Mojave (Moj)	Southwestern Great Basin including Owens-Amargosa-Mojave River system (Mono Lake shared with Lahontan Area) (Reheis et al. 2002a; Knott et al. 2008)
Gila (Gil)	Gila River basin with Ríos Sonoyta and Concepción (Echelle 2008; Hedrick and Hurt 2012) and Colorado River delta (Minckley and Marsh 2009)
Cabeza de Vaca (CdV)	Endorheic basin of Pluvial Lake Cabeza de Vaca (Strain 1971; Mack et al. 2006) with Mimbres River and Ríos Casas Grandes, San Pedro, Santa María, and Santa Clara, Lagunas Babícora, de Guzmán, de Santa María, El Barreal, and Ojo del Diablo; includes Ríos Mátape, Mayo, Sonora, and Yaqui, Pacific Slope (Schönhuth et al. 2014; Abadía-Cardoso et al. 2015)
Conchos (Con)	Río Conchos basin with Alamito and Terlingua Creeks (Hoagstrom et al. 2014); includes Ríos Fuerte, Mocorito, and Sinaloa, Pacific Slope (Schönhuth et al. 2014; Abadía-Cardoso et al. 2015)
Rift (Rif)	Rio Grande basin portion of Rio Grande Rift, San Juan Mountains to Mesilla Bolsón (Mack and Giles 2004) with Pecos and South Canadian headwaters in Sangre de Cristo Mountains (Rogers et al. 2014; Galindo et al. 2016)
Tularosa (T)	Endorheic Hueco Bolsón–Tularosa Basin (Sheng et al. 2001; Mack et al. 2006)
Parras (Par)	Parras Basin with endorheic Ríos Aguanaval and Nazas and through-flowing Ríos Culiacán, Piaxtla, San Lorenzo, and Tunal (Mezquital), Pacific Slope (Schönhuth et al. 2014; Abadía-Cardoso et al. 2015)
Pánuco (Pán)	Desert (i.e., northwestern) portion of Río Pánuco basin with La Media Luna and Ríos El Salto, Gallinas, Santa Maria, Tamesí, and Verde
Coahuila (Coa)	Downstream-most Río Bravo–Rio Grande (Hoagstrom et al. 2014) with Ríos Salado and San Juan, Bolsón de Cuatro Ciénegas, El Potosí, and Sandia; also included are separate Ríos San Fernando and Soto la Marina
Devils (Dev)	Edwards Plateau (i.e., Devils River Uplift) section of Río Bravo–Rio Grande basin with lower Pecos River and Independence Creek, Devils River, and smaller tributaries from Chisos Mountains (i.e., Mariscal Canyon) to Río Salado confluence (i.e., Maravillas, Pinto, San Felipe, Sycamore, and Tornillo Creeks, and Ríos de Nava–la Compuerta, Escondido, La Pinta Piedra, San Diego, and San Rodrigo) (Hoagstrom et al. 2014)
Llano (Lla)	Edwards Plateau (i.e., Llano Uplift) outside of Río Bravo-Rio Grande basin, with upper Colorado, Nueces, San Antonio, and Guadalupe Rivers and Brazos River tributaries extending west onto the plateau; includes Balcones Fault Zone (Hoagstrom et al. 2014)
Capitan (Cap)	Middle Pecos River basin in Roswell and Delaware basins (i.e., Capitan Basins) (Hoagstrom et al. 2014; Osborne et al. 2016), with Ríos Hondo, Felix, and Peñasco, Bitter, Salt, and Tularosa Creeks, Black River and Three Rivers, Comanche, Diamond Y, Leon, Phantom, Sandia, and San Solomon Springs
Texas Rolling Plains (TRP)	Upper Brazos and Red River basins (Madole et al. 1991) draining southwestern Staked Plain (i.e., Llano Estacado) (Caran and Baumgardner 1990)
Ogallala (Oga)	Arkansas, Kansas, and Platte River basins on Great Plains (i.e., Ogallala Aquifer)
Northern Great Plains (NGP)	Great Plains from Niobrara River basin to Mackenzie River delta
Northern Great Basin–Colorado Plateau subdivision	Pacific Ocean tributaries north of Sierra Nevada and Cape Mendocino
Bidahochi (Bid)	Colorado River basin upstream of Grand Wash Fault (Colorado Plateau), named for Miocene Bidahochi Formation (Chapin 2008; Spencer et al. 2008)
Bouse (Bou)	Blythe, Bristol, Cottonwood, Havasu, Las Vegas, and Mojave basins (i.e., Bouse Embayment), including Colorado River Valley from Boulder Canyon to Chocolate Mountains, Bill Williams River (Roskowski et al. 2010)
Hualapai (Hua)	Colorado River basin between Grand Wash Fault and Boulder Canyon including Meadow Valley Wash, Muddy River (i.e., Pluvial White River system), and Virgin River, named for Miocene Lake Hualapai (Roskowski et al. 2010; Crossey et al. 2015)
Railroad Valley (Rr)	Railroad Lake basin with Railroad, Reveille, Warm Springs, Hot Creek, and Fish Lake valleys (Hubbs and Miller 1948)
Bonneville (Bon)	Bonneville Basin (Reheis et al. 2014) excluding Bear River basin upstream from Cache Valley (Smith et al. 2002; Loxterman and Keeley 2012)

Area of endemism	Description
Glenn's Ferry (GlF)	Snake River basin upstream from Hells Canyon (i.e., Pluvial Lake Glenn's Ferry basin), with Malheur, Owyhee, Powder, upper Snake, Wood, and upper Yellowstone Rivers (Missouri River basin), and Bear River upstream of Cache Valley (Smith et al. 2002; Loxterman and Keeley 2012)
Solitary (So)	Four endorheic basins, between Bonneville and Lahontan basins, including Butte, Goshute, Ruby, and Steptoe valleys (i.e., pluvial lakes Waring, Franklin, Gale, and Steptoe) (Houston et al. 2015)
Lahontan (Lah)	Pluvial Lake Lahontan basin with Carson, Humboldt, Truckee, and Walker Rivers, Pluvial Lake Clover (Clover and Independence valleys), and Pluvial Lake Newark (Newark Valley; Reheis et al. 2002b, 2014) (Mono Lake shared with Mojave Area) (Reheis et al. 2002a)
Pit	Pit River basin with Eagle and Goose Lakes, Surprise and Warner valleys, Chewaucan (Summer Lake, Lake Abert), and Fort Rock (Silver Lake) basins (Hubbs and Miller 1948; Negrini 2002)
Klamath (Kla)	Klamath River basin upstream of Klamath Falls with basin of ancient Lake Modoc (Hubbs and Miller 1948; Negrini 2002)
Catlow (Cat)	Catlow and Guano valleys (Hubbs and Miller 1948; Harris 2000)
Harney (Har)	Harney Basin with Donner and Blitzen and Silvies Rivers and Harney and Malheur Lakes (Hubbs and Miller 1948; Negrini 2002)
Alvord (Alv)	Alvord Basin with Alvord and Borax Lakes (Hubbs and Miller 1948; Harris 2000)

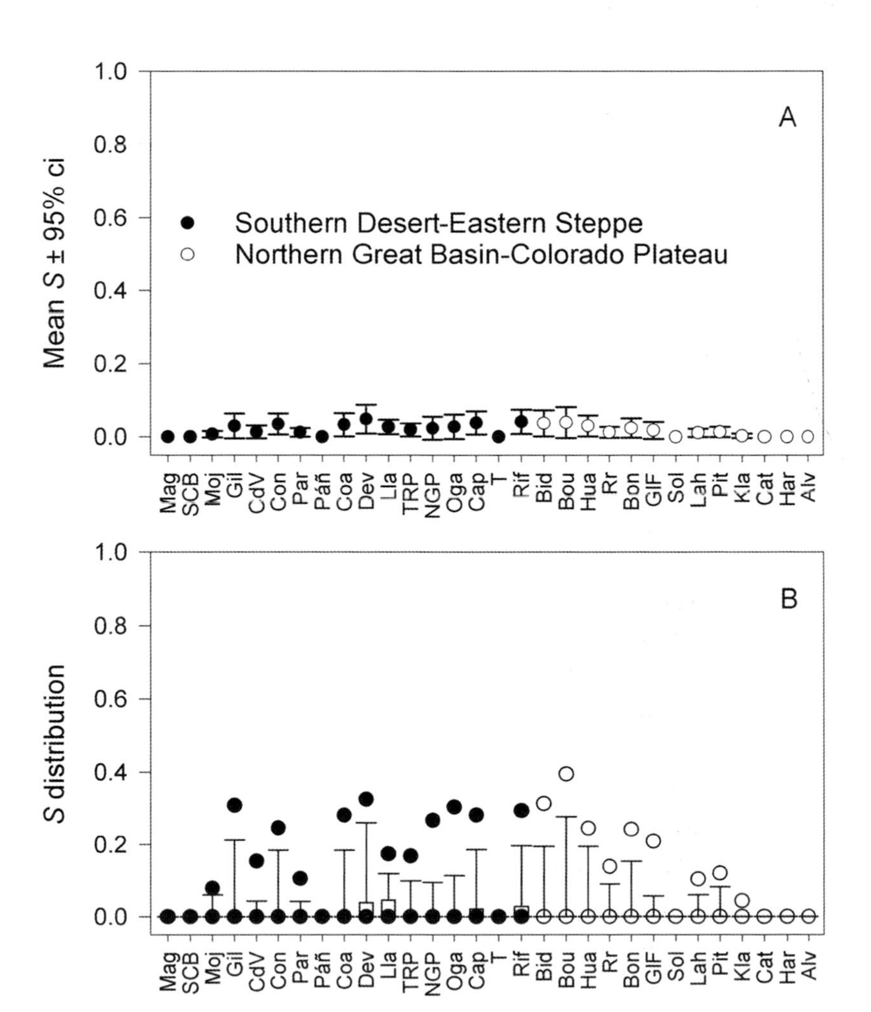

Fig. 3.2 Sørensen's coefficient of similarity (*S*; 0 = no similarity, 1 = total similarity) pairwise comparisons for each of 30 areas of endemism with the other 29 areas of endemism. Symbol colors indicate regional subdivisions. *A*, mean *S* with 95% confidence intervals for all pairwise comparisons (*n* = 29 for each area). *B*, boxplots of *S* for all pairwise comparisons by area of endemism (*n* = 29 for each area). Due to low similarities, 5th, 10th, 25th, and 50th percentiles all = 0. In four cases, the 75th percentile is visible as the upper extension of the box. Upper whiskers = 90th percentiles, upper points = 95th percentiles.

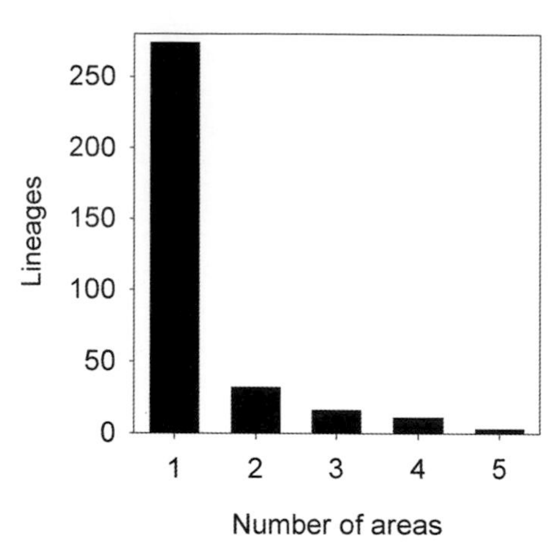

Fig. 3.3 Number of fish lineages endemic to the North American desert with distributions in 1, 2, 3, 4, or 5 areas of endemism (N = 336). No lineage occupied more than 5 areas.

(Devils, Llano, Capitan, Rift), all of which shared faunal exchanges in the Pleistocene (Osborne et al. 2016), have a 75th percentile of pairwise similarities greater than 0. Even these are less than 0.05. Within this region, *M. aestivalis, N. jemezanus,* and *Ictalurus lupus* each occupy five areas of endemism, more than any other desert fish lineage. These are riverine fishes, suited for inter-area dispersal (Bean et al. 2011; Page and Burr 2011; competition and hybridization with nonnative *Ictalurus punctatus* now restrict *I. lupus* to headwater streams, McClure-Baker et al. 2010).

CLADISTIC PATTERNS

Modern molecular methods and accrual of molecular data in open-access "banks" (e.g., GenBank, www.ncbi.nlm.nih.gov/genbank/) facilitate organization of fish lineages into evolutionary clades—monophyletic groups that contain all, and only, the descendants of a common ancestor. Cladistic placement is available for 337 of the 338 desert-endemic lineages (all except the extinct *Stypodon signifer*). On the basis of these data, desert-endemic lineages have been placed in 53 clades belonging to 15 families (table 3.2). Many clades produced multiple endemic lineages (mean = 6 ± 8.1 SD lineages per clade).

Clade reconstruction facilitates estimates of the timing of lineage diversification (Near et al. 2005). Estimates available for 38 clades indicate Early Miocene initiation of endemism for 1 clade, Middle Miocene for

Table 3.2 Summary of phylogenetic clades including fishes endemic to the North American desert.

Clade	Endemism initiated[a]	Lineages	Areas: Total (By subdivision)[b]	References
PETROMYZONTIDAE				
Entosphenus	NE	2	2 (0,2)	Lorion et al. 2000; Docker 2006
CYPRINIDAE				
Plagopterin	Late Miocene	9	5 (1,4)	Bufalino and Mayden 2010; Schönhuth et al. 2012b
Western chub-pikeminnow	Late Miocene	36	22 (9,13)	Schönhuth et al. 2014
Mylocheilus	Late Miocene	3	3 (0,3)	Bufalino and Mayden 2010; Houston et al. 2010
Campostoma	Pliocene	2	3 (3,0)	Bufalino and Mayden 2010; Domínguez-Domínguez et al. 2011
Tiaroga	Middle Miocene	1	1 (1,0)	Martin and Bonett 2015
Exoglossum	Pliocene	16	11 (3,8)	Bufalino and Mayden 2010; Smith et al. 2017
Dionda	Middle Miocene	12	6 (6,0)	Schönhuth et al. 2012a; Martin and Bonett 2015
Platygobio clade	Early Miocene	6	9 (9,0)	Bufalino and Mayden 2010; Martin and Bonett 2015
Algansea-Agosia	Middle Miocene	2	4 (3,1)	Schönhuth et al. 2012b; Pérez-Rodríguez et al. 2015
Notropis	Late Miocene	5	8 (8,0)	Martin and Bonett 2015; Conway and Kim 2016
Yuriria	Pliocene	2	2 (2,0)	Martin and Bonett 2015; Pérez-Rodríguez et al. 2015
Codoma-Cyprinella	Middle Miocene	24	9 (9,0)	Schönhuth and Mayden 2010; Osborne et al. 2016
"*Notropis*" *longirostris* clade	Late Miocene	1	1 (1,0)	Martin and Bonett 2015
Hybognathus	Middle Miocene	4	9 (9,0)	Moyer et al. 2009; Martin and Bonett 2015
Alburnops	Middle Miocene	9	7 (7,0)	Martin and Bonett 2015

Clade	Endemism initiated[a]	Lineages	Areas: Total (By subdivision)[b]	References
CATOSTOMIDAE				
Cycleptinae	NE	1	4 (4,0)	Buth and Mayden 2001; Doosey et al. 2010
Catostomus	Late Miocene	20	14 (5,9)	Doosey et al. 2010; Unmack et al. 2014
Moxostoma	Late Miocene	5	7 (7,0)	Clements et al. 2012; Pérez-Rodríguez et al. 2016
Pantosteus	Late Miocene	13	13 (7,6)	McPhee et al. 2008; Unmack et al. 2014
ICTALURIDAE				
Troglobites	Late Miocene	4	4 (4,0)	Arce-H. et al. 2017
Ictalurus	Late Miocene	5	8 (8,0)	Arce-H. et al. 2017
SALMONIDAE				
Oncorhynchus clarkii group	Pliocene	10	6 (2,5)	Wilson and Turner 2009; Sağlam et al. 2017
Oncorhynchus mykiss group	Quaternary	13	9 (5,4)	Currens et al. 2009; Crête-Lafrenière et al. 2012
Prosopium	Quaternary	8	5 (1,4)	Miller 2006; Crête-Lafrenière et al. 2012
ATHERINOPSIDAE				
Atherinella	NE	1	1 (1,0)	Miller et al. 2005; Bloom et al. 2009
Chirostoma	NE	1	1 (1,0)	Miller et al. 2005; Bloom et al. 2009
GOODEIDAE				
Empetrichthyinae	Middle Miocene	10	3 (1,2)	Pérez-Rodríguez et al. 2015; Jimenez et al. 2017
Characodontini	Middle Miocene	3	1 (1,0)	Webb et al. 2004; Pérez-Rodríguez et al. 2015
Girardinichthyini	Late Miocene	1	1 (1,0)	Webb et al. 2004; Pérez-Rodríguez et al. 2015
Chapalichthyini	Late Miocene	1	1 (1,0)	Webb et al. 2004; Pérez-Rodríguez et al. 2015
FUNDULIDAE				
Fundulus	NE	1	1 (1,0)	Ghedotti and Davis 2013
Wileyichthys	NE	1	1 (1,0)	Ghedotti and Davis 2013
Plancterus	NE	2	6 (6,0)	Ghedotti and Davis 2013
Lucania	NE	1	1 (1,0)	Ghedotti and Davis 2013
CYPRINODONTIDAE				
Cualac	Late Miocene	1	1 (1,0)	Echelle et al. 2005
Megupsilon	Late Miocene	1	1 (1,0)	Echelle et al. 2005
Cyprinodon	Late Miocene	39	10 (10,0)	Echelle et al. 2005; Echelle 2008
POECILIIDAE				
Poecilia	Middle Miocene	1	1 (1,0)	Meredith et al. 2011
Gambusia	NE	16	7 (7,0)	Rauchenberger 1989; Echelle et al. 2013
Poeciliopsis	NE	4	3 (3,0)	Mateos et al. 2002; Hedrick and Hurt 2012
Xiphophorus	NE	10	2 (2,0)	Kang et al. 2013
COTTIDAE				
Cottopsis	NE	4	2 (0,2)	Kinziger et al. 2005; Baumsteiger et al. 2012; Smith and Busby 2014
Uranidea	NE	3	3 (0,3)	Kinziger et al. 2005; Smith and Busby 2014
Cottus greenei clade	NE	2	1 (0,1)	Kinziger et al. 2005
CENTRARCHIDAE				
Lepomis	Quaternary	1	1 (1,0)	Near et al. 2005; Coghill et al. 2013

Clade	Endemism initiated[a]	Lineages	Areas: Total (By subdivision)[b]	References
Micropterus	Late Miocene	1	1 (1,0)	Near et al. 2003
PERCIDAE				
Percina	Quaternary	1	1 (1,0)	Near and Benard 2004
Microperca	Pliocene	1	1 (1,0)	Kelly et al. 2012; Echelle et al. 2015
Oligocephalus	Late Miocene	9	8 (8,0)	Kelly et al. 2012; Bossu et al. 2013
CICHLIDAE				
Herichthys	Pliocene	2	2 (2,0)	Hulsey et al. 2010; de la Maza-Benignos et al. 2015
Nosferatu	Pliocene	5	1 (1,0)	Hulsey et al. 2010; de la Maza-Benignos et al. 2015
GOBIESOCIDAE				
Gobiesox	NE	1	1 (1,0)	Espinosa-Pérez and Castro-Aguirre 1996
Count		337	236 (170,67)	

Note: One lineage has not been linked to a phylogenetic clade: Cyprinidae, *Stypodon signifer.*

[a]Refers to estimates of the first divergence of endemics to the desert region, not an estimate for all lineage members. NE = not estimated.

[b]Areas in parentheses indicate number of areas occupied in each subdivision of the desert region: Southern Desert–Eastern Steppe, Northern Great Basin–Colorado Plateau.

9, Late Miocene for 16, Pliocene for 8, and Quaternary for 4. Some clades that began to diversify in the Miocene continued to do so thereafter (e.g., *Exoglossum,* Smith et al. 2017; *Moxostoma,* Pérez-Rodríguez et al. 2016; and *Cyprinodon,* Echelle 2008).

Synchronous divergence among lineages is supported in specific cases, as with *Gila* and *Pantosteus* regarding dispersal from the Bidahochi to the Rift Area 9–10 million years ago (Spencer et al. 2008). However, overall comparative estimates for timing of lineage origins reveal asynchronous divergence of endemic lineages over about 23 million years. This extended period corresponds with fragmentation of river systems via aridity and tectonism (Chapin 2008; Galloway et al. 2011). Persistent desert habitats consistently influenced evolution of fishes throughout this period (sensu Gámez et al. 2017). Desert-compatible adaptations such as (1) tolerance of high turbidity, salinity, and temperatures, (2) boom-bust population dynamics, and (3) behaviors to avoid inhospitable conditions are evident in many desert fishes (Deacon and Minckley 1974).

BIOGEOGRAPHIC SUBDIVISIONS

A major boundary divides the 30 areas of endemism into two subdivisions: (1) rivers flowing to the Southern California Bight, Gulf of California, Gulf of Mexico, Hudson Bay, or Arctic Ocean (Southern Desert–Eastern Steppe); and (2) rivers flowing to the Pacific Ocean north of the Sierra Nevada (Northern Great Basin–Colorado Plateau). Prolonged hydrographic separation of drainages north and west of Grand Canyon from those south and east explains this subdivision (Spencer et al. 2008). As noted above, this separation ended during the Miocene-Pliocene transition, about 5 million years ago (Roskowski et al. 2010; Crossey et al. 2015).

This subdivision boundary could be the most distinct for North American fishes. Only 10 of 336 endemic lineages (3%) occur in both subdivisions. Two events explain these exceptions. First, some endemics of the Bidahochi Area of the Northern Great Basin–Colorado Plateau (*Gila elegans, Gila robusta, P. lucius, Catostomus latipinnis,* and *X. texanus*) colonized the Hualapai, Bouse, and Gila Areas of the Southern Desert–Eastern Steppe following the formation of Grand Canyon (Spencer et al. 2008). Some endemics in the Gila Area of the Southern Desert–Eastern Steppe (*Agosia chrysogaster, Plagopterus argentissimus, Catostomus insignis, Pantosteus clarkii*) used this corridor in reverse. Second, *Siphateles obesus* reached the Mojave Area of the Southern Desert–Eastern Steppe from the Lahontan Area of the Northern Great Basin–Colorado Plateau (Harris 2000), presumably through the ancestor to Mono Lake, California—Pluvial Lake Russell—

which alternately connected with both basins (Reheis et al. 2002a).

Trans-subdivision distributions are also few among clades. Only 9 of 53 clades (17%) are represented by lineages in each subdivision. These include (1) clades with access through Grand Canyon (western chub–pikeminnow, *Exoglossum*, *Catostomus*, *Pantosteus*; Spencer et al. 2008), (2) clades with marine dispersal (*Oncorhynchus mykiss*; Abadía-Cardoso et al. 2015), and (3) clades with headwater dispersal (*Oncorhynchus clarkii*; Rogers et al. 2014).

Endemic richness differs between subdivisions. Seventeen areas of endemism in the Southern Desert–Eastern Steppe subdivision harbor 243 desert-endemic lineages, whereas 13 areas of endemism in the Northern Great Basin–Colorado Plateau subdivision harbor only 103. Potential explanations for this discrepancy are debated, but include (1) greater geographic isolation of Northern Great Basin–Colorado Plateau areas from climate refuges in the Gulf of Mexico basin (Smith et al. 2010; Griffiths 2015), (2) higher rates of extinction due to less geologic stability within the Northern Great Basin–Colorado Plateau from the Miocene to the present (Smith et al. 2010), (3) higher rates of dispersal and extinction in the Northern Great Basin–Colorado Plateau due to climate flux in the Quaternary (Smith et al. 2010; Leprieur et al. 2011), (4) reduced ecosystem productivity in the Northern Great Basin–Colorado Plateau (Oberdorff et al. 2011), or (5) reduced thermal equability in the Northern Great Basin–Colorado Plateau (Griffiths et al. 2014). The higher number of areas of endemism in the Southern Desert–Eastern Steppe also increases potential for inter-area diversification.

Southern Desert–Eastern Steppe Subdivision

In the Southern Desert–Eastern Steppe subdivision (table 3.3), dispersal barriers extending the length of the Rio Grande Rift confined lineages east of the Continental Divide, with two local exceptions: (1) stream capture transferred Rift Area *Pantosteus* cf. *plebeius* to the Bidahochi Area, where they hybridized with *P. discobolus* (hybrids = *P. discobolus jarrovii*; Unmack et al. 2014), and (2) stream capture transferred Cabeza de Vaca Area *P. plebeius* to the Gila Area (Sapillo Creek; McPhee et al. 2008). Crossing of the Continental Divide was more common south of the Rift, as in *C. ornatum* (Schonhuth et al. 2011), *Codoma ornata* (Schönhuth et al. 2015), and *Moxostoma* (Pérez-Rodríguez et al. 2016), which explains faunal relatedness between areas of the southern Gulf of Mexico and Gulf of California drainages.

Table 3.3. Distribution of desert-endemic fishes among Southern Desert–Eastern Steppe areas of endemism, with niche affiliations. Abbreviations correspond to table 3.1.

Lineage	Area of endemism																	Niche[a]		
	Mag	SCB	Moj	Gil	CdV	Con	Rif	T	Par	Pán	Coa	Dev	Lla	Cap	TRP	Oga	NGP	Hw	Al	Sp
CYPRINIDAE																				
Creek chub–plagopterin																				
Meda fulgida				X															+	
Plagopterus argentissimus[b]				X															+	
Western minnow																				
Gila brevicauda					X													+		
G. conspersa									X									+	+	
G. diatenia				X															+	
G. elegans[b]				X															+	
G. eremica					X														+	+
G. intermedia				X															+	+
G. minacae					X	X			X									+	+	

Lineage	Area of endemism																	Niche[a]		
	Mag	SCB	Moj	Gil	CdV	Con	Rif	T	Par	Pán	Coa	Dev	Lla	Cap	TRP	Oga	NGP	Hw	Al	Sp
G. modesta-pandora[29]							X				X			X				+		+
G. nigra				X														+		+
G. nigrescens					X													+	+	
G. orcuttii		X																+	+	
G. pulchra						X												+	+	
G. purpurea					X															+
G. robusta[b]				X														+	+	
G. sp. Tunal[29]									X									+	+	
Ptychocheilus lucius[b]				X															+	
Siphateles mohavensis[16]			X																+	
S. obesus[b,16]			X																+	
Campostoma																				
Campostoma ornatum[11,27]					X	X												+	+	
C. cf. *ornatum* Nazas-Piaxtla[11,27]									X									+	+	
Tiaroga																				
Rhinichthys cobitis				X															+	
Exoglossum																				
Rhinichthys cf. *osculus* Amargosa[31]			X															+	+	
R. cf. *osculus* Whitmore[31]			X																	+
R. cf. *osculus* Gila[31]				X														+	+	
R. cf. *osculus* Los Angeles[31]		X																+		
Dionda																				
Dionda argentosa												X								+
D. diaboli											X	X							+	
D. episcopa[28]														X						+
D. cf. *episcopa* upper Pecos[28]														X						+
D. flavipinnis[28]													X							+
D. cf. *flavipinnis* Colorado[28]													X							+
D. melanops[28]											X								+	
D. nigrotaeniata[28]													X							+
D. serena[28]													X							+
D. sp. Conchos[28]						X													+	
D. sp. Tunal[28]									X										+	
D. texensis[28]													X							+

Lineage	Area of endemism																Niche[a]		
	Mag	SCB	Moj	Gil	CdV	Con	Rif	T	Par	Pán	Coa	Dev	Lla	Cap	TRP	Oga	NGP	Hw	Al	Sp
Platygobio clade																				
Macrhybopsis aestivalis						X	X				X	X		X					+	
M. australis															X				+	
M. gelida																X	X		+	
M. meeki																	X		+	
M. tetranema																X			+	
Platygobio gracilis							X									X	X	+	+	
Algansea–Agosia																				
Agosia chrysogaster[b]				X															+	
A. cf. *chrysogaster* Yaqui-Mayo-Fuerte-Sinaloa					X	X													+	
Notropis																				
Notropis amabilis[8]											X	X	X	X					+	+
Notropis girardi																X			+	
N. jemezanus						X	X				X	X		X					+	
N. megalops[8,34]						X					X	X								+
N. oxyrhynchus															X				+	
Yuriria																				
Notropis aulidion†									X										+	
N. calabazas[18]										X									+	
Codoma-Cyprinella																				
Codoma ornata[31]						X			X									+	+	
C. cf. *ornata* upper Conchos[31]					X	X												+	+	
C. cf. *ornata* Nazas[31]									X									+	+	
C. cf. *ornata* Tunal[31]									X									+	+	
Cyprinella alvarezdelvillari									X											+
C. bocagrande					X															+
C. forlonensis[26]										X									+	
C. formosa					X														+	
C. garmani									X										+	+
C. lepida													X							+
C. cf. *lepida* Nueces[26]													X							+
C. "lutrensis" blairi												X								+
C. "lutrensis" Brazos[22]															X				+	

Lineage	Area of endemism																Niche[a]		
	Mag	SCB	Moj	Gil	CdV	Con	Rif	T	Par	Pán	Coa	Dev	Lla	Cap	TRP	Oga	NGP	Hw	Al	Sp
C. "lutrensis" Capitan-Chihuahua-Colorado[22]						X	X					X							+	
C. "lutrensis" Red[22]															X				+	
C. panarcys						X													+	
C. proserpina											X	X								+
C. rutila											X								+	
C. xanthicara											X									+
Tampichthys catostomops										X									+	
T. dichromus										X									+	
T. mandibularis										X									+	
T. rasconis										X									+	
"Notropis" longirostris clade																				
Notropis chihuahua						X													+	+
Hybognathus																				
Hybognathus amarus							X				X	X		X					+	
H. argyritis																X	X		+	
H. placitus													X		X	X	X		+	
Notropis nazas									X										+	
Alburnops																				
Notropis aguirrepequenoi											X								+	
N. bairdi															X				+	
N. braytoni						X					X	X							+	
N. buccula															X				+	
N. orca†						X	X				X	X							+	
N. saladonis†											X								+	
N. simus pecosensis														X					+	
N. s. simus†							X												+	
N. tropicus										X									+	
Unclassified																				
Stypodon signifer†									X											+
CATOSTOMIDAE																				
Cycleptinae																				
Cycleptus cf. *elongatus* Rio Grande[4]						X	X				X	X							+	

Lineage	Area of endemism																	Niche[a]		
	Mag	SCB	Moj	Gil	CdV	Con	Rif	T	Par	Pán	Coa	Dev	Lla	Cap	TRP	Oga	NGP	Hw	Al	Sp
Catostomus																				
Catostomus bernardini					X	X			X									+	+	
C. cahita					X													+		
C. fumeiventris			X																+	
C. insignis[b]				X															+	
C. latipinnis[b]				X															+	
C. leopoldi					X													+		
C. wigginsi					X														+	+
Xyrauchen texanus[b,6,12]				X															+	
Moxostoma																				
Moxostoma albidum[23]											X	X							+	
M. cf. *albidum* Soto la Marina[23]											X								+	
M. congestum[23]							X						X	X					+	
M. milleri[23]									X										+	
M. sp. Río Conchos[23]						X													+	
Pantosteus																				
Pantosteus clarkii[b,32]				X															+	
P. jordani[32]																	X	+	+	
P. nebuliferus									X										+	
P. cf. *plebeius* Rift[19]							X											+		
P. plebeius Cabeza de Vaca[19]				X	X															
P. cf. *plebeius* Conchos[32]						X														
P. cf. *plebeius* Mezquital[32]									X									+		
P. santaanae[32]		X																	+	
ICTALURIDAE																				
Ictalurus																				
Ictalurus lupus									X		X	X	X	X				+	+	+
I. cf. *lupus* Río Conchos						X													+	+
I. mexicanus										X								+	+	
I. pricei[5]					X													+	+	
I. cf. *pricei* Río Sinaloa[5]									X									+	+	
Troglobites																				
Prietella lundbergi										X										+

Lineage	Area of endemism																Niche[a]		
	Mag	SCB	Moj	Gil	CdV	Con	Rif	T	Par	Pán	Coa	Dev	Lla	Cap	TRP	Oga	NGP	Hw	Al	Sp
P. phreatophila											X	X								+
Satan eurystomus													X							+
Trogloglanis pattersoni													X							+
SALMONIDAE																				
Oncorhynchus clarkii																				
Oncorhynchus clarkii macdonaldi[20,24]																X		+		
Oncorhynchus clarkii stomias[20,24]																X		+		
O. clarkii virginalis[20,24]							X											+		
Oncorhynchus mykiss																				
Oncorhynchus apache				X														+		
O. chrysogaster[1]						X			X									+		
O. cf. *chrysogaster* Verde-Molino[1]						X												+		
O. gilae				X														+		
O. mykiss nelsoni[2]		X																+		
O. cf. *mykiss* North Sierra Madre Occidental[1]					X	X												+		
O. cf. *mykiss* Piaxtla-San Lorenzo[1]									X									+		
Prosopium																				
Prosopium williamsoni Missouri[17]																	X	+		
Atherinopsidae																				
Atherinella																				
Atherinella elegans						X													+	+
Chirostoma																				
Chirostoma mezquital									X										+	+
GOODEIDAE																				
Crenichthys-Empetrichthys																				
Empetrichthys latos concavus†			X																	+
E. l. latos			X																	+
E. l. pahrump†			X																	+

Lineage	Area of endemism																Niche[a]		
	Mag	SCB	Moj	Gil	CdV	Con	Rif	T	Par	Pán	Coa	Dev	Lla	Cap	TRP	Oga	NGP	Hw	Al	Sp
E. merriami†			X																	+
Characodontini																				
Characodon audax									X											+
C. garmani†									X											+
C. lateralis									X											+
Girardinichthyini																				
Ataeniobius toweri										X									+	+
Chapalichthyini																				
Xenoophorus captivus										X										+
FUNDULIDAE																				
Fundulus																				
Fundulus philpisteri[15]											X									+
Wileyichthys																				
Fundulus lima	X																			+
Plancterus																				
Fundulus kansae																X	X		+	
F. zebrinus												X	X	X	X				+	
Lucania																				
Lucania interioris											X									+
CYPRINODONTIDAE																				
Cualac																				
Cualac tessellatus										X										+
Megupsilon																				
Megupsilon aporus†											X									+
Cyprinodon																				
Cyprinodon albivelis					X														+	+
C. alvarezi†											X									+
C. arcuatus†				X															+	+
C. atrorus											X								+	
C. bifasciatus											X									+
C. bobmilleri											X									+
C. bovinus														X						+
C. ceciliae†											X									+
C. diabolis			X																	+
C. elegans														X						+
C. eremus				X															+	+

Lineage	Area of endemism																	Niche[a]		
	Mag	SCB	Moj	Gil	CdV	Con	Rif	T	Par	Pán	Coa	Dev	Lla	Cap	TRP	Oga	NGP	Hw	Al	Sp
C. eximius						X						X							+	
C. fontinalis					X															+
C. inmemoriam†											X									+
C. julimes						X														+
C. latifasciatus†									X											+
C. longidorsalis†											X									+
C. macrolepis						X														+
C. macularius				X															+	+
C. meeki†									X										+	+
C. nazas									X										+	
C. nevadensis amargosae			X																	+
C. n. calidae†			X																	+
C. n. mionectes			X																	+
C. n. nevadensis			X																	+
C. n. pectoralis			X																	+
C. n. shoshone			X																	+
C. pachycephalus						X														+
C. pecosensis														X					+	+
C. pisteri					X														+	+
C. radiosus			X																+	+
C. rubrofluviatilis[13]															X				+	
C. cf. *rubrofluviatilis* Brazos[13]															X				+	
C. salinus milleri			X																	+
C. s. salinus			X																	+
C. salvadori						X													+	
C. sp. Aguanaval[13]									X										+	
C. tularosa								X												+
C. veronicae†											X									+
POECILIIDAE																				
Poecilia																				
Poecilia latipunctata										X										+
Gambusia																				
Gambusia alvarezi						X														+
G. amistadensis†												X								+
G. aurata										X									+	
G. gaigei												X								+
G. geiseri													X							+
G. georgei†													X							+
G. heterochir													X							+

Lineage	Area of endemism																	Niche[a]		
	Mag	SCB	Moj	Gil	CdV	Con	Rif	T	Par	Pán	Coa	Dev	Lla	Cap	TRP	Oga	NGP	Hw	Al	Sp
G. hurtadoi						X														+
G. krumholzi[14]												X							+	+
G. longispinis											X								+	
G. marshi											X									+
G. nobilis														X						+
G. senilis					X	X						X								+
G. speciosa											X	X							+	
G. vittata										X									+	+
G. zarskei[21]						X														+
Poeciliopsis																				
Poeciliopsis lucida						X													+	
P. monacha					X	X												+		
P. occidentalis				X															+	
P. sonoriensis					X														+	
Xiphophorus																				
Xiphophorus continens										X								+		
X. couchianus											X									+
X. gordoni											X									+
X. meyeri											X									+
X. montezumae										X										+
X. multilineatus										X										+
X. nezahualcoyotl										X										+
X. nigrensis										X										+
X. pygmaeus										X										+
X. xiphidium											X								+	+
CENTRARCHIDAE																				
Lepomis																				
Lepomis cf. *megalotis* Cuatro Ciénegas[7]											X									+
Micropterus																				
Micropterus treculii													X							+
PERCIDAE																				
Percina																				
Percina carbonaria													X							+
Microperca																				
Etheostoma fonticola													X							+

Lineage	Area of endemism																	Niche[a]		
	Mag	SCB	Moj	Gil	CdV	Con	Rif	T	Par	Pán	Coa	Dev	Lla	Cap	TRP	Oga	NGP	Hw	Al	Sp
Oligocephalus																				
Etheostoma australe						X													+	
E. grahami											X	X							+	+
E. lepidum Balcones[33]													X							+
E. lepidum Concho-San Saba[33]													X							+
E. lepidum Capitan[33]														X						+
E. lugoi											X									+
E. pottsi						X			X									+		
E. pulchellum[3]													X		X	X			+	+
E. segrex											X									+
CICHLIDAE																				
Herichthys																				
Herichthys minckleyi											X									+
H. tamasopoensis										X								+		
Nosferatu																				
Nosferatu bartoni[9,10]										X										+
N. labridens[9,10]										X										+
N. pame[9,10]										X								+		
N. pratinus[9,10]										X									+	
N. steindachneri[9,10]										X								+		
Gobiesocidae																				
Gobiesox juniperoserrai[†25]	X																			+
Count	2	4	19	22	23	37	12	1	30	29	43	24	23	16	11	10	8	42	122	118

Sources: [1]Abadía-Cardoso et al. 2015; [2]Abadía-Cardoso et al. 2016; [3]Bossu et al. 2013; [4]Buth and Mayden 2001; [5]Castañeda-Rivera et al. 2014; [6]Chen and Mayden 2012; [7]Coghill et al. 2013; [8]Conway and Kim 2016; [9]De la Maza-Benignos and Lozano-Vilano 2013; [10]De la Maza-Benignos et al. 2015; [11]Domínguez-Domínguez et al. 2011; [12]Doosey et al. 2010; [13]Echelle et al. 2005; [14]Echelle et al. 2013; [15]García-Ramírez et al. 2006; [16]Harris 2000; [17]Miller 2006; [18]Lyons and Mercado-Silva 2004; [19]McPhee et al. 2008; [20]Metcalf et al. 2012; [21]Meyer et al. 2010; [22]Osborne et al. 2016; [23]Pérez-Rodríguez et al. 2016; [24]Rogers et al. 2014; [25]Ruiz-Campos et al. 2014; [26]Schönhuth and Mayden 2010; [27]Schönhuth et al. 2011; [28]Schönhuth et al. 2012a; [29]Schönhuth et al. 2014; [30]Schönhuth et al. 2015; [31]Smith et al. 2017; [32]Unmack et al. 2014; [33]D. J. MacGuigan and T. J. Near, personal communication July 2015; [34]K. W. Conway, personal communication, February 2017.

Note: Except where other citations are provided, species names and distributions follow Miller et al. 2005; Minckley and Marsh 2009; Moyle 2002; Page and Burr 2011; Page et al. 2013.

[a]Niches: Headwater (Hw) includes steep streams, rivers, and lakes dominated by coarse substrates, low ambient temperatures, and low productivity; Spring (Sp) includes habitats dominated by groundwater inflows with groundwater-dependent temperatures; Alluvial (Al) includes lower-gradient streams, rivers, and lakes with mixed substrates, seasonal ambient temperatures, and relatively high productivity.

[b]Also native in Northern Great Basin–Colorado Plateau areas (see table 3.4).

[†]Extinct.

Northern Great Basin–Colorado Plateau Subdivision

Northern Great Basin–Colorado Plateau areas shared prehistoric connections to the Pacific Ocean via routes passing north of the Sierra Nevada (table 3.4). Colorado Plateau (i.e., Bidahochi Area) endemics diverged from northwestern relatives when the nascent upper Colorado River basin became endorheic (Smith et al. 2002; Spencer et al. 2008). As already discussed, this area later became integrated with the lower Colorado River basin via Grand Canyon.

Table 3.4 Distribution of desert-endemic fishes among Northern Great Basin–Colorado Plateau areas of endemism, with niche affiliations. Abbreviations correspond to table 3.1.

Lineage	Area of endemism													Niche[a]		
	Bid	Bou	Hua	Rr	Bon	GlF	Sol	Lah	Pit	Kla	Cat	Har	Alv	Hw	Al	Sp
PETROMYZONTIDAE																
Entosphenus																
Entosphenus lethophagus									X	X				+		
E. minimus										X				+		
CYPRINIDAE																
Creek chub–plagopterin																
Lepidomeda albivallis			X													+
L. aliciae					X										+	
L. altivelis[†]			X													+
L. copei						X									+	
L. mollispinis mollispinis			X												+	
L. m. pratensis			X													+
L. vittata	X													+	+	
Plagopterus argentissimus[b]			X												+	
Western minnow																
Eremichthys acros								X								+
Gila atraria[10]					X	X									+	
G. atraria Bear-Snake[10]						X									+	
G. coerulea										X					+	
G. cypha	X	X													+	
G. elegans[b]	X	X													+	
G. jordani			X													+
G. robusta[b]	X	X												+	+	
G. seminuda			X												+	
Hesperoleucas symmetricus mitrulus[1]									X					+	+	
Moapa coriacea			X													+
Ptychocheilus lucius[b]	X	X	X												+	
Relictus solitarius							X									+
Siphateles. alvordensis-boraxobius[8]													X			+

Lineage	Area of endemism													Niche[a]		
	Bid	Bou	Hua	Rr	Bon	GlF	Sol	Lah	Pit	Kla	Cat	Har	Alv	Hw	Al	Sp
S. bicolor[8]										X				+	+	
S. cf. *bicolor* Silver Lake[8]									X					+	+	
S. columbianus[8]												X		+	+	
S. eurysomas[8]											X			+	+	
S. isolatus-newarkensis[8]								X						+	+	
S. obesus[b,8]				X				X	X					+	+	
S. thalassinus[8]									X					+	+	
Mylocheilus																
Iotichthys phlegethontis					X										+	+
Richardsonius balteatus Bonneville-Snake[9]					X	X									+	
R. egregius								X							+	
Exoglossum																
Rhinichthys deaconi[†]		X														+
R. cf. *osculus* Oregon Lakes[19]									X					+		
R. cf. *osculus* Harney[19]												X				+
R. cf. *osculus* Salmon Falls Creek[19]						X								+		
R. cf. *osculus* Thousand Springs[19]					X									+		
R. cf. *osculus* Snake Valley[19]					X									+		
R. cf. *osculus* Lahontan[19]						X		X						+	+	
R. cf. *osculus* Clover[19]								X								+
R. cf. *osculus* East Walker[19]								X						+		
R. cf. *osculus* Hualapai[19]			X		X									+	+	
R. cf. *osculus* Plateau[19]	X	X	X		X									+	+	
R. cf. *osculus* Little Colorado[19]	X													+		
Algansea-Agosia																
Agosia chrysogaster[b]		X													+	
CATOSTOMIDAE																
Catostomus																
Catostomus ardens[4,16]					X									+	+	
C. cf. *ardens* Bear-Snake-Weber-Provo[16]					X	X								+	+	
C. insignis[b]		X												+	+	
C. latipinnis[b]	X	X	X												+	
C. cf. *latipinnis* Little Colorado	X														+	
C. microps									X					+		
C. snyderi[5]										X					+	
C. tahoensis								X						+	+	

Lineage	Area of endemism													Niche[a]		
	Bid	Bou	Hua	Rr	Bon	GlF	Sol	Lah	Pit	Kla	Cat	Har	Alv	Hw	Al	Sp
C. warnerensis												X			+	
Chasmistes brevirostris[3,5,7,13]										X					+	
C. cujus[3,6]								X							+	
C. liorus[2,3,6]					X										+	
C. muriei†[4,6]						X									+	
Deltistes luxatus[3,6,13]										X					+	
Xyrauchen texanus[b,3,6]	X	X	X												+	
Pantosteus																
Pantosteus clarkii[b,21]		X	X												+	
P. discobolus discobolus[21]	X														+	
P. discobolus jarrovii[21]	X													+		
P. lahontan[21]								X						+	+	
P. platyrhynchus[21]	X				X	X								+	+	
P. virescens[21]					X	X									+	
SALMONIDAE																
Oncorhynchus clarkii																
Oncorhynchus clarkii alvordensis-henshawi[12]								X						+		
O. clarkii seleniris[18]								X						+		
O. clarkii behnkei-bouvieri-utah[12]						X								+		
O. clarkii pleuriticus[14,17]	X													+		
O. clarkii Colorado-Gunnison River[14,17]	X													+		
O. clarkii Great Basin[12]			X		X	X								+		
O. clarkii San Juan River[14,17]	X													+		
Oncorhynchus mykiss																
Oncorhynchus mykiss cf. *stonei* Sacramento[5]									X					+		
O. mykiss cf. *gairdneri* Harney[5]												X		+		
O. mykiss newberrii[5]										X				+		
O. cf. *mykiss* Fort Rock[5]									X					+		
O. cf. *mykiss* Klamath[5]										X				+		
O. cf. *mykiss* Catlow[5]											X			+		
Prosopium																
Prosopium abyssicola						X								+		
P. gemmifer						X								+		
P. spilonotus						X								+		
P. williamsoni Big Lost[15]						X								+		
P. williamsoni Big Wood[15]						X								+		
P. williamsoni Bonneville-Upper Snake-Green[15]	X				X	X								+		

Lineage	Area of endemism													Niche[a]		
	Bid	Bou	Hua	Rr	Bon	GlF	Sol	Lah	Pit	Kla	Cat	Har	Alv	Hw	Al	Sp
P. williamsoni Lahontan[15]								X						+		
GOODEIDAE																
Empetrichthyinae																
Crenichthys baileyi albivallis[22]			X													+
C. b. baileyi[22]			X													+
C. b. grandis[22]			X													+
C. b. moapae[22]			X													+
C. b. thermophilus[22]			X													+
C. nevadae				X												+
COTTIDAE																
Cottopsis																
Cottopsis asperrimus[11,20]									X					+		+
C. klamathensis klamathensis[11,20]										X				+		
C. princeps[11,20]										X				+		
C. tenuis[11,20]										X				+		
Uranidea																
Uranidea bendirei[11,20]												X		+		
U. echinatus[†11,20]					X									+		
U. extensus[11,20]						X									+	
Cottus greenei clade																
Cottus greenei[11,20]						X										+
C. leiopomus[11,20]						X								+		
Count	17	11	20	2	16	21	1	13	10	12	2	5	1	43	47	20

Sources: [1]Aguilar and Jones 2009; [2]Belk and Schaalje 2016; [3]Chen and Mayden 2012; [4]Cole et al. 2008; [5]Currens et al. 2009; [6]Doosey et al. 2010; [7]Dowling et al. 2016; [8]Harris 2000; [9]Houston et al. 2014; [10]Johnson 2002; [11]Kinziger et al. 2005; [12]Loxterman and Keeley 2012; [13]Markle et al. 2005; [14]Metcalf et al. 2012; [15]Miller 2006; [16]Mock et al. 2006; [17]Rogers et al. 2014; [18]Sağlam et al. 2017; [19]Smith et al. 2017; [20]Smith and Busby 2014; [21]Unmack et al. 2014; [22]Williams and Wilde 1981.

Note: Except where other citations are provided, species names and distributions follow Behnke 2002; Moyle 2002; De la Maza-Benignos 2009; Minckley and Marsh 2009; Page and Burr 2011; Page et al. 2013.

[a]Niches: Headwater (Hw) includes steep streams, rivers, and lakes dominated by coarse substrates, low ambient temperatures, and low productivity; Spring (Sp) includes habitats dominated by groundwater inflows with groundwater-dependent temperatures; Alluvial (Al) includes lower-gradient streams, rivers, and lakes with mixed substrates, seasonal ambient temperatures, and relatively high productivity.

[b]Also native in Southern Desert–Eastern Steppe areas (see table 3.3).

[†]Extinct.

ECOLOGICAL SEGREGATION

Desert-endemic fishes have three primary niches, each supporting substantial biodiversity within the desert region (Minckley 1991; Minckley and Marsh 2009): (1) rocky, sediment-poor headwaters, typically low-order mountain streams and rivers; (2) sediment-rich alluvial valleys with rivers and streams of diverse size; and (3) springs, which can vary widely in size and geographic setting across the desert region (Deacon and Minckley 1974; Hubbs 1995). Some endemic lineages are habitat-specific, whereas others are not (see tables 3.3–3.4). Alluvial habitats support the most lineages (122 in the Southern Desert–Eastern Steppe subdivision, 48 in the Northern Great Basin–Colorado Plateau subdivision), springs the second-most (119 and 20, respectively), and headwaters the third-most (47 and 56, respectively).

DESERT-ENDEMIC EVOLUTION

Faunal dissimilarity among areas of endemism indicates that most lineage diversity arose independently. The following conceptual model of faunal assembly (fig. 3.4) employs the concept of ecological "faunal filtering" (Jackson and Harvey 1989; Moyle 2002), but represents filters as nonhierarchical and interactive (Tonn et al. 1990; Jackson et al. 2001). This is because the evolution of endemism is a two- or three-step process that reticulates, meaning that an incipient endemic lineage can begin the divergence process at any time, not necessarily in synchrony with other lineages.

Step 1: Colonization

Founders of a desert-endemic lineage create a pool of colonists either by occupying an area of endemism

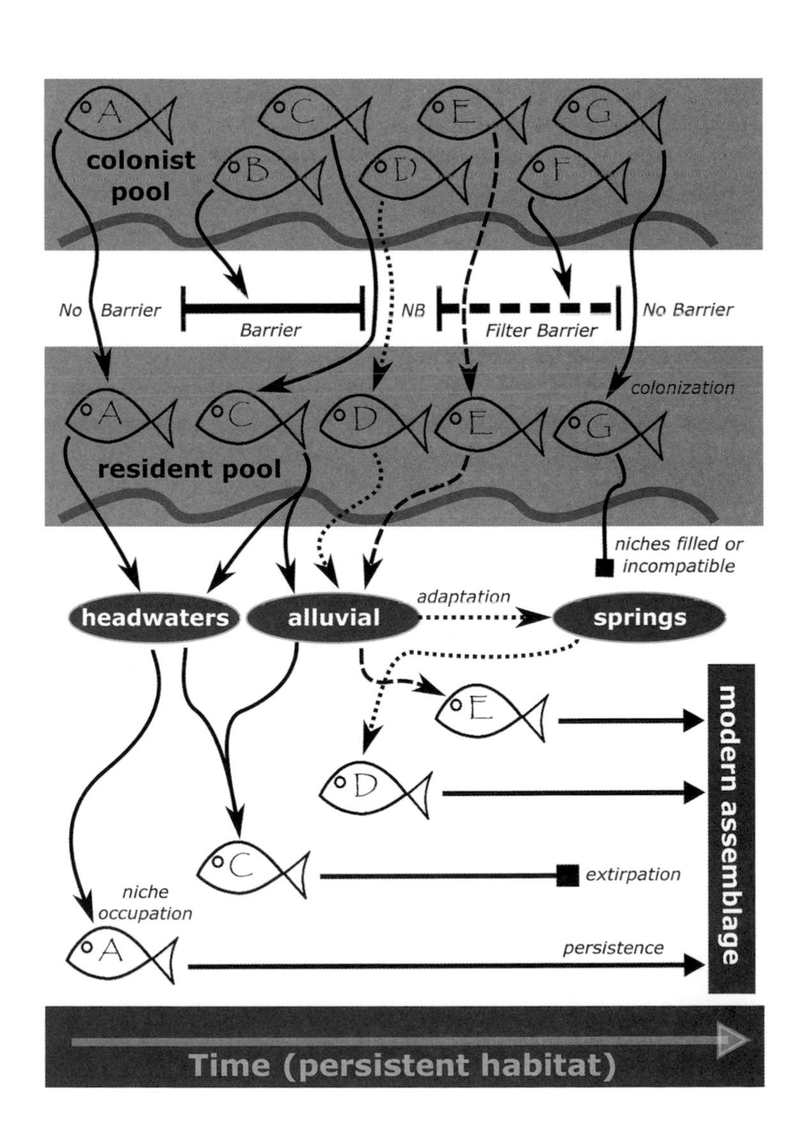

Fig. 3.4 Dynamic filtering of endemic fish faunas in areas of endemism of the North American desert (see table 3.1 and fig. 3.1). Events decrease in age from left to right. Pool of colonists = lineages with access. Persistent populations form the pool of residents. Endemic assemblages include divergent residents of long-term persistence within available niches (headwaters, alluvial, springs), regardless of time of colonization. Examples: *A*, one or more headwater specialists gain residency early in the history of an area and persist; *B*, a barrier precludes colonization; *C*, one or more alluvial-headwater generalists colonize, reside for an extended period, but are eventually extirpated; *D*, one or more alluvial specialists colonize and reside, later adapting to spring-fed habitat, where they persist; *E*, one or more alluvial specialists colonize across a filter barrier (i.e., a large valley that excludes headwater specialists) and persist; *F*, a filter barrier blocks one or more headwater specialists; *G*, colonization succeeds, but residency fails.

prior to desertification and persisting thereafter or by colonizing from outside the desert via dispersal corridors. Because clades have differing histories, their phylogeny represents a dynamic faunal filter. For example, *Cyprinodon* invaded certain areas of endemism accessible from the Gulf of Mexico basin to the southeast (Echelle et al. 2005). In contrast, *Oncorhynchus clarkii* invaded different areas of endemism that could be accessed from Pacific drainages on the northwest (Behnke 2002).

Phylogeny is also an integrative faunal filter that consolidates heritable traits like physiological tolerance, dispersal ability, dispersal tendency, and time of existence. Thus, the distribution of salmonids, for example, reflects their Nearctic affiliation and Pleistocene origin (Wilson and Turner 2009), coldwater affinity (Knouft and Page 2011), and high dispersal ability (Griffiths 2017). The contrasting distributions of goodeids reflect their Neotropical affiliation, warm-water affinity, Miocene origin, and localized life histories that reduce dispersal ability and tendency (Pérez-Rodríguez et al. 2015).

Step 2: Residency

Once they are present, dispersal barriers and habitat-specific adaptations can isolate colonists. Populations that join a pool of residents may persist through evolutionary adaptation, occupation of stable habitats, ability to track habitat across the landscape, or (most likely) some combination of these factors (Minckley et al. 1986). Lineages of residents may colonize adjacent areas if and when suitable dispersal corridors emerge.

Step 3: Faunal Assembly

An assemblage accrues in an area of endemism as multiple lineages colonize it. The reticulate nature of the model (see fig. 3.4) indicates that colonization can be periodic, depending on climate and hydrography. New colonists join residents and, if they persist, eventually become residents themselves (albeit of more recent origin). For example, although the fauna of the Railroad Area (i.e., Railroad Valley) includes just two endemics, *Siphateles obesus* and *Crenichthys nevadae*, they have different times of origin and routes of colonization (Hubbs and Miller 1948). That is, the two-lineage endemic fauna was assembled in two separate events.

It is also possible for representatives from the same clade to colonize an area repeatedly. This adds lineage diversity if lineages remain distinct. For instance, the presence of endemic congeners from two clades in the Devils Area belies a history of alternating isolation and interconnection. In *Dionda*, *D. diaboli* is older and *D. argentosa* younger (Schönhuth et al. 2012a). In *Cyprinella*, *C. proserpina* is older and *C. "lutrensis"* younger (Schönhuth and Mayden 2010). Presumably, the older lineages colonized the Devils Area at an earlier opportunity sometime in the Miocene, preceding a Late Miocene to Pliocene period of isolation (Hoagstrom et al. 2014). The younger lineages colonized the region at a later time, when colonization again became possible (Osborne et al. 2016).

In situ isolation and divergence within an area of endemism can increase the diversity of endemic lineages. In the desert region, this occurred via at least four models of gene flow (Meffe and Vrijenhoek 1988; Hubbs 1995; Langerhans et al. 2007; Hughes et al. 2009): (1) Death Valley model: habitat fragmentation severs gene flow; (2) stream-hierarchy model: drainage configuration mediates gene flow; (3) headwater model: headwater-restricted gene flow; and (4) spring-isolation model: spring-restricted gene flow. The Death Valley model pertains where environmental harshness strongly fragments a population such that remnant populations have complex, idiosyncratic structure independent of geographic proximity, as among Mojave Area *Cyprinodon* (Duvernell and Turner 1998; Echelle 2008). The stream-hierarchy model pertains for populations with genetic structure corresponding to geographic proximity, as in *Pantosteus discobolus* of the Colorado River basin (Hopken et al. 2013), *Lepidomeda copei* (Blakney et al. 2014), *G. robusta* (Dowling et al. 2015), and *Meda fulgida*, *Rhinichthys cobitis*, *Rhinichthys osculus*, and *Pantosteus clarkii* (Pilger et al. 2017). The headwater model pertains to populations restricted to headwater habitats, like *Oncorhynchus* (Campbell et al. 2011; Metcalf et al. 2012) and *Gila pandora* (Galindo et al. 2016). Similarly, the spring-isolation model pertains to populations restricted to spring-fed habitat, like certain *Cyprinella* and *Dionda* (Carson et al. 2014), *Cyprinodon* (Echelle and Echelle 1994; Tobler and Carson 2010), and *Gambusia* (Davis et al. 2006; Hubbs 2014).

Following the terminology of Cowman and Bellwood (2013), each area of endemism is a "center of survival" that has allowed ancient endemic lineages to persist through tens of thousands to millions of years. Areas with diversification in situ also became "centers of origin," nurturing their own lineage diversity. Areas that accrued diversity over time through either colonization or in situ lineage diversification became "centers of accumulation" for endemic diversity. For example, the Pánuco Area is a center of survival for 29 area-specific lineages. Their estimated ages of origin indicate that this has been an area of lineage accumulation for over 11 million years. Intra-area diversification in *Tampichthys* (Schönhuth et al. 2008), *Xiphophorus* (Kang et al. 2013), and *Nosferatu* (de la Maza-Benignos et al. 2015) indicates that the Pánuco Area also is a center of origin.

CONSERVATION

This inventory shows collective biodiversity within the desert region to be the sum of unique contributions from 30 areas of endemism. Persistence of suitable habitat is critical in all areas, making water-resource conservation paramount (Andersen and Deacon 2001; Ruhí et al. 2016). Further, endemics may exhibit adaptations to specific ecological processes that must also be conserved (e.g., Hoagstrom and Turner 2015).

Desert-endemic faunas were assembled over millions of years. Hence the statement, "Irreplaceable life forms are in our collective hands, and they are far too precious to drop" (Pister 1999). These lineages have long relied on ecological refugia that, coupled with their own adaptations, confer population resistance and resilience (Smith et al. 2002; Blakney et al. 2014). Human impacts have made this an insecure legacy through ecosystem degradation (Mueller and Marsh 2002; Miyazono et al. 2015), fragmentation (Fagan et al. 2002; Miyazono and Taylor 2013), and introductions of invasive species (Minckley 1991; Olden et al. 2006). As a guide for conservation, a desert fish habitat ethic—modeled after the Land Ethic of Aldo Leopold (Pister 1992; Piccolo 2017)—is proposed: *A thing is right when it tends to preserve ecosystem processes that support endemic fishes and protect them from negative interactions with nonnatives.*

Imperilment and extinction of desert-endemic fishes are often symptoms of unsustainable water use (Contreras-Balderas and Lozano-Vilano 1994; Deacon et al. 2007). Extreme cases involve outright habitat destruction (Miller 1984; Contreras-Balderas and Lozano-Vilano 1996). Awareness of threats can prompt establishment of captive populations to forestall extinction (Finger et al. 2013; Osborne et al. 2013), but it seems unlikely that available resources will support propagation efforts for 336 endemic lineages. Further, wild populations require functioning ecosystems (Pister and Unkel 1989; Keppel et al. 2012). Hence, the most parsimonious means for comprehensive conservation is preservation and restoration of natural aquatic habitats (Williams et al. 2011; Lozano-Vilano and de la Maza-Benignos 2017) as needed by each endemic lineage.

ACKNOWLEDGMENTS

Mauricio de la Maza-Benignos and Peter Unmack made helpful editorial suggestions. Matthew Mayfield skillfully and generously prepared figure 3.1.

DEDICATION

To Phil Pister, other founders of the Desert Fishes Council, and all DFC members, past, present, and future: you inspire us.

REFERENCES

Abadía-Cardoso, A., J. C. Garza, R. L. Mayden, and F. J. García de León. 2015. Genetic structure of Pacific trout at the extreme southern end of their native range. *PLoS ONE* 10: e0141775.

Abadía-Cardoso, A., D. E. Pearse, S. Jacobson, J. Marshall, D. Dalrymple, F. Kawasaki, G. Ruiz-Campos, and J. C. Garza. 2016. Population genetic structure and ancestry of steelhead/rainbow trout (*Oncorhynchus mykiss*) at the extreme southern edge of their range in North America. *Conservation Genetics* 17: 675–89.

Aguilar, A., and W. J. Jones. 2009. Nuclear and mitochondrial diversification in two native California minnows: Insights into taxonomic identity and regional phylogeography. *Molecular Phylogenetics and Evolution* 51: 373–81.

Andersen, M. E., and J. E. Deacon. 2001. Population size of Devils Hole pupfish (*Cyprinodon diabolis*) correlates with water level. *Copeia* 2001: 224–28.

Arce-H., M., J. G. Lundberg, and M. A. O'Leary. 2017. Phylogeny of the North American catfish family Ictaluridae (Teleostei: Siluriformes) combining morphology, genes and fossils. *Cladistics* 33: 406–28.

Baumsteiger, J., A. P. Kinziger, and A. Aguilar. 2012. Life history and biogeographic diversification of an endemic western North American freshwater fish clade using a comparative species tree approach. *Molecular Phylogenetics and Evolution* 65: 940–52.

Bean, P. T., J. T. Jackson, D. J. McHenry, T. H. Bonner, and M. R. J. Forstner. 2011. Rediscovery of the headwater catfish *Ictalurus lupus* (Ictaluridae) in a western Gulf-Slope drainage. *Southwestern Naturalist* 56: 285–89.

Behnke, R. J. 2002. *Trout and Salmon of North America*. New York: Free Press.

Belk, M. C., and G. B. Schaalje. 2016. Multivariate heritability of shape in June sucker (*Chasmistes liorus*) and Utah sucker (*Catostomus ardens*): Shape as a functional trait for discriminating closely related species. *Development Genes and Evolution* 226: 197–207.

Blakney, J. R., J. L. Loxterman, and E. R. Keeley. 2014. Range-wide comparisons of northern leatherside chub populations reveal historical and contemporary patterns of genetic variation. *Conservation Genetics* 15: 757–70.

Bloom, D. D., K. R. Piller, J. Lyons, N. Mercado-Silva, and M. Medina-Nava. 2009. Systematics and biogeography of the silverside tribe Menidiini (Teleostomi: Atherinopsidae) based on the mitochondrial ND2 gene. *Copeia* 2009: 408–17.

Bossu, C. M., J. M. Beaulieu, P. A. Ceas, and T. J. Near. 2013. Explicit tests of paleodrainage connections of southeastern North America and the historical biogeography of orangethroat darters (Percidae: *Etheostoma*: *Ceasia*). *Molecular Ecology* 22: 5397–417.

Bufalino, A. P., and R. L. Mayden. 2010. Phylogenetic evaluation of North American Leuciscidae (Actinopterygii: Cypriniformes: Cyprinoidea) as inferred from analyses of mitochondrial and nuclear DNA sequences. *Systematics and Biodiversity* 8: 493–505.

Buth, D. G., and R. L. Mayden. 2001. Allozymic and isozymic evidence for polytypy in the North American catostomid genus *Cycleptus*. *Copeia* 2001: 899–906.

Campbell, M. R., C. C. Kozfkay, K. A. Meyer, M. S. Powell, and R. N. Williams. 2011. Historical influences of volcanism and glaciation in shaping mitochondrial DNA variation and distribution in Yellowstone cutthroat trout across its native range. *Transactions of the American Fisheries Society* 140: 91–107.

Caran, S. C., and R. W. Baumgardner, Jr. 1990. Quaternary stratigraphy and paleoenvironments of the Texas Rolling Plains. *Geological Society of America Bulletin* 102: 768–85.

Carson, E. W., A. H. Hanna, G. P. Garrett, R. J. Edwards, and J. R. Gold. 2014. Conservation genetics of cyprinid fishes in the upper Nueces River basin in central Texas. *Southwestern Naturalist* 59: 1–8.

Castañeda-Rivera, M., J. M. Grijalva-Chon, L. E. Gutiérrez-Millán, G. Ruiz-Campos, and A. Varela-Romero. 2014. Analysis of the *Ictalurus pricei* complex (Teleostei: Ictaluridae) in northwest Mexico based on mitochondrial DNA. *Southwestern Naturalist* 59: 434–38.

Chapin, C. E. 2008. Interplay of oceanographic and paleoclimate events with tectonism during middle to late Miocene sedimentation across the southwestern USA. *Geosphere* 4: 976–91.

Chen, W.-J., and R. L. Mayden. 2012. Phylogeny of suckers (Teleostei: Cypriniformes: Catostomidae): Further evidence of relationships provided by the single-copy nuclear gene IRBP2. *Zootaxa* 3586: 195–210.

Clements, M. D., H. L. Bart, Jr., and D. L. Hurley. 2012. A different perspective on the phylogenetic relationships of the Moxostomatini (Cypriniformes: Catostomidae) based on cytochrome-*b* and growth hormone intron sequences. *Molecular Phylogenetics and Evolution* 63: 159–67.

Coghill, L. M., C. D. Hulsey, J. Chaves-Campos, F. J. García de León, and S. G. Johnson. 2013. Phylogeography and conservation genetics of a distinct lineage of sunfish in the Cuatro Ciénegas Valley of Mexico. *PLoS ONE* 8: e77013.

Cole, D. D., K. E. Mock, B. L. Cardall, and T. A. Crowl. 2008. Morphological and genetic structuring in the Utah Lake sucker complex. *Molecular Ecology* 17: 5189–204.

Conway, K. W., and D. Kim. 2016. Redescription of the Texas shiner *Notropis amabilis* from the southwestern United States and northern Mexico with the reinstatement of *N. megalops* (Teleostei: Cyprinidae). *Ichthyological Exploration of Freshwaters* 26: 305–40.

Contreras-Balderas, S. 1969. Perspectivas de la ictiofauna en las zonas áridas del norte de México. *ICASALS Publication* 3: 293–304.

Contreras-Balderas, S., and M. L. Lozano-Vilano. 1994. Water, endangered fishes, and development perspectives in arid lands of Mexico. *Conservation Biology* 8: 379–87.

Contreras-Balderas, S., and M. L. Lozano-Vilano. 1996. Extinction of most Sandia and Potosí valleys (Nuevo León, Mexico) endemic pupfishes, crayfishes and snails. *Ichthyological Exploration of Freshwaters* 7: 33–40.

Cowman, P. F., and D. R. Bellwood. 2013. The historical biogeography of coral reef fishes: Global patterns of origination and dispersal. *Journal of Biogeography* 40: 209–24.

Crête-Lafrenière, A., L. K. Weir, and L. Bernatchez. 2012. Framing the Salmonidae family phylogenetic portrait: A more complete picture from increased taxon sampling. *PLoS ONE* 7: e46662.

Crossey, L. C., K. E. Karlstrom, R. Dorsey, J. Pearce, E. Wan, L. S. Beard, Y. Asmerom, V. Polyak, R. S. Crow, A. Cohen, J. Bright, and M. E. Pecha. 2015. Importance of groundwater in propagating downward integration of the 6–5 ma Colorado River system: Geochemistry of springs, travertines, and lacustrine carbonates of the Grand Canyon region over the past 12 ma. *Geosphere* 11: 660–82.

Culp, J. M., T. D. Prowse, and E. A. Luiker. 2005. Mackenzie River basin. In A. C. Benke and C. E. Cushing, eds., *Rivers of North America*, 804–50. Amsterdam: Elsevier Academic Press.

Currens, K. P., C. B. Schreck, and H. W. Li. 2009. Evolutionary ecology of redband trout. *Transactions of the American Fisheries Society* 138: 797–817.

Davis, S. K., A. A. Echelle, and R. A. Van Den Bussche. 2006. Lack of cytonuclear genetic introgression despite long-term hybridization and backcrossing between two poeciliid fishes (*Gambusia heterochir* and *G. affinis*). *Copeia* 2006: 351–59.

Deacon, J. E., and W. L. Minckley. 1974. Desert fishes. In G. W. Brown, Jr., ed., *Desert Biology*, vol. 2, 385–488. New York: Academic Press.

Deacon, J. E., A. E. Williams, C. Deacon Williams, and J. E. Williams. 2007. Fueling population growth in Las Vegas: How large-scale groundwater withdrawal could burn regional biodiversity. *BioScience* 57: 688–98.

De la Maza-Benignos, M. 2009. *Los Peces del Río Conchos*. Alianza WWF-FGRA y Gobierno del Estado de Chihuahua.

De la Maza-Benignos, M., and M. L. Lozano-Vilano. 2013. Description of three new species of the genus *Herichthys* (Perciformes: Cichlidae) from eastern Mexico, with redescription of *H. labridens, H. steindachneri*, and *H. pantostictus*. *Zootaxa* 3734: 101–29.

De la Maza-Benignos, M., C. P. Ornelas-García, M. L. Lozano-Vilano, M. E. García-Ramírez, and I. Doadrio. 2015. Phylogeographic analysis of genus *Herichthys* (Perciformes: Cichlidae), with descriptions of *Nosferatu* new genus and *H. tephua* n. sp. *Hydrobiologia* 748: 201–31.

Docker, M. F. 2006. Bill Beamish's contributions to lamprey research and recent advances in the field. *Guelph Ichthyology Reviews* 7: 1–52.

Domínguez-Domínguez, O., M. Vila, R. Pérez-Rodríguez, N. Remón, and I. Doadrio. 2011. Complex evolutionary history of the Mexican stoneroller *Campostoma ornatum* Girard, 1856 (Actinopterygii: Cyprinidae). *BMC Evolutionary Biology* 11: 153.

Doosey, M. H., H. L. Bart, Jr., K. Saitoh, and M. Miya. 2010. Phylogenetic relationships of catostomid fishes (Actinopterygii: Cypriniformes), based on mitochondrial ND4/ND5 gene sequences. *Molecular Phylogenetics and Evolution* 54: 1028–34.

Dowling, T. E., C. D. Anderson, P. C. Marsh, and M. S. Rosenberg. 2015. Population structure in the Roundtail Chub (*Gila robusta* complex) of the Gila River basin as determined by microsatellites: Evolutionary and conservation implications. *PLoS ONE* 10: e0139832.

Dowling, T. E., D. F. Markle, G. J. Tranah, E. W. Carson, D. W. Wagman, and B. P. May. 2016. Introgressive hybridization and the evolution of lake-adapted catostomid fishes. *PLoS ONE* 11: e0149884.

Duvernell, D. D., and B. J. Turner. 1998. Evolutionary genetics of Death Valley pupfish populations: Mitochondrial DNA sequence variation and population structure. *Molecular Ecology* 7: 279–88.

Echelle, A. A. 2008. The western North American pupfish clade (Cyprinodontidae: *Cyprinodon*): Mitochondrial DNA divergence and drainage history. *Geological Society of America Special Papers* 439: 27–38.

Echelle, A. A., E. W. Carson, A. F. Echelle, R. A. Van Den Bussche, T. E. Dowling, and A. Meyer. 2005. Historical biogeography of the New-World pupfish genus *Cyprinodon* (Teleostei: Cyprinodontidae). *Copeia* 2005: 320–39.

Echelle, A. A., M. L. Lozano-Vilano, S. Baker, W. D. Wilson, A. F. Echelle, G. P. Garrett, and R. J. Edwards. 2013. Conservation genetics of *Gambusia krumholzi* (Teleostei: Poeciliidae) with assessment of the species status of *G. clarkhubbsi* and hybridization with *G. speciosa*. *Copeia* 2013: 72–79.

Echelle, A. A., M. R. Schwemm, N. J. Lang, J. S. Baker, R. M. Wood, T. J. Near, and W. L. Fisher. 2015. Molecular systematics of the least darter (Percidae: *Etheostoma microperca*): Historical biogeography and conservation implications. *Copeia* 103: 87–98.

Echelle, A. F., and A. A. Echelle. 1994. Assessment of genetic introgression between two pupfish species, *Cyprinodon elegans* and *C. variegatus* (Cyprinodontidae), after more than 20 years of secondary contact. *Copeia* 1994: 590–97.

Espinosa-Pérez, H., and J. L. Castro-Aguirre. 1996. A new freshwater clingfish (Pisces: Gobiesocidae) from Baja California Sur, México. *Bulletin of the Southern California Academy of Science* 95: 120–26.

Fagan, W. F., P. J. Unmack, C. Burgess, and W. L. Minckley. 2002. Rarity, fragmentation, and extinction risk in desert fishes. *Ecology* 83: 3250–56.

Finger, A. J., S. Parmenter, and B. P. May. 2013. Conservation of the Owens pupfish: Genetic effects of multiple translocations and extirpations. *Transactions of the American Fisheries Society* 142: 1430–43.

Galindo, R., W. D. Wilson, and C. A. Caldwell. 2016. Geographic distribution of genetic diversity in populations of Rio Grande chub *Gila pandora*. *Conservation Genetics* 17: 1081–91.

Galloway, W. E., T. L. Whiteaker, and P. Ganey-Curry. 2011. History of Cenozoic North American drainage basin evolution, sediment yield, and accumulation in the Gulf of Mexico basin. *Geosphere* 7: 938–73.

Gámez, N., S. S. Nihei, E. Scheinvar, and J. J. Morrone. 2017. A temporally dynamic approach for cladistic biogeography and the processes underlying the biogeographic patterns of North American deserts. *Journal of Zoological Systematics and Evolutionary Research* 55: 11–18.

García-Ramírez, M. E., S. Contreras-Balderas, and M. L. Lozano-Vilano. 2006. *Fundulus philpisteri* sp. nov. (Teleostei: Fundulidae) from the Río San Fernando basin, Nuevo León, México. In M. L. Lozano-Vilano and A. J. Contreras-Balderas, eds., *Studies of North American Desert Fishes in Honor of E. P (Phil) Pister, Conservationist*, 13–19. Monterrey: Universidad Autónoma de Nuevo León.

Ghedotti, M. J., and M. P. Davis. 2013. Phylogeny, classification, and evolution of salinity tolerance of the North American topminnows and killifishes, family Fundulidae (Teleostei: Cyprinodontiformes). *Fieldiana: Life and Earth Sciences* 7: 1–65.

Griffiths, D. 2015. Connectivity and vagility determine spatial richness gradients and diversification of freshwater fish in North American and Europe. *Biological Journal of the Linnean Society* 116: 773–86.

Griffiths, D. 2017. Connectivity and vagility determine beta diversity and nestedness in North American and European freshwater fish. *Journal of Biogeography* 44: 1723–33.

Griffiths, D., C. McGonigle, and R. Quinn. 2014. Climate and species richness patterns of freshwater fish in North America and Europe. *Journal of Biogeography* 41: 452–63.

Harris, P. M. 2000. Systematic studies of the genus *Siphateles* (Ostariophysi: Cyprinidae) from western North America. PhD diss., Oregon State University.

Hedrick, P. W., and C. R. Hurt. 2012. Conservation genetics and evolution in an endangered species: Research in Sonoran topminnows. *Evolutionary Applications* 5: 806–19.

Hoagstrom, C. W., and T. F. Turner. 2015. Recruitment ecology of pelagic-broadcast spawning minnows: Paradigms from the ocean advance science and conservation of an imperiled freshwater fauna. *Fish and Fisheries* 16: 282–99.

Hoagstrom, C. W., V. Ung, and K. Taylor. 2014. Miocene rivers and taxon cycles clarify the comparative biogeography of North American highland fishes. *Journal of Biogeography* 41: 644–58.

Hopken, M. W., M. R. Douglas, and M. E. Douglas. 2013. Stream hierarchy defines riverscape genetics of a North American desert fish. *Molecular Ecology* 22: 956–71.

Houston, D. D., R. P. Evans, and D. K. Shiozawa. 2015. Pluvial drainage patterns and Holocene desiccation influenced the genetic architecture of relict dace, *Relictus solitarius* (Teleostei: Cyprinidae). *PLoS ONE* 10: e0128433.

Houston, D. D., D. K. Shiozawa, and B. R. Riddle. 2010. Phylogenetic relationships of the western North American cyprinid genus *Richardsonius* with an overview of phylogeographic structure. *Molecular Phylogenetics and Evolution* 55: 259–73.

Houston, D. D., D. K. Shiozawa, B. T. Smith, and B. R. Riddle. 2014. Investigating the effects of Pleistocene events on genetic divergence within *Richardsonius balteatus*, a widely distributed western North American minnow. *BMC Evolutionary Biology* 14: 111.

Hubbs, C. 1995. Springs and spring runs as unique aquatic systems. *Copeia* 1995: 989–91.

Hubbs, C. 2014. Differences in spring versus stream fish assemblages. In C. A. Hoyt and J. Karges, eds., *Proceedings of the Sixth Symposium on the Natural Resources of the Chihuahuan Desert Region*, 376–95. Fort Davis, TX: Chihuahuan Desert Research Institute.

Hubbs, C. L. 1940. Fishes of the desert. *Biologist* 22: 61–69.

Hubbs, C. L., and R. R. Miller. 1948. The zoological evidence: Correlation between fish distribution and hydrographic history in the desert basins of western United States. In E. Blackwelder, *The Great Basin, with Emphasis on Glacial and Postglacial Times*, University of Utah Biological Series vol. 10, no. 7, 17–166. Salt Lake City: University of Utah.

Hughes, J. M., D. J. Schmidt, and D. S. Finn. 2009. Genes in streams: Using DNA to understand the movement of freshwater fauna and their riverine habitat. *BioScience* 59: 573–83.

Hulsey, C. D., P. R. Hollingsworth, Jr., and J. A. Fordyce. 2010. Temporal diversification of Central American cichlids. *BMC Evolutionary Biology* 10: 279.

Jackson, D. A., and H. H. Harvey. 1989. Biogeographic associations in fish assemblages: Local vs. regional processes. *Ecology* 70: 1472–85.

Jackson, D. A., P. R. Peres-Neto, and J. D. Olden. 2001. What controls who is where in freshwater fish communities—the roles of biotic, abiotic, and spatial factors. *Canadian Journal of Fisheries and Aquatic Sciences* 58: 157–70.

Jimenez, M., S. C. Goodchild, C. A. Stockwell, and S. C. Lema. 2017. Characterization and phylogenetic analysis of complete mitochondrial genomes for two desert cyprinodontoid fishes, *Empetrichthys latos* and *Crenichthys baileyi*. *Gene* 626: 163–72.

Johnson, J. B. 2002. Evolution after the flood: Phylogeography of the desert fish Utah chub. *Evolution* 56: 948–60.

Kang, J. H., M. Schartl, R. B. Walter, and A. Meyer. 2013. Comprehensive phylogenetic analysis of all species of swordtails and platies (Pisces: genus *Xiphophorus*) uncovers a hybrid origin of a swordtail fish, *Xiphophorus monticolus*, and demonstrates that the sexually selected sword originated in the ancestral lineage of the genus, but was lost again secondarily. *BMC Evolutionary Biology* 13: 25.

Kelly, N. B., T. J. Near, and S. H. Alonzo. 2012. Diversification of egg-deposition behaviors and the evolution of male parental care in darters (Teleostei: Percidae: Etheostomatinae). *Journal of Evolutionary Biology* 25: 836–46.

Keppel, G., K. P. Van Niel, G. W. Wardell-Johnson, C. J. Yates, M. Byrne, L. Mucina, A. G. T. Schut, S. D. Hopper, and S. E. Franklin. 2012. Refugia: Identifying and understanding safe havens for biodiversity under climate change. *Global Ecology and Biogeography* 21: 393–404.

Kinziger, A. P., R. M. Wood, and D. A. Neely. 2005. Molecular systematics of the genus *Cottus* (Scorpaeniformes: Cottidae). *Copeia* 2005: 303–11.

Knott, J. R., M. N. Machette, R. E. Klinger, A. M. Sarna-Wojcicki, J. C. Liddicoat, J. C. Tinsley III, B. T. David, and V. M. Ebbs. 2008. Reconstructing late Pliocene to middle Pleistocene Death Valley lakes and river systems as a test of pupfish (Cyprinodontidae) dispersal hypotheses. *Geological Society of America Special Papers* 439: 1–26.

Knouft, J. H., and L. M. Page. 2011. Assessment of the relationships of geographic variation in species richness to climate and landscape variables within and among lineages of North American freshwater fishes. *Journal of Biogeography* 38: 2259–69.

Langerhans, R. B., M. E. Gifford, and E. O. Joseph. 2007. Ecological speciation in *Gambusia* fishes. *Evolution* 61: 2056–74.

Legendre, P., and L. Legendre. 1998. *Numerical Ecology*. 2nd English ed. Amsterdam: Elsevier.

Leprieur, F., P. A. Tedesco, B. Hugueny, O. Beauchard, H. H. Dürr, S. Brosse, and T. Oberdorff. 2011. Partitioning global patterns of freshwater fish beta diversity reveals contrasting signatures of past climate changes. *Ecology Letters* 14: 325–34.

Lorion, C. M., D. F. Markle, S. B. Reid, and M. F. Docker. 2000. Redescription of the presumed-extinct Miller Lake lamprey, *Lampetra minima*. *Copeia* 2000: 1019–28.

Loxterman, J. L., and E. R. Keeley. 2012. Watershed boundaries and geographic isolation: Patterns of diversification in cutthroat trout from western North America. *BMC Evolutionary Biology* 12: 38.

Lozano-Vilano, M. L., and M. de la Maza-Benignos. 2017. Diversity and status of Mexican killifishes. *Journal of Fish Biology* 90: 3–38.

Lyons, J., and N. Mercado-Silva. 2004. *Notropis calabazas* (Teleostei; Cyprinidae): New species from the Río Pánuco basin of central México. *Copeia* 2004: 868–75.

Mack, G. H., and K. A. Giles. 2004. *The Geology of New Mexico, a Geologic History*. Special Publication 11. Socorro: New Mexico Geological Society.

Mack, G. H., W. R. Seager, M. R. Leeder, M. Perez-Arlucca, and S. L. Salyards. 2006. Pliocene and Quaternary history of the Rio Grande, the axial river of the southern Rio Grande rift, New Mexico, USA. *Earth-Science Reviews* 79: 141–62.

Madole, R. F., C. R. Ferring, M. J. Guccione, S. A. Hall, W. C. Johnson, and C. J. Sorensen. 1991. Quaternary geology of the Osage Plains and Interior Highlands. In R. B. Morrison, ed., *Quaternary Nonglacial Geology: Conterminous U.S.*, 503–46. Boulder, CO: Geological Society of America.

Markle, D. F., M. R. Cavalluzzi, and D. C. Simon. 2005. Morphology and taxonomy of Klamath Basin suckers (Catostomidae). *Western North American Naturalist* 65: 473–89.

Martin, S. D., and R. M. Bonett. 2015. Biogeography and divergent patterns of body size disparification in North American minnows. *Molecular Phylogenetics and Evolution* 93: 17–28.

Mateos, M., O. I. Sanjur, and R. C. Vrijenhoek. 2002. Historical biogeography of the livebearing fish genus *Poeciliopsis* (Poeciliidae: Cyprinodontiformes). *Evolution* 56: 972–84.

McClure-Baker, S. A., A. A. Echelle, R. A. Van den Bussche, A. F. Echelle, D. A. Hendrickson, and G. P. Garrett. 2010. Genetic status of headwater catfish in Texas and New Mexico: A perspective from mtDNA and morphology. *Transactions of the American Fisheries Society* 139: 1780–91.

McPhail, J. D., and C. C. Lindsey. 1970. *Freshwater Fishes of Northwestern Canada and Alaska*. Fisheries Research Board of Canada Bulletin 173.

McPhee, M. V., M. J. Osborne, and T. F. Turner. 2008. Genetic diversity, population structure, and demographic history of the Rio Grande sucker, *Catostomus* (*Pantosteus*) *plebeius*, in New Mexico. *Copeia* 2008: 191–99.

Meffe, G. K., and R. C. Vrijenhoek. 1988. Conservation genetics in the management of desert fishes. *Conservation Biology* 2: 157–69.

Meredith, R. W., M. N. Pires, D. N. Reznick, and M. S. Springer. 2011. Molecular phylogenetic relationships and the coevolution of placentotrophy and superfetation in *Poecilia* (Poeciliidae: Cyprinodontiformes). *Molecular Phylogenetics and Evolution* 59: 148–57.

Metcalf, J. L., S. L. Stowell, C. M. Kennedy, K. B. Rogers, D. McDonald, J. Epp, K. Keepers, A. Cooper, J. J. Austin, and A. P. Martin. 2012. Historical stocking data and 19th century DNA reveal human-induced changes to native diversity and distribution of cutthroat trout. *Molecular Ecology* 21: 5194–207.

Meyer, M. K., S. Schories, and M. Schartl. 2010. Description of *Gambusia zarskei* sp. nov.—a new poeciliid fish from the upper Rio Conchos system, Chihuahua, Mexico (Teleostei: Cyprinodontiformes: Poeciliidae). *Vertebrate Zoology* 60: 11–18.

Miller, B. A. 2006. The phylogeography of *Prosopium* in western North America. MSc thesis, Brigham Young University.

Miller, R. R. 1946. The need for ichthyological surveys of the major rivers of western North America. *Science* 104: 517–19.

Miller, R. R. 1961. Man and the changing fish fauna of the American Southwest. *Papers of the Michigan Academy of Science, Arts, and Letters* 46: 365–404.

Miller, R. R. 1984. *Rhinichthys deaconi*, a new species of dace (Pisces: Cyprinidae) from southern Nevada. *Occasional Papers of the Museum of Zoology, University of Michigan* 707: 1–21.

Miller, R. R., and A. A. Echelle. 1975. *Cyprinodon tularosa*, a new cyprinodontid fish from the Tularosa Basin, New Mexico. *Southwestern Naturalist* 19: 365–77.

Miller, R. R., W. L. Minckley, and S. M. Norris. 2005. *Freshwater fishes of México*. Chicago: University of Chicago Press.

Minckley, W. L. 1991. *Native Fishes of Arid Lands: A Dwindling Resource of the Desert Southwest*. USDA Forest Service, Rocky Mountain Forest and Range Experiment Station, General Technical Report RM-206: 1–45.

Minckley, W. L., and J. E. Deacon. 1968. Southwestern fishes and the enigma of "endangered species." *Science* 159: 1424–32.

Minckley, W. L., D. A. Hendrickson, and C. E. Bond. 1986. Geography of western North American freshwater fishes: Description and relationships to intracontinental tectonism. In C. H. Hocutt and E. O. Wiley, eds., *The Zoogeography of North American Freshwater Fishes*, 519–613. New York: John Wiley & Sons.

Minckley, W. L., and P. C. Marsh. 2009. *Inland Fishes of the Greater Southwest: Chronicle of a Vanishing Biota*. Tucson: University of Arizona Press.

Miyazono, S., R. Patiño, and C. M. Taylor. 2015. Desertification, salinization, and biotic homogenization in a dryland river ecosystem. *Journal of the Total Environment* 511: 444–53.

Miyazono, S., and C. M. Taylor. 2013. Effects of habitat size and isolation on species immigration-extinction dynamics and community nestedness in a desert river system. *Freshwater Biology* 58: 1303–12.

Mock, K. E., R. P. Evans, M. Crawford, B. L. Cardall, S. U. Janecke, and M. P. Miller. 2006. Rangewide molecular structuring in the Utah sucker (*Catostomus ardens*). *Molecular Ecology* 15: 2223–38.

Moyer, G. R., R. K. Remington, and T. F. Turner. 2009. Incongruent gene trees, complex evolutionary processes, and the phylogeny of a group of North American minnows (*Hybognathus* Agassiz 1855). *Molecular Phylogenetics and Evolution* 50: 514–25.

Moyle, P. B. 2002. *Inland Fishes of California*. Revised and expanded. Berkeley: University of California Press.

Mueller, G. A., and P. C. Marsh. 2002. *Lost: A Desert River and Its Native Fishes: A Historical Perspective of the Lower Colorado River*. US Geological Survey Information and Technology Report USGS/BRD/ITR-2002-0010.

Near, T. J., and M. F. Benard. 2004. Rapid allopatric speciation in logperch darters (Percidae: *Percina*). *Evolution* 58: 2798–808.

Near, T. J., D. I. Bolnick, and P. C. Wainwright. 2005. Fossil calibrations and molecular divergence time estimates in centrarchid fishes (Teleostei: Centrarchidae). *Evolution* 59: 1768–882.

Near, T. J., T. W. Kassler, J. B. Koppelman, C. B. Dillman, and D. P. Philipp. 2003. Speciation in North American black basses, *Micropterus* (Actinopterygii: Centrarchidae). *Evolution* 57: 1610–21.

Negrini, R. M. 2002. Pluvial lake sizes in the northwestern Great Basin throughout the Quaternary period. In R. Hershler, D. B. Madsen, and D. R. Currey, eds., *Great Basin Aquatic Systems History*, 11–52. Smithsonian Contributions to the Earth Sciences, 33.

Nelson, J. S., and M. J. Paetz. 1992. *The Fishes of Alberta*. 2nd ed. Edmonton: University of Alberta Press.

Oberdorff, T., P. A. Tedesco, B. Hugueny, F. Leprieur, O. Beauchard, S. Brosse, and H. H. Dürr. 2011. Global and regional patterns in riverine fish species richness: A review. *International Journal of Ecology* 2011: 967631.

Olden, J. D., N. L. Poff, and K. R. Bestgen. 2006. Life-history strategies predict fish invasions and extirpations in the Colorado River basin. *Ecological Monographs* 76: 25–40.

Osborne, M. J., T. A. Diver, C. W. Hoagstrom, and T. F. Turner. 2016. Biogeography of "*Cyprinella lutrensis*": Intensive genetic sampling from the Pecos River "melting pot" reveals a dynamic history and phylogenetic complexity. *Biological Journal of the Linnean Society* 117: 264–84.

Osborne, M. J., T. L. Perez, C. S. Altenbach, and T. F. Turner. 2013. Genetic analysis of captive spawning strategies for the endangered Rio Grande Silvery Minnow. *Journal of Heredity* 104: 437–46.

Page, L. M., and B. M. Burr. 2011. *Peterson Field Guide to Freshwater Fishes of North America North of Mexico*. 2nd ed. Boston: Houghton Mifflin Harcourt.

Page, L. M., H. Espinosa-Pérez, L. T. Findley, C. R. Gilbert, R. N. Lea, N. E. Mandrak, R. L. Mayden, and J. S. Nelson. 2013. *Common and Scientific Names of Fishes from the United States, Canada, and Mexico*. 7th ed. Special Publication 34. Bethesda, MD: American Fisheries Society.

Parenti, L. R., and M. C. Ebach. 2009. *Comparative Biogeography, Discovering and Classifying Biogeographical Patterns of a Dynamic Earth*. Los Angeles: University of California Press.

Parsons, A. J., and A. D. Abrahams. 2009. Geomorphology of desert environments. In A. J. Parsons and A. D. Abrahams, eds., *Geomorphology of Desert Environments*, 2nd ed., 3–7. Berlin: Springer.

Pérez-Rodríguez, R., O. Domínguez-Domínguez, I. Doadrio, E. Cuevas-García, and G. Pérez-Ponce de León. 2015. Comparative historical biogeography of three groups of Nearctic freshwater fishes across central Mexico. *Journal of Fish Biology* 86: 993–1015.

Pérez-Rodríguez, R., O. Domínguez-Domínguez, A. F. Mar-Silva, I. Doadrio, and G. Pérez-Ponce de León. 2016. The historical biogeography of the southern group of the sucker genus *Moxostoma* (Teleostei: Catostomidae) and the colonization of central Mexico. *Zoological Journal of the Linnean Society* 177: 633–47.

Piccolo, J. J. 2017. The Land Ethic and conservation of native salmonids. *Ecology of Freshwater Fish* 26: 160–64.

Pilger, T. J., K. B. Gido, D. L. Propst, J. E. Whitney, and T. F. Turner. 2017. River network architecture, genetic effective size, and distributional patterns predict differences in genetic structure across species in a dryland stream fish community. *Molecular Ecology* 26: 2687–97.

Pister, E. P. 1990. Desert fishes: An interdisciplinary approach to endangered species conservation in North America. *Journal of Fish Biology* 37 (supplement A): 183–87.

Pister, E. P. 1991. The Desert Fishes Council: Catalyst for change. In W. L. Minckley and J. E. Deacon, eds., *Battle against Extinction: Native Fish Management in the American West*, 55–68. Tucson: University of Arizona Press.

Pister, E. P. 1992. Ethical considerations in conservation of biodiversity. *Transactions of the North American Wildlife and Natural Resources Conference* 57: 355–64.

Pister, E. P. 1999. Professional obligations in the conservation of fishes. *Environmental Biology of Fishes* 55: 13–20.

Pister, E. P., and W. C. Unkel. 1989. Conservation of western wetlands and their fishes. In R. R. Sharitz and J. W. Gibbons, eds., *Freshwater Wetlands and Wildlife*, 475–85. Springfield, VA: National Technical Information Service.

Rauchenberger, M. 1989. Systematics and biogeography of the genus *Gambusia* (Cyprinodontiformes: Poeciliidae). *American Museum Novitates* 2951: 1–74.

Reheis, M. C., K. D. Adams, C. G. Oviatt, and S. N. Bacon. 2014. Pluvial lakes in the Great Basin of the western United States—a view from the outcrop. *Quaternary Science Reviews* 97: 33–57.

Reheis, M. C., A. M. Sarna-Wojcicki, R. L. Reynolds, C. A. Repenning, and M. D. Mifflin. 2002b. Pliocene to middle Pleistocene lakes in the western Great Basin: Ages and connections. In R. Hershler, D. B. Madsen, and D. R. Currey, eds., *Great Basin Aquatic Systems History*, 53–108. Smithsonian Contributions to the Earth Sciences, 33.

Reheis, M. C., S. Stine, and A. M. Sarna-Wojcicki. 2002a. Drainage reversals in Mono Basin during the late Pliocene and Pleistocene. *Geological Society of America Bulletin* 114: 991–1006.

Robison, H. W., and T. M. Buchanan. 1988. *Fishes of Arkansas*. Fayetteville: University of Arkansas Press.

Rogers, K. B., K. R. Bestgen, and J. Epp. 2014. Using genetic diversity to inform conservation efforts for native cutthroat trout of the southern Rocky Mountains. *Wild Trout Symposium* 11: 218–28.

Roskowski, J. A., P. J. Patchett, J. E. Spencer, P. A. Pearthree, D. L. Dettman, J. E. Faulds, and A. C. Reynolds. 2010. A late Miocene–early Pliocene chain of lakes fed by the Colorado River: Evidence from Sr, C, and O isotopes of the Bouse Formation and related units between Grand Canyon and the Gulf of California. *Geological Society of America Bulletin* 122: 1625–36.

Ruhí, A., J. D. Olden, and J. L. Sabo. 2016. Declining streamflow induces collapse and replacement of native fish in the American Southwest. *Frontiers in Ecology and the Environment* 14: 465–72.

Ruiz-Campos, G., F. Camarena-Rosales, A. González-Acosta, A. M., Maeda-Martínez, F. J. García de León, A. Varela-Romero, and A. Andreu-Soler. 2014. Estatus actual de conservación de seis especies de peces dulceacuícolas de la península de Baja California, México. *Revista Mexicana de Biodiversidad* 85: 1235–48.

Ruiz-Campos, G., J. L. Castro-Aguirre, S. Contreras-Balderas, M. L. Lozano-Vilano, A. F. González-Acosta, and S. Sánchez-González. 2002. An annotated distributional checklist of the freshwater fish from Baja California Sur, México. *Reviews in Fish Biology and Fisheries* 12: 143–55.

Sağlam, İ. K., D. J. Prince, M. Meek, O. A. Ali, M. R. Miller, M. Peacock, H. Neville, A. Goodbla, C. Mellison, W. Somer, B. May, and A. J. Finger. 2017. Genomic analysis reveals genetic distinctiveness of the Paiute cutthroat trout (*Oncorhynchus clarkii seleniris*). *Transactions of the American Fisheries Society* 146: 1291–302.

Schönhuth, S., M. J. Blum, M. L. Lozano-Vilano, D. A. Neely, A. Varela-Romero, H. Espinosa, A. Perdices, and R. L. Mayden. 2011. Inter-basin exchange and repeated headwater capture across the Sierra Madre Occidental inferred from the phylogeography of Mexican stonerollers. *Journal of Biogeography* 38: 1406–21.

Schönhuth, S., I. Doadrio, O. Domínguez-Domínguez, D. M. Hillis, and R. L. Mayden. 2008. Molecular evolution of southern North American Cyprinidae (Actinopterygii), with the description of the new genus *Tampichthys* from central Mexico. *Molecular Phylogenetics and Evolution* 47: 729–56.

Schönhuth, S., D. M. Hillis, D. A. Neely, M. L. Lozano-Vilano, A. Perdices, and R. L. Mayden. 2012a. Phylogeny, diversity, and species delimitation of the North American round-nosed minnows (Teleostei: *Dionda*), as inferred from mitochondrial and nuclear DNA sequences. *Molecular Phylogenetics and Evolution* 62: 427–46.

Schönhuth, S., M. L. Lozano-Vilano, A. Perdices, H. Espinosa, and R. L. Mayden. 2015. Phylogeny, genetic diversity and phylogeography of the genus *Codoma* (Teleostei, Cyprinidae). *Zoologica Scripta* 44: 11–28.

Schönhuth, S., and R. L. Mayden. 2010. Phylogenetic relationships in the genus *Cyprinella* (Actinopterygii: Cyprinidae) based on mitochondrial and nuclear gene sequences. *Molecular Phylogenetics and Evolution* 55: 77–98.

Schönhuth, S., A. Perdices, M. L. Lozano-Vilano, F. J. García de León, H. Espinosa, and R. L. Mayden. 2014. Phylogenetic relationships of North American western chubs of the genus *Gila* (Cyprinidae, Teleostei), with emphasis on southern species. *Molecular Phylogenetics and Evolution* 70: 210–30.

Schönhuth, S., D. K. Shiozawa, T. E. Dowling, and R. L. Mayden. 2012b. Molecular systematics of western North American cyprinids (Cypriniformes: Cyprinidae). *Zootaxa* 3586: 281–303.

Sheng, Z., R. E. Mace, and M. P. Fahy. 2001. The Hueco Bolson: An aquifer at the crossroads. In R. E. Mace, W. F. Mullican III, and E. S. Angle, eds., *Aquifers of West Texas*, 66–75. Texas Water Development Board Report no. 356.

Smith, G. R., C. Badgley, T. P. Eiting, and P. S. Larson. 2010. Species diversity gradients in relation to geological history in North American freshwater fishes. *Evolutionary Ecology Research* 12: 693–726.

Smith, G. R., J. Chow, P. J. Unmack, D. F. Markle, and T. E. Dowling. 2017. Fishes of the Mio-Pliocene western Snake River Plain and vicinity. II. Evolution of the *Rhinichthys osculus* complex (Teleostei: Cyprinidae) in western North America. *Miscellaneous Publications of the Museum of Zoology, University of Michigan* 204: 45–83.

Smith, G. R., T. E. Dowling, K. W. Gobalet, T. Lugaski, D. K. Shiozawa, and R. P. Evans. 2002. Biogeography and timing of evolutionary events among Great Basin fishes. In R. Hershler, D. B. Madsen, and D. R. Currey, eds., *Great Basin Aquatic Systems History*, 175–234. Smithsonian Contributions to the Earth Sciences, 33.

Smith, M. L., and R. R. Miller. 1986. The evolution of the Rio Grande basin as inferred from its fish fauna. In C. H. Hocutt and E. O. Wiley, eds., *The Zoogeography of North American Freshwater Fishes*, 457–85. New York: John Wiley & Sons.

Smith, W. L., and M. S. Busby. 2014. Phylogeny and taxonomy of sculpins, sandfishes, and snailfishes (Perciformes: Cottoidei) with comments on the phylogenetic significance of their early-life-history specializations. *Molecular Phylogenetics and Evolution* 79: 332–52.

Spencer, J. E., G. R. Smith, and T. E. Dowling. 2008. Middle to late Cenozoic geology, hydrography, and fish evolution in the American Southwest. *Geological Society of America Special Papers* 439: 279–99.

Stewart, K. W., and D. A. Watkinson. 2004. *The Freshwater Fishes of Manitoba*. Winnipeg: University of Manitoba Press.

Strain, W. S. 1971. Late Cenozoic bolson integration in the Chihuahua tectonic belt. *West Texas Geological Society Publication* 71: 167–73.

Thomas, D. S. G. 2011. Arid environments: Their nature and extent. In D. S. G. Thomas, ed., *Arid Zone Geomorphology: Process, Form and Change in Drylands*, 3rd ed., 3–16. Oxford: Wiley-Blackwell.

Tobler, M., and E. W. Carson. 2010. Environmental variation, hybridization, and phenotypic diversification in Cuatro Ciénegas pupfishes. *Journal of Evolutionary Biology* 23: 1475–89.

Tonn, W. M., J. J. Magnuson, M. Rask, and J. Toivonen. 1990. Intercontinental comparison of small-lake fish assemblages: The balance between local and regional processes. *American Naturalist* 136: 345–75.

Unmack, P. J., T. E. Dowling, N. J. Laitinen, C. L. Secor, R. L. Mayden, D. K. Shiozawa, and G. R. Smith. 2014. Influence of introgression and geological processes on phylogenetic relationships of western North American mountain suckers (*Pantosteus*, Catostomidae). *PLoS ONE* 9: e900061.

Webb, S. A., J. A. Graves, C. Macias-Garcia, A. E. Magurran, D. Ó. Foighil, and M. G. Ritchie. 2004. Molecular phylogeny of the livebearing Goodeidae (Cyprinodontiformes). *Molecular Phylogenetics and Evolution* 30: 527–44.

Williams, J. E., and G. R. Wilde. 1981. Taxonomic status and morphology of isolated populations of the White River springfish, *Crenichthys baileyi* (Cyprinodontidae). *Southwestern Naturalist* 25: 485–503.

Williams, J. E., R. N. Williams, R. F. Thurow, L. Elwell, D. P. Philipp, F. A. Harris, J. L. Kershner, P. J. Martinez, D. Miller, G. H. Reeves, C. A. Frissell, and J. R. Sedell. 2011. Native fish conservation areas: A vision for larger-scale conservation of native fish communities. *Fisheries* 36: 267–77.

Wilson, W. D., and T. F. Turner. 2009. Phylogenetic analysis of Pacific cutthroat trout (*Oncorhynchus clarki* ssp.: Salmonidae) based on partial mtDNA ND4 sequences: A closer look at the highly fragmented inland species. *Molecular Phylogenetics and Evolution* 52: 406–15.

4 Living with Aliens

Nonnative Fishes in the American Southwest

Peter B. Moyle

Alien fishes and invertebrates dominate most of the aquatic ecosystems in the American Southwest (Olden and Poff 2005; Gido et al. 2013). They are not going away and are likely to increase in distribution and abundance because of climate change, increased demand for water, and continued introductions and movement by people. This chapter is an essay dealing with the ensuing problems for native fishes. It does not deal extensively with the broad issues of invasive species and their impacts, which are covered in numerous publications (e.g., Fuller et al. 1999; Moyle and Marchetti 2006). This chapter also does not include a catalog of alien species and their impacts in the Southwest. Instead, its goal is to address the basic problem: How do we live with alien species when they have so many negative effects on native species?

First, what is an alien (nonnative) species? I follow Blackburn et al. (2014, 4) who define an alien species as "a species moved by human activities beyond the limits of its native geographic range into an area in which it does not naturally occur. The movement allows the species to overcome fundamental biogeographic barriers to its natural dispersal. Common synonyms are exotic, introduced, nonindigenous, or non-native." When a species spreads widely and does demonstrable harm, it is labeled as invasive.

The basic problem, of course, is that most alien fishes have desirable features, such as being large, edible, and sporty, and so have a strong constituency for maintaining their populations. We often forget the huge efforts that were made to bring alien species to the Southwest. Common carp *Cyprinus carpio* were carried from Europe to California in barrels in ships that went around Cape Horn. Carp, trout, bass, sunfish, and catfish were transported throughout the West in special railroad cars, with agency biologists meeting the trains to deliver the fish to distant locations by truck, horse, and backpack.[1] At that time, most native fishes in the Southwest were regarded by decision makers as "rough fish" or "trash fish," even if they were valued for food and fertilizer by local people.

There was some shift in cultural values in the United States after the Endangered Species Act of 1973 placed a high value on obscure native fishes, especially those headed toward extinction, as this volume demonstrates. Such fishes often compete directly with alien game fishes for habitat, especially in water released from dams. And most people who angle still prefer to catch the aliens. Alien-fish catchers still move fish around, despite prohibitions against doing so. In California, for example, northern pike *Esox lucius* were illegally introduced twice and eradicated at great expense (Moyle 2002).

Alien fishes are especially successful in the Southwest because almost every stream of size is dammed. Reservoirs behind dams favor a wide array of aliens, such as common carp, centrarchid basses, and catfishes, while excluding most natives. And alien fishes also colonize reaches upstream and downstream from the reservoirs. Reservoirs are aquatic sores on the landscape that infect every waterway to which they connect with alien creatures.

What all this means is that, like it or not, we have to live with alien species that have varying impacts on native fishes. The question then becomes, Can we live with them and still have thriving native fishes? The answer, naturally, is "It depends." Unfortunately, it depends on many factors, including idiosyncratic traits of both native and alien species and how much effort is devoted to conservation of natives and their habitats. But battling aliens seems to be the favored, if failing, strategy, as discussed in the rest of this essay. The essay is mostly couched in military terms because that is standard practice when discussing alien invaders.[2]

BATTLING ALIENS: ARE WE LOSING THE WAR?

The predecessor of this book is entitled *Battle against Extinction*, and alien invasions are one front in this battle (Minckley and Deacon 1991). Alien species are a leading cause of declines of native fishes, in conjunction with habitat loss and change. As Arthington et al. (2016) point out, however, extinctions of freshwater fishes have been surprisingly rare. The IUCN has documented, so far, "only" 69 global extinctions of freshwater fishes (Darwall and Freyhof 2016). But many species are in the queue leading to extinction, especially in the Southwest, and are likely to be "saved" only by extreme measures, such as breeding in conservation hatcheries (Baumsteiger and Moyle 2017). In addition, climate change, with its increase in water temperatures and changes in precipitation patterns, is likely to favor alien fishes throughout the western United States (Moyle et al. 2013).

The global shift in fish faunas toward alien species is resulting in biotic homogenization (Rahel 2002; Olden et al. 2004). This homogenization, in turn, reflects convergence of aquatic habitats, especially riverine habitats dominated by reservoirs (Poff et al. 2007). In the Southwest, as in other arid regions, the dominant fishes in most rivers are of diverse origins, including common carp (Eurasia) and one or more species of "black bass" (*Micropterus*, eastern USA), sunfish (*Lepomis*, eastern USA), mosquitofish (*Gambusia*, southeastern USA), tilapias (*Oreochromis*, Africa), bullhead catfishes (*Ameiurus*, eastern USA), and trout (Salmonidae, multiple origins) (Rahel 2002; see also regional sources). Usually, there are some native fishes that persist despite altered conditions; in the Southwest, they are typically suckers (*Catostomus*), which have flexible life histories and physiological and behavioral mechanisms for surviving adverse conditions. But a major contributor to homogenization is loss of native species, often local endemics, following invasions of widespread alien species (Light and Moyle 2015).[3]

The forward movement of this juggernaut indicates that we are losing the battle against extinction in the Southwest and against the seizure of aquatic ecosystems by alien invaders. It suggests that we need to develop alternative management strategies for our aquatic

1. A book that documents the incredible efforts to bring alien fishes to California alone is Dill and Cordone (1997).

2. This tradition is partly due to Charles Elton (1958), who uses military metaphors in his classic book *The Ecology of Invasions of Animals and Plants*. Larson (2005) argues that using militaristic terms in invasion biology is counterproductive.

3. Baumsteiger and Moyle (2017) document that determining extinction in freshwater fishes is quite difficult, much less determining exact causes.

faunas, especially endemic fishes. At present, the best strategies mostly involve intensively managed refuges, fortresses we can defend against invaders.[4]

BETRAYAL? NATIVE SPECIES AS INVADERS

As ecosystems change in response to anthropogenic disturbance and alien invasions, native fishes have three alternatives: (1) extirpation, (2) integration (adaptation), or (3) invasion, especially of new habitats. Extirpation is the unfortunate pathway of native fishes that cannot adapt quickly to the one-two punch of habitat change and alien invaders. Integration means living in ecosystems with alien species. In California, for example, Sacramento sucker *C. occidentalis*, prickly sculpin *Cottus asper*, hitch *Lavinia exilicauda*, and rainbow trout *Oncorhynchus mykiss* are native species that have successfully joined reservoir fish assemblages. Native invaders are native fishes that have become abundant in altered habitats, either within their historical native range or adjacent to it, and often have negative effects on other natives.

In the Pecos River, New Mexico, Hoagstrom et al. (2010) documented the complete replacement of the Rio Grande silvery minnow *Hybognathus amarus* by the nonnative plains minnow *H. placitus* in a 158 km reach of river over 10 years. The two species are similar in morphology and behavior, but the plains minnow is widespread, while the endangered Rio Grande silvery minnow is endemic to the Rio Grande basin. This replacement of endemic species with more widespread similar species seems to be a general problem in streams of the Rio Grande watershed, especially where streams have been altered (Hoagstrom et al. 2010). Likewise, Bonner and Wilde (2000) found that decreasing flows in the Canadian River downstream of a Texas reservoir resulted in invasion of native fishes from tributaries, which largely replaced the native big-river fauna.

As the saying goes, "Nature abhors a vacuum,"[5] so it should not be surprising that native fishes will invade a new or highly disturbed habitat in absence of better-adapted alien invaders or that some natives actually thrive in such habitats. Examples of native invaders are relatively uncommon compared with outside invaders, suggesting that they rarely meet the criteria needed to be a successful invader or human symbiont (Moyle and Marchetti 2006).

A potential pathway for a native invader is through translocation: introduction into new, often fishless, habitats for the purposes of conservation (Lema et al., chap. 22, this volume). Translocation is a fairly common practice for pupfishes (*Cyprinodon*) and other small fishes in the Southwest, as well as for native trout (Fausch et al. 2009). There is likely to be increased consideration of translocation as a response to global warming, where survival of a species may depend on moving it to cooler water (at higher elevation or latitude). The dilemma is that such fishes may assume the characteristics of an invasive species in the new habitat, where they may endanger endemic invertebrates and amphibians.

A special case of native invasion involves the reintroduction of native fishes into native habitats that have been severely altered, usually through continuous plantings of hatchery fish. This practice is common with rainbow trout, for example, throughout its native range. It is a strategy currently being used to place razorback suckers *Xyrauchen texanus* back in the Colorado River (Bestgen et al., chap. 21, this volume). Whether the resulting fish population represents the species or a simulacrum of it is an open question (Baumsteiger and Moyle 2017).

GETTING ALONG: SYMBIOSIS

Understanding each species' biological relationship to humans can save managers a lot of trouble. It is also important to understand that such relationships can change as societal values change. The interactions of humans and alien species can be regarded as symbioses, usually expressed as mutualism, commensalism, or parasitism (or some other negative interaction). For alien game fishes, the symbiotic relationship is mutualism, where both species benefit. Trout and bass species enjoy global distributions, thanks to enthusiastic movement by people who wanted fisheries for cultur-

4. The need for refuges (preserves) was an important theme of the predecessor of this book; see Moyle and Sato (1991) and Williams (1991) and section 4 of this volume. Closs et al. (2016) devote considerable space in their book to the difficulties of fish conservation.

5. This proverb is attributed to Aristotle. It was used in an ecological and evolutionary context by naturalist Joseph Grinnell in a 1924 essay in *Ecology* [5(3):225–29].

ally desirable fishes in their local waters. In many areas, however, the relationship has turned negative as harm to newly valued native fishes and ecosystems is realized. Large sums are now spent in the Southwest using piscicides and other means to eradicate alien trout in favor of native species.

For some introductions, the relationship is commensal, where the species benefits with no known harm to humans or ecosystems. For example, bigscale logperch *Percina macrolepida* was introduced as a hitchhiker in a shipment of largemouth bass into California. It quickly spread throughout the Central Valley and Southern California, inhabiting, with other alien fishes, highly altered habitats such as leveed sloughs, reservoirs, and rice paddies. There have been no obvious impacts on native fishes despite its abundance (Moyle 2002). Ironically, it is considered a threatened species in its native New Mexico.

Relationships where the species benefits but humans, or at least human-dominated ecosystems, are harmed are many, although often equivocal. One person's pest is another person's—or animal's—food. For example, Mississippi silverside *Menidia audens* was introduced into California for insect control. It spread like a disease through California's aqueduct system to both natural and artificial systems. There it is doing harm by preying on or competing with native fishes such as delta smelt *Hypomesus transpacificus*, but also benefits herons and terns because it serves as prey (Moyle 2002; PBM, pers. obs).

TAKING ACTION: WHEN TO FIGHT

The best defense is offense when it comes to alien invasions: there is no substitute for prevention of species introductions. State and federal agencies have had some success in preventing invasions, but lack sufficient resources to engage in the monitoring and education efforts needed for effective prevention. On a large scale, funding is more likely to be available for emergency eradication than for systematic prevention campaigns, despite the difficulty of doing battle after an invasion has taken place. However, there are clearly some alien species and situations that respond to strong management actions, while there are others that do not. This means that some kind of decision-making process is needed to determine the best use of limited resources.

The choices for setting management priorities, especially for newly established aliens, can be labeled as (1) eradication, (2) control, and (3) acceptance (doing nothing). Using a "triage" system like this is a bit simplistic, but the alternatives are either not to set priorities systematically or to develop a complex prioritization formula, such as that of Blackburn et al. (2014). Wiens (2016) points out the many problems of the latter approach, including lack of sufficient information to make such a formula work for taking quick, cost-effective action.

Britton et al. (2011) present a decision tree for assessing the risks involved in taking action to deal with a fish invasion, including risks to the environment and to other species caused by the management action. The tree provides a systematic framework for decision making. This tool was successfully applied retrospectively to three fish introductions in Great Britain, using a quantitative risk analysis tool (Fish Invasiveness Scoring Kit, or FISK) (Vilizzi and Copp 2013). It has the additional advantage of being able to predict potential impacts of likely invaders before they invade. One case history in Britton et al. (2011) is fathead minnow *Pimephales promelas*, which is frequently introduced into waters of the Southwest by its use as bait. The minnow was deemed a low-risk species, except possibly to amphibians, but its presence in just two ponds made eradication justified and practical.

Eradication

Eradication, or at least containment, is typically aimed at either new invaders where the invasion is taking place or at established populations of harmful aliens in a confined area. There are many examples of successful eradication of alien trout from mountain lakes and small streams to benefit either native trout or amphibians. Diverse species of trout were planted in "barren" mountain lakes throughout the West, where they changed the lake ecosystems and, in particular, wiped out frogs and salamanders that depended on the lakes. In the Sierra Nevada, California, eradication of trout from lakes using gillnets resulted in recovery of native frogs (*Rana* spp.) in many locations (Knapp et al. 2007).

A more complex case is that of Fossil Creek, Arizona. It had flows regulated by a dam and was invaded by alien fishes, so five native fishes were in danger of extirpation. Eradication of the alien fishes, along with improved flows due to dam decommissioning, resulted in rapid recovery of native fish populations. Removal of alien fish was determined the major reason for success (Marks et al. 2010).

Perhaps the biggest eradication program currently proposed is the use of cyprinid herpesvirus-3 to eradicate common carp from the Murray-Darling River system in Australia (McColl et al. 2014). As far as is known, the virus affects only common carp. McColl et al. (2014) point out that success of the virus program will depend on using other methods for control as well in an integrated pest management program. If this eradication program is successful, no doubt similar efforts will be attempted wherever common carp have been introduced.

It is worth emphasizing that eradication programs have a long history, with many failures, mostly because it is extremely difficult to kill every fish in a treated system. This difficulty increases with the size of the water body being treated. Most infamous was the attempt to eradicate common carp and other aliens, as well as native minnows and suckers, from the Green River in Utah in 1962, prior to the completion of Flaming Gorge Dam. The operation contributed to the endangerment of a number of native fishes, but had little long-term effect on alien fishes, which thrived in the altered flow regime. Common carp, especially, have resiliency because of their longevity, mobility, physiological tolerances, and high fecundity; they were back and reproducing in the poisoned section of the river by 1964 (Holden 1991).

Overall, an alien fish species considered for eradication should have (1) a population confined to an isolated area, (2) a high likelihood of being harmful if allowed to spread from the introduction site, and (3) a high probability of actually being eradicated by the methods being considered. The eradication program should also have no demonstrable environmental harm, such as also eradicating endemic invertebrates or having undesirable long-term ecosystem consequences. Kopf et al. (2017) note that for long-established species being considered for eradication, the potential impact of loss of that species from the ecosystem also needs to be considered; they propose twelve questions that should be answered before an eradication program proceeds.

Control

Control is often an alternative where eradication is not possible, especially for long-established species that cause problems for fisheries or endangered species. For example, many refuges for native salmonids have been created by building barriers across a stream and then eradicating all alien trout above the barrier before repatriating the native trout (Lusardi et al. 2015). Control is justified if (1) the alien species cannot be eradicated, (2) the species is known to be harmful, especially to native fishes, (3) control measures will be sufficiently effective to reduce the negative effects of the species, (4) control measures will not have irreversible environmental consequences, and (5) control is cost-effective, because it is likely to be continuously needed.

Perhaps the best-known control program in the Southwest is the large-scale effort to reduce populations of rainbow trout *Oncorhynchus mykiss* and other alien species in the Colorado River below Glen Canyon Dam to improve conditions for the endangered humpback chub *Gila cypha* and other native species. The initial four years of the removal program by electrofishing reduced alien species by about half, with a concomitant increase in native species abundance (Coggins et al. 2011). However, the change in fauna coincided with a drought, which also reduced habitat for trout. Nevertheless, the control program was deemed successful enough so that it is being applied in other areas in the Colorado River watershed as well, such as Bright Angel Creek (https://www.nps.gov/grca/learn/nature/trout-reduction.htm).

One method for control of alien fishes increasingly being considered for regulated streams in the Southwest is using a natural flow regime (Poff et al. 1997), which mimics the flow regime under which the native fishes evolved, and thus gives them an advantage. Especially important is having a period of high spring flow for native fish spawning (Gido and Propst 2012; Gido et al. 2013). Ruhí et al. (2015, 2016) show that reduction of variation in streamflow benefits alien species in southwestern streams while extirpating native fishes. In Putah Creek, California, a natural flow regime was instituted that included elevated spring flows and minimum year-round flows (Kiernan et al. 2012). The flow regime and cooler temperatures dramatically shifted the fish fauna to favor native species, although the shift was much less downstream, where large pools (from gravel mining) and warmer water favored aliens such as largemouth bass, sunfishes, and common carp. However, the responses of alien fishes to natural flow regimes can be variable. Gido et al. (2013), for example, found that alien but river-adapted smallmouth bass *M. dolomieu* often responded positively to an unregulated, presumably natural, flow regime. These studies also show the benefits of an ecosystem-based approach (as opposed to a single-species approach) to discouraging alien fishes. However, in Grand Canyon, experimental high-flow releases from Glen Canyon Dam to the Colorado River had little impact on fishes (Valdez et al. 2001). Apparently, the flood flows were not extreme enough in magnitude and duration to push out alien fishes or restructure the habitat in ways that favored native fishes.

Control programs, like eradication programs, are usually aimed at single species in one limited geographic area or watershed, although they are likely to affect other species as well (e.g., rainbow trout control in Grand Canyon). Once undertaken, control programs need continuous action and monitoring to make sure they are actually working. Thus, the electrofishing program that removes alien fishes from the Colorado River has to be either continuous or periodically repeated to have an impact. Managers also have to be aware that evolution (natural selection) is working against them. Alien species with high fecundities or reproductive rates have the potential to adapt rapidly to control measures, especially those involving removal of individuals, much as fish populations respond to fisheries with changes in size, age at reproduction, and other characteristics.

The most workable control programs seem to be those that suppress a suite of alien species while favoring the natives with other actions. In Putah Creek, a flow regime designed to favor native fishes was successful, but most (though not all) alien fishes were still present in low numbers after more than 10 years of a restored natural flow regime (Kiernan et al. 2012). In Martis Creek, California, a dammed stream with a largely unregulated flow regime, the relative abundance of native and alien fishes fluctuated in response to changes in the flow regime, which were sufficient among years to let alien and native species coexist (Kiernan and Moyle 2012). In contrast, Propst et al. (2008) found that a natural flow regime in the relatively undisturbed Gila River, New Mexico, was inadequate by itself to keep alien fishes at bay; alien fishes chronically invaded during periods of low flow. Propst et al. (2008) consequently recommended a control program for alien fishes, along with maintaining a natural flow regime, to keep the river dominated by native fishes. When a control program was instituted for three major predators, at least one species of native fish had a positive response to the removal, although invasion of another alien predator may have negated some of the positive aspects (Propst et al. 2015). However, unusual natural flows apparently eliminated most nonnative predators from the river, at least temporarily (D. Propst, pers. comm.).

Acceptance

Acceptance of alien species as part of a local ecosystem is very hard for those engaged in native fish conservation. But for better or worse, alien fishes are integrated into southwestern aquatic ecosystems, often coexisting with native species, though often at the cost of extirpations. Their degree of integration seems to be partly related to number of years since the original invasion took place, which is major reason for fighting invasions in the early stages. The degree of integration also depends on how much the current habitat differs from the historical habitat (e.g., reservoir vs. river).

The novel aquatic ecosystems that consequently dominate southwestern waters (1) contain mixtures of native and alien species, including plants, fish, amphibians, and invertebrates, (2) occur in highly altered (by people) physical environments, (3) include people as an integral part of the system, and (4) have mostly developed in the past century (Moyle 2013; Murcia et al. 2014). Novel ecosystems "have a tendency to self-organize and manifest novel qualities without intensive human management" (Hobbs et al. 2013, 58). The self-organizing part of the definition is actually one of the more remarkable traits of novel aquatic ecosystems. In these ecosystems, species of diverse geographic origins seem to "get along," at least in the short periods in which they have been studied (Moyle 2013; Light and Moyle 2015). In fact, in the Southwest, so many waters support novel ecosystems, as defined above, that they can hardly be considered novel.

Extinctions of native fishes, of course, have happened in the organizational process, but usually at least a few native species are integrated into most of the systems. This is true even of the highly altered and highly invaded (62 fish species) lower Colorado River, where a handful of native species are managing to persist (Olden et al. 2006), albeit sometimes assisted by artificial propagation and other extreme measures.

Most alien fishes are so well adapted to novel conditions that they are going to continue to thrive in southwestern ecosystems indefinitely. Many are likely to become even more abundant because of their adaptations for living in human-dominated ecosystems, especially under rapid climate change (Moyle et al. 2011). Acceptance of this reality has major implications for management of native fishes because we have to understand how we can shape novel ecosystems to favor natives. As indicated, Putah Creek provides a rare example of this kind of management (Kiernan et al. 2012).

RECONCILIATION ECOLOGY: LIVING WITH NEW REALITIES

Most efforts to conserve native aquatic species focus on "natural" systems that can be set aside as preserves (preservation) or on altered systems that can be restored to more natural conditions (restoration). Alternatively, aquatic systems can be managed to favor single species, as is required by state and federal endangered species acts in the United States, and as has traditionally been done in fisheries and wildlife management. Unfortunately, most ecosystems today require considerable human intervention if they are to support desirable, usually native, species. Rosenzweig (2003, 7) addresses this dilemma through reconciliation ecology, defined as the "science of inventing, establishing, and maintaining new habitats to conserve species diversity in places where people live, work, and play." Reconciliation ecology acknowledges that humans dominate all ecosystems on the planet, which gives us the responsibility to determine what we want these integrated ecosystems to look like and what species we want them to contain. Closely related to the concept of reconciliation ecology is the idea of novel (no-analog) ecosystems (Moyle 2013). Here I have emphasized that novel aquatic ecosystems virtually always contain mixtures of native and nonnative species. While I have mainly discussed fishes, alien invertebrates, such as the Japanese clam and various crayfish species (e.g., *Orconectes virilis*), as well as alien aquatic plants (e.g., *Potamogeton crispus*), are usually also present in southwestern streams and rivers. Diverse and aggressive alien species tend to dominate the most disturbed environments. Unfortunately, a majority of stream reaches in the Southwest have regulated flows, have altered channels, and contain alien species. Dams at least can allow these waters to be managed as novel ecosystems that "share" water with cities and farms. The general tools available to managers are flow- and temperature-regime manipulations, habitat management, and alien species control. A flow regime that discourages aliens and favors natives is most likely to work if accompanied by large-scale habitat improvements, such as re-creation of meanders and pools and reconnection of floodplains. Where justified, control programs for aliens (more than just fishes) are needed. The Lower Colorado River Multi-species Conservation Program uses some of these ideas (https://www.lcrmscp.gov/fish/fish_augmentation.html). Reconciliation ecology therefore involves creating novel ecosystems that are highly managed with distinct goals in mind, such as maintaining large populations of rare native fishes.

CONCLUSIONS

Alien fishes will always be with us and will continue to threaten native fishes. *Battle against Extinction* is still being waged, but reconciliation ecology may provide a more realistic vision for the future. Unfortunately, we are currently losing the war against the seizure of aquatic ecosystems by alien species, one skirmish at a time. Consequently, we must develop alternative management strategies for aquatic biotas, especially endemic fishes. Scarce resources should be concentrated in places where they can really make a difference. Victories may depend heavily on developing managed refuges, fortresses we can defend against invaders when we cannot develop novel ecosystems that support desired species.

Most aquatic ecosystems in the western United States today are far from pristine, and they require considerable human manipulation to support native species. The pathway through reconciliation ecology at least acknowledges that people increasingly dominate aquatic ecosystems. Recognizing this fact leaves us with the responsibility to determine what we want these ecosystems to look like and what species we want them to contain, now and in the future, preferably through manipulation of flows and habitat. Direct control and eradication of aliens, however, have to remain as important tools for native fish conservation. Sometimes reconciliation is possible only after a battle against invasive aliens has been won; then we can work for coexistence of the rest of the biota, deciding who dies and who lives. Playing God will require a lot of humility.

ACKNOWLEDGMENTS

I deeply appreciate the comments of Kurt Fausch and John Wiens on drafts of this manuscript.

REFERENCES

Arthington, A. H., N. K. Dulvey, W. Gladstone, and I. J. Winfield. 2016. Fish conservation in freshwater and marine realms: Status, threats, and management. *Aquatic Conservation: Marine and Freshwater Ecosystems* 26: 838–57.

Baumsteiger, J., and P. B. Moyle. 2017. Assessing extinction. *BioScience* 67: 357–66.

Blackburn, T. M., F. Essl, T. Evans, P. E. Hulme, J. M. Jeschke, I. Kühn, S. Kumschick, Z. Marková, A. Mrugała, W. Nentwig, and J. Pergl. 2014. A unified classification of alien species based on the magnitude of their environmental impacts. *PLoS Biology* 12 (5): e1001850.

Bonner, T. H., and G. R. Wilde. 2000. Changes in the Canadian River fish assemblage associated with reservoir construction. *Journal of Freshwater Ecology* 15: 189–98.

Britton, J. R., G. H. Copp, M. Brazier, and G. D. Davies. 2011. A modular assessment tool for managing introduced fishes according to risks of species and their populations, and impacts of management actions. *Biological Invasions* 13: 2847–60.

Closs, G. P., M. Krkosek, and J. D. Olden, eds. 2016. *Conservation of Freshwater Fishes*. Cambridge: Cambridge University Press.

Coggins, L. G., Jr., M. D. Yard, and W. E. Pine III. 2011. Nonnative fish control in the Colorado River in Grand Canyon, Arizona: An effective program or serendipitous timing? *Transactions of the American Fisheries Society* 140: 456–70.

Darwall, W. R. T., and J. Freyhof. 2016. Lost fishes, who is counting? The extent of the threat to freshwater fish biodiversity. In G. P. Closs, M. Krkosek, and J. D. Olden, eds., *Conservation of Freshwater Fishes*, 1–36. Cambridge: Cambridge University Press.

Dill, W. A., and A. J. Cordone. 1997. *History and Status of Introduced Fishes in California, 1871–1996*. California Department of Fish and Game Fish Bulletin 178.

Elton, C. 1958. The Ecology of Invasions by Plants and Animals. London: Methuen.

Fausch, K. D., B. E. Rieman, J. B. Dunham, M. K. Young, and D. P. Peterson. 2009. Invasion versus isolation: Trade-offs in managing native salmonids with barriers to upstream movement. *Conservation Biology* 23: 859–70.

Fuller, P. L., L. G. Nico, and J. D. Williams. 1999. *Nonindigenous Fishes Introduced into Inland Waters of the United States*. Special Publication 27. Bethesda, MD: American Fisheries Society.

Gido, K. B., and D. L. Propst. 2012. Long-term dynamics of native and nonnative fishes in the San Juan River, New Mexico and Utah, under a partially managed flow regime. *Transactions of the American Fisheries Society* 141: 645–59.

Gido, K. B., D. L. Propst, J. D. Olden, and K. R. Bestgen. 2013. Multidecadal responses of native and introduced fishes to natural and altered flow regimes in the American Southwest. *Canadian Journal of Fisheries and Aquatic Sciences* 70: 554–64.

Hoagstrom, C. W., N. D. Zymonas, S. R. Davenport, D. L. Propst, and J. E. Brooks. 2010. Rapid species replacements between fishes of the North American plains: A case history from the Pecos River. *Aquatic Invasions* 5: 141–53.

Hobbs, R. J., E. S. Higgs, and C. Hall. 2013. *Novel Ecosystems: Intervening in the New Ecological World Order*. West Sussex, UK: John Wiley & Sons.

Holden, P. B. 1991. Ghosts of the Green River: Impacts of the Green River poisoning on management of native fishes. In W. L. Minckley and J. E. Deacon, eds., *Battle against Extinction: Native Fish Management in the American West*, 43–53. Tucson: University of Arizona Press.

Kiernan, J. D., and P. B. Moyle. 2012. Flows, droughts, and aliens: Factors affecting the fish assemblage in a Sierra Nevada, California, stream. *Ecological Applications* 22: 1146–61.

Kiernan, J. D., P. B. Moyle, and P. K. Crain. 2012. Restoring native fish assemblages to a regulated California stream using the natural flow regime concept. *Ecological Applications* 22: 1472–82.

Knapp, R. A., D. M. Boiano, and V. T. Vredenburg. 2007. Removal of nonnative fish results in population expansion of a declining amphibian (mountain yellow-legged frog, *Rana muscosa*). *Biological Conservation* 135: 11–20.

Kopf, R. K., D. G. Nimmo, P. Humphries, L. J. Baumgartner, M. Bode, N. R. Bond, A. E. Byrom, J. Cucherousset, R. P. Keller, A. J. King, H. M. McGinness, P. B. Moyle, and J. D. Olden. 2017. Confronting the risks of large-scale invasive species control. *Nature Ecology and Evolution* 1. https://doi.org/10.1038/s41559-017-0172.

Larson, B. M. H. 2005. The war of the roses: Demilitarizing invasion biology. *Frontiers in Ecology and the Environment* 3: 495–500.

Light, T., and P. B. Moyle. 2015. Assembly rules and novel assemblages in aquatic ecosystems. In J. Canning-Clode, ed., *Biological Invasions in Changing Ecosystems: Vectors, Ecological Impacts, Management, and Predictions*, 432–57. Warsaw/Berlin: De Gruyter Open.

Lusardi, R. A., M. R. Stephens, P. B. Moyle, C. L. McGuire, and J. M. Hull. 2015. Threat evolution: Negative feedbacks between management action and species recovery in threatened trout (Salmonidae). *Reviews in Fish Biology and Fisheries* 25: 521–35.

Marks, J. C., G. A. Haden, M. O'Neill, and C. Pace. 2010. Effects of flow restoration and exotic species removal on recovery of native fish: Lessons from a dam decommissioning. *Restoration Ecology* 18: 934–43.

McColl, K. A., B. D. Cooke, and A. Sunarto. 2014. Viral biocontrol of invasive vertebrates: Lessons from the past applied to cyprinid herpesvirus-3 and carp (*Cyprinus carpio*) control in Australia. *Biological Control* 72: 109–17.

Minckley. W. L., and J. E. Deacon, editors. 1991. *Battle against Extinction: Native Fish Management in the American West.* Tucson: University of Arizona Press.

Moyle, P. B. 2002. *Inland Fishes of California*. Revised and expanded. Berkeley: University of California Press.

Moyle, P. B. 2013. Novel aquatic ecosystems: The new reality for streams in California and other Mediterranean climate regions. *River Research and Applications* 30: 1335–44.

Moyle, P. B., J. V. E. Katz, and R. M. Quiñones. 2011. Rapid decline of California's native inland fishes: A status assessment. *Biological Conservation* 144: 2414–23.

Moyle, P. B., J. D. Kiernan, P. K. Crain, and R. M. Quiñones. 2013. Climate change vulnerability of native and alien freshwater fishes of California: A systematic assessment approach. *PLoS ONE*. http://dx.plos.org/10.1371/journal.pone.0063883.

Moyle, P. B., and M. P. Marchetti. 2006. Predicting invasion success: Freshwater fishes in California as a model. *BioScience* 56: 515–24.

Moyle, P. B., and G. M. Sato. 1991. On the design of preserves to protect native fishes. In W. L. Minckley and J. E. Deacon, eds., *Battle against Extinction: Native Fish Management in the American West*.155–69. Tucson: University of Arizona Press.

Murcia, C., J. Aronson, G. H. Kattan, D. Moreno-Mateos, K. Dixon, and D. Simberloff. 2014. A critique of the "novel ecosystem" concept. *Trends in Ecology and Evolution* 29: 548–53.

Olden, J. D., and N. L. Poff. 2005. Long-term trends of native and non-native fish faunas in the American Southwest. *Animal Biodiversity and Conservation* 28: 75–89.

Olden, J. D., N. L. Poff, and K. R. Bestgen. 2006. Life-history strategies predict fish invasions and extirpations in the Colorado River basin. *Ecological Monographs* 76: 25–40.

Olden, J. D., N. L. Poff, M. R. Douglas, M. E. Douglas, and K. D. Fausch. 2004. Ecological and evolutionary consequences of biotic homogenization. *Trends in Ecology and Evolution* 19: 18–24.

Poff, N. L., J. D. Allan, M. B. Bain, J. R. Karr, K. L. Prestegaard, B. D. Richter, R. E. Sparks, and J. C. Stromberg. 1997. The natural flow regime. *BioScience* 47: 769–84.

Poff, N. L., J. D. Olden, D. M. Merritt, and D. M. Pepin. 2007. Homogenization of regional river dynamics by dams and global biodiversity implications. *Proceedings of the National Academy of Sciences* 104: 5732–37.

Propst, D. L., K. B. Gido, and J. A. Stefferud. 2008. Natural flow regimes, nonnative fishes, and native fish persistence in arid-land river systems. *Ecological Applications* 18: 1236–52.

Propst, D. L., K. B. Gido, J. E. Whitney, E. I. Gilbert, T. J. Pilger, A. M. Monié, Y. M. Paroz, J. M. Wick, J. A. Monzingo, and D. M. Myers. 2015. Efficacy of mechanically removing nonnative predators from a desert stream. *River Research and Applications* 31: 692–703.

Rahel, F. J. 2002. Homogenization of freshwater faunas. *Annual Review of Ecology and Systematics* 2002: 291–315.

Rosenzweig, M. L. 2003. *Win-Win Ecology: How the Earth's Species Can Survive in the Midst of Human Enterprise*. Oxford: Oxford University Press.

Ruhí, A., E. E. Holmes, J. N. Rinne, and J. L. Sabo. 2015. Anomalous droughts, not invasion, decrease persistence of native fishes in a desert river. *Global Change Biology* 21: 1482–96.

Ruhí, A., J. D. Olden, and J. L. Sabo. 2016. Declining streamflow induces collapse and replacement of native fish in the American Southwest. *Frontiers in Ecology and Evolution* 14: 465–72.

Valdez, R. A., T. L. Hoffnagle, C. C. McIvor, T. McKinney, and W. C. Leibfried. 2001. Effects of a test flood on fishes of the Colorado River in Grand Canyon, Arizona. *Ecological Applications* 11: 686–700.

Vilizzi, L., and G. H. Copp. 2013. Application of FISK, an invasiveness screening tool for non-native freshwater fishes, in the Murray-Darling Basin (Southeastern Australia). *Risk Analysis* 33: 1432–40.

Wiens, J. A. 2016. *Ecological Challenges and Conservation Conundrums: Essays and Reflections for a Changing World*. Chichester, UK: Wiley & Sons.

Williams, J. E. 1991. Preserves and refuges for native western fishes: History and management. In W. L. Minckley and J. E. Deacon, eds., *Battle against Extinction: Native Fish Management in the American West*, 171–90. Tucson: University of Arizona Press.

5 Current Conservation Status of Some Freshwater Species and Their Habitats in México

María de Lourdes Lozano-Vilano,
Armando J. Contreras-Balderas,
Gorgonio Ruiz-Campos,
and María Elena García-Ramírez

Since the work of Contreras-Balderas (1991) and Contreras-Balderas et al. (2003), no updated comprehensive information has been available regarding the conservation status of threatened freshwater fish species in México at regional or broader scales. However, since that time several authors have reported on the status of certain fishes that inhabit the inland waters of México. Two of the most recent studies of Mexican fishes were done by Ruiz-Campos et al. (2014a) and Lozano-Vilano and de la Maza-Benignos (2017), who documented the status of six species in the Baja California peninsula and of the Mexican killifish species, respectively. In this chapter, we summarize the pertinent literature that has been generated on the conservation status of some fishes in México (table 5.1) and include current population monitoring data and observations of other ichthyologists.

Several lists describing the conservation status of freshwater fishes in North America have been published (Deacon 1979; Williams et al. 1989; Jelks et al. 2008). These lists created standardized criteria for the definition of status of species crossing different regions and political boundaries that are useful for conservation and management programs. The Norma Oficial Mexicana list of Mexican Species at Risk (NOM-059-SEMARNAT-2010) (DOF 2010) now includes 187 freshwater fishes, after information from new research and publications is added. The list was also supplemented with additional information collected during sampling site visits in México from 1991 through 2017. Some Mexican fishes in freshwater habitats are protected indirectly via protection of water resources for human consumption. Direct protections come from natural protected areas such as the federal Biome Reserve and National Parks; state natural protected areas; and local natural areas important for municipalities.

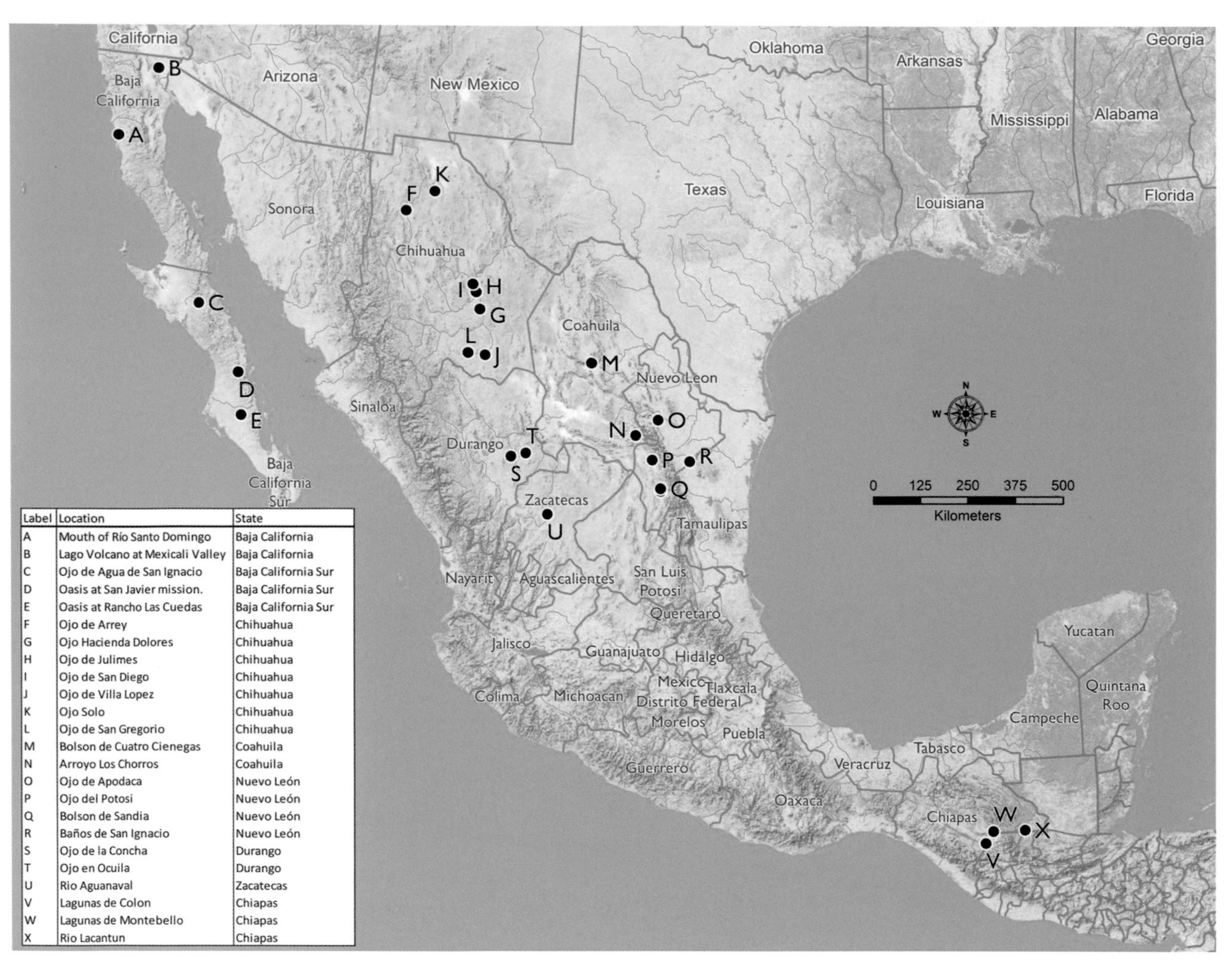

Label	Location	State
A	Mouth of Río Santo Domingo	Baja California
B	Lago Volcano at Mexicali Valley	Baja California
C	Ojo de Agua de San Ignacio	Baja California Sur
D	Oasis at San Javier mission.	Baja California Sur
E	Oasis at Rancho Las Cuedas	Baja California Sur
F	Ojo de Arrey	Chihuahua
G	Ojo Hacienda Dolores	Chihuahua
H	Ojo de Julimes	Chihuahua
I	Ojo de San Diego	Chihuahua
J	Ojo de Villa Lopez	Chihuahua
K	Ojo Solo	Chihuahua
L	Ojo de San Gregorio	Chihuahua
M	Bolson de Cuatro Cienegas	Coahuila
N	Arroyo Los Chorros	Coahuila
O	Ojo de Apodaca	Nuevo León
P	Ojo del Potosi	Nuevo León
Q	Bolson de Sandia	Nuevo León
R	Baños de San Ignacio	Nuevo León
S	Ojo de la Concha	Durango
T	Ojo en Ocuila	Durango
U	Rio Aguanaval	Zacatecas
V	Lagunas de Colon	Chiapas
W	Lagunas de Montebello	Chiapas
X	Rio Lacantun	Chiapas

Fig. 5.1 Locations of sampling sites and areas discussed in this chapter.

In this chapter, we make recommendations for the legal protection of freshwater fishes in different locations and states of México; precise locality information and site descriptions are also presented (fig. 5.1).

HABITATS AND FISHES

Baja California

Mouth of Río Santo Domingo

This location was formerly called the mouth of the Río San Ramón, and until 1997, it supported one of the few remnant populations of threespine stickleback *Gasterosteus aculeatus* in northwestern Baja California (Ruiz-Campos et al. 2000). Myers (1930) first reported threespine stickleback there, and its presence was verified in 1996 (Ruiz-Campos et al. 2000). However, subsequent sampling between 2000 and 2013 has not detected the species, and it is probably extirpated (Ruiz-Campos et al. 2014a). Excessive pumping of water in the lower part of this river for agricultural irrigation in the towns of Vicente Guerrero and Camalú has caused the river to dry, which, when coupled with increased extraction of sand, has altered habitat for this anadromous fish.

Lago Volcano at Mexicali Valley

This lake, situated at the base of Cerro Prieto volcano, represents one of the historical sites of the desert pupfish *Cyprinodon macularius* in the lower Río Colorado basin of México (Follett 1960). The lake has expanded to an area of 6 km² due to input of residual water from the cooling system of the Cerro Prieto Geothermal Station (Varela-Romero et al. 2003). The desert pup-

Table 5.1. Extinct native fishes of México. Names follow Page et al. (2013). Status information is the best available based on literature sources and recent surveys completed by the authors.

FAMILY Species	Common name	Status[a]	Habitat association
CYPRINIDAE			
Notropis amecae	Ameca shiner	Extinct	Small stream
N. orca	Phantom shiner	Extinct	Large river
N. simus simus	Rio Grande bluntnose shiner	Extinct	Large river
Stypodon signifer	Stumptooth minnow	Extinct	Springs
CYPRINODONTIDAE			
Cyprinodon alvarezi	Potosí pupfish	Extinct (C)	Springs
C. ceciliae	La Presita pupfish	Extinct	Springs
C. inmemoriam	La Trinidad pupfish	Extinct	Springs
C. latifasciatus	Parras pupfish	Extinct	Springs
C. longidorsalis	Charco Palma pupfish	Extinct (C)	Springs
C. veronicae	Charco Azul pupfish	Extinct (C)	Springs
Undescribed *Cyprinodon*	Ojo de Villa Lopez[b]	Extinct	Springs
Undescribed *Cyprinodon*	Ojo en San Pedro Ocuila[b]	Extinct	Springs
Megupsilon aporus	Catarina pupfish	Extinct	springs
GOODEIDAE			
Skiffia francesae	Golden skiffia	Extinct (C)	Springs
POECILIIDAE			
Xiphophorus couchianus	Monterrey platyfish	Extinct (C)	Springs

[a](C) indicates that captive populations may exist.
[b]Locality.

fish population has been stable in abundance over the last 20 years, based on monitoring since 1996 (Varela-Romero et al. 2003; Ruiz-Campos et al. 2014a; Ruiz-Campos 2016). Population abundance and the proportion of males to females is stable due to lack of competition with other fishes, which are limited by salinity levels greater than 35 ppt (Ruiz-Campos et al. 2013).

Baja California Sur

Ojo de Agua de San Ignacio

This spring is the type locality of the Baja California killifish *Fundulus lima*, one of two freshwater fishes endemic to Baja California Sur. Spring water has been used for irrigation of orchards and domestic purposes in the town of San Ignacio since the establishment of Jesuit missionaries. In the 1970s, common carp *Cyprinus carpio* and American bullfrog *Lithobates (Rana) catesbeiana* were introduced to promote rural aquaculture. In the 1980s, two fishes of Neotropical affinity, guppy *Poecilia reticulata* and green swordtail *Xiphophorus hellerii*, were introduced by aquarists. In 1996, a local resident stocked a fourth nonnative fish, redbelly tilapia *Tilapia* cf. *zillii*, in this oasis. In just a few years, this African cichlid was the dominant fish in the Río San Ignacio basin (Ruiz-Campos et al. 2014a), with relative abundance as high as 98% at many of the sites monitored between 2002 and 2017 (Ruiz-Campos et al. 2006; Ruiz-Campos 2014b; G. Ruiz-Campos, pers. obs.). Nonnative western mosquitofish *Gambusia affinis* has also recently been detected in the spring, albeit in low abundance. Abundance of *F. lima* in the spring site has varied substantially from 2002 to 2017, ranging from almost absent up to 3.1 individuals per minnow trap

Fig. 5.2 Oasis at Rancho Las Cuedas in Santa Rita basin (Sierra de la Giganta), municipality of Comondú, Baja California Sur, México, showing changes in habitat conditions for native killifish during three sampling events: *(Top)*, April 16, 1998; *(Middle)*, April 18, 2002; *(Bottom)*, January 21, 2005. (Photos by Gorgonio Ruiz-Campos.)

per hour (Ruiz-Campos et al. 2014a; G. Ruiz-Campos, pers. obs.).

Oasis at San Javier Mission

At this site, situated on the western slope of the Sierra de la Giganta at an elevation of 463 m above sea level, *F. lima* was first reported by Follett (1960). In 1977, additional *F. lima* specimens were collected, and no other fish species were detected (Ruiz-Campos 2012). Subsequent intensive sampling at this same locality (1998, 2004, and 2010) failed to detect endemic *F. lima*, but abundant nonnative *T.* cf. *zillii* and *P. reticulata* were found (Ruiz-Campos et al. 2014c).

Oasis at Rancho Las Cuedas

This site is in the Río San Luis basin on the western slope of the Sierra de la Giganta, at an elevation of 173 meters above sea level (Ruiz-Campos et al. 2014c). Sampling in 1998 (fig. 5.2*A*) detected *F. lima* ($n = 20$) in relatively high abundance compared with nonnative *T.* cf. *zillii* ($n = 8$). However, only four years later (2002, fig. 5.2*B*), *T.* cf. *zillii* was dominant ($n = 39$) and only two *F. lima* were found. Sampling two years later (2006, fig. 5.2*C*) detected no *F. lima*, and it is presumed extirpated (Ruiz-Campos 2012).

Chihuahua

Ojo de Arrey

The spring has been dry since 1999, and the local population of whitefin pupfish *Cyprinodon albivelis* has been extirpated (Lozano-Vilano and de la Maza-Benignos 2017).

Ojo Hacienda Dolores

This site, which is used for recreation, was last visited in 2012. Presence of nonnative Mozambique tilapia *Oreochromis mossambicus* was noted in 2004, but abundance of native largescale pupfish *Cyprinodon macrolepis* was high in 2012.

Ojo de Julimes

Minckley and Minckley (1986) noted an undescribed

Cyprinodon sp. at this location, which was later described as Julimes pupfish *Cyprinodon julimes* by de la Maza-Benignos and Vela-Valladares (2009). The spring is thermal and much reduced in flow, but government and nongovernmental associations declared the spring a protected area. *Cyprinodon julimes* was proposed as a species in danger of extinction in the list of Mexican Species at Risk in NOM-059-SEMARNAT (de la Maza-Benignos et al. 2012, 2014a, b, c). This area has a program for sustainable development of water, which is unusual for the Chihuahuan Desert and may improve prospects for survival of the Julimes pupfish (de la Maza-Benignos et al. 2014b).

Ojo de San Diego

This site was last visited in 2012. The resident population of bighead pupfish *Cyprinodon pachycephalus* is critically threatened. Swimming pools for human recreation were recently developed, which is a strong stressor on this species.

Ojo de Villa Lopez

This large spring was reported by Contreras-Balderas (1991) to support two undescribed fishes, a *Cyprinodon* sp. and a *Gambusia* sp. In 2005, those two species were found in addition to six other native fishes. Sampling in 2012 detected only three native fishes (*Gambusia* sp., red shiner *Cyprinella lutrensis*, and Chihuahua shiner *Notropis chihuahua*), and two nonnative species (*O. mossambicus* and bluegill *Lepomis macrochirus*). The *Cyprinodon* sp. has not been detected since 2006, and we believe it is extinct (see table 5.1), possibly because of drying in intervening years.

Ojo Solo

Ojo Solo was once part of a five-spring complex in Ejido Rancho Nuevo, Villa Ahumada, in Bolsón de los Muertos in northwestern Chihuahua (Ojo de Carbonera, Ojo de las Varas, Ojo el Medio, Ojo Solo, and Ojo del Apache). Only Ojo Solo remains and has been reduced to 30% of its former area. Ojo Solo supports endemic Carbonera pupfish *Cyprinodon fontinalis* and largemouth shiner *Cyprinella bocagrande* (Lozano-Vilano and de la Maza-Benignos 2017), and each was at risk of extinction when the last remaining natural spring habitat dried in January 2017 (de la Maza-Benignos 2017). Pronatura Noreste created a natural refuge habitat in a nearby desert spring, Ojo Caliente, and in 2014 began translocation of pupfish, Chihuahuan dwarf crayfish *Cambarellus chihuahuae*, and shiners to the new location (Carson et al. 2015), where they survive.

Ojo de San Gregorio

This small spring supports yellowfin gambusia *Gambusia alvarezi*. Since 2006, the owners of the spring have constructed a pool downstream, which is a refuge, and we observed thousands of *G. alvarezi* during our last visit (2006). We consider the population stable.

Coahuila

Bolsón de Cuatro Ciénegas

This basin has been impacted by extraction of water for nearby villages, agriculture, and recreation, which has caused desiccation of many springs in the valley. The fishes are also threatened by nonnative jewel cichlid *Hemichromis guttatus*, which is distributed in Río Mezquitez and in lagoons and pozas in the basin. The most important of these is Poza Churince, where we removed 47,218 *H. guttatus* from 1999 to 2005. Unfortunately, we did not eliminate the species, and in our last visit (2014), we removed an additional 3,756 *H. guttatus*. As a result, Minckley's cichlid *Herichthys minckleyi* is functionally extinct, as only old adults (8–10 years old) persisted and juveniles, larvae, and nests were absent. In addition, Cuatro Ciénegas pupfish *Cyprinodon bifasciatus*, Cuatro Ciénegas shiner *Cyprinella xanthicara*, longear sunfish *Lepomis* cf. *megalotis*, largemouth bass *Micropterus salmoides*, and roundnose minnow *Dionda* cf. *episcopa* are all extirpated or extinct. Only robust gambusia *Gambusia marshi* remains. In Poza El Anteojo we eliminated *H. guttatus* following removal of 19,071 individuals from 2000 to 2002 (Lozano-Vilano et al. 2006).

Arroyo Los Chorros

This site in the San Juan basin, Coahuila, is the type locality of Salinas chub *Gila modesta*. The habitat is degraded due to channelization of the spring, and in 2004 only parts of it remained, but the species persists (fig.

Fig. 5.3 Los Chorros, San Juan basin, Coahuila, showing changes in habitat conditions for the type locality of *Gila modesta: (Left)*, July 26, 2004; *(Right)*, May 15, 2016. (Photos by María de Lourdes Lozano-Vilano.)

5.3). Hybrids of nonnative *Xiphophorus variatus* and *X. maculatus* (variable platyfish × southern platyfish) have been introduced. The spring is also close to a road, which puts it at additional risk from pollutants.

Nuevo León

Ojo de Apodaca

This spring is part of a complex of springs in the basin of Río Santa Catarina in the municipalities of Santa Catarina, Monterrey, Guadalupe, Apodaca, and Juarez. Each municipality historically had one or more springs that supported populations of Monterrey platyfish *Xiphophorus couchianus*, but by 1991, all of these populations were extirpated except for that in Ojo de Apodaca, the last refuge of this species (Contreras-Balderas 1991). Our recent sampling (2014) indicated that it, too, was destroyed, so all natural populations of the species are extinct. There is a captive population at the Texas State University, San Marcos.

Ojo del Potosí

Two endemic species, the Potosí pupfish *Cyprinodon alvarezi* and Catarina pupfish *Megupsilon aporus*, were found in this spring. After 1994, the spring nearly dried (Contreras-Balderas and Lozano-Vilano 1996) due to water extraction for irrigation. By 1996, the spring dried completely, and the two pupfish species are extinct in the wild.

There are some captive populations of *C. alvarezi* held by aquarists, with 362 individuals in five different locations (M. Arroyo, pers. comm.). *Megupsilon aporus* is gone from this locality and is apparently extinct because no captive populations exist (M. de la Maza-Benignos and M. Arroyo, pers. comm.).

Bolsón de Sandia

This area supported several springs, each with one endemic pupfish. These species included La Presita pupfish *Cyprinodon ceciliae*, La Trinidad pupfish *C. inmemoriam*, La Palma pupfish *C. longidorsalis*, and Charco Azul pupfish *C. veronicae*. All the springs are now dry due to overexploitation for irrigation, and the pupfishes are extinct in the wild. Lozano-Vilano and de la Maza-Benignos (2017) indicated that *Cyprinodon alvarezi* and *C. longidorsalis* exist in aquarium populations in Europe and the United States (M. Arroyo, pers. comm.).

Fig. 5.4 Baños de San Ignacio, San Fernando basin, Linares, Nuevo León, showing changes in habitat conditions for the type locality of *Cyprinodon bobmilleri* and *Fundulus philpisteri*: *(Top)*, July 23, 2004; *(Bottom)*, March 16, 2013. (Photos by María de Lourdes Lozano-Vilano.)

Cyprinodon inmemoriam has been extinct since October 1985, before its formal description, and no captive populations exist. The spring that supported *C. ceciliae* has dried completely since 1991, and because there are no captive populations, the species is extinct. Miller et al. (2005) indicated that *C. alvarezi*, *C. veronicae*, *C. longidorsalis*, and *Megupsilon aporus* are in captivity, but now the status of these species in different locations in México and the United States is less certain. We know conservation aquarists hold *C. veronicae* (510 individuals in three locations) and *C. longidorsalis* (353 individuals in three locations) (M. Arroyo, pers. comm.).

Baños de San Ignacio

This large thermal spring/swamp in the San Fernando basin, Linares, Nuevo León, is about 20 m wide and 1.4 m deep, and is rich in sulfur compounds (Lozano-Vilano and Contreras-Balderas 1999; García-Ramírez et al. 2002). The endemics, San Ignacio pupfish *Cyprinodon bobmilleri* and conservationist killifish *Fundulus philpisteri*, coexist with other native species, including Mexican tetra *Astyanax mexicanus*, Tex-Mex gambusia *Gambusia speciosa*, Amazon molly *Poecilia formosa*, and Texas cichlid *Herichthys* cf. *cyanoguttatus*. Survival of these fishes is threatened by low water levels in dry seasons, and during our last visit (2013), the swamp was nearly dry and flow of the spring stream was greatly diminished (fig. 5.4).

Durango

Ojo de la Concha

The Nazas pupfish *Cyprinodon nazas*, reported at this site by Miller (1976), may be extirpated (A. Maeda-

Martínez, pers. comm.), along with the Tepehuán shiner *Cyprinella alvarezdelvillari* (Contreras-Balderas and Lozano-Vilano 1994). The introduction of nonnative blue tilapia *Oreochromis aureus* and porthole livebearer *Poeciliopsis gracilis* occurred in 2002. Both were abundant until 2008. After the property owners opened the floodgates of the dam and emptied most of the water, most tilapia and porthole livebearers were removed, but the status of native fishes is not known.

Ojo en San Pedro de Ocuila

The undescribed *Cyprinodon* sp. inhabiting this spring was reduced in abundance by 1988. It was found with two established nonnatives, goldfish *Carassius auratus* and *O. mossambicus*. The local townspeople constructed a concrete pool to collect the spring water, and since that time (1988), the *Cyprinodon* has not been seen and is presumed extinct.

Zacatecas

Río Aguanaval

Collections in 1971 documented *C. nazas* in Rancho Grande and Río Grande. Subsequent sampling in 2006–2007 indicated the species was not present and is likely extirpated.

Chiapas

Lagunas de Colón

The area was last visited in 2014. We noted that natural conditions had been altered by construction of pools for swimming. Native fishes known from here include largelip killifish *Profundulus labialis*, pale catfish *Rhamdia guatemalensis*, Macabí tetra *Brycon guatemalensis*, Chiapa de Corzo cichlid *Chiapaheros grammodes*, and redhead cichlid *Paraneetroplus synspilus*.

Lagunas de Montebello

The area was last visited in 2014, when we noted increased human populations and the town increasingly encroaching on the lagoons. Native fishes known from here include largelip killifish *P. labialis*, upper Grijalva livebearer *Poeciliopsis hnilickai*, stippled gambusia *Gambusia sexradiata*, northern checkmark cichlid *Theraps intermedius*, and green swordtail *Xiphophorus hellerii*. We also documented nonnative common carp *Cyprinus carpio* and unidentified tilapia *Oreochromis* sp., and observed *M. salmoides*. Rainbow trout *Oncorhynchus mykiss* was noted in some lagunas in our first visit in 1978, and it is unknown if more nonnative species are present.

Río Lacantun

In 2004–2007, Lozano-Vilano et al. (2007) conducted a fish inventory to determine the condition of the commercial fishery and other fishes in Río Lacantun. Later (2010–2013), Lozano-Vilano et al. (2013) observed a decline in the number and abundance of native species and increased abundance of nonnative species. Currently, there are 52 native species and 6 nonnative species. Five native species found here are protected under Mexican law, including Lacandon sea catfish *Potamarius nelsoni*, pale catfish, northern checkmark cichlid (Special Concern), Chiapas cichlid *Torichthys socolofi* (threatened), and Chiapas catfish *Lacantunia enigmatica* (endangered). Nonnative species were grass carp *Ctenopharyngodon idella*, *Cyprinus carpio*, vermiculated sailfin catfish *Pterygoplichthys disjunctivus*, Amazon sailfin catfish *P. pardalis*, *O. mossambicus*, and *O. aureus*. Because this river is in the Reserva de la Biosfera de Montes Azules, it is important to understand the condition of the fish community. The five main problems we found affecting native fishes were (1) increased human populations; (2) agricultural and cattle production practices that led to reduction or elimination of riparian vegetation; (3) river pollution from agrochemicals, including DDT; (4) fishery overharvest; and (5) presence of nonnative species.

CONCLUSIONS

The future of native fishes in México, particularly those in small springs in the Chihuahuan and Sonoran Deserts, is not bright. In the list of Mexican Species at Risk, 64 species of freshwater fishes have endangered status, and while helpful federal legislation has been passed in support of habitat protection and native fishes over the last 20 years, it has not been sufficient to prevent the extinction of several species. Furthermore, the status of captive populations that are extinct in the wild is tenuous and their future uncertain. Minimally, we need more research describing the distribution, taxonomy, ecology, population dynamics, and genetic structure of these fishes to better understand declines in their populations and habitat. It is inconceivable that the human population in México has no awareness of the magnitude of the problems associated with alteration and destruction of aquatic environments. This is especially unfortunate because fishes may be a barometer of the integrity or deterioration of our ecosystems and thus of our future. This is because fishes respond quickly to changes in the environment, are excellent monitors of water quality, and measure the suitability of the environment for humans. Therefore, it is fundamentally necessary that governments and agencies, both national and international, establish new rules for the care and management of water, fishes, and habitats. Sustainable water resources are urgently needed, and conservation must be performed in a holistic context by considering the hydrologic basin as the unit of management. It is necessary to develop programs to maintain or increase environmental flows, and to restore critical habitats, for all declining and rare species, regardless of listing status.

REFERENCES

Carson, E. W., C. Pedraza-Lara, M. L. Lozano-Vilano, G. A. Rodríguez-Almaráz, I. Banda Villanueva, L. A. Sepúlveda-Hernández, L. Vela-Valladares, A. Cantú-Garza, and M. de la Maza-Benignos. 2015. The rediscovery and precarious status of Chihuahuan dwarf crayfish *Cambarellus chihuahuae*. *Occasional Papers of the Museum of Southwestern Biology* 12: 1–7.

Contreras-Balderas, S. 1991. Conservation of Mexican freshwater fishes: Some protected sites and species, and recent federal legislation. In W. L. Minckley and J. E. Deacon, eds., *Battle against Extinction: Native Fish Management in the American West*, 191–97. Tucson: University of Arizona Press.

Contreras-Balderas, S., P. Almada-Villela, M. L. Lozano-Vilano, and M. E. García-Ramírez. 2003. Freshwater fish at risk or extinct in Mexico. *Reviews in Fish Biology and Fisheries* 12: 241–51.

Contreras-Balderas, S., and M. L. Lozano-Vilano. 1994. *Cyprinella alvarezdelvillari*, a new cyprinid fish from Rio Nazas of Mexico, with a Key to the *Lepida* Clade. *Copeia* 1994: 897–906.

Contreras-Balderas, S., and M. L. Lozano-Vilano. 1996. Extinction of most Sandia and Potosí Valleys (Nuevo León, México) endemic pupfishes, crayfishes and snails. *Ichthyological Exploration of Freshwaters* 7: 33–40.

Deacon, J. E. 1979. Endangered and threatened fishes of the West. *Great Basin Naturalist Memoirs* 3: 41–64.

De la Maza-Benignos, M., J. A. Rodriguez-Pineda, A. de la Mora-Covarrubias, E. W. Carson, M. Quiñonez-Martínez, P. Lavín-Murcio, L. Vela-Valladares, M. L. Lozano-Vilano, H. Parra-Gallo, H. A. Macías-Duarte, T. Lebgue-Keleng, E. Pando-Pando, M. Pando-Pando, M. Andazola-González, A. Anchondo-Najera, G. Quintana-Martínez, I. A. Banda-Villanueva, H. J. Ibarrola-Reyes, and J. Zapata-López. 2012. *Planes de Manejo y Programa de Monitoreo de Signos Vitales para las Áreas de Manantiales de la UMA El Pandeño; y San Diego de Alcalá en el Desierto Chihuahuense*. Vol. 1. Pronatura Noreste, A.C. (editor). Amigos del Pandeño, A.C.

De la Maza-Benignos, M., and L. Vela-Valladares. 2009. *Cyprinodon julimes*. In M. de la Maza-Benignos, *Los Peces del Rio Conchos*, 185–89. Chihuahua: Alianza WWF-FGRA y Gobierno del Estado de Chihuahua.

De la Maza-Benignos, M., L. Vela-Valladares, E. W. Carson, and M. L. Lozano-Vilano. 2014c. Plan de acción para la conservación de los peces endémicos del estado de Chihuahua tales como el pez cachorrito de Julimes (*Cyprinodon julimes*). In M. de la Maza-Benignos, N. Gonzalez-Hernandez, I. Banda-Villanueva, and L. Vela-Valladares (Compiladores), *Plan de acción para la conservación y recuperación de especies de fauna silvestre prioritaria en el estado de Chihuahua*, 109–13. Pronatura Noreste, A.C. y Gobierno del Estado de Chihuahua, México.

De la Maza-Benignos, M., L. Vela-Valladares, and M. L. Lozano-Vilano. 2014a. El pez *Cyprinodon julimes* y su hábitat. In A. Cruz-Angón, ed., *La biodiversidad en Chihuahua: Estudio de Estado*, 493–94. México: CONABIO.

De la Maza-Benignos, M., L. Vela-Valladares, M. L. Lozano-Vilano, M. E. García-Ramírez, J. Zapata-Lopez, A. J. Contreras-Balderas, and E. W. Carson. 2014b. The potential of holistic approaches to conservation of desert springs: A case study of El Pandeño Spring and its microendemic pupfish *Cyprinodon julimes* in the Chihuahuan Desert at Julimes, Chihuahua, Mexico. In M. de la Maza-Benignos, M. L. Lozano-Vilano, and E. W. Carson, eds., *Conservation of Desert Wetlands and Their Biotas*, 1–45. Special Publications, vol. 1. Albuquerque: Museum of Southwestern Biology, Pronatura Noreste, and Universidad Autónoma de Nuevo León.

De la Maza-Benignos, M. 2017. Sobre la extinción del hábitat natural del cachorrito de Carbonera. Pronatura Noreste, A.C. https://www.youtube.com/watch?v=Pw-5keHlgzM.

DOF. 2010. NORMA Oficial Mexicana NOM-059-SEMARNAT-2010, Protección ambiental—Especies nativas de México de flora y fauna silvestres—Categorías de riesgo y especificaciones para su inclusión, exclusión o cambio—Lista de especies en riesgo. Diario Oficial de la Federación, 30 de diciembre de 2010 (Segunda Sección). http://www.pronaturane.org/sblock/admin/images/Planes%20de%20Manejo%20y%20Programa%20de%20Monitoreo%20Vol%20%201%20mayo%202012%20final.pdf.

Follett, W. I. 1960. The freshwater fishes: Their origins and affinities. *Systematic Zoology* 9: 212–32.

García-Ramírez, M. E., S. Contreras-Balderas, and M. L. Lozano-Vilano. 2002. *Fundulus philpisteri* sp. nov. (Teleostei: Fundulidae) from the Río San Fernando basin, Nuevo León, México. In *Studies of North American Desert Fishes in Honor of P. (Phil) Pister, Conservationist*, 13–19. Monterrey: Universidad Autónoma de Nuevo León.

Jelks, H. L., S. J. Walsh, N. M. Burkhead, S. Contreras-Balderas, E. Díaz-Pardo, D. A. Hendrickson, J. Lyons, N. E. Mandrak,

F. Mccormick, J. S. Nelson, S. P. Platania, B. A. Porter, C. B. Renaud, J. J. Schmitter-Soto, E. B. Taylor, and M. L. Warren, Jr. 2008. Conservation status of imperiled North American freshwater and diadromous fishes. *Fisheries* 33: 372–407.

Lozano-Vilano, M. L., and S. Contreras-Balderas. 1999. *Cyprinodon bobmilleri:* New species of pupfish from Nuevo León, México (Pisces: Cyprinodontidae). *Copeia* 1999: 382–87.

Lozano-Vilano, M. L., A. J. Contreras-Balderas, and M. E. García-Ramírez. 2006. Eradication of spotted jewelfish, *Hemichromis guttatus,* from Poza San Jose del Anteojo, Cuatro Ciénegas Bolsón, Coahuila, Mexico. *Southwestern Naturalist* 51: 554–56.

Lozano-Vilano, M. L., S. Contreras-Balderas, and M. E. García-Ramírez. 2007. *Ordenamiento de la actividad pesquera en la ribera del Rio Lacantun de la Reserva de la Biosfera de Montes Azules.* Monterrey: Centro Interdisciplinario de Biodiversidad y Ambiente (CeIBA) y Universidad Autónoma de Nuevo León.

Lozano-Vilano, M. L., S. Contreras-Balderas, and M. E. García-Ramírez. 2013. *Ordenamiento de la actividad pesquera en la ribera del Rio Lacantun de la Reserva de la Biosfera de Montes Azules. Informe.* Monterrey: Centro Interdisciplinario de Biodiversidad y Ambiente (CeIBA) y Universidad Autónoma de Nuevo León.

Lozano-Vilano, M. L., and M. de la Maza-Benignos. 2017. Diversity and status of Mexican killifishes. *Journal of Fish Biology* 90: 3–38.

Miller, R. R. 1976. Four new species of the group *Cyprinodon* from México, with a key to the *C. eximius* complex. *Bulletin of the Southern California Academy of Sciences* 75: 68–75.

Miller, R. R., W. L. Minckley, and S. M. Norris. 2005. *Freshwater Fishes of México.* Chicago: University of Chicago Press.

Minckley, W. L., and C. O. Minckley. 1986. *Cyprinodon pachycephalus,* a new species of pupfish (Cyprinodontidae) from the Chihuahuan Desert of northern Mexico. *Copeia* 1986: 184–92.

Myers, G. S. 1930. The killifish of San Ignacio and the stickleback of San Ramon, lower California. *Proceedings of the California Academy of Sciences* 19: 95–104.

Page, L. M., H. Espinosa-Pérez, L. T. Findley, C. R. Gilbert, R. N. Lea, N. E. Mandrak, R. L. Mayden, and J. S. Nelson. 2013. *Common and Scientific Names of Fishes from the United States, Canada, and Mexico.* 7th ed. Special Publication 34. Bethesda, MD: American Fisheries Society.

Ruiz-Campos, G. 2012. *Catálogo de Peces Dulceacuícolas de Baja California Sur.* México, D.F.: Instituto Nacional de Ecología, SEMARNAT.

Ruiz-Campos, G. 2016. *Manejo Poblacional y de Hábitat del Pez Cachorrito del Desierto* (Cyprinodon macularius) *en la Laguna de Evaporación del Campo Geotérmico de Cerro Prieto, Mexicali, Baja California, México.* Informe técnico final del proyecto AA-018TOQ034-E24-2016, Comisión Federal de Electricidad.

Ruiz-Campos, G., A. Andreu-Soler, and A. Varela-Romero. 2013. Condition status of the endangered desert pupfish, *Cyprinodon macularius* Baird and Girard, 1853, in the Lower Colorado River basin (Mexico). *Journal of Applied Ichthyology* 29: 555–61.

Ruiz-Campos, G., A. Andreu-Soler, M. R. Vidal-Abarca Gutiérrez, J. Delgadillo-Rodríguez, M. L. Suárez-Alonso, C. González-Abraham, and V. H. Luja. 2014c. *Catálogo de humedales dulceacuícolas de Baja California Sur.* México, D.F.: Instituto Nacional de Ecología y Cambio Climático, SEMARNAT.

Ruiz-Campos, G., F. Camarena-Rosales, S. Contreras-Balderas, C. A. Reyes-Valdez, J. de la Cruz-Agüero, and E. Torres-Balcazar. 2006. Distribution and abundance of the endangered killifish, *Fundulus lima* (Teleostei: Fundulidae), in oases of central Baja California Peninsula, México. *Southwestern Naturalist* 51: 502–9.

Ruiz-Campos, G., F. Camarena-Rosales, A. F. González-Acosta, A. M. Maeda-Martínez, F. J. García de León, A. Varela-Romero, and A. Andreu-Soler. 2014a. Estatus actual de conservación de seis especies de peces dulceacuícolas de la Península de Baja California, México. *Revista Mexicana de Biodiversidad* 85: 1235–48.

Ruiz-Campos, G., S. Contreras-Balderas, M. L. Lozano-Vilano, S. González-Guzmán, and J. Alaníz-García. 2000. Ecological and distributional status of the continental fishes of northwestern Baja California, Mexico. *Bulletin of the Southern California Academy of Sciences* 99: 59–90.

Ruiz-Campos, G., A. Varela-Romero, S. Sánchez-González, F. Camarena-Rosales, A. Maeda-Martínez, A. F. González-Acosta, A. Andreu-Soler, E. Campos-González, and J. Delgadillo-Rodríguez. 2014b. Peces invasores del noroeste de México. In R. E. Mendoza-Alfaro and P. Kolef, eds., *Especies Acuáticas Invasoras en México,* 375–400. México D.F.: CONABIO.

Varela-Romero, A., G. Ruiz-Campos, L. M. Yepiz-Velazquez, and J. Alaníz-García. 2003. Distribution, habitat, and conservation status of desert pupfish (*Cyprinodon macularius*) in the lower Colorado River basin, Mexico. *Reviews in Fish Biology and Fisheries* 12: 157–65.

Williams, J. E., J. E. Johnson, D. A. Hendrickson, S. Contreras-Balderas, J. D. Williams, M. Navarro-Mendoza, D. E. McAllister, and J. E. Deacon. 1989. Fishes of North America endangered, threatened, or special concern. *Fisheries* 14: 2–20.

6 Ghosts of Our Making

Extinct Aquatic Species of the North American Desert Region

Jack E. Williams and Donald W. Sada

The life of a water-dependent species in the desert can be tough. Across North America, 39% of fish species are at some level of risk of extinction (Jelks et al. 2008), but in desert regions of the continent, the percentage of native fishes at risk is higher. Fishes in this desert region are characterized by lower species richness, but higher endemism than those in other regions, and those species that occupy a more restricted range are subject to greater threats from water withdrawals and introductions of nonnative species. According to the Arizona Game and Fish Department (2017), of the 36 native fishes in Arizona, 20 are listed as endangered or threatened pursuant to the Endangered Species Act, one is already extinct, and many of the remaining fishes are listed as sensitive or declining. Although invertebrates are generally less well known than fishes, the risk for aquatic invertebrates in desert systems appears high as well. Springsnails, for instance, are a diverse but imperiled group in arid lands (Hershler et al. 2014a).

Across North America, extinctions of freshwater fishes have accelerated since 1950 as cities have boomed and dams and other landscape-scale habitat modifications have become more common (Burkhead 2012). Aquatic habitats in desert regions have been greatly impacted by human activity. Riparian areas along desert springs and streams have been overgrazed by livestock, flows have been diverted to quench the ever-growing thirst of expanding cities, and nonnative predators and competitors have been widely introduced. With threats so high, it should not be surprising that the number of extinctions during the past century has increased dramatically (Miller et al. 1989; Burkhead 2012).

Extinction is certain to occur for aquatic species if their water supply is removed. Deserts are defined by scarcity of water, yet these regions are home to some of the largest and fastest-growing cities, including Las Vegas, Phoenix, Tucson, and El Paso–Juarez. The Las Vegas region obtains water from two sources, groundwater and the Colorado River at Lake Mead. A surge in groundwater pumping during the 1950s eliminated spring habitat for the Las Vegas dace *Rhinichthys deaconi*, causing its extinction (Miller et al. 1989), and plans by the Southern Nevada Water Authority to tap into regional aquifers to supplement the declining water levels in Lake Mead could signal a broader wave of extinction for spring-dependent species (Deacon et al. 2007). In Texas, 63 of 281 significant springs had gone dry due to diversions and groundwater withdrawal by the 1970s, probably causing extinctions among aquatic invertebrate groups before their presence could be documented (Unmack and Minckley 2008).

Whether conservation efforts will hold the line on extinctions in the face of growing demand for freshwater in North America's desert regions is a central question for society. To date, the effectiveness of conservation efforts has been mixed, the rate of extinctions is increasing, and the future survival of remaining aquatic species in the desert is far from certain. To paraphrase the famous line of George Santayana, if we do not learn from our mistakes of the past, we are doomed to repeat them in the future.

This chapter reports on some of society's more enduring mistakes: human-caused extinctions of species. Its focus is on recent (since 1900) extinctions of fishes and aquatic invertebrates from North American deserts and nearby regions. In many cases, extinctions have become known only through the determination of early scientists to search out rare species and to document their passing as thoroughly as was possible at the time. In the 1930s, 1940s, and 1950s, Carl Hubbs and Robert Miller surveyed the desert for fishes, collecting specimens, determining their taxonomic status, describing new species, and recording notes on their conservation and threats (Hubbs and Miller 1948; Miller 1948). In the 1960s, 1970s, and 1980s, Salvador Contreras-Balderas, W. L. Minckley, James Deacon, Clark Hubbs, and others expanded on these early efforts. Without the hard work of these pioneering conservation biologists, not only would our knowledge of rare species be less, but the number of extinct species would undoubtedly be much greater.

Despite the determined efforts of this small but dedicated cadre of scientists, knowledge of many desert fishes and invertebrates is surprisingly scant. Although some of the larger river systems and prominent spring systems and their species are well known and frequently monitored, other habitats have been surveyed infrequently, if at all. In many areas where fishes are surveyed, associated populations of invertebrates are ignored, in part because of bias toward funding work on vertebrate species, but also because taxonomic experts on many invertebrate taxa are scarce. The number of scientific surveys of some of the major spring systems in Coahuila or Nuevo León can be counted on one hand. For example, Contreras-Balderas and Lozano-Vilano (1996) recount infrequent surveys of major spring systems of northern México and describe how the loss of entire aquatic communities went practically unnoticed. Unmack and Minckley (2008) expressed similar concerns for large spring systems in Nevada and Texas that were pumped dry before surveys of invertebrate communities were made.

The geographic scope of this chapter extends from the Great Basin of the western United States southward through the Mojave, Sonoran, and Chihuahuan Deserts to as far south as the Mexican states of Zacatecas, San Luis Potosí, and Jalisco. For some of the extinct species discussed herein, desperate conservation measures were employed, but ultimately failed. But for many others, scientists simply had no knowledge of the species and the threats to their existence until it was too late. There is no doubt that additional extinctions have occurred but have gone unnoticed and unrecorded. Where springs and desert wetlands were pumped dry or diverted into pipes as cities and agriculture grew, subsequent scientific surveys yield dry spring holes with no trace of the species that once lived there.

Knowledge of what is extinct and when its extinction occurred may be clouded by taxonomic uncertainty. The springsnail *Pyrgulopsis* (known in scientific shorthand as "pyrg") in Longstreet Spring in Ash Meadows, Nevada, is a case in point. The snail was first collected in the late 1940s or early 1950s and was considered by the original collector to represent a new but undescribed species. Springsnails are a diverse group with many species in the Ash Meadows area, so it is not unreasonable to presume that the individuals in Longstreet Spring were distinct. By the 1970s, groundwater pumping had dried Longstreet Spring, and the springsnail was declared extinct (Norment 2014); however, no formal description of the Longstreet springsnail as a distinct species was ever completed. The Longstreet pyrg *Pyrgulopsis* sp. is an example of a population that became extinct before taxonomists were able to conclude whether the population should be described as a separate species or lumped with other nearby populations (R. Hershler, pers. comm., May 31, 2017). Regardless of its taxonomic status, the desiccation of Longstreet Spring and the loss of its springsnail are tragic and highlight how the pace of habitat destruction often races ahead of our scientific understanding.

While the slow pace of taxonomic studies may limit determination of what are distinct species, new genetic techniques are revealing previously unknown species. Genetic techniques sometimes reveal "cryptic" species that appear morphologically identical to other populations but are identified as distinct by molecular studies that uncover genetic differences among populations. Springsnails, which have been characterized as "undersplit taxonomically" (Hershler et al. 2014a), are probably in need of further genetic studies to identify potential cryptic species.

Although the conservation trend for most aquatic species in desert regions is toward continued imperilment and loss of populations, occasionally a species previously believed to be extinct is found persisting. Such was the case when a 2012 survey rediscovered the Chihuahuan dwarf crayfish *Cambarellus chihuahuae* (Carson et al. 2015). Although most of the springs from which this crayfish was known have dried due to excessive groundwater extraction, one spring in a more isolated part of the basin maintained surface-water flows and a crayfish population. Nonetheless, this crayfish remains critically endangered. Similarly, the Nevada Department of Wildlife (J. Elliot, pers. comm.) reports that the Independence Valley tui chub *Siphateles bicolor isolate* persists, despite earlier claims that the subspecies was extinct (Miller et al. 1989).

For all the reasons described above—paucity of scientific surveys, uncertain taxonomies, cryptic species, and rediscoveries—any list of extinct species should be considered more provisional than the title of this chapter might suggest. With that caveat, the actual number of extinct species undoubtedly exceeds the list provided herein.

The list of extinct fishes and aquatic invertebrates presented herein was compiled from a variety of sources (see references in table 6.1). The works of Robert Miller and colleagues provide some of the best records of extinct species in the United States and México (Miller 1948; Miller et al. 1989, 2005). For knowledge of extinct aquatic invertebrates, Robert Hershler provided critical information (in personal communications as well as papers cited in the references).

[Box 6.1] The Nomenclature of Extinction

Extinct: A species or subspecies is considered extinct when the last individual of that taxonomic group has died.

Functionally extinct: A species or subspecies is considered functionally extinct when population numbers or suitable habitat shrink to such levels that remaining populations lose viability.

Extinct in the wild: A species or subspecies is extinct in the wild when the only known living members of that taxonomic group survive in aquaria, hatcheries, or other forms of captivity.

Extirpated: A species or subspecies is considered extirpated from a specific habitat area when all members of that taxonomic group cease to exist in that area, but may still survive elsewhere.

EXTINCTION OF AQUATIC BIOTA IN NORTH AMERICAN DESERTS

Thirty-two fish taxa and 23 aquatic invertebrate taxa were determined to be extinct within the desert and adjacent regions of the United States and México (table 6.1). Additionally, five fishes (three pupfishes, one goodeid, and one poeciliid) are considered extinct in the wild (table 6.2). The five fishes listed as extinct in the wild are believed to survive in aquaria, but they must be considered critically endangered. Miller et al. (2005) reported that Catarina pupfish *Megupsilon aporus* is being cultured in México and the United States after groundwater pumping and introduction of largemouth bass *Micropterus salmoides* caused extirpation of all wild populations near El Potosí in Nuevo León, but Lozano-Vilano et al. (chap. 5, this volume) consider it to be extinct.

Rare species are notoriously difficult to maintain in captivity for long periods, especially species like

Table 6.1. Recently extinct fish and aquatic invertebrate taxa from North American deserts and adjacent areas. Extinction dates are approximate or indicated as unknown. Numbers correspond to the locations mapped on fig. 6.1.

Common name	Scientific name	Historical range	Extinction date	Sources
FISHES				
Family Catostomidae				
1. Snake River sucker	*Chasmistes muriei*	Upper Snake River, Wyoming	1928	Miller et al. 1989
Family Cottidae				
2. Utah Lake sculpin	*Cottus echinatus*	Utah Lake, Utah	1930	Miller et al. 1989
Family Cyprinidae				
3. Maravillas red shiner	*Cyprinella lutrensis blairi*	Big Bend, Texas	1954	Miller et al. 1989
4. Carpa de Río Tunal, Río Tunal shiner	*Dionda* sp.	Río del Tunal, Durango	Unknown	Minckley and Miller 2005
5. Thicktail chub	*Gila crassicauda*	Central Valley, California	1957	Miller et al. 1989
6. Pahranagat spinedace	*Lepidomeda altivelis*	Pahranagat Valley, Nevada	1940	Miller et al. 1989
7. Ameca shiner	*Notropis amecae*	Río Ameca, Jalisco	1970	Miller et al. 1989; Lozano-Vilano et al., chap. 5, this volume
8. Carpita de Durango, Durango shiner	*Notropis aulidion*	Río del Tunal, Durango	1965	Miller et al. 1989, 2005
9. Phantom shiner	*Notropis orca*	Rio Grande, New Mexico, Texas, Chihuahua, Tamaulipas	1964	Miller et al. 1989
10. Carpita de Salado, Salado shiner	*Notropis saladonis*	Río Salado, Nuevo León and Coahuila	1970	Miller et al. 2005
11. Rio Grande bluntnose shiner	*Notropis simus simus*	Rio Grande, New Mexico, Texas, Coahuila	1964	Miller et al. 1989
12. Clear Lake splittail	*Pogonichthys ciscoides*	Clear Lake, California	1970	Miller et al. 1989
13. Las Vegas dace	*Rhinichthys deaconi*	Las Vegas Valley, Nevada	1955	Miller et al. 1989
14. Grass Valley speckled dace	*Rhinichthys osculus reliquus*	Grass Valley, Nevada	1950	Miller et al. 1989
15. High Rock Spring tui chub	*Siphateles bicolor* ssp.	High Rock Spring, California	1989	Moyle et al. 1995
16. Stumptooth minnow	*Stypodon signifer*	Parras Valley, Coahuila	1950	Miller et al. 2005
Family Cyprinodontidae				
17. La Presita pupfish	*Cyprinodon ceciliae*	Bolsón de Sandia, Nuevo León	1990	Contreras-Balderas and Lozano-Vilano 1996; Miller et al. 2005
18. La Trinidad pupfish	*Cyprinodon inmemoriam*	Bolsón de Sandia, Nuevo León	1985	Contreras-Balderas and Lozano-Vilano 1996; Lozano-Vilano et al., chap. 5, this volume
19. Parras pupfish	*Cyprinodon latifasciatus*	Parras Valley, Coahuila	1930	Miller et al. 1989; 2005
20. Tecopa pupfish	*Cyprinodon nevadensis calidae*	Tecopa, California	1971	Miller et al. 1989
21. Monkey Spring pupfish	*Cyprinodon* sp.	Monkey Spring, Arizona	1971	Minckley et al. 1991
22. Ojo de Villa Lopez pupfish	*Cyprinodon* sp.	Ojo de Villa Lopez, Chihuahua	2006	Lozano-Vilano et al., chap. 5, this volume
23. Ojo en Ocuila pupfish	*Cyprinodon* sp.	Ojo en Ocuila, Durango	ca. 1990	Lozano-Vilano et al., chap. 5, this volume
24. Catarina pupfish	*Megupsilon aporus*	El Potosí, Nuevo León	ca. 2000	Miller et al. 2005; Lozano-Vilano et al., chap. 5, this volume

Common name	Scientific name	Historical range	Extinction date	Sources
Family Goodeidae				
25. Parras characodon, Mexcalpique de Parras	*Characodon garmani*	Parras Valley, Coahuila	1900	Miller et al. 1989; 2005
26. Ash Meadows poolfish	*Empetrichthys merriami*	Ash Meadows, Nevada	1948	Miller et al. 1989
27. Pahrump Ranch poolfish	*Empetrichthys latos pahrump*	Pahrump Valley, Nevada	1958	Miller et al. 1989
28. Raycraft Ranch poolfish	*Empetrichthys latos concavus*	Pahrump Valley, Nevada	1960	Miller et al. 1989
Family Poeciliidae				
29. Amistad gambusia	*Gambusia amistadensis*	Goodenough Spring, Texas	1973	Peden 1973; Miller et al. 1989
30. San Marcos gambusia	*Gambusia georgei*	San Marcos River, Texas	1983	Miller et al. 1989
Family Salmonidae				
31. Alvord cutthroat trout	*Oncorhynchus clarkii alvordensis*	Alvord Basin, Oregon, Nevada	1940	Miller et al. 1989
32. Yellowfin cutthroat trout	*Oncorhynchus clarkii macdonaldi*	Twin Lakes, Colorado	1910	Behnke 2002
AQUATIC INVERTEBRATES				
Family Astacidae				
33. Sooty crayfish	*Pacifastacus nigrescens*	San Francisco, California	Prior to 1980	Taylor et al. 2007
Family Cambaridae				
34. Unnamed crayfish	*Cambarellus alvarezi*	Ejido El Potosí, Nuevo León	ca. 1994	Alvarez et al. 2010
35. Charco Azul crayfish	*Cambarellus* sp.	Bolsón de Sandia, Nuevo León	1990	Contreras-Balderas and Lozano-Vilano 1996
Family Cochliopidae				
36. Phantom Lake springsnail	*Juturnia brunei*	Phantom Lake, Texas	ca. 1984	Hershler et al. 2014b
37. Hertlein's tryonia	*Tryonia hertleini*	Spring near terminus of Río Casas Grandes, Chihuahua	1975	Hershler 2001
38. Julimes tryonia	*Tryonia julimesensis*	Springs near Julimes, Chihuahua	Between 1991 and 2001	Hershler et al. 2011
39. Oasis Ranch tryonia	*Tryonia oasiensis*	Caroline Spring, Texas	2009	Hershler et al. 2011
40. Santa Rosa tryonia	*Tryonia santarosae*	Ojo de Santa Rosa, Chihuahua	Between 1973 and 1990	Hershler et al. 2014b
41. Shi-Kuei's tryonia	*Tryonia shikuei*	Ojo de Federico and Ojo de San Juan, Chihuahua	Mid-1980s	Hershler et al. 2014b
42. Unnamed tryonia	*Tryonia* sp.	Spring near La Laguna, Chihuahua	Between 1971 and 1990	Hershler et al. 2014b
43. Panaca Big Spring tryonia	*Tryonia* sp.	Panaca Big Spring, Nevada	1945	D. W. Sada, unpublished data
Family Dytiscidae				
44. Mono Lake Hygrotus diving beetle	*Hygrotus artus*	Mono Lake, California	Unknown	Evans and Hogue 2006

Common name	Scientific name	Historical range	Extinction date	Sources
Family Elmidae				
45. Stephan's riffle beetle	*Heterelmis stephani*	Madera Canyon, Arizona	1993	Brown 1972
Family Hydrobiidae				
46. Unnamed pyrg	*Pyrgulopsis brandi*	Springs at Las Palomas, Chihuahua	1970s	Hershler 1994
47. Nevada pyrg	*Pyrgulopsis nevadensis*	Pyramid and Walker Lakes, Nevada	Unknown	D. W. Sada, field notes
48. Fish Lake pyrg	*Pyrgulopsis ruinosa*	Fish Lake, Nevada	1995	D. W. Sada, field notes
49. Longstreet pyrg	*Pyrgulopsis* sp.	Longstreet Spring, Nevada	1970s	R. Hershler, pers. comm.
Family Lymnaeidae				
50. Fish Springs marshsnail	*Stagnicola pilsbryi*	Fish Springs, Utah	1970	Cordeiro and Perez 2012a
51. Thickshell pondsnail	*Stagnicola utahensis*	Utah Lake, Utah	Late 1930s or early 1940s	Cordeiro and Perez 2012b
Family Physidae				
52. Fish Lake physa	*Physa microstriata*	Fish Lake, Utah	Between 1929 and 1989	Bogan 2000
Family Unionidae				
53. Rio Grande monkeyface	*Rotundaria couchiana*	Rio Grande, New Mexico, Texas, Chihuahua	ca. 1900	Campbell and Lydeard 2012
Family Valvatidae				
54. Beltran's valvata snail	*Valvata beltrani*	Charco Azul, Nuevo León	1980s	Contreras-Balderas and Lozano-Vilano 1996
55. Unnamed valvata snail	*Valvata* sp.	Charco Azul, Nuevo León	1980s	Contreras-Balderas and Lozano-Vilano 1996

Table 6.2. Aquatic taxa from North American deserts that are extinct in the wild. Numbers correspond to the locations mapped on fig. 6.1.

Common name	Scientific name	Historical range	Sources
Family Cyprinodontidae			
56. Charco Palma pupfish	*Cyprinodon longidorsalis*	Charco La Palma, Nuevo León	Contreras-Balderas and Lozano-Vilano 1996; Minckley and Miller 2005
57. Charco Azul pupfish	*Cyprinodon veronicae*	Bolsón de Sandia, Nuevo León	Contreras-Balderas and Lozano-Vilano 1996; Minckley and Miller 2005
58. Potosí pupfish	*Cyprinodon alvarezi*	El Potosí, Nuevo León	Minckley and Miller 2005
Family Goodeidae			
59. Golden skiffia	*Skiffia francesae*	Río Teuchitlan, Jalisco	Miller et al. 2005; Lozano-Vilano et al., chap. 5, this volume
Family Poeciliidae			
60. Monterrey platyfish	*Xiphophorus couchianus*	Río Santa Cararina, Nuevo León	Lozano-Vilano et al., chap. 5, this volume

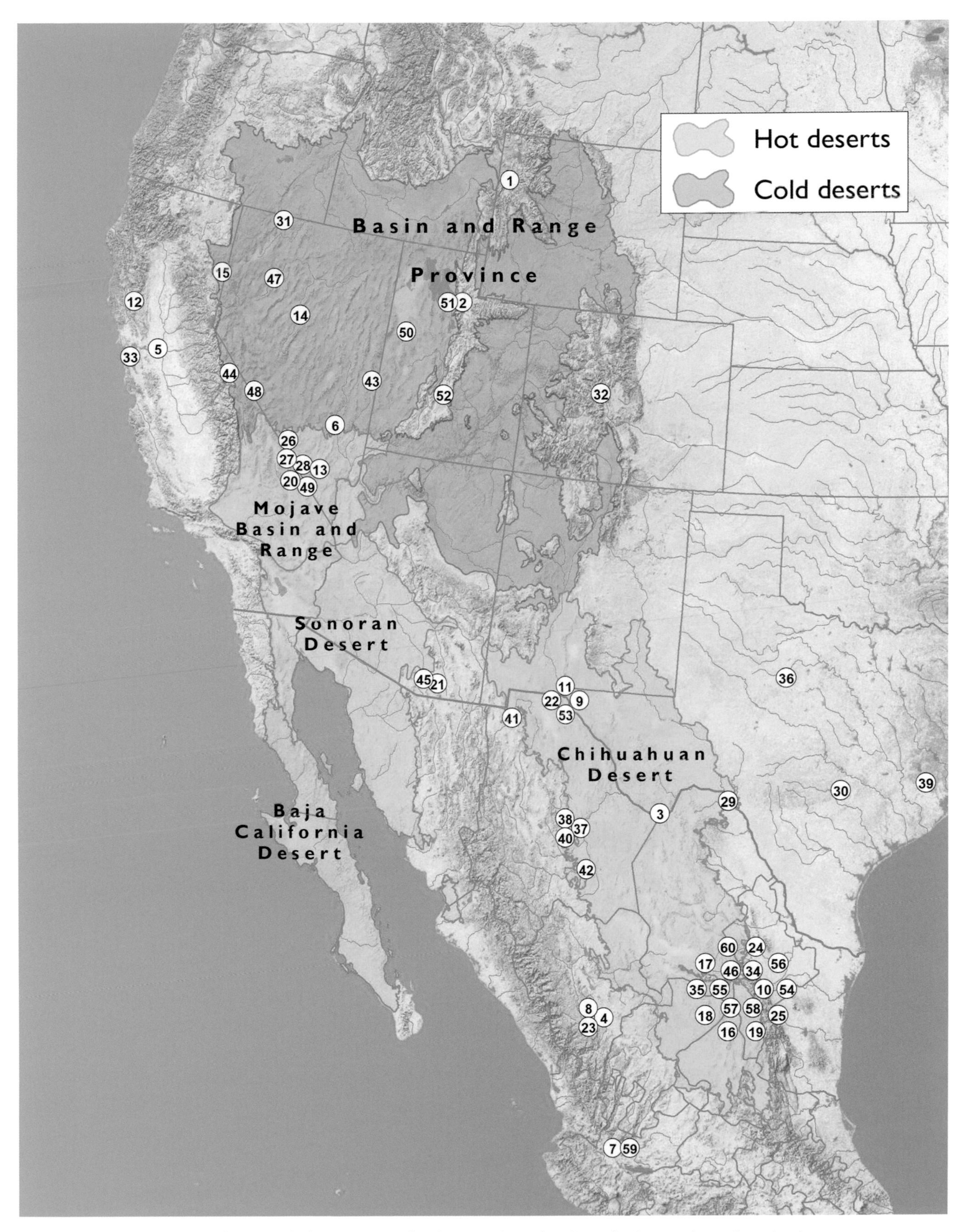

Fig. 6.1 Map of North American deserts and adjacent regions showing approximate locations of extinct species and species that are extinct in the wild. For wide-ranging taxa, an approximate midpoint is shown. See tables 6.1 and 6.2 for cross-referencing of the numbers shown on the map with species lists.

pupfish that have short lifespans. Eventually, it is more likely than not that some failure of electricity or water supplies, or some sort of accident or human error, will result in the loss of these artificially maintained populations. If they do persist for many generations, genetic problems are likely to arise due to inbreeding and small population sizes. Unless native habitats are restored and the species repatriated from captive stocks into their historical ranges, their likelihood of survival will decrease each year. Not only is it difficult to maintain these species in captivity over the long term, but novel selection pressures in aquaria and the short generation times of many of these species will result in genetic drift away from their natural evolutionary trajectory. For these reasons, conservationists rightly focus on the protection and restoration of natural habitats, especially for species that are endemic to a single spring or spring system, as the most prudent way of ensuring the future existence of naturally rare species, rather than relying on translocation of populations into artificial habitats (Williams et al. 2005).

As noted earlier, the list of extinct species is likely to be incomplete, especially for invertebrate taxa, such as springsnails, riffle beetles (genera *Heterelmis, Microcylloepus*), amphipods (genus *Hyalella*), and naucorids (genus *Ambrysus*) that did not receive attention from scientific collectors until recent decades. Even today, our knowledge of locally endemic aquatic invertebrate taxa is woefully inadequate in more remote desert regions.

Extinct wetland-dependent amphibians and mammals are beyond the scope of this chapter, but these losses nonetheless reinforce the degree to which some desert wetland areas have been degraded. The Vegas Valley leopard frog *Lithobates fisheri* was known only from wetlands in Nevada's Las Vegas Valley, some of which also supported the extinct Las Vegas dace. Today, springs supporting these wetlands have been pumped dry to maintain the burgeoning population in the metropolitan Las Vegas region, which now supports more than 2 million people.

The distribution of extinct species, and of those that are extinct in the wild, shows clusters of extinctions located in the Chihuahuan and Mojave Deserts (fig. 6.1). These clusters are consistent with areas of high aquatic species diversity and local endemism, such as the Ash Meadows–Death Valley region, the Rio Grande basin, and the Chihuahuan spring complexes in México. While many extinct species are narrow endemics, some recently extinct species historically occurred in larger river systems, such as the Rio Grande, and their loss demonstrates the significant habitat and community changes that have occurred during the past century in major southwestern rivers.

OBITUARIES OF AQUATIC SPECIES

Few records mark the passing of most species. The following are selected case studies of extinct species that contain a more complete historical record than other such species. Understanding the process of extinction through these case studies should help us prevent extinctions in the future. Additional obituaries for extinct desert fishes can be found in Miller et al. (1989), Minckley et al. (1991) and Norment (2014). Whether well known or not, the passing of any species should be marked and remembered. As E. O. Wilson remarked in *The Diversity of Life* (1992, 32), "Every scrap of biological diversity is priceless, to be learned and cherished, and never to be surrendered without a struggle."

Amistad Gambusia *Gambusia amistadensis*

At one time, Goodenough Spring, located on the Texas side of the Rio Grande, was the third largest spring in Texas (fig. 6.2). This spring was the only natural habitat for the Amistad gambusia, which is now extinct.

Large dam construction and the associated impoundment of rivers is a primary cause of population loss and habitat fragmentation for many native fishes of the Southwest, but seldom are the impacts severe enough to cause the extinction of a species. It was not until construction of Amistad Dam was completed in 1968 and water was first impounded behind it that this fish came to the attention of scientists as a potentially distinct species. Collections of the native *Gambusia* were made in April and August of 1968; the latter specimens were collected from among flooded prickly-pear cactus as the reservoir's waters rose (Peden 1973).

Recognizing the potential loss of a species, scientists from the University of Texas at Austin established stocks of the *Gambusia* at the university. When it was formally described as a new species in 1973, no specimen could be located in reservoir waters around Goodenough Spring, and the species was considered to be extinct in the wild (Peden 1973).

Fig. 6.2 Goodenough Spring, Val Verde County, Texas, 1941. This spring is the only known natural habitat of the Amistad gambusia, but it now lies buried below nearly 50 m of water in Amistad Reservoir. (Photo courtesy of the Louis Leurig family.)

The fate of the captive stocks was well documented by Hubbs and Jensen (1984). Initially, scientists examined the potential to introduce the Amistad gambusia into other nearby springs, but they were all inhabited by western mosquitofish *Gambusia affinis*, which readily hybridizes with other *Gambusia* species. In 1974, stocks of the Amistad gambusia were moved from the University of Texas to Dexter National Fish Hatchery in New Mexico in the hopes of maintaining a population there. But routine population monitoring in 1979 found that the stocks had become contaminated by *G. affinis* and that the Amistad gambusia had apparently been eliminated by hybridization. Species of *Gambusia* are notoriously hard to distinguish without microscopic examination of the male gonopodium, so it is understandable how routine handling of the fish would not detect an introduced *Gambusia* or their hybrids. When the contamination was discovered at Dexter, it was found that one of the habitats at the university had dried and that the other had also been contaminated by *G. affinis* (Hubbs and Jensen 1984).

Prior to the construction of Amistad Dam, *Gambusia affinis* was known from the Rio Grande, including where the downstream outflow of Goodenough Spring entered the Rio Grande. Apparently, the distinct environment at Goodenough Spring favored the Amistad gambusia and excluded the more common *G. affinis*. Ultimately, hybridization with a nonnative congener eliminated the last stocks of the Amistad gambusia, which illustrates the importance of maintaining the unique habitats of single-spring endemic species and the hazards of attempting to maintain stocks of different, but related, species in artificial environments.

Nevada Pyrg *Pyrgulopsis nevadensis*

The Nevada pyrg is one of the few lentic members of the genus *Pyrgulopsis* in the western United States (Hershler 1994). It was described in 1883 from material dredged from Pyramid and Walker Lakes in the western Lahontan Basin by I. C. Russell in 1882 (Stearns 1883). It is known only from these terminal lakes, but live specimens have not been seen since its description, and it is believed extinct due to its absence from collections made during recent dredging and limnological studies (Robertson 1978; Galat et al. 1981). Large windrows of empty shells now characterize shorelines of both lakes, indicating that it must have been abundant at one time. All other western Lahontan Basin *Pyrgulopsis* occupy springs. Extensive surveys of springs in the region over the past 25 years have found only springsnails (*Fluminicola turbiniformis, F. dalli, P. longiglans*, and *P. gibba*) (Hershler 1998, 1999).

Little is known about Nevada pyrg life history or habitat requirements, which makes it difficult to determine the causes of its extinction. However, inflow into both lakes was dramatically reduced before the turn of the twentieth century by upstream diversion for agriculture and municipalities. Continued declines in water level over the following decades also changed water chemistry. Although there is no information about the tolerance of the Nevada pyrg to altered lake conditions, it was probably extirpated from both lakes by toxic conditions, particularly in the hypolimnion. These impacts also extirpated lentic populations of Lahontan cutthroat trout *Oncorhynchus clarkii henshawi* (federally listed as threatened) and Tahoe sucker *Catostomus tahoensis*, and justified listing of cui-ui *Chasmistes cujus*, which is endemic to Pyramid Lake, as endangered (Scoppettone and Vinyard 1981; Coffin and Cowan 1995). From 1882 to 1996, anthropogenic diversions lowered the surface elevation of Walker Lake by about 50 m, decreased its volume from 11.1 to 2.7 km^3, and increased its salinity from 2,500 to 13,300 mg/L. Increasing salinity and other factors relating to hypolimnetic anoxia created conditions that stressed tui chubs *Siphateles bicolor* and extirpated the cutthroat trout, even hatchery stocks (La Rivers 1962; Beutel et al. 2013). These changes are less than what occurred 9,000 to 5,000 years ago, when the lake dried following the natural diversion of the Walker River into the Carson River basin (Adams 2007). Survival of pyrgs in Walker Lake through this period is unknown, but could be attributable to its persistence in springs that are below current lake levels or to its reintroduction by birds (primarily white pelicans *Pelicanus erythrorhynchos*) that nest on Pyramid Lake islands, and feed there and at Walker Lake.

Julimes Tryonia *Tryonia julimesensis*

Julimes tryonia is one species of a diverse group of narrowly endemic springsnails from the Chihuahuan Desert of Chihuahua and Texas (Hershler et al. 2011). Hershler et al. (2011) described 13 species of *Tryonia* from isolated spring systems in Chihuahua, Durango, and Texas, including *Tryonia julimesensis*, as part of a springsnail species flock in the Río Conchos drainage. Julimes tryonia is known only from unnamed springs along the Río Conchos south-southeast of Julimes, Chihuahua.

Most springsnails in North America's deserts are crenophiles (Hershler 1998; Polhemus and Polhemus 2002). Like most crenophilic macroinvertebrates, springsnails are most abundant near spring sources where the environment (e.g., temperature, discharge, dissolved oxygen concentration, and pH) is more constant (e.g., McCabe 1998). Their abundance incrementally decreases downstream as environmental variability increases, and their distribution terminates where variability exceeds their tolerance. Members of the genus *Tryonia* are small organisms with shells typically only about 2 mm in width, but their abundance near spring sources can be high, with individuals appearing as a nearly solid carpet of shells along the springbrook.

When first discovered in 1991, the Julimes tryonia was abundant in the main spring runs within a series of hot springs along the east side of the Río Conchos (Hershler et al. 2011). At that time, the species was also found in a small rheocrene just west of the main springs. In both habitats, springsnails were collected from 44°C water, among the warmest water known to support these species (Hershler et al. 2011). When the habitats were resurveyed in 2001, the springs had been dredged, and the species could not be located. Hershler et al. (2011) believed the species became extinct sometime between 1991 and 2001.

Springs and marshes within the Chihuahuan Desert are rapidly disappearing as urban development and agriculture increase the rates of groundwater withdrawal, surface habitat disturbance, and spring failures. Within the Río Conchos drainage, Edwards et al. (2003) reported that surveys conducted in 1994–1995 found that the native fish fauna was relatively intact when compared with collections made during the 1950s, but they warned that the increasing spread of high-capacity pumps and pipelines would degrade aquatic habitats in the region. Hershler et al. (2011) described their efforts to discover and describe the springsnail fauna of the region prior to its loss by water extraction as a "race against time." Unless sustainable water use practices are implemented, additional aquatic species in Chihuahua and elsewhere are likely to follow the Julimes tryonia into extinction.

Tecopa Pupfish *Cyprinodon nevadensis calidae*

The Tecopa pupfish has the dubious distinction of being the first species to be removed from the US list of endangered and threatened wildlife because it became extinct. This pupfish was included on Appendix D, United States List of Endangered Native Fish and Wildlife, which was published in the *Federal Register* on October 13, 1970, before the passage of the Endangered Species Act in 1973 (USFWS 1970). The final rule removing the pupfish from the endangered species list was published in the *Federal Register* on January 15, 1982. Ironically, the species may have been extinct when it was listed as endangered in October 1970 or soon thereafter.

The first known collections of Tecopa pupfish are from 1942 and include specimens collected a few hundred meters downstream of North Tecopa Hot Spring and South Tecopa Hot Spring, both tributaries of the Amargosa River in California (Miller 1948). Specimens of the Tecopa pupfish were collected on March 28, 1954, and on February 2, 1970, from the outflows of the hot springs, but surveys conducted in 1972 and 1977 in the outflows of the Tecopa Hot Springs, as well as other nearby springs, were unable to locate the subspecies and confirmed its extinction (Miller et al. 1989).

The natural range of the Tecopa pupfish was confined to the hot outflows below each of the springs, which were separated by only about 10 m. Type specimens for this species were collected from 40°C water, among the hottest water temperatures recorded for any fish species (Miller 1948). Miller (1948) reported most Tecopa pupfish in the outflow of South Tecopa Hot Spring as occurring 115–320 m downstream of a small concrete barrier below a bathhouse at the spring head. Tecopa pupfish were fairly common in water of 36.5°C and abundant in water from 32°C to 36°C (Miller et al. 1989). The Amargosa pupfish *C. n. amargosae* occurred farther downstream and in the Amargosa River. Hybrids between the two closely related pupfishes probably occurred in cooler downstream waters (Miller et al. 1989).

The cause of the Tecopa pupfish's extinction is undoubtedly related to development of the hot springs and outflows into bathhouses and soaking pools. Rudimentary bathhouses were already present at the head pool of both North and South Tecopa Hot Springs when Miller visited the site on September 26, 1942 (Miller 1948, plate XII). The outflows of both North and South Tecopa Hot Springs were channelized and combined in 1965, which probably increased flow and reduced suitable habitat (USFWS 1982). Introduction of western mosquitofish and bluegill *Lepomis macrochirus* also may have contributed to the decline (USFWS 1982).

Extinction of the Tecopa pupfish should not be attributed to failure of the Endangered Species Act, as the subspecies became extinct before its passage in 1973. It is unknown whether the owners of Tecopa Hot Springs were aware of the unique subspecies of pupfish present on their property as the hot springs were developed during the 1950s–1970s.

Poolfishes of the Genus *Empetrichthys*

In 1948, the Museum of Zoology at the University of Michigan published what would become a classic in desert fish literature: *The Cyprinodont Fishes of the Death Valley System of Eastern California and Southwestern Nevada* by Robert Rush Miller. In this treatise, which constituted the bulk of his doctoral dissertation from the University of Michigan, Miller described the results of numerous field trips that he took with his father, Ralph R. Miller, and his wife, Frances Hubbs Miller, to the Death Valley region between 1936 and 1942, which netted more than 10,000 fish specimens for the Museum of Zoology. From these collections, Miller (1948) would describe an amazing 10 new species and subspecies of desert fishes, including a new species *Empetrichthys latos* and three new subspecies within this species: Pahrump poolfish *E. l. latos*, Pahrump Ranch poolfish *E. l. pahrump*, and Raycraft Ranch poolfish *E. l. concavus*. Miller (1948) also detailed his collection of 22 specimens of the Ash Meadows poolfish *E. merriami*, which was originally described by Charles Gilbert in 1893 (fig. 6.3).

The three subspecies of *E. latos* were collected from springs in Pahrump Valley, Nevada. The Pahrump poolfish was restricted to the main spring pool at Manse Ranch (Manse Spring), where poolfish must have been common, given the large number of fish collected in 1938 (Miller 1948). About 10 km northwest of Manse Ranch, there were two main springs on Pahrump Ranch. Poolfish collected from a marsh downstream of one of the springs would be named the Pahrump

Fig. 6.3 Illustration of the extinct Ash Meadows poolfish as it appeared in David Starr Jordan's *Fishes* (1907). The "enormously enlarged" (Jordan 1907) pharyngeal teeth of this species are illustrated at the upper right.

Ranch poolfish by Miller. Miller (1948) noted that poolfish were common in the downstream marsh. but absent in the other spring at Pahrump Ranch. The latter spring had been dredged in 1941, and in 1942, Miller found only a few nonnative common carp *Cyprinus carpio* in the degraded spring habitat. Approximately 0.8 km north of Pahrump Ranch was Raycraft Ranch, where the third subspecies of poolfish was located. The poolfish on Raycraft Ranch were collected from a large spring-fed pool. When Robert and Frances Miller surveyed the waters of Raycraft Ranch in 1942, poolfish were not common, but introduced common carp were present (Miller 1948).

Just to the west of Pahrump Valley near the California border lies a series of springs and wetlands known as Ash Meadows. From 1936 to 1942, Miller collected extensively throughout Ash Meadows, but found only 22 specimens of the Ash Meadows poolfish in five springs. Miller was a determined and thorough collector, and if poolfish had been more common, it is likely he would have collected more (he collected 3,861 pupfish from Ash Meadows during this same time). According to Miller (1948), only three poolfish were collected in a 1930 survey by Myers and Wales. As their name implies, poolfish prefer large spring-fed pools to shallow spring outflows, so their habitat was more restricted than that of the pupfish in Ash Meadows. Poolfish may have been rare because of competition from pupfish; poolfish were historically abundant in Pahrump Valley spring pools and marshes where no other fish were native. Surveys conducted in the 1930s preceded the spring alteration, water diversions, and exotic species introductions that occurred in later years.

The last known Ash Meadows poolfish was collected from Big Spring (Deep Spring of Miller 1948) on September 7, 1948, by Hildemann and Kopec (Miller et al. 1989). Efforts to collect poolfish in 1953 and subsequent years failed. Crayfish *Procambarus clarkii* and bullfrogs were widely introduced into Ash Meadows springs in the 1930s and probably preyed on the poolfish. Western mosquitofish, sailfin mollies *Poecilia latipinna*, and shortfin mollies *P. mexicana* were also introduced to and became common in these springs. Together, these nonnative species may have sealed the fate of poolfish in Ash Meadows.

To the east in Pahrump Valley, the combination of spring dredging, water diversions, and groundwater pumping, along with introductions of common carp, goldfish *Carassius auratus*, and bullfrogs, ultimately proved fatal for the Pahrump Ranch poolfish and Raycraft Ranch poolfish. La Rivers (1962) reported that poolfish were present at Pahrump Ranch during 1958, but that common carp and bullfrogs were abundant. Extinction of the Pahrump Ranch poolfish probably occurred shortly thereafter when one of the springs failed because of excessive groundwater pumping (Miller et al. 1989). A similar fate was reported for poolfish at Raycraft Ranch. Poolfish were last collected from Raycraft Ranch in 1953, when common carp and bullfrogs were present (Miller et al. 1989). By the 1960s, the spring pool at Manse Ranch still sustained poolfish, but this habitat was changing rapidly as groundwater was being removed by pumping. It should be noted that these springs were not small. Historical records indicate that the spring pool at Manse Ranch was quite large, with an outflow of about 10,220 l/

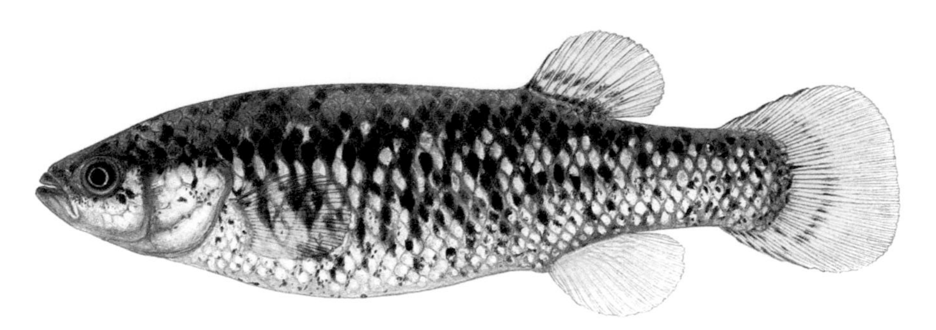

Fig. 6.4 Pahrump poolfish *Empetrichthys latos latos*, the only surviving member of the poolfish genus *Empetrichthys*, now known only in transplanted populations. (Artwork courtesy of Joe Tomelleri.)

min (Norment 2014). The springs probably harbored unique aquatic invertebrates that were also eliminated when the springs dried (Unmack and Minckley 2008).

Poolfish persisted in Manse Spring through the 1960s and early 1970s, and conservationists made considerable efforts to save this last remaining member of the genus *Empetrichthys* from extinction (fig. 6.4). Jim Deacon and Josh Williams (2010) described efforts conducted during the 1960s to remove invasive goldfish from Manse Spring, including the spectacular but ultimately unsuccessful use of dynamite. Facing the potential loss of the population due to the impact of goldfish and continued groundwater pumping, state and federal resource agency staff transplanted the endangered poolfish into other locations in Nevada. Two transplants failed, but poolfish persist at Corn Creek Springs on the Desert National Wildlife Refuge just north of Las Vegas, at Shoshone Ponds Natural Area in northern Nevada, and in a pond on Spring Mountain Ranch State Park. Any remaining poolfish were extirpated from Manse Spring when the spring failed in August 1975 because of excessive groundwater pumping (Deacon and Williams 2010).

Alvord Cutthroat Trout, *Oncorhynchus clarkii alvordensis*

In 1934, Carl Hubbs collected trout from two Alvord Basin streams: Virgin Creek, in northern Nevada, and nearby Trout Creek, across the border in Oregon. The specimens collected from Trout Creek appeared to be hybrids with introduced rainbow trout *Oncorhynchus mykiss*, but 83 trout collected on August 2, 1934, from Virgin Creek appeared to be pure cutthroat trout. These Virgin Creek fish would become the type specimens for the *alvordensis* subspecies, as noted then by Hubbs and published in Robert Behnke's (2002) classic *Trout and Salmon of North America*. Hubbs also collected 11 rainbow trout from Virgin Creek and noted that 6,000 rainbow trout fingerlings had been stocked there in 1933. Hybridization between the native cutthroat trout and introduced rainbow trout was evident in material collected from Virgin Creek during the 1950s and again in the 1970s, resulting in the belief that pure *alvordensis* were extinct (Williams and Bond 1983). Basibranchial teeth, a diagnostic feature of cutthroat trout, were absent in fish samples from the 1950s (Williams and Bond 1983). Genetic analyses of trout collected from Virgin Creek from 1984 to 1986 revealed large proportions of rainbow trout alleles (Tol and French 1988).

Introductions of rainbow trout have resulted in hybridization with various subspecies of cutthroat trout throughout the West. Genetic introgression, which is a common result of hybrid matings and backcrosses, results in a broad mixture of genetic material between the native and introduced trout and a morphological pattern in which individuals more closely resemble either one parent trout or the other.

The possibility of finding remaining genetically pure or nearly pure Alvord cutthroat trout has long intrigued the angling and conservation communities. In 1986, about 25 trout that phenotypically resembled the native Alvord strain were collected from Virgin Creek and transplanted into Jackson Creek in the Lahontan Basin in Nevada. Behnke (2002) reported that the Jackson Creek transplant failed, but rumors persist that some fish survived in a remote canyon area (J. Tomelleri, pers. comm.). In 2005, Behnke penned an article in *Trout* magazine entitled "Ivory-Billed Trout," where he claimed that early records indicated that trout were moved from the Alvord Basin into a nearby Oregon stream in 1928, prior to rainbow stocking, raising the possibility that some pure Alvord cutthroat trout might be present in the high desert of Oregon. Specimens conforming to Hubbs's original descriptions of Alvord cutthroat trout were subsequently found in Guano

Fig. 6.5 Drawing of a cutthroat trout taken from Guano Creek, Oregon, that is phenotypically similar to *O. clarkii alvordensis* as described by Carl Hubbs. (Illustration from photo taken by David Kortum. Artwork courtesy of Joe Tomelleri.)

Creek, raising the possibility that the native trout might have persisted in that stream (fig. 6.5). Unfortunately, that stream has also been stocked with rainbow trout, increasing the likelihood that these specimens were also of hybrid origin.

Despite the high likelihood that genetically pure Alvord cutthroat trout are extinct, the idea of preserving some remnants of the subspecies—some sort of phenotypic recovery—remains attractive to trout enthusiasts.

WHY EXTINCTION MATTERS

Scientists who track species endangerment and loss agree that we have entered a "sixth mass extinction" and that this one is caused by humans. By comparing the extinction rates of various vertebrate groups (mammals, birds, fishes, amphibians, and reptiles) from 1500 to the present with those from 1900 to the present, we find that the extinction rate during the past century has been 8 to 100 times greater than the rate from the earlier period (Ceballos et al. 2015). Ceballos and colleagues (2015) estimated that the current mass extinction episode is unprecedented in human history and unlike anything experienced on Earth for the past 65 million years.

Clearly, the trend toward increasing extinctions is unsustainable, and if such losses continue for future human generations, we will lose the benefits of a biologically rich environment and the many ecosystem services that diverse natural communities provide. There is a growing scientific consensus that more diverse communities tend to be more stable and resistant to disturbance than communities built around fewer species or populations (Tilman et al. 1996; Ptacnik et al. 2008). Biologically rich systems recover from disturbances that would destabilize ecosystems composed of few species. This ability of diverse natural systems to maintain their function and productivity in the face of rapid environmental change has been termed the "portfolio effect" (Figge 2004) and is the likely reason why certain commercial and recreational fisheries sustain themselves over time despite environmental change while others collapse (Carlson and Satterthwaite 2011).

In this assessment of extinct desert species, diversions of water supplies and introductions of nonnative species are the primary drivers of extinction. Narrow endemics, such as species that occur in a single spring, appear to be the most vulnerable to extinction. Excessive groundwater and surface-water withdrawals for municipal and agricultural uses have dried many desert spring systems. Loss of surface flows indicates that water is being withdrawn faster than it is replaced by precipitation and groundwater recharge, and that withdrawal at this rate, if continued, will eventually lead to aquifer failure and the inability of the region to support human life without importing water from distant sources. This unsustainable use of water resources and the resulting extinctions of species provide a clear call for immediate changes in management of water supplies if desert ecosystems and human communities are to be sustained.

Increases in the numbers of endangered and extinct species are a clear cautionary signal about the future of humans in arid lands. Where will the water come from to sustain the cities of Phoenix, Tucson, El Paso–Juarez, and Las Vegas when groundwater supplies have been depleted and the flows of the Colorado and Rio Grande diminish due to overuse and climate change? In their assessment of the potential impacts of the Southern Nevada Water Authority's plans to increase mining of groundwater aquifers to support growth in Las Vegas, Deacon and colleagues (2007) noted that "we must acknowledge limits to water availability as we strive to strike a balance between human water demand and the needs of natural systems and future generations." If limits to growth are not acknowledged, a future of ecological crisis and loss of opportunities for future generations can be assured.

Governments have long recognized the inherent value of natural places and endangered species by protecting areas as national parks and monuments and by passing legislation such as the US Endangered Species Act. Yet, as society grows increasingly focused on short-term gains at the expense of future quality of life, will our children and grandchildren grow up in a world rich in biological resources, or in a more depauperate one where people might one day lament what used to be and what might have been? Alfred Russel Wallace acknowledged the importance of current choices for future generations in the following passage from an 1863 report in the *Journal of the Royal Geographical Society* that perhaps is more relevant today than ever.

> [The naturalist] looks upon every species of animal and plant now living as the individual letters which go to make up one of the volumes of our Earth's history; and, as a few lost letters may make a sentence unintelligible, so the extinction of the numerous forms of life which the progress of cultivation invariably entails will necessarily render obscure this invaluable record of the past. It is, therefore, an important object [to preserve them].
>
> If this is not done, future ages will certainly look back upon us as a people so immersed in the pursuit of wealth as to be blind to higher considerations. They will charge us with having culpably allowed the destruction of some of those records of Creation which we had it in our power to preserve; and, while professing to regard every living thing as the direct handiwork and best evidence of a Creator, yet, with a strange inconsistency, seeing many of them perish irrecoverably from the face of the Earth, uncared for and unknown.

REFERENCES

Adams, K. D. 2007. Late Holocene sedimentary environments and lake-level fluctuations at Walker Lake, Nevada, USA. *Geological Society of America Bulletin* 119: 126–39.

Alvarez, F., M. López-Mejía, and C. Pedraza-Lara. 2010. *Cambarellus alvarezi*. *IUCN Red List of Threatened Species*. e.T153825A4550209. http://dx.doi.org/10.2305/IUCN.UK.2010-3.RLTS.T153825A4550209.en.

Arizona Game and Fish Department. 2017. *Native Fish Management and Conservation*. http://www.azgfd.gov/w_c/nongameandendangeredwildlifeprogram/nativefish.shtml.

Behnke, R. J. 2002. *Trout and Salmon of North America*. New York: Free Press.

Behnke, R. J. 2005. Ivory-billed trout. *Trout* 47: 56–58.

Beutel, M. W., A. J. Horne, J. C. Roth, and N. J. Barratt. 2013. Limnological effects of anthropogenic desiccation of a large, saline lake, Walker Lake, Nevada. *Hydrobiologia* 466: 91–105.

Bogan, A. E. 2000. *Physella microstriata*. *IUCN Red List of Threatened Species*. e.T17240A6884311. http://dx.doi.org/10.2305/IUCN.UK.2000.RLTS.T17240A6884311.en.

Brown, H. P. 1972. Synopsis of the genus *Heterelmis* Sharp in the United States, with description of a new species from Arizona (Coleoptera, Dropoidae, Elmidae). *Entomological News* 83: 229–38.

Burkhead, N. M. 2012. Extinction rates in North American freshwater fishes, 1900–2010. *BioScience* 62: 798–808.

Campbell, D., and C. Lydeard. 2012. The genera of *Pleurobemini* (Bivalvia: Unionidae: Ambleminae). *American Malacological Bulletin* 30: 19–38.

Carlson, S. M., and W. H. Satterthwaite. 2011. Weakened portfolio effect in a collapsed salmon population complex. *Canadian Journal of Fisheries and Aquatic Sciences* 68: 1579–89.

Carson, E. W., C. Pedraza-Lara, M. L. Lozano-Vilano, G. A. Rodríguez-Almaráz, I. Banda-Villanueva, L. A. Sepúlveda-Hernández, L. Vela-Valladares, A. Cantú-Garza, and M. de la Maza-Benignos. 2015. The rediscovery and precarious status of Chihuahuan dwarf crayfish, *Cambarellus chihuahuae*. *Occasional Papers of the Museum of Southwestern Biology* 12: 1–7.

Ceballos, G., P. R. Ehrlich, A. D. Barnosky, A. Garcia, R. M. Pringle, and T. M. Palmer. 2015. Accelerated modern human-induced species losses: Entering the sixth mass extinction. *Science Advances* 1: e1400253.

Coffin, P. D., and W. F. Cowan. 1995. *Lahontan Cutthroat Trout (*Oncorhynchus clarkii henshawi*) Recovery Plan*. Portland, OR: US Fish and Wildlife Service.

Contreras-Balderas, S., and M. L. Lozano-Vilano. 1996. Extinction of most Sandia and Potosí valleys (Nuevo León, México) endemic pupfishes, crayfishes and snails. *Ichthyological Exploration of Freshwaters* 7: 33–40.

Cordeiro, J., and K. Perez. 2012. *Stagnicola pilsbryi*. *IUCN Red List of Threatened Species*. www.iucnredlist.org/details/177724/0. Accessed June 13, 2017.

Cordeiro, J., and K. Perez. 2012. *Stagnicola utahensis*. *IUCN Red List of Threatened Species*. www.iucnredlist.org/details/20706/0. Accessed June 13, 2017.

Deacon, J. E., and J. E. Williams. 2010. Retrospective evaluation of the effects of human disturbance and goldfish introduction on endangered Pahrump poolfish. *Western North American Naturalist* 70: 425–36.

Deacon, J. E., A. E. Williams, C. Deacon Williams, and J. E. Williams. 2007. Fueling population growth in Las Vegas: How large-scale groundwater withdrawal could burn regional biodiversity. *BioScience* 57: 688–98.

Edwards, R. J., G. P. Garrett, and E. Marsh-Matthews. 2003. Fish assemblages of the Rio Conchos Basin, Mexico, with emphasis on their conservation and status. In G. P. Garrett and N. L. Allan, eds., *Aquatic Fauna of the Northern Chihuahuan Desert*, 75–89. Museum of Texas Tech University, Special Publication no. 46.

Evans, A. V., and J. N. Hogue. 2006. *Field Guide to the Beetles of California*. Berkeley: University of California Press.

Figge, F. 2004. Bio-Folio: Applying portfolio theory to biodiversity. *Biological Conservation* 13: 827–49.

Galat, D. L., E. L. Lider, S. Vigg, and S. R. Robertson. 1981. Limnology of a large, deep, North American terminal lake, Pyramid Lake, Nevada, U.S.A. *Hydrobiologia* 82: 281–317.

Gilbert, C. H. 1893. Report on the fishes of the Death Valley Expedition collected in Southern California and Nevada in 1891, with descriptions of new species. *North American Fauna* 7: 229–34.

Hershler, R. 1994. A review of the North American freshwater snail genus *Pyrgulopsis* (Hydrobiidae). *Smithsonian Contributions to Zoology* 554: 1–115.

Hershler, R. 1998. A systematic review of the hydrobiid snails (Gastropoda: Rissooidea) of the Great Basin, western United States. Part I. Genus *Pyrgulopsis*. *Veliger* 41: 1–132.

Hershler, R. 1999. A systematic review of the hydrobiid snails (Gastropoda: Rissooidea) of the Great Basin, western United States. Part II. Genera *Colligyrus*, *Eremopyrgus*, *Fluminicola*, *Pristinicola*, and *Tryonia*. *Veliger* 42: 306–77.

Hershler, R. 2001. Systematics of the North and Central American aquatic snail genus *Tryonia* (Rissooidea: Hydrobiidae). *Smithsonian Contributions to Zoology* no. 612.

Hershler, R., H.-P. Liu, and J. Howard. 2014a. Springsnails: A new conservation focus in western North America. *BioScience* 64: 693–700.

Hershler, R., J. J. Landye, H.-P. Liu, M. de la Maza-Benignos, P. Ornelas, and E. W. Carson. 2014b. New species and records of Chihuahuan desert springsnails, with a new combination for *Tryonia brunei*. *Western North American Naturalist* 74: 47–65.

Hershler, R., H.-P. Liu, and J. J. Landye. 2011. New species and records of springsnails (Caenogastropoda: Cochliopidae: *Tryonia*) from the Chihuahuan Desert (Mexico and United States), an imperiled biodiversity hotspot. *Zootaxa* 3001: 1–32.

Hubbs, C., and B. L. Jensen. 1984. Extinction of *Gambusia amistadensis*, an endangered fish. *Copeia* 1984: 529–30.

Hubbs, C. L., and R. R. Miller. 1948. The zoological evidence: Correlation between fish distribution and hydrographic history in the desert basins of western United States. In E. Blackwelder, *The Great Basin, with Emphasis on Glacial and Postglacial Times*, University of Utah Biological Series vol. 10, no. 7, 17–166. Salt Lake City: University of Utah.

Jelks, H. L., S. J. Walsh, N. M. Burkhead, S. Contreras-Balderas, E. Diaz-Pardo, D. A. Hendrickson, J. Lyons, N. E. Mandrake, F. McCormick, J. S. Nelson, S. P. Platania, B. A. Porter, C. B. Renaud, J. J. Schmitter-Soto, E. B. Taylor, and M. L. Warren, Jr. 2008. Conservation status of imperiled North American freshwater and diadromous fishes. *Fisheries* 33: 372–407.

Jordan, D. S. 1907. *Fishes*. New York: Henry Holt.

La Rivers, I. 1962. *Fishes and Fisheries of Nevada*. Carson City: Nevada State Fish and Game Commission.

McCabe, D. J. 1998. Biological communities in springbrooks. In L. Botosaneau, ed., *The Biology of Springs and Springbrooks*, 221–28. Studies in Crenobiology. Leiden, The Netherlands: Backhuys Publishers.

Miller, R. R. 1948. *The Cyprinodont Fishes of the Death Valley System of Eastern California and Southwestern Nevada*. Miscellaneous Publications of the Museum of Zoology, University of Michigan, 68. Ann Arbor: University of Michigan Press.

Miller, R. R., W. L. Minckley, and S. M. Norris. 2005. *Freshwater Fishes of México*. Chicago: University of Chicago Press.

Miller, R. R., J. D. Williams, and J. E. Williams. 1989. Extinctions of North American fishes during the past century. *Fisheries* 14: 22–38.

Minckley, W. L., G. K. Meffe, and D. L. Soltz. 1991. Conservation and management of short-lived fishes: The cyprinodontoids. In W. L. Minckley and J. E. Deacon, eds., *Battle against Extinction: Native Fish Management in the American West*, 247–82. Tucson: University of Arizona Press.

Minckley, W. L., and R. R. Miller. 2005. Extirpation, extinction, and conservation. In R. R. Miller, W. L. Minckley, and S. M. Norris, *Freshwater Fishes of México*, 60–62. Chicago: University of Chicago Press.

Moyle, P. B., R. M. Yoshiyama, J. E. Williams, and E. D. Wikramanayake. 1995. *Fish Species of Special Concern in California*. 2nd ed. Rancho Cordova, CA: California Department of Fish and Game.

Norment, C. 2014. *Relicts of a Beautiful Sea: Survival, Extinction, and Conservation in a Desert World*. Chapel Hill: University of North Carolina Press.

Peden, A. E. 1973. Virtual extinction of *Gambusia amistadensis* n. sp., a poeciliid fish from Texas. *Copeia* 1973: 210–21.

Polhemus, D. A., and J. T. Polhemus. 2002. Basin and ranges: The biogeography of aquatic true bugs (Insecta: Heteroptera) in the Great Basin. In R. Hershler, D. B. Madsen, and D. R. Currey, eds., *Great Basin Aquatic Systems History*, 235–54. Smithsonian Contributions to the Earth Sciences, 33.

Ptacnik, R., A. G. Solimini, T. Andersen, T. Tamminen, P. Brettum, L. Lepisto, E. Willen, and S. Rekolainen. 2008. Diversity predicts stability and resource use efficiency in natural phytoplankton communities. *Proceedings of the National Academy of Sciences* 105: 5134–38.

Robertson, S. R. 1978. The distribution and relative abundance of benthic macroinvertebrates in Pyramid Lake, Nevada. MA thesis, University of Nevada.

Scoppettone, G. G., and G. Vinyard. 1981. Life history and management of four endangered lacustrine suckers. In W. L. Minckley and J. E. Deacon, eds., *Battle against Extinction: Native Fish Management in the American West*, 359–78. Tucson: University of Arizona Press.

Stearns, R. E. C. 1883. Description of a new hydrobiinoid gastropod from mountain lakes of the Sierra Nevada, with remarks on allied species and the physiographical features of said region. *Proceedings of the Academy of Natural Sciences of Philadelphia* 35: 171–76.

Taylor, C. A., G. A. Schuster, J. E. Cooper, R. J. DiStefano, A. G. Eversole, P. Hamr, H. H. Hobbs III, H. W. Robison, C. E. Skelton, and R. F. Thoma. 2007. A reassessment of the conservation status of crayfishes of the United States and Canada after 10+ years of increased awareness. *Fisheries* 32: 372–89.

Tilman, D., D. Wedin, and J. Knops. 1996. Productivity and sustainability influenced by biodiversity in grassland ecosystems. *Nature* 379: 718–20.

Tol, D., and J. French. 1988. Status of a hybridized population of Alvord cutthroat trout from Virgin Creek, Nevada. In R. E. Gresswell, ed., *Status and Management of Interior Stocks of Cutthroat Trout*, 116–20. American Fisheries Society Symposium, no. 4.

Unmack, P. J., and W. L. Minckley. 2008. The demise of desert springs. In L. E. Stevens and V. J. Meretsky, eds., *Aridland Springs in North America: Ecology and Conservation*, 11–34. Tucson: University of Arizona Press.

USFWS (US Fish and Wildlife Service). 1970. Title 50—Wildlife and Fisheries. *Federal Register* 35: 16047–48.

USFWS (US Fish and Wildlife Service). 1982. Endangered and threatened wildlife and plants: Deregulation of the Tecopa pupfish. *Federal Register* 47: 2317–19.

Wallace, A. R. 1863. On the physical geography of the Malay Archipelago. *Journal of the Royal Geographical Society* 30: 172–77.

Williams, J. E., and C. E. Bond. 1983. Status and life history notes on the native fishes of the Alvord Basin, Oregon and Nevada. *Great Basin Naturalist* 43: 409–20.

Williams, J. E., C. A. Macdonald, C. Deacon Williams, H. Weeks, G. Lampman, and D. W. Sada. 2005. Prospects of recovering endemic fishes pursuant to the US Endangered Species Act. *Fisheries* 30: 24–29.

Wilson, E. O. 1992. *The Diversity of Life*. Cambridge, MA: Harvard University Press.

∴Monuments to the Dam Age

SECTION 2

Racing to Collapse

7 Running on Empty

Southwestern Water Supplies and Climate Change

Bradley H. Udall

Two iconic Southwestern rivers, the Colorado River and the Rio Grande, are notable for enormous anthropogenic changes that began in the late nineteenth century and continue to this day. These changes began with the construction of massive water infrastructure in the early twentieth century to support first agricultural and later, growing municipal uses. Elephant Butte Dam on the Rio Grande was finished by 1916, and Hoover Dam had closed on the Colorado River by 1935, each representing the first major water storage project on its river. As the twentieth century proceeded, high population growth in southwestern cities and new agricultural water demands from federal water projects drove increasing diversions and consumptive use, all facilitated by enormous new infrastructure projects. In both basins, annual consumptive use ultimately grew to equal annual inflow, with major impacts, especially to natural flow–dependent ecosystems. Since about 1990, the Colorado River has failed to reach the Sea of Cortez (Pitt 2001), and the Rio Grande has dried downstream of El Paso, Texas, and in other reaches as well, for decades (Scurlock 1998; Thomson 2011).

Human management of these rivers grew in complexity as our ability to move water via infrastructure increased. Water development was initially supported by state laws, interstate agreements, and federal laws and funding. Later, state and federal environmental laws and interstate lawsuits substantially modified these original arrangements. In addition to agricultural and municipal users, Native Americans began to obtain their treaty water rights around 1980, although many rights are still unresolved (Stern 2015). This convoluted mix of water users, water laws, and sophisticated infrastructure led to complicated operating rules and agreements, many of which now constrain possible solutions to water shortages and conservation actions.

Infrastructure played an especially large role in all of these changes, with dams and large delivery canals as key components. By allowing for significant water storage over years, dams provided new reliability for water users in a geographic area with uncertain precipitation, and they prevented harmful floods. However, they also changed within- and among-year flow patterns, facilitated large new consumptive uses, cooled water temperatures, removed sediment, and allowed river channelization to occur (Poff et al. 1997). These and other changes drove declines in native aquatic and riparian species in each basin that led to federal listing of several taxa as endangered. In the latter part of the twentieth century, the introduction of nonnative aquatic and riparian species added another stressor to these systems and complicated the recovery of endangered species. Finally, as the twenty-first century unfolds, climate change has begun to affect these rivers through long-term unprecedented drought, earlier spring runoff, and reductions in snowpack (MacDonald 2010; Seager et al. 2013; Dettinger et al. 2015; Udall and Overpeck 2017). For better or worse, all of these changes have fundamentally altered these two rivers, with no return to pre–European settlement conditions now possible.

THE COLORADO RIVER

The Colorado River flows for over 2,300 kilometers as it drains parts of seven states and two nations in the American Southwest with a total area of 637,000 km^2 (Kammerer 1990) (fig. 7.1). Its waters irrigate over 2 million hectares (5 million acres) of land inside and outside the basin, and serve 40 million Americans in every major southwestern city, both inside and outside the basin proper (USBR 2012). The river's waters were originally allocated under the 1922 Colorado River Compact, which split the river into a Lower Basin (California, Nevada, and Arizona) and an Upper Basin (Colorado, Utah, New Mexico, Wyoming) (Hundley 2009). A later 1944 international treaty set aside 1.85 billion cubic meters (B m^3) per year (1.5 million acre-feet [maf] per year) for México, considered to be the most senior right on the river (Hundley 1966). Agriculture consumes over 80% of the total water use in the basin (Cohen et al. 2013; Colorado River Basin Stakeholders 2015), a figure consistent with water use in arid-land rivers around the globe. Additional laws, agreements under the National Environmental Policy Act (NEPA), international treaties, and Supreme Court decrees have added to the original agreements (Meyers 1966) and constitute what is called "The Law of the River."

Upper Basin agriculture is mainly alfalfa and pasture for cattle. The Upper Basin climate, with a few exceptions near Grand Junction and Montrose, Colorado, does not support other crops due to the relatively short growing season and cool temperatures. Notably, alfalfa is highly consumptive of water (Glennon and Culp 2012; Robbins 2014) in both the Upper and Lower basins, consuming almost 1 B m^3 (800,000 acre-feet [af]) and 2.7 B m^3 (2.2 maf) annually, respectively (Cohen et al. 2013). In the Upper Basin, this high consumptive use is due mainly to the vast acreage of the crop, and in the Lower Basin it is due to high temperatures, a year-round growing season, and significant acreage. Lower Basin agriculture is extremely varied and includes many winter vegetables, wheat, and cotton, in addition to alfalfa (Cohen et al. 2013). The Imperial Valley in California and the Yuma area in Arizona, both near the border with México, have year-round growing seasons and provide much of the nation's high-value winter produce. The Imperial Irrigation District serves almost 202,500 ha (500,000 acres) of farmland with annual diversions in excess of 3.0 B m^3 (2.5 maf), making it the nation's largest agricultural diverter (Imperial Irrigation District 2004). Yuma supports approximately 71,000 ha (175,000 acres) of farmland with over 1.1 million cubic meters (M m^3) (875,000 acre-feet [af]) of annual diversions (USBR 1981; Sauder 2009; Noble 2015). Both areas have senior water rights arising from farming that began around 1900.

Reflecting the aridity of the southwestern region, most of the cities served by Colorado River water are either beyond the basin's boundaries (Los Angeles, San Diego, Denver, Salt Lake City, Albuquerque) or significantly uphill from their supply points (Las Vegas, Phoenix, Tucson). The Colorado River provides roughly 25% of the greater Los Angeles supply, 50% of Colorado Front Range and Albuquerque supplies, and about 123 M m^3 (100,000 af) for the Salt Lake City area. Las Vegas is 90% dependent on Colorado River water and recently spent $1.5 billion building a new pumping plant and intake into the very bottom of Lake Mead to ensure deliveries no matter how low the reservoir. During the recent 2011–2017 California drought,

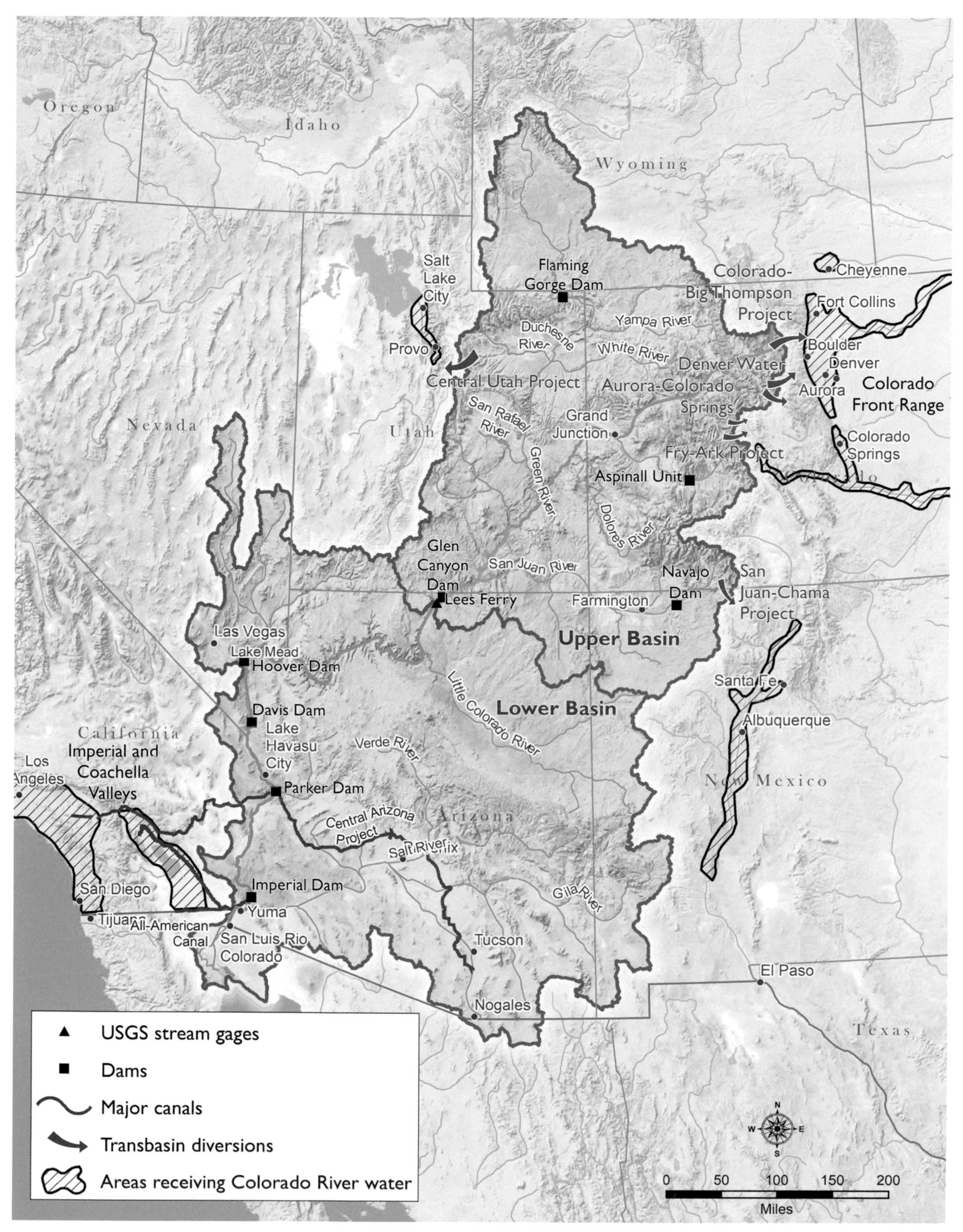

Fig. 7.1 The Colorado River basin and surrounding areas. Major tributaries, delivery canals, dams, trans-basin diversions, and areas of use out of the basin are noted. The Lees Ferry USGS gage just downstream of Glen Canyon Dam, which divides the river into the Upper Basin and Lower Basin, is indicated.

Colorado River supplies were crucial for Los Angeles because the California State Water Project, its third source after local supplies, had no or very low water allocations in many of those years. Without Colorado River water, many southwestern cities would face substantial shortages.

The river is fed primarily by winter snowpack from the Rocky Mountains, with roughly 15% of the total basin area (mostly in the Upper Basin) generating 85% of the flow. The twentieth-century mean annual flow at Lees Ferry, the dividing line between Upper and Lower Basins, was approximately 18.5 B m^3 (15 maf), with half of that amount allocated to the Lower Basin and half to the Upper Basin under the Colorado River Compact. A small but important part of the basin's runoff (~9.25 M m^3, 750,000 af) enters the river below Lees Ferry in Grand Canyon (Hely 1969). Unfortunately, the twentieth century is now known to have been anomalously wet, especially at the time of compact negotiations, and we also now know that multi-decadal megadroughts substantially more severe than those in the twentieth century have occurred several times during the past 2,000 years (Woodhouse et al. 2006; Meko et al. 2007). Climate change is now expected to exacerbate droughts, increase the risk of multi-decadal drought, and lower the mean flow (Ault et al. 2014, 2016; Cook et al. 2015; Udall and Overpeck 2017).

The river system contains over 200 storage facilities, with most storage in 15 US Bureau of Reclamation (USBR) reservoirs. The basin's significant vertical drop, combined with mountains and canyons, provided numerous high-quality dam sites. Hoover Dam, completed in 1935, and Glen Canyon Dam, completed in 1963, together provide over 61.7 B m^3 (50 maf) of storage in the nation's two largest reservoirs, Lakes Mead and Powell. An additional 12.3 B m^3 (10 maf) is stored in Upper Basin reservoirs, mostly in Colorado and Wyoming (USBR 1981). Total basin reservoir storage is four times the annual flow, which is a very large amount relative to most other major rivers in the world. This storage is useful to buffer supplies during droughts, but it is also difficult to refill when demands are roughly equivalent to supplies (USBR 2012), a state of affairs that began in approximately 2003. From 2000 to 2004, the volume of Lakes Mead and Powell declined by 50%, and since that time they have not refilled.

A number of environmental concerns challenge basin natural resources (Pitt et al. 2000; Pitt 2001; Adler 2007; Cohen 2014). Dams and other infrastructure on the river have blocked fish movements, restricted sediment transport, and changed the timing and temperature of flows. Diversions take the entire flow of the river, such that since about 1990, the river no longer reaches the Sea of Cortez. In addition, introduced nonnative fishes, whose presence is facilitated by reservoirs, threaten endemic fishes. In the Upper Basin, there are federally listed endangered fish in multiple tributaries, including the San Juan, the Gunnison, the Colorado main stem, and the Green River and its larger tributaries. There are recovery programs under the Endangered Species Act in place (Adler 2007), largely funded from hydropower sales, which have been reduced during the drought. In the Lower Basin, the Lower Colorado River Multi-species Conservation Program began in 2005 (US Department of the Interior 2005). It will provide about $650 million over 50 years to conserve 26 species, 6 threatened and/or endangered and 20 that are not federally listed (Adler 2007).

The Salton Sea is California's largest lake by surface area. It was created by accident in 1905–1907 when the Colorado River broke out of its banks and flowed entirely into the Salton Sink (Hundley 2009). Since that time, over 12.3 B m^3/yr (1 maf/yr) in agricultural return flows from the Imperial Irrigation District have kept it alive, although it has shrunk in size from about 130,000 ha (500 sq mi) in 1907 to about 90,650 ha (350 sq mi) now. It is a very shallow lake, with large portions just a meter or so deep. The Salton Sea is a crucial stop along the Pacific flyway, with over 400 species of birds using the lake for sustenance (Cohen 2014). Since about 2000, efficiency improvements in the Imperial Irrigation District and water transfers to Southern California have reduced the lake's inflows, and in the next few years, it will drop precipitously due to transfers and additional efficiency improvements in the district. This will make the lake hypersaline, killing the fish and further shrinking the surface area. Exposure of more lake-bed soils will create substantial air quality problems from windblown dust (Cohen 2014). The region already has some of worst air quality in the nation; Imperial County, for example, has the highest rates of asthma in California. In 2003, as part of a water transfer agreement with San Diego, the state pledged funding to mitigate these problems, and in 2007, the state released a $9 billion mitigation plan. The Pacific Institute, a water-focused nongovernmental organization, put the costs of inaction at $30 billion to $70 billion (Cohen 2014). Only in 2017, under pressure from a number of groups, did the state commit to spending $380 million to begin to address some of the health and environmental issues.

In addition to these large-scale built-in or developing challenges, significant Native American tribal water rights still need to be determined for multiple federally created reservations (MacDonnell 2016). Under the Winters Doctrine, these tribes are entitled to reserved water rights (Royster 2006; Shurts 2000). Only some of these tribal rights have been fully quantified, with most tribes still seeking to finalize their rights residing in Arizona. Supplies for tribal rights come from the states in which the tribes live, despite the federal nature of the obligation (Royster 1994). Arizona has by now allocated all of its most senior Central Arizona Project (CAP) water right for tribes, and remaining settlements will have to come from its lower-priority CAP supplies that are most likely to be curtailed under the terms of the 1968 agreement as basin supplies contract (Bark 2006; Weldon and McKnight 2007).

As large as the environmental and tribal issues are, they are overshadowed by the intersection of three other problems: systematic overuse, the inflexible legal rules by which the river is managed, and climate change, manifested most visibly as unprecedented long-term drought, all of which threaten the reliability of existing water uses. These three enormous, interrelated, and complicated problems provide the material for the rest of this section.

Overuse has two manifestations, one in the Upper Basin and one in the Lower Basin. In the Lower Basin, the passage of the 1968 Colorado River Basin Project Act created a long-term built-in water deficit, known as the structural deficit. This 14.8 B m^3/yr (1.2 maf/yr) imbalance is the difference between the water legally available to the Lower Basin states and México (11.1 B m^3, 9.0 maf) and the amount actually used by the three states and México (12.6 B m^3/yr, 10.2 maf/yr) (Collum and McCann 2014). The 1968 act authorized the construction of the Central Arizona Project, a 2 B m^3/yr (1.6 maf/yr), 238 km (336-mile) uphill diversion from the Colorado River near Lake Havasu City to Phoenix, Tucson, and agricultural areas in central Arizona (Johnson 1977). Of this amount, approximately 1.5 B m^3/yr (1.2 maf/yr) is dependent on the Upper Basin states not using their full allocation. This use is allowed by the Colorado River Compact's Article III, Section (e), but would have to cease if the Upper Basin were to use more water or if climate change or long-term drought were to reduce flows.

At the time the 1968 act was signed, it was known that over time this "extra" water would decline due to increasing use in the Upper Basin (Tipton and Kalmbach, Inc. 1965; Johnson 1977). The only unknown was when the Upper Basin would use its full allocation, thus depriving the Lower Basin of the water. Importantly, however, the 1968 act did not envision 2 million residents in Las Vegas being 90% reliant on Lake Mead for their water. In addition, loss of flows due to climate change was not a consideration, and such reductions in flow will only speed the day of reckoning in the Lower Basin. Indeed, climate change reductions could make the built-in Lower Basin deficit worse if declines were to occur in the limited but critical 9.25 M m^3/yr (750,000 af/yr) inflows into Grand Canyon below Lees Ferry or the additional 9.25 M m^3/yr (750,000 af/yr) that enters below Hoover Dam. Flow reductions in this reach would have to be met by reductions in Lower Basin demand, unlike reductions in flows upstream of Lees Ferry, which would be met by reductions in Upper Basin use, at least according to one common interpretation of the compact and the delivery obligation of the Upper Basin (see below).

The unprecedented drought that began in 2000 in the Colorado River basin, resulting in nearly 20% annual reductions in flows and a 50% loss of Lake Mead and Lake Powell reservoir storage, has made the built-in deficit obvious. Facing sharp storage declines at the turn of the twenty-first century, the states acted quickly, and by 2007, they had agreed to new operational rules for the reservoirs and had defined shortage amounts based on Lake Mead elevations (US Department of the Interior 2007). However, only 740,000 M m^3 (600,000 af) of the 1.48 B m^3 (1.2 maf) built-in deficit was covered by the agreement. With continuing drought conditions, in 2010 and again from 2014 to 2017, Lake Mead dropped to within just a few feet of the first shortage trigger point established by the 2007 rules. After years of negotiations (Wines 2014a, b; Davis 2016; James 2016), the Lower Basin states in 2019 agreed to a temporary solution through 2026 to solve this difficult problem (Jacobs 2019). The agreement is currently awaiting congressional authorization (Jaspers 2019).

Recent studies have explored the chances that Lakes Mead and Powell will empty with the structural deficit in place combined with reductions in inflows under climate change; all reach the conclusion that this is a very likely outcome under current management practices sometime before 2050 and perhaps as soon as 2020 (Barnett and Pierce 2008; Barsugli et al. 2009). This is not an especially surprising finding, as whenever demands outpace supplies, reservoirs, no matter how big, will empty at some point.

Although the structural deficit has not yet been fully resolved, positive steps have been taken on other overuse factors. The 2007 agreement allowed, for the first time, carryover storage in Lake Mead, which encourages conservation. A number of conservation efforts have also propped up the reservoir in recent years, keeping it from hitting the shortage trigger point. Minute 319, agreed to in 2012, allowed México to store water in Lake Mead and also provided for an environmental "pulse flow," the first of its kind, in the dewatered reach in México from Morales Dam to the ocean. Minute 323, signed in late 2017, replaces the now expired Minute 319. It allows for the continued storage of water in Lake Mead by México, provides for shared cutbacks in time of drought, and provides funding and water for habitat restoration in México over the next decade.

One proposal promoted by an environmental nongovernmental organization, known as "Fill Mead First," argues that combining water from Lakes Mead and Powell into just one reservoir, Mead, would save water by reducing evaporative surface area while simultaneously opening up portions of flooded Glen Canyon for recreation. A recent analysis, however, suggests that the water savings would be illusory, and that the new, smaller Lake Powell would be difficult to manage due to infrastructure water release constraints at Glen Canyon Dam when Lake Powell's elevation is below its penstock intakes (Schmidt et al. 2016).

The Upper Basin faces a different, but related, overuse problem. Under Article III, Section (d) of the 1922 compact, the Upper Basin states agreed not to reduce the flows at Lees Ferry below 92.5 B m^3 (75 maf) in any given 10-year period (Colorado River Compact 1922). This agreement was reached as an imperfect way to limit Upper Basin consumptive use and avoid hoarding (Hundley 2009). When the compact was negotiated, it was never anticipated that this limit would be reached because the supply was thought to exceed the amount allocated (Hundley 2009). The flows, however, have proved to be less than originally thought, thus providing the Upper Basin with an uncertain, continually varying amount of water. This situation makes planning for future Upper Basin development difficult, and it also creates the possibility of a "compact call," the forced curtailment of Upper Basin use to meet the requirements of Article III, Section (d).

The Upper Basin's 10-year running delivery obligation would probably become untenable in a basin with significantly reduced supply and with a vast disparity in water use between the Lower (about 12.4–13.6 B m^3/yr, 10–11 maf/yr) and Upper (5.6 B m^3/yr, 4.5 maf/yr) Basins, an outcome not anticipated by the compact, which allocated equal shares to the two basins. The Upper Basin delivery obligation was never intended to be a risk allocation plan in a basin with uncertain supplies (Hundley 2009). A compact call would precipitate very difficult intra- and interstate discussions, with equity, economics, and the environment all key issues. Only one state, Colorado, has seriously studied the issue (MWH Americas 2012, 2014).

Climate change is the final element of the three major problems in the basin. In general, multi-model median climate change projections show decreasing water availability in the Colorado River basin as it warms, although the complete model range is large and runs from gains to losses. These projections have consistently indicated that southern parts of the basin are likely to face precipitation declines and enhanced droughts, in contrast to the northernmost parts of the basin, which may experience smaller precipitation declines or even increases (Cayan et al. 2013; Melillo et al. 2014; Polade et al. 2014; Ayers et al. 2016). Temperatures are, of course, projected to warm throughout the basin, and evaporative (and transpiration) demands are thus generally expected to increase.

In recent years, a number of papers have tied increased temperatures to declines in runoff efficiency (Woodhouse et al. 2016; Lehner et al. 2017; Udall and Overpeck 2017) and to future large temperature-induced flow declines of 20% by midcentury and 35% by the century's end. While additional precipitation may partially counteract these losses, offsetting the entire temperature-induced loss in 2050 would require as much precipitation as occurred in the wettest decade of the twentieth century, and by 2100 would require almost twice that amount. Scientists have taken to calling twenty-first century droughts "hot droughts" because the loss of flows is due, at least in part, to higher temperatures, rather than just reductions in precipitation, as was the case in the twentieth century.

Furthermore, dust blowing in from the deserts south and west of the Rocky Mountains darkens the snowpack, which thus absorbs more solar energy. This dust deposition may advance runoff timing by three weeks and decrease runoff quantity by 5% (Painter et al. 2010). Much of this blowing dust results from land use disturbances such as construction, grazing, and off-road vehicles in the region. Dust has been tied to aridity, which in turn is directly tied to climate change. Under a future heavy dust scenario, a minor additional 1% loss of flow and a significant additional three-week advance in runoff timing are projected (Deems et al. 2013).

The current 20-year drought, which has a clear temperature-induced flow reduction component, has shown no signs of ending. With additional warming certain and no long-term trend showing hope for increased precipitation, the temperature-induced flow decline seems likely to continue. The basin states and México have made progress in addressing some of the most pressing issues of the drought, including overuse and the structural deficit. However, if the drought continues to reduce flows, additional difficult and painful steps will be needed to bring the basin into balance.

THE RIO GRANDE

The Rio Grande runs 3,060 kilometers from its US headwaters in Colorado and New Mexico to the Gulf of Mexico, with the final 1,900 km from El Paso, Texas, forming the international border with México (Hill 1974). The river's headwaters in Colorado's eastern San Juan Mountains drain a small area compared with the Colorado River (fig. 7.2), and consequently those headwaters produce a comparatively small amount of flow, about 1.2 B m^3/yr (1 maf/yr), from snowpack. Snowpack and subsequent flows are also highly variable from year to year (Gutzler 2011). Near these headwaters, the Rio Chama, the largest US tributary to the Rio Grande, contributes another approximately 493 M m^3/yr (400,000 af/yr) from a drainage that also includes Colorado's southern San Juan Mountains (Thomson 2011). In some years, a pronounced summer monsoon provides significant precipitation, but the resulting flows are irregular, sediment laden, unregulated, and generally do not increase reservoir storage. The flows in the upper Rio Grande are thus only about 10% of the total flow in the Colorado River.

This discussion focuses on the river upstream of Fort Quitman, Texas, just south of El Paso and Ciudad Juarez, México. The river is often dry at Fort Quitman due to upstream water extractions and regains flow only at its confluence with México's Río Conchos, which usually provides all of the water for the lower river (Paddock 2001; Schmandt 2002). The Rio Grande upstream from Fort Quitman, sometimes known as the upper Rio Grande, comprises three distinct agricultural segments: Colorado's San Luis Valley, managed by Colorado's state engineer; the middle Rio Grande, in New Mexico, from near Cochiti Reservoir to Elephant Butte Reservoir, managed by the Middle Rio Grande Conservancy District; and the Paso del Norte region, from Elephant Butte Reservoir to Fort Quitman, also mostly in New Mexico but extending into Texas, and managed by the Elephant Butte Irrigation District and El Paso County Water Improvement District #1. The US Bureau of Reclamation and the US Army Corps of Engineers also play important water management roles throughout the Upper Rio Grande.

The river is governed by a 1906 international treaty, a three-state Rio Grande Compact signed in 1939, and the 1944 international treaty that also covers the Colorado River. The compact was designed to protect senior agricultural water rights both in Colorado and near El Paso. Under the compact, the upper two sections have annual requirements for water delivery to river sections downstream that vary according to input flows at index gages, with increasing percentages of the input flows required at output gages as the input flows increase. The Rio Grande Compact has been the source of much interstate litigation among the three signatory states (Paddock 2001). Recent unresolved litigation involves the consequences of groundwater pumping on a large scale near the New Mexico–Texas border in the Mesilla Valley, something not envisioned by the compact. The 1906 treaty requires a relatively small 74 M m^3 (60,000 af) delivery to México near El Paso, while the 1944 treaty focuses on México's delivery requirement for the Río Conchos.

The Rio Grande is unique in the West in that it has four forms of water rights, two of which are unique to New Mexico. Along with state water rights acquired under prior appropriation and federal tribal rights under

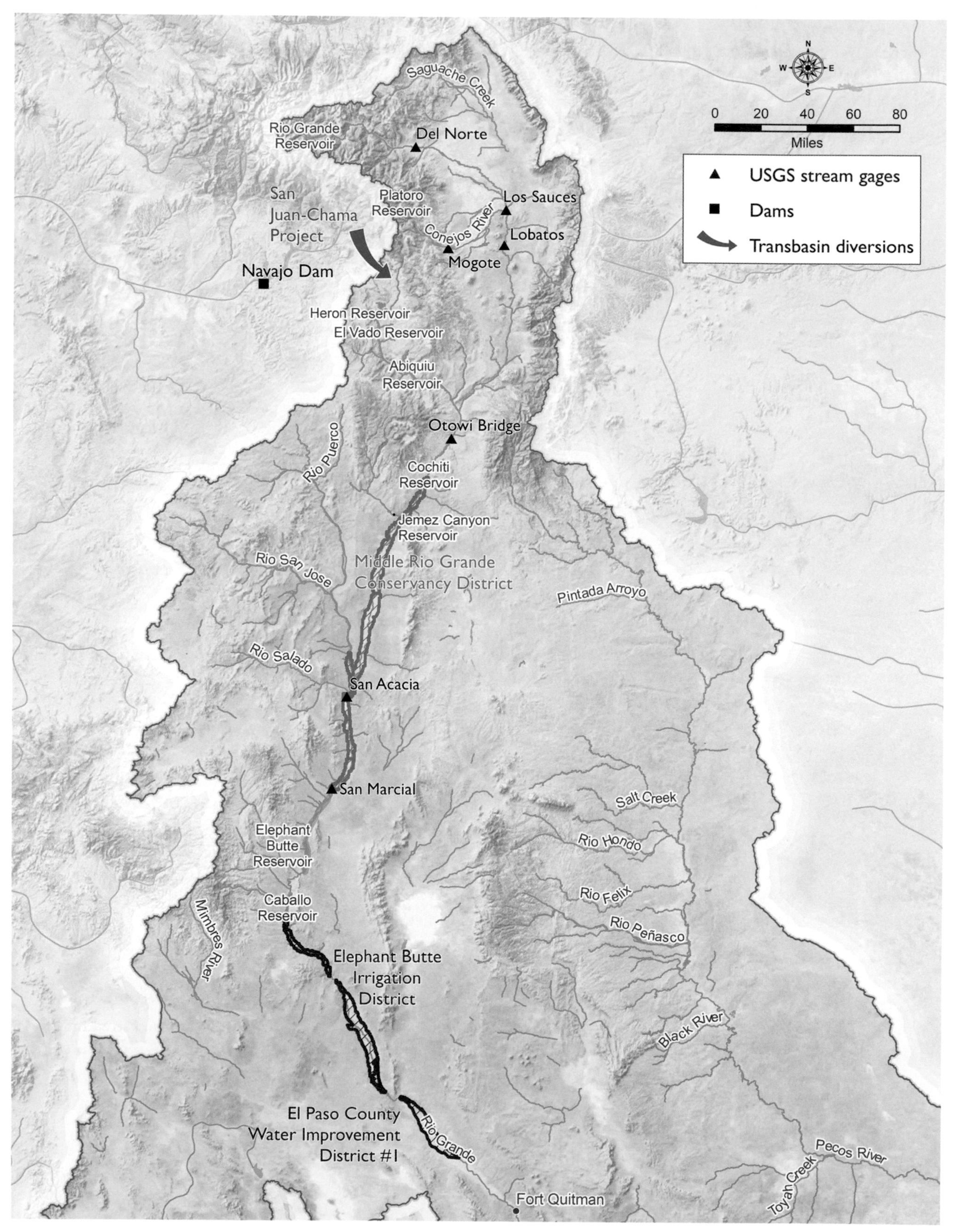

Fig. 7.2 The Rio Grande basin and surrounding areas. Major tributaries, dams, and the San Juan–Chama diversion from the Colorado River are noted. Key agricultural areas are hatched, and important USGS gages for Rio Grande Compact administration are indicated.

the Winters Doctrine, there are also Spanish acequia rights and Pueblo rights. Acequias are communal agricultural ditches with shared water rights (Crawford 1988; Rivera 2011). Pueblo rights pre-date the existence of the United States and even México. They were acknowledged by the Spanish prior to the creation of México and are now considered to date from "time immemorial," unlike federal tribal reserved rights, which carry the date of the establishment of the reservation.

Three large cities—Albuquerque, population 555,000; El Paso, population 672,000; and Ciudad Juarez, population 1.3 million—rely on the Rio Grande for large fractions of their water supplies, although they also rely on heavy groundwater pumping. In the past there were significant aquifer overdrafts, but in recent years these have stabilized. However, as with other regions around the West, agriculture is the dominant water use. The US Bureau of Reclamation operates four federal projects in the basin: the Rio Grande Project, the Middle Rio Grande Project, the Closed Basin Project, and the San Juan–Chama Project, all upstream of Texas.

The Rio Grande Project was approved in 1905, and by 1916, the largest reservoir on the river, Elephant Butte, was providing 2.47 B m^3 (2 maf) of storage to service project lands (Littlefield 2008) in the Paso del Norte region and beyond. Water from this reservoir is delivered by Elephant Butte Irrigation District and El Paso County Water Improvement District #1 to farmers in New Mexico and Texas. The Rio Grande Project services 72,900 ha (180,000 acres) of US land and another 10,125 ha (25,000 acres) in México (USBR 1981). The major crops irrigated are cotton, alfalfa, pecans, vegetables, and grain. In 1938, a smaller 419 M m^3 (340,000 af) reservoir, Caballo, was constructed downstream of Elephant Butte to provide flood control storage lost at Elephant Butte due to siltation and to allow for wintertime power production at Elephant Butte without loss of water.

The US Bureau of Reclamation's Middle Rio Grande Project, approved in the 1950s, involved the rehabilitation of an existing regional irrigation system, the Middle Rio Grande Conservancy District. The USBR channelized the Rio Grande in this river section, creating numerous environmental problems. Approximately 40,500 ha (100,000 acres) are irrigated by the project. Alfalfa, barley, wheat, oats, corn, fruits, and vegetables are the principal crops grown (USBR 1981). A major reservoir on the Rio Chama, El Vado, with about 247 M m^3 (200,000 af) of storage, was built in the 1930s to service these lands (Mann 2007).

Still farther upstream, in Colorado's San Luis Valley, agriculture and irrigation developed in the mid- to late 1800s, prior to the 1902 federal Reclamation Act (Littlefield 2008). The Closed Basin Project pumps approximately 18 M m^3/yr (15,000 af/yr) of groundwater into the Rio Grande from what was once thought to be a closed basin nearby. It was completed in the 1970s to provide some of Colorado's Rio Grande Compact obligation, thus allowing additional use of Rio Grande water that would otherwise be needed for downstream compact compliance. The San Luis Valley is home to the oldest water rights in Colorado and some of the oldest in the basin, dating to the 1850s. Management in the San Luis Valley is by the Colorado State Engineer and the local Division Engineer (Vandiver 2011).

An important trans-basin project in New Mexico connecting the Colorado River to the Rio Grande, the San Juan–Chama Project, was completed in 1978 (ABCWUA 2016; USBR 1981; Whipple 2007; Flanigan and Haas 2008). This project diverts approximately 123 m m^3/yr (100,000 af/yr) from the San Juan River through a lengthy tunnel into the upper reaches of the Rio Chama, where the waters are captured by a large 493 M m^3 (400,000 af) reservoir, Heron, on what was once a small tributary. Rio Chama water is contracted for by the city of Albuquerque (59.4 M m^3/yr, 48,200 af/yr), the Middle Rio Grande Conservancy District (25.8 M m^3/yr, 20,900 af/yr), the Jicarilla Apache Tribe (8.0 M m^3/yr, 6,500 af/yr), and the city of Santa Fe (9.9 M m^3/yr, 5,600 af/yr), and about 6.2 M m^3 (5,000 af) is used by a number of small towns and an irrigation district.

In the early twenty-first-century drought, San Juan–Chama water was used by the Bureau of Reclamation as an emergency supply to prevent downstream Rio Grande drying so as to supply habitat for the Rio Grande silvery minnow *Hybognathus amarus* (Katz 2007). This use was highly controversial and was later litigated. Federal law now specifically disallows such use. In 2008, the City of Albuquerque completed its $400 million diversion works, which allowed it to use its San Juan–Chama supplies, thereby reducing its reliance on local groundwater (Stomp 2014). San Juan–Chama water now comprises about half of its annual use, with the remainder mostly coming from groundwater pumping. Aquifers in the Albuquerque area have rebounded significantly since use of the San Juan–Chama supplies began (ABCWUA 2016).

The US Army Corp of Engineers also operates several flood control reservoirs, including Abiquiu on the Rio Chama just upstream of its confluence with the Rio Grande, Cochiti on the mainstem, and smaller, mostly dry sediment and flood control reservoirs on tributaries at Galisteo and Jemez Canyon. In recent years, Abiquiu Reservoir has played an important role in regulating flow and also has an important conservation pool (~246.7 M m^3, 200,000 af) (Kelly et al. 2007). Since its completion in 1973, Cochiti Reservoir has prevented large floods from harming Albuquerque and other downstream areas, but this protection has come at great environmental and social cost. The displacement of the Cochiti Pueblo from its historical lands to build the reservoir raises substantial environmental justice issues (Pecos 2007).

Like nearly all western rivers, the Rio Grande has been heavily modified by the activities of humans. Channelization and on-stream reservoir construction have led to changes in sediment loading and capture, changes in the seasonal and annual hydrograph, and increases in salinity (Phillips et al. 2011). By 1962, over 115,000 inexpensive 4.9 m high "jetty jacks" had been placed along the river for over 160 km, channelizing it (Grassel 2002). Most of them remain in place, although they are obscured by vegetation and sediment. The reach below Cochiti Dam, one of the 20 largest earthfill dams in the world, has been extensively studied because of the dam's profound impacts on the river (Richard 2001; Richard and Julien 2003; Richard et al. 2005). In this reach, sediment storage in Cochiti Reservoir has resulted in a sediment-starved river, and therefore in a downcut and straightened river channel. Conversely, in reaches below Albuquerque after the Middle Rio Grande diversions occur, the river carries too little flow to transport the available sediment, creating other problems including the need for emergency dredging to prevent out-of-channel flows (USBR 2013a). In the 1950s, USBR built an approximately 128 km "low flow" conveyance canal to move water to Elephant Butte Reservoir from near San Acacia to facilitate the required Rio Grande Compact deliveries to the reservoir (Phillips et al. 2011). Prior to construction of the conveyance canal, some of the deliveries were not making it to Elephant Butte. After the canal was built, the main channel was often deprived of water because the canal was below the level of the river.

Because of limited water supplies and extensive habitat alteration, the Rio Grande has been the scene of one of the iconic Endangered Species Act battles in the West. In 1994, the Rio Grande silvery minnow was federally listed as endangered. At the time, the fish was found only in the reach between Cochiti Dam and Elephant Butte Reservoir, a mere 5% of its historical range from northern New Mexico to the Gulf of Mexico (Bestgen and Platania 1991). Severe habitat loss, channelization, blockage of fish movement, too much and too little sediment, and increased salinity contributed to its decline (Cowley 2006). A severe drought began in 1996, and that same year the diversion of the entire river brought about a large minnow kill. The USBR began using stored water from its San Juan–Chama Project near the headwaters of the river to benefit the minnow against the protests of the project's contracted users (Katz 2007).

Beginning in 1999, complicated legal disputes ensued over a proposed recovery plan, critical habitat designation, biological opinions, and the legality of using San Juan–Chama water for the minnow (Katz 2007; Kelly and McKean 2011; DuMars 2012). Proposed solutions for Rio Grande silvery minnow recovery included removal of Cochiti Dam, thus re-enabling minnow access to a river segment that never historically dried, providing more water for the river, controlling sedimentation in river uplands, "naturalizing" irrigation drains to mimic habitat, moving levees farther back to allow ecosystem services to occur in the natural floodplain, and enacting strict water conservation measures (Cowley 2006).

In late 2016, the US Fish and Wildlife Service issued a long-awaited biological opinion (BO) affecting the river (USFWS 2016). The 15-year opinion, which replaced the expired but extended 2003 opinion, was a non-jeopardy decision on federal and state water activities as they impact the Rio Grande silvery minnow, the southwestern willow flycatcher *Empidonax traillii extimus*, and the yellow-billed cuckoo *Coccyzus americanus* from the Colorado–New Mexico state line to Elephant Butte Reservoir. It requires 86 conservation measures. The BO focuses on improving fish densities rather than mandating seasonal flow targets as the previous opinion did. Without water, however, the fish obviously cannot survive. One key requirement is that fish passageways around existing diversion dams be constructed to allow better fish movement (Bestgen et al. 2010).

At least three studies have investigated future flows of the Rio Grande under the influence of climate change (Hurd and Coonrod 2012; USBR 2013b; Elias et al. 2015). The 2013 USBR study indicated that by 2100, flows available for irrigation use in the uppermost reach in Colorado's San Luis Valley could decline by 25%. Divertible flows in the Middle Rio Grande were projected to decline by 35%, in large part because the compact allows Colorado to use more flow at lower flow levels, requiring less delivery to New Mexico. Downstream of Elephant Butte Reservoir, flows could decline by 50%. These forecasted declines are USBR's worst-case climate change modeled flow outcomes scenario in the United States and reflect the small size of the basin, the small size of its primary runoff-generating snow-covered areas, and its southerly position, which places it within the zone where climate change is most likely to entail significant precipitation declines and substantial increases in temperatures.

Hurd and Coonrod (2012), using three older-generation climate models from the set used by USBR, evaluated the economic impacts of climate change. All of the models showed declines in flow, the worst verging on half of the current flow. Agriculture would bear 80% or more of the economic losses. Elias et al. (2015) used models from a subset of older and newer climate models, pre-selecting four models simulating hot/dry, hot/wet, warm/wet, and warm/dry climates. As such, the study found a variety of outcomes, from less to more flow. It should be noted that even the slightly wetter, newer climate models are in general agreement that drying is the most likely outcome for the basin, even if a few models show increased wetness. The Elias et al. (2015) study was not a multi-model ensemble, the technique generally used to evaluate model agreement.

A recent study documented an unprecedented decline in runoff efficiency (the proportion of snowpack that becomes river flow) in the Rio Grande basin since 1980 as temperatures have warmed (Lehner et al. 2017). The decreasing runoff efficiency from 1980 to 2015 was unprecedented in the context of the last 445 years as reconstructed from paleoclimate proxies. The study found that very low runoff efficiencies are 2.5 to 3 times more likely when temperatures are high. This finding, along with other recent studies in the Colorado River basin (Woodhouse et al. 2016; Udall and Overpeck 2017), strongly suggests that the basin will undergo snowmelt flow reductions as it warms. A climate change–enhanced monsoon may reduce these losses, but the monsoons have been unpredictable and highly variable from year to year, and precipitation is mostly not storable in reservoirs. Furthermore, climate models currently do a poor job of modeling this type of precipitation, and hence it is very difficult to assess its likelihood.

CONCLUDING THOUGHTS

The Rio Grande and the Colorado River are the two most important rivers in the American Southwest. Both were profoundly and irreversibly changed by humans during the twentieth century to provide water for human endeavors. In the twenty-first century, both rivers and their basins will undergo more major changes as anthropogenic warming continues to modify the water cycle through changes in the amount, timing, form, temperature, and quality of water supplies. Human and natural systems will suffer additional impacts because of these ongoing changes. Although there are no universal fixes that will restore these rivers to their natural state while providing humans with ample supplies for cities and agriculture, there are a number of actions that can be taken to minimize additional harm and perhaps incrementally restore some river reaches. These actions rest on a fundamental truth: It is unlikely that the current southwestern drought in the Colorado River and Rio Grande basins will end, because we have entered a new period of declining flows due to increased temperatures. Yes, we may have short periods with above-average flows, and in the case of the Colorado River, the northern portions may at some point regularly see more flow. But on the whole, we should expect less water in these rivers as climate change unfolds in the twenty-first century.

Consistent reductions in our water supplies point toward a number of actions. First, any plan to increase diversions and consumption is folly. In the Colorado River system, additional Upper Basin diversions will at best take water away from existing Lower Basin uses, creating additional inter-basin stress, and at worst will be costly blunders that will reduce water supply reliability. Instead, we need to be thinking about how to equitably deal with less water for all users. Basin water managers will need to find new ways of sharing water among agriculture, the environment, and municipalities, using economic incentives to drive desirable water saving behavior, including payments to compensate willing sellers, whenever possible.

Agricultural users and the environment will probably be hit the hardest, as they have smaller constituencies and fewer resources. Crop switching, deficit irrigation, and rotational fallowing should all be used to reduce agricultural demand while minimizing the economic impact of less water. Water reuse and a strong, lasting water conservation ethic in cities need to become the norm. Southwestern cities should continue their very successful strategy over the last two decades of pushing down total water use while undergoing significant growth. In some cases, as a last resort, we may need to appeal to voluntary shared sacrifice to manage our way through severe shortage—nineteenth-century water law, with its absolute winners and losers, is a poor solution for the twenty-first century, where the total loss of water by some users so that others can have a full supply can create economic and environmental disasters.

We will need to use our existing infrastructure wisely to minimize risk, rather than solely as big buckets to maximize consumptive use. Recent, compensated, voluntary efforts to prop up storage in the Colorado River have avoided water shortages at little cost and provide one successful model for effective water conservation. We should look at new ways of re-operating reservoirs to support river restoration and to save water wherever possible.

Some of the upper tributaries in the Colorado River and Rio Grande are reasonably healthy; we should avoid degrading these systems any further. Burgeoning efforts in the Colorado River delta to restore ecosystems have been modestly successful, and with the signing of Minute 323, such efforts will continue. These efforts should be supported by all water users, as all bear some responsibility for the overall system degradation. We should also continue to resolve long-standing Indian water rights claims. Tribes control nearly 20% of all water in the Colorado River, and future solutions must acknowledge their water rights and needs. USBR's Colorado River basin study has successfully brought stakeholders together to generate shared system-wide visions that overcome typical battle lines. With its input and output index gages, the Rio Grande Compact unfortunately encourages the management of three entirely separate river systems: the Colorado portion, the Middle Rio Grande near Albuquerque, and the segment below Elephant Butte Dam. Discussions rarely occur across these boundaries, unlike the Colorado, where nearly everyone wants a say on everything. Such system-wide thinking in the Rio Grande could also promote new ideas and solutions, including novel reservoir operations.

Water problems in the Southwest are here to stay. These problems represent a fundamental governance challenge to all who inhabit the Southwest. How do we best manage a resource that is key to our lives and the lives of many creatures, that promotes and supports our economic well-being, and that provides our sense of place? Each locality reliant on these great rivers has its own unique needs and stakeholders, and has an upriver community from whom they receive water, and a downstream community to whom they send water. The Colorado River and the Rio Grande, like all rivers, tie us together in critical ways. Only through shared governance will we derive solutions that work for the broadest set of citizens and for the environment upon which we depend.

REFERENCES

ABCWUA (Albuquerque Bernalillo County Water Utility Authority). 2016. *Water 2120: Securing Our Water Future.*

Adler, R. W. 2007. *Restoring Colorado River Ecosystems: A Troubled Sense of Immensity*. Washington, DC: Island Press.

Ault, T. R., J. E. Cole, J. T. Overpeck, G. T. Pederson, and D. M. Meko. 2014. Assessing the risk of persistent drought using climate model simulations and paleoclimate data. *Journal of Climate* 27: 7529–49.

Ault, T. R., J. S. Mankin, B. I. Cook, and J. E. Smerdon. 2016. Relative impacts of mitigation, temperature, and precipitation on 21st-century megadrought risk in the American Southwest. *Science Advances* 2 (10): e1600873.

Ayers, J., D. L. Ficklin, I. T. Stewart, and M. Strunk. 2016. Comparison of CMIP3 and CMIP5 projected hydrologic conditions over the upper Colorado River basin. *International Journal of Climatology* 36: 3807–18.

Bark, Rosalind H. 2006. Water reallocation by settlement: Who wins, who loses, who pays? Bepress Legal Series, 1454.

Barnett, T. P., and D. W. Pierce. 2008. When will Lake Mead go dry? *Water Resources Research* 44: W03201. https://doi.org/10.1029/2007WR006704.

Barsugli, J. J., K. Nowak, B. Rajagopalan, J. R. Prairie, and B. Harding. 2009. Comment on When will Lake Mead go dry? by T. P. Barnett and D. W. Pierce. *Water Resources Research* 45. https://doi.org/10.1029/2008WR007627.

Bestgen, K. R., B. Mefford, J. M. Bundy, C. D. Walford, and R. I. Compton. 2010. Swimming performance and fishway model passage success of Rio Grande silvery minnow. *Transactions of the American Fisheries Society* 139: 433–48.

Bestgen, K. R., and S. P. Platania. 1991. Status and conservation of the Rio Grande silvery minnow, *Hybognathus amarus*. *Southwestern Naturalist* 36: 225–32.

Cayan, D., M. Tyree, K. E. Kunkel, C. L. Castro, A. Gershunov, J. Barsugli, A. Ray, J. Overpeck, M. Anderson, J. Russell, B. Rajagopalan, I. Rangwala, and P. Duffy. 2013. Future climate: Projected average. In G. Garfin, A. A. Jardine, R. Meredith, M. Black, and S. LeRoy, eds., *Assessment of Climate Change in the Southwest United States: A Report Prepared for the National Climate Assessment*, 101–25. Washington, DC: Island Press.

Cohen, M. 2014. *Hazard's Toll: The Costs of Inaction at the Salton Sea*. Pacific Institute for Studies in Development, Environment and Security. http://bibpurl.oclc.org/web/74566 http://pacinst.org/publication/hazards-toll/.

Cohen, M., J. Christian-Smith, and J. Berggren. 2013. *Water to Supply the Land*. http://www.pacinst.org/wp-content/uploads/2013/05/pacinst-crb-ag.pdf.

Collum, C., and T. McCann. 2014. *Central Arizona Project Board Report Agenda Number 9*. Phoenix: Central Arizona Project. http://www.cap-az.com/documents/meetings/05-01-2014/9.%20Colorado%20River%20Report%20May%201%20Board.pdf.

Colorado River Basin Stakeholders. 2015. *Colorado River Basin Stakeholders Moving Forward to Address Challenges Identified in the Colorado River Basin Water Supply and Demand Study, Phase 1 Report*. Washington, DC: US Bureau of Reclamation. https://www.usbr.gov/lc/region/programs/crbstudy/MovingForward/Phase1Report/fullreport.pdf.

Colorado River Compact. 1922. https://www.usbr.gov/lc/region/g1000/pdfiles/crcompct.pdf.

Cook, B. I., T. R. Ault, and J. E. Smerdon. 2015. Unprecedented 21st century drought risk in the American Southwest and Central Plains. *Science Advances* 1 (1): e1400082–e1400082.

Cowley, D. E. 2006. Strategies for ecological restoration of the Middle Rio Grande in New Mexico and recovery of the endangered Rio Grande silvery minnow. *Reviews in Fisheries Science* 14: 169–86.

Crawford, S. 1988. *Mayordomo: Chronicle of an Acequia in Northern New Mexico*. 1st ed. Albuquerque: University of New Mexico Press.

Davis, Tony. 2016. Big CAP cuts coming as 3-State Water Agreement nears. *Arizona Daily Star*, April 23, 2016. http://tucson.com/news/local/big-cap-cuts-coming-as-state-water-agreement-nears/article_876e3aa6-6cf0-53ec-bd0c-95be8c6468ae.html.

Deems, J. S., T. H. Painter, J. J. Barsugli, J. Belnap, and B. Udall. 2013. Combined impacts of current and future dust deposition and regional warming on Colorado River basin snow dynamics and hydrology. *Hydrology and Earth System Sciences* 17: 4401–13.

Dettinger, M., B. Udall, and A. Georgakakos. 2015. Western water and climate change. *Ecological Applications* 25: 2069–93.

DuMars, C. 2012. The Middle Rio Grande minnow wars. In C. T. Ortega Klett, ed., *One Hundred Years of Water Wars in New Mexico 1912–2012*, 123–40. Santa Fe: Sunstone Press.

Elias, E. H., A. Rango, C. M. Steele, J. F. Mejia, and R. Smith. 2015. Assessing climate change impacts on water availability of snowmelt-dominated basins of the upper Rio Grande basin. *Journal of Hydrology: Regional Studies* 3 (March): 525–46.

Flanigan, K. G., and A. I. Haas. 2008. Impact of full beneficial use of San Juan–Chama Project water by the City of Albuquerque on New Mexico's Rio Grande Compact obligations. *Natural Resources Journal* 48: 371.

Glennon, P., and R. Culp. 2012. Parched in the West but shipping water to China, bale by bale. *Wall Street Journal*, October 5, 2012, Opinion. http://online.wsj.com/news/articles/SB10000872396390444517304577653432417208116.

Grassel, K. 2002. *Taking Out the Jacks: Issues of Jetty Jack Removal in Bosque and River Restoration Planning*. http://dspace.unm.edu/handle/1928/10490.

Gutzler, D. S. 2011. Climate and drought in New Mexico. In D. Brookshire, H. Gupta, and O. P. Matthews, eds., *Water Policy in New Mexico*, 56–70. RFF Press Water Policy Series. Washington, DC: RFF Press.

Hely, A. G. 1969. *Lower Colorado River Water Supply—Its Magnitude and Distribution*. USGS Professional Paper Numbered Series 486-D. http://pubs.er.usgs.gov/publication/pp486D.

Hill, R. A. 1974. Development of the Rio Grande Compact of 1938. *Natural Resources Journal* 14: 163–200.

Hundley, N., Jr. 1966. *Dividing the Waters: A Century of Controversy between the United States and Mexico*. 1st ed. Berkeley: University of California Press.

Hundley, N., Jr. 2009. *Water and the West: The Colorado River Compact and the Politics of Water in the American West*. 2nd ed. Berkeley: University of California Press.

Hurd, B., and J. Coonrod. 2012. Hydro-economic consequences of climate change in the upper Rio Grande. *Climate Research* 53: 103–18.

Imperial Irrigation District. 2004. Quantification Settlement Agreement: Imperial Irrigation District/San Diego County Water Authority Water Conservation and Transfer Agreement Annual Implementation Agreement. http://www.iid.com/water/library/qsa-water-transfer/qsa-annual-reports.

Jacobs, J. 2019. States reach drought pact as largest user protests. *E&E News*, March 20, 2019. https://www.eenews.net/stories/1060127745.

James, I. 2016. California weighs sharing "pain" of Colorado River cuts. *Desert Sun*, April 26, 2016. http://www.desertsun.com/story/news/environment/2016/04/26/california-weighs-sharing-pain-colorado-river-cuts/83510014/.

Jaspers, B. 2019. Bipartisan drought plan legislation coming to Congress "very soon." Arizona Public Media, March 28, 2019. https://www.azpm.org/p/home-articles-news/2019/3/28/148689-bipartisan-drought-plan-legislation-coming-to-congress-very-soon/.

Johnson, R. 1977. *The Central Arizona Project, 1918–1968*. Tucson: University of Arizona Press.

Kammerer, J. C. 1990. *Largest Rivers in the United States*. US Geological Survey Open-File Report 87-242.

Katz, L. 2007. History of the minnow litigation and its implications for the future of reservoir operations on the Rio Grande. *Natural Resources Journal* 47: 675–91.

Kelly, S., I. Augusten, J. Mann, and L. Katz. 2007. History of the Rio Grande reservoirs in New Mexico: Legislation and litigation. *Natural Resources Journal* 47: 525–613.

Kelly, S., and S. McKean. 2011. The Rio Grande silvery minnow: Eleven years of litigation. *Water Matters* 8. http://uttoncenter.unm.edu/pdfs/Silvery_Minnow_litigation.pdf.

Lehner, F., E. R. Wahl, A. W. Wood, D. B. Blatchford, and D. Llewellyn. 2017. Assessing recent declines in upper Rio Grande River runoff efficiency from a paleoclimate perspective: Rio Grande declines in runoff efficiency. *Geophysical Research Letters* 44: 4124–33. https://doi.org/10.1002/2017GL073253.

Littlefield, D. R. 2008. *Conflict on the Rio Grande: Water and the Law, 1879–1939*. 1st ed. Norman: University of Oklahoma Press.

MacDonald, G. M. 2010. Water, climate change, and sustainability in the Southwest. *Proceedings of the National Academy of Sciences* 107: 21256–62.

MacDonnell, L. 2016. *Tribes and Water in the Colorado River Basin*. Colorado River Research Group. 6 p. https://www.coloradoriverresearchgroup.org/publications.html.

Mann, J. 2007. A reservoir runs through it: A legislative and administrative history of the Six Pueblos' right to store prior and paramount water at El Vado. *Natural Resources Journal* 47: 733–68.

Meko, D. M., C. A. Woodhouse, C. A. Baisan, T. Knight, J. J. Lukas, M. K. Hughes, and M. W. Salzer. 2007. Medieval drought in the upper Colorado River basin. *Geophysical Research Letters* 34. http://onlinelibrary.wiley.com/doi/10.1029/2007GL029988/full.

Melillo, J. M., T. T. C. Richmond, and G. W. Yohe. 2014. *Climate Change Impacts in the United States: The Third National Climate Assessment*. US Global Change Research Program. http://admin.globalchange.gov/sites/globalchange/files/Ch_0a_FrontMatter_ThirdNCA_GovtReviewDraft_Nov_22_2013_clean.pdf.

Meyers, C. J. 1966. The Colorado River. *Stanford Law Review* 19: 1–75.

MWH Americas, Inc. 2012. Colorado River Water Bank Feasibility Study: Phase 1.

MWH Americas, Inc. 2014. Colorado River Water Bank Feasibility Study: Phase 2.

Noble, Wade. 2015. A case study in efficiency—Agriculture and water use in the Yuma, Arizona area. Yuma County Agriculture Water Coalition. http://nebula.wsimg.com/5967b5f521b14d1d8497cb59d22a2909?AccessKeyId=D824E441302DA8098207&disposition=0&alloworigin=1.

Paddock, William A. 2001. The Rio Grande Compact of 1938. *University of Denver Water Law Review* 5: 1.

Painter, T. H., J. S. Deems, J. Belnap, A. F. Hamlet, C. C. Landry, and B. Udall. 2010. Response of Colorado River runoff to dust radiative forcing in snow. *Proceedings of the National Academy of Sciences* 107: 17125–30.

Pecos, R. 2007. The history of Cochiti Lake from the Pueblo perspective. *Natural Resources Journal* 47: 639–52.

Phillips, F., E. Hall, and M. Black. 2011. *Reining In the Rio Grande: People, Land, and Water*. Albuquerque: University of New Mexico Press.

Pitt, J. 2001. Can we restore the Colorado River delta? *Journal of Arid Environments* 49: 211–20.

Pitt, J., D. F. Luecke, M. J. Cohen, and E. P. Glenn. 2000. Two nations, one river: Managing ecosystem conservation in the Colorado River delta. *Natural Resources Journal* 40: 819–64.

Poff, N. L., J. D. Allan, M. B. Bain, J. R. Karr, K. L. Prestegaard, B. D. Richter, R. E. Sparks, and J. C. Stromberg. 1997. The natural flow regime. *BioScience* 47: 769–84.

Polade, S. D., D. W. Pierce, D. R. Cayan, A. Gershunov, and M. D. Dettinger. 2014. The key role of dry days in changing regional climate and precipitation regimes. *Scientific Reports* 4 (March). https://doi.org/10.1038/srep04364.

Richard, G. A. 2001. Quantification and prediction of lateral channel adjustments downstream from Cochiti Dam, Rio Grande, NM. PhD diss., Colorado State University. http://www.fws.gov/bhg/Literature/GAR-Dissertation.pdf.

Richard, G. A., and P. Julien. 2003. Dam impacts on and restoration of an alluvial river—Rio Grande, New Mexico. *International Journal of Sediment Research* 18 (2): 89–96. http://www.treesearch.fs.fed.us/pubs/28475.

Richard, G. A., P. Y. Julien, and D. C. Baird. 2005. Statistical analysis of lateral migration of the Rio Grande, New Mexico. *Geomorphology* 71: 139–55.

Rivera, J. A. 2011. The historical role of acequias and agriculture in New Mexico. In D. Brookshire, H. Gupta, and O. P. Matthews, eds., *Water Policy in New Mexico*. Washington, DC: RFF Press.

Robbins, T. 2014. In time of drought, U.S. West's alfalfa exports are criticized. NPR.org, August 12, 2014. http://www.npr.org/2014/08/12/339753108/in-time-of-drought-arizona-s-alpha-exports-criticized.

Royster, J. 1994. A primer on Indian water rights: More questions than answers. *Tulsa Law Journal* 30: 61.

Royster, J. 2006. Indian water and the federal trust: Some proposals for federal action. *Natural Resources Journal* 46: 375–98.

Sauder, Robert A. 2009. *The Yuma Reclamation Project: Irrigation, Indian Allotment, and Settlement along the Lower Colorado River*. Reno: University of Nevada Press.

Schmandt, J. 2002. Bi-national water issues in the Rio Grande/Río Bravo basin. *Water Policy* 4: 137–55.

Schmidt, J. C., M. Kraft, D. Tuzlak, and A. Walker. 2016. *Fill Mead First: A Technical Assessment*. Logan: Center for Colorado River Studies, Utah State University.

Scurlock, D. 1998. *From the Rio to the Sierra: An Environmental History of the Middle Rio Grande Basin*. General Technical Report RMRS-GTR-5. Fort Collins, CO: USDA Rocky Mountain Research Station. http://books.google.com/books/about/Statistical_theory_and_methodology_in_sc.html?id=XU1MAAAAMAAJ.

Seager, R., M. Ting, C. Li, N. Naik, B. Cook, and J. Nakamura. 2013. Projections of declining surface-water availability for the southwestern United States. *Nature Climate Change* 3: 482–86.

Shurts, J. 2000. *Indian Reserved Water Rights: The Winters Doctrine in Its Social and Legal Context*. 1st ed. Norman: University of Oklahoma Press.

Stern, C. 2015. *Indian Water Rights Settlements*. Washington, DC: Congressional Research Service. http://nationalaglawcenter.org/wp-content/uploads/assets/crs/R44148.pdf.

Stomp, J. 2014. Albuquerque's water resources management: Integrated strategy meets area challenges. *The Water Report*, no. 121, 1–9.

Thomson, B. 2011. Water resources in New Mexico. In D. Brookshire, H. Gupta, and O. P. Matthews, eds., *Water Policy in New Mexico*, 25–55. Washington, DC: RFF Press.

Tipton and Kalmbach, Inc. 1965. *Water Supplies of the Colorado River*. Denver: Upper Colorado River Commission.

Udall, B., and J. Overpeck. 2017. The twenty-first-century Colorado River hot drought and implications for the future. *Water Resources Research* 53: 2404–18.

USBR (US Bureau of Reclamation). 1981. *Project Data*. US Department of the Interior.

USBR (US Bureau of Reclamation). 2012. *Colorado River Basin Water Supply and Demand Study*. Denver: US Department of the Interior, US Bureau of Reclamation, Technical Services Center.

USBR (US Bureau of Reclamation). 2013a. *River Maintenance Program—San Marcial Delta Water Conveyance Channel Maintenance Project Biological Assessment*. Albuquerque: US Bureau of Reclamation.

USBR (US Bureau of Reclamation). 2013b. *West-Wide Climate Risk Assessment: Upper Rio Grande Impact Assessment*. Albuquerque: US Bureau of Reclamation.

US Department of the Interior. 2005. *Record of Decision Lower Colorado River Multi-species Conservation Plan*. https://www.lcrmscp.gov/publications/rec_of_dec_apr05.pdf.

US Department of the Interior. 2007. *Record of Decision Colorado River Interim Guidelines for Lower Basin Shortages and the Coordinated Operations for Lake Powell and Lake Mead*. Washington, DC: Department of the Interior. http://www.usbr.gov/lc/region/programs/strategies/RecordofDecision.pdf.

USFWS (US Fish and Wildlife Service). 2016. *Final Biological and Conference Opinion for Bureau of Reclamation, Bureau of Indian Affairs, and Non-Federal Water Management and Maintenance Activities on the Middle Rio Grande, New Mexico, December*. https://www.fws.gov/southwest/es/NewMexico/documents/BO/2013-0033_MRG_BiOp_Final.pdf.

Vandiver, S. E. 2011. The administration of the Rio Grande Compact in Colorado. In K. O. Verburg, *Colorado River Documents 2008*. Denver: US Bureau of Reclamation.

Weldon, J. B., Jr., and L. M. McKnight. 2007. Future Indian water settlements in Arizona: The race to the bottom of the waterhole. *Arizona Law Review* 49: 441–67.

Whipple, J. 2007. *San Juan–Chama Project Water Supply*. Santa Fe: New Mexico Interstate Stream Commission.

Wines, M. 2014a. Colorado River drought forces a painful reckoning for states. *New York Times*, January 6, 2014.

Wines, M. 2014b. Arizona cities could face cutbacks in water from Colorado River, officials say. *New York Times*, June 17, 2014. http://www.nytimes.com/2014/06/18/us/arizona-cities-could-face-cutbacks-in-water-from-colorado-river-officials-say.html.

Woodhouse, C. A., S. T. Gray, and D. M. Meko. 2006. Updated streamflow reconstructions for the upper Colorado River basin. *Water Resources Research* 42. https://doi.org/10.1029/2005WR004455.

Woodhouse, C. A., G. T. Pederson, K. Morino, S. A. McAfee, and G. J. McCabe. 2016. Increasing influence of air temperature on upper Colorado River streamflow. *Geophysical Research Letters* 43, January. https://doi.org/10.1002/2015GL067613.

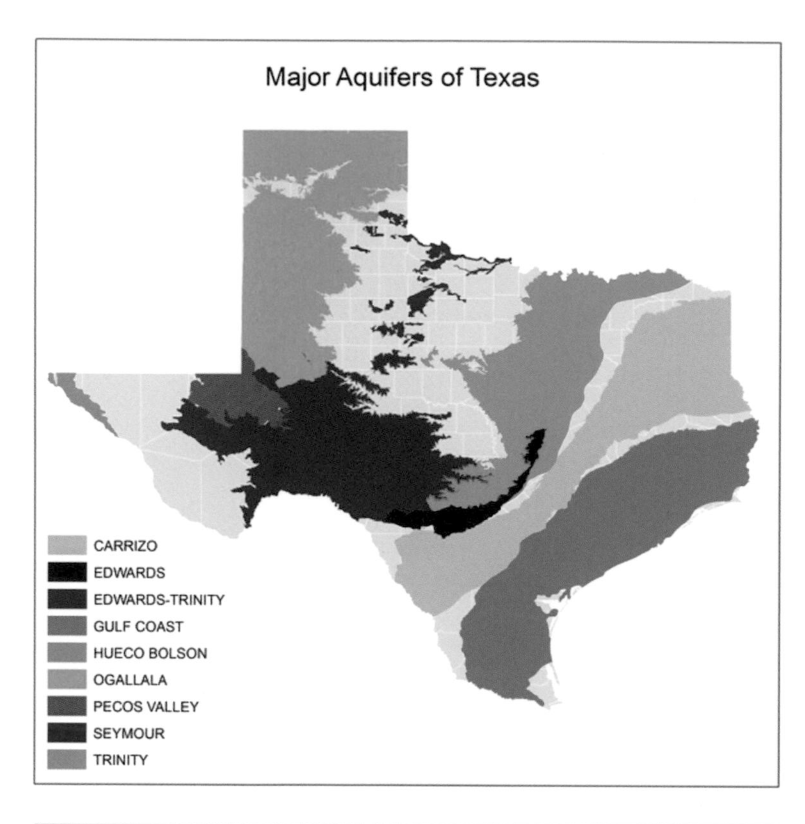

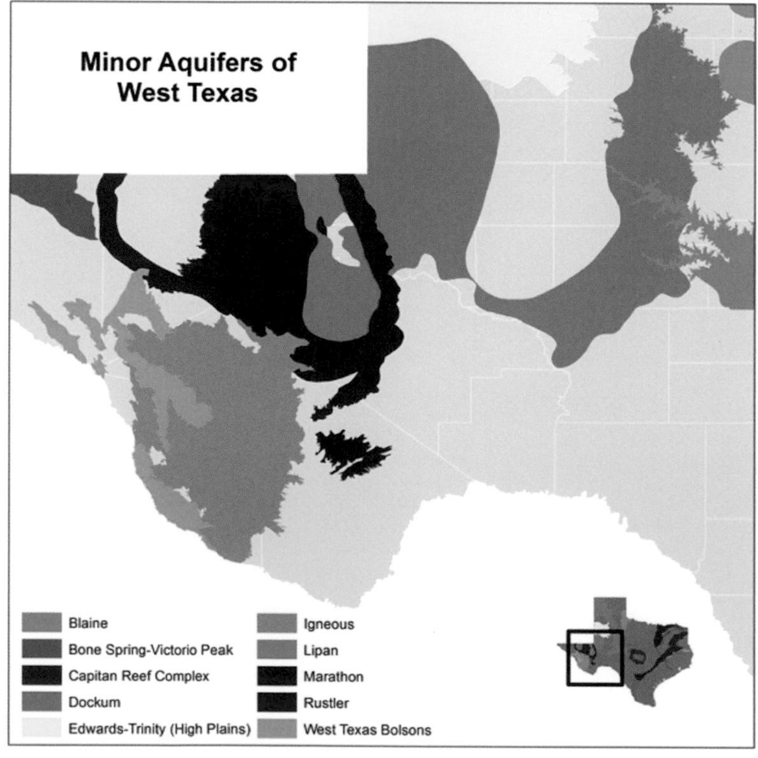

Fig. 8.1 Major and minor aquifers of the Chihuahuan Desert of Texas. (Source: Texas Water Development Board, https://www.twdb.texas.gov/groundwater/aquifer/.)

8 Mining Hidden Waters

Groundwater Depletion, Aquatic Habitat Degradation, and Loss of Fish Diversity in the Chihuahuan Desert Ecoregion of Texas

Gary P. Garrett, Megan G. Bean, Robert J. Edwards, and Dean A. Hendrickson

Desert ecosystems are fragile and slow to recover from perturbations, and some changes may be irreparable. Exploitation of limited resources, particularly groundwater pumping, has degraded natural systems in the Chihuahuan Desert ecoregion of Texas, caused degradation of aquatic habitats, and resulted in extirpation and extinction of species and, ultimately, losses of entire ecosystems (Garrett and Edwards 2001). Deep trenching of streams by overgrazing-induced erosion has resulted in lowered water tables and further desiccation of watersheds. This contributes to spring failure, ciénega drying, and the transformation of previously flowing streams into dry arroyos (Minckley et al. 1991).

The Chihuahuan Desert of Texas includes the most remote watersheds in the state, and they contain a wide variety of habitats with many uniquely adapted plants and animals. In addition to flows from the Rio Grande, Río Conchos, and Pecos River, three major aquifers (Hueco-Mesilla bolsóns, Pecos Valley, and Edwards–Trinity Plateau) and several minor aquifers of varying water quality, yields, and geology (Bone Spring–Victorio Peak, Capitan Reef, Dockum, Igneous, Marathon, Rustler, and West Texas bolsóns) provide water to the area (fig. 8.1) (Mace 2001). The individual groundwater basins are connected through regional flow systems in fractured, karstic carbonate rocks. These fracture trends also connect the major recharge and spring areas (Sharp et al. 2003).

These aquatic habitats have undergone substantial anthropogenic modifications, including degraded water quality, diversion of surface water, groundwater depletion and associated reduction in spring discharge and instream flows, channelization, bank instability and erosion, sedimentation, impoundment, and extensive introduction of nonnative species (Edwards et al. 2002). Some of the most heavily impacted habitats are desert springs and associated ciénegas. These aquatic habitats were seldom damaged on purpose; put simply, water is rare in the desert, and people exploit it for a variety of reasons (Garrett and Edwards 2001). Although streamflow in Texas is governed by the "prior appropriation" doctrine, which states that it is publicly owned and that permits are therefore required for use of surface water, little to no protection exists for groundwater because it is governed by the "rule of capture," which states that groundwater belongs to the landowner. This unfortunate dichotomy allows a landowner to pump from an aquifer to the extent that not only damages spring flow, ciénegas, streams, and rivers. but also allows the "taking" of the neighbor's property by lowering the water table and even drying up wells, which oddly contradicts the intended sacrosanct nature of property rights in Texas.

The springs, ciénegas, creeks, and rivers of the Chihuahuan Desert ecoregion of Texas now hardly resemble their natural state (table 8.1, fig. 8.2), formerly characterized by springs supporting large ciénegas and watercourses lined with gallery forests and diverse riparian zones. Springs and their associated ciénegas provided habitat for a wide variety of plants and animals, some of which are endemic to these systems (Hendrickson and Minckley 1984). Ciénegas harbor not only unique species, but entire communities of interacting organisms that depend on these fragile habitats for survival. Desert fishes are particularly dependent on these systems.

Table 8.1. Top five Chihuahuan Desert springs in Texas, with historical flow rates. Comanche and Phantom Lake springs no longer flow, San Solomon Springs have been reduced to 720 L/s. The filling of Amistad Reservoir affected both Goodenough Springs (inundated) and San Felipe Springs (increased flow from additional aquifer head pressure).

Spring	Flow (L/s)
Goodenough Springs	3,900
San Felipe Springs	2,600
Comanche Springs	1,100
San Solomon Springs	990
Phantom Lake Springs	330

Source: Brune 1981.

Habitat modifications, in addition to introductions of nonnative species and extinctions of native species, may cause irreversible damage to these ecosystems. Under these circumstances, periodic droughts are even more devastating, as they provoke increased groundwater pumping for agricultural and municipal uses. Such extreme conditions put stress on fish community equilibrium, with more tolerant species gaining a competitive and numerical advantage. Tributary creeks tend to be impacted more severely, yet are critical to the breeding and rearing of the young of many native species. Changes have been gradual and long-term, some taking place since the mid-1800s (Miller 1961), but their effects have been compounded over time and are now dramatic. What remains of unmodified habitats and intact flora and fauna needs careful management if they are to be preserved. Many of the fishes of the area could serve well as biological indicators of the overall integrity of these desert aquatic ecosystems, and studies of aquatic communities can provide valuable baseline data for future management decisions and actions affecting the larger, binational Chihuahuan Desert ecoregion. By involving individuals and local governments, the likelihood of achieving long-term benefits for natural resources, as well as public health and quality of life, is enhanced (Garrett and Edwards 2001). While perturbations such as pollution, reduced groundwater, and dam construction are theoretically fixable, recovery to a pristine state is unlikely.

Although many data gaps exist, what is known is somewhat grim. Of the more than 100 moderate (2.8–28 L/s) and major (>28 L/s) historical springs in the Chihuahuan Desert ecoregion of Texas, 50% are no longer extant. Early records, some as far back as 1583, mention expansive ciénegas and abundant fishes (Brune 1981). Very likely, some species were extinct before being discovered. Others have persisted, but just barely. Approximately half of the native fishes are threatened with extinction or are already extirpated or extinct (Hubbs 1990). Likely extinctions in this area include Maravillas red shiner *Cyprinella lutrensis blairi*, phantom shiner *Notropis orca*, Rio Grande bluntnose shiner *Notropis simus simus*, and Amistad gambusia *Gambusia amistadensis* (Miller et al. 1989; Hubbs et al. 2008). Regional extirpations include shovelnose sturgeon *Scaphirhynchus platorynchus*, Rio Grande silvery minnow *Hybognathus amarus*, Chihuahua catfish *Ictalurus* sp., Rio Grande cutthroat trout *Oncorhynchus*

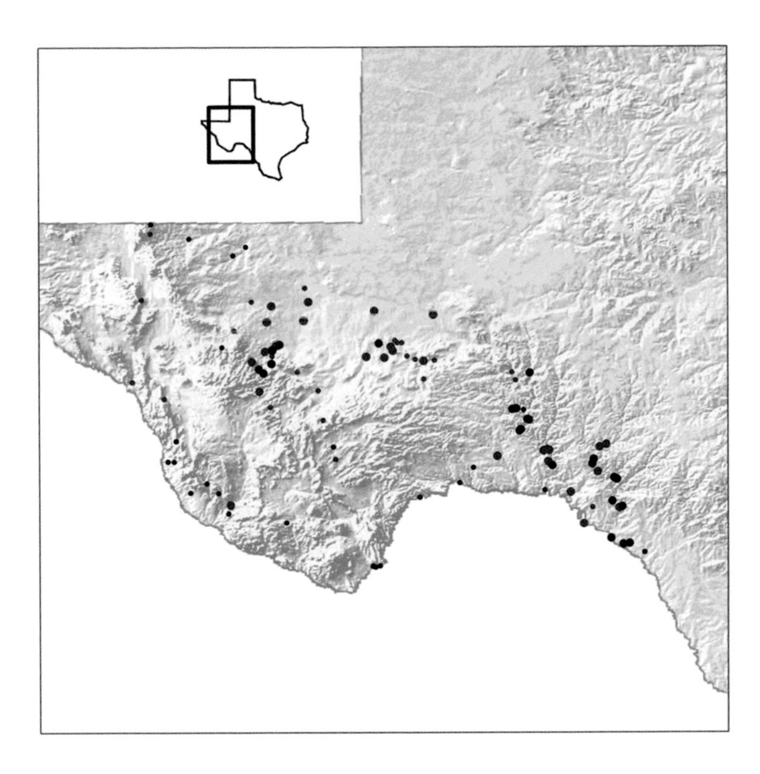

Fig. 8.2 Chihuahuan Desert springs in Texas: major springs (large points, >28 L/s) and moderate springs (small points, 2.8–28 L/s) prior to 1950 (all points), no longer extant (red points), and currently extant (black points). (Sources: Brune 1975, 1981, and Springs Online, http://springsdata.org/search/index.php.)

clarkii virginalis, and blotched gambusia *Gambusia senilis* (Bestgen and Platania 1990; Hubbs et al. 2008). Endemic species other than fishes are being lost as well (e.g., Rio Grande monkeyface *Quadrula couchiana*, false spike *Quincuncina mitchelli*, and Mexican fawnsfoot *Truncilla cognata*; Howells and Garrett 1995). Left unchecked, this trend of species extirpation and extinction is likely to continue.

BIG BEND REACH OF THE RIO GRANDE

The Big Bend reach of the Rio Grande is functionally composed of two distinct segments (above and below Mariscal Canyon), each characterized by unique attributes of base flow, sediment movement, and water quality. These differences are primarily due to reduced base flow and water quality in the upper segment and considerable spring-flow inputs and improved water quality in the lower segment. Due to human alterations, the Rio Grande between El Paso and the confluence with the Río Conchos is functionally an intermittent stream, though its pristine state had continuous flow sufficient to provide for the movement of large, main-channel fishes as far upstream as Albuquerque (Miller 1977). In the lower segment, the extensive Edwards–Trinity Plateau Aquifer in Texas and the Cerro Colorado–La Partida and Serranías del Burro Aquifers in Coahuila sustain base flows (Bennett and Urbanczyk 2014), keeping habitat largely intact and supporting a high diversity of native aquatic species (Bennett et al. 2014).

The springs in the area of Big Bend National Park were once more numerous and had greater discharges than at present. During the late nineteenth and early twentieth centuries, extensive farming, ranching, and mining operations in the area caused many of the springs to dry (Brune 1981). One of the larger springs in the Big Bend reach is Boquillas Hot Springs, adjacent to the Rio Grande. Approximately 1 km to the west of these springs was another spring system in a marshy area known as Graham Ranch Springs. The federally endangered Big Bend gambusia *Gambusia gaigei* originally inhabited both of these springs. The population in Boquillas Hot Springs became extinct when the springs failed in 1954 (Hubbs and Brodrick 1963), but the population in the Graham Ranch Springs (now called Spring 4) persisted. This population was subsequently extirpated by a renovation effort to remove introduced sport fishes from the springs. Prior to the renovation, many individuals were relocated to various isolated springs in Big Bend National Park and aquaria in Austin, Texas. Only three individuals from the Austin aquarium population survived, and two new refuges were established near Spring 4, called the Spring 1 Refuge and the Clark Hubbs Refuge (USFWS 2012).

Water management practices have eliminated the natural hydrograph and reduced overall streamflow in the Big Bend reach of the Rio Grande (Schmidt et al. 2003). Introduction and expansion of nonnative riparian plant species has altered stream channel morphology, sediment dynamics, aquatic habitats, and riparian vegetative cover (Dean and Schmidt 2011). Specifically, giant reed *Arundo donax* and salt cedar *Tamarix* spp. affect channel sediment retention, aquatic habitat, and riparian communities by covering up and eliminating backwaters and side channels, diminishing channel conveyance capacity, and increasing flooding frequency (Dean and Schmidt 2011; Garrett and Edwards 2014). These impacts have modified the once wide and shallow channel into one that is narrower and deeper. Additional threats to native fishes and their habitats in the Big Bend reach include further alteration of spring flows by groundwater extraction (Donnelly 2007), deteriorating aquatic habitat (Heard et al. 2012), elevated concentrations of mercury, arsenic, and selenium in fishes (Schmitt et al. 2005), water quality deterioration (Sandoval-Solís et al. 2010; Bennett et al. 2012), and climate change (Ingol-Blanco 2011).

The changes described above appear to be the main contributors to the extinctions and extirpations of native fishes in the Big Bend reach. One of the most dramatic examples is the federally endangered Rio Grande silvery minnow. This species originally had one of the widest distributions of any species in the Rio Grande drainage, ranging from northern New Mexico to the mouth of the river on the Gulf Coast as well as in the Pecos River. The species is now extirpated from the Pecos River and Rio Grande everywhere except for a small segment in New Mexico (USFWS 2010). An effort to repatriate the Rio Grande silvery minnow to part of its previous range in the Big Bend reach was initiated in the first decade of the twenty-first century. Following federal designation of the population as experimental, repatriations began in December 2008, and additional individuals have been stocked annually. While limited reproduction has occurred, sufficient recruitment of individuals for a self-sustaining population has not been documented (Edwards and Garrett 2013).

Other imperiled fishes in the Big Bend reach of the Rio Grande include American eel *Anguilla rostrata*, Conchos roundnose minnow *Dionda* sp., speckled chub *Macrhybopsis aestivalis*, Tamaulipas shiner *Notropis braytoni*, Rio Grande shiner *Notropis jemezanus*, longnose dace *Rhinichthys cataractae*, Mexican redhorse *Moxostoma austrinum*, Rio Grande blue catfish *Ictalurus furcatus* ssp., and headwater catfish *Ictalurus lupus*, as well as four state threatened species: Mexican stoneroller *Campostoma ornatum*, Chihuahua shiner *Notropis chihuahua*, Rio Grande blue sucker *Cycleptus* sp., and Conchos pupfish *Cyprinodon eximius* (Edwards et al. 2002; Hanna et al. 2013; Hendrickson and Cohen 2015).

BALMORHEA SPRINGS COMPLEX

Balmorhea Springs Complex is considered the largest and most important of the remaining desert spring systems in the Chihuahuan Desert ecoregion of Texas (Karges 2014). This group of springs occurs on the Toyah Creek valley floor near the base of the Davis and Barrilla Mountains. Phantom Lake, San Solomon, and Giffin Springs are artesian, emanating from a regional flow system of northwest–southeast-oriented solution channels in the Cretaceous limestone formation that connects individual groundwater basins with local recharge from the Davis Mountains (Sharp et al. 2003; Uliana et al. 2007). This regional flow ultimately becomes inflow to the Edwards–Trinity Plateau Aquifer (Bumgarner et al. 2012). Saragosa, East Sandia, and West Sandia Springs are gravity springs fed by a relatively shallow aquifer in the sand and gravel beds covering the Toyah Creek valley (Saragosa and West Sandia Springs no longer flow). The system is recharged by local rainfall and runoff from the Davis Mountains (White et al. 1941; Ashworth et al. 1997).

Historically, this complex of springs created an extensive, interconnected network of ciénegas fed by cumulative spring discharges of approximately 1,500 L/s (White et al. 1941). This dynamic mosaic persisted as isolated subunits during droughts and reconnected during high spring flows and low-level flooding (Karges 2014). This network of ciénegas has since been drained, and spring flows have been diverted to a system of canals to irrigate fields for agriculture (fig. 8.3). By 2001, combined flows averaged approximately 960 L/s (Texas Water Development Board 2005), and all that remains today for fish habitat are artificial refuge ciénegas and irrigation canals (Garrett 2003).

Imperiled fishes of this springs complex include

the federally endangered Pecos gambusia *Gambusia nobilis* and Comanche Springs pupfish *Cyprinodon elegans*, as well as roundnose minnow *Dionda episcopa* and headwater catfish. In addition to these fishes, other imperiled species include federally endangered diminutive amphipod *Gammarus hyalleloides*, Phantom Cave snail *Pyrgulopsis texana*, Phantom springsnail *Tryonia cheatumi*, and the federally threatened Pecos sunflower *Helianthus paradoxus*.

Due to a long-term pattern of increasing aridity in the region, spring flows were already declining prior to the development of extensive irrigation in the late 1940s (Sharp et al. 2003). However, the declines have increased dramatically in the past 50 years, and in addition, aquifers are now threatened by oil and gas development.

Recent discoveries of oil and gas and related plans for extensive hydraulic fracturing ("fracking") in the area pose numerous risks for the aquifers, springs, and ciénegas. According to the Apache Corporation (2016), it has leased mineral rights to more than 140,000 ha around and north of Balmorhea in order to exploit the "Alpine High" oil and gas discovery, which is estimated to contain 2 trillion cubic meters of natural gas and 3 billion barrels of oil. The most immediate environmental threat from this endeavor is the large de-

Fig. 8.3 Phantom Lake, 1938 (*top*) and 2000 (*bottom*). (1938 photo by Carl Hubbs, used by permission of Clark Hubbs, 2000 photo by Robert J. Edwards.)

mand for water for fracking and the resultant impact on groundwater resources and spring flows. The fracking process in this region typically requires 1,000–10,000 m^3 of water per well (Gallegos et al. 2015). With 2,000–3,000 oil and gas wells planned in the area, an estimated 2,000,000–30,000,000 m^3 of water may be withdrawn from the aquifers. Chemical and isotopic data reveal that most of the aquifer waters date to the Pleistocene (Sharp et al. 2003); therefore, substantial water extraction along the aquifer flow paths would probably reduce groundwater levels, and recharge would be insufficient to maintain spring flows (Texas Water Development Board 2005). Other risks include contamination from surface spills and the potential for fracking fluids and fracking wastewater to migrate from the new wellbores, as well as from previously abandoned wells, into the aquifers (Myers 2016).

Phantom Lake Spring (see fig. 8.3) ceased flowing in 1999 (Scudday 2003), and a small spring pool is now artificially maintained by a pump that accesses the aquifer through the spring source, Phantom Cave (Lewis et al. 2013). Some protection for the remainder of the Balmorhea Springs Complex is provided by Texas Parks and Wildlife Department, which owns the 18 ha Balmorhea State Park where San Solomon Springs, San Solomon Ciénega Refuge, and Clark Hubbs Ciénega Refuge are located. In addition, The Nature Conservancy owns and manages 100 ha of land that contains East Sandia and West Sandia Springs (Karges 2014).

PECOS RIVER WATERSHED

Agricultural and municipal water diversions have greatly diminished water quantity in the upper reach of the Pecos River in Texas (immediately downstream of Red Bluff Reservoir) and increased salinity to near that of seawater. The high salinity created conditions that allowed the golden alga (*Prymnesium parvum*), a species of algae previously known to occur in estuarine systems, to become established and repeatedly form toxic blooms deadly to fish. These conditions resulted in the loss of many fish species from this reach of the Pecos River (Rhodes and Hubbs 1992). Independence Creek is the major tributary to the lower reach of the Pecos River, and spring flows from there account for about a 40% increase in water volume in the river. The spring inputs from Independence Creek, as well as other springs, reduce sodium concentrations in the Pecos River by approximately 50% (Upper Rio Grande Basin and Bay Expert Science Team 2012).

The major existing and former springs in the Pecos River watershed are Comanche, Leon, and Diamond-Y Springs. As Gunnar Brune (1981, 357) so eloquently noted, "Failure of Comanche Springs was probably the most spectacular example in Texas of man's abuse of nature." Flowing at 1,200–1,900 L/s, this complex of springs was one of the largest in Texas, but flow completely ceased in 1962 due to aquifer pumping for irrigation from a well field up-gradient of the springs during the drought of the 1950s (Mace 2001). The outflow from the springs along Comanche Creek supported a vast ciénega, approximately 25 km long. The drying of the springs was not only an ecological disaster, but also had severe impacts on the more than 100 farmers who had, since the 1860s, depended on the waters flowing from Comanche Springs and the ciénega for irrigation of approximately 2,500 ha of cropland (Brune 1981).

In an attempt to establish its water rights and protect the springs, the local water district sued the small number of pumpers west of Fort Stockton whose actions had severely diminished the spring flows. The pumpers prevailed in the lawsuit by basing their defense on a 1904 case that had established the concept of "rule of capture." The rule of capture allowed well owners to pump as much water as desired, regardless of impacts on the aquifer. This case was also the one in which the Texas Supreme Court had determined that "because the existence, origin, movement, and course of such water, and causes that govern and direct their movements, are so secret, occult, and concealed . . . an attempt to administer any set of legal rules in respect to them would be involved in hopeless uncertainty, and would, therefore, be practically impossible" (Potter 2004: 2) and therefore, that it was impossible to provide any protections for the springs and the organisms that depended on them (including humans). Ultimately, the flows of Comanche Springs ceased, the ciénega dried up, the native flora and fauna disappeared, the surface irrigators lost their farms, and their land reverted to desert (Garrett 2003). Thus, in a matter of only a few years, farming in central Pecos County shifted from an economy based on free, free-flowing spring water to one based on expensive pumping of groundwater. Even today, during wet years, the springs can flow a small amount until the next growing season, when the pumps come back on (Scudday 2003).

Known extirpations caused by the aquifer drawdown and drying of Comanche Springs include populations of *Ictalurus lupus, Gambusia nobilis* and *Cyprinodon elegans* (Hubbs and Springer 1957; Hubbs 2003; Fishes of Texas Project database, www.fishesoftexas.org).

Leon Springs, up-gradient in the same aquifer as Comanche Springs, was also modified to provide irrigation for farming. Originally the springs were deep and up to 30 m in diameter, and they supported a large ciénega that extended for many kilometers downstream (Brune 1981). During the 1920s, a stone and earth dam created Lake Leon (Scudday 2003), which backed water up to, or over, Leon Springs. Unfortunately, this perturbation probably led to the extirpation of both *Gambusia nobilis* and *Cyprinodon bovinus*, as none were collected by Carl Hubbs in his 1938 survey of this type locality for the Leon Springs pupfish (Hubbs 1980; Minckley et al. 1991). The same groundwater pumping that led to the demise of Comanche Springs in 1961 also dried Leon Springs in 1958 (Brune 1981).

Although Diamond-Y Springs were not as large as Leon Springs, they continue to flow and provide habitat for the federally endangered *Gambusia nobilis* and *Cyprinodon bovinus* as well as the plains killifish *Fundulus zebrinus*, rainwater killifish *Lucania parva*, and western mosquitofish *Gambusia affinis*, and the nonnative largespring gambusia *G. geiseri* and green sunfish *Lepomis cyanellus*; however, the flow is greatly reduced from historical levels (Scudday 2003). It is fortunate for the Diamond-Y Springs ecosystem that these waters are not derived from the same aquifer as Comanche and Leon Springs, but from a mixing between Rustler Aquifer waters and local rainfall (Sharp et al. 2003). Other rare species in this system include the federally endangered invertebrates Diamond tryonia *Pseudotryonia adamantina*, Gonzales tryonia *Tryonia circumstriata*, Pecos amphipod *Gammarus pecos*, and Pecos assiminea *Assiminea pecos*, and the federally threatened Pecos sunflower *Helianthus paradoxus*. Some degree of protection is afforded the inhabitants of the ciénega at Diamond-Y Springs in that The Nature Conservancy owns 1,600 ha that encompass it and is committed to its maintenance and perpetuation. Unfortunately, while state water law would allow The Nature Conservancy to pump as much of the water that it "owns" beneath its land as it would like, that law does not allow it to protect this "owned" water. Additionally, the Diamond-Y Springs are adjacent to an active oil and gas extraction field. Working wells are within 100 m of surface water, a natural gas refinery is 30 m upslope from the spring, and old brine pits are just a few meters away.

DEVILS RIVER WATERSHED

The springs of the Devils River and surrounding area are fed by the Edwards–Trinity Plateau Aquifer, which produces the largest number of springs in Texas, with 46 occurring in Val Verde County alone, as well as the third (Goodenough Springs) and fourth (San Felipe Springs) largest springs in the state (Brune 1981). Goodenough Springs, now covered by Amistad Reservoir, still maintain a significant discharge under the lake surface (Ashworth and Stein 2005). *Gambusia amistadensis* was endemic to the headsprings and the 1.3 km spring run downstream to its confluence with the Rio Grande (Peden 1973), but inundation by the reservoir resulted in its extinction.

Devils River is situated in an ecological transition zone at the confluence of three ecoregions (Chihuahuan Desert, Edwards Plateau, and Southern Texas Plains) and as a result supports a high level of biodiversity and endemism. The biodiversity is well represented by the fishes of this region, which include the federally endangered Mexican blindcat *Prietella phreatophila*, the federally threatened Devils River minnow *Dionda diaboli*, and the state threatened proserpine shiner *Cyprinella proserpina*, spotfin gambusia *Gambusia krumholzi*, blotched gambusia, Conchos pupfish, and Rio Grande darter *Etheostoma grahami*, as well as the Manantial roundnose minnow *Dionda argentosa*, Tamaulipas shiner, Rio Grande shiner, West Texas shiner *Notropis megalops*, longlip jumprock *Moxostoma albidum*, headwater catfish, and Rio Grande largemouth bass *Micropterus salmoides nuecensis*.

Although most of the Devils River flows through private property, several conservation areas and initiatives exist within the basin. The Texas Parks and Wildlife Department currently protects 15,000 ha in the Devils River State Natural Area. In addition, The Nature Conservancy owns and manages the 1,900 ha Dolan Falls Preserve, and a total of 63,000 ha of private and public lands are currently under conservation easements (Garrett et al. 2014).

The primary threat to aquatic organisms in the Devils River watershed is groundwater extraction from the Edwards–Trinity Plateau Aquifer and the resultant reduction or loss of spring flows. This threat may be exacerbated by an anticipated increase in hydraulic fracturing in the Wolfcamp Shale formation, which extends into the northern portion of Val Verde County. A recent US Geological Survey estimate (Gaswirth et al. 2016) suggested that it will be the largest continuous oil and gas deposit ever discovered in the United States, containing 20 billion barrels of oil and 450 billion cubic meters of natural gas.

The status of the federally endangered Mexican blindcat is especially relevant to aquifer issues and may be directly impacted by massive groundwater withdrawals. A natural population of Mexican blindcat, previously known only from México, was recently (2016) documented in a cave in Val Verde County, Texas (Texas Natural History Collections catalog #60413, #60414, and #60415). It is likely that the species previously occurred at least throughout the subterranean environments of the northern Coahuila, México, and Val Verde County, Texas, portions of the Edwards–Trinity Aquifer, moving via aquifer interconnections beneath the Rio Grande. These aquatic cave environments are at risk from many factors, including aquifer depletion and groundwater pollution. Agricultural, industrial, and municipal activities in local recharge areas potentially threaten aquifer water quality. Commercial interbasin transfers of groundwater in this area, which have recently been proposed, could have detrimental effects on overall water quantity and could obstruct access to habitats by precluding fish movements through subterranean passages.

HOPE FOR THE FUTURE

Although dire scenarios abound throughout the region, there are glimmers of hope and possibility. Many landowners are coming to understand the stakes involved and the potential losses they could incur from excessive water mining and the associated environmental degradation. Current and planned conservation efforts in Texas focus on addressing landscape and aquatic resource conservation through private landowners. Private landowner involvement is always an important component of natural resource conservation, but in a state like Texas, where more than 95% of the land is privately owned, it is essential for any meaningful progress in managing desert aquatic habitats and associated groundwater systems.

Conservation agencies, organizations, and local stakeholder/landowner groups have prioritized private land conservation as a means to achieve long-term conservation and restoration goals. These types of partnerships effectively promote awareness and stewardship of habitats and organize landowner involvement in local conservation projects. Conservation agencies and organizations such as the Texas Parks and Wildlife Department, the US Fish and Wildlife Service Partners Program, the Natural Resource Conservation Service, The Nature Conservancy, the Desert Fish Habitat Partnership, local groundwater conservation districts, and the Rio Grande Joint Venture work with private landowners to provide conservation best-management practices and support on-the-ground projects to maintain or restore habitats to support functional ecosystems.

The Texas Parks and Wildlife Department, in particular, is working with the University of Texas and the Siglo Group to develop Native Fish Conservation Areas (NFCAs) in Texas, and several are planned for the Chihuahuan Desert region (fig. 8.4). The NFCA approach, which represents an ecologically focused conservation prioritization of watershed segments that serve as native fish "strongholds," will raise awareness of aquifer–spring–ciénega–stream connections and provide for effective implementation of conservation activities. It designates areas that can adequately support maintenance of processes that create habitat complexity, protect all life stages, and support long-term persistence of priority species, and it provides a framework for their sustainable management over time (Williams et al. 2011). These NFCAs can then function as priority areas for conservation investments to promote integrated, holistic conservation strategies that enable the long-term persistence of freshwater biodiversity. Addressing proximate threats to fish species, such as spring flow declines, reduced streamflow, loss of natural flow regimes, siltation/sediment imbalance, reduced water quality, habitat fragmentation, barriers to movement, habitat loss, and nonnative species impacts can also provide favorable outcomes for a broad array of both aquatic and terrestrial species. Concomitant benefits of watershed conservation actions ultimately yield healthier, more resilient ecosystems.

Grassroots stakeholder groups have also formed to address local watershed issues and advocate for private lands and the landscape. Initially, many of these groups were formed when landowners and stakeholders want-

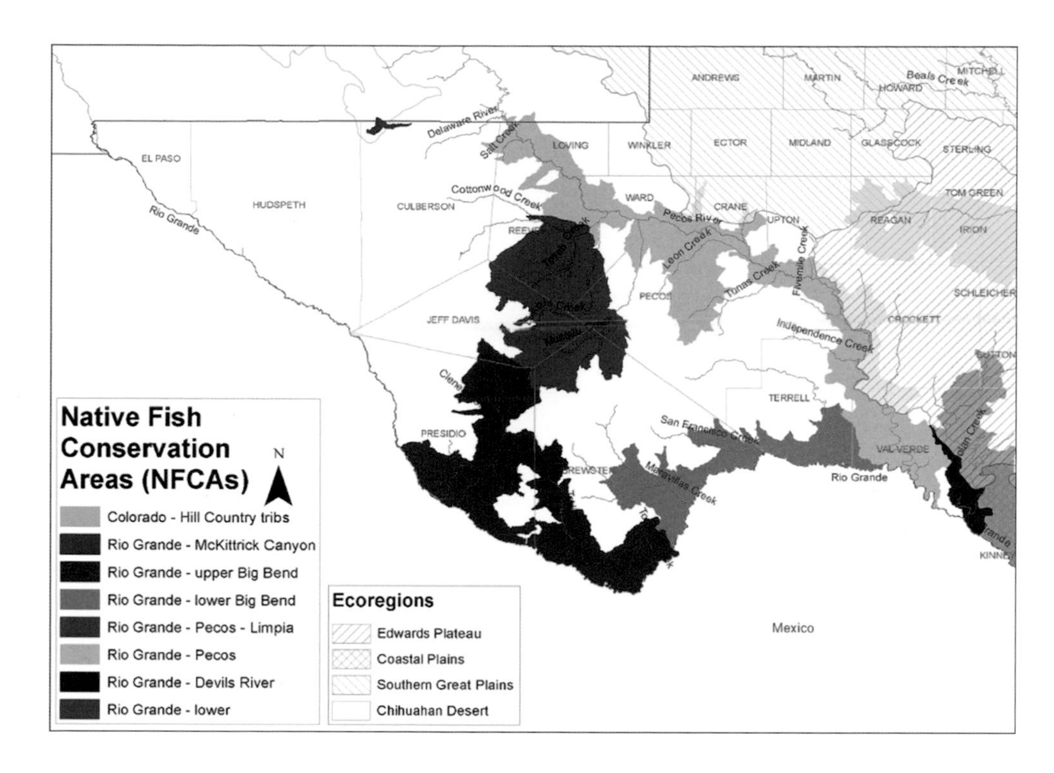

Fig. 8.4 Native Fish Conservation Areas in the Chihuahuan Desert of Texas.

ed to have an organized and more effective advocacy voice about specific watershed issues. Several issues, including groundwater extraction, oil and gas exploration and development, pipeline construction through eminent domain, and increased landscape fragmentation provided the impetus for the formation of groups such as the Devils River Conservancy and Big Bend Conservation Alliance. These groups built networks of passionate landowners and stakeholders, voiced cohesive opposition or support of issues, and now use their network to educate other landowners about broader conservation and stewardship issues and ethics. Expanding from issue-specific missions to broader conservation goals and objectives has allowed these organizations to connect landowners with conservation agencies and organizations, increasing their landscape conservation impact. By connecting landowners (who might otherwise be skeptical) with state and federal conservation agencies, they are increasing opportunities for conservation professionals to provide those landowners with technical guidance and recommendations about land management practices and aquatic species conservation. Continued commitments, collaborative conservation efforts, and watershed management planning by conservation agencies, local alliances, and private landowners will be critical to the long-term success of desert aquatic habitat and species conservation.

REFERENCES

Apache Corporation. 2016. Apache Corporation Discovers Significant New Resource Play in Southern Delaware Basin. http://investor.apachecorp.com/releasedetail.cfm?ReleaseID=988060. Accessed February 23, 2017.

Ashworth, J. B., D. B. Coker, and W. Tschirhart. 1997. *Evaluation of Diminished Spring Flows in the Toyah Creek Valley, Texas*. Texas Water Development Board Open File Report.

Ashworth, J. B., and W. G. Stein. 2005. *Springs of Kinney and Val Verde Counties*. Prepared for Plateau Regional Water Planning Group.

Bennett, J., B. Brauch, and K. Urbanczyk. 2012. Estimating ground water contribution of the Edwards–Trinity Plateau Aquifer to the Big Bend reach of the Rio Grande, Texas. *Geological Society of America Abstracts with Programs* 44: 2.

Bennett, J., M. Briggs, and S. Sandoval-Solís. 2014. Rio Grande–Río Bravo River Corridor. In M. D. Wesson, C. Hallmich, J. Bennett, C. Sifuentes Lugo, A. Garcia, A. M. Roberson, J. Karges, and G. P. Garrett, eds., *Conservation Assessment for the Big Bend–Río Bravo Region: A Binational Collaborative Approach to Conservation*, 21–23. Montreal, QC: Commission for Environmental Cooperation.

Bennett, J., and K. Urbanczyk. 2014. Springs of the Lower Canyons. In M. D. Wesson, C. Hallmich, J. Bennett, C. Sifuentes Lugo, A. Garcia, A. M. Roberson, J. Karges, and G. P. Garrett, eds., *Conservation Assessment for the Big Bend–Río Bravo Region: A Binational Collaborative Approach to Conservation*, 23–24. Montreal, QC: Commission for Environmental Cooperation.

Bestgen, K. R., and S. P. Platania. 1990. Extirpation of *Notropis simus simus* (Cope) and *Notropis orca* Woolman (Pisces: Cyprinidae) from the Rio Grande in New Mexico, with notes

on their life history. *Occasional Papers of the Museum of Southwestern Biology* 6: 1–8.
Brune, G. 1975. *Major and Historical Springs of Texas.* Texas Water Development Board Report 189: 1–94.
Brune, G. 1981. *Springs of Texas.* Vol. I. Fort Worth: Branch-Smith.
Bumgarner, J. R., G. P. Stanton, A. P. Teeple, J. V. Thomas, N. A. Houston, J. D. Payne, and M. Musgrove. 2012. *A Conceptual Model of the Hydrogeologic Framework, Geochemistry, and Groundwater-Flow System of the Edwards-Trinity and Related Aquifers in the Pecos County Region, Texas.* US Geological Survey Scientific Investigations Report 2012-5124.
Dean, D. J., and J. C. Schmidt. 2011. The role of feedback mechanisms in historic channel changes of the lower Rio Grande in the Big Bend region. *Geomorphology* 126: 333–49.
Donnelly, A. C. A. 2007. *Groundwater Availability Run.* Texas Water Development Board Report no. 06-16.
Edwards, R. J., and G. P. Garrett. 2013. *Biological Monitoring of the Repatriation Efforts for the Endangered Rio Grande Silvery Minnow* (Hybognathus amarus) *in Texas.* Final Report, Texas State Wildlife Grants, TPWD contract #186743 (E-91-1).
Edwards, R. J., G. P. Garrett, and E. Marsh-Matthews. 2002. Conservation and status of the fish communities inhabiting the Río Conchos basin and middle Rio Grande, México and U.S.A. *Reviews in Fish Biology and Fisheries* 12: 119–32.
Gallegos, T. J., B. A. Varela, S. S. Haines, and M. A. Engle. 2015. Hydraulic fracturing water use variability in the United States and potential environmental implications. *Water Resources Research* 51: 5839–45.
Garrett, G. P. 2003. Innovative approaches to recover endangered species. In G. P. Garrett and N. L. Allan, eds., *Aquatic Fauna of the Northern Chihuahuan Desert,* 151–60. Museum of Texas Tech University, Special Publication no. 46.
Garrett, G. P., and R. J. Edwards. 2001. Regional ecology and environmental issues. In R. E. Mace, W. F. Mullican III, and E. S. Angle, eds., *Aquifers of West Texas,* 56–65. Texas Water Development Board Report no. 356.
Garrett, G. P., and R. J. Edwards. 2014. Changes in fish populations in the Lower Canyons of the Rio Grande. In C. A. Hoyt and J. Karges, eds., *Proceedings of the Sixth Symposium on the Natural Resources of the Chihuahuan Desert Region,* 396–408. Fort Davis, TX: Chihuahuan Desert Research Institute. http://hdl.handle.net/2152/62996.
Garrett, G. P., J. Karges, and E. Verdecchia. 2014. Devils River. In M. D. Wesson, C. Hallmich, J. Bennett, C. Sifuentes Lugo, A. Garcia, A. M. Roberson, J. Karges, and G. P. Garrett, eds., *Conservation Assessment for the Big Bend–Río Bravo Region: A Binational Collaborative Approach to Conservation,* 29–30. Montreal, QC: Commission for Environmental Cooperation.
Gaswirth, S. B., K. R. Marra, P. G. Lillis, T. J. Mercier, H. M. Leathers-Miller, C. J. Schenk, T. R. Klett, P. A. Le, M. E. Tennyson, S. J. Hawkins, M. E. Brownfield, J. K. Pitman, and T. M. Finn. 2016. *Assessment of Undiscovered Continuous Oil Resources in the Wolfcamp Shale of the Midland Basin, Permian Basin Province, Texas, 2016.* US Geological Survey Fact Sheet 2016-3092.
Hanna, A. H., K. W. Conway, E. W. Carson, G. P. Garrett, and J. R. Gold. 2013. Conservation genetics of an undescribed species of *Dionda* (Teleostei: Cyprinidae) in the Rio Grande drainage in western Texas. *Southwestern Naturalist* 58: 35–40.
Heard, T. C., J. S. Perkin, and T. H. Bonner. 2012. Intra-annual variation in fish communities and habitat associations in a Chihuahuan Desert reach of the Rio Grande/Río Bravo Del Norte. *Western North American Naturalist* 72: 1–15.
Hendrickson, D. A., and A. E. Cohen. 2015. Fishes of Texas Project Database (Version 2.0). https://doi.org/10.17603/C3WC70.
Hendrickson, D. A., and W. L. Minckley. 1984. Ciénegas: Vanishing climax communities of the American Southwest. *Desert Plants* 6: 131–74.
Howells, R. G., and G. P. Garrett. 1995. Freshwater mussel surveys of Rio Grande tributaries in Chihuahua, Mexico. *Triannual Unionid Report* 8: 10.
Hubbs, C. 1980. Solution to the *C. bovinus* problem: Eradication of a pupfish genome. *Proceedings of the Desert Fishes Council* 10: 9–18.
Hubbs, C. 1990. Declining fishes of the Chihuahuan Desert. In A. M. Powell, R. R. Hollander, J. C. Barlow, W. B. McGillivray, and D. J. Schmidly, eds., *Third Symposium on Resources of the Chihuahuan Desert Region, United States and Mexico,* 89–96. Fort Davis, TX: Chihuahuan Desert Research Institute.
Hubbs, C. 2003. Spring-endemic *Gambusia* of the Chihuahuan Desert. In G. P. Garrett, and N. L. Allan, eds., *Aquatic Fauna of the Northern Chihuahuan Desert,* 127–34. Museum of Texas Tech University, Special Publication no. 46.
Hubbs, C., and H. J. Brodrick. 1963. Current abundance of *Gambusia gaigei,* an endangered fish species. *Southwestern Naturalist* 8: 46–48.
Hubbs, C., R. J. Edwards, and G. P. Garrett. 2008. An annotated checklist of the freshwater fishes of Texas, with keys to identification of species. *Texas Journal of Science.* http://www.texasacademyofscience.org/.
Hubbs, C., and V. G. Springer. 1957. A revision of the *Gambusia nobilis* species group, with descriptions of three new species, and notes on their variation, ecology and evolution. *Texas Journal of Science* 9: 279–327.
Ingol-Blanco, E. 2011. Modeling climate change impacts on hydrology and water resources: Case study Río Conchos Basin. PhD diss., University of Texas at Austin.
Karges, J. 2014. Balmorhea Springs Complex. In M. D. Wesson, C. Hallmich, J. Bennett, C. Sifuentes Lugo, A. Garcia, A. M. Roberson, J. Karges, and G. P. Garrett, eds., *Conservation Assessment for the Big Bend–Río Bravo Region: A Binational Collaborative Approach to Conservation,* 34–35. Montreal, QC: Commission for Environmental Cooperation.
Lewis, R. H., N. L. Allan, S. B. Stoops, G. P. Garrett, C. W. Kroll, J. West, and R. Deaton. 2013. Status of the endangered Pecos Gambusia (*Gambusia nobilis*) and Comanche Springs Pupfish (*Cyprinodon elegans*) in Phantom Lake Spring, Texas. *Southwestern Naturalist* 58: 234–38.
Mace, R. E. 2001. Aquifers of West Texas: An overview. In R. E. Mace, W. F. Mullican III, and E. S. Angle, eds., *Aquifers of West Texas,* 1–16. Texas Water Development Board Report no. 356.
Miller, R. R. 1961. Man and the changing fish fauna of the American Southwest. *Papers of the Michigan Academy of Science, Arts, and Letters* 46: 365–404.

Miller, R. R. 1977. Composition and derivation of the native fish fauna of the Chihuahuan Desert region. In R. H. Wauer and D. H. Riskind, eds., *Transactions of the Symposium on the Biological Resources of the Chihuahuan Desert Region, United States and Mexico*, 365–82. US National Park Service Transactions and Proceedings, no. 3.

Miller, R. R., J. D. Williams, and J. E. Williams. 1989. Extinctions of North American fishes during the past century. *Fisheries* 14: 22–38.

Minckley, W. L., G. K. Meffe, and D. L. Soltz. 1991. Conservation and management of short-lived fishes: The cyprinodontoids. In W. L. Minckley and J. E. Deacon, eds., *Battle against Extinction: Native Fish Management in the American West*, 247–82. Tucson: University of Arizona Press.

Myers, T. 2016. *A Preliminary Analysis of the Risks to the Balmorhea Springs Complex Posed by Unconventional Oil & Gas Development*. Final report prepared for Earthworks. www.earthworksaction.org/files/publications/BalmorheaSpringsAssessmentFINAL.pdf.

Peden, A. E. 1973. Virtual extinction of *Gambusia amistadensis* n. sp., a poeciliid fish from Texas. *Copeia* 1973: 210–21.

Potter, H. G. 2004. History and evolution of the Rule of Capture. In W. F. Mullican III and S. Schwartz, eds., *100 Years of Rule of Capture: From East to Groundwater Management*, 1–10. Texas Water Development Board Report no. 361.

Rhodes, K., and C. Hubbs. 1992. Recovery of Pecos River fishes from a red tide fish kill. *Southwestern Naturalist* 37: 178–87.

Sandoval-Solís, S., B. Reith, and D. C. McKinney. 2010. *Hydrologic Analysis Before and After Reservoir Alteration at the Big Bend Reach, Rio Grande/Río Bravo*. Center for Research in Water Resources Online Report 10-06, University of Texas at Austin. www.crwr.utexas.edu/reports/2010/rpt10-6.shtml.

Schmidt, J. C., B. L. Everitt, and G. A. Richard. 2003. Hydrology and geomorphology of the Rio Grande and implications for river rehabilitation. In G. P. Garrett and N. L. Allan, eds., *Aquatic Fauna of the Northern Chihuahuan Desert*, 25–46. Museum of Texas Tech University, Special Publication no. 46.

Schmitt, C. J., J. E. Hinck, V. S. Blazer, N. D. Denslow, G. M. Dethloff, T. M. Bartish, J. J. Coyle, and D. E. Tillitt. 2005. Environmental contaminants and biomarker responses in fish from the Rio Grande and its U.S. tributaries: Spatial and temporal trends. *Science of the Total Environment* 350: 161–93.

Scudday, J. F. 2003. My favorite old fishing holes in West Texas: Where did they go? In G. P. Garrett and N. L. Allan, eds., *Aquatic Fauna of the Northern Chihuahuan Desert*, 135–40. Museum of Texas Tech University, Special Publication no.46.

Sharp, J. M., Jr., R. Boghici, and M. Uliana. 2003. Groundwater systems feeding the springs of West Texas. In G. P. Garrett and N. L. Allan, eds., *Aquatic Fauna of the Northern Chihuahuan Desert*, 1–11. Museum of Texas Tech University, Special Publication no. 46.

Texas Water Development Board. 2005. *Diminished Spring Flows in the San Solomon Springs System, Trans-Pecos, Texas*. Report to the Texas Parks and Wildlife Department. Section 6, Endangered Species Grant Number WER69, Study Number 84312.

Uliana, M. M., J. L. Banner, and J. M. Sharp, Jr. 2007. Regional groundwater flow paths in Trans-Pecos, Texas, inferred from oxygen, hydrogen, and strontium isotopes. *Journal of Hydrology* 334: 334–46.

Upper Rio Grande Basin and Bay Expert Science Team. 2012. *Environmental Flows Recommendations Report*. Final Submission to the Environmental Flows Advisory Group, Rio Grande Basin and Bay Area Stakeholders Committee and Texas Commission on Environmental Quality.

USFWS (US Fish and Wildlife Service). 2010. *Rio Grande Silvery Minnow Recovery Plan* (Hybognathus amarus), First Revision. Albuquerque: US Fish and Wildlife Service.

USFWS (US Fish and Wildlife Service). 2012. *Big Bend Gambusia,* Gambusia gaigei, *5-Year Review, Summary and Evaluation*. Austin, TX: US Fish and Wildlife Service.

White, W., H. Gale, and S. Nye. 1941. *Geology and Ground-Water Resources of the Balmorhea Area, Western Texas*. Geological Survey Water-Supply Paper 849-C. US Department of the Interior.

Williams, J. E., R. N. Williams, R. F. Thurow, L. Elwell, D. P. Philipp, F. A. Harris, J. L. Kershner, P. J. Martinez, D. Miller, G. H. Reeves, C. A. Frissell, and J. R. Sedell. 2011. Native Fish Conservation Areas: A vision for large-scale conservation of native fish communities. *Fisheries* 36: 267–77.

9 Southwestern Fish and Aquatic Systems: The Climate Challenge

Jonathan T. Overpeck and Scott A. Bonar

The southwestern United States is home to a diverse assemblage of native fishes, many found nowhere else. However, their continued existence is at increasing peril, as indicated by the high percentages of native fish species currently listed as federally threatened or endangered in states across the region (e.g., Arizona 56%, New Mexico 20%, and Nevada 49%; https://www.azgfd.com/wildlife/nativefish/; http://heritage.nv.gov/node/227; https://www.fws.gov/endangered/species/index.html). The Southwest is also a region characterized by large environmental, ecological, and topographic gradients, extending from the lowest and driest deserts of the United States to some of its highest mountains. Moreover, it is a region that is already being affected by climate change, particularly in the form of a strong regional manifestation of global warming, with a regional pervasiveness and rates of change that are outstripping most other parts of the continent. If left unchecked, climate change could far exceed the rates and magnitudes of change that have characterized the region over the last millions of years. The future impacts of continued climate change in the Southwest will be dramatic, and many water resources and fish populations will probably be at grave risk under unmitigated climate change.

Our goal is to provide an overview of the climate challenge facing the Southwest. This challenge, which has already begun, features the interaction of natural climate variability with human-driven change. Although our primary focus is on the region centered on Arizona, New Mexico, Colorado, Utah, Nevada, Southern California, and western Texas, much of what we cover is relevant to the broader semiarid areas of the American West and northern México. We examine the types of impacts the future climate will have on fish populations and their environments and end with solution options that emphasize the need for a multipronged approach. Climate adaptation strategies in the form of improved management and conservation are essential, just as are broader societal interventions to reduce the causes of climate change (i.e., emissions of greenhouse gases).

CLIMATE AND THE SOUTHWEST

Climate change is not new to the North American Southwest, and the rich record of past change, from meteorological instruments and paleoclimate proxy sources such as tree rings, cave formations, and lake sediments, is typically understood by climate scientists to represent the natural range of climate variability (Garfin et al. 2013). The largest changes in recent Earth history are the natural astronomically driven swings between glacial and interglacial periods, whereas smaller-scale natural processes such as the El Niño–Southern Oscillation, droughts, and summer monsoon variability have meant a steady alternation of wet and dry periods ranging from years to decades (Garfin et al. 2013; Routson et al. 2016). The fishes of the Southwest, and the environments upon which they depend, evolved in this envelope of natural variability (Settele et al. 2014).

As important and palpable as the influences of the natural climate system are for the Southwest, they are now being overwhelmed by the impacts of human-caused global climate change (Stocker et al. 2013; Garfin et al. 2013). Like most of the planet, the Southwest has been warming steadily over the last hundred years, with short periods of warm and cold variation superimposed on the now clear trend of ever-increasing temperatures. Global warming since 1900 is now in excess of 1°C (fig. 9.1), and warming in the Southwest and in the headwaters of the Colorado River has been comparable (Udall and Overpeck 2017). Much evidence now indicates that the ongoing period of global and regional warming cannot be ascribed to natural causes alone. This evidence includes the following observations: (1) Warming has been strongly attributed or "fingerprinted" primarily to the increasing concentrations of greenhouse gases in the atmosphere, and especially increases in carbon dioxide produced by burning fossil fuels. (2) Study of natural cycles and change suggests that the Northern Hemisphere would still be slowly cooling in the absence of the greenhouse gas forcing. (3) No natural cycle or mode of variation has been found that can explain the recent warming. (4) Changes in solar irradiation, which can be measured precisely from space, are unable to explain the warming. (5) The pattern of temperature change, which shows the lower atmosphere (troposphere) warming and the upper atmosphere (stratosphere) cooling, provides more evidence against possible solar warming of climate. (6) Basic chemistry and physics dictate that increases in greenhouse gas concentrations in the atmosphere, even in trace amounts, must cause the Earth's climate system to trap more heat (Stocker et al. 2013).

What can be expected for the future? Climate scientists have strong confidence that climate warming will

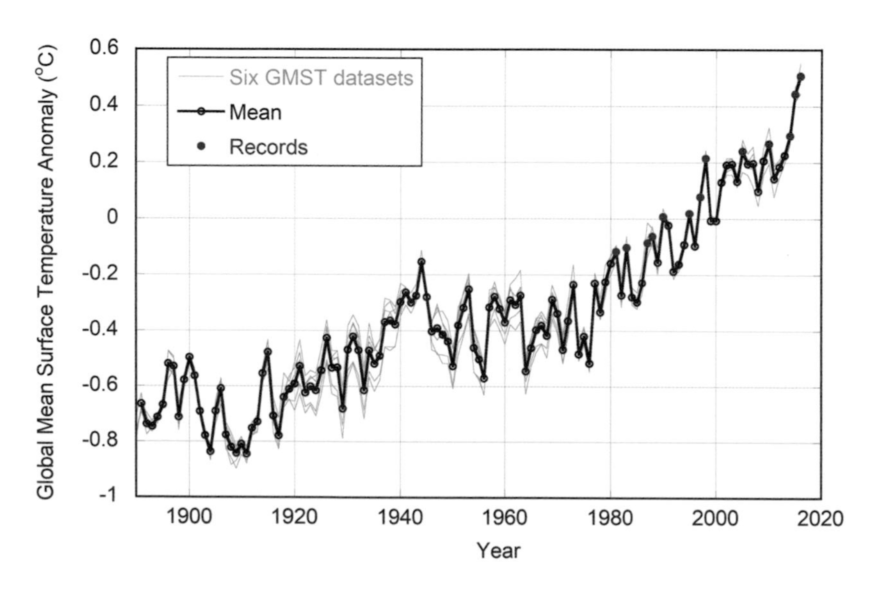

Fig. 9.1 Mean and range of six global mean surface temperature records, showing the steady rise in global temperatures over the last ca. 120 years (anomalies relative to the 1981–2010 mean). The red dots show the steady progression of record-breaking warm years since 1980, as well as the unprecedented 0.24°C jump in global surface temperatures during 2014–2016. Decade-scale periods of slower and faster warming are primarily the result of climate variability interacting with the more dominant human-forced warming trend. (From Yin et al. 2018.)

continue as long as carbon dioxide and other greenhouse gases are added to the atmosphere (Stocker et al. 2013). If left unchecked, the rate and magnitude of warming will probably exceed anything seen in millions of years (Masson-Delmotte et al. 2013); that is, in the absence of curbing nearly all greenhouse gas emissions, both the rate and magnitude of future warming (>5°C warmer than current temperatures in 100 years) will probably exceed anything seen since the fish now present in the Southwest first evolved.

Although the rate of global atmospheric warming briefly slowed early in the twenty-first century, as it has done multiple times as anthropogenic warming interacts with natural climate variability, there is no sign that the overall postindustrial rate of global warming has slowed. Years since 2000 hold most of the top-ten records for warmest year to date, and 2014, 2015, and 2016 have shattered the global record, combining for an unprecedented 0.24°C rise in global temperature in just three years (Yin et al. 2018).

With current technology, anthropogenic global warming and associated climate change are largely irreversible and can only be plateaued at a specific level of change. Over 90% of the heat trapped by the human-driven rise in atmospheric greenhouse gas concentrations is taken up by the oceans, and the record of ocean warming shows no sign of slowing over the last 40 years (Stocker et al. 2013). Even if human emissions of greenhouse gases were to slow and then cease, much of the carbon dioxide already added to the atmosphere would remain there for decades, centuries, or even millennia, continuing to warm the planet. Furthermore, even if greenhouse gas concentrations in the atmosphere were slowly to return to levels closer to preindustrial conditions, the ocean would compensate by releasing stored anthropogenic heat, keeping the planet largely in its new human-warmed climate state for at least a millennium (Solomon et al. 2009).

The largest uncertainty in how the Southwest's climate will change in the future stems not from climate model uncertainty (box 9.1), but from an inability to predict how human behavior will change, including how emissions of greenhouse gases will be slowed. Because of this uncertainty, climate scientists assess how the Earth's climate will change as a function of different greenhouse gas emission scenarios. If business-as-usual emissions prevail, southwestern climate changes will clearly be unprecedented in the evolutionary history of fishes and other wildlife in the region. In contrast, if global emission reduction targets, such as that agreed to in Paris by almost all nations in 2015, are met, warming of the planet, and the Southwest, may be limited to just about double what has been seen so far. However, even this "best-case" scenario means a new, hotter southwestern climate that could be unprecedented in many locations and would be irreversible for centuries to come. Moreover, this best-case scenario requires that the nations of the world meet their agreed-to greenhouse gas reduction commitments in an unwavering manner. If a major carbon emitter like the United States fails to meet its agreed-to goals, then the Southwest will be even hotter for centuries to come. The costs of inaction will be high, and not limited to unprecedented warmth alone.

Temperature will not be the only factor involved in a changing climate. Climate science indicates that many other changes will occur, scaling in magnitude and rate of change with the rate and magnitude of mean annual temperature change (Stocker et al. 2013). These changes will include mean annual and cool-season precipitation declines in the southern reaches of the Southwest, declining snowpack amounts and durations, increasing heat wave severity and duration, droughts that are much hotter and more dominated by temperature than in the past, and declining streamflows (Overpeck 2013; Overpeck et al. 2013; Cook et al. 2015; Lehner et al. 2017; Udall and Overpeck 2017). These trends are ongoing and are all projected to continue into the future. The greatest uncertainty is how large the changes will be, which is primarily a function of future greenhouse gas emissions. The more emissions, the greater the warming, and the greater the magnitude of all of the above trends.

Less confidence exists as to the degree to which a number of other climate changes may occur. For example, the current consensus among climate scientists is that the amount of summer monsoon precipitation will not change, but that it might shift later into the summer and fall (Cook and Seager 2013). Thus, the hottest, driest time of the year that occurs before the monsoon season will continue to become hotter, drier, and longer (Weiss et al. 2009). Another change that has been predicted, as well as observed in many regions globally (Stocker et al. 2013), is precipitation becoming more episodic and intense. Although models indicate that the Southwest's cool-season precipitation should intensify

[Box 9.1] Why Are Climate Models Needed and Why Should We Trust Them?

There are many reasons why computer models have been a foundation of science for years, and the case for using them to model the Earth's climate is especially strong. There is no precedent from Earth history for the future we are facing. Although there have been multiple periods in geologic history when atmospheric carbon dioxide levels were higher than they are today (and in each case, these periods were warmer than today; Masson-Delmotte et al. 2013), other critical features of the planet, including the locations of oceans, continents, ice sheets, and mountains, were not the same as today. For this reason, we need to rely on global climate models to assess how climate will change in the future, as well as to understand many details of that future that are not available from study of past climates (Masson-Delmotte 2013; Stocker et al. 2013). Just as models that allow airplanes to be flown and landed in almost any conditions on autopilot must be carefully constructed and tested, global climate models have been tested and improved a great deal since the 1970s, when they first started to see serious use for understanding nuclear winter and climate change. With each passing decade, the models have had to meet ever more rigorous tests, including tests of their ability to simulate realistic average past and present climates. Climate models are similar to weather models, but instead of being used to simulate weather on a particular day of the year, climate models are designed to simulate average weather conditions (i.e., climate) over longer periods of time well into the future.

We know that climate models have utility because they undergo an ever-growing, extensive, and sophisticated range of tests. To be useful, a climate model needs to prove that it can simulate many aspects of the modern climate, including natural variability in the climate system as well as the evolution of climate over the last 150 years or longer. A famous test occurred after Mount Pinatubo erupted late in 1991, in the largest volcanic eruption of the twentieth century (Hansen et al. 1992). Scientists at NASA used their then state-of-the-art climate model to project how the planet's climate would respond to the huge eruption over the next several years. The climate model indicated that the climate would first cool rapidly due to the volcanic solid and liquid particles (aerosols) blown into the stratosphere, then gradually warm again as the volcanic aerosols fell out of the stratosphere, with the ocean acting as a flywheel to spread the response over several years. Even though the NASA model was crude compared with those available today, it did a remarkable job of getting the projection right. Since then, volcanic eruptions have become one of the many ways climate models are tested.

Climate models are also routinely tested against paleoclimate data describing climates of the last 1,000 years, the last 21,000 years (since the peak of the last Ice Age, when the planet was colder than today by about the same amount it could warm in the next 100 years if greenhouse gas emissions are not curbed), and a host of other ancient periods when the climate of the Earth was dramatically different, and often warmer, than today. Climate models are able to simulate many aspects of these ancient climates with considerable accuracy, giving us additional confidence that we can trust the same models to simulate many elements of the future climate (Masson-Delmotte 2013). The level of agreement among the approximately 40 state-of-the-art global climate and "Earth System" models developed around the world also helps indicate the amount of confidence we can place in specific projections. Finally, given that significant climate change has already occurred, comparison of each model's projections with the Earth's observed climate change provides a further test to show which projections can be accepted with confidence and how models can be improved. As models continue to be developed and tested, climate scientists have greater confidence in some climate projections (e.g., warming and the influence of this warming) and less in others (e.g., the regional specifics of how precipitation amounts will change). However, climate models continue to demonstrate that they are valuable tools for understanding the changes that we can expect in the future.

(Dominguez et al. 2012), it is still unclear if this is occurring (Gershunov et al. 2013). However, if this signal does emerge from the noise of natural variability, larger and more frequent flood events can be expected. This observed change in the global precipitation regime is a direct result of the atmosphere's ability to hold more water as it warms; more water is thus available to precipitate when conditions are right. This is also the reason scientists are confident that evaporation and evapotranspiration demand will increase as the climate warms (Vano et al. 2014).

Many headwaters of southwestern rivers (e.g., the Colorado River and Rio Grande) are in mountainous areas in the northern reaches of the Southwest. Future precipitation patterns in these regions, particularly as related to snowpack, are uncertain. Some climate models project significant increases in precipitation in the Rocky Mountain headwaters with continued warming, while others project decreases; to date, empirical evidence (gage data) suggests that no or little change has occurred, which matches projections of another subset of global models (Udall and Overpeck 2017). This means that rivers of the Southwest are likely to become increasingly dominated by temperature-driven reductions in flows, despite periods of above-normal snowpack that occasionally compensate for temperature-driven reductions. Warming drives up evaporation, evapotranspiration, sublimation of snow, and water use by plants as the growing season lengthens and tree lines advance upward, all at the expense of river flow (Goulden and Bales 2014; Udall and Overpeck 2017). Thus, rivers of the region are getting warmer and flowing less, and this trend could lead to drastically less water for fish (fig. 9.2), just as warming and drying will affect springs and lakes.

The previous effects discussed, however, do not describe the full extent of the climate challenge. The Southwest is one of the most drought-prone regions

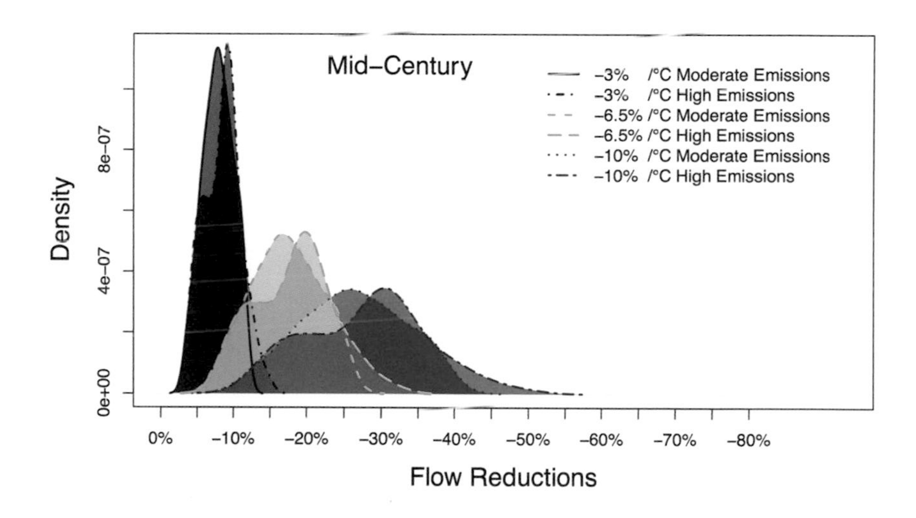

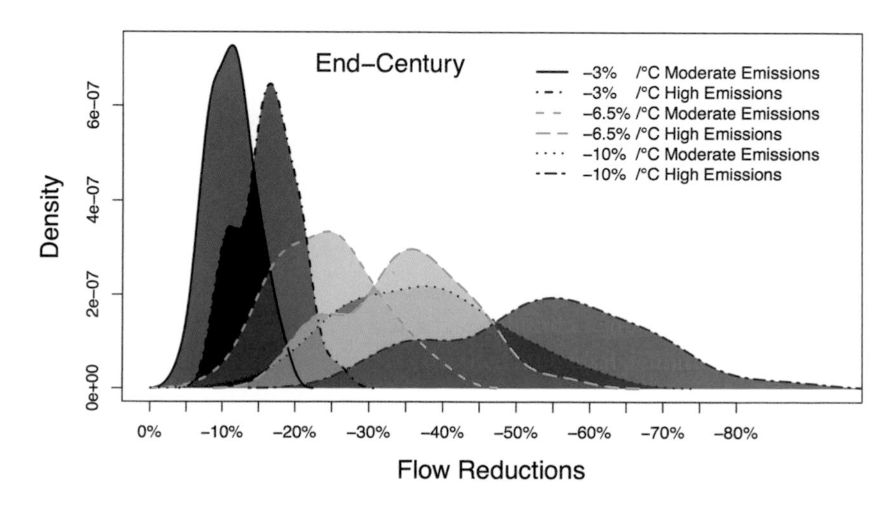

Fig. 9.2 Projections of flow (relative to average flow in the twentieth century) of the upper Colorado River basin under a warming climate for the mid-twenty-first century (*top*) and the end of the century (*bottom*). The six distributions in each plot are probability density functions constructed using the hundreds of downscaled World Climate Research Programme Coupled Model Intercomparison Project (CMIP) phase 3 and phase 5 climate model temperature projections for the upper basin stored by the US Bureau of Reclamation, for two different greenhouse gas emission scenarios (moderate and high) and three different flow sensitivities to warming (3%, 6.5%, and 10%). Like a bell-shaped "normal" curve, each probability density function gives the relative likelihood of future flow reductions of a certain magnitude; the peak in each curve shows the most likely flow reduction and the tails give the range of possible flow reductions. Greenhouse gas emission reductions will be felt more by the end of the century than by midcentury, and the fact that temperature-driven flow reductions have already probably reached 10% suggests that the red curves (high flow sensitivity to warming) are the most realistic. Flow reductions caused by drought would be in addition to those shown here. (From Udall and Overpeck 2017.)

on the planet. The annually dated record from tree rings indicates that multi-decadal megadroughts have occurred regularly in the Southwest, and that droughts up to 50 years long with only one year of above-normal precipitation have taken place in the last 2,000 years (Woodhouse and Overpeck 1998; Routson et al. 2011). Warming only exacerbates already serious megadrought risks, and it is increasingly clear that the biggest river in the region, the Colorado, could see flow reductions during future hot megadroughts of 50%–75% unless human-caused climate change is stopped (Vano et al. 2014; Udall and Overpeck 2017). Smaller water bodies could fare much worse.

Groundwater is a critical driver of surface-water supplies, and climate change is likely to continue lowering groundwater levels where there is ongoing recharge—hotter, drier climates mean less water getting into the ground. However, overuse may be a bigger threat to groundwater than climate change in many parts of the Southwest (Famiglietti et al. 2011; Scanlon et al. 2015). The same climate science that suggests there will be decades-long periods of drought suggests that there will be comparable periods of above-normal precipitation superimposed on the warming and drying trend.

The health and survivability of aquatic habitats are linked to riparian and upland terrestrial vegetation, and there are clear signs that climate changes in the Southwest—most notably warming and hot drought—are also causing profound changes to this vegetation (Overpeck 2013). These changes include widespread forest and woodland tree mortality (Adams et al. 2017) and a large increase in severe wildfires (Kitzberger et al. 2017). Most recently, soil crusts of the region have been found to be at risk (Rutherford et al. 2017). All of this evidence suggests more large episodes of erosion and sediment loading of aquatic systems as climate change continues.

What confidence can be ascribed to these predictions? The agreement between observed and predicted change provides high levels of confidence in some aspects of the Southwest's future climate challenge. We know with confidence that both mean and extreme temperatures will continue to rise, lending confidence to other aspects of the Southwest's climate future. Other changes, most notably a possible change in cool-season precipitation in southwestern stream headwaters and changes in the critical summer monsoon, are more uncertain. Furthermore, as the Earth's past climate record has shown, unexpected changes can occur at unpredictable accelerated and abrupt rates (Alley et al. 2003; Overpeck and Cole 2006). For example, the current meltdown of the Arctic is proceeding at a much faster rate than was projected two decades ago (National Research Council 2013), and some hydrologic systems appear to have gone through abrupt shifts in their mean state, suddenly becoming much drier or wetter (Alley et al. 2003; Overpeck and Cole 2006). These examples suggest that the changing climate of the Southwest may have significant surprises in store for local climate, hydrology, fish, wildlife, and humans, and this fact needs to be taken into consideration by managers (National Research Council 2013; Overpeck 2014).

The impacts of climate change in the Southwest are combining with increased human use of both groundwater resources and surface flows to put southwestern aquatic systems at even greater risk. Under any scenario, human activities are the primary forces behind changes in the quality of the Southwest's aquatic habitat. The opportunity to curb these forces (e.g., unsustainable groundwater pumping and surface-water diversions as well as greenhouse gases emissions) to conserve fish populations and other wildlife exists, but it is uncertain that humans will assume the responsibility.

POTENTIAL IMPACTS OF CLIMATE CHANGE ON SOUTHWESTERN FISHES AND THEIR ENVIRONMENTS

A climate that is changing as fast as projected will dramatically impact aquatic ecosystems in North American deserts, putting numerous species at increased risk of extirpation or extinction. Here we focus on effects of climate change on desert fish communities; however, climate change is also affecting most stream and riparian plants and animals, both terrestrial and aquatic. These effects are wide-ranging because the importance of riparian environments in arid lands is considerably disproportionate to their area. For example, in Arizona, only 0.4% of the state's total area is riparian (Zaimes 2007), yet 80% of Arizona vertebrate species spend some portion of their life cycle in riparian zones (Zaimes 2007, calculated from data in Hubbard 1977). The magnitude of the importance of riparian areas is similar in other arid regions (Brinson et al. 1981; Free et al. 2015).

Major effects of climate change on aquatic organisms living in desert waters can result from increased water temperatures, diminished water volume, the changing nature of extremes in temperature and precipitation, the seasonality of the changes, and the influences on aquatic systems due to climate impacts on terrestrial vegetation. Climate impacts on terrestrial vegetation through drought and warming can also drive extreme wildfires, violent flooding, and substantial erosion. These factors affect fishes in many ways (Paukert et al. 2016), and the degree of effect depends on the species involved, the type of water body it lives in, and the area in which it is located.

Nonnative fishes are common in the arid West (Allard 1978; Fuller et al. 1999). In response to demands from European settlers, the US Fish Commission conducted massive fish transplantation programs by rail and road during the late 1800s and early 1900s (Leonard 1979). Currently, one in every four fishes in western streams is nonnative (Schade and Bonar 2005). Five western states, mostly those of the arid Intermountain West, have nonnative fishes in over 60% of their total stream kilometers. Climate change will directly affect economically important recreational fishes (many of which are nonnative), other nonnative fishes, and interactions among native and nonnative fishes.

As the climate warms, so will surface waters (Mohseni and Stefan 1999). Surface waters experience day and night temperature fluctuations similar to air temperature fluctuations. The speed of change in surface-water temperature is affected by the volume of water, and smaller water bodies will change temperature more rapidly than larger ones. Groundwater influences this dynamic (Caissie and Luce 2017). Groundwater temperature is related to the surrounding air temperature, but less so than that of surface water. At depths up to about 25 m, groundwater temperature tracks mean annual air temperature (Heath 1983). Below 25 m, groundwater temperature is affected by both surface air temperature and geothermal influences. Climate change will affect organisms living both in surface waters and in caves and springs fed by groundwater sources.

Warmer water temperatures will stress fishes and other aquatic organisms that evolved under cooler climates. Ectotherms such as fish have optimal temperatures at which they carry out their life's activities, and temperatures that deviate from the optimum can affect various physiological processes (Whitney et al. 2016). Higher temperatures can affect a fish's neuroendocrine processes (Chadwick et al. 2015), metabolic processes (Fry 1947), immune defenses against parasites and pathogens (Bowden 2008), and various stages of reproduction (Whitney et al. 2016). Optimum temperatures for fishes vary depending on their activities; for example, temperatures optimal for adult feeding and growth can be different from those required for reproduction.

Many desert fishes are not tolerant of high temperatures (Carveth et al. 2006). The Cottonball Marsh pupfish *Cyprinodon salinus milleri* of Death Valley, which is tolerant of extreme thermal conditions, is an exception. This fish lives in a strange landscape consisting of salt-rimmed spring-fed streams on the floor of Death Valley that can reach temperatures over 38°C. Carveth et al. (2006) tested the relative tolerances of 18 species of native and nonnative fishes found in Arizona and obtained surprising results (fig. 9.3). They found that some small-bodied fishes, such as Gila topminnow *Poeciliopsis occidentalis* and other pupfishes, are resistant to high temperatures, having evolved in some of warmest water temperatures of North American deserts. Conversely, most other Arizona desert fish species are not nearly as tolerant. Roundtail chub *Gila robusta*, spikedace *Meda fulgida*, desert sucker *Catostomus clarkii*, loach minnow *Rhinichthys cobitis*, and razorback sucker *Xyrauchen texanus* have intermediate tolerances, similar to those of temperate zone species from the eastern United States. Some arid-land salmonids, such as Apache trout *Oncorhynchus apache* (Lee and Rinne 1980; Recsetar et al. 2014), Rio Grande cutthroat trout *Oncorhynchus clarkii virginalis* (Recsetar et al. 2012; Zeigler et al. 2013), and Gila trout *Oncorhynchus gilae* (Lee and Rinne 1980), are quite sensitive to high temperatures. Their tolerances are more similar to those of salmonid species living in the Pacific Northwest than to those of desert species living below the Colorado Plateau only 150 km away.

The relative thermal tolerances of fishes and other aquatic species are important in determining which species will be most successful when co-occupied streams warm in a changing climate. In Arizona's high-elevation West Fork Black River, Apache trout inhabit upper, forested cool canyons, where waters are well below their chronic lethal thermal tolerances. However, nonnative virile crayfish *Orconectes virilis* inhabit lower meadow sections of the stream in sunny areas where

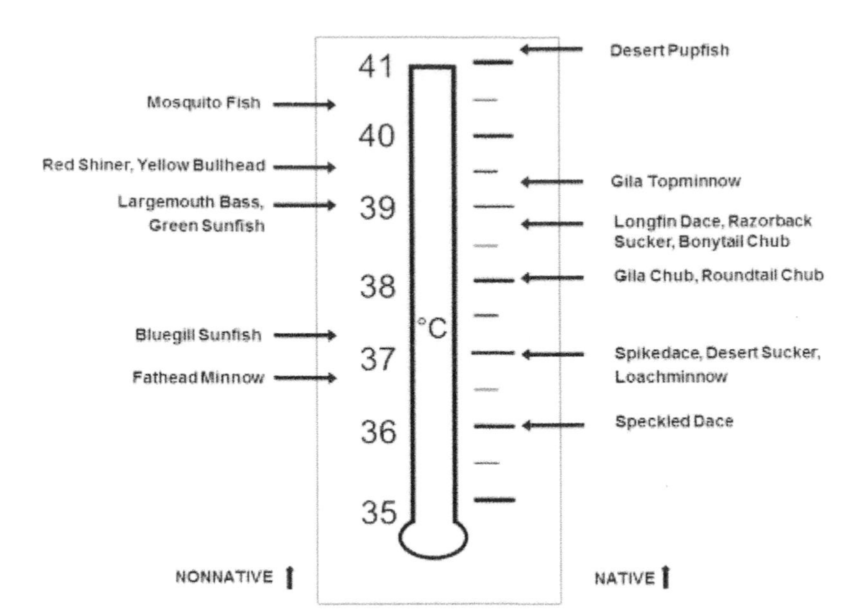

Fig. 9.3 Lethal thermal maxima of common Arizona fishes acclimated at 25°C. (From Carveth et al. 2006).

water temperatures are warmer—above the tolerances of the Apache trout (Petre and Bonar 2017).

Water temperatures and other natural cues, such as flow changes and photoperiod, govern fish spawning events (Falke et al. 2010). Alteration of natural conditions, including the onset times of critical temperatures or the lengths of temperature regimes, by climate change may affect fish spawning in terms of timing and in other ways we do not yet understand. For example, because spawning of predator and prey can be linked, alteration of temperature may affect predator-prey relationships (Davies et al. 1982).

Climate change will result in higher temperatures, but its effect on the Southwest's precipitation is less clear, particularly with respect to changes in its means, extremes, and seasonality. Nevertheless, higher air and water temperatures will increase evaporation, and this, along with multiple other factors, including the projected larger human population in the Southwest, will reduce available groundwater and surface water. The resulting decrease in streamflow will shrink the habitat available for fishes, fragment habitats so fish cannot move from one portion of a water body to another (Fagan et al. 2005), and affect water quality (Whitney et al. 2016).

The typical flow pattern of an unregulated southwestern desert stream is illustrated by Aravaipa Creek (fig. 9.4). In winter and early spring, small flood events, steady rains, and snowmelt (depending on year) fill streams and clean fine sediment from the substrate (Propst and Gido 2004; Yarnell et al. 2015). In late spring, many native fishes spawn on the descending limb of the hydrograph. In early to mid-summer (June and July), streams experience their summer base flows. Summer base flow is often limiting for aquatic organisms because the wetted area of streams is lowest, and water temperatures are warmest, at that time. During late summer, monsoonal rainstorms occur, and flash floods often result. These floods create new channels, provide nutrients, and displace nonnative fishes, preferentially those nonnative species that have not evolved in these flashy stream systems (Minckley and Meffe 1987). These floods stop during the fall, which is a relatively calm time before the gentler rain and snow events start again in the late fall.

Diversion of surface water and groundwater extraction for cities, agriculture, and other uses has dramatically reduced surface flows throughout the American Southwest. Streams that formerly flowed, such as the lower Salt, lower Gila, Santa Cruz, and San Pedro, now flow just a fraction of their former lengths, if at all. Reductions in summer base flow affect the amount of habitat available to different species to grow, feed, and escape predators. Different fish species favor different types of mesohabitat. Reduction in streamflow first dries riffles and leaves pool habitat. Many nonnative fishes commonly found in southwestern streams evolved in slow-moving large river systems of the East, where pool habitat was more common. Thus, low flows and diminished glides and riffles can result in condi-

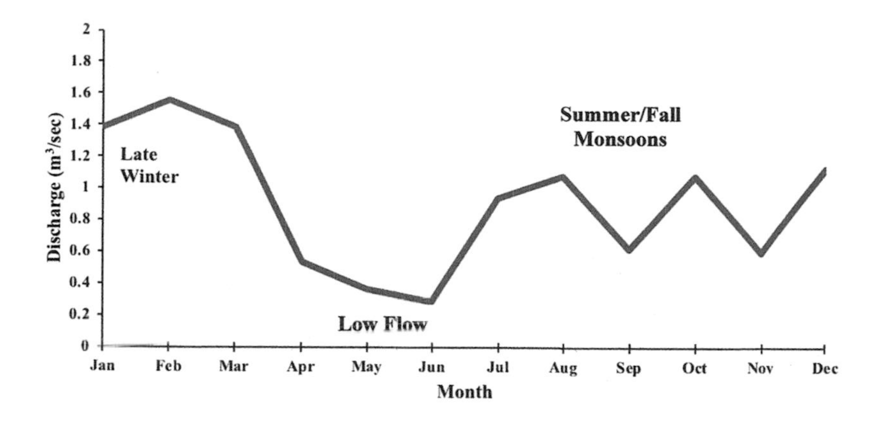

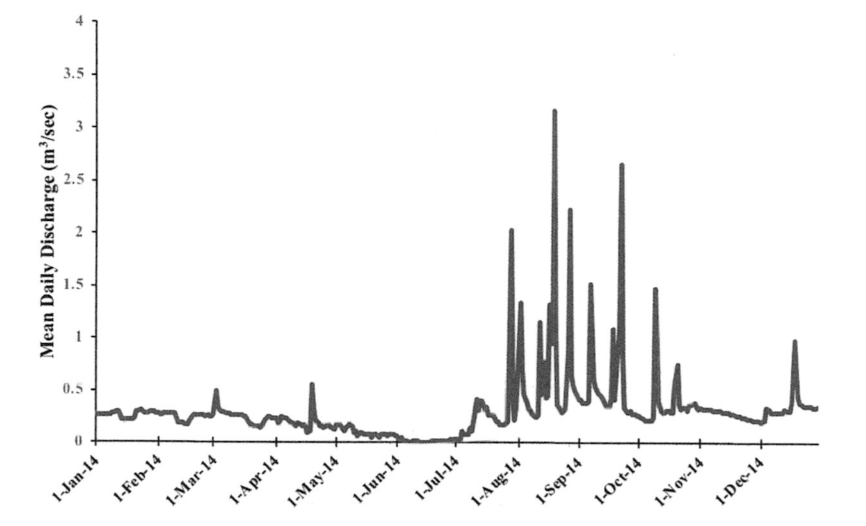

Fig. 9.4 Hydrographs of Aravaipa Creek, Arizona, from USGS stream gage (USGS 09473000), showing mean monthly discharge during 1931–2015 (*top*) and discharge for one year, 2014 (*bottom*), demonstrating the degree of short-duration flooding in the system during late summer and early fall. (Top plot adapted by authors from data provided by USGS Current Water Data for the Nation, https://waterdata.usgs.gov/nwis/rt; bottom plot is a hydrograph derived from data from same source.)

tions that favor many common nonnative species.

Changes to other parts of the hydrograph can also affect the ecology of fishes found in streams. Loach minnow spawns in cobble substrate clean of fine sediments. Cleaning of fines occurs as flows diminish following spring peaks. Nonnative fishes, because they did not evolve with the flash flooding experienced by western fishes, are often differentially displaced by floods, especially in canyon-bound systems (Minckley and Meffe 1987). Even subtle components of flow may be important in the life history of a species. This is illustrated by Moapa dace *Moapa coriacea*, which lives in small spring systems in southern Nevada. Individuals brought into captivity failed to spawn under various manipulations until a combination of flow across the surface and flow directed into cobble substrate resulted in successful spawning (Ruggirello et al. 2015). Many, perhaps all, desert fish species require a rather specific mix of physical stimulants to spawn. Climate change that alters storm frequency or events that affect the amount or timing of water flow can be expected to alter this delicate balance.

Decreases in the amount of suitable habitat for fishes increase the fragmentation of populations on a landscape scale. Jaeger et al. (2014) predicted that by the mid-twenty-first century, stream connectivity for native fishes in an Arizona river will decline by 6%–9% within a year and 12%–18% during spring spawning months. Those species of desert fish with the most fragmented distributions are more than five times as likely to suffer local extirpations as those with continuous distributions (Fagan et al. 2005).

The effects of a hotter, drier climate on terrestrial plant communities in watersheds along streams are likely to increase water quality problems. Already forests are dying at rapid rates due to desiccation and insect infestations (Adams et al. 2017) and thus becoming increasing susceptible to wildfire. Wildfire removes canopies from streamside vegetation that cools streams and filters sediments entering streams. Levels of suspended sediment can be as high as 700,000 mg/L in streams following fires (Rinne 1996). Although species are susceptible to different levels of sediment, high suspended sediment levels such as those after a wildfire increase mortality in most species. For example, LC50 of Yaqui chub *Gila purpurea* fry was 8,372 mg/L suspended sediment (for a 12-hour period). For embryos, it was 3,977 mg/L (Clark-Barkalow and Bonar 2015).

Suspended sediment is not the only water quality factor that affects streams in a drying, warming climate. Salt levels in many southwestern streams and lakes are already very high. Lowering water levels increases concentrations of salt, which affects the ionic balance of fishes (Whitney et al. 2016). Higgins and Wilde (2005) found that following Dust Bowl droughts, fish assemblages in prairie streams shifted to more salt-tolerant species.

Clearly, climate change will affect southwestern streams in many ways, including alteration of flow, temperature, and other aspects of water quality. The specifics of many of these changes are unknown. It is probable, however, that these changes will have negative consequences for native fishes.

SOUTHWESTERN FISHES AND THEIR CLIMATE CHALLENGE: TOWARD SOLUTIONS

Tools available for conserving desert fishes and their environments under threat of climate change include social, legal, and site-specific management techniques as well as region-wide plans. The effectiveness of these different options will vary depending on their feasibility under given circumstances, which in turn is related to the degree of public support and the funds available to implement an option.

Social science techniques help increase public willingness to address the overall problem of climate change. Lack of acceptance of the causes and importance of climate change by some segments of the public continues to limit action. Approximately 40% of Americans still think that current climate change is simply one of the Earth's natural cycles (Gallup 2018a), as opposed to 97% of climate studies, which state humans are causing climate change (Cook et al. 2013). The last six Gallup polls show that only 1% to 3% of the American public believe environmental issues are among our most important (Gallup 2018b). Therefore, if climate change is to be slowed, more efforts must be made to help the general public understand the gravity of continued climate change, prioritize the issue, and act to support programs and officials that regard human-caused climate change as a serious threat to both humans and the resources on which they depend.

Climate change can be minimized only by public and official commitment to slowing greenhouse gas emissions. Changing public attitudes has been historically in the realm of psychologists, social scientists, marketers, and business professionals; however, biologists can also employ improved communication skills to promote change. Social science techniques such as conflict resolution and verbal judo (Bonar 2007; Bonar and Fraidenburg 2010); persuasion techniques (Bonar 2007; Cialdini 2008), and negotiation (Fisher et al. 1991; Susskind et al. 1999) all have been used to great effect to encourage action on conservation issues and are increasingly being seen as important tools for biologists. The public should be aware that failure to act to slow emissions has the potential to result in centuries of warming and other deleterious impacts.

The magnitude of the climate change issue can discourage individual fisheries biologists and others from tackling it. However, throughout history, motivated, seemingly ordinary natural resource managers have successfully led efforts to reduce the effects of anthropogenic change over large regions. Past examples include Hugh H. Bennett, who started the Soil Conservation Service, which fought poor soil management across the United States during the 1930s Dust Bowl (Helms 2010); Rachel Carson, an editor for the US Fish and Wildlife Service, who cemented the nation's attention on the effect of pesticides on wildlife (Carson 1962); and Edwin "Phil" Pister, a former California Department of Fish and Game biologist, who was one of the founders of the Desert Fishes Council and is known worldwide for his promotion of environmental ethics and desert fish conservation (Fraidenburg 2007). Scientists, managers, and others, including those whose work is mainly in the aquatic realm, should never lose sight of the larger picture: reducing emissions of greenhouse gases, at the level of the state, region, nation, and globe, is the ultimate solution to human-caused climate change, and the ability to make an overall difference is within their reach.

Nonetheless, because the oceans act as a sink for carbon dioxide and heat, no matter how vigorously we act to reduce greenhouse gas emissions, the climate will not be returning to historical conditions for perhaps millennia, barring any miraculous technological breakthroughs. Therefore, we will also invariably have to manage—or adapt to—a changing future (Stein et al. 2013). The degree to which we will need to adapt will depend on how much climate change we allow to occur.

Climate change impacts are broad and widespread. A daunting task is identifying which sites should be prioritized for protection. Sometimes, common sense will help identify sites and species that require immediate action. For example, does the species occur across a large range? Are similar, less fragmented streams found in cooler environments that contain the same species (e.g., farther north in the Northern Hemisphere)? Protecting an individual lake, with the overall goal of cutthroat trout conservation, may not be that important, because native cutthroat trout occupy a wide distribution across North America. However, a site containing a rare pupfish, springsnail, or other endemic species could be prioritized for protection, simply because the species is found nowhere else. Public interest, which can be mustered for the protection of particular sites or species, and which translates to funds available for protection, will necessarily influence decisions.

Sites with appropriate conditions for species of interest that can serve as refuges can be identified, protected, and connected (Bush et al. 2014; Morelli et al. 2016). For example, headwater streams can harbor fishes that require lower temperatures. Streams can be connected by ensuring that flow and temperatures remain sufficient for fish movement among different stream segments. Where ecosystems cannot be preserved or connected, slowing rates of climate and other changes to allow the ecosystem to adapt as well as possible may be an option.

Use of existing laws—such as instream water rights, Endangered Species Act protections, and federal reserved water rights—and further legal protections will be needed to maintain streamflow because of water demand from an increasing human population and increased evaporation due to warming temperatures. In some states, securing legal protection for streamflow can be challenging under existing laws because it is often incumbent on biologists, hydrologists, and other managers to prove to the court, using strong arguments and supporting data, that fish populations need specified minimum amounts of water to maintain particular life processes (e.g., *Winters v. United States* [1908]; Arizona Revised Statutes § 45-152.01), whereas in other states, instream flows can be purchased from willing sellers without requirements for proof of need.

Cooling of streams can be accomplished by maintaining flow and providing shade. Where streams are small and well mixed, pools, unless they are considerably large and deep, do not provide effective temperature refuges (Bilby 1984; Nielsen et al. 1994; Bonar and Petre 2015). Stream temperature models (e.g., Bartholow 1989; Isaak et al. 2010) coupled with knowledge of the temperature tolerances and preferences of various southwestern aquatic organisms (e.g., Lee and Rinne 1980; Widmer et al. 2006; Carveth et al. 2007; Recsetar and Bonar 2013; Zeigler et al. 2013) can guide management strategies. For example, Price (2013) used the USGS stream temperature model SSTemp (Bartholow 1989) to calculate numbers of trees of various species to add to cool southwestern stream segments by specific numbers of degrees to meet goals set by the thermal tolerances of the native fishes at the location.

Fish will be stressed by reduced streamflow and higher water temperatures. Therefore, other stressors, such as nonnative fishes (Coggins et al. 2011; Propst et al. 2015) or wildfires (Schultz et al. 2012), should be reduced. The Four Forest Restoration Initiative Stakeholders Group (2017) helps allow stressed fishes to survive in a stream reach.

Historically, site-specific fisheries management (e.g., management of regulations on one lake or management of water temperatures in one stream segment) has been sufficient. Yet because climate change is a global phenomenon that occurs across political boundaries, management typically now must be coordinated at larger landscape scales. This requires governments of various regions to work together to implement joint plans (Stein et al. 2013; Lynch et al. 2014). These planning processes employ a variety of procedures and models to help jurisdictions work together to maximize benefits across a region. Projected changes in land use and climate-induced shifts are used to identify individual sites and species where management would provide the most benefit (Bond et al. 2014; Bush et al. 2014; Guse et al. 2015). Plans can also specify the best time to act based on the projected chronology of change in specific areas (Werners et al. 2015). Sites or species can be prioritized by their vulnerability (e.g., their exposure, sensitivity, or adaptive capacity) (Glick et al. 2011; Case et al. 2015).

Plans that include climate change action in their efforts to restore southwestern fishes and their habitats are generally new, and most are in the early stages of implementation. Examples of large-scale plans being implemented include California WaterFix (California Department of Water Resources and US Bureau of Reclamation 2016; California Natural Resources Agency 2017a) and EcoRestore (California Natural Resources Agency 2017b), which are designed to reroute water and restore habitat in the Sacramento–San Joaquin River delta. These efforts schedule removal of water at times and places to minimize saltwater intrusion into the freshwater system, reduce entrainment of fishes, and maximize aquatic habitat connectivity. Managing watersheds to prevent catastrophic fires and massive flooding beyond that of historical conditions is also under way. The Four Forest Restoration Initiative, a plan focusing on habitat resiliency, is being adopted in the inland Southwest (Schultz et al. 2012; Four Forest Restoration Initiative Stakeholders Group 2017). This plan discusses means to manage and increase the resiliency of over 970,000 mainly forested hectares across Arizona and New Mexico to help prevent catastrophic wildfires and floods. Objectives include maintaining old-growth stands of timber and improving fish and wildlife habitat while clearing fuels that contribute to high-intensity fires.

Recent US Fish and Wildlife Service species recovery plans include climate change as a factor and plan for it (e.g., USFWS 2015, 2017); however, other USFWS plans written earlier do not include climate change as a threat (e.g., USFWS 1990a, b). If climate change planning is to be included in species protection, many of these older plans need to be updated.

In isolated water bodies, where natural dispersal is not possible, assisted migration to water bodies where more favorable conditions exist might be an option; however, it is a topic of contentious debate. Those in opposition to assisted migration reference the great damage that transplanted nonnative species have caused (Ricciardi and Simberloff 2009). Others suggest that on a case-by-case basis, with careful planning, assisted migration of a species beyond its current range may be a necessary, if reluctantly used, option (Schlaepfer et al. 2009; Schwartz et al. 2009), and may fill an undone ecosystem service by replacing an earlier species that could not survive the location with one that can (Lunt et al. 2013).

Successful decisions regarding "safe" refuges and assisted migration destinations will become increasingly challenging with growing levels of climate change because projecting some consequences of climate change is difficult. Although we know with great certainty that warming will continue as long as emissions of greenhouse gases continue, there is inherently less confidence in projecting the likelihood of highly unusual (i.e., several standard deviations from the mean) extreme events. Examples include heatwaves of duration and severity that greatly exceed expectations, extreme flood events, or unprecedented hot droughts, which have the potential to last many decades in the Southwest. These and other types of poorly understood extremes happened in the past and could potentially overwhelm designated refuges or designated "safe havens" for at-risk fish populations. Thus, to be effective under continued climate change, contingency plans (and funding to implement them) and adequate redundancy must be incorporated into management plans to provide backup options if the unexpected occurs. This is particularly true for fish species that become limited to just a small number of extant populations.

Managing some species in the wild may be impossible because of limited time, money, or public interest. Therefore, continued development and maintenance of captive populations is important. Seed banks have been established in the Norwegian Arctic to preserve plants that might disappear from wild populations (Westengen et al. 2013). Captive populations of highly endangered terrestrial and aquatic species have been developed and maintained in zoos, in aquaria, and at artificial refuges. Zoo and aquarium populations provide a source of organisms if wild habitat is restored, plus a valuable educational opportunity for those who visit the aquarium or zoo. A population of Devils Hole pupfish has been developed and maintained in a mesocosm aquarium facility adjacent to Devils Hole to provide fish in case of catastrophe (Feuerbacher et al. 2016). Thirty-three refuge populations of the desert pupfish complex *Cyprinodon macularius* and *Cyprinodon eremus* have been maintained in pools at private and public parks, zoos, and aquaria (Koike et al. 2008). Maintenance of populations in zoos, aquaria and other artificial refuge sites, although not an optimal solution, may be the only way to maintain some species.

CONCLUSION

The climate of the Southwest is changing as dramatically as that of any region in North America, and many of the changes are well understood. Warming, as well as other climate shifts related to warming, is clearly attributable to human emissions of greenhouse gases, primarily due to burning of fossil fuels. Given the major impacts climate change is having on aquatic and upland systems in the region, the only truly safe option for fishes of the region is to ensure that climate change does not continue. Given that this option is unlikely in the short term, all agencies and other players committed to fish conservation need to help increase public concern for reducing greenhouse gas emissions, while simultaneously planning for climate impacts across large landscapes with new and updated fish and habitat recovery plans and periodic status reviews that include climate change impacts. Legal approaches to protecting streamflow, groundwater, and springs need to be strengthened significantly, just as new on-the-ground methods to connect suitable habitats, cool streams, and manage watersheds must be considered, along with improved strategies for assisted migration. Adaptive management methods that monitor waters and other critical elements of the environment must be deployed aggressively so that management actions can be adjusted as needed. Refuges for those fishes that are most likely to be extirpated by climate change need to be developed. Perhaps most importantly, scientists and managers alike need to engage with the public more effectively to communicate the climate challenges to both fish and humankind, and the importance of immediate action.

REFERENCES

Adams, H. D., G. A. Barron-Gafford, R. L. Minor, A. A. Gardea, L. P. Bentley, D. J. Law, D. D. Breshears, N. G. McDowell, and T. E. Huxman. 2017. Temperature response surfaces for mortality risk of tree species with future drought. *Environmental Research Letters* 12: 115014.

Allard, D. C. 1978. *Spencer Fullerton Baird and the U.S. Fish Commission*. New York: Arno Press.

Alley, R. B., J. Marotzke, W. D. Nordhaus, J. T. Overpeck, D. M. Peteet, R. A. Pielke, Jr., R. T. Pierrehumbert, P. B. Rhines, T. F. Stocker, L. D. Talley, and J. M. Wallace. 2003. Abrupt climate change. *Science* 299: 2005–10.

Bartholow, J. M. 1989. *Stream Temperature Investigations: Field and Analytic Methods*. Instream Flow Information Paper no. 13. US Fish and Wildlife Service Biological Report 89 (17).

Bilby, R. E. 1984. Characteristics and frequency of cool-water areas in a western Washington stream. *Journal of Freshwater Ecology* 2: 593–602.

Bonar, S. A. 2007. *The Conservation Professional's Guide to Working with People*. Washington, DC: Island Press.

Bonar, S. A., and M. Fraidenburg. 2010. Communication and conflict resolution in fisheries management. In W. A. Hubert and M. C. Quist, eds., *Inland Fisheries Management in North America*, 3rd ed., 157–84. Bethesda, MD: American Fisheries Society.

Bonar, S. A., and S. J. Petre. 2015. Ground-based thermal imaging of stream surface temperatures: Technique and evaluation. *North American Journal of Fisheries Management* 35: 1209–18.

Bond, N. R., J. R. Thomson, and P. Reich. 2014. Incorporating climate change in conservation planning for freshwater fishes. *Diversity and Distributions* 20: 931–42.

Bowden, T. J. 2008. Modulation of the immune system of fish by their environment. *Fish and Shellfish Immunology* 25: 373–83.

Brinson, M. M., B. L. Swift, R. C. Plantico, and J. S. Barclay. 1981. *Riparian Ecosystems: Their Ecology and Status*. Washington, DC: US Fish and Wildlife Service Biological Services Program.

Bush, A., V. Hermoso, S. Linke, D. Nipperess, E. Turak, and L. Hughes. 2014. Freshwater conservation planning under climate change: Demonstrating proactive approaches for Australian Odonata. *Journal of Applied Ecology* 51: 1273–81.

Caissie, D., and C. H. Luce. 2017. Quantifying stream bed advection and conduction heat fluxes. *Water Resources Research* 53: 1595–624.

California Department of Water Resources and US Bureau of Reclamation. 2016. *Final Environmental Impact Report/Environmental Impact Statement for the Bay Delta Conservation Plan/California WaterFix—Volume I*. Final EIR/EIS for the BDCP/California WaterFix. December. (DOE/EIS-0515.) (ICF 00139.14.) Sacramento: ICF International.

California Natural Resources Agency. 2017a. *California Waterfix is Alternative 4a*. California Natural Resources Agency, Sacramento. https://www.californiawaterfix.com/. (October 2017).

California Natural Resources Agency. 2017b. *California EcoRestore: A Stronger Delta Ecosystem*. California Natural Resources Agency, Sacramento. http://resources.ca.gov/ecorestore/ (October 2017).

Carson, R. 1962. *Silent Spring*. Boston: Houghton Mifflin.

Carveth, C. J., A. M. Widmer, and S. A. Bonar. 2006. Comparison of upper thermal tolerances of native and nonnative fish species in Arizona. *Transactions of the American Fisheries Society* 135: 1433–40.

Carveth, C. J., A. M. Widmer, S. A. Bonar, and J. R. Simms. 2007. An examination of the effects of chronic static and fluctuating temperature on the growth and survival of spikedace, *Meda fulgida*, with implications for management. *Journal of Thermal Biology* 32: 102–8.

Case, M. J., J. J. Lawler, and J. A. Tomasevic. 2015. Relative sensitivity to climate change of species in northwestern North America. *Biological Conservation* 187: 127–33.

Chadwick, J. G., K. H. Nislow, and S. D. McCormick. 2015. Thermal onset of cellular and endocrine stress responses correspond to ecological limits in Brook Trout, an iconic cold-water fish. *Conservation Physiology* 3: cov017.

Cialdini, R. B. 2008. *Influence: Science and Practice*. 5th ed. Boston: Allyn and Bacon.

Clark-Barkalow, S. L., and S. A. Bonar. 2015. Effects of suspended sediment on survival of Yaqui Chub, an endangered US/Mexico borderlands cyprinid. *Transactions of the American Fisheries Society* 144: 345–51.

Coggins, L. G., Jr., M. D. Yard, and W. E. Pine III. 2011. Nonnative fish control in the Colorado River in Grand Canyon, Arizona: An effective program or serendipitous timing? *Transactions of the American Fisheries Society* 140: 456–70.

Cook, B. I., T. R. Ault, and J. E. Smerdon. 2015. Unprecedented 21st century drought risk in the American Southwest and Central Plains. *Science Advances* 1: e1400082.

Cook, B. I., and R. Seager. 2013. The response of the North American Monsoon to increased greenhouse gas forcing. *Journal of Climate* 118: 1690–99.

Cook, J., D. Nuccitelli, S. A. Green, M. Richardson, B. Winkler, R. Painting, R. Way, P. Jacobs, and A. Skuce. 2013. Quantifying the consensus on anthropogenic global warming in the scientific literature. *Environmental Research Letters* 8. https://doi.org/10.1088/1748-9326/8/2/024024.

Davies, W. D., W. L. Shelton, and S. P. Malvestuto. 1982. Prey-dependent recruitment of largemouth bass: A conceptual model. *Fisheries* 7: 12–15.

Dominguez, F., E. Rivera, D. P. Lettenmaier, and C. L. Castro. 2012. Changes in winter precipitation extremes for the western United States under a warmer climate as simulated by regional climate models. *Geophysical Research Letters* 39. https://doi.org/10.1029/2011GL050762.

Fagan, W. F., C. Aumann, C. M. Kennedy, and P. J. Unmack. 2005. Rarity, fragmentation, and the scale dependence of extinction risk in desert fishes. *Ecology* 86: 34–41.

Falke, J. A., K. D. Fausch, K. R. Bestgen, and L. L. Bailey. 2010. Spawning phenology and habitat use in a Great Plains, USA, stream fish assemblage: An occupancy estimation approach. *Canadian Journal of Fisheries and Aquatic Sciences* 67: 1942–56.

Famiglietti, J. S., M. Lo, S. L. Ho, J. Bethune, K. J. Anderson, T. H. Syed, S. C. Swenson, C. R. de Linage, and M. Rodell. 2011. Satellites measure recent rates of groundwater depletion in California's Central Valley. *Geophysical Research Letters* 38. https://doi.org/10.1029/2010GL046442.

Feuerbacher, O., S. A. Bonar, and P. J. Barrett. 2016. Design and testing of a mesocosm-scale habitat for culturing the endangered Devils Hole pupfish. *North American Journal of Aquaculture* 78: 259–69.

Fisher, R., W. Ury, and B. Patton. 1991. *Getting to Yes: Negotiating Agreement Without Giving In*. New York: Penguin Books.

Four Forest Restoration Initiative Stakeholders Group. 2017. *The Four Forest Restoration Initiative* (October 2017). http://www.4fri.org/.

Fraidenburg, M. E. 2007. *Intelligent Courage: Natural Resource Careers That Make a Difference*. Malabar, FL: Krieger Publishing.

Free, C., G. S. Baxter, C. R. Dickman, A. Lisle, and L. K. P. Leung. 2015. Diversity and community composition of vertebrates in desert river habitats. *PLoS ONE* 10 (12): e0144258.

Fry, F. E. J. 1947. Effects of the environment on animal activity. *Publications of the Ontario Fisheries Research Laboratory* 68: 1–63.

Fuller, P. L., L. G. Nico, and J. D. Williams. 1999. *Nonindigenous Fishes Introduced into Inland Waters of the United States*. Special Publication 27. Bethesda, MD: American Fisheries Society.

Gallup. 2018a. Global warming concern steady despite some partisan shifts. http://news.gallup.com/poll/231530/global-warming-concern-steady-despite-partisan-shifts.aspx.

Gallup. 2018b. In Depth: Topic A to Z. Most Important Problem. http://news.gallup.com/poll/1675/most-important-problem.aspx.

Garfin, G. A., A Jardine, R. Merideth, M. Black, and S. LeRoy, eds. 2013. *Assessment of Climate Change in the Southwest United States: A Report Prepared for the National Climate Assessment*. Washington, DC: Island Press.

Gershunov, A., B. Rajagopalan, J. Overpeck, K. Guirguis, D. Cayan, M. Hughes, M. Dettinger, C. Castro, R. Schwartz, M. Anderson, A. Ray, J. Barsugli, T. Cavazos, and M. Alexander. 2013. Future climate: Projected extremes. In G. A. Garfin, A. Jardine, R. Merideth, M. Black, and S. LeRoy, eds., *Assessment of Climate Change in the Southwest United States: A Report Prepared for the National Climate Assessment*,126–47. Washington, DC: Island Press.

Glick, P., B. A. Stein, and N. A. Edelson, eds. 2011. *Scanning the Conservation Horizon: A Guide to Climate Change Vulnerability Assessment*. Washington, DC: National Wildlife Federation.

Goulden, M. L., and R. C. Bales. 2014. Mountain runoff vulnerability to increased evapotranspiration with vegetation expansion. *Proceedings of the National Academy of Sciences* 111: 14071–75.

Guse, B., J. Kail, J. Radinger, M. Schröder, J. Kiesel, D. Hering, C. Wolter, and N. Fohrer. 2015. Eco-hydrologic model cascades: Simulating land use and climate change impacts on hydrology, hydraulics and habitats for fish and macroinvertebrates. *Science of the Total Environment* 533: 542–56.

Hansen, J., A. Lacis, R. Ruedy, and M. Sato. 1992. Potential climate impact of Mount Pinatubo eruption. *Geophysical Research Letters* 19: 215–18.

Heath, R. C. 1983. *Basic Ground-Water Hydrology*. Water Supply Paper 2220. Washington, DC: US Geological Survey.

Helms, D. 2010. Hugh Hammond Bennett and the creation of the Soil Conservation Service. *Journal of Soil and Water Conservation* 65: 37A–47A.

Higgins, C. L., and G. R. Wilde. 2005. The role of salinity in structuring fish assemblages in a prairie stream system. *Hydrobiologia* 549: 197–203.

Hubbard, J. P. 1977. Importance of riparian ecosystems: Biotic considerations. In R. R. Johnson and D. A. Jones, eds., *Importance, Preservation and Management of Riparian Habitat*. USDA Forest Service General Technical Report RM-GTR-43.

Isaak, D. J., C. H. Luce, B. E. Rieman, D. E. Nagel, E. E. Peterson, D. L. Horan, S. Parkes, and G. L. Chandler. 2010. Effects of climate change and wildfire on stream temperatures and salmonid thermal habitat in a mountain river network. *Ecological Applications* 20: 1350–71.

Jaeger, K. L., J. D. Olden, and N. A. Pelland. 2014. Climate change poised to threaten hydrologic connectivity and endemic fishes in dryland streams. *Proceedings of the National Academy of Sciences* 111: 13894–99.

Kitzberger, T., D. A. Falk, A. L. Westerling, and T. W. Swetnam.

2017. Direct and indirect climate controls predict heterogeneous early-mid 21st century wildfire burned area across western and boreal North America. *PLoS ONE* 12: e0188486.

Koike, H., A. A. Echelle, D. Loftis, and R. A. Van Den Bussche. 2008. Microsatellite DNA analysis of success in conserving genetic diversity after 33 years of refuge management for the desert pupfish complex. *Animal Conservation* 11: 321–29.

Lee, R. M., and J. N. Rinne. 1980. Critical thermal maxima of five trout species in the southwestern United States. *Transactions of the American Fisheries Society* 109: 632–35.

Lehner, F. E., R. Wahl, A. W. Wood, D. B. Blatchford, and D. Llewellyn. 2017. Assessing recent declines in Upper Rio Grande runoff efficiency from a paleoclimate perspective. *Geophysical Research Letters* 44: 4124–33. https://doi.org/10.1002/2017GL073253.

Leonard, J. R. 1979. *The Fish Car Era*. Washington, DC: US Fish and Wildlife Service.

Lunt, I. D., M. Byrne, J. J. Hellmann, N. J. Mitchell, S. T. Garnett, M. W. Hayward, T. G. Martin, E. McDonald-Madden, S. E. Williams, and K. K. Zander. 2013. Using assisted colonization to conserve biodiversity and restore ecosystem function under climate change. *Biological Conservation* 157: 172–77.

Lynch, A. J., E. Varela-Acevedo, and W. W. Taylor. 2014. The need for decision-support tools for a changing climate: Application to inland fisheries management. *Fisheries Management and Ecology* 22: 14–24.

Masson-Delmotte, V., M. Schulz, A. Abe-Ouchi, J. Beer, A. Ganopolski, J. F. González Rouco, E. Jansen, K. Lambeck, J. Luterbacher, T. Naish, T. Osborn, B. Otto-Bliesner, T. Quinn, R. Ramesh, M. Rojas, X. Shao, and A. Timmermann. 2013. Information from Paleoclimate Archives. In T. F. Stocker, D. Qin, G.-K. Plattner, M. Tignor, S. K. Allen, J. Boschung, A. Nauels, Y. Xia, V. Bex, and P. M. Midgley, eds., *Climate Change 2013: The Physical Science Basis. Contribution of Working Group I to the Fifth Assessment Report of the Intergovernmental Panel on Climate Change*. Cambridge: Cambridge University Press.

Minckley, W. L., and G. K. Meffe. 1987. Differential selection by flooding in stream-fish communities of the arid American Southwest. In W. J. Matthews and D. C. Heines, eds., *Community and Evolutionary Ecology of North American Stream Fishes*, 93–104. Norman: University of Oklahoma Press.

Mohseni, O., and H. G. Stefan. 1999. Stream temperature/air temperature relationship: A physical interpretation. *Journal of Hydrology* 218: 128–41.

Morelli, T. L., C. Daly, S. Z. Dobrowski, D. M. Dulen, J. L. Ebersole, S. T. Jackson, J. D. Lundquist, C. I. Millar, S. P. Maher, W. B. Monahan, K. R. Nydick, K. T. Redmond, S. C. Sawyer, S. Stock, and S. R. Beissinger. 2016. Managing climate change refugia for climate adaptation. *PLoS ONE* 11 (8): e0159909.

National Research Council. 2013. *Abrupt Impacts of Climate Change: Anticipating Surprises*. Washington, DC: National Academies Press.

Nielsen, J. L., T. E. Lisle, and V. Ozaki. 1994. Thermally stratified pools and their use by steelhead in northern California streams. *Transactions of the American Fisheries Society* 123: 613–26.

Overpeck, J. T. 2013. The challenge of hot drought. *Nature* 503: 350–51.

Overpeck, J. T. 2014. The challenge of biodiversity adaptation under climate change. In J. P. Palutikof, S. L. Boulter, J. Barnett, and D. Rissik, eds., *Applied Studies in Climate Adaptation*, 61–67. Oxford: Wiley. https://doi.org/10.1002/9781118845028.ch8.

Overpeck, J. T., and J. E. Cole. 2006. Abrupt change in the Earth's climate system. *Annual Review of Environment and Resources* 31: 1–31.

Overpeck, J., G. Garfin, A. Jardine, D. Busch, D. Cayan, M. Dettinger, E. Fleishman, A. Gershunov, G. MacDonald, K. R. Travis, and B. H. Udall. 2013. Summary for decision makers. In G. A. Garfin, R. M. Jardine, M. Black, and S. LeRoy, eds. *Assessment of Climate Change in the Southwest United States: A Report Prepared for the National Climate Assessment*, 1–20. Washington, DC: Island Press.

Paukert, C. P., A. J. Lynch, and J. E. Whitney. 2016. Effects of climate change on North American inland fishes: Introduction to the special issue. *Fisheries* 41: 329–30.

Petre, S. J., and S. A. Bonar. 2017. Determination of habitat requirements for Apache Trout. *Transactions of the American Fisheries Society* 146: 1–15.

Price, J. E. 2013. Potential methods to cool streams containing Apache trout in the White Mountains of Arizona and implications for climate change. MS thesis, University of Arizona.

Propst, D. L., and K. B. Gido. 2004. Responses of native and nonnative fishes to natural flow regime mimicry in the San Juan River. *Transactions of the American Fisheries Society* 133: 922–31.

Propst, D. L., K. B. Gido, J. E. Whitney, E. I. Gilbert, T. J. Pilger, A. M. Monié, Y. M. Paroz, J. M. Wick, J. A. Monzingo, and D. M. Myers. 2015. Efficacy of mechanically removing nonnative predators from a desert stream. *River Research and Applications* 31: 692–703.

Recsetar, M. S., and S. A. Bonar. 2013. Survival of Apache trout eggs and alevins under static and fluctuating temperature regimes. *Transactions of the American Fisheries Society* 142: 373–79.

Recestar, M. S., S. A. Bonar, and O. G. Feuerbacher. 2014. Growth and survival of Apache Trout under static and fluctuating temperature regimes. *Transactions of the American Fisheries Society* 143: 1247–54.

Recsetar, M. S., M. P. Zeigler, D. L. Ward, S. A. Bonar, and C. A. Caldwell. 2012. Relationship between fish size and thermal tolerance. *Transactions of the American Fisheries Society* 141: 1433–38.

Ricciardi, A., and D. Simberloff. 2009. Assisted colonization is not a viable conservation strategy. *Trends in Ecology and Evolution* 24: 248–53.

Rinne, J. N. 1996. Short-term effects of wildfire on fishes and aquatic macroinvertebrates in the southwestern United States. *North American Journal of Fisheries Management* 16: 653–58.

Routson, C. C., C. A. Woodhouse, and J. T. Overpeck. 2011. Second century megadrought in the Rio Grande headwaters, Colorado: How unusual was medieval drought? *Geophysical Research Letters* 38, L22703. https://doi.org/10.1029/2011GL050015.

Routson, C. C., C. A. Woodhouse, J. T. Overpeck, J. L. Betancourt, and N. P. McKay. 2016. Teleconnected ocean forcing of western North American droughts and pluvials during the last millennium. *Quaternary Science Reviews* 146: 238–50.

Ruggirello, J. E., S. A. Bonar, O. G. Feuerbacher, L. H. Simons, and C. Powers. 2015. *Spawning Ecology and Captive Husbandry of Endangered Moapa Dace*. Tucson: Arizona Cooperative Fish and Wildlife Research Unit Fisheries Research Report 12–15.

Rutherford, W. A., T. H. Painter, S. Ferrenberg, J. Belnap, G. S. Okin, C. Flagg, and S. C. Reed. 2017. Albedo feedbacks to future climate via climate change impacts on dryland biocrusts. *Scientific Reports* 7. https://doi.org/10.1038/srep44188.

Scanlon, B. R., Z. Zhang, R. C. Reedy, D. R. Pool, H. Save, D. Long, J. Chen, D. M. Wolock, B. D. Conway, and D. Winester. 2015. Hydrologic implications of GRACE satellite data in the Colorado River basin. *Water Resources Research*. https://doi.org/10.1002/2015WR018090.

Schade, C. B., and S. A. Bonar. 2005. Distribution and abundance of nonnative fishes in streams of the western United States. *North American Journal of Fisheries Management* 25: 1386–94.

Schlaepfer, M. A., W. D. Helenbrook, K. B. Searing, and K. T. Shoemaker. 2009. Assisted colonization: Evaluating contrasting management actions (and values) in the face of uncertainty. *Trends in Ecology and Evolution* 24: 471–72.

Schultz, C. A., T. Jedd, and R. D. Beam. 2012. The collaborative forest landscape restoration program: A history and overview of the first projects. *Journal of Forestry* 110: 381–91.

Schwartz, M. W., J. J. Hellmann, and J. S. McLachlan. 2009. The precautionary principle in managed relocation is misguided advice. *Trends in Ecology and Evolution* 24: 474.

Settele, J., R. Scholes, R. Betts, S. Bunn, P. Leadley, D. Nepstad, J. T. Overpeck, and M. A. Taboada. 2014. Terrestrial and inland water systems. In C. B. Field, V. R. Barros, D. J. Dokken, K. J. Mach, M. D. Mastrandrea, T. E. Bilir, M. Chatterjee, K. L. Ebi, Y. O. Estrada, R. C. Genova, B. Girma, E. S. Kissel, A. N. Levy, S. MacCracken, P. R. Mastrandrea, and L. L. White, eds., *Climate Change (2014). Impacts, Adaptation, and Vulnerability. Part A: Global and Sectoral Aspects. Contribution of Working Group II to the Fifth Assessment Report of the Intergovernmental Panel on Climate Change*, 271–359. Cambridge: Cambridge University Press.

Solomon, S., G.-K. Plattner, R. Knutti, and P. Friedlingstein. 2009. Irreversible climate change due to carbon dioxide emissions. *Proceedings of the National Academy of Sciences* 106: 1704–9.

Stein, B. A., A. Staudt, M. S. Cross, N. S. Dubois, C. Enquist, R. Griffis, L. J. Hansen, J. J. Hellmann, J. J. Lawler, E. J. Nelson, and A. Pairis. 2013. Preparing for and managing change: Climate adaptation for biodiversity and ecosystems. *Frontiers in Ecology and the Environment* 11: 502–10.

Stocker, T. F., D. Qin, G.-K. Plattner, M. Tignor, S. K. Allen, J. Boschung, A. Nauels, Y. Xia, V. Bex and P. M. Midgley, eds. 2013. *Climate Change 2013: The Physical Science Basis. Contribution of Working Group I to the Fifth Assessment Report of the Intergovernmental Panel on Climate Change*. Cambridge: Cambridge University Press.

Susskind, L., P. Levy, and J. Thomas-Larmer. 1999. *Negotiating Environmental Agreements: How to Avoid Escalating Confrontation, Needless Costs and Unnecessary Litigation*. Washington, DC: Island Press.

Udall, B., and J. Overpeck. 2017. The twenty-first-century Colorado River hot drought and implications for the future. *Water Resources Research* 53: 2404–18.

USFWS (US Fish and Wildlife Service). 1990a. *Spikedace Recovery Plan*. Albuquerque: US Fish and Wildlife Service.

USFWS (US Fish and Wildlife Service). 1990b. *Recovery Plan for the Endangered and Threatened Species of Ash Meadows, Nevada*. Portland: US Fish and Wildlife Service.

USFWS (US Fish and Wildlife Service). 2015. *Recovery Plan for the Kendall Warm Springs Dace* (Rhinichthys osculus thermalis). Revision. Cheyenne: US Fish and Wildlife Service.

USFWS (US Fish and Wildlife Service). 2017. *Recovery Plan for the Santa Ana Sucker*. Sacramento: US Fish and Wildlife Service.

Vano, J. A., B. Udall, D. R. Cayan, J. T. Overpeck, L. D. Brekke, T. Das, H. C. Hartmann, H. G. Hidalgo, M. Hoerling, G. J. McCabe, K. Morino, R. S. Webb, K. Werner, and D. P. Lettenmaier. 2014. Understanding uncertainties in future Colorado River streamflow. *Bulletin of the American Meteorology Society*: 95: 59–78.

Weiss, J. L., C. L. Castro, and J. T. Overpeck. 2009. The changing character of climate, drought, and the seasons in the southwestern U.S.A. *Journal of Climate* 22: 5918–32.

Werners, S. E., E. van Slobbe, T. Bölscher, A. Oost, S. Pfenninger, G. Trombi, M. Bindi, and M. Moriondo. 2015. Turning points in climate change adaptation. *Ecology and Society* 20: 3. http://dx.doi.org/10.5751/ES-07403-200403.

Westengen, O. T., S. Jeppson, and L. Guarino. 2013. Global ex-situ crop diversity conservation and the Svalbard Global Seed Vault: Assessing the current status. *PLoS ONE* 8 (5): e64146.

Whitney, J. E., R. Al-Chokhachy, D. B. Bunnell, C. A. Caldwell, S. J. Cooke, E. J. Eliason, M. Rogers, A. J. Lynch, and C. P. Paukert. 2016. Physiological basis of climate change impacts on North American inland fishes. *Fisheries* 41: 332–45.

Widmer, A. M., C. J. Carveth, S. A. Bonar, and J. R. Simms. 2006. Upper temperature tolerance of loach minnow under acute, chronic, and fluctuating thermal regimes. *Transactions of the American Fisheries Society* 135: 755–62.

Woodhouse, C. A., and J. T. Overpeck. 1998. 2000 years of drought variability in the central United States. *Bulletin of the American Meteorological Society* 79: 2693–714.

Yarnell, S. M., G. E. Petts, J. C. Schmidt, A. A. Whipple, E. E. Beller, C. N. Dahm, P. Goodwin, and J. H. Viers. 2015. Functional flows in modified riverscapes: Hydrographs, habitats and opportunities. *BioScience* 65: 963–72.

Yin, J., J. T. Overpeck, C. Peyser, and R. Stouffer. 2018. Big jump of record warm global mean surface temperature in 2014–2016 related to unusually large oceanic heat releases. *Geophysical Research Letters* 45: 1069–78. https://doi.org/10.1002/2017GL076500.

Zaimes, G. N. 2007. *Understanding Arizona's Riparian Areas*. Report AZ1432. Tucson: Arizona Cooperative Extension, University of Arizona.

Zeigler, M. P., S. F. Brinkman, C. A. Caldwell, A. S. Todd, M. S. Recsetar, and S. A. Bonar. 2013. Upper thermal tolerances of Rio Grande cutthroat trout under constant and fluctuating temperatures. *Transactions of the American Fisheries Society* 142: 1395–405.

10 Novel Drought Regimes Restructure Aquatic Invertebrate Communities in Arid-Land Streams

Kate S. Boersma and David A. Lytle

Aquatic invertebrates inhabiting arid-land streams possess diverse adaptations to withstand droughts and floods. In the western United States, seasonal precipitation regimes of alternating flooding and drying have existed for much of the current interglacial period (Betancourt et al. 1990; Holmgren et al. 2003). Consequently, arid-land aquatic invertebrates have evolved diverse phenotypes (fig. 10.1) that represent adaptations to hydrologic variability (Williams 1985) and allow them to withstand, and even take advantage of, climate extremes (Tonkin et al. 2018). Aquatic invertebrates can often survive harsh environmental conditions that would be fatal to other arid-land aquatic organisms (Boulton and Lake 2008), including fishes.

However, anthropogenic water withdrawals and aridification are increasing the frequency, severity, and duration of droughts (Balling and Goodrich 2010; Cayan et al. 2010). Predictable seasonal droughts are being replaced by more stochastic supraseasonal drought regimes (Lake 2003). Human population growth in the region's cities and concomitant increases in water withdrawals for human use decrease water availability for aquatic ecosystems (Deacon et al. 2007; Vörösmarty et al. 2010). Additionally, earlier onset of summer and higher mean summer temperatures lead to earlier snowmelt (Seager et al. 2007; Barnett et al. 2008), higher evaporative potential (Seager et al. 2013), and declining surface water (Cook et al. 2014). Climate variability is predicted to increase over the next century as rainstorms become more intense and droughts more extreme (Ruff et al. 2011; Cook et al. 2015).

Arid-land aquatic invertebrates are adapted to withstand severe droughts, and many species persist through, or recover rapidly from, seasonal stream drying (Boersma et al. 2014a). However, novel drought regimes associated

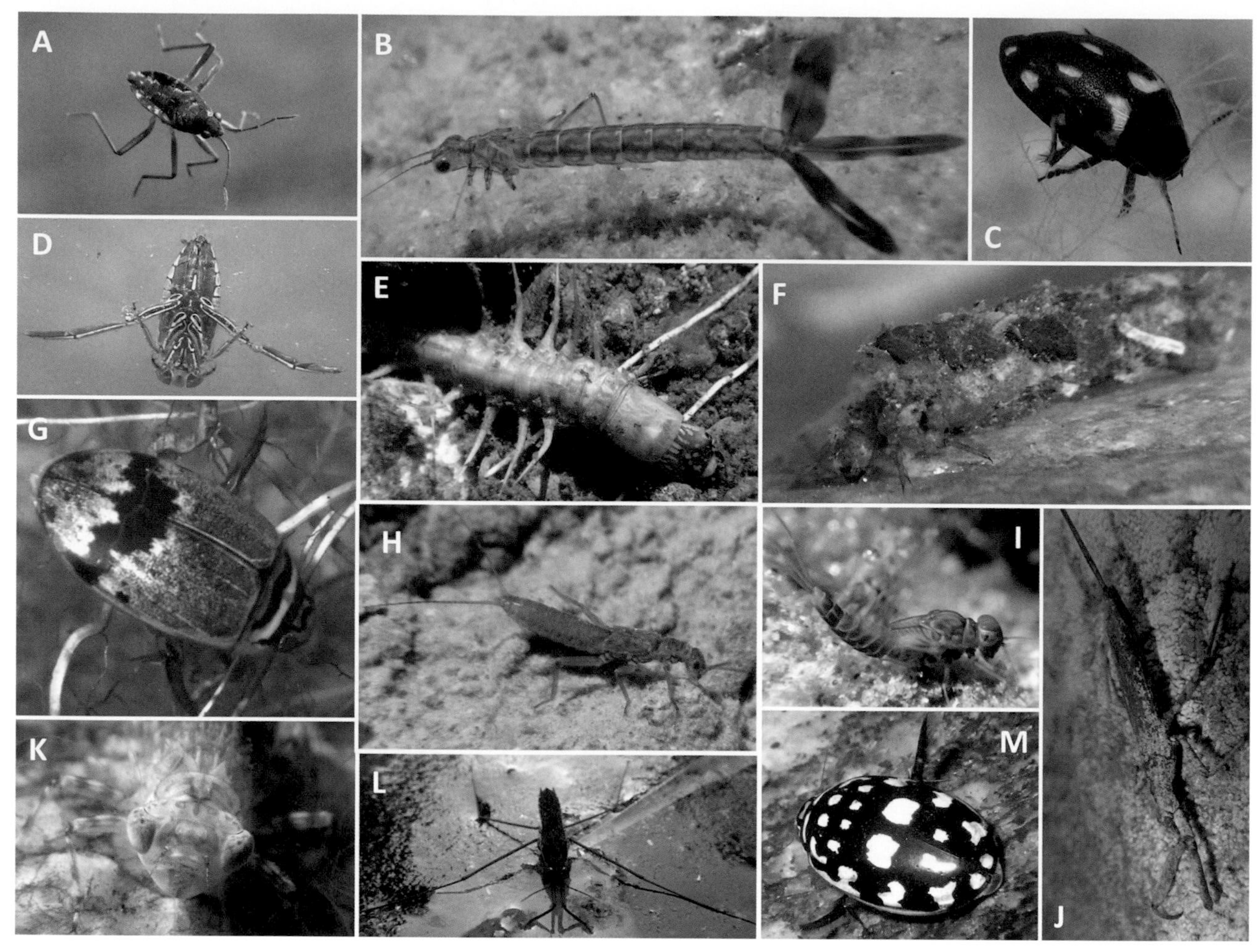

Fig. 10.1 Common aquatic invertebrate species in fragmented intermittent streams of the American Southwest and northwestern México: *A*, water strider (*Microvelia* sp.); *B*, great spreadwing damselfly (*Archilestes grandis*); *C*, diving beetle (*Stictotarsus corvinus*); *D*, redbacked backswimmer (*Notonecta lobata*); *E*, fishfly (*Neohermes filicornis*); *F*, caddisfly (*Hesperophylax* sp.); *G*, diving beetle (*Thermonectus nigrofasciatus*); *H*, Arizona snowfly (*Mesocapnia arizonensis*); *I*, mayfly (*Callibaetis* sp.); *J*, water scorpion (*Curicta pronotata*); *K*, red rock skimmer (*Paltothemis lineatipes*); *L*, common water strider (*Aquarius remigis*); *M*, sunburst diving beetle (*Thermonectus marmoratus*). (Photo credits: M. T. Bogan.)

with increasing aridity can be catastrophic disturbances that restructure aquatic invertebrate communities (Sponseller et al. 2010; Bogan et al. 2015). The disappearance of vulnerable taxa during catastrophic droughts can have cascading effects on aquatic ecosystems and generate novel community trajectories. These indirect effects of droughts make it difficult to ensure conservation of aquatic ecosystems. Effective ecosystem conservation requires intimate a priori knowledge of how drying disturbance affects functionally important species and how changes in local populations of those species propagate through the rest of the ecosystem.

Quantifying the effects of droughts on ecosystem functioning is challenging, but functional-trait analyses provide a promising approach. Invertebrate functional traits describe both how individual species respond to droughts (in the case of traits such as dispersal capacity and desiccation tolerance) and how they contribute to ecosystem functioning (in the case of traits such as body size, trophic level, and diet). Analyses of functional diversity combine functional-trait information on the organisms in a community and thus can yield insights on the effects of droughts that would not be apparent when studying diversity based on species identity alone.

In this chapter, we begin by providing a general overview of those aquatic invertebrate functional traits that increase the ability of local populations to persist through drought. We then profile an arid-land stream in southeastern Arizona, French Joe Canyon, and discuss invertebrate responses to its novel drought regimes. We highlight two particularly vulnerable species: the giant water bug (*Abedus herberti*) and the sycamore caddisfly (*Phylloicus mexicanus*), to demonstrate the potential ecosystem-level consequences of local extinctions of functionally important taxa. Then, we examine the functional diversity of arid-land aquatic invertebrate communities as a way to assess species vulnerability to drought and predict the consequences of local species extinctions. We end with a discussion of management strategies for invertebrate conservation. As will become apparent, adaptive traits that buffer invertebrate communities against predictable seasonal drought disturbances are often inadequate to ensure persistence in the face of emerging novel hydrologic disturbance regimes.

ADAPTATIONS TO HYDROLOGIC EXTREMES

Droughts cause a suite of abiotic changes to stream habitats, including decreases in water level, dissolved oxygen, and overall habitat area as well as increases in conductivity and temperature, all of which exert physiological stress on organisms (Lake 2003). Thus, species richness typically decreases with decreasing surface-water permanence (Bogan et al. 2013). Nevertheless, even temporary aquatic habitats that rewet only once every few years can harbor intermittent-habitat specialists (Cover et al. 2015; Bogan 2017).

Entire assemblages of aquatic invertebrates can demonstrate high resistance and resilience to predictable drying disturbances. Here we define *resistance* as the capacity to withstand drought stress in situ and *resilience* as the capacity for extirpated populations to rapidly recolonize or repopulate following cessation of drought conditions (Lake 2013). Boersma et al. (2014a) conducted a drying manipulation experiment in which they exposed such assemblages to different evaporation treatments, ranging from control pools with stable habitat conditions to experimental pools with near-complete desiccation. They found no differences in species richness or composition among drying treatments, even though the most severe treatment contained only small pools of water (<1 cm depth) and patches of damp sediment. Woodward et al. (2015) also observed near-complete recovery from seasonal drying in streams that retained some limited flow during the dry season. The high resistance of aquatic invertebrate assemblages suggests that these arid-land stream communities are adapted to seasonal droughts, so long as some minute quantity of aquatic habitat remains as a refuge. Due to the importance of these refuge habitats for community persistence, we follow the aquatic invertebrate literature convention and define habitats that contain any surface water year-round as *perennial* and those that dry completely for portions of the year as *intermittent*.

A history of exposure to drought exerts selective pressure and yields functional traits that provide drought resistance and resilience, including behavioral, physiological, and life history adaptations. Many aquatic insects, such as dragonflies and aquatic beetles, can follow gradients of moisture (Lytle et al. 2008) and humidity (Rebora et al. 2007) to avoid desiccation. Some desert-adapted crayfish (e.g., *Geocharax*) can create burrows that capture water and serve as dry-season refuges (Stubbington et al. 2017). Particularly hardy taxa like dobsonflies (*Neohermes filicornis*, Cover et al. 2015) and winter stoneflies (*Mesocapnia arizonensis*, Bogan 2017) burrow deep within the sediment and remain buried for years until flow returns. Many species of aquatic beetles and true bugs are present in water as winged adults and can fly to escape drying streams or ponds (Chester and Robson 2011). Thus, organisms create or seek out habitats to match their environmental tolerances.

Physiological adaptations in arid-land aquatic invertebrates provide higher tolerance to environmental extremes than in invertebrates inhabiting temperate habitats (Boulton and Lake 2008). Many desert-adapted taxa have modified physiological mechanisms such as heat-shock proteins (Clark et al. 2012), ammonium excretion (Ganser et al. 2015), and energy metabolism (Luo et al. 2014), which allow for increased maximum thermal limits (Stewart et al. 2013) or even complete desiccation tolerance (Sota and Mogi 1992).

Life history adaptations allow invertebrate species that live in temporary aquatic habitats to synchronize their life cycles with seasonal drying and rewetting events and survive dry periods as terrestrial or drought-resistant adults, larvae, or eggs. For insects that split their lives between an aquatic larval phase and an aerial adult phase (e.g., dragonflies, Odonata; caddisflies, Trichoptera; and true flies, Diptera), development and metamorphosis to terrestrial life stages is often accelerated to precede drying of intermittent habitats (De Block and Stoks 2005; Drummond et al. 2015). Individuals can then complete their terrestrial phase when aquatic habitat is absent and lay eggs that will hatch once water returns. Other species use periodic drying as an environmental cue to trigger aestivation (Wickson et al. 2012), diapause (Bogan 2017), or pupation (Cover et al. 2015).

With their many adaptive traits conferring high biological resistance and resilience, it may appear as though arid-land aquatic invertebrate communities are impervious to the effects of drought. However, unprecedented multi-year, supraseasonal droughts lack the predictable seasonality of annual dry/wet cycles (Barnett et al. 2008; Cayan et al. 2010), generating novel environmental conditions to which even some desert-adapted aquatic invertebrates are vulnerable (Lake 2013).

FRENCH JOE CANYON

Is there a threshold beyond which drought regimes exceed the capacity of aquatic food webs to recover? Answering this question requires site- and species-specific knowledge. Herein, we present a case study from a stream in southeastern Arizona and profile the vulnerability of two functionally important species: an apex predator and a large-bodied detritivore. These profiles demonstrate how novel climate cycles and elimination of functionally important taxa can trigger trophic cascades that restructure entire aquatic invertebrate communities.

French Joe Canyon provides a characteristic example of changing hydrology resulting from supraseasonal drought and the subsequent development of novel aquatic invertebrate community structure (Bogan and Lytle 2011). This canyon, located in the Whetstone Mountains in southeastern Arizona, once housed a small perennial stream—albeit one that contracted to just a few fragmented pools during summer. A long prehistory of surface water remained evident in its extensive travertine deposits and lush maidenhair fern glen. Records from early explorers and the presence of flightless aquatic taxa also suggested this stream had been historically permanent at least for centuries. Severe drought from 1999 through 2005 ended this legacy when, in June 2005, the stream dried completely. This drying event marked the transition from a perennial stream with at least some year-round surface water to an intermittent one that dries completely and rewets periodically.

Fig. 10.2 Giant water bug (*Abedus herberti*) adult. (Photo credit: M. T. Bogan.)

Bogan and Lytle (2011) sampled aquatic invertebrates in French Joe Canyon for four years before and four years after the transition to intermittent flow. Remarkably, they found no differences in total species richness before and after the change in hydrology (i.e., the number of species observed did not change). However, community composition shifted dramatically. Other studies suggest that once aquatic habitats dry and rewet, dispersal capacity largely determines the potential for recolonization and recovery (Bogan et al. 2015; Phillipsen et al. 2015). This pattern was apparent in French Joe Canyon as well. Following the transition to intermittency, highly vagile colonists—those able to disperse more than 10 km from aquatic habitats in adjacent canyons—replaced flightless, dispersal-limited taxa. Among the local extinctions observed were two functionally important taxa: an apex predator (the giant water bug) and a large-bodied detritivore (the sycamore caddisfly).

The Giant Water Bug

The giant water bug (*Abedus herberti*, Hemiptera: Belostomatidae; fig. 10.2) is a large (25–40 mm adult length), long-lived (up to three years) predatory insect that inhabits perennial streams of Arizona and Sonora (Menke 1960, 1977). It respires atmospheric oxygen and must surface regularly to breathe. The giant water bug can survive out of the water for up to 24 hours (C. Goforth, pers. comm.) and even longer in humid environments (M. T. Bogan, pers. comm.), but it requires water for all stages of reproduction and development

(Smith 1979). It is flightless, however, and found exclusively in perennial aquatic habitats. These characteristics, combined with its long lifespan, limited dispersal capacity, and large body size, make the giant water bug vulnerable to unprecedented drought disturbances, and several local extinctions have been observed in southeastern Arizona over the past two decades (KSB and DAL, unpublished data; Bogan and Lytle 2011; historical records in Menke 1960).

The fate of the giant water bug in French Joe Canyon and elsewhere is of particular ecological interest because it is the apex predator in perennial pools. It is a sit-and-wait predator that uses raptorial forelimbs and piercing mouthparts to capture prey, inject digestive enzymes, and ingest the liquefied tissue. This feeding biology allows it to consume prey items that are larger than itself, including both invertebrates and vertebrates. Apex predators are functionally important to maintain food-web stability, and their extirpation can initiate trophic cascades (e.g., Paine 1966; Borrvall and Ebenman 2006). In streams, disappearance of apex predators can affect entire food webs, from macroinvertebrates (Rodríguez-Lozano et al. 2015) to diatoms (Koetsier 2005).

Boersma et al. (2014b) documented the consequences of the extirpation of the giant water bug for aquatic communities with a manipulative experiment. They created experimental invertebrate assemblages that mimicked natural communities found in French Joe Canyon and removed giant water bugs from half of the assemblages. Apex-predator removal triggered a trophic cascade: the abundance of large mesopredators increased from approximately 8 to approximately 24 individuals in assemblages with the giant water bug removed. These mesopredators were mostly beetles (e.g., *Rhantus*) and true bugs (e.g., *Microvelia*) that could aerially colonize from elsewhere (e.g., fig. 10.1A, B, C, D, E, G, J, K, L, M). They benefited from the ecological vacancy created by apex-predator removal (Boersma et al. 2014b), a phenomenon known as mesopredator release (Ritchie and Johnson 2009). In this case, the dynamics of resistance and resilience facilitated mesopredator release because the apex predator and mesopredators differed in their capacity to survive and recolonize following the transition to intermittency (Bogan and Boersma 2012). That is, novel communities created by unprecedented drought regimes were limited to taxa that either resisted periodic desiccation or exhibited resilience through recolonization (fig. 10.3).

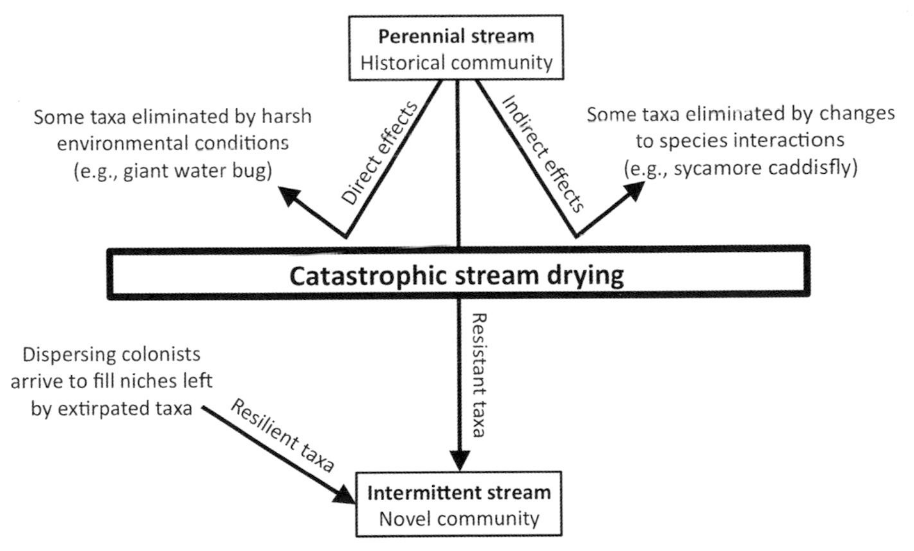

Fig. 10.3 Effects of catastrophic stream drying on arid-land aquatic invertebrate communities. *Top center:* Perennial arid-land streams harbor diverse aquatic invertebrate communities, even if surface water fragments into isolated pools for portions of the year. *Center:* Entire-stream drying acts as a catastrophic disturbance that eliminates vulnerable species (deflected lines), either directly through the harsh environment (such as the giant water bug *Abedus herberti*), or indirectly through cascading effects of the extirpation of other vulnerable taxa (such as the sycamore caddisfly, *Phylloicus mexicanus*). *Bottom center:* Catastrophic drying converts perennial streams to intermittent streams with no perennial refuge habitats and produces novel communities comprising (*1*) resistant taxa able to survive direct and indirect effects of drought, such as some stoneflies and true flies (vertical line), and (*2*) long-distance dispersers that colonize rapidly from perennial habitats of other canyons, such as some beetles and true bugs (diagonal line). These communities are functionally depauperate, missing the top-down effects of the apex predator (giant water bug) and the bottom-up effects of the large-bodied shredder (sycamore caddisfly) that breaks down leaf litter into fine particulate organic matter.

Inferred dispersal and recolonization potential in giant water bugs

Dispersal capacity is a functional trait of fundamental importance in arid-land aquatic conservation because it facilitates recolonization following disturbance. The loss of giant-water-bug populations from French Joe Canyon and other streams in the region is particularly concerning given their limited dispersal capacity and low likelihood of recolonization, as documented by a series of genetic studies on population structure. Finn et al. (2007) used mitochondrial DNA to compare giant-water-bug populations within and among mountain ranges in southeastern Arizona, including the French Joe Canyon population, to test several theories of population connectivity. They found that populations are genetically distinct, even when separated by only a few kilometers, suggesting that many populations have persisted in isolation since the last glacial period (ca. 10,000 BCE), when a cooler and wetter climate facilitated dispersal and allowed greater stream and population connectivity. Phillipsen and Lytle (2013) further compared genetic distance among isolated giant-water-bug populations and found that populations are so isolated that deep genetic divergences occur between populations separated by only a handful of kilometers, probably due to genetic drift acting on population structure over time.

When overland dispersal occurs, it is probably in conjunction with a highly adaptive behavioral trait—rainfall response behavior (RRB)—a remarkable flood-escape mechanism in which individuals respond to sustained heavy rainfall hitting the surface of stream pools by crawling out of and away from the stream (Lytle 1999; Lytle and Smith 2004). Individuals exhibiting RRB typically return to their stream of origin within 24 hours post-flood (Lytle 1999). However, because giant water bugs travel laterally and upslope away from the stream during RRB, it is possible that individuals occasionally crawl over low headwater divides and encounter streams on the opposite side of the mountain range, which would be consistent with the genetic structure described above (Finn et al. 2007; Phillipsen and Lytle 2013). Behavioral experiments have documented that RRB in individual giant-water-bug populations, including the one extirpated from French Joe Canyon, is adapted to the highly localized patterns of flash flooding that are specific to each canyon: individuals from flood-prone canyons respond more quickly to rainfall as a cue that flooding is imminent than do individuals from canyons where flooding is uncommon (Lytle et al. 2008). This observation suggests that the unique character of each giant-water-bug population represents local adaptation as well as genetic drift.

Less is known about drought-induced dispersal and the capacity of extirpated populations for recovery following extreme drought. Rare long-distance dispersal events in aquatic invertebrates may occur during stream contraction when the few remaining refuge pools dry and individuals are forced to disperse in an attempt to avoid desiccation. Drought-escape behavior (DEB) has been observed in several other arid-land aquatic invertebrates (Smith 1973; Velasco and Millan 1998; Lytle et al. 2008), and there is evidence that giant water bugs also exhibit DEB. Boersma and Lytle (2014) discovered a single male giant water bug walking downslope in a dry streambed in High Creek, Galiuro Mountains, Arizona. It was 110 m downstream from the nearest water source, which was in the process of drying completely. It is possible that rare dispersal events like this one promote gene flow among populations that have been historically isolated. However, given the limited extent of perennial aquatic habitat during summer, it seems unlikely that DEB could result in successful colonization with enough regularity to rescue imperiled populations. In the Boersma and Lytle (2014) example, there was no remaining perennial surface water for the dispersing individual to locate anywhere in the High Creek drainage. Thus, the large distances among potential source populations and increasing aridification in the region may make DEB a poor or inviable mechanism for recolonization in remote streams such as French Joe Canyon or High Creek.

Implications for currently imperiled populations

As of November 2017, giant water bugs had not returned to French Joe Canyon following their extirpation in 2005 (M. T. Bogan, pers. comm.). However, observations of recolonization following extirpation of other populations provide a glimmer of hope. Ash Creek is a fragmented stream in the Galiuro Mountains, Arizona, that contained a healthy giant-water-bug population from 2000 to 2002 (KSB and DAL, unpublished data). In 2003, perennial water in the drainage decreased to two small pools (<5 m^2 wetted surface area), and the species disappeared. Twice-yearly sampling of the limited remaining aquatic habitat from 2003 to 2008 failed to detect giant water bugs (Bogan 2012). Then, in 2009, two males were encountered. Giant water bugs are conspicuous, especially in such a small habitat area, which makes it unlikely they were present from 2003 to 2008 but undetected. Perhaps these two individuals dispersed from the nearest potential source population 4 km west in Redfield Canyon. Unfortunately, two males cannot reestablish a population. Sampling in 2010 located only one male, and in 2011 none were detected (M. T. Bogan, pers. comm.). Subsequent visits confirmed that this population was again extirpated. Aside from Ash Creek and French Joe Canyon, other population extirpations in the region are suspected, including High Creek (KSB and DAL, unpublished data) and an unspecified drainage in the Baboquivari Mountains (historical records in Menke 1960).

The Sycamore Caddisfly

Aquatic apex predators like the giant water bug are known to be vulnerable to the effects of climate change (Estes et al. 2011), but recent evidence suggests that other large-bodied aquatic organisms are also disproportionately at risk (Daufresne et al. 2009). The sycamore caddisfly (Trichoptera: Calamoceratidae; fig. 10.4) is another species that was extirpated from French Joe Canyon during the transition to intermittency in 2005. It is a large (22 mm as larvae), univoltine caddisfly with larvae that use sycamore leaves as case-building material and as a primary food source. This species is one of the few detritivores in forested arid-land streams that shreds coarse particulate organic matter into fine particulate organic matter, increasing availability of carbon for the rest of the detritivore food web. Sycamore caddisflies can be numerically abundant in small headwater streams (e.g., 80 individuals/m^2; Lytle 2000), and were common in French Joe Canyon prior to the transition to intermittency.

Adaptations to flash flooding

In arid regions of the southwestern United States and northwestern México, sycamore caddisflies inhabit perennial streams that are exposed to powerful late-

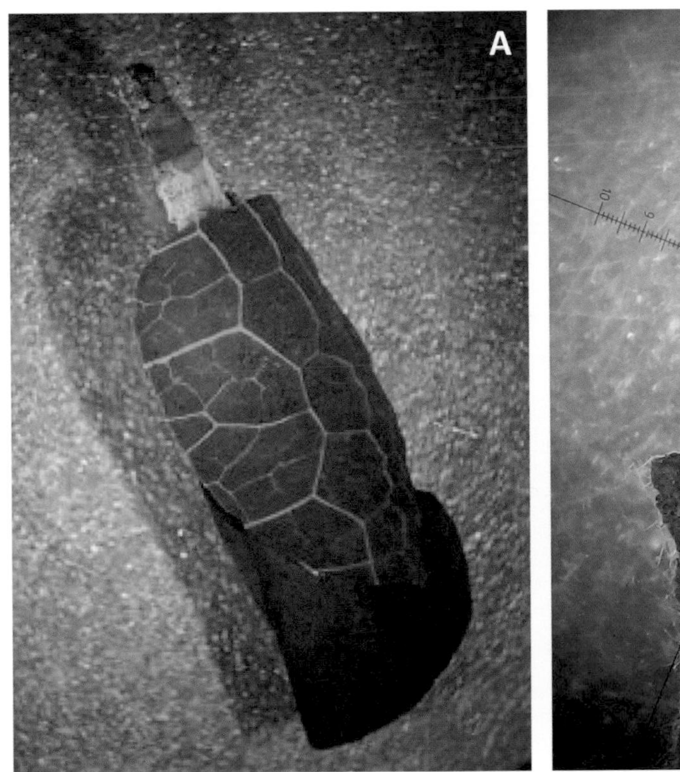

Fig. 10.4 Sycamore caddisfly (*Phylloicus mexicanus*) larva. (*A*) Larvae construct cases made of the leaves of the sycamore tree (*Platanus wrightii*). (Photo credit: K. S. Boersma.) (*B*) Detail of larva in a sycamore leaf case. (Photo credit: R. Van Driesche.)

summer monsoons and stream-scouring flash floods like those found in French Joe Canyon. Larvae that remain in streams during a flash flood suffer 96% mortality (Lytle 2000). Consequently, the sycamore caddisfly has evolved an elegant mechanism to escape flash flooding: it synchronizes its life history timing to emerge from the larval aquatic stage into the terrestrial adult stage before onset of the monsoon season (Lytle 2003). However, individuals face a trade-off in timing of emergence: they can either emerge earlier to better avoid catastrophic floods (albeit at a smaller body size and with lower fecundity), or later at a larger body size with greater fecundity (but with more risk of mortality by flooding) (Lytle 2002). Thus, individuals within populations exhibit a range of emergence times as a bet-hedging strategy. This adaptation confers the ability to survive annual cycles of flash flooding even if the timing of floods varies, but it does not protect the caddisfly from the effects of supraseasonal drought.

When Bogan and Lytle (2011) recorded catastrophic drying at French Joe Canyon in 2005, the sycamore caddisfly disappeared from the site, and it had not returned as of November 2017 (M. T. Bogan, pers. comm.). Caddisfly functional traits reflect this low resistance and resilience to intermittency: (1) their long lifespan and slow development means that they require surface water for much of the year in order to successfully reproduce, and (2) adults are weak fliers and have limited dispersal capacity to recolonize previously occupied habitats following extirpation (Wiggins 1977). Thus, the disappearance of perennial surface water in French Joe Canyon removed not only the apex predator that regulated populations of other aquatic insects, but also a large-bodied detritivore responsible for shredding leaf litter.

Subsequent manipulative experiments further suggested that sycamore caddisflies are negatively impacted by local extinctions of the giant water bug (Boersma et al. 2014b), possibly due to the latter species' ability to suppress mesopredators that prey upon the caddisfly. We hypothesize that caddisfly extirpation may have been caused by increasing predation pressure from

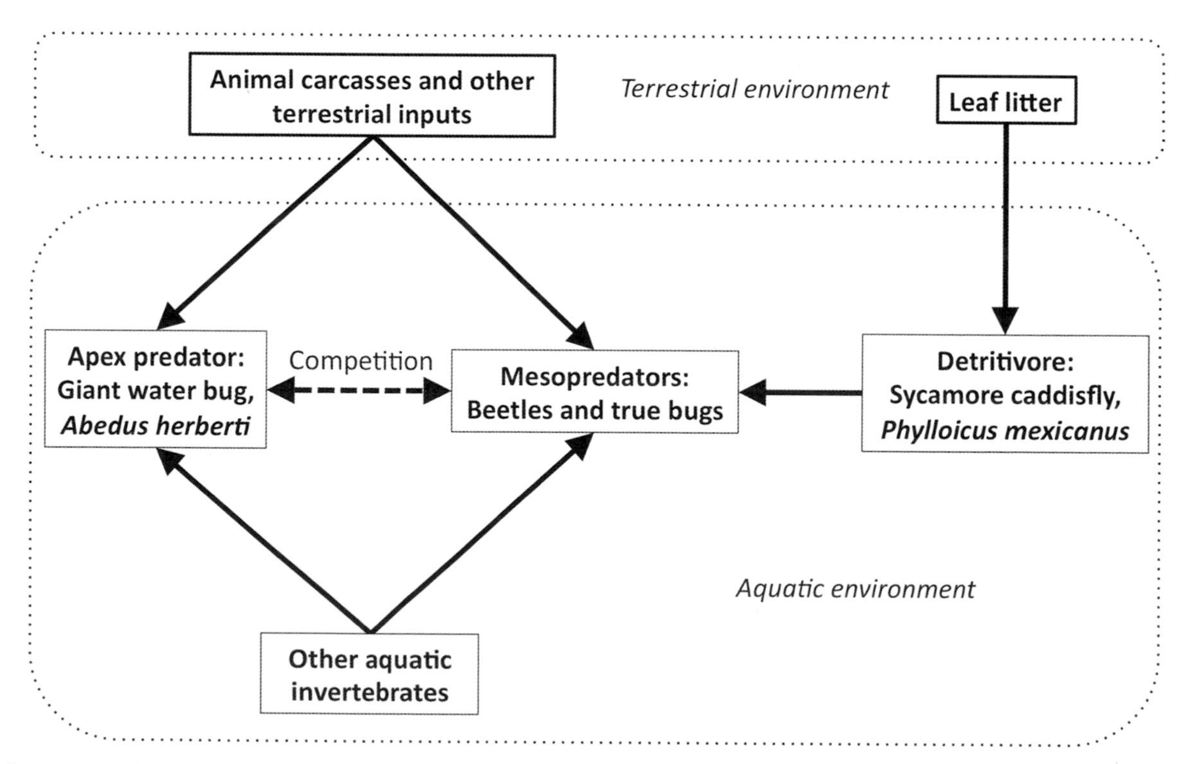

Fig. 10.5 Hypothesized trophic interactions in a simplified French Joe Canyon food web. The terrestrial environment (dotted lines, above) contributes animal and plant matter to the aquatic environment (dotted lines, below). The giant water bug and mesopredators compete (dashed arrow) to consume other aquatic invertebrates and terrestrial inputs (solid arrows). The sycamore caddisfly consumes leaf litter from the terrestrial environment and is consumed or otherwise negatively impacted by aquatic mesopredators (solid arrows). When the giant water bug is locally extirpated, this releases mesopredators from competition and may increase predation on the sycamore caddisfly (mesopredator release).

mesopredators released from competition when the giant water bug was extirpated (mesopredator release, fig. 10.5). However, this mechanism remains untested and awaits future controlled experiments that illustrate the direct interaction between mesopredators and the sycamore caddisfly.

As a shredding species that integrates terrestrial-leaf resources into aquatic food webs, the sycamore caddisfly plays an important functional role by increasing carbon availability within the ecosystem (e.g., Wallace and Webster 1996). While apex predator extirpation modifies top-down control of aquatic food webs, the extirpation of detritivores like the sycamore caddisfly modifies bottom-up control, meaning that the food web in French Joe Canyon was disrupted from basal and apex trophic levels, concurrently. The combination of top-down and bottom-up food-web alterations may explain the radical changes in community structure following the shift from perennial to intermittent conditions (see fig. 10.3).

FUNCTIONAL DIVERSITY AND DROUGHT

As the case study demonstrates, it is not just aquatic *biodiversity* that is at risk with climate change, but also aquatic ecosystem *functioning*. In aquatic ecosystems worldwide, decreasing functional diversity has been associated with negative impacts on ecosystem functioning (Dewson et al. 2007; Ledger et al. 2011) and on provisioning of ecosystem services (Olden et al. 2008; Strecker et al. 2011). Ecological functions such as decomposition, nutrient cycling, and biomass production are impacted by drought (Cadotte et al. 2011), and functional traits provide clues to how aquatic invertebrate responses to drought mediate these changes.

Recently developed databases provide information on the functional traits of aquatic invertebrates from arid-land streams of the southwestern United States and northwestern México (available in Schriever et al. 2015; Boersma et al. 2014a; Cañedo-Argüelles et al. 2016) and enable analyses of how functional diversity responds to drought (Cañedo-Argüelles et al. 2015; Boersma et al. 2016). These studies reveal that functional diversity may be more sensitive to drought than taxonomic diversity.

If species with unique ecological roles are replaced by species with roles that are already present, the total number of species (taxonomic richness) remains unchanged but functional diversity decreases, potentially diminishing ecosystem functioning (Carmona et al. 2012). In French Joe Canyon, functional richness decreased following the transition from perennial to intermittent hydrology (Boersma et al. 2016), even though taxonomic richness did not change (Bogan and Lytle 2011). This observation is alarming because it suggests that even if perennial habitat returns to an intermittent creek, the ecosystem services provided by extirpated taxa (e.g., shredding of coarse particulate organic matter) may not.

Despite a growing understanding of how functional traits like dispersal capacity, lifespan, and desiccation tolerance are associated with particular species' vulnerability to supraseasonal droughts, less is known about how changes in community-wide functional diversity will affect ecosystem functioning under new drought regimes. For example, we can infer that the disappearance of a large-bodied shredder will impact leaf litter decomposition, but until we quantify leaf litter consumption and excretion rates in sycamore caddisflies and other detritivorous taxa, we are unable to predict the extent of the negative consequences of these extirpations. Future studies linking drought-induced reductions in functional diversity with ecosystem attributes, such as decomposition and nutrient cycling, are necessary to describe the consequences of novel drought regimes in detail.

CONSERVATION OF AQUATIC INVERTEBRATE COMMUNITIES

Genetic isolation, degradation and loss of aquatic habitats, increasing environmental extremes, and trophic collapse are threats that arid-land aquatic invertebrates share with desert fishes and other aquatic organisms (Mims et al. 2014). In many cases, similar conservation strategies can be applied to vertebrates and invertebrates. Here we make recommendations inferred from the research presented in this chapter.

Recognizing Vulnerable Taxa

As our case study demonstrates, aquatic invertebrates are not all equivalent in their capacity to withstand and recover from novel drought disturbances. Specific trait combinations, such as large body size, limited dispersal capacity, long generation time, and low tolerance to environmental extremes, can often identify taxa that are disproportionately vulnerable. Nor are invertebrates equivalent in their contributions to ecosystem functioning. In aquatic communities comprising dozens of species, it may be necessary to prioritize conservation of functionally important taxa—those that disproportionately contribute to ecosystem functioning and the stability of food webs. As shown here, apex predators and large-bodied shredders are more vulnerable to drought, functionally more important, and less likely to rapidly recolonize intermittent habitats than are species filling other functional roles. Thus, their conservation should be a top priority to preserve functional diversity in isolated aquatic habitats of arid-land stream ecosystems.

Refuge Habitats

How might this conservation occur? Arid-land aquatic invertebrate communities exhibit high resistance and resilience to the extreme fluctuations in precipitation and declines in water quality that occur during droughts (Lake 2003). The "tipping point" and transition to novel community states appear to occur with catastrophic, whole-stream drying that makes formerly perennial habitats become intermittent. Creating artificial pools in drought-prone catchments may help counteract this effect and maintain vulnerable populations (Carini et al. 2006), and may also provide stopover habitats to enable recolonization after drying (e.g., Mims et al. 2016). In fact, irrigation pools and reservoirs designed for agriculture and livestock use can serve as effective refuge habitats for many aquatic species (Hazell et al. 2001; Abellán et al. 2006), although care must be taken to avoid harboring nonnative species.

Salvage and Reintroduction

Focusing conservation efforts around target taxa may increase the overlap between invertebrate and fish conservation strategies. During catastrophic drying events, desert fish conservationists often enact rescue operations that include salvage of threatened populations and assisted recolonization following rewetting (Minckley and Deacon 1991; Lema et al., chap. 22, this volume). Salvage and reintroduction have also been used to successfully reestablish populations of the threatened Ash Meadows naucorid (*Ambrysus amargosus*), a flightless predatory invertebrate endemic to Ash Meadows National Wildlife Refuge, Nevada. Similar strategies may be useful with functionally important taxa like the giant water bug and sycamore caddisfly. However, it remains challenging to reestablish intact food webs so that target taxa can recolonize functioning ecosystems instead of depauperate, functionally impaired communities that might not support them. Introducing freshwater fishes to habitats without prey is a primary cause of reintroduction failure (Cochran-Biederman et al. 2015), and similar outcomes can be expected with freshwater invertebrates if food-web structure is not considered during reintroduction. Furthermore, as discussed earlier, populations of poorly dispersing species such as giant water bugs possess unique genetic structures, behaviors, and morphologies. In this case, it may never be possible to fully replace the diversity that is lost when a unique population becomes locally extinct.

Ecosystem-Wide Conservation

Due to the central role that invertebrates play in maintaining food-web stability and ecosystem functioning, we advocate an ecosystem-wide approach to conservation and restoration of arid-land aquatic habitats that takes aquatic invertebrates into account along with other taxa. Conservation efforts intended to preserve the physical and biological integrity of an entire habitat, including its hydrologic properties, are more likely to conserve a diverse assemblage of aquatic organisms than single-species conservation efforts alone. The ecological roles of all native species, including the diverse spectrum of aquatic invertebrates, need to be understood, recognized, and conserved.

REFERENCES

Abellán, P., D. Sanchéz-Fernández, A. Millán, F. Botella, J. A. Sánchez-Zapata, and A. Giménez. 2006. Irrigation pools as macroinvertebrate habitat in a semi-arid agricultural landscape (SE Spain). *Journal of Arid Environments* 67: 255–69.

Balling, R. C., and G. B. Goodrich. 2010. Increasing drought in the American Southwest? A continental perspective using a spatial analytical evaluation of recent trends. *Physical Geography* 31: 293–306.

Barnett, T. P., D. W. Pierce, H. G. Hidalgo, C. Bonfils, B. D. Santer, T. Das, G. Bala, A. W. Wood, T. Nozawa, A. A. Mirin, D. R. Cayan, and M. D. Dettinger. 2008. Human-induced changes in the hydrology of the western United States. *Science* 319: 1080–83.

Betancourt, J. L., T. R. Van Devender, and P. S. Martin. 1990. *Packrat Middens: The Last 40,000 Years of Biotic Change*. Tucson: University of Arizona Press.

Boersma, K. S., M. T. Bogan, B. A. Henrichs, and D. A. Lytle. 2014a. Invertebrate assemblages of pools in arid-land streams have high functional redundancy and are resistant to severe drying. *Freshwater Biology* 59: 491–501.

Boersma, K. S., M. T. Bogan, B. A. Henrichs, and D. A. Lytle. 2014b. Top predator removals have consistent effects on large species despite high environmental variability. *Oikos* 123: 807–16.

Boersma, K. S., L. E. Dee, S. J. Miller, M. T. Bogan, D. A. Lytle, and A. I. Gitelman. 2016. Linking multidimensional functional diversity to quantitative methods: A graphical hypothesis-evaluation framework. *Ecology* 97: 583–93.

Boersma, K. S., and D. A. Lytle. 2014. Overland dispersal and drought-escape behavior in a flightless aquatic insect, *Abedus herberti* (Hemiptera: Belostomatidae). *Southwestern Naturalist* 59: 301–2.

Bogan, M. T. 2012. Drought, dispersal and community dynamics in arid-land streams. PhD diss., Oregon State University.

Bogan, M. T. 2017. Hurry up and wait: Life cycle and distribution of an intermittent stream specialist (*Mesocapnia arizonensis*). *Freshwater Science* 36: 805–15.

Bogan, M. T., and K. S. Boersma. 2012. Aerial dispersal of aquatic invertebrates along and away from arid-land streams. *Freshwater Science* 31: 1131–44.

Bogan, M. T., K. S. Boersma, and D. A. Lytle. 2015. Resistance and resilience of invertebrate communities to seasonal and supraseasonal drought in arid-land headwater streams. *Freshwater Biology* 60: 2547–58.

Bogan, M. T., O. Gutiérrez-Ruacho, A. Alvarado-Castro, and D. A. Lytle. 2013. Habitat type and permanence determine local aquatic invertebrate community structure in the Madrean Sky Islands. In G. J. Gottfried, P. F. Folliott, B. S. Gebow, L. G. Eskew, and L. C. Collins, eds., *Merging Science and Management in a Rapidly Changing World: Biodiversity and Management of the Madrean Archipelago III*, 277–82. US Forest Service Proceedings RMRS-P-67. Fort Collins, CO: USDA Forest Service.

Bogan, M. T., and D. A. Lytle. 2011. Severe drought drives novel community trajectories in desert stream pools. *Freshwater Biology* 56: 2070–81.

Borrvall, C., and B. Ebenman. 2006. Early onset of secondary extinctions in ecological communities following the loss of top predators. *Ecology Letters* 9: 435–42.

Boulton, A. J., and P. S. Lake. 2008. Effects of drought on stream insects and its ecological consequences. In J. Lancaster and R. A. Briers, eds., *Aquatic Insects: Challenges to Populations*, 81–102. London: CABI Publishing.

Cadotte, M. W., K. Carscadden, and N. Mirotchnick. 2011. Beyond species: Functional diversity and the maintenance of ecological processes and services. *Journal of Applied Ecology* 48: 1079–87.

Cañedo-Argüelles, M., K. S. Boersma, M. T. Bogan, J. D. Olden, I. Phillipsen, T. A. Schriever, and D. A. Lytle. 2015. Dispersal strength determines meta-community structure in a dendritic riverine network. *Journal of Biogeography* 42: 778–90.

Cañedo-Argüelles, M., M. T. Bogan, D. A. Lytle, and N. Prat. 2016. Are Chironomidae (Diptera) good indicators of water scarcity? Dryland streams as a case study. *Ecological Indicators* 71: 155–62.

Carini, G., J. M. Hughes, and S. E. Bunn. 2006. The role of waterholes as "refugia" in sustaining genetic diversity and variation of two freshwater species in dryland river systems (Western Queensland, Australia). *Freshwater Biology* 51: 1434–46.

Carmona, C. P., F. M. Azcárate, F. de Bello, H. S. Ollero, J. Lepš, and B. Peco. 2012. Taxonomical and functional diversity turnover in mediterranean grasslands: Interactions between grazing, habitat type and rainfall. *Journal of Applied Ecology* 49: 1084–93.

Cayan, D. R., T. Das, D. W. Pierce, T. P. Barnett, M. Tyree, and A. Gershunov. 2010. Future dryness in the Southwest US and the hydrology of the early 21st century drought. *Proceedings of the National Academy of Sciences* 107: 21271–76.

Chester, E. T., and B. J. Robson. 2011. Drought refuges, spatial scale and recolonisation by invertebrates in non-perennial streams. *Freshwater Biology* 56: 2094–104.

Clark, M. S., N. Y. Denekamp, M. A. S. Thorne, R. Reinhardt, M. Drungowski, M. W. Albrecht, S. Klages, A. Beck, M. Kube, and E. Lubzens. 2012. Long-term survival of hydrated resting eggs from *Brachionus plicatilis*. *PLoS ONE* 7: e29365.

Cochran-Biederman, J. L., K. E. Wyman, W. E. French, and G. L. Loppnow. 2015. Identifying correlates of success and failure of native freshwater fish reintroductions. *Conservation Biology* 29: 175–86.

Cook, B. I., T. R. Ault, and J. E. Smerdon. 2015. Unprecedented 21st century drought risk in the American Southwest and Central Plains. *Science Advances* 1: e1400082.

Cook, B. I., J. E. Smerdon, R. Seager, and S. Coats. 2014. Global warming and 21st century drying. *Climate Dynamics* 43: 2607–27.

Cover, M. R., J. H. Seo, and V. H. Resh. 2015. Life history, burrowing behavior, and distribution of *Neohermes filicornis* (Megaloptera: Corydalidae), a long-lived aquatic insect in intermittent streams. *Western North American Naturalist* 75: 474–90.

Daufresne, M., K. Lengfellner, and U. Sommer. 2009. Global warming benefits the small in aquatic ecosystems. *Proceedings of the National Academy of Sciences* 106: 12788–93.

Deacon, J. E., A. E. Williams, C. Deacon Williams, and J. E. Williams. 2007. Fueling population growth in Las Vegas: How large-scale groundwater withdrawal could burn regional biodiversity. *BioScience* 57: 688–98.

De Block, M., and R. Stoks. 2005. Pond drying and hatching date shape the tradeoff between age and size at emergence in a damselfly. *Oikos* 108: 485–94.

Dewson, Z. S., A. B. W. James, and R. G. Death. 2007. Stream ecosystem functioning under reduced flow conditions. *Ecological Applications* 17: 1797–808.

Drummond, L. R., A. R. McIntosh, and S. T. Larned. 2015. Invertebrate community dynamics and insect emergence in response to pool drying in a temporary river. *Freshwater Biology* 60: 1596–612.

Estes, J. A., J. Terborgh, J. S. Brashares, M. E. Power, J. Berger, W. J. Bond, S. R. Carpenter, T. E. Essington, R. D. Holt, J. B. C. Jackson, R. J. Marquis, L. Oksanen, T. Oksanen, R. T.

Paine, E. K. Pikitch, W. J. Ripple, S. A. Sandin, M. Scheffer, T. W. Schoener, J. B. Shurin, A. R. E. Sinclair, M. E. Soulé, R. Virtanen, and D. A. Wardle. 2011. Trophic downgrading of Planet Earth. *Science* 333: 301–6.

Finn, D. S., M. S. Blouin, and D. A. Lytle. 2007. Population genetic structure reveals terrestrial affinities for a headwater stream insect. *Freshwater Biology* 52: 1881–97.

Ganser, A. M., T. J. Newton, and R. J. Haro. 2015. Effects of elevated water temperature on physiological responses in adult freshwater mussels. *Freshwater Biology* 60: 1705–16.

Hazell, D., R. Cunningham, D. Lindenmayer, B. Mackey, and W. Osborne. 2001. Use of farm dams as frog habitat in an Australian agricultural landscape: Factors affecting species richness and distribution. *Biological Conservation* 102: 155–69.

Holmgren, C. A., M. C. Penalba, K. A. Rylander, and J. L. Betancourt. 2003. A 16,000 ^{14}C yr B.P. packrat midden series from the USA-Mexico Borderlands. *Quaternary Research* 60: 319–29.

Koetsier, P. 2005. Response of a stream diatom community to top predator manipulations. *Aquatic Sciences* 67: 517–27.

Lake, P. S. 2003. Ecological effects of perturbation by drought in flowing waters. *Freshwater Biology* 48: 1161–72.

Lake, P. S. 2013. Resistance, resilience and restoration. *Ecological Management and Restoration* 14: 20–24.

Ledger, M. E., F. K. Edwards, L. E. Brown, A. M. Milner, and G. Woodward. 2011. Impact of simulated drought on ecosystem biomass production: An experimental test in stream mesocosms. *Global Change Biology* 17: 2288–97.

Luo, Y., C. Li, A. G. Landis, G. Wang, J. Stoeckel, and E. Peatman. 2014. Transcriptomic profiling of differential responses to drought in two freshwater mussel species, the giant floater *Pyganodon grandis* and the pondhorn *Uniomerus tetralasmus*. *PLoS ONE* 9: e89481.

Lytle, D. A. 1999. Use of rainfall cues by *Abedus herberti* (Hemiptera: Belostomatidae): A mechanism for avoiding flash floods. *Journal of Insect Behavior* 12: 1–12.

Lytle, D. A. 2000. Biotic and abiotic effects of flash flooding in a montane desert stream. *Archiv für Hydrobiologie* 150: 85–100.

Lytle, D. A. 2002. Flash floods and aquatic insect life-history evolution: Evaluation of multiple models. *Ecology* 83: 370–85.

Lytle, D. A. 2003. Reconstructing long-term flood regimes with rainfall data: Effects of flood timing on caddisfly populations. *Southwestern Naturalist* 48: 36–42.

Lytle, D. A., J. D. Olden, and L. E. McMullen. 2008. Drought-escape behaviors of aquatic insects may be adaptations to highly variable flow regimes characteristic of desert rivers. *Southwestern Naturalist* 53: 399–402.

Lytle, D. A., and R. Smith. 2004. Exaptation and flash flood escape in the giant water bugs. *Journal of Insect Behavior* 17: 169–78.

Menke, A. S. 1960. A taxonomic study of the genus *Abedus* Stal (Hemiptera, Belostomatidae). *University of California Publications in Entomology* 16: 393–440.

Menke, A. S. 1977. Synonymical notes and new distribution records in *Abedus* (Hemiptera, Belostomatidae). *Southwestern Naturalist* 22: 115–23.

Mims, M. C., L. Hauser, C. S. Goldberg, and J. D. Olden. 2016. Genetic differentiation, isolation-by-distance, and metapopulation dynamics of the Arizona treefrog (*Hyla wrightorum*) in an isolated portion of its range. *PLoS ONE* 11: e0160655.

Mims, M. C., I. C. Phillipsen, D. A. Lytle, E. E. H. Kirk, and J. D. Olden. 2014. Ecological strategies predict associations between aquatic and genetic connectivity for dryland amphibians. *Ecology* 96: 1371–82.

Minckley, W., and J. Deacon. 1991. *Battle against Extinction: Native Fish Management in the American West.* Tucson: University of Arizona Press.

Olden, J. D., N. L. Poff, and K. R. Bestgen. 2008. Trait synergisms and the rarity, extirpation, and extinction risk of desert fishes. *Ecology* 89: 847–56.

Paine, R. T. 1966. Food web complexity and species diversity. *American Naturalist* 100: 65–75.

Phillipsen, I. C., E. H. Kirk, M. T. Bogan, M. C. Mims, J. D. Olden, and D. A. Lytle. 2015. Dispersal ability and habitat requirements determine landscape-level genetic patterns in desert aquatic insects. *Molecular Ecology* 24: 54–69.

Phillipsen, I. C., and D. A. Lytle. 2013. Aquatic insects in a sea of desert: Population genetic structure is shaped by limited dispersal in a naturally fragmented landscape. *Ecography* 36: 731–43.

Rebora, M., S. Piersanti, G. Salerno, E. Conti, and E. Gaino. 2007. Water deprivation tolerance and humidity response in a larval dragonfly: A possible adaptation for survival in drying ponds. *Physiological Entomology* 32: 121–26.

Ritchie, E. G., and C. N. Johnson. 2009. Predator interactions, mesopredator release and biodiversity conservation. *Ecology Letters* 12: 982–98.

Rodríguez-Lozano, P., I. Verkaik, M. Rieradevall, and N. Prat. 2015. Small but powerful: Top predator local extinction affects ecosystem structure and function in an intermittent stream. *PLoS ONE* 10: e0117630.

Ruff, T. W., Y. Kushnir, and R. Seager. 2011. Comparing twentieth- and twenty-first-century patterns of interannual precipitation variability over the western United States and northern Mexico. *Journal of Hydrometeorology* 13: 366–78.

Schriever, T. A., M. T. Bogan, K. S. Boersma, M. Cañedo-Argüelles, K. L. Jaeger, J. D. Olden, and D. A. Lytle. 2015. Hydrology shapes taxonomic and functional structure of desert stream invertebrate communities. *Freshwater Science* 34: 399–409.

Seager, R., M. Ting, I. Held, Y. Kushnir, J. Lu, G. Vecchi, H.-P. Huang, N. Harnik, A. Leetmaa, N.-C. Lau, C. Li, J. Velez, and N. Naik. 2007. Model projections of an imminent transition to a more arid climate in southwestern North America. *Science* 316: 1181–84.

Seager, R., M. Ting, C. Li, N. Naik, B. Cook, J. Nakamura, and H. Liu. 2013. Projections of declining surface-water availability for the southwestern United States. *Nature Climate Change* 3: 482–86.

Smith, R. L. 1973. Aspects of the biology of three species of the genus *Rhantus* (Coleoptera: Dytiscidae) with special reference to the acoustical behavior of two. *Canadian Entomologist* 105: 909–19.

Smith, R. L. 1979. Repeated copulation and sperm precedence—paternity assurance for a male brooding water bug. *Science* 205: 1029–31.

Sota, T., and M. Mogi. 1992. Interspecific variation in desiccation survival time of *Aedes* (Stegomyia) mosquito eggs is correlated with habitat and egg size. *Oecologia* 90: 353–58.

Sponseller, R. A., N. B. Grimm, A. J. Boulton, and J. L. Sabo. 2010. Responses of macroinvertebrate communities to long-term flow variability in a Sonoran Desert stream. *Global Change Biology* 16: 2891–900.

Stewart, B. A., P. G. Close, P. A. Cook, and P. M. Davies. 2013. Upper thermal tolerances of key taxonomic groups of stream invertebrates. *Hydrobiologia* 718: 131–40.

Strecker, A. L., J. D. Olden, J. B. Whittier, and C. P. Paukert. 2011. Defining conservation priorities for freshwater fishes according to taxonomic, functional, and phylogenetic diversity. *Ecological Applications* 21: 3002–13.

Stubbington, R., M. T. Bogan, N. Bonada, A. J. Boulton, T. Datry, C. Leigh, and R. Vander Vorste. 2017. The biota of intermittent rivers and ephemeral streams: Aquatic invertebrates. In T. Datry, N. Bonada, and A. J. Boulton, eds., *Intermittent Rivers and Ephemeral Streams: Ecology and Management*, 217–44. London: Academic Press.

Tonkin, J. D., D. M. Merritt, J. D. Olden, L. V. Reynolds, and D. A. Lytle. 2018. Flow regime alteration degrades ecological networks in riparian ecosystems. *Nature Ecology and Evolution* 2: 86–93.

Velasco, J., and A. Millan. 1998. Insect dispersal in a drying desert stream: Effects of temperature and water loss. *Southwestern Naturalist* 43: 80–87.

Vörösmarty, C. J., P. B. McIntyre, M. O. Gessner, D. Dudgeon, A. Prusevich, P. Green, S. Glidden, S. E. Bunn, C. A. Sullivan, C. R. Liermann, and P. M. Davies. 2010. Global threats to human water security and river biodiversity. *Nature* 467: 555–61.

Wallace, J. B., and J. R. Webster. 1996. The role of macroinvertebrates in stream ecosystem function. *Annual Review of Entomology* 41: 115–39.

Wickson, S., E. T. Chester, and B. J. Robson. 2012. Aestivation provides flexible mechanisms for survival of stream drying in a larval trichopteran (Leptoceridae). *Marine and Freshwater Research* 63: 821.

Wiggins, G. B. 1977. *Larvae of the North American caddisfly genera (Trichoptera)*. Toronto: University of Toronto Press.

Williams, W. D. 1985. Biotic adaptations in temporary lentic waters, with special reference to those in semi-arid and arid regions. *Hydrobiologia* 125: 85–110.

Woodward, G., N. Bonada, H. B. Feeley, and P. S. Giller. 2015. Resilience of a stream community to extreme climatic events and long-term recovery from a catastrophic flood. *Freshwater Biology* 60: 2497–510.

11 The Exotic Dilemma

Lessons Learned from Efforts to Recover Native Colorado River Basin Fishes

Brandon Albrecht, Ron Kegerries, Ron Rogers, and Paul Holden

Invasions of nonnative species have long impacted desert aquatic ecosystems and their native fishes, and the history of attempted control of nonnatives continues to be written. Nonnatives threaten native desert fishes at all life stages through (1) direct predation of eggs, larvae, juveniles, and adults; (2) habitat alteration (e.g., removal of vegetation); (3) competition and displacement; and (4) hybridization (Rinne 1996; Wind 2004; Valdez and Muth 2005). Relatively aggressive nonnative fishes (e.g., centrarchids) can have a predatory or competitive advantage over native species that did not evolve with comparable levels of competitive or predatory pressures (Johnson et al. 2011). Further, habitat modifications in desert aquatic ecosystems have set the stage for the influx of many nonnative taxa (Carlson and Muth 1989; Minckley and Deacon 1991), and ever-increasing human demands for water, exacerbated by persistent drought conditions, are causing desert aquatic habitats to dwindle (Nash and Gleick 1991; Christensen et al. 2004; Ault et al. 2016).

In smaller streams and isolated water bodies, nonnative control has meant attempting to eliminate every individual. Active-removal techniques include application of piscicides (ironically, the same tool once used to remove natives), mechanical removal of all individuals (Mueller 2005; Coggins and Yard 2010), and mechanical removal of specific nonnative predators or competitors (Marsh et al. 2015; Propst et al. 2015). Where complete removal is possible, native fishes are maintained in nonnative-free environments (Minckley et al. 2003; Mueller 2006).

In larger systems, where complete elimination has proved impossible, managers have sought ways for nonnative and native fish species to coexist by reducing the stress of nonnatives on natives. Ecological techniques that

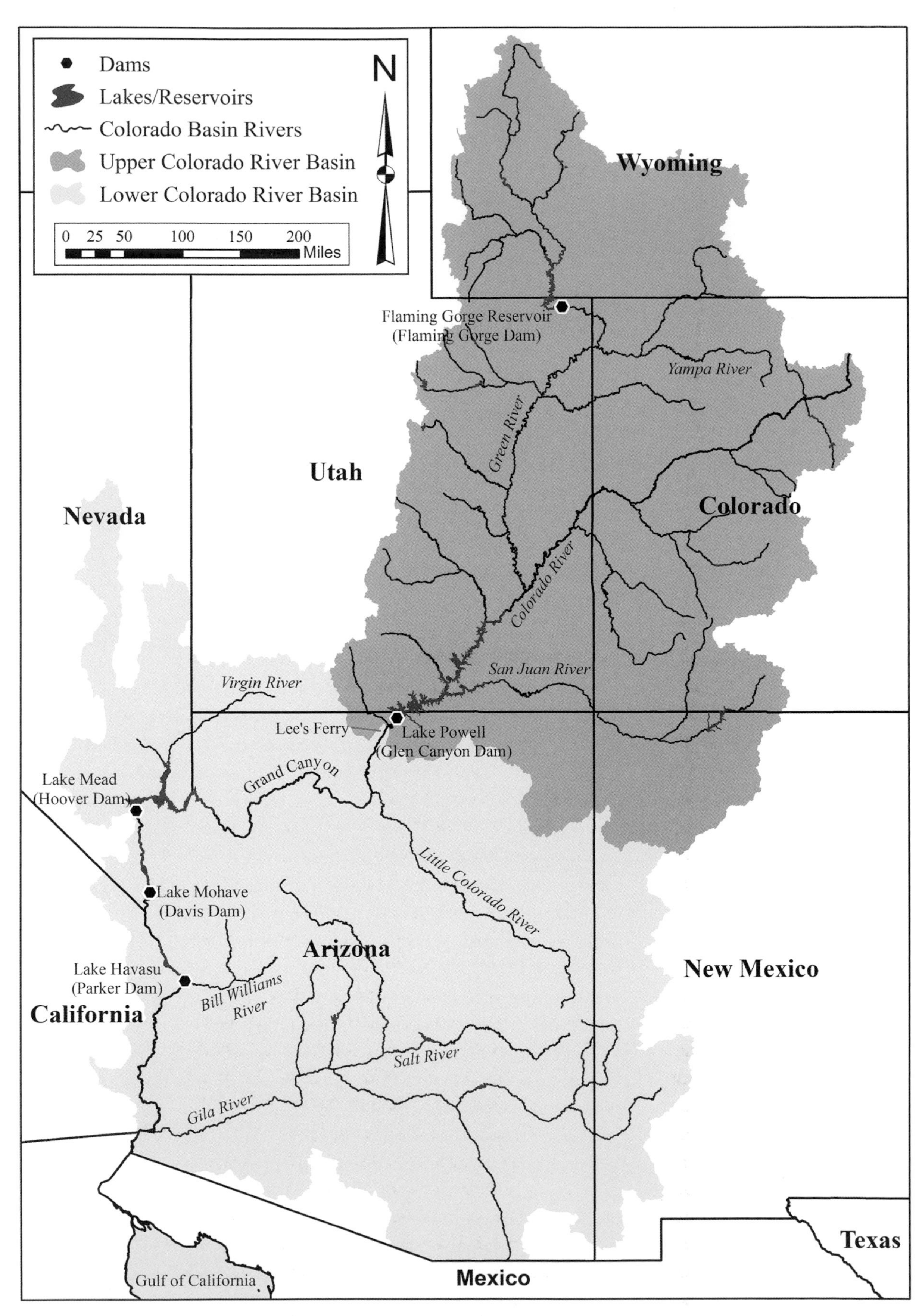

Fig. 11.1 Map of the Colorado River basin. (Map courtesy Bio-West, Inc.)

promote coexistence of natives with nonnatives include manipulation of streamflows to disadvantage nonnative fishes (Propst et al. 2008; LaGory et al. 2012), improvement of habitat diversity and complexity (Poff et al. 1997; Propst and Gido 2004), and protection of existing large, complex habitats from degradation (Propst et al. 2008; Haak and Williams 2012).

Efforts to manage nonnative fishes are ongoing, but different viewpoints on nonnative fish management are present among biologists and managers. These viewpoints range from beliefs that recovery and conservation of native species hinge on the removal of nonnative species to suggestions that habitat ultimately dictates the impacts of nonnative species. Nevertheless, there is still a need for effective tools and approaches that function in these ever-changing environments. This chapter describes the nonnative fish dilemma, and highlights research successes and failures, in light of the ever-changing attitudes toward nonnative fishes within the Colorado River basin.

COLORADO RIVER BASIN

For the purposes of this chapter, the Colorado River and its tributaries above Lees Ferry, near Glen Canyon Dam, are defined as the upper Colorado River basin (UCRB), whereas the Colorado River and its tributaries below Lees Ferry are defined as the lower Colorado River basin (LCRB) (fig. 11.1). Nonnatives currently dominate the fish communities of both basins, with over 60 species present (table 11.1).

Nonnative fishes were first introduced into the Colorado River basin in the early 1900s, and for many decades the basin was managed in favor of nonnatives (Rinne 1996). For example, in 1962, native fishes were purposely removed via chemical piscicides to promote and manage resources for nonnative sport fishes (Holden 1991). This management action gained much public attention, helping prompt eventual passage of the Endangered Species Act of 1973 (Pister 1991). Nevertheless, the effects of nonnatives, along with dams and flow manipulations, continued to negatively affect native fishes (Minckley and Deacon 1991). Combined effects of these alterations resulted in widespread declines in native species (Rinne 1996). As described by Holden (1991, 54), "The major difference between the situation in 1962 [Green River piscicide treatment] and now [1991] is that many native fishes are now protected by law, and state and federal agencies are working to learn more about them and how best to manage them."

Because of its relatively low native fish diversity, high degree of endemism, and (in many cases) highly altered habitats, the Colorado River basin is particularly susceptible to nonnative fish invasion and proliferation (Miller 1961; Hawkins and Nesler 1991; Valdez

Fig. 11.2 A particularly nonnative-fish-laden trammel net haul from the upper Colorado River basin (*top*), and a trammel net full of nonnative fishes from the lower Colorado River basin (*bottom*). (Photos courtesy of BIO-WEST, Inc.)

and Muth 2005). Water development beginning in the early 1900s has resulted in impoundments and diversions throughout the Colorado River basin (Carlson and Muth 1989; Olden and Poff 2005). These projects altered riverine habitats, creating large warmwater reservoirs (Olden and Poff 2005; Valdez and Muth 2005), which are optimal habitats for some nonnative warmwater species, many of which were introduced as forage and sport fish. Through time, many of these nonnatives spread throughout the basin, both up- and downstream of reservoirs, to occupy habitats important for native fishes (Olden and Poff 2005; Albrecht et al. 2017) (fig. 11.2).

ACTIVE APPROACHES, DILEMMAS, AND POTENTIAL LESSONS LEARNED

Upper Colorado River Basin (UCRB)

Nonnative predators and competitors (see table 11.1) emerged in the UCRB as a major threat approximately 25 years ago, and their impacts appear to have increased under persistent drought conditions (UCREFRP 2014). Although all nonnative fish species are problematic, research has identified northern pike *Esox lucius*, smallmouth bass *Micropterus dolomieu*, and walleye *Sander vitreus* as the most problematic species in riverine portions of the UCRB because of their high abundance, overlapping habitat use with natives, and ability to consume most native fishes at multiple life stages (Johnson et al. 2008; UCREFRP 2014).

The Upper Colorado River Endangered Fish Recovery Program (UCREFRP) is working to reduce the numbers of nonnative fishes through a variety of methods and projects. Comprehensive efforts were strategically considered, expanded, refined, and improved over the past decade. These efforts focused on actively removing nonnative fishes from riverine habitats, preventing their escape from upstream reservoirs, revising stocking guidelines, changing harvest regulations, and conducting public outreach (UCREFRP 2014). In-river removal, which is the primary threat-reduction action taken, has been evaluated by several researchers. Although numbers of smallmouth bass and other nonnatives can be reduced through electrofishing or seining, these methods do not necessarily generate a positive response from natives (Bestgen et al. 2007; Skorupski et al. 2012; Breton et al. 2014). Furthermore, removal efforts may need to continue as long as source populations exist (Breton et al. 2014), and it may be that the abundance of adult native fishes is often insufficient to produce enough larvae to elicit a positive response to nonnative removal (Bestgen et al. 2007).

In other studies, removal of northern pike was ineffective, with more individuals annually recruited than removed, indicating that either greater or more targeted removal efforts are needed (Zelasko et al. 2016). Walleye removal efforts are also ongoing, but without noticeable declines in adult catch rates (Michaud et al. 2016). Thus, researchers have concluded that reducing in-river reproduction and emigration from off-channel or other external sources would be necessary for controlling nonnative fish populations (Breton et al. 2014; Zelasko et al. 2016). Additionally, the UCREFRP has recently focused on the installation of escape-prevention devices at upstream reservoirs, such as Elkhead, Rifle Gap, Starvation, and Ridgway, as well as the eradication of illegally introduced populations, such as those at Paonia, Miramonte, and Red Fleet Reservoirs (UCREFRP 2014). Over the past 15 years, nonnative fish management in the UCREFRP has been adaptive, and it continues to improve toward this end. However, much of the success is measured by the number of nonnative fish removed rather than the response of the native fish community.

San Juan River

In the San Juan River sub-basin of the UCRB, removal of large-bodied nonnative fishes is thought necessary to increase survival of native fish species (Brooks et al. 2000). Channel catfish *Ictalurus punctatus* is perceived as the most problematic nonnative fish species (Hines 2016), but removal efforts have yet to demonstrate a strong response by native fish (Franssen et al. 2014). For 15 years, these efforts focused primarily on reducing channel catfish and common carp *Cyprinus carpio*. A decline in common carp relative abundance suggests that removal efforts affected that species, but channel catfish abundance estimates remain relatively stable (Duran 2016). Meanwhile, the populations of native razorback sucker *Xyrauchen texanus* and Colorado pikeminnow *Ptychocheilus lucius* are both sustained by stocking efforts due to little natural recruitment. Although it is unlikely that channel catfish and common carp will be eradicated, the justification for continuing

Table 11.1 Nonnative fishes in the upper Colorado River basin (UCRB) and lower Colorado River basin (LCRB).

FAMILIES AND SPECIES	BASIN PRESENCE
Catostomidae (suckers)	
Bigmouth buffalo, *Ictiobus cyprinellus*	LCRB
Utah sucker, *Catostomus ardens*	UCRB
Longnose sucker, *C. catostomus*	UCRB
White sucker, *C. commersonii*	UCRB
Centrarchidae (sunfishes)	
Rock bass, *Ambloplites rupestris*	LCRB
Green sunfish, *Lepomis cyanellus*	UCRB/LCRB
Bluegill, *L. macrochirus*	UCRB/LCRB
Redear sunfish, *L. microlophus*	LCRB
Warmouth, *L. gulosus*	LCRB
Smallmouth bass, *Micropterus dolomieu*	UCRB/LCRB
Largemouth bass, *M. salmoides*	UCRB/LCRB
White crappie, *Pomoxis annularis*	UCRB/LCRB
Black crappie, *P. nigromaculatus*	UCRB/LCRB
Cichlidae (cichlids)	
Blue tilapia, *Oreochromis aureus*	LCRB
Mozambique tilapia, *O. mossambicus*	LCRB
Redbelly tilapia, *O. zillii*	LCRB
Clupeidae (herrings)	
Gizzard shad, *Dorosoma cepedianum*	UCRB/LCRB
Threadfin shad, *D. petenense*	UCRB/LCRB
Cyprinidae (minnows)	
Goldfish, *Carassius auratus*	UCRB/LCRB
Grass carp, *Ctenopharyngodon idella*	UCRB/LCRB
Red shiner, *Cyprinella lutrensis*	UCRB/LCRB
Common carp, *Cyprinus carpio*	UCRB/LCRB
Utah chub, *Gila atraria*	UCRB
Brassy minnow, *Hybognathus hankinsoni*	UCRB
Plains minnow, *H. placitus*	UCRB
Golden shiner, *Notemigonus crysoleucas*	UCRB/LCRB
Sand shiner, *Notropis stramineus*	UCRB
Fathead minnow, *Pimephales promelas*	UCRB/LCRB
Bullhead minnow, *P. vigilax*	UCRB
Longnose dace, *Rhinichthys cataractae*	UCRB
Redside shiner, *Richardsonius balteatus*	UCRB
Creek chub, *Semotilus atromaculatus*	UCRB
Leatherside chub, *Snyderichthys copei*	UCRB
Cyprinodontidae (killifishes and pupfishes)	
Plains topminnow, *Fundulus sciadicus*	UCRB/LCRB

FAMILIES AND SPECIES	BASIN PRESENCE
Plains killifish, *F. zebrinus*	UCRB/LCRB
Rainwater killifish, *Lucania parva*	UCRB
Esocidae (pikes)	
Northern pike, *Esox lucius*	UCRB/LCRB
Gadidae (cods)	
Burbot, *Lota lota*	UCRB
Gasterosteidae (sticklebacks)	
Brook stickleback, *Culaea inconstans*	UCRB
Ictaluridae (catfishes)	
Black bullhead, *Ameiurus melas*	UCRB/LCRB
Yellow bullhead, *A. natalis*	UCRB/LCRB
Brown bullhead, *A. nebulosus*	UCRB
Flathead catfish, *Pylodictis olivaris*	UCRB/LCRB
Channel catfish, *Ictalurus punctatus*	UCRB/LCRB
Loricariidae (suckermouth armored catfishes)	
Sailfin catfish, *Pterygoplichthys disjunctivus*	LCRB
Moronidae (temperate basses)	
Yellow bass, *Morone mississippiensis*	LCRB
White bass, *M. chrysops*	UCRB
Striped bass, *M. saxatilis*	UCRB/LCRB
Percidae (perches)	
Iowa darter, *Etheostoma exile*	UCRB
Johnny darter, *E. nigrum*	UCRB
Yellow perch, *Perca flavescens*	UCRB
Walleye, *Sander vitreus*	UCRB/LCRB
Poeciliidae (livebearers)	
Sailfin molly, *Poecilia latipinna*	LCRB
Mexican molly, *P. mexicana*	LCRB
Platyfish, *Xiphophorus* spp.	LCRB
Western mosquitofish, *Gambusia affinis*	UCRB/LCRB
Salmonidae (trout and salmon)	
Arctic grayling, *Thymallus arcticus*	LCRB
Yellowstone cutthroat trout, *Oncorhynchus clarkii bouvieri*	UCRB
Greenback cutthroat trout, *O. c. stomias*	UCRB
Coho salmon, *O. kisutch*	UCRB
Rainbow trout, *O. mykiss*	UCRB/LCRB
Kokanee, *O. nerka*	UCRB
Brown trout, *Salmo trutta*	UCRB/LCRB
Brook trout, *Salvelinus fontinalis*	UCRB
Lake trout, *S. namaycush*	UCRB

Sources: Modified from Marsh and Minckley 1985, Minckley et al. 2003, Valdez and Muth 2005, Pitts 2008, Pool et al. 2010, and UCREFRP 2014.

their removal is the hope that the number of nonnative individuals can be reduced sufficiently to reduce predation and competition with native fish to a level that increases post-stocking success (Duran 2016). Overall, the channel catfish population has proved resilient to mechanical removal, sustaining its large-scale demographic impacts on native fishes (Franssen et al. 2014).

Lower Colorado River Basin (LCRB)

Mainstem Lower Colorado River

The Lower Colorado River was once a large, dynamic, meandering braided river and floodplain that included abundant isolated pools, oxbow lakes, and backwater habitats, with distinct river segments separated by narrow canyons and high-gradient reaches (Mueller and Marsh 2002). The historical river delta was a vast area with complex braided-channel features and nursery habitat, occupied by a thriving native fish community. However, in less than 90 years, this section of river became heavily managed, with mainstem dams, diversions, levees, and canals designed to store over five years of average annual inflow for generating electricity, irrigating crops, and supplying domestic uses. At least 20 reservoirs now inundate about 1,750 km^2 of floodplain habitat and largely prevent the river from reaching the Gulf of California.

While the LCRB historically had few predatory fishes, many predatory nonnatives have been introduced since the 1930s as food and sport resources (as described for the UCRB above) or as biological controls for pests, such as western mosquitofish *Gambusia affinis* for mosquito abatement (Rinne 1996; Mueller and Marsh 2002). Stabilized habitat conditions in reservoirs and reservoir tailwaters probably fueled fisheries managers' desire to address public interests in sport fishing and pest control and gave nonnative fish species opportunities to establish extensive populations, overwhelm native species, and become the dominant aquatic fauna in the LCRB.

Habitat changes and nonnative fishes are obstacles for native fish conservation and recovery under current conditions in the LCRB (Minckley et al. 2003), especially below Lake Mead. All life stages of native fish are vulnerable to predation, but larvae and juveniles are particularly susceptible (Marsh et al. 2015; Ehlo et al. 2017). It has been suggested that removal of nonnative fishes from the LCRB may allow native fish stocks to persist in, if not re-inhabit, their former ranges (Mueller and Marsh 2002). However, the complete removal of nonnative fish is probably impossible for several reasons: (1) current technology is cost-prohibitive, (2) current habitat conditions favor nonnatives, and (3) returning the river to pre-1930s conditions is infeasible (Rinne 1996; Mueller and Marsh 2002; Mueller 2005). A more likely scenario is increased demand for LCRB resources (e.g., water, electricity, recreation) as the human population increases. Further, anglers desiring fishing opportunities rely on state agencies to provide fishable populations of popular nonnatives (Mueller and Marsh 2002), creating conflict within state agencies (charged with providing sport fisheries and, at the same time, protecting native species) as well as between state agencies focused on management for recreation versus federal agencies focused on endangered species protection (Clarkson et al. 2005).

An important strategy for managing native fish species conservation and recovery in the LCRB has been to isolate the most vulnerable life stages in off-channel refuge habitats from which nonnatives have been removed, later repatriating adults reared in these refuges into mainstem habitats (Marsh et al. 2015). Presumably, adults have a higher likelihood of survival, but even the largest adults may be susceptible to nonnative fish predation (Karam and Marsh 2010; Marsh et al. 2015). Nonetheless, this strategy can allow for some survival and recruitment and help preserve genetic diversity (Minckley et al. 2003; Marsh et al. 2015). This strategy has been tested, with mixed results, in isolated backwaters surrounding Lake Mohave (e.g., Yuma and Davis Coves), Cibola High Levee Pond, and Imperial Ponds, as well as at other locations. Wild-spawned larvae have also been reared to adulthood in hatcheries and restocked into the wild.

The success of both razorback sucker and bonytail *Gila elegans* recruitment at Cibola High Levee Pond supports the hypothesis that it may be possible to develop self-sustaining populations of native fish in isolated habitats for purposes of repatriation (Minckley et al. 2003; Mueller 2006, 2007). Successful reproduction and recruitment were first observed in 1998 (Marsh 2000). By 2003, catch data suggested that the fish community was 99.9% native and contained both razorback sucker and bonytail (G. Mueller, pers. comm.). Unfortunately, nonnative largemouth bass *Micropterus salmoides* were observed in the pond in 2004, and the abundance of native fishes quickly diminished. Efforts at the Imperial Ponds and Yuma Cove were also ultimately unsuccessful due to failure to entirely eliminate nonnative fishes (Marsh et al. 2015). These observations create a clear and poignant model of nonnative effects on native populations.

Smaller backwaters around Lake Mohave drain in the winter as water levels in Lake Mohave decline, which effectively eliminates nonnative fish that could contaminate the backwaters in the following year (Marsh et al. 2015). This habitat feature could be valuable for isolating the most vulnerable life stages of native fishes. At Davis Cove, larvae have been successfully reared to juveniles, and observations about habitat use, survival, and population dynamics have been made. Repatriation of adults has provided more stable demographics and higher genetic diversity within the Lake Mohave brood stock. Despite the advantages of these isolated habitats for native fish conservation and recovery, nonnative fishes are pervasive and frequently invade as a result of floods, failed screen systems, human stocking, and possible transport by birds and other wildlife (Minckley et al. 2003; Marsh et al. 2015), which makes this method of recovery unsustainable without long-term, possibly permanent, human intervention.

Ultimately, removal and isolation of nonnative fishes is labor intensive, costly, and probably unsustainable (e.g., Mueller 2005). Regardless of the challenges and difficulties nonnative fishes present to native fishes, natives in the LCRB have had some successes. The river reach between Davis Dam and Parker Dam has been home to a restored population of flannelmouth sucker *Catostomus latipinnis*, and now more razorback suckers can be found in that same reach than in Lake Mohave (Mueller and Wydoski 2004; Best and Lantow 2012). Additionally, razorback suckers have been routinely located after they were stocked near the Bill Williams River area of Lake Havasu (Ehlo et al. 2016). These successes appear to have a common theme: habitat complexity benefits natives.

Virgin River

The Virgin River sub-basin of the LCRB supports two federally listed endangered fish species—woundfin *Plagopterus argentissimus* and Virgin River chub *Gila seminuda*—and presents an example of eradication efforts on a smaller tributary to the Colorado River. When agricultural and municipal water demands increased in this system in the early twentieth century, water diversions became common and permanent. Additionally, Hoover Dam (completed in 1935) allowed Lake Mead to inundate 80 km of the lower Virgin River. Reductions in water flow, lack of suitable habitat, and the invasion of nonnative fishes from Lake Mead and other off-channel sources led to a dramatic decline in woundfin and Virgin River chub. Early monitoring efforts documented the decline of native fishes but failed to address effective recovery actions (Holden et al. 2005).

In 1996, efforts began to eradicate nonnative fishes from the Virgin River, with red shiner *Cyprinella lutrensis* as the target species. This was accomplished by installing barriers and staging efforts from upstream to downstream, reach by reach, to remove red shiners with chemical piscicide after some native fishes were evacuated and held in off-channel facilities (Rehm and Fridell 2008). Although red shiner was successfully eliminated downstream to the Utah-Arizona border, the response of native fishes, particularly woundfin, was marginal through 2014, probably due to intermittent streamflows and elevated water temperatures (M. Schijf, pers. comm.). Happily, thereafter, woundfin captures upstream of the Utah-Arizona border barrier increased from about 2,500 in 2015 to 9,000 in 2016 (M. Schijf, pers. comm.). Increased instream flows, more favorable runoff patterns, and presumably, increased turbidity levels led to greater woundfin recruitment and survival (M. Schijf, pers. comm.; Ward et al. 2016). Although this effort is an example of successful nonnative fish removal, it is unclear if the positive trend will persist, and it is likely that removal of red shiner alone will not be sufficient to recover or enhance native fish populations. Rather, the combination of reduced nonnative fish predation and competition with both natural and management-induced improvements in habitat quantity and quality is paramount to ensure recovery and conservation.

Native trout

Native trout of the LCRB also face challenges from nonnative fishes. As elsewhere, hybridization, predation, and competition threaten native trout (Propst et al. 1992; Rinne and Janisch 1995; Finlayson et al. 2005), and these threats compound other stressors such as habitat degradation and climate change (Williams et al. 2015). However, in smaller streams, especially in headwater areas, removal and exclusion of nonnative trout species have demonstrated some success (Rinne and Janisch 1995; Propst and Stefferud 1997). Additionally, the management of both Apache trout *Oncorhynchus apache* and Gila trout *O. gilae* could potentially enhance sport-fishing opportunities while simultaneously recovering native species (Rinne and Janisch 1995). For example, habitat improvements and nonnative fish removal efforts with the aim of reestablishing Apache trout within their historical range have succeeded and been supported by the general public and sport-fishing community.

PASSIVE APPROACHES, DILEMMAS, AND POTENTIAL LESSONS LEARNED

Reservoir Inflows, Complex Habitats, and Resilient Native Fishes

In contrast to the active techniques described above, another approach that appears to have worked in specific cases has been to simply leave things alone. This passive approach demonstrates the resiliency and plasticity of certain native fish species that, in some locations and under specific conditions, can survive and recruit despite the presence of nonnative species, without active intervention.

An example occurs at tributary inflow areas of and the inflows of some mainstem Colorado River impoundments. These inflow areas provide habitat that is relatively similar to the historical braided river channel, with its complex in- and off-channel habitats, that was once more common to the Colorado River (fig. 11.3). Currently, impoundments have altered many of the low-gradient segments of the Colorado River (Mueller and Marsh 2002; Marsh et al. 2015). The increased cover, in the form of vegetation and turbidity, found in these more dynamic habitats probably protects native fishes from nonnative predators (Ward et al. 2016; Albrecht et al. 2017; Kegerries et al. 2017). Recruitment of wild-origin razorback suckers has been documented nearly annually in inflow areas of Lake Mead since the 1970s, and based on data from recent captures of unmarked fish, it is possible that recruitment also occurs in Lake Powell near the Colorado River and San Juan River inflows (Albrecht et al. 2010, 2017).

Thus, reservoir inflows may be functioning examples of how, given appropriate habitat and cover features (vegetation and turbidity), native species can maintain recruitment and persist despite habitat modifications and competition and predation pressure from thriving nonnative fishes (Albrecht et al. 2010, 2017). Complex inflow-area habitats may be a key to future conservation and recovery efforts not only for razorback sucker, but also for other native species, due to their similarity to historical habitats that many native species are well adapted to.

While the razorback sucker is a good example of an endangered fish species that demonstrates sufficient plasticity and resiliency to do well in these complex la-

Fig. 11.3 Channel complexity at the Colorado River inflow area, Lake Mead. (Photo courtesy of BIO-WEST, Inc.)

custrine or fluvial habitats despite nonnative fish predators and competitors, bonytail, humpback chub *Gila cypha*, Colorado pikeminnow, flannelmouth sucker, bluehead sucker *Catostomus discobolus*, and other native fishes have also been found within either the lentic or lotic portions of the complex inflow habitats at Lakes Powell and Mead (Albrecht et al. 2014, 2017). In fact, there has recently been an increase in humpback chub distribution and abundance in the lower Grand Canyon, and once-rare captures near Lake Mead are increasingly common (Kegerries et al. 2016). Combinations of cover and complex and dynamic habitats are important components to consider for future habitat management and restoration efforts (Albrecht et al. 2017; Kegerries et al. 2017).

Further, the natural flow regime concept underscores the importance of dynamic, complex habitats to maintaining ecological integrity, implying that flow-regime restoration will benefit native fishes while disadvantaging nonnative species (Poff et al. 1997; Propst and Gido 2004). For example, Propst et al. (2008) found that a natural flow regime, in a location with comparatively little anthropogenic modification and coupled with active management and removal of nonnative fishes, worked to conserve the native fish assemblage in the upper Gila River drainage. Similarly, maintenance and restoration of a more natural flow regime in the San Juan River helped native fish recruitment. Collectively, studies like these lend credence to the notion that complex habitats created by natural river flows are important for native fish conservation.

CONCLUSIONS

Although nonnative fishes continue to threaten native fish populations, one hopeful aspect of the situation is that fisheries managers have developed multiple techniques to prevent further deterioration of native fish populations. Mueller (2005) reviewed predatory fish removal efforts and highlighted their successes in headwater streams, isolated ponds, and springs for salmonids and spring-dwelling fishes. Also, persistent and emerging complex habitats appear to have conservation potential. It may be effective to let native fishes demonstrate their plasticity and resiliency where natural or restored habitats and flow regimes disadvantage nonnative species to the benefit of natives. Other emerging techniques include development and use of triploids, "daughterless" carp, the "Trojan Y chromosome" strategy, and other genetic manipulations that might increase control over intentionally stocked nonnative species (Allen and Wattendorf 1987; Teem and Gutierrez 2014). The future will hopefully bring additional tools and strategies.

However, native fishes have often been unjustifiably characterized as poor baitfish and sport fish, which has devalued them in the eyes of many, furthering the popularity of nonnative fish species (Mueller 2005). As transportation and access to lakes and streams improves, the public will probably continue to desire well-known, nonnative sport and food fishes. If so, then the nonnative fish presence may exceed the ability of biologists and resource managers to reduce the threat, especially because source populations are often untouched and, in some cases, bolstered.

Many biologists believe that nonnative fishes are particularly harmful to native species when habitat conditions are stable, modified, and/or simplified. If that is true, then simultaneous nonnative fish control, preservation and restoration of habitat complexity, and flow-regime restoration may best advantage native species. In extreme cases, the only viable option may be to isolate native fish populations, or certain life stages, from nonnatives.

Managers may sometimes be faced with conducting continued mechanical-removal programs to ameliorate nonnative threats (see the UCRB example above). In the most advanced cases, a combination of hatchery propagation, stocking into the wild, piscicide and mechanical removal of nonnatives, habitat renovation, and flow-regime restoration may all be necessary (see the Virgin River example above).

The nonnative fish dilemma in the Colorado River basin has been historically addressed through nonnative fish removal. For example, the mechanical removal of nonnative fish, with its hefty price tag—over $4,000,000 since 2001 in the San Juan River alone (Franssen et al. 2014)—has arguably hindered implementation of a more ecosystem-driven approach. Is a better understanding of ecosystem biology, with improved insight into using habitats in their present-day state to perhaps achieve a broader and more complex ecosystem to more fully benefit native species, needed? Should biologists and managers think bigger than simply removing one nonnative species at a time? Could greater efforts geared toward enhancing segment- and basin-scale factors (e.g., flow regimes) be useful to this end? Are such ideas even realistic possibilities that society would tolerate?

While millions of dollars have been allocated to reducing nonnative fish impacts on native species, fewer dollars have been allocated to provide education, outreach, and public-and-agency collaboration programs, even though these approaches may stand to be the ultimate and most sustainable solutions to curbing future introductions of new, novel, and problematic nonnative species. For example, in 2016, roughly 1% of the UCREFRP's budget was spent on education and outreach. The San Juan River Basin Recovery Implementation Program (SJRIP) similarly allocated 1% of its 2016 budget to education and outreach (UCREFRP and SJRIP 2016). Conversely, the proportion of funds dedicated to some of the removal techniques described above ranged from 16% to 20%.

A shift in the mentality of fisheries biologists and managers may be needed. Perhaps the most lasting approach, with the best potential to stem the tide of ever-increasing nonnative species, is through public education and outreach programs that teach the value of native species, rather than increased voltage for mechanical removal (i.e., electrofishing) or more widespread application of piscicides. This may be possible because, according to the National Park Service, over 305 million people (a number almost equal to the entire US population) visited US national parks in 2015 (Olson 2016), which demonstrates a societal value of protecting wild animals and ecosystems. Over two decades ago, Rinne (1996, 149, 156) described the effects of introduced fishes on native fishes of the southwestern United States and concluded that

> presently the native fish fauna is endangered. Future conservation efforts must be innovative, vigilant, and include (1) research into the mechanisms of interactions between native and introduced fishes in the state, (2) conservation and restoration of habitats for native species in an ecosystem or river basin concept, (3) incorporation of a value system for native fishes, and (4) stringent regulations for importation of nonnative fishes. . . .
>
> . . . Through education, native fishes should be demonstrated to be of value to our society in the same manner that a rationale was provided for introduction of nonnative species.

The nonnative fish dilemma is an ongoing battle. Just because an action can be implemented does not ensure it is the best practice given available resources and ecological context. Also, the scale of nonnative fish removal efforts, and the ability to incorporate them with other habitat- and flow-related restorations, should be considered before implementation. The biggest questions may now be related to how to measure the success of removal and restoration efforts. Does a simple reduction in numbers of nonnative fish equal success? Are the most feasible methods to enhance native fish populations being used? One could argue that hypothesis-based monitoring and research to examine all life stages (e.g., Bestgen et al. 2011; LaGory et al. 2012) are made more applicable by focusing on flow regimes and habitats that give native fish an advantage.

ACKNOWLEDGMENTS

The authors thank Sandra Livingston Turner, Chadd VanZanten, and peer reviewers for reviewing previous versions of this work, and Harrison Mohn for mapping assistance and editorial review.

REFERENCES

Albrecht, B., P. B. Holden, R. Kegerries, and M. E. Golden. 2010. Razorback sucker recruitment in Lake Mead, Nevada-Arizona, why here? *Lake and Reservoir Management* 26: 336–44.

Albrecht, B., R. Kegerries, J. M. Barkstedt, W. H. Brandenburg, A. L. Barkalow, S. P. Platania, M. McKinstry, B. Healy, J. Stolberg, and Z. Shattuck. 2014. *Razorback Sucker* Xyrauchen texanus *Research and Monitoring in the Colorado River Inflow Area of Lake Mead and the Lower Grand Canyon, Arizona and Nevada*. Final Report. Salt Lake City: US Bureau of Reclamation, Upper Colorado Region.

Albrecht, B., H. E. Mohn, R. Kegerries, M. C. McKinstry, R. Rogers, T. Francis, B. Hines, J. Stolberg, D. Ryden, D. Elverud, B. Schleicher, K. Creighton, B. Healy, and B. Senger. 2017. Use of inflow areas in two Colorado River basin reservoirs by the endangered razorback sucker (*Xyrauchen texanus*). *Western North American Naturalist* 77: 500-14.

Allen, S. K., Jr., and R. J. Wattendorf. 1987. Triploid Grass Carp: Status and management implications. *Fisheries* 12: 20–24.

Ault, T. R., J. S. Mankin, B. I. Cook, and J. E. Smerdon. 2016. Relative impacts of mitigation, temperature, and precipitation on 21st-century megadrought risk in the American Southwest. *Science Advances* 2: e1600873.

Best, E., and J. Lantow. 2012. *Investigations of Flannelmouth Sucker Habitat Use, Preference, and Recruitment Downstream of Davis Dam in the Lower Colorado River, 2006–2010*. Boulder City, NV: Bureau of Reclamation, Lower Colorado River Multi-species Conservation Program.

Bestgen, K. R., C. D. Walford, A. A. Hill, and J. A. Hawkins. 2007. *Native Fish Response to Removal of Non-native Predator Fish in the Yampa River, Colorado*. Final report. Denver: Recovery Implementation Program for Endangered Fishes in the Upper Colorado River Basin, Project Number 140.

Bestgen, K. R., G. B. Haines, and A. A. Hill. 2011. *Synthesis of Flood Plain Wetland Information: Timing of Razorback Sucker Reproduction in the Green River, Utah, Related to Stream Flow, Water Temperature, and Flood Plain Wetland Availability*. Final report. Denver: US Fish and Wildlife Service, Upper Colorado River Endangered Fish Recovery Program.

Breton, A. R., D. L. Winkelman, J. Hawkins, and K. R. Bestgen. 2014. *Population Trends of Smallmouth Bass in the Upper Colorado River Basin with an Evaluation of Removal Effects*. Final Report. Denver: US Fish and Wildlife Service, Upper Colorado River Endangered Fish Recovery Program.

Brooks, J. E., M. J. Buntjer, and J. R. Smith. 2000. *Non-native Species Interactions: Management Implications to Aid in Recovery of the Colorado Pikeminnow* Ptychocheilus lucius *and Razorback Sucker* Xyrauchen texanus *in the San Juan River, CO-NM-UT*. Albuquerque: San Juan River Basin Recovery Implementation Program and US Fish and Wildlife Service.

Carlson, C. A., and R. T. Muth. 1989. The Colorado River: Lifeline of the American Southwest. *Canadian Special Publication of Fisheries and Aquatic Sciences* 106: 220–39.

Christensen, N. S., A. W. Wood, N. Voisin, D. P. Lettenmaier, and R. N. Palmer. 2004. The effects of climate change on the hydrology and water resources of the Colorado River basin. *Climate Change* 62: 337–63.

Clarkson, R. W., P. C. Marsh, S. E. Stefferud, and J. A. Stefferud. 2005. Conflicts between native fish and nonnative sport fish management in the Southwestern United States. *Fisheries* 30: 20–27.

Coggins, L. G., Jr., and M. D. Yard. 2010. *Mechanical Removal of Nonnative Fish in the Colorado River within Grand Canyon*. US Geological Survey Open-File Report 2010-5135: 227–34.

Duran, B. R. 2016. *Endangered Fish Monitoring and Nonnative Species Monitoring and Control in the Upper/Middle San Juan River: 2015*. Albuquerque: US Fish and Wildlife Service, San Juan River Basin Recovery Implementation Program.

Ehlo, C. A., B. R. Kesner, and P. C. Marsh. 2016. *Comparative Survival of Repatriated Razorback Suckers in Lower Colorado River Reach 3, 2015 Annual Report*. Boulder City: US Bureau of Reclamation, Lower Colorado River Multi-species Conservation Program.

Ehlo, C. A., M. J. Saltzgiver, T. E. Dowling, P. C. Marsh, and B. R. Kesner. 2017. Use of molecular techniques to confirm nonnative fish predation on razorback sucker larvae in Lake Mohave, Arizona and Nevada. *Transactions of the American Fisheries Society* 146: 201–5.

Finlayson, B., W. Somer, D. Duffield, D. Propst, C. Mellison, T. Pettengill, H. Sexauer, T. Nesler, S. Gurtin, J. Elliot, F. Partridge, and D. Skaar. 2005. Native inland trout restoration on national forests in the Western United States: Time for improvement? *Fisheries* 30: 10–19.

Franssen, N. R., J. E. Davis, D. W. Ryden, and K. B. Gido. 2014. Fish community responses to mechanical removal of nonnative fishes in a large southwestern river. *Fisheries* 39: 352–63.

Haak, A. L., and J. E. Williams. 2012. Spreading the risk: Native trout management in a warmer and less-certain future. *North American Journal of Fisheries Management* 32: 387–401.

Hawkins, J. A., and T. P. Nesler. 1991. *Nonnative Fishes of the Upper Colorado River Basin: An Issue Paper*. Lakewood: US Fish and Wildlife Service, Upper Colorado River Endangered Fish Recovery Program.

Hines, B. 2016. *Endangered Fish Monitoring and Nonnative Fish Control in the Lower San Juan River 2015*. Albuquerque: US Fish and Wildlife Service, San Juan River Basin Recovery Implementation Program.

Holden, P. B. 1991. Ghosts of the Green River: Impacts of Green River poisoning on management of native fishes. In W. L. Minckley and J. E. Deacon, eds., *Battle against Extinction: Native Fish Management in the American West*, 43–54. Tucson: University of Arizona Press.

Holden, P. B., J. E. Deacon, and M. E. Golden. 2005. Historical changes in fishes of the Virgin-Moapa River system: Continuing decline of a unique native fauna. *American Fisheries Society Symposium* 45: 99–114.

Johnson, B. M., P. J. Martinez, J. A. Hawkins, and K. R. Bestgen. 2008. Ranking predatory threats to nonnative fishes in the Yampa River, Colorado, via bioenergetics modeling. *North American Journal of Fisheries Management* 28: 1941–53.

Johnson, J. E., M. G. Pardew, and M. M. Lyttle. 2011. Predator recognition and avoidance by larval razorback sucker and northern hog sucker. *Transactions of the American Fisheries Society* 122: 1139–45.

Karam, A. P., and P. C. Marsh. 2010. Predation of adult razorback sucker and bonytail by striped bass in Lake Mohave, Arizona-Nevada. *Western North American Naturalist* 70: 117–20.

Kegerries, R. B., B. Albrecht, E. I. Gilbert, W. H. Brandenburg, A. L. Barkalow, H. Mohn, R. Rogers, M. McKinstry, B. Healy, J. Stolberg, E. Omana Smith, and M. Edwards. 2016. *Razorback Sucker* Xyrauchen texanus *Research and Monitoring in the Colorado River Inflow Area of Lake Mead and the Lower Grand Canyon, Arizona and Nevada*. Logan, UT: BIO-WEST, Inc.

Kegerries, R. B., B. C. Albrecht, E. I. Gilbert, W. H. Brandenburg, A. L. Barkalow, M. C. McKinstry, H. E. Mohn, B. D. Healy, J. R. Stolberg, E. C. Omana Smith, C. B. Nelson, and R. J. Rogers. 2017. Occurrence and reproduction by razorback sucker (*Xyrauchen texanus*) in the Grand Canyon, Arizona. *Southwestern Naturalist* 62: 227-32.

LaGory, K., T. Chart, K. R. Bestgen, J. Wilhite, S. Capron, D. Speas, H. Hermansen, K. McAbee, J. Mohrman, M. Trammell, and B. Albrecht. 2012. *Study Plan to Examine the Effects of Using Larval Razorback Sucker Occurrence in the Green River as a Trigger for Flaming Gorge Dam Peak Releases*. Denver: US Fish and Wildlife Service, Upper Colorado River Endangered Fish Recovery Program.

Marsh, P. C. 2000. *Fish Population Status and Evaluation in the Cibola High Levee Pond*. Boulder City, NV: US Bureau of Reclamation.

Marsh, P. C., and W. L. Minckley. 1985. *Aquatic Resources of the Yuma Division, Lower Colorado River*. Final Report. Boulder City, NV: US Bureau of Reclamation.

Marsh, P. C., T. E. Dowling, B. R. Kesner, T. F. Turner, and W. L. Minckley. 2015. Conservation to stem imminent extinction: The fight to save razorback sucker *Xyrauchen texanus* in Lake Mohave and its implications for species recovery. *Copeia* 2015: 141–56.

Michaud, C., T. Francis, R. C. Schelly, M. T. Jones, and E. Kluender. 2016. *Evaluation of Walleye Removal in the Upper Colorado River basin.* Lakewood, CO: US Fish and Wildlife Service, Upper Colorado River Endangered Fish Recovery Program.

Miller, R. R. 1961. Man and the changing fish fauna of the American Southwest. *Papers of the Michigan Academy of Science, Arts, and Letters* 46: 365-404.

Minckley, W. L., and J. E. Deacon, eds, 1991. *Battle against Extinction: Native Fish Management in the American West.* Tucson: University of Arizona Press.

Minckley, W. L., P. C. Marsh, J. E. Deacon, T. E. Dowling, P. E. Hedrick, W. J. Matthews, and G. Mueller. 2003. A conservation plan for lower Colorado River native fishes. *BioScience* 53: 219–34.

Mueller, G. A. 2005. Predatory fish removal and native fish recovery in the Colorado River mainstem: What have we learned? *Fisheries* 30: 10–19.

Mueller, G. A. 2006. *Ecology of Bonytail and Razorback Sucker and the Role of Off-Channel Habitats in Their Recovery.* US Geological Survey Scientific Investigations Report 2006-5056.

Mueller, G. A. 2007. *Native Fish Sanctuary Project—Development Phase, 2007.* US Geological Survey Open-File Report 2008-1126.

Mueller, G. A., and P. C. Marsh. 2002. *Lost: A Desert River and Its Native Fishes: A Historical Perspective of the Lower Colorado River.* US Geological Survey Information and Technology Report USGS/BRD/ITR-2002-0010.

Mueller, G. A., and R. Wydoski. 2004. Reintroduction of the flannelmouth sucker in the lower Colorado River. *North American Journal of Fisheries Management* 24: 41–46.

Nash, L. L., and P. H. Gleick. 1991. Sensitivity of streamflow in the Colorado basin to climatic changes. *Journal of Hydrology* 125: 221–41.

Olden, J. D., and N. L. Poff. 2005. Long-term trends of native and non-native fish faunas in the American Southwest. *Animal Biodiversity and Conservation* 28: 75–89.

Olson, J. 2016. America's national parks: Record number of visitors in 2015. Press release, National Park Service, January 27.

Pister, E. P. 1991. The Desert Fishes Council: Catalyst for change. In W. L. Minckley and J. E. Deacon, eds., *Battle against Extinction: Native Fish Management in the American West,* 55–68. Tucson: University of Arizona Press.

Pitts, K. L. 2008. Assessing threats to native fishes of the lower Colorado River basin. MSc thesis, Kansas State University.

Poff, N. L., D. Allan, M. B. Bain, J. R. Karr, K. L. Prestegaard, B. D. Richter, R. E. Sparks, and J. C. Stromberg. 1997. The natural flow regime. *BioScience* 47: 769-83.

Pool, T. K., J. D. Olden, J. B. Whittier, and C. P. Paukert. 2010. Environmental drivers of fish functional diversity and composition in the lower Colorado River basin. *Canadian Journal of Fish and Aquatic Science* 67: 1791-807.

Propst, D. L., and K. B. Gido. 2004. Responses of native and nonnative fishes to natural flow regime mimicry in the San Juan River. *Transactions of the American Fisheries Society* 133: 922-31.

Propst, D. L., K. B. Gido, and J. A. Stefferud. 2008. Natural flow regimes, nonnative fishes, and native fish persistence in arid-land river systems. *Ecological Applications* 18: 1236-52.

Propst, D. L., K. B. Gido, J. E. Whitney, E. I. Gilbert, T. J. Pilger, A. M. Monié, Y. M. Paroz, J. M. Wick, J. A. Monzingo, and D. M. Myers. 2015. Efficacy of mechanically removing nonnative predators from a desert stream. *River Research and Applications* 31: 692–703.

Propst, D. L., J. A. Stefferud, and P. R. Turner. 1992. Conservation and status of Gila trout, *Oncorhynchus gilae. Southwestern Naturalist* 37: 117–25.

Propst, D. L., and J. A. Stefferud. 1997. Population dynamics of Gila trout in the Gila River drainage of the south-western United States. *Journal of Fish Biology* 51: 1137–54.

Rehm, A. H., and R. A. Fridell. 2008. *Virgin River Basin Red Shiner Eradication Program.* Publication Number 09-02. Salt Lake City: Utah Division of Wildlife Resources.

Rinne, J. N. 1996. The effects of introduced fishes on native fishes: Arizona, Southwestern United States. In *Protection of Aquatic Diversity, Proceedings of the World Fisheries Conference, Theme* 3, 149–59. New Delhi: Oxford and IBH Publishing.

Rinne, J. N., and J. Janisch. 1995. Coldwater fish stocking and native fish in Arizona: Past, present and future. *American Fisheries Society Symposium* 15: 397–406.

Skorupski, J. A., M. J. Breen, and L. Monroe. 2012. *Native Fish Response to Nonnative Fish Removal from 2005–2008 in the Middle Green River, Utah.* Lakewood, CO: US Fish and Wildlife Service, Upper Colorado River Basin Endangered Fish Recovery Program.

Teem, J. L., and J. B. Gutierrez. 2014. Combining the Trojan Y chromosome and daughterless carp eradication strategies. *Biological Invasions* 16: 1231-40.

UCREFRP (Upper Colorado River Endangered Fish Recovery Program). 2014. *Upper Colorado River Basin Nonnative and Invasive Aquatic Species Prevention and Control Strategy.* Lakewood, CO: US Fish and Wildlife Service, Upper Colorado River Basin Endangered Fish Recovery Program.

UCREFRP and SJRIP (Upper Colorado River Endangered Fish Recovery Program and San Juan River Basin Recovery Implementation Program). 2016. *2015–2016 Program Highlights.* Denver: US Fish and Wildlife Service.

Valdez, R. A., and R. T. Muth. 2005. Ecology and conservation of native fishes in the upper Colorado River basin. *American Fisheries Society Symposium* 45: 157–204.

Ward, D. L., R. Morton-Starner, and B. Vaage. 2016. Effects of turbidity on predation vulnerability of juvenile humpback chub to rainbow trout and brown trout. *Journal of Fish and Wildlife Management* 7: 205–12.

Williams, J. E., H. M. Neville, A. L. Haak, W. T. Colyer, S. J. Wenger, and S. Bradshaw. 2015. Climate change adaptation and restoration of western trout streams: Opportunities and strategies. *Fisheries* 40: 304–17.

Wind, E. 2004. *Effects of Non-native Predators on Aquatic Ecosystems.* Victoria: British Columbia Ministry of Environment, Wildlife Bulletin no. B-123.

Zelasko, K. A., K. R. Bestgen, J. A. Hawkins, and G. C. White. 2016. Evaluation of a long-term predator removal program: Abundance and population dynamics of invasive northern pike in the Yampa River, Colorado. *Transactions of the American Fisheries Society* 145: 1153–70.

::Necessity

SECTION 3

Improving the Odds

12 Applying Endangered Species Act Protections to Desert Fishes

Assessment and Opportunities

Matthew E. Andersen
and James E. Brooks

The Endangered Species Act (ESA) of 1973, as amended, is the single most important piece of American federal legislation intended to conserve species, including desert fishes (Clark 2013; Malcom and Li 2015). Conserving species protects biodiversity, so in this respect the ESA is the American equivalent of what in some countries is the national biodiversity law (Medaglia et al. 2014; Lowell and Kelly 2016) or rare species conservation law (Rudd et al. 2016). The lead federal agency for administration and enforcement of the ESA in freshwater habitats is the US Fish and Wildlife Service (USFWS). Both American and foreign species may be listed under the ESA because USFWS and associated agencies recognize that activities of Americans and foreign nationals traveling into and out of the United States can have impacts on species beyond our country's borders.

Since publication of *Battle against Extinction: Native Fish Management in the American West* (Minckley and Deacon 1991), implementation strategies under the ESA have broadened, largely in response to increasing opposition to ESA regulatory actions and associated political and bureaucratic actions.

[Box 12.1]

The purpose of the Endangered Species Act as described in Section 2: "The purposes of the Act are to provide a means whereby the ecosystems upon which endangered species and threatened species depend may be conserved."

Herein we review the original administration of the ESA regulatory process and describe changes in ESA implementation since the early 1990s. We also discuss how these changes have affected on-the-ground recovery efforts and what, if any, effects they have had on recovery success.

HISTORICAL APPLICATION OF ESA TO DESERT FISHES

Preceded by the Endangered Species Preservation Act of 1966 and the Endangered Species Conservation Act of 1969, the ESA was one of a suite of environmental laws enacted in the early 1970s (Lowell and Kelly 2016). These statutes were written in response to growing recognition by American society that post–World War II economic resource development was negatively affecting the environment, often to the point that human health and well-being were degraded. In the 1970s, this shared national narrative of a need to recover and protect the environment was powerful (Sarewitz 2004).

The ESA has been invoked to protect many threatened or rare desert fishes. There remain, however, species identified by the International Union for Conservation of Nature (IUCN) as being under one or more threats to persistence that have not been afforded ESA protection (table 12.1). The fact that some species subject to threats are not listed under the ESA indicates the administrative and financial burden the ESA imposes and the increased public and political resistance to additional listings. In some cases, these species have been protected through substantive conservation actions by states (Clark 2013; Wilson 2015), universities (Bezzerides and Bestgen 2002), nongovernmental organizations (NGOs) (Dauwalter et al. 2011), and others (Scott et al. 2010; Evans et al. 2013), thus allowing USFWS to find them sufficiently protected for the immediate future.

The ESA's original emphasis was to place restrictions and prohibitions on federal agencies and their actions with respect to the rarest and most vulnerable species. The law requires that federal agencies ensure that their actions are not likely to jeopardize the continued existence of a listed species or result in the destruction or adverse modification of habitat deemed critical for a listed species (Section 7 of the ESA). The venue for that assurance, for freshwater and most terrestrial species, is consultation with USFWS. The formal Section 7 consultation process concludes with a biological opinion (BO) that provides a more or less comprehensive evaluation of an agency's proposed project and identifies any changes necessary to conserve the species and allow the project to proceed. The BO may conclude that the project causes jeopardy to the continued survival of a species (a jeopardy opinion), that it does not (a non-jeopardy opinion), or that it would cause adverse modification of the species' critical habitat. Negotiations between USFWS and the federal agency proposing or approving the project then attempt to reach a compromise decision that modifies the project to remove the adverse effects, or allows the project to go forward while addressing its species and habitat impacts through conservation measures. This section of the ESA has generated considerable controversy, with opponents arguing that the ESA's consultation requirements negatively and unfairly affect economic interests (Seasholes 2007), and that it is too heavy a "hammer" (fig. 12.1).

Recent authors have documented modern examples of the integration of habitat protection provisions into natural resource management (Sheridan 2007;

Table 12.1 Comparison of the number of desert fish species documented as facing identifiable threats with the numbers that are receiving protections and actions. Of the fish species listed by IUCN and USFWS, there are 38 in common; IUCN lists 10 species not listed by USFWS under the ESA, and USFWS lists 12 not listed by IUCN. Thus the 38 fish species addressed by conservation programs are a mix of listed and non-listed species.

Under threat	IUCN listed	USFWS listed	Listed by both IUCN and USFWS	Subject to a formal conservation program
126	48	50	38	38

Sources: Desert Fishes Council Species Tracking List, http://www.desertfishes.org//species-tracking/species-page, accessed November 27, 2017; International Union for Conservation of Nature (IUCN) *Red List*, http://www.iucnredlist.org, accessed November 27, 2017; US Fish and Wildlife Service, Environmental Conservation Online System, http://ecos.fws.gov/ecp, accessed November 27, 2017.

Carroll et al. 2010) and development planning (Jonas et al. 2013). Groups seeking to limit development in the American West have used its habitat protection provisions as a basis for suing USFWS to restrict development and thereby protect species (Baier 2016). But analyses of decisions formulated through the Section 7 process have demonstrated that "no project was stopped or extensively altered due to USFWS finding jeopardy or adverse modification" (Malcom and Li 2015, 15846). Very few Section 7 consultations result in outright rejection of the proposed action (Owen 2012).

The ESA's designation of habitat as a critical component of species conservation was a key factor in the effort to conserve one of the first desert fish listed (Deacon and Deacon Williams 1991) in the case of Devils Hole, in Nye County, Nevada, the unique, remote habitat that is home to the Devils Hole pupfish *Cyprinodon diabolis* (Hausner et al. 2014). Though the case was ultimately resolved at the US Supreme Court (Deacon and Deacon Williams 1991; Williams and Propst, chap. 1, this volume), the ESA was invoked to stop water withdrawals shown to have a significant negative impact on the Devils Hole pupfish population (Andersen and Deacon 2001; Hausner et al. 2014). Similar threats to species with very limited habitats occur elsewhere in the arid American West. Persistence of the Borax Lake chub *Gila boraxobius*, for example, could also be dramatically influenced by groundwater withdrawal in areas distant from its habitat (Williams et al. 2005).

In the eighteenth, nineteenth, and early twentieth centuries, aridity and the lack of transportation infrastructure in the western United States limited practical opportunities for land development. Much western land was too arid for conventional European-style agriculture, and extractive industries had not overcome the huge challenge of transporting their products to market from remote areas. In the late twentieth century, residents of the western states increasingly bristled at the land ownership and management roles the federal government had assumed (Lowell and Kelly 2016) as human populations, federally subsidized water delivery projects (Reisner 1986), new techniques for drilling into deep aquifers, federally subsidized transportation projects, and demand for rare earth elements became more widespread. During the 1980s, this resentment spawned the Sagebrush Rebellion, which encouraged western governors and county officials to use legal maneuvers to impede or stop federal conservation actions in their states. The ESA is among a suite of laws establishing federal management authority over natural resources that governors and citizens cite as being objectionable (SPLC 2001; Tetreault 2014). This strong state and local resistance to federal natural resource regulations has been a consideration as USFWS decides how best to use the ESA to protect species while also seeking to cooperate with state and local officials and communities.

Fig. 12.1 The Endangered Species Act is sometimes called a hammer for implementing species and habitat protections. (Photo by James Brooks.)

Actions by Congress to limit the ESA's regulatory impacts on resource development activities have been nearly continuous since 1973 and have increased in recent years (Pombo 2005; Union of Concerned Scientists 2016). Most of these actions have been designed to weaken the regulatory mechanisms of the ESA, and most have been unsuccessful. In response to congressional pressure, and with the growing realization that ESA implementation was being stalled by local action, the Clinton administration, under Secretary of the Interior Bruce Babbitt, implemented a series of policy changes in 1994 (USFWS 1994a, b, c, d, e, f). The primary objectives of these policy changes were (1) increasing the transparency of the ESA process, (2) improving the quality of the science used in ESA decisions, and (3) increasing participation in the ESA process by state governments and private landowners (Bear 1996). More recently, the costs of ESA compliance have proved to be objectionable to some members of Congress, who are seeking to include requirements for assessing economic impacts within the language of the ESA (Olson et al. 2017), which now requires that a species' ecological status be independently assessed using the "best available science."

FLEXIBLE APPROACHES TO ESA IMPLEMENTATION

The original application of the ESA to species protection was through strict regulation, reflecting a philosophy of making any errors in favor of species conservation rather than resource development. This strict enforcement of ESA provisions, which was deemed necessary to prevent continuing or future species endangerment, has been described as a classic example of the command and control approach to environmental regulation (Housein 2002). Conflicts with land users rose as the number of species protected under the ESA increased; in response, public and political opposition to the ESA increased substantially.

Beginning in the early 1980s, USFWS made numerous changes to increase the flexibility of its implementation of ESA protections not only for species listed under the ESA, but for "candidate species" and those formally proposed for listing. The flexible actions it authorized included incidental take permitting and habitat conservation planning to allow federal and non-federal landowners opportunities to contribute to species conservation while continuing, in agreed-upon fashion, existing land and water use practices. In general, the increased flexibility of ESA implementation has increased landowner cooperation and participation in species conservation. Whether it has been equally effective in promoting species recovery is uncertain (Li and Male 2013; Eubanks 2015).[6]

Candidate Conservation Agreements (CCAs) and CCAs with Assurances (CCAAs) were developed and implemented in the mid-1990s to provide protections for candidate and other at-risk, unlisted species. They were derived from policy formulated by USFWS, not by amendment to the ESA. They include provisions that allow all private and public landowners to participate. For species already listed, Section 10 of the ESA was modified to include the use of Habitat Conservation Plans (HCP) and Safe Harbor Agreements (SHA) (table 12.2).

Candidate Conservation Agreements

Candidate species are those for which there is adequate biological and threat information to support listing under the ESA. The concept of Candidate Conservation Agreements (CCAs) was adopted by USFWS in 1985 in response to a growing need to provide protection to the large number of species that potentially warranted listing but were unlikely to receive the full protection of the ESA. Implementation of a CCA presumes that proactive conservation measures will preclude the ultimate need to list a species. The CCA is a voluntary agreement between USFWS and private property owners and/or federal land management agencies in which the parties agree to identify and implement conservation measures that would reduce threats to the relevant species.

A prime example relevant to the conservation of desert fishes is the multi-party CCA for Pecos pupfish *Cyprinodon pecosensis* (USFWS 1999) entered into by representatives of the states of New Mexico and Texas, the Bureau of Land Management (BLM), and the USFWS. The conservation efforts identified in this CCA were primarily related to securing habitats to prevent the introduction and spread of a hybridizing congener, sheepshead minnow *C. variegatus*, which had been introduced into the lower Pecos River in Texas and had spread upstream to near Carlsbad, New Mexico. State agencies implemented additional rules regarding live baitfish use, and land management agencies were required to isolate and secure off-channel Pecos River pupfish habitats where sheepshead minnow did not occur to prevent invasion.

6. Objective assessments of the relative recovery success of species protected by strict versus flexible ESA implementation have not been conducted. Such assessments would be difficult, if not impossible, because of a lack of meaningful control and treatment groups.

Table 12.2. Comparison of the characteristics of types of conservation agreements among partners. These modifications to ESA implementation allow greater flexibility and support for landowner participation in species and habitat protection activities. Take of a species is incidental take as determined through Section 7 procedures.

Conservation agreement type	Year implemented	Landowner status	Species ESA status	Take allowed	Major provisions
Candidate Conservation Agreement (CCA)	1983	Private or public	Candidate or proposed	No	Identify and implement conservation measures; does not require additional measures if species status does not improve, but no assurances if species is listed
Candidate Conservation Agreement with Assurances (CCAA)	1995	Non-federal	Candidate or Proposed	Yes	Landowner not required to conduct additional conservation activities if species is listed; Enhancement of Survival Permit is issued that identifies allowable take
Habitat Conservation Plan	1982	Private	Listed	Yes	Authorizes take of a listed species; does not authorize activities that result in take
Safe Harbor Agreement	1995	Non-federal	Listed	Yes	Environmental baseline established at beginning of process; allows return to baseline at end of agreement
Recovery Program	1969	Federal and non-federal	Listed	Yes	Dual-goal programs intended to recover species while all legal water development continues, often in consultation with multi-agency recovery team
Conservation Agreement	1995	Federal and non-federal	At-risk or candidate	No	State-led conservation actions for at-risk species to reduce probability of listing

Candidate Conservation Agreements with Assurances

Candidate Conservation Agreements with Assurances (CCAAs) are nearly identical in intent to CCAs, with the exception that incidental take of a species is allowed through an Enhancement of Survival Permit. If the species protected under the CCAA is ultimately listed despite the actions taken under the CCAA, no additional restrictions will be imposed on non-federal landowners. Uses of federal lands may be restricted in the case of listing.

A CCAA was issued for the Page springsnail *Pyrgulopsis morrisoni* to address its conservation needs in springs in the Verde River valley of central Arizona (USFWS 2009). This hydrobiid snail is endemic to a series of springs in the Oak Creek drainage located on private lands, on National Park Service lands, and within and adjacent to a state-owned fish hatchery. The purpose of this CCAA was to assess the status of the species, identify and remove threats, increase the species' distribution and abundance, and restore spring habitats. Activities required of landowners under the CCAA included providing access to springs on private lands for monitoring of the species and its introduction into suitable habitats, and making alterations in hatchery management activities consistent with conservation needs of the species.

Habitat Conservation Plans

In 1982, Congress authorized procedures to grant a permit, under Section 10 of the ESA, allowing incidental take of a species in the course of resource development, provided that the landowner develops an approved Habitat Conservation Plan (HCP). The plan describes potential impacts to the species and how they will be either minimized or mitigated, and includes a detailed description of planned activities and funding. HCPs for desert fishes vary considerably in scale.

The HCP for the Lower Colorado River Multi-species Conservation Program (LCRMSCP 2004) covers a variety of listed and unlisted plant and animal species along the economically important lower Colorado River ecosystem. Water management activities, including diversions, returns, and hydropower generation, are included in the HCP, and conservation measures for listed and unlisted species alike are identified. The permit applicants covered under the LCRMSCP include 23 entities from the states of Arizona, California, and Nevada. The most prominent species addressed by the program is the desert tortoise *Gopherus agassizii*. The principal fish species covered are humpback chub *Gila cypha*, bonytail *G. elegans*, and razorback sucker *Xyrauchen texanus*.

On a smaller scale, the HCP for the El Coronado Ranch in southeast Arizona applies to specific activities on a private ranch. There is a single ESA Section 10 permit applicant, but signatories to the plan also include the US Forest Service and the Arizona Game and Fish Department (USFWS 1998). This HCP includes watershed and riparian habitat improvement and species introductions for three fishes, the federally protected Yaqui chub *G. purpurea*, an undescribed Yaqui form of longfin dace *Agosia chrysogaster*, and Yaqui catfish *Ictalurus pricei*.

Safe Harbor Agreements

Safe Harbor Agreements (SHAs) were initially implemented in 1995, and the final rule was published in 1999. Under an SHA, non-federal property owners may voluntarily design and implement management activities to maintain and/or improve habitats for the benefit of species listed under the ESA. These property owners will not be subject to increased regulatory oversight if the activities attract those species to areas they did not previously inhabit or increase their abundance and distribution. Baseline conditions (for species and/or habitat) are established at the beginning of the agreement, and property owners may return to those conditions at the end of the agreement period.

The state of Nevada obtained SHAs to address the concerns of private property owners regarding conservation of Lahontan cutthroat trout *Oncorhynchus clarkii henshawi* in the Quinn, Blackrock, and upper Humboldt River basins (USFWS 2006b). Beginning in 2004, "umbrella" SHAs were established by USFWS and Nevada to cover these private owners. The SHA identified a total of 53 streams that held Lahontan cutthroat trout, only 6 of which did not flow through private lands. Applying the SHA to private lands was intended to greatly increase the amount of stream habitat available for Lahontan cutthroat trout conservation as individual landowner conservation actions were negotiated and implemented. These SHAs did not establish a specific baseline or set of baselines, recognizing that each unique situation would be addressed separately, covered by an individual cooperative agreement, and based on species abundance and distribution and/or habitat conditions.

In 2004, on the middle Rio Grande in central New Mexico, USFWS established an SHA with the Pueblo of Santa Ana for conservation activities for Rio Grande silvery minnow *Hybognathus amarus*, southwestern willow flycatcher *Empidonax traillii extimus*, and bald eagle *Haliaeetus leucocephalus* (USFWS 2004). Under this SHA, the Pueblo implemented management activities in the riverine corridor to maintain and/or improve habitats for the three species. For Rio Grande silvery minnow, habitat-based activities included the installation of gradient reduction structures to mitigate channel incision and abandonment of the surrounding floodplain in order to improve and diversify aquatic habitats. The baseline established for Rio Grande silvery minnow was based on the presence of the species in the Pueblo portion of the Rio Grande. That baseline was determined to be one Rio Grande silvery minnow, which, as presented on page 8 of the SHA, "indicates that the Pueblo reaches of the Rio Grande . . . are occupied." That baseline was subjected to minimal scrutiny due to sovereign tribal trust issues and tribal control of all technical information collected within Pueblo boundaries.

RECOVERY PROGRAMS

Current ESA policies allow for the establishment of multi-year, multi-stakeholder recovery programs for listed species, but their costs are often high. Resource developers and users that are members of a formal recovery program must continue to adhere to ESA legal requirements, especially those of Section 7, typically associated with approval for federally funded or authorized projects being contingent on participation by federal and non-federal entities in the recovery program (e.g., USFWS 1987, 1991).

Among the recovery programs established for desert fishes, the Upper Colorado River Endangered Fish Recovery Program (UCREFRP) is well known and widely assessed. The upper basin encompasses the Colorado River and its tributaries upstream of Lees Ferry, Arizona. The UCREFRP shares many resources with, and is often assessed with, the San Juan River Basin Recovery Implementation Program (SJRIP) of Colorado, New Mexico, and Utah (GAO 2006).

First established as the replacement for a native fish research program in 1988 (Minckley et al. 2003), the UCREFRP was convened by the USFWS with the intention of recovering four endangered species of big-river fishes—humpback chub, bonytail, Colorado pikeminnow, and razorback sucker—while at the same time allowing legally authorized water development to occur (giving the developers "operational certainty") in the upper Colorado River basin. The twin, coequal goals of the UCREFRP, conservation and development, have high potential to be contradictory, and have therefore drawn criticism from some (e.g., Brower et al. 2001), while other reviewers have focused on the concurrent successes and challenges of the program (e.g., Lochhead 1996).

For USFWS, state fish and wildlife agencies, academics, and NGOs concerned with species and the environment, the UCREFRP has been a substantial source of funding for studies that have increased the available data on these long-lived, difficult-to-observe fish species (e.g., Johnson et al. 2008; Bottcher et al. 2013; Webber et al. 2013). Some of these data have been used to support contradictory viewpoints (Sarewitz 2004). The largest UCREFRP-funded projects, however, have been for infrastructure, typically a target of spending in such programs, where tangible hard structures are readily visible monuments to conservation progress and the funding contributed by cooperators, in contrast to the often slower, less tangible march of scientific investigation.

The UCREFRP and the SJRIP require payments from states and developers in exchange for operational certainty, which has dramatically increased federal purchasing power. The US Government Accounting Office (GAO) estimated that between 1992 and 2006, for example, the UCREFRP and the SJRIP spent $34.6 million, of which $1.6 million came from the USFWS budget (GAO 2006). For the fiscal years 1989–2013, the UCREFRP expended over $239 million. In this 25-year period, the average annual expenditure was more than $11 million (Tart 2014). Fish recovery programs were not the only targets of USFWS recovery spending, however; the USFWS estimate for all recovery spending on endangered species between 1989 and 2004, for example, was $11.1 billion (Langpap and Kerkvliet 2010). As states and developers continue to fund scientific and management research efforts, the mounting cost becomes an issue of increasing concern. If the prior rates of UCREFRP spending were to continue to the scheduled 2023 closure of the program, it could be on track to spend an additional $100 million or more. External observers question, sometimes loudly, whether endangered fish in the upper Colorado River basin are worth more than $330 million. Those concerned with the conservation of this unique fauna and the preservation of habitats that continue to provide many environmental and societal benefits argue that such ESA expenditures are reasonable, especially in exchange for the increased operational certainty provided by participation in the recovery program (Langpap and Kerkvliet 2010; Loomis and Ballweber 2012; Tart 2014). Many other fish in the arid American West could benefit from additional research and conservation support (see table 12.1), but have not received the concerted effort afforded the UCREFRP focal species, mostly because of the broad-scale demands on Colorado River water.

The combined historical and new knowledge developed by the UCREFRP cooperators has not identified a means by which active management will not be necessary to maintain the focal species in their current habitat, altered as it is from its native state. A suite of actions appears to be necessary in perpetuity, especially the removal of nonnative predatory species (Mueller 2005), but also including hatchery rearing programs and management of flows (via dam releases) and backwater (off-channel) habitat. Never fully achieving recovery, but rather avoiding extinction with the support of associated perpetual management, is a common legacy

[Box 12.2]

Currently available evidence suggests that complete recovery of native desert fish species, to the point where conservation actions are no longer required, is unlikely for most species.

for western fish and other recovery programs (Mueller 2005; Williams et al. 2005; Kodric-Brown and Brown 2007).

Conservation Agreements

State and local resistance to federal regulations and associated costs has spawned additional recovery and conservation programs for fishes modeled, in part, on the UCREFRP and SJRIP. These programs include the June Sucker Recovery Implementation Program, the Virgin River Fishes Recovery Program, the Glen Canyon Dam Adaptive Management Program, and several species-specific programs. They are based on Conservation Agreements, a new, alternative class of agreements among federal, state, and business partners to seek locally or regionally developed conservation alternatives. These agreements allow for greater flexibility than may be possible under strict ESA enforcement. Their provisions reduce the need for federal permitting for operations, impose no penalties for incidental take (accidental fish mortalities), and expand the ability of individual states, utilities, water authorities, and landowners to negotiate land and water use. Some habitat disturbances necessary to maintain open water, such as clearing of vegetation (Kodric-Brown and Brown 2007), require fewer stakeholders to negotiate, and can be implemented more quickly, under the Conservation Agreement approach than if the species were listed under the ESA (Wilson 2015).

Recent studies suggested that three fishes native to the Colorado River drainage, but not the subject of any formal recovery program, were at risk (Rees et al. 2005a, b; Bezzerides and Bestgen 2002; Ptacek et al. 2005): roundtail chub *Gila robusta*, bluehead sucker *Pantosteus discobolus*, and flannelmouth sucker *Catostomus latipinnis* (hereinafter the "three species"). To address these conservation concerns, the Colorado River basin states of Arizona, Colorado, Nevada, New Mexico, Utah, and Wyoming entered into a range-wide Conservation Agreement (UDWR 2006) and included the three species in their state Wildlife Action Plans and other conservation activities (Carman 2007; Walters et al. 2015).

Despite well-intentioned, ongoing efforts by state and local agencies and academics, the forecast for the three species is not encouraging in much of their native range (Bestgen et al. 2011; Webber et al. 2013; Walters et al. 2015), with a few exceptions (Van Haverbeke et al. 2013). Habitat modifications and reductions, competition with nonnative fish species, and hybridizations continue to pose risks for the three species, although distinct populations may be able to adapt to locally altered conditions and the presence of nonnative fishes (Melis et al. 2010; Cathcart et al. 2015). In Arizona, the USFWS initiated consultation on a proposal to list the roundtail chub as threatened under the ESA in 2015, but, on the basis of a revised taxonomy that combined three nominative species into one, USFWS is currently (2018) reconsidering whether or not to list this species. More time will be needed, beyond 2018, to determine whether the three-species Conservation Agreement is ultimately successful.

If Conservation Agreements are to effectively maintain or improve a species' status, and thereby remove the need to list the species under ESA authority, monitoring of their effectiveness (Clark 2013) is necessary. Some examples of species that appear to have been sufficiently conserved by Conservation Agreement actions include Virgin spinedace *Lepidomeda mollispinis*, Colorado River cutthroat trout *Oncorhynchus clarkii pleuriticus*, Bonneville cutthroat trout *O. c. utah* (first petitioned for listing in 1979), and least chub *Iotichthys phlegethontis* (first listed as a candidate species in 1972) (Wilson 2015).

BEST AVAILABLE SCIENCE

One of the factors distinguishing the ESA from other environmental laws in the United States and elsewhere is the requirement that USFWS use the best available science when making listing and certain other decisions (Lowell and Kelly 2016). As with other aspects of the act, this language has resulted in different, and sometimes contradictory, interpretations.

Compliance with this provision of the ESA has required all parties, including government agencies, developers, academics, NGOs, and others, to increase their scientific sophistication and rigor. Though some programs have put substantial effort into attempting to use shared, consistent data (e.g., Melis et al. 2006), differing worldviews, differing interpretations of data, and sometimes outright different and contradictory facts have clashed as sophisticated, educated people, all looking at similar or related data through different personal, cultural, and political lenses, have disagreed over how science should be used and interpreted (Sarewitz 2004).

There are many examples of such science controversies surrounding environmental protection, as illustrated by Sarewitz (2004). For desert fishes, there may be no better example than that of the middle Rio Grande in New Mexico. The issuance of a Section 7 decision regarding water management and the conservation of the endangered Rio Grande silvery minnow (USFWS 2003) resulted in the development of the Middle Rio Grande Endangered Species Collaborative Program, which has been consumed by science controversy. The controversy has centered on the use of a particular sampling method (seining) and metric (catch per unit effort, CPUE) for assessing the species' status and response to management actions. Debate on the program's use of science has resulted in three formal reviews conducted by panels of external experts (PAP 2005; Atkins and CP Callahan, Inc. 2012; Hubert et al. 2016). The first two reviews (PAP 2005; Atkins and CP Callahan, Inc. 2012) were disputed by a vocal and forceful minority of program participants, which resulted in the conduct and acceptance of the final review (Hubert et al. 2016). The findings of the third review panel again upheld the sampling and analytical methods used and provided recommendations to assist in improving sampling design and use of the CPUE metric. Regardless, the controversy did not abate.

A parallel fourth review effort was completed after the Hubert et al. (2016) review in an attempt to provide an adaptive management framework for use in Middle Rio Grande water operations and environmental compliance (Noon et al. 2017). This review, like the previous three, was conducted by a panel of qualified experts, whose conclusions disagreed in part with that of Hubert et al. (2016). Thus, controversy over the science used to integrate water management and endangered species conservation primarily served to further confound decision makers' efforts to conclude well-founded deliberations.

The collaborative program failed to minimize controversy by addressing the fundamental needs of both sides of the issue at its initiation, as Sarewitz (2004) and Walters (2007) had advised it to do. The initial disagreement over the use of CPUE began on the Pecos River in the 1990s after a similar Section 7 consultation on the effects of federal water management on the federally protected Pecos bluntnose shiner *Notropis simus pecosensis* (USFWS 2006a) and carried over to the middle Rio Grande. An initial research and data-gathering program for the Pecos was conducted between 1992 and 1997, as required by the consultation, and was followed by a series of population and habitat monitoring projects to describe species status in response to reservoir-controlled flows in occupied habitat (Hoagstrom et al. 2007, 2008). New Mexico water management agencies, concerned with ensuring legally required interstate water deliveries to Texas, employed private consultants to critique the work of Hoagstrom and colleagues. The consultants' reports derided the Hoagstrom et al. (2007, 2008) data used by USFWS as "fatally flawed" and inadequate for detection of the species (Widmer et al. 2010), a finding consistent with the interests of their employers. Archdeacon and Davenport (2013) responded in defense of their methodology and data quality, while Widmer et al. (2013) provided the final consultants' response. The principal result of this debate was that, while there may be multiple ways to consider a dataset, the original collection methods and data analyses presented by Hoagstrom et al. (2007, 2008) were found to be appropriate.

Thus, species managers and water managers found themselves in conflict over what the science says about water and fish. For both the middle Rio Grande and Pecos River examples, agency managers did not start the program with a full understanding of the needs of both sides of the issue so as to avoid, or at least minimize, political controversy over the science (Sarewitz 2004). What was not addressed by Sarewitz (2004) is the deliberate role that private consultants can play in perpetuating a science controversy (Lightman 2016; McGarity and Wagner 2008). An appropriate analogy to their role is the historical opposition by the tobacco industry to empirical evidence of the relationship between smoking tobacco and lung cancer (Mukherjee 2010). These examples from New Mexico, however, contrast with other fish conservation programs in the Southwest, such as the UCREFRP, the Glen Canyon Dam Adaptive Management Program, and the Lower Colorado River Multi-species Conservation Program, which have not experienced controversies to such a degree. These programs are not perfect, and disagreements still erupt, but their initial development has allowed for continual conversation and shared assessment of models and recommended management actions without continually disruptive science controversy. More upfront conversations among participants at the outset of a program are needed to agree on a shared perspective and on how collected data will be used (Walters 2007).

ASSESSING DIFFERENT APPROACHES TO ESA IMPLEMENTATION

The flexibility of the ESA, appended and modified in response to successes and failures since its inception in 1973, raises questions regarding the best approach to preventing extinction and achieving species recovery. Does strict enforcement of the ESA improve recovery efforts and successes? Do modified interpretations of ESA authority result in improved recovery efforts and successes? If the answer to the second question is yes, then are there adequate checks and balances in place to ensure compliance with the intent of the ESA? Regardless of the answers to these questions, the authority of the ESA to provide the impetus to protect listed species is clear, as in the case of Candidate Conservation Agreements. Further, while the ESA has been criticized for not removing species from protected status, implementation of the law has prevented the extinction of numerous species (Taylor et al. 2005; Williams et al. 2005; Evans et al. 2016). But few ESA-listed species have been recovered. Until the Oregon chub *Oregonichthys crameri* was delisted in 2015, no fish species had been removed from the ESA because of its human-assisted recovery.

Improvements to ESA implementation have been made since 1982 with the inclusion of permitting for incidental take and Habitat Conservation Plans, and subsequently with a variety of policy changes in the early 1990s. Enforcement of the ESA continues to generate controversy, and various authors have suggested ways to lessen that controversy by removing the ambiguity of terms used in regulatory actions (Clark 2013), particularly within BOs produced during Section 7 consultations (Malcom and Li 2015). Gibbs and Currie (2012) believe that it is difficult to assess the effectiveness of the ESA when species recovery data are insufficient, and they highlight the need for continued, additional high-quality data and associated analyses. Regardless, several evaluations have demonstrated that a species listing under the ESA, along with development and implementation of a recovery plan and designation of critical habitat, contributes to improved species status (Male and Bean 2005; Taylor et al. 2005; Gibbs and Currie 2012).

The ESA has been a powerful tool for protecting desert fish species, though fewer fish species than terrestrial species have been recovered. This contrast directly reflects the habitat where these animals are found. In nearly all cases, it is impossible to find more water for fish habitat; in the case of the American West, the situation is becoming increasingly dire as droughts become more frequent and severe (Seager et al. 2012), and as human populations put more demand on limited water resources. The ESA has been critically important for driving constructive engagement of diverse stakeholders, but if or when water shortages become more critical, there will be increasingly fewer resources for which to negotiate, and it seems unlikely that water for fish will take precedence over human demands.

IMPROVING ESA IMPLEMENTATION

Use of the "best available science" is critical to successful implementation of ESA requirements. The policy makers, resource managers, and various publics involved in natural resource management are largely nonscientists, so if scientists are to be relevant to ESA implementation, they must do more than communicating technical results and conclusions (Sarewitz 2004). Sarewitz (2004) suggests that if decision makers are not to be confounded by "competing or confrontational science," agency administrators and policy makers must engage scientists early in the process. Early coordination serves to avoid perpetuating lingering disagreements borne out of misunderstandings and lack of knowledge.

To further understanding and appreciation of the role of science, Sullivan et al. (2006) provided an overview of the problems of communicating scientific works to nonscientists and suggested a general strategy for overcoming much of the opposition to technical data so that sound resource management decisions can be made. Dahlstrom (2014) suggested that "storytelling" through narrative formats, particularly through mass media, would encourage increased comprehension and interest by nonscientists and result in more engagement. Thus, a shared narrative among those on both sides of environmental controversies and those tasked with species conservation is critical for success.

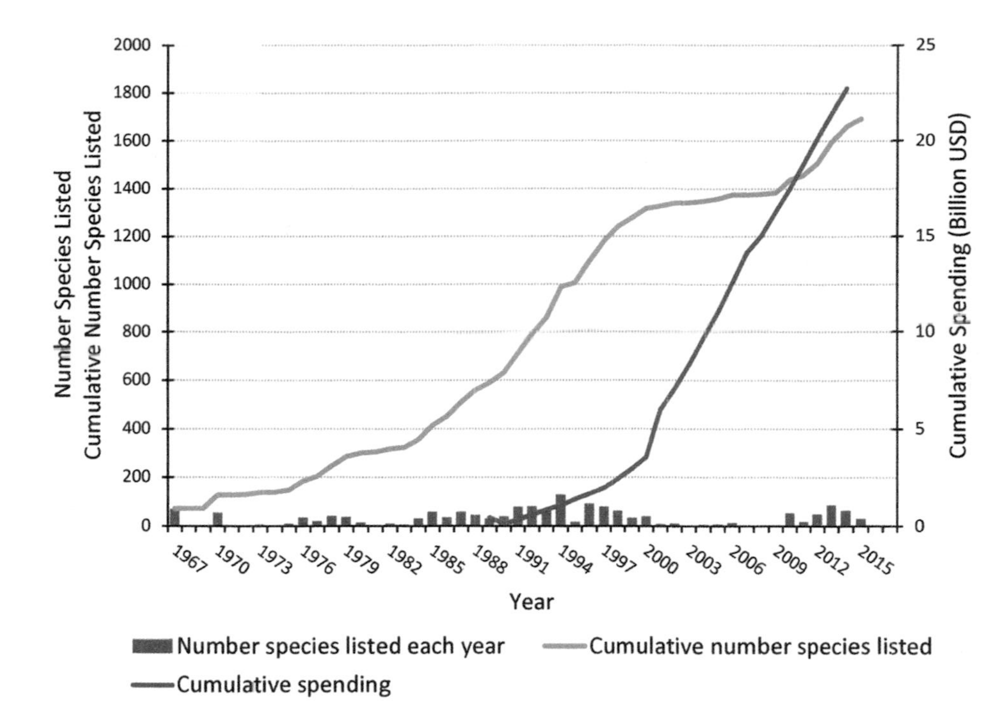

Fig. 12.2 Graph of cumulative count of listed species plotted against cumulative spending for ESA implementation.

Policy changes and additions that have improved science, increased transparency, and supported non-federal agency participation in the ESA regulatory process are integral to the conservation of listed species (Bear 1996; Evans et al. 2016). Most criticisms of the ESA cited in this chapter involve the lack of communication and transparency on the part of federal agencies, particularly the USFWS, and the lack of involvement of non-federal entities in the development of ESA decisions. All stakeholders potentially affected by ESA decisions have a responsibility to engage actively with one another. Along with continual improvements to ESA implementation, there is a concomitant need for two-way communication and cooperation. This need is particularly relevant to the more contentious issues surrounding listing decisions, recovery planning, including critical habitat designation, and Section 7 consultations.

FUNDING ESA IMPLEMENTATION

Paramount to improving the implementation of ESA efforts to list, protect, and recover species is adequate funding. It is not difficult to find assessments of available USFWS and other federal funding sources for species recovery that point out the woefully inadequate funding of recovery efforts (Restani and Marzluff 2002; Gerber 2016; Greenwald et al. 2016). Most, though not all, recovery plans identify not only recovery actions and priorities, but also the funding that will be required for their implementation. Estimates of the amounts required to adequately fund species recovery efforts in 2016 ranged from $1.21 billion per year (Gerber 2016) to $2.3 billion per year (Greenwald et al. 2016). These figures exceed the 2016 budget justification for recovery funding for USFWS—$78 million—by 15–29 times (USDA, Budget Justifications and Performance Information, Fiscal Year 2016). The number of species listed and the amount spent on ESA implementation continue to increase, but the demand for funds far exceeds the funds available, and the gap is widened as the cumulative number of listed species increases[7] (fig. 12.2). A full understanding of the political realities of ESA recovery funding in today's world does not support an optimistic view of meeting these funding challenges in the future.

7. Regardless of how species priorities are developed, use of a consistent priority review and development process would significantly increase the likelihood of more equitable funding. Use of a ranking system, or one that similarly incorporates and prioritizes threat, recoverability, and taxonomic status in a listing process, would help to address all concerns either for or against the species listing. A solid biological basis and a review system that considers all sides serves to more equitably disperse limited available recovery funding in these times of politically inflicted austerity.

CONCLUSIONS

The legislative history of the ESA demonstrates that there was a publicly supported commitment by the federal government to recognize and list the most imperiled species and an associated direction that neither the federal government, nor any of its agencies, should contribute to the extinction of those species. Though written in concise language with strict focus, the ESA has had landmark conservation impacts, evolving along the way into a relatively flexible statute. It was written with the expectation that federal agencies would work toward species recovery and required that their actions be reviewed to ensure that species are not pushed into extinction.

In the years since the ESA's passage, permitting for incidental take, support for conservation actions, and rewards for contributing toward recovery have been expanded to include non-federal entities. Most of these modifications were accomplished as policies formulated by USFWS. For such a short piece of legislation with such narrowness of focus and clarity of intent, the ESA has survived this stretching and bending and a plethora of reinterpretations to meet these new demands quite well. It still sets an admirable goal: that species will not be allowed to become extinct due to human actions.

Managers of fish species listed under the ESA strike a delicate balance among three management priorities: recovery, flexibility, and funding. Sufficient, case-specific emphasis on each of these priorities is necessary for success, but is challenging to achieve. If there is too little or too much emphasis on any priority, then progress on the others is negatively affected. Current economic pressures suggest that the current, historically high levels of funding are not sustainable. If complete management flexibility is maintained, including flexibility to develop resources, then recovery opportunities diminish or are lost as populations and habitats are negatively affected. If managers or politicians seek to reduce funding to modest levels or eliminate it completely, then, similarly, recovery opportunities are diminished or lost, and flexibility is reduced. All recovery actions, recovery programs, and conservation actions strive for a balance among these considerations in their own particular circumstances. As managers, governments, stakeholders, and economic interests consider allowances for economic development under the various ESA alternatives, serious negotiators will evaluate how much environmental loss for the sake of cumulative, short-term economic gains can be permitted before long-term habitat resiliency is lost. Sufficient contiguous habitat must be available, particularly considering the increasing impacts of climate change, for habitats to be resilient and available to support species in the long term.

The US National Climate Assessment predicts less precipitation and more droughts in the American Southwest in the future (Melillo et al. 2014). Such changes will increase the potential for conflicts over water between human society and natural systems (Jaeger et al. 2014). Though USFWS and its partners have employed ESA protections to maintain many rare desert fish species, a changing climate will put new, perhaps untenable, pressures on the status quo (Hopken et al. 2013; Ruhí et al. 2016). Rare species with small populations in tenuous habitats are not likely to fare well, and severe water shortages will lengthen the list of aquatic ecosystems that are tenuous habitats for aquatic species (Hausner et al. 2014; Gangloff et al. 2016). Though federal institutions in the United States have historically supported the ESA and other environmental protection legislation, such legal support may not be forthcoming in the future (Olson et al. 2017).

Continuing to implement the ESA in a flexible manner will be required in the future if there is to be any hope of maintaining the vulnerable ichthyofauna of the American West. Congress was responding to habitat threats stemming from unbridled economic development and use of resources when it passed the act in 1973, but it did not foresee today's dramatic climate changes and the possibility that those habitats, especially aquatic habitats, might not exist in the future. Ongoing conservation actions, such as management of flows and habitats and nonnative fish removal, are likely to remain necessary for maintaining species. All parties engaged in these actions should agree to share in the effort to build sound science that improves our understanding of species and their life history needs as well as the effectiveness of management actions. Managers should be flexible enough to make modest sacrifices, if necessary, to preserve a shared biodiversity legacy. In the case of desert fishes, preserving this unique ichthyofauna will contribute to the preservation of critical water supplies for both humans and ecosystems.

Those who seek to conserve desert fishes will need to be aware that more data do not always result in more conservation. Audiences will not always be moved by more facts. A shared conceptual model of the systems in question can be difficult to create, but is critical for advancing management and the associated science. This conceptual model is a narrative that requires all parties to listen to one another, remaining open-minded about different perspectives, and that encompasses as many concerns as possible. There is a strong need to incorporate social and economic information into this model. Social and economic information should not change the natural science findings, but these three kinds of information should be considered collectively by decision makers. Shared perspectives from broad stakeholder communities will increase broad-based support for conservation actions, and the ichthyofauna of the American West will need this broad-based support to survive into the twenty-second century.

ACKNOWLEDGMENTS

The authors are grateful for the constructive critiques from Scott Vanderkooi, Jennifer Fowler-Propst, and Gail Kobetich that improved this chapter. Jakob Drozd provided critically important assistance by researching species listing information and by preparing figure 12.2. Erica Lam helped format tables.

REFERENCES

Andersen, M. E., and J. E. Deacon. 2001. Population size of Devils Hole pupfish (*Cyprinodon diabolis*) correlates with water level. *Copeia* 2001: 224–28.

Archdeacon, T. P., and S. R. Davenport. 2013. Comment: Detection and population estimation for small-bodied fishes in a sand-bed river. *North American Journal of Fisheries Management* 33: 446–52.

Atkins and CP Callahan, Inc. 2012. *Scientific Review Report (DRAFT), Rio Grande Silvery Minnow, Population Estimate and Population Monitoring*. Albuquerque: US Bureau of Reclamation.

Baier, L. E. 2016. *Inside the Equal Access to Justice Act: Environmental Litigation and the Crippling Battle over America's Lands, Endangered Species, and Critical Habitats*. Lanham, MD: Rowan and Littlefield.

Bear, D. 1996. Reform of the Endangered Species Act: Overview of administrative reforms [congressional hearing material submitted by Bruce E. Babbitt, Secretary, Department of the Interior]. In *Biodiversity Protection: Implementation and Reform of the Endangered Species Act (Summer Conference, June 9–12)*. http://scholar.law.colorado.edu/biodiversity-protection-implementation-and-reform-endangered-species-act/18.

Bestgen, K. R., P. Budy, and W. J. Miller. 2011. *Status and Trends of Flannelmouth Sucker* Catostomus latipinnis, *Bluehead Sucker* Catostomus discobolus, *and Roundtail Chub* Gila robusta, *in the Dolores River, Colorado, and Opportunities for Population Improvement: Phase II Report*. Larval Fish Laboratory Contribution 166 and Intermountain Center for River Rehabilitation and Restoration 2011(2): 1–55.

Bezzerides, N., and K. Bestgen. 2002. *Status Review of Roundtail Chub* Gila robusta, *Flannelmouth Sucker* Catostomus latipinnis, *and Bluehead Sucker* Catostomus discobolus *in the Colorado River Basin*. Larval Fish Laboratory Contribution 118. Fort Collins: Colorado State University.

Bottcher, J. L., T. E. Walsworth, G. P. Thiede, P. Budy, and D. W. Speas. 2013. Frequent usage of tributaries by the endangered fishes of the upper Colorado River basin: Observations from the San Rafael River, Utah. *North American Journal of Fisheries Management* 33: 585–94.

Brower, A., C. Reedy, and J. Yelin-Kefer. 2001. Consensus versus conservation in the Upper Colorado River Basin Recovery Implementation Program. *Conservation Biology* 15: 1001–7.

Carman, S. M. 2007. *Bluehead Sucker* Catostomus discobolus *and Flannelmouth Sucker* Catostomus latipinnis *Conservation Strategy*. Santa Fe: New Mexico Department of Game and Fish.

Carroll, C., J. A. Vucetich, M. P. Nelson, D. J. Rohlf, and M. K. Phillips. 2010. Geography and recovery under the US Endangered Species Act. *Conservation Biology* 24: 395–403.

Cathcart, C. N., K. B. Gido, and M. C. McKinstry. 2015. Fish community distributions and movements in two tributaries of the San Juan River, USA. *Transactions of the American Fisheries Society* 144: 1013–28.

Clark, J. R. 2013. The Endangered Species Act at 40: Opportunities for improvements. *BioScience* 63: 924–25.

Dahlstrom, M. F. 2014. Using narratives and storytelling to communicate science with nonexpert audiences. *Proceedings of the National Academy of Sciences* 111: 13614–20.

Dauwalter, D. C., J. S. Sanderson, J. E. Williams, and J. R. Sedell. 2011. Identification and implementation of Native Fish Conservation Areas in the upper Colorado River basin. *Fisheries* 36: 278–88.

Deacon, J. E., and C. Deacon Williams. 1991. Ash Meadows and the legacy of the Devils Hole pupfish. In W. L. Minckley and J. E. Deacon, eds., *Battle against Extinction: Native Fish Management in the American West*, 69–87. Tucson: University of Arizona Press.

Eubanks, W. S. II. 2015. Subverting Congress' intent: The recent misapplication of Section 10 of the Endangered Species Act and its consequent impacts on sensitive wildlife and habitat. *Environmental Affairs Legal Review* 42(2): 259.

Evans., D. M., J. P. Che-Castaldo, D. Crouse, F. W. Davis, R. Epanchin-Niell, C. H. Flather, R. K. Frohlich, D. D. Goble, Y.-W. Li, T. D. Male, L. L. Master, M. P. Moskwik, M. C. Neel, B. R. Noon, C. Parmesan, M. W. Schwartz, J. M. Scott, and B. K. Williams. 2016. *Species Recovery in the United States: Increasing the Effectiveness of the Endangered Species Act*. Issues in Ecology, Report no. 20, Ecological Society of America.

Evans, D. M., D. D. Goble, and J. M. Scott. 2013. New priorities as the Endangered Species Act turns 40. *Frontiers in Ecology and the Environment* 11: 510.

Gangloff, M. M., G. J. Edgar, and B. Wilson. 2016. Imperiled species in aquatic ecosystems: Emerging threats, management and future prognoses. *Aquatic Conservation Marine and Freshwater Ecosystems* 26: 858–71.

GAO (United States Government Accountability Office). 2006. Endangered Species: Many factors may affect the length of time to recover select species. 06-730. Washington, DC.

Gerber, L. R. 2016. Conservation triage or injurious neglect in endangered species recovery. *Proceedings of the National Academy of Sciences* 113: 3563–66.

Gibbs, K. E., and D. J. Currie. 2012. Protecting endangered species: Do the main legislative tools work? *PLoS ONE* 7: e35730.

Greenwald, N., B. Hartl, L. Mehrhoff, and J. Pang. 2016. *Shortchanged: Funding Needed to Save America's Most Endangered Species*. Tucson: Center for Biological Diversity.

Hausner, M. B., K. P. Wilson, D. B. Gaines, F. Suarez, G. G. Scoppettone, and S. W. Tyler. 2014. Life in a fishbowl: Prospects for the endangered Devils Hole pupfish (*Cyprinodon diabolis*) in a changing climate. *Water Resources Research* 50: 7020–34.

Hoagstrom, C. W., J. E. Brooks, and S. R. Davenport. 2007. Recent habitat association and the historical decline of *Notropis simus pecosensis*. *River Research and Applications* 23: 1–15.

Hoagstrom, C. W., J. E. Brooks, and S. R. Davenport. 2008. Spatiotemporal population trends of *Notropis simus* in relation to habitat conditions and the annual flow regime of the Pecos River, 1992–2005. *Copeia* 2008: 5–15.

Hopken, M. W., M. R. Douglas, and M. E. Douglas. 2013. Stream hierarchy defines riverscape genetics of a North American desert fish. *Molecular Ecology* 22: 956–71.

Housein, J. G. 2002. Endangered species and Safe Harbor Agreements: How should they be used? MS thesis, Virginia Polytechnic Institute and State University.

Hubert, W., M. Fabrizio, R. Hughes, and M. Cusack. 2016. Summary of Findings by the External Expert Panelists: Rio Grande Silvery Minnow Population Monitoring Workshop. Isleta Casino and Resort, December 8–10, 2015. Albuquerque: US Bureau of Reclamation, Area Office.

Jaeger, K. L., J. D. Olden, and N. A. Pelland. 2014. Climate change poised to threaten hydrologic connectivity and endemic fishes in dryland streams. *Proceedings of the National Academy of Sciences* 111: 13894–99.

Johnson, B. M., P. J. Martinez, J. A. Hawkins, and K. R. Bestgen. 2008. Ranking predatory threats by nonnative fishes in the Yampa River, Colorado, via bioenergetics modeling. *North American Journal of Fisheries Management* 28: 1941–53.

Jonas, A. E. G., S. Pincetl, and J. Sullivan. 2013. Endangered neoliberal suburbanism? The use of the federal Endangered Species Act as a growth management tool in Southern California. *Urban Studies* 50: 2311–31.

Kodric-Brown, A., and J. H. Brown. 2007. Native fishes, exotic mammals, and the conservation of desert springs. *Frontiers in Ecology and Environment* 5: 549–53.

Langpap, C., and J. Kerkvliet. 2010. Allocating conservation resources under the Endangered Species Act. *American Journal of Agricultural Economics* 92: 110–24.

LCRMSCP (Lower Colorado River Multi-species Conservation Program). 2004. *Lower Colorado River Multi-species Conservation Program*. Vol. 2, *Habitat Conservation Plan*. Final. December 17. (J&S 00450.00.) Sacramento, CA. https://lcrmscp.gov/publications/hcp_volii_dec04.pdf.

Li, Y.-W., and T. Male. 2013. *Protecting Unlisted Species: Assessing and Improving Candidate Conservation Agreements with Assurances*. Washington, DC: Defenders of Wildlife.

Lightman, B., ed. 2016. *A Companion to the History of Science*. New York: John Wiley & Sons.

Lochhead, J. S. 1996. *Upper Colorado River Fish: A Recovery Program That Is Working—Myth or Reality?* Biodiversity Protection: Implementation and Reform of the Endangered Species Act (Summer Conference, June 9–12).

Loomis, J., and J. A. Ballweber. 2012. A policy analysis of the collaborative Upper Colorado River Basin Endangered Fish Recovery Program: Cost savings or cost shifting? *Natural Resources Journal* 52: 337–62.

Lowell, N., and R. P. Kelly. 2016. Evaluating agency use of "best available science" under the United State Endangered Species Act. *Biological Conservation* 196: 53–59.

Malcom, J. W., and Y.-W. Li. 2015. Data contradict common perceptions about a controversial provision of the US Endangered Species Act. *Proceedings of the National Academy of Sciences* 112: 15844–49.

Male, T. D., and M. J. Bean. 2005. Measuring progress in US endangered species conservation. *Ecology Letters* 8: 986–92.

McGarity, T. O., and W. Wagner. 2008. *Bending Science: How Special Interests Corrupt Public Health Research*. Cambridge, MA: Harvard University Press.

Medaglia, J. C., F.-K. Phillips, F. Perron-Welch, J. Rohe, and R. Jiménez-Aybar. 2014. *Biodiversity Legislation Study; A Review of Biodiversity Legislation in 8 Countries*. London, Hamburg, and Montreal: Global Legislators' Organization (GLOBE), World Future Council, and Centre for International Sustainable Development Law.

Melillo, J. M., T. C. Richmond, and G. W. Yohe, eds. 2014. *Climate Change Impacts in the United States: The Third National Climate Assessment*. US Global Change Research Program.

Melis, T. S., J. F. Hamill, L. G. Coggins, Jr., G. E. Bennett, P. E. Grams, T. A. Kennedy, D. M. Kubly, and B. E. Ralston, eds. 2010. *Proceedings of the Colorado River Basin Science and Resource Management Symposium, November 18–20, 2008, Scottsdale, Arizona*. US Geological Survey Scientific Investigations Report 2010-5135.

Melis, T. S., S. J. D. Martell, L. G. Coggins, Jr., W. E. Pine III, and M. E. Andersen. 2006. Adaptive management of the Colorado River Ecosystem below Glen Canyon Dam, Arizona: Using science and modeling to resolve uncertainty in river management. *Proceedings of the American Water Resources Association Specialty Conference on Adaptive Management and Water Infrastructure*, June 26–28, 2006, Missoula, MT.

Minckley, W. L., and J. E. Deacon, eds. 1991. *Battle against Extinction: Native Fish Management in the American West*. Tucson: University of Arizona Press.

Minckley, W. L., P. C. Marsh, J. E. Deacon, T. E. Dowling, P. W. Hedrick, W. J. Matthews, and G. Mueller. 2003. A conservation plan for native fishes of the lower Colorado River. *BioScience* 53: 219–34.

Mueller, G. A. 2005. Predatory fish removal and native fish recovery in the Colorado River mainstem: What have we learned? *Fisheries* 30: 10–19.

Mukherjee, S. 2010. *The Emperor of All Maladies*. New York: Scribner, Simon and Schuster.

Noon, B., D. Hankin, T. Dunne, and G. Grossman. 2017. *Independent Science Panel Findings Report: Rio Grande Silvery Minnow Key Scientific Uncertainties and Study Recommendations.* Prepared for the US Army Corps of Engineers, Albuquerque District on Behalf of the Middle Rio Grande Endangered Species Collaborative Program. Prepared by Geosystems Analysis, Inc. Albuquerque, NM. June 2017. Contract no. W912PP-15-C-0008. http://riograndem.wikispaces.com/RGSM+Panel+Workshop.

Olson, P. G., M. C. Burgess, L. B. Gohmert, and B. Babin. 2017. A bill to amend the Endangered Species Act of 1973 to require review of the economic cost of adding a species to the list of endangered species or threatened species, and for other purposes. United States House of Representatives Resolution 717 (H.R. 717). https://www.congress.gov/bill/115th-congress/house-bill/717/text.

Owen, D. 2012. Critical habitat and the challenge of regulating small harms. *Florida Law Review* 64: 141–99.

PAP (Rio Grande Silvery Minnow Program Advisory Panel). 2005. *Review of Research, Monitoring, and Management of Rio Grande Silvery Minnow in the Middle Rio Grande.* Draft report submitted to Middle Rio Grande Endangered Species Act Collaborative Program, US Fish and Wildlife Service, Albuquerque, NM.

Pombo, R. W. 2005. Implementation of the Endangered Species Act of 1973. Report to the House Committee on Resources, 109th Congress, May 2005. http://www.eswr.com/docs/505/esaimplrprtresources.pdf.

Ptacek, J. A., D. E. Rees, and W. J. Miller. 2005. *Bluehead Sucker* (Catostomus discobolus)*—A Technical Conservation Assessment.* Fort Collins, CO: USDA Forest Service.

Rees, D. E., J. A. Ptacek, R. J. Carr, and W. J. Miller. 2005a. *Flannelmouth Sucker* (Catostomus latipinnis)*—A Technical Conservation Assessment.* Fort Collins, CO: USDA Forest Service.

Rees, D. E., J. A. Ptacek, and W. J. Miller. 2005b. *Roundtail Chub* (Gila robusta robusta)*—A Technical Conservation Assessment.* Fort Collins, CO: USDA Forest Service.

Reisner, M. 1986. *Cadillac Desert: The American West and Its Disappearing Water.* New York: Viking Penguin.

Restani, M., and J. M. Marzluff. 2002. Funding extinction? Biological needs and political realities in the allocation of resources to endangered species recovery. *BioScience* 52: 169–77.

Rudd, M. A., S. Andres, and M. Kilfoil. 2016. Non-use economic values for little-known aquatic species at risk: Comparing choice experiment results from surveys focused on species, guilds, and ecosystems. *Environmental Management* 58: 476–90.

Ruhí, A., J. D. Olden, and J. L. Sabo. 2016. Declining streamflow induces collapse and replacement of native fish in the American Southwest. *Frontiers in Ecology and Environment* 36: 487–532.

Sarewitz, D. 2004. How science makes environmental controversies worse. *Environmental and Science Policy* 7: 385–403.

Scott, J. M., D. D. Goble, A. M. Haines, J. A. Wiens, and M. C. Neel. 2010. Conservation-reliant species and the future of conservation. *Conservation Letters* 3: 91–97.

Seager, R., M. Ting, C. Li, N Naik, B. Cook, J. Nakamura, and H. Liu. 2012. Projections of declining surface-water availability for the southwestern United States. *Nature Climate Change.* https://doi.org/10.1038/NCLIMATE1787.

Seasholes, B. 2007. *Bad for Species, Bad for People: What's Wrong with the Endangered Species Act and How to Fix It.* National Center for Policy Analysis Report no. 303. Dallas, Texas.

Sheridan, T. E. 2007. Embattled ranchers, endangered species, and urban sprawl: The political ecology of the new American West. *Annual Review of Anthropology* 36: 121–38.

SPLC (Southern Poverty Law Center). 2001. Conflict in Klamath. *Intelligence Report.* Montgomery, AL. https://www.splcenter.org/fighting-hate/intelligence-report/2001/conflict-klamath.

Sullivan, P. J., J. M. Acheson, P. L. Angermeier, T. Faast, J. Flemma, C. M. Jones, E. E. Knudsen, T. J. Minello, D. H. Secor, R. Wunderlich, and B. A. Zanetell. 2006. Defining and implementing best available science for fisheries and environmental science, policy, and management. *Fisheries* 31: 460–65.

Tart, J. K. 2014. The Upper Colorado River Endangered Fish Recovery Program: Twenty-five years in the making. *University of Denver Water Law Review*, January 16.

Taylor, M. F. J., K. F. Suckling, and J. J. Rachlinski. 2005. The effectiveness of the Endangered Species Act: A quantitative analysis. *BioScience* 55: 360–67.

Tetreault, S. 2014. BLM chief: Lawbreakers in Bundy confrontation will be held accountable. *Las Vegas Review-Journal*, May 12, 2014. http://www.reviewjournal.com/news/bundy-blm/blm-chief-lawbreakers-bundy-confrontation-will-be-held-accountable.

UDWR (Utah Division of Wildlife Resources). 2006. *Range-Wide Conservation Agreement and Strategy for Roundtail Chub* Gila robusta *Bluehead Sucker* Catostomus discobolus *and Flannelmouth Sucker* Catostomus latipinnis. Prepared for Colorado River Fish and Wildlife Council. Publication no. 06-18. Salt Lake City: Utah Department of Natural Resources.

Union of Concerned Scientists. 2016. *The Endangered Endangered Species Act.* Cambridge, MA. https://www.ucsusa.org/center-science-and-democracy/preserving-science-based-safeguards/the-endangered-endangered-species-act.

USFWS (US Fish and Wildlife Service). 1987. *Final Recovery Implementation Program for Endangered Fish Species in the Upper Colorado River Basin.* Denver: US Department of the Interior Fish and Wildlife Service.

USFWS (US Fish and Wildlife Service). 1991. *Final Biological Opinion for the Animas–La Plata Project, Colorado and New Mexico.* Denver: US Department of the Interior Fish and Wildlife Service.

USFWS (US Fish and Wildlife Service). 1994a. Endangered and threatened wildlife and plants: Notice of interagency cooperative policy for peer review in Endangered Species Act activities. *Federal Register* 59 (126): 34270.

USFWS (US Fish and Wildlife Service). 1994b. Endangered and threatened wildlife and plants: Notice of interagency cooperative policy on information standards under the Endangered Species Act activities. *Federal Register* 59 (126): 34271.

USFWS (US Fish and Wildlife Service). 1994c. Endangered and threatened wildlife and plants: Notice of interagency cooperative policy for Endangered Species Act Section 9 prohibitions. *Federal Register* 59 (126): 34272.

USFWS (US Fish and Wildlife Service). 1994d. Endangered and threatened wildlife and plants: Notice of interagency cooperative policy on recovery plan participation and implementation under the Endangered Species Act activities. *Federal Register* 59 (126): 34272–73.

USFWS (US Fish and Wildlife Service). 1994e. Endangered and threatened wildlife and plants: Notice of interagency cooperative policy for the ecosystem approach to the Endangered Species Act activities. *Federal Register* 59 (126): 34273–74.

USFWS (US Fish and Wildlife Service). 1994f. Endangered and threatened wildlife and plants: Notice of interagency cooperative policy regarding the role of state agencies in Endangered Species Act activities. *Federal Register* 59 (126): 34274–75.

USFWS (US Fish and Wildlife Service). 1998. *Implementing Agreement, El Coronado Ranch, West Turkey Creek, Cochise County, Arizona: Habitat Conservation Plan to Establish a Program for the Conservation of Yaqui Chub, Yaqui Form of Longfin Dace, and Yaqui Catfish.* Phoenix: US Fish and Wildlife Service.

USFWS (US Fish and Wildlife Service). 1999. *Conservation Agreement for the Pecos Pupfish between and among Texas Parks and Wildlife Department, New Mexico Department of Game and Fish, New Mexico State Parks, US Bureau of Land Management and US Fish and Wildlife Service.* Albuquerque: US Fish and Wildlife Service.

USFWS (US Fish and Wildlife Service). 2003. *Biological and Conference Opinions on the Effects of Actions Associated with the Programmatic Biological Assessment of Bureau of Reclamation's Water and River Maintenance Operations, Army Corps of Engineers' Flood Control Operation and Related Non-federal actions on the Middle Rio Grande, New Mexico.* Cons. no. 2-22-03-F-0129. Albuquerque: US Fish and Wildlife Service.

USFWS (US Fish and Wildlife Service). 2004. *Intra-Service Biological Opinion Regarding the Proposed Issuance of an Incidental Take Permit (TE-035920-0) and Approval of a Safe Harbor Agreement for the Rio Grande Silvery Minnow, Southwestern Willow Flycatcher, and Bald Eagle to the Pueblo of Santa Ana in New Mexico.* Cons. no. 2-22-04-F-369. Albuquerque: US Fish and Wildlife Service. https://www.fws.gov/southwest/es/NewMexico/documents/BO/2004-0369%20Safe%20Harbor%20Agreement%20Pueblo%20of%20Santa%20Ana%20with%20clarification%20memo.pdf.

USFWS (US Fish and Wildlife Service). 2006a. *Biological Opinion for the Bureau of Reclamation's Proposed Carlsbad Project Water Operations and Water Supply Conservation, 2006–2016 (May 18, 2006).* Cons. no. 22420-2006-F-0096. Albuquerque: US Fish and Wildlife Service. https://www.fws.gov/southwest/es/NewMexico/documents/BO/2006-0096_10_year_BO_Pecos_River_Operations.pdf.

USFWS (US Fish and Wildlife Service). 2006b. *Safe Harbor Agreement for Voluntary Enhancement/Restoration Activities Benefiting Lahontan Cutthroat Trout on Non-federal Lands within the Upper Humboldt River Distinct Population Segment Area.* Reno: US Fish and Wildlife Service. https://www.fws.gov/nevada/protected_species/fish/documents/lct/lct_sha_final_23june2006.pdf.

USFWS (US Fish and Wildlife Service). 2009. Candidate conservation agreement with assurances for the Page springsnail (*Pyrgulopsis morrisoni*). Phoenix: US Fish and Wildlife Service.

Van Haverbeke, D. R., D. M. Stone, L. G. Coggins, Jr., and M. J. Pillow. 2013. Long-term monitoring of an endangered desert fish and factors influencing population dynamics. *Journal of Fish and Wildlife Management* 4: 163–77.

Walters, A. W., K. R. Sherrill, and J. R. Hickey. 2015. Three-species fish assemblage. In N. B. Carr and C. P. Melcher, eds., *Wyoming Basin Rapid Ecoregional Assessment*, 529–57. US Geological Survey Open-File Report 2015-1155.

Walters, C. J. 2007. Is adaptive management helping to solve fisheries problems? *Ambio* 36: 304–7.

Webber, P. A., K. R. Bestgen, and G. B. Haines. 2013. Tributary spawning by endangered Colorado River basin fishes in the White River. *North American Journal of Fisheries Management* 33: 1166–71.

Widmer, A. M., L. L. Burckhardt, J. W. Kehmeier, E. J. Gonzales, C. N. Medley, and R. A. Valdez. 2010. Detection and population estimation for small-bodied fishes in a sand-bed river. *North American Journal of Fisheries Management* 30: 1553–70.

Williams, J. E., C. A. Macdonald, C. Deacon Williams, H. Weeks, G. Lampman, and D. W. Sada. 2005. Prospects for recovering endemic fishes pursuant to the US Endangered Species Act. *Fisheries* 30: 24–29.

Wilson, K. W. 2015. Conservation agreements and strategies: Utah's proactive approach to conserving native aquatic species. Abstract, Desert Fishes Council Annual Meeting.

13 The Value of Specimen Collections for Conserving Biodiversity

*Adam E. Cohen,
Dean A. Hendrickson,
and Gary P. Garrett*

Conservation biology is a multifaceted discipline with the primary objective of protection and perpetuation of entire natural communities and ecosystems (Soulé 1985). As such, it requires defensible and ideally verifiable information about community composition over time. Specimens held in natural history museums are the preferred form of evidence documenting historical populations because they persist in collections for hundreds of years or more. They can be examined along with original notes and labels for verification of specimen identification, location, and date of collection (Pauly 1995; Saenz-Arroyo et al. 2005; Lister 2011). Until humans began to preserve specimens and associated data, observations were non-verifiable and thus easily dismissed when found to conflict with other sources or opinions.

Since widespread acceptance in the late 1700s of the Linnean binomial system of nomenclature, taxonomists have been compiling collections of organisms. Initial efforts focused on use of specimens for taxon descriptions. Later, expanding collections became useful to those studying morphology in the context of evolution and ecology. As those disciplines began to focus on variation within and among species, requiring more specimens for comparisons, contributions to collections increased and, concurrently, long-term preservation methods improved. Centuries later, those specimens would become useful for understanding changes in biogeographic patterns over time. By the mid-1900s, collectors, including many cited in this volume, were certainly aware of a link between collections and conservation, as species losses were noticed and collections were needed to document the natural histories of species they thought might go extinct or be locally extirpated.

[Box 13.1] Glossary of Museum Collection Terms Used in This Chapter

Collecting event: Collection effort at a single location and date, typically resulting in multiple lots.

Museum lot: A collection of physical objects assigned to a single catalog number. Can vary by discipline, but in ichthyology it usually refers to a jar of specimens of the same species collected during a single collecting event.

Museum record: A row in a spreadsheet or database with data fields related to a museum lot (e.g., date, location, identification, and collector).

Occurrence record: A row in a spreadsheet or database with at least the data fields that would allow the mapping of an individual organism's location in time and space and its identity.

Specimen: Physical organism, which can be a single individual or part of an individual.

The data held in specimens are infinitely vast because specimens hold physical evidence down to the molecular level. The contents of their guts, the parasites they host, even the chemicals held in their tissues are informative regarding the environment from which they came. But more recently, inventory data, traditionally held in variable formats in handwritten ledgers and used primarily as an internal resource for keeping track of specimens, have been recognized as having value to a broad audience.

Recent digitization and databasing efforts in most collections have made their inventories readily available in an easily shared and computer-readable format that is more durable and accessible than the original paper labels and ledgers. Widespread use of database software and the internet has made museum data publicly available via institutional websites and global biodiversity data aggregating services such as the Global Biodiversity Information Facility (GBIF), VertNet, FishNet2, and iDigBio. Researchers and natural resource managers are finding these data valuable for basic and applied research relevant to conservation, evolution, and ecology.

The data held by museums are constantly being subjected to quality control as researchers examine documents and specimens. Museum data can thus be considered a "living" resource that is improving and growing with time. Despite this, many specimens have not been examined since first being cataloged in collections, and many institutions do not have the resources needed to implement quality control and standardize their data. As should be expected for data with long and diverse administrative histories, these data often contain errors that are difficult or impossible for users to ferret out and correct. Consequently, large-scale data standardization (Chapman 2005) and specimen examination efforts are often needed before they can be effectively used (Maldonado et al. 2015). Regional collaborative projects, employing appropriate taxonomic expertise and consultation with those who collected the specimens, are key to correcting and providing quality-controlled data needed by the public, managers, and researchers. This chapter describes how the regional, quality-controlled, and methodically curated Fishes of Texas Project (Hendrickson and Cohen 2015) and its database have made use of specimens to aid conservation decision making in Texas. The Fishes of Texas Project provides an example of what could be done for other regions and/or taxonomic groups.

THE FISHES OF TEXAS PROJECT

The Fishes of Texas Project (FoTX; www.fishesoftexas.org), first funded in 2006, is a collaboration between the University of Texas at Austin's Biodiversity Collections (Texas Natural History Collections, TNHC) and the Texas Parks and Wildlife Department (TPWD). It builds on a strong, more than half-century-long legacy of Texas fish collection by Dr. Clark Hubbs and his associates as well as many others at other institutions. Its primary goal is to provide researchers, resource managers, and the general public with a comprehensive and authoritative source of information on the freshwater fishes of Texas, which constitute one of the most diverse ichthyofaunas in western North America. The published information on this distinctive fauna is scattered over many, often obscure and difficult to obtain datasets, scientific papers, and agency reports. Many of the journals these data are held in are hard to find or require fees to access, and many reports are not published digitally and exist only as paper copies that are also difficult to find. Even when these publications are obtained, the raw occurrence data (location, date, and species) are often not provided, and when they are provided they are not in a format, such as a spreadsheet or database, amenable to merging with other datasets to facilitate analysis. The need for a clearinghouse of such information for use by scientists, resource managers, and the lay public is apparent.

The FoTX's primary resource is its large and comprehensive database of more than 125,000 specimen-based fish occurrence records dating back to 1851 and covering all species known from the entire extent of Texas river basins within and beyond Texas's borders. Data were complied from 42 museums, primarily in the United States and México, including small regional collections with scientifically valuable specimens that would otherwise not be available because their data are unavailable online. The data include 649 of Texas's freshwater, marine, and estuarine fish species, including known extinct and extirpated species (i.e., Maravillas red shiner *Cyprinella lutrensis blairi*, Rio Grande silvery minnow *Hybognathus amarus*, phantom shiner *Notropis orca*, Pecos bluntnose shiner *Notropis simus pecosensis*, Rio Grande bluntnose shiner *Notropis simus simus*, Amistad gambusia *Gambusia amistadensis*, San Marcos gambusia *Gambusia georgei*, and blotched gambusia *Gambusia senilis*), excepting only Rio Grande cutthroat trout *Oncorhynchus clarkii virginalis*, for which specimens were never collected in Texas. It also documents specimen-vouchered occurrences of all the state's rare and endangered species and most (many were never vouchered as specimens) of the nonnative species recorded.

Obtaining these data directly from the contributing collections and doing the necessary record cleaning would not be feasible for most researchers with limited time and resources. Finding these data would require knowledge of their existence and separate inquiries to each institution. Once they were compiled, further investments in data processing would be required to update taxonomy, standardize dates, geo-reference localities, fill in data holes when possible, and perform the especially time-intensive task of verification of outlier records via inspection of specimens and original documentation. Now these data, with these improvements completed, are conveniently available online in a single digital database.

DATA, SPECIMENS, AND LABELS FOR DATA CLEANING

Because museum specimen records are sometimes incomplete or have errors (Chapman 2005; Maldonado et al. 2015), an effort to evaluate the FoTX database for likely errors and make corrections when it was possible to do so confidently was critical. An advantage of the large dataset was that specimen records themselves assisted in error detection and synonymy of locality and collector names. By coalescing records into logical groups, such as those with the same or similar coordinates, collectors, and dates, it was possible to evaluate other record fields to identify errors and correct them. Furthermore, if several independent records were linked to the same collecting event, then data present for one record, but lacking for others, could be applied to all records in the event. For example, a *Micropterus salmoides* largemouth bass specimen, collected by William H. Emory in 1854 (also described in Girard 1859) at "Minneville Rio, Tex," was uninformative regarding the history of the species because the location could not be found in modern gazetteers. The same was true for a record of another species held at a different museum, with matching date and collector and with a similar location, "Minnewitla Riv. Tex." A close phonetic match, "Manahuilla Creek," was found in gazetteers that also happened to run adjacent to "Minnehulla Church." That location fit with the route of the collector as indicated by other independent records (Girard 1859), so "Manahuilla Creek" was applied to all records in the event. Variants of Native American locality names are common. Original names were translated phonetically by early collectors onto labels that were perhaps further changed by transcription errors as data were copied into ledgers. Adding to the confusion, place names also evolved, making matching records to correct locations sometimes impossible. Especially for old records, or for rare species, efforts to match records are worthwhile. This example is important because it is near the edge of the native distribution of largemouth bass in Texas, and the next collection from Manahuilla Creek was not made until 122 years later—the next largemouth bass was collected there 164 years later!

Through geo-referencing and mapping, it was determined that 6,236 records (5% of the database) were spatial or temporal outliers, or otherwise represented specimens likely to be incorrectly identified, and thus required examination. Of these, 2,884 (~46% of the records flagged for examination) were the result of errors in identification; the remainder were correctly identified initially and thus represented temporal or spatial range extensions, or edge of range occurrences elucidating range peripheries.

In addition to museum specimen records, the FoTX website includes related field notes, museum documentation, and associated imagery when provided by data donors or collectors. Data transcription errors are often discovered via comparisons of original labels and field notes with collections' ledgers and databases. Even the specimens themselves sometimes yielded clues to their true provenance via cut marks, clipped fins, or measurements when compared with separate reports such as collectors' field notes and published literature.

This process of methodical and often tedious specimen and documentation examination has substantially redefined the accepted ranges of many species (both native and nonnative), and has even led to the discovery of three species not reported in recent literature documenting Texas's fauna (Hubbs et al. 2008; Page and Burr 2011). Highland stoneroller *Campostoma spadiceum* was thought to be restricted to Oklahoma and Arkansas, but redetermination of previously misidentified specimens collected in 1953 and 1986 from Aiken Creek in Bowie County verified its presence in Texas, and FoTX sampling efforts in 2012–2014 confirmed its persistence. Although overlooked (Hubbs et al. 2008; Page and Burr 2011), longlip jumprock *Moxostoma albidum* was described as occurring in Texas (Robins and Raney 1957; Doosey et al. 2010; Pérez-Rodríguez et al. 2016), but no Texas specimens were in the FoTX database initially. Using characters provided by Robins and Raney (1957) and methodical reexamination of specimens, FoTX added 15 records of this species. Black redhorse *Moxostoma duquesnei* was thought to range only as far south as the Texas-Oklahoma border, but reidentification of a purported gray redhorse *M. congestum* collected in 1998 documented its presence in Texas. This process is ongoing, and selection of specimens for examination is now in part determined by database user comments and suggestions.

DATA-DIRECTED PRIORITIZED SAMPLING

Although Texas has large areas lacking perennial streams, where fish are not likely to exist, the assembled data show that the state's fish fauna is more poorly sampled than initially anticipated. Eighteen (~7%) of the state's 254 counties remain unsampled, and many counties are undersampled, with 66 having fewer than 10 collecting events. Only 39 counties have more than 100 collecting events. Similarly, temporal coverage of sampling is skewed. The first Texas specimens curated in a formal museum were collected in 1851, and by 1900 there were 294 collecting events. By 1920, the total had climbed to 336, and to 1,930 by 1950. An era of more intense sampling followed, and by 2011, there were 14,682 collecting events from Texas, obtained by thousands of collectors.

It is not possible to go back in time to collect and fill in temporal gaps, but new specimen data will continue to be added from contributing museums from sometimes long-held backlogs of specimens awaiting cataloging. Occasional "sleeper" specimens are discovered in jars of multiple specimens cataloged as one taxon.

The Fishes of Texas Project initiated a sampling program in 2008 to fill spatial gaps in the database, adding 6,739 museum lots from 539 locations. These efforts updated distributions for most native and nonnative freshwater and estuarine species statewide. They included 276 species, one of which (bigeye shiner *Notropis boops*) was not previously known to occur in the state and another (*Campostoma spadiceum*) collected shortly after having been discovered in Texas via specimen examinations. Range extensions were documented for at least nine native species (brook silverside *Labidesthes sicculus*, golden topminnow *Fundulus chrysotus*, Gulf killifish *F. grandis*, saltmarsh topminnow *F. jenkinsi*, least killifish *Heterandria formosa*, Amazon molly *Poecilia formosa*, sheepshead minnow *Cyprinodon variegatus*, Rio Grande cichlid *Herichthys cyanoguttatus*, and naked goby *Gobiosoma bosc*) and six nonnative species (grass carp *Ctenopharyngodon idella*, armadillo del rio *Hypostomus* sp., vermiculated sailfin catfish *Pterygoplichthys disjunctivus*, bluefin killifish *Lucania goodei*, variable platyfish *Xiphophorus variatus*, and blue tilapia *Oreochromis aureus*). Gulf pipefish *Syngnathus scovelli*, an estuarine species, was documented in a freshwater reservoir, indicating that the species can complete its life cycle there (Martin et al. 2013).

NATIVE RANGES AND RANGE CHANGES THROUGH TIME

The Fishes of Texas Project data have supported the changing of previously accepted native ranges for some species in important ways. For example, the federally endangered sharpnose shiner *Notropis oxyrhynchus* was believed to have been introduced to and then extirpated from the Colorado River (Hubbs et al. 2008). Examination of collections of Colorado River cyprinids from before 1960, however, yielded five sharpnose shiner records from 1884 to 1955, thus supporting its native status there (USFWS 2014). Failure to collect the species from the Colorado River since 1955, despite substantial efforts, suggests it is probably extirpated. In addition, smalleye shiner *N. buccula* is documented in the Colorado River by a single specimen collected in 1952 (TNHC2635). The native status of sharpnose shiner in the Colorado River lends credibility to *N. buccula* being native as well, especially as these two are commonly syntopic in the Brazos River. Smalleye shiner is often confused with other species, and it would not be surprising to discover specimens of it in a jar with a similar species.

Examination of museum specimens previously identified as weed shiner *Notropis texanus* from the Brazos River basin determined that the species never occurred there. This finding was surprising, since it contradicts all but one (Lee et al. 1980) of the published maps and range descriptions for the species since 1953 (Knapp 1953; Conner and Suttkus 1986; Thomas et al. 2007; Hubbs et al. 2008; Page and Burr 2011). Museum specimens collected pre-1950 support the species being native to the basins on either side of the Brazos (Colorado and Trinity Rivers), but no specimen-based evidence exists for it being present in the Brazos basin. The separate populations, presumably lacking gene flow, may merit additional morphological and genetic assessments. Similarly, work is being conducted on defining historical ranges for three morphologically similar *Moxostoma* species in the Rio Grande (*M. albidum*, Mexican redhorse *M. austrinum*, and *M. congestum*) with the aid of genotyped specimens (Clements 2008). These examples highlight the value of vouchering specimens, especially those that are difficult to identify.

Historical FoTX data and expert opinion were used to make determinations of native ranges for all Texas freshwater fishes (at the resolution of United States Geological Survey HUC8; USGS 2017). Reflecting the scientific process of advancing new knowledge, those native ranges often differ from what has been published in other sources (Knapp 1953; Conner and Suttkus 1986; Thomas et al. 2007; Page and Burr 2011; Maxwell 2012) as a result of new insights provided by the FoTX's comprehensive specimen-verified data. These native range polygons are now visible on the website, and occurrence records can be compared with them to assess whether they represent native or nonnative occurrences.

The quality-controlled data revealed various patterns of range contraction over the past 50 to 100 years for species including *Notropis buccula*, Rio Grande shiner *N. jemezanus*, *N. oxyrhynchus*, and headwater catfish *Ictalurus lupus*. But for at least 16 species, there is a reduction trend in which the southwestern edges of ranges are shifting to the northeast. These species include bowfin *Amia calva*, western creek chubsucker *Erimyzon claviformis*, lake chubsucker *E. sucetta*, pallid shiner *Hybopsis amnis*, emerald shiner *Notropis atherinoides*, ironcolor shiner *N. chalybaeus*, suckermouth minnow *Phenacobius mirabilis*, swamp darter *Etheostoma fusiforme*, slough darter *E. gracile*, cypress darter *E. proeliare*, and blackside darter *Percina maculata*. Some species reveal the same trend through the presence of relict populations southwest of their larger northeasterly range, such as silver chub *Macrhybopsis storeriana*, creek chub *Semotilus atromaculatus*, spotted sucker *Minytrema melanops*, greenthroat darter *Etheostoma lepidum*, goldstripe darter *E. parvipinne*, and river darter *Percina shumardi*.

Also detected via FoTX data are range expansions for many other species, including *Labidesthes sicculus*, inland silverside *Menidia beryllina*, *Herichthys cyanoguttatus*, and numerous nonnatives. *Fundulus grandis* and *Cyprinodon variegatus* are salt tolerant and are invading the upper reaches of the Rio Grande, Colorado, Brazos, and Red Rivers (only *F. grandis* in Red) as declining spring flow is increasing salinity. Others, such as *Fundulus chrysotus*, *Heterandria formosa* and *Poecilia formosa*, are likely to be using artificial canals to move along the coast.

NATURAL HISTORY AND HISTORICAL DISTRIBUTIONS OF NEWLY DESCRIBED SPECIES

Modern ichthyological research is replete with examples of genetic analyses resulting in discovery or confirmation of what researchers considered to be distinct forms, and new species are often split from another. Recently, four cyprinid species in Texas were split into many new taxa. Roundnose minnow *Dionda episcopa* is now seven species (Schönhuth et al. 2012), speckled chub *Macrhybopsis aestivalis* is now five species (Eisenhour 2004), red shiner *Cyprinella lutrensis* is now nine species (Richardson and Gold 1995) and Texas shiner *Notropis amabilis* is now two species (Conway and Kim 2016). Once new species are formally described, typically with the aid of museum specimens, and if the descriptions include morphological diagnoses, it is possible to go to the specimen record to reidentify specimens, determine historical ranges of the newly described species, and assess how their ranges might be changing and/or their populations expanding or declining. Often very little is known about the newly described species because there was little scientific rationale to motivate research and collection. Once a subset of what was thought to be a single, widespread species is determined to be a new species, typically with a much smaller range, there is often a critical need for information that could inform conservation actions. Museum specimens are useful in such cases and may be the only way to retrieve historical natural history information about the species, including gut contents, parasites, and demographic attributes.

Museum specimens collected long before DNA was discovered have now been used in important scientific discoveries (De Bruyn et al. 2011; Metcalf et al. 2012), and because of progress on using formalin-preserved specimens in genetics research, a potentially large archive is increasingly available to researchers (Hykin et al. 2015; Ruane and Austin 2017). These advances could perhaps be employed for extirpated or extinct Texas fishes that are known only from specimens.

Modeling and Conservation Network Planning

For most Texas fishes, FoTX's database, with its large number of collections distributed across the state's diverse landscape, has enabled creation of species distribution models (SDMs) that use occurrence data and environmental variables as input for niche modeling programs such as MaxEnt (Phillips and Dudík 2008). Estimates of occurrence probabilities for species have been the foundation for conservation efforts in Texas and are especially valuable when private property limits access to sometimes large unsampled stream reaches. Comparison of modeled occurrence probabilities with fish surveys in the James River and Barton Creek, Texas, demonstrated that model estimates are reliable predictors of the fish community (Labay et al. 2011) and those estimates have since been used in conservation network planning (Ciarleglio et al. 2009; Moilanen et al. 2011; Hendrickson and Labay 2014). Conservation network planning software programs aim to define target areas where conservation efforts can be most effective in terms defined by the user (e.g., number of species, number of natives, land use, and distance from human populations). In Texas, model predictions derived from the FoTX data defined 21 Native Fish Conservation Areas (http://nativefishconservation.org/initiatives/texas-native-fish-conservation-network/) that TPWD is now using to prioritize statewide freshwater aquatic ecosystem conservation efforts. The SDMs may also be useful in evaluating stream condition because they can be used to establish a theoretical historical baseline community for comparison with modern collections. This method has the potential to replace the controversial index site comparisons that are currently standard practice in bioassessment (Labay et al. 2015).

Use by Agencies

Publicly available databases expand opportunities for education and research, often beyond what can be anticipated and tracked after data are extracted from the site. Tracking via Google Analytics indicated use of the FoTX database by 6,200 unique individuals in 7,900 sessions over a recent six-month period (April to September 2017). The target audience has always been natural resource professionals and researchers, including staff at TPWD, who use the data for environmental permitting recommendations and conservation efforts, including updating the list of Species of Greatest Conservation Need used in allocation of conservation funds. Collaborations with TPWD have also included use of the data to update its Natural Diversity Database (NDD) and its species conservation assessments. Because the NDD feeds NatureServe's database, and it in turn is used by the International Union for Conservation of Nature (IUCN), the FoTX data are helping to define state, national, and global rankings used by researchers for large-scale assessments and to inform conservation actions at the state and federal level.

CONCLUSIONS

Museum specimens and associated data are a sound and verifiable basis for research and conservation decision making because specimens and related data collected long ago can be reexamined and verified. Such data are particularly valuable in an era when facts are frequently questioned by politicians and a dubious public, as they create a reliable basis for sound science, especially in litigious settings.

Increased deposition of specimens in museums should be encouraged. Collections made today, even of common and widespread species, will accommodate future research needs that cannot easily be predicted. A foundation of modern science is repeatability, and there is no better way to repeat any study than by use of the same samples, or a portion of the samples, used in the study being repeated. Vouchering all specimens is not always possible or even desirable when endangered species are involved, but quality control assessments are possible if researchers archive representative specimens. Given the remarkable past value of specimens in various research applications, it is difficult to envision all their uses in the future, but there is no doubt they have great potential for supporting conservation efforts.

The work done to develop and normalize the data in the Fishes of Texas Project took many years of collaborative work by a group of regional faunal experts in academia and government. Their expertise ensured foundational work that will save future researchers intensive additional work so that they may focus on additional data improvements or focused tasks. Additionally, the project establishes a foundational dataset upon which a community of researchers can build scientific conclusions. By repatriating cleaned data to their original specimen-holding institutions, the project's efforts improve data held at institutions all over the world. Widespread use of flawed occurrence data could be greatly reduced if other academic institutions, government agencies, and NGOs were to develop similar collaborative projects.

REFERENCES

Chapman, A. D. 2005. Principles and methods of data cleaning—Primary species and species-occurrence data, version 1.0. Report for the Global Biodiversity Information Facility, Copenhagen. http://www.gbif.org/document/80528.

Ciarleglio, M., J. W. Barnes, and S. Sarkar. 2009. ConsNet: New software for the selection of conservation area networks with spatial and multi-criteria analyses. *Ecography* 32: 205–9.

Clements, M. D. 2008. Molecular systematics and biogeography of the Moxostomatini (Cypriniformes: Catostomidae), with special reference to taxa in Texas and Mexico. PhD. diss., Tulane University.

Conner, J. V., and R. D. Suttkus. 1986. Zoogeography of freshwater fishes of the Western Gulf Slope of North America. In C. H. Hocutt and E. O. Wiley, eds., *Zoogeography of North American Freshwater Fishes*, 413–56. New York: John Wiley & Sons.

Conway, K. W., and D. Kim. 2016. Redescription of the Texas Shiner *Notropis amabilis* from the southwestern United States and Northern Mexico with the reinstatement of *N. megalops* (Teleostei: Cyprinidae). *Ichthyological Exploration of Freshwaters* 26: 305–40.

De Bruyn, M., L. R. Parenti, and G. R. Carvalho. 2011. Successful extraction of DNA from archived alcohol-fixed white-eye fish specimens using an ancient DNA protocol. *Journal of Fish Biology* 78: 2074–79.

Doosey, M. H., H. L. Bart, Jr., K. Saitoh, and M. Miya. 2010. Phylogenetic relationships of catostomid fishes (Actinopterygii: Cypriniformes) based on mitochondrial ND4/ND5 gene sequences. *Molecular Phylogenetics and Evolution* 54: 1028–34.

Eisenhour, D. J. 2004. Systematics, variation, and speciation of the *Macrhybopsis aestivalis* complex west of the Mississippi River. *Bulletin of the Alabama Museum of Natural History* 23: 9–48.

Girard, C. F. 1859. Ichthyology of the boundary. *U.S. and Mexican Boundary Survey Report* 2: 1–85.

Hendrickson, D. A., and A. E. Cohen. 2015. Fishes of Texas Project Database (Version 2.0). https://doi.org/10.17603/C3WC70.

Hendrickson, D. A., and B. J. Labay. 2014. *Conservation Assessment and Mapping Products for GPLCC Priority Fish Taxa*. Final contract report to US Fish and Wildlife Service, Great Plains Landscape Conservation Cooperative. Austin: University of Texas at Austin. http://hdl.handle.net/2152/27744.

Hubbs, C., R. J. Edwards, and G. P. Garrett. 2008. An annotated checklist of the freshwater fishes of Texas, with keys to identification of species. 2nd ed. *Texas Journal of Science Supplement (July)*: 2–87. https://doi.org/10.15781/T22Z13563.

Hykin, S. M., K. Bi, and J. A. McGuire. 2015. Fixing formalin: A method to recover genomic-scale DNA sequence data from formalin-fixed museum specimens using high-throughput sequencing. *PLoS ONE* 10 (10): 1–16. e0141579.

Knapp, F. T. 1953. *Fishes Found in the Freshwaters of Texas*. Brunswick, GA: Ragland Studio and Litho Printing Co.

Labay, B. J., A. E. Cohen, B. Sissel, D. A. Hendrickson, F. D. Martin, and S. Sarkar. 2011. Assessing historical fish community composition using surveys, historical collection data, and species distribution models. *PLoS ONE* 6 (9): e25145.

Labay, B. J., D. A. Hendrickson, A. E. Cohen, T. H. Bonner, R. S. King, L. J. Kleinsasser, G. W. Linam, and K. O. Winemiller. 2015. Can species distribution models aid bioassessment when reference sites are lacking? Tests based on freshwater fishes. *Environmental Management* 56: 835–46.

Lee, D. S., C. R. Gilbert, C. H. Hocutt, R. E. Jenkins, D. E. McAllister, and J. R. Stauffer, Jr. 1980. *Atlas of North American Freshwater Fishes*. Raleigh: North Carolina State Museum of Natural History.

Lister, A. M. 2011. Natural history collections as sources of long-term datasets. *Trends in Ecology and Evolution* 26: 153–54.

Maldonado, C., C. I. Molina, A. Zizka, C. Persson, C. M. Taylor, J. Albán, E. Chilquillo, N. Rønsted, and A. Antonelli. 2015. Estimating species diversity and distribution in the era of big data: To what extent can we trust public databases? *Global Ecology and Biogeography* 24: 973–84.

Martin, F. D., A. E. Cohen, B. J. Labay, M. J. Casarez, and D. A. Hendrickson. 2013. Apparent persistence of a landlocked population of Gulf pipefish, *Syngnathus scovelli*. *Southwestern Naturalist* 58: 376–78.

Maxwell, R. J. 2012. Patterns of endemism and species richness of fishes of the Western Gulf Slope. MS thesis, Texas State University.

Metcalf, J. L., S. L. Stowell, C. M. Kennedy, K. B. Rogers, D. McDonald, J. Epp, K. Keepers, A. Cooper, J. J. Austin, and A. P. Martin. 2012. Historical stocking data and 19th century DNA reveal human-induced changes to native diversity and distribution of cutthroat trout. *Molecular Ecology* 21: 5194–207.

Moilanen, A., A. Arponen, J. Leppänen, L. Meller, and H. Kujala. 2011. Zonation: Spatial conservation planning framework and software version 3.0 user manual. University of Helsinki Research Portal, University of Helsinki. https://tuhat.helsinki.fi/portal/en/publications/zonation-spatial-co(a1006241-f66a-46e1-8105-eff0f0c494e4).html.

Page, L. M., and B. M. Burr. 2011. *Peterson Field Guide to Freshwater Fishes of North America North of Mexico*. Boston: Houghton Mifflin Harcourt.

Pauly, D. 1995. Anecdotes and the shifting baseline syndrome of fisheries. *Trends in Ecology and Evolution* 10: 430.

Pérez-Rodríguez, R., O. Domínguez-Domínguez, A. F. Mar-Silva, I. Doadrio, and G. Pérez-Ponce de León. 2016. The historical biogeography of the southern group of the sucker genus *Moxostoma* (Teleostei: Catostomidae) and the colonization of central Mexico. *Zoological Journal of the Linnean Society* 177: 633–47.

Phillips, S. J., and M. Dudík. 2008. Modeling of species distributions with Maxent: New extensions and a comprehensive evaluation. *Ecography* 31: 161–75.

Richardson, L. R., and J. R. Gold. 1995. Evolution of the *Cyprinella lutrensis* species-complex. II. Systematics and biogeography of the Edwards Plateau shiner, *Cyprinella lepida*. *Copeia* 1995: 28–37.

Robins, C. R., and E. C. Raney. 1957. The systematic status of the suckers of the genus *Moxostoma* from Texas, New Mexico and Mexico. *Tulane Studies in Zoology* 5: 291–381.

Ruane, S., and C. C. Austin. 2017. Phylogenomics using formalin-fixed and 100+ year-old intractable natural history specimens. *Molecular Ecology Resources* 17: 1003–8.

Saenz-Arroyo, A., C. Roberts, J. Torre, M. Carino-Olvera, and R. Enriquez-Andrade. 2005. Rapidly shifting environmental baselines among fishers of the Gulf of California. *Proceedings of the Royal Society B* 272: 1957–62.

Schönhuth, S., D. M. Hillis, D. A. Neely, M. L. Lozano-Vilano, A. Perdices, and R. L. Mayden. 2012. Phylogeny, diversity, and species delimitation of the North American round-nosed minnows (Teleostei: *Dionda*), as inferred from mitochondrial and nuclear DNA sequences. *Molecular Phylogenetics and Evolution* 62: 427–46.

Soulé, M. E. 1985. What is conservation biology? *BioScience* 35: 727–34.

Thomas, C., T. H. Bonner, and B. G. Whiteside. 2007. *Freshwater Fishes of Texas: A Field Guide*. College Station: Texas A&M University Press.

USFWS (US Fish and Wildlife Service). 2014. *Species Status Assessment Report for the Sharpnose Shiner* (Notropis oxyrhynchus) *and Smalleye Shiner* (N. buccula). Species Status Assessment Reports. Arlington, TX: US Fish and Wildlife Service. https://catalog.data.gov/dataset/species-status-assessment-report-for-the-sharpnose-shiner-notropis-oxyrhynchus-and-smalleye.

USGS (US Geological Survey). 2017. The watershed boundary dataset (WBD). US Geological Survey. ftp://rockyftp.cr.usgs.gov/vdelivery/Datasets/Staged/Hydrography/WBD/.

14

Conservation Genetics of Desert Fishes in the Genomics Age

Thomas F. Turner,
Thomas E. Dowling,
Trevor J. Krabbenhoft,
Megan J. Osborne,
and Tyler J. Pilger

FROM PROTEINS TO GENOMES IN CONSERVATION GENETICS

Much has been written about the utility of genetic approaches in the conservation of desert fishes, including Echelle's (1991) chapter in *Battle against Extinction*. Echelle's review focused primarily on analysis of variation in enzymes (called allozymes). Protein electrophoresis was used to characterize allelic variation at allozyme gene loci, and this technique, coupled with population genetic theory from the "Modern Synthesis," built the foundations of our understanding of genetic variation within and among natural populations. Concurrently, recombinant DNA technology was adopted into the field of population genetics, which led to the use of restriction fragment length polymorphism (RFLP) analysis and related approaches to characterize genetic variation in nucleotide sequences rather than in proteins.

Since then, advances in molecular genetics have occurred at a remarkable pace. A foundational innovation was the polymerase chain reaction (PCR), which can be used for copying and sequencing specific DNA fragments in virtually any laboratory. Widespread availability of PCR led to cost-effective DNA (i.e., Sanger) sequencing of individual genes and analysis of DNA microsatellites. Now, massively parallel or "next-generation" (NextGen) DNA sequencing generates data for the entire genome in a single assay, and thousands of genes are routinely analyzed on a population level. How will these advances change the ways in which genetic data and theory are applied to conservation problems, and what new insights might these technologies provide?

Despite the rapidity of technological advances, the basic questions and goals of conservation genetics have changed little since Echelle's chapter was published. Identification of distinct taxonomic units and of the evolutionary relationships of lineages remain first steps in conservation planning. Detection and mitigation of introgressive hybridization of native and nonnative species still motivates conservation and management efforts. Likewise, characterizing and interpreting genetic variation within and among populations in a species remains a central focus. What has changed is the ability to quantitatively evaluate local genetic adaptation, especially in polygenic traits (Lynch 1996), and physiological scope for coping with environmental variation through ecological genomic analysis. There is renewed appreciation for the value of time series for identifying trajectories of genetic response to conservation and management actions (Schwartz et al. 2007; Dowling et al. 2014). The new technology will inform long-standing problems of conservation genetics. Big datasets and associated bioinformatics approaches are projected to open new doors to predictive frameworks. We use examples from the published literature to illustrate these possibilities and some of the limitations of next-generation DNA and RNA sequencing technology for addressing old and new conservation issues.

THE TOOLBOX OF CONSERVATION GENETICS AND GENOMICS

Technical jargon and acronyms are barriers to understanding how molecular genetics contributes to desert fish conservation. In this section, a brief overview of commonly used techniques is provided, and the efficacy of each approach for answering questions in conservation science is considered. This section is not exhaustive, but is meant to provide a concise reference for readers less familiar with molecular genetics. More detailed and complete reviews and protocols can be found in Allendorf et al. (2010), Wit et al. (2012), and Benestan et al. (2016).

Sanger sequencing involves amplifying individual gene loci via PCR and targeted *primers* (short, complementary DNA sequences that flank genes or regions of interest and initiate amplification), followed by DNA sequencing of fragments up to 500–1,000 base pairs (bp) in length. Sanger sequencing has been a mainstay of conservation genetics since the early 1990s because it is cost-effective and broadly applicable across taxa. Likewise, modes and rates of mutation at the DNA sequence level are relatively well understood, and powerful and widely disseminated analytical software is available to estimate phylogenies and differentiation among populations. Sanger sequencing is used when rapid genetic assessment is needed, or for taxonomic identification of unknown samples (e.g., DNA barcoding and eDNA; see below).

Genetic variation at *microsatellite loci* is based on differences in fragment lengths found in short, repeated, and often hypervariable regions of DNA spread ubiquitously throughout the genome. Once specific loci are identified and amplified by PCR, they can be characterized using an automated sequencer, yielding insight into allelic (fragment-length) variation across hundreds or thousands of individuals for loci with many alleles.

Targeted SNP genotyping is like the microsatellite technique in that the genotyping of many individuals is the goal, but rather than focusing on repeat-sequence length variation, this method characterizes variation in point mutations (base substitutions) across individuals. The variable sites are termed *single-nucleotide polymorphisms* (SNPs). Targeted SNP genotyping is usually done in one of two ways: with NextGen DNA sequencing applied to PCR products (e.g., GT-seq; Campbell et al. 2015), or by quantitative PCR with allele-specific fluorescent probes (e.g., Taqman and Fluidigm SNP assays). Both methods require primers designed to target specific regions amplified by PCR.

Restriction site–associated DNA sequencing (**RAD-seq**) is a whole-genome sub-sampling method that allows for sequencing thousands of genomic regions with relatively high throughput (i.e., hundreds of individuals in a single sequencing run). Genomic regions are isolated using restriction enzymes that digest genomic DNA into smaller fragments. Samples are then individually labeled (by attaching unique DNA sequence identifiers) and pooled for NextGen sequencing. The resulting SNP genotypes are compared among individuals (see above).

RNA sequencing (**RNA-seq**) is a method for estimating genome-wide gene expression in a tissue sample. Transcribed RNA is isolated from tissue and converted to more stable DNA by enzymatic reverse transcription. Unique DNA sequence identifiers are attached to DNA of individual samples. Samples are then pooled and sequenced on a NextGen apparatus. The relative abundance of distinct transcript sequences is a proxy for gene expression when assay conditions are properly controlled. RNA-seq is widely used in ecological genomics as a means to understand how organisms interact with their environment at the genetic level. In addition, SNPs can be identified and genotyped using RNA-seq. Because RNA-seq characterizes only the subset of the genome that is transcribed (e.g., ~5%), it is an excellent method for identifying parts of the genome under selection.

Sequence capture is a method that uses RNA or DNA probes to enrich a genomic DNA sample for a predetermined subset of loci. The probes can be designed to be complementary to any part of a genome, such as protein-coding genes or noncoding regions. Sequence capture is more expensive than RADseq, particularly when large numbers of probes are used, but generally cheaper than whole-genome resequencing.

Whole-genome resequencing relies on NextGen sequencing of genomic DNA. Enormous amounts of data are generated, but the dataset consists of millions of relatively short fragments of DNA (<250 bp) that are mapped and arranged into a genome by bioinformatic computation. When fragments are short, genome assembly is much less precise, especially for organisms whose genome architecture is unknown. Long-read (i.e., DNA fragments >10,000 bp) sequencing substantially improves the quality of genome assemblies by spanning large parts of genomes. Other important considerations include the trade-off between sequencing depth (i.e., the number of sequences obtained for a specific genomic region) and number of individuals studied.

Environmental DNA **(e-DNA)** analysis aims to identify the presence and, in some cases, quantify the abundance of target DNA sequences from environmental samples, such as filtrate from water. The approach involves isolation and amplification of DNA (typically mtDNA) from the sample using specialized DNA-binding membranes.

Epigenetic profiling identifies genome-wide changes in DNA structure other than base-pair substitutions. Epigenetic modifications to DNA, such as methylation of individual nucleotides, can reduce gene expression and thus affect organismal performance.

CONSERVATION GENETICS AND GENOMICS IN PRACTICE

Conservation genetics questions are hierarchical, beginning most broadly with an understanding of when and how particular evolutionary lineages (i.e., metapopulations, distinct population segments, species) became distinct from close relatives. Once a focal lineage is identified, the goal is to understand the ecological and evolutionary factors that regulate its genetic diversity in space and time. These factors include abundance, distribution, population connectivity, introgression, mutation, and artificial (e.g., human-imposed) and natural selection. Such understanding motivates management and recovery actions that could include nonnative species removal, population augmentation, assisted migration, and other on-the-ground activities. Molecular approaches can then be used to monitor genetic variation to evaluate and modify management and recovery actions so as to maximize genetic and ecological diversity and enhance probabilities of lineage persistence.

Phylogenetics and Lineage Identity

DNA sequence–based methods are most commonly used for characterizing deeper evolutionary relationships among populations and species. For decades, Sanger sequencing was the method of choice, yielding many significant papers on desert fishes (e.g., Houston et al. 2014; Unmack et al. 2014). However, single-gene sequencing is now largely being replaced by NextGen sequencing, which yields DNA sequences for thousands to tens of thousands of loci. More data potentially provide more accurate and precise estimates of phylogeographic and phylogenetic relationships (Narum et al. 2013; Ellegren 2014). Loci obtained from RADseq are randomly scattered around the genome and are assumed to represent all linkage groups (Peterson et al. 2012; Amores et al. 2014); therefore, these methods can detect adaptive as well as neutral genetic variants (e.g., Hohenlohe et al. 2013). NextGen sequencing allows researchers to use multiple loci for generating phylogenetic trees, with the (perhaps naïve) expectation that concatenation of all loci could accurately recover the evolutionary history of organisms of interest (Brito and Edwards 2009, but see below).

With this increased resolution come costs and potential consequences frequently not considered when initiating a NextGen sequencing project. While tens of thousands of loci yield greater resolution, RADseq also leads to more false positives and genotyping errors (Fourcade et al. 2013; Mastretta-Yanes et al. 2015), requiring greater consideration of the statistical ramifications of results. In addition, this quantity of information requires extensive and rigorous bioinformatic analyses (Mastretta-Yanes et al. 2015; Paris et al. 2017) to ensure that the data are sufficient and appropriate for addressing specific questions. For example, RADseq is subject to experimental error and other artifacts that can lead to missing data, genotyping errors, and poor coverage of the genome for some individuals (Arnold et al. 2013; Gautier et al. 2013), although methodological refinements have alleviated some of these issues (e.g., Ali et al. 2016; Hoffberg et al. 2016). Adequate coverage of the genome is critical because poor coverage can result in missing data and genotyping errors (Nielsen et al. 2011; Alex-Buerkle and Gompert 2013).

A comparison of two RADseq studies on Devils Hole pupfish *Cyprinodon diabolis* illustrates how differences in depth of sequence coverage, extent of missing data, model selection, and choice of bioinformatic filtering criteria can lead to radically different conclusions. One study concluded that Devils Hole was colonized between 103 and 800 years ago, and that colonization was followed by genetic assimilation (Martin et al. 2016). A second study placed the date of colonization around 60,000 years ago (Sağlam et al. 2016), which is consistent with what is known about the geologic history of the region (Winograd et al. 1992). Notably, Martin et al. (2016) analyzed loci with up to 90% missing data and used a very high mutation rate (e.g., ~1,000 times faster than that of humans) to account for their conclusions. Both Martin et al. (2016) and Sağlam et al. (2016) expressed high levels of confidence in their (mutually exclusive) results. Because critical decisions about whether and how to implement assisted gene flow or hatchery programs depend on the evolutionary provenance of Devils Hole pupfish, independent scrutiny of the original data, including critical evaluation of how missing data affected the conclusions, must be conducted prior to the development and implementation of a genetic management plan.

Even with adequate genomic coverage and stringent computational filtering, NextGen sequencing data may not recover the correct evolutionary history for a focal lineage for any of several reasons: (1) insufficient data to resolve relationships (Funk and Omland 2003), (2) incomplete and random sorting of ancestral polymorphism (Brito and Edwards 2009), or (3) introgressive hybridization (Arnold 2016). Each of these problems yields different expected outcomes, and therefore it is possible to discriminate among them. If there are too few characters to recover individual gene trees, the majority of gene trees will be unresolved. This is especially problematic for RADseq as there are typically only one or a few biallelic SNPs per RAD locus, and in this case, phylogenetic relationships can be obtained only through consensus across all variable loci. Even if sufficient characters can be generated in RADseq to resolve relationships, individual gene trees may differ due to incomplete lineage sorting, which complicates the consensus methods that are frequently relied upon in phylogeny reconstruction. Concatenation of SNPs obtained from the small DNA fragments that are usually obtained in NextGen sequencing necessitates combining data with different genealogical histories to estimate phylogenetic relationships. In worst cases, analysis of concatenated SNPs across many loci can give an incorrect phylogenetic tree with very high statistical support (Tonini et al. 2015). These problems can be addressed computationally (Chifman and Kubatko 2014), or with DNA sequencing methods that generate longer sequence fragments, as these often contain more variable positions per gene and make it more likely that individual gene trees can be recovered and statistically contrasted.

Introgressive hybridization produces the most severe violation of the assumptions of phylogenetic analysis, as genes do not share a common genealogical history, resulting in a "phylogenetic web" instead of a resolved tree. Introgressive hybridization is prevalent in many native western fishes (e.g., minnows, Gerber et al. 2001; pupfish, Carson et al. 2012; suckers, Dowling et al. 2016; trout, Kovach et al. 2016), and divergence in the face of gene exchange (Arnold 2016) results in individual genes exhibiting different evolutionary histories (Harrison 2012; Harrison and Larson 2016). Discordance will be especially evident for traits that exhibit unusual patterns of inheritance (e.g., mitochondrial DNA, Y chromosomes; Funk and Omland 2003) or are adaptive (Hohenlohe et al. 2013). Given that most RAD sequences are neutral, the consensus tree may be biased and mask discordance. Again, newer technologies that generate longer sequences might allow for contrasts among gene trees to identify such conflicts.

Hybridization is especially problematic for conservation biologists, as they need to consider the potential impact of introgression in various contexts (Dowling et al. 1992a, b). When introgression results from human actions (e.g., widespread stocking of rainbow trout), the products of hybridization should be considered detrimental, as they could result in the swamping of locally adapted taxa (e.g., Propst et al., chap. 20, this volume; Smith and Stearley 2018). Conversely, introgression can also occur naturally as part of the evolutionary process (Arnold 2016), enhancing evolutionary potential through the transfer of adaptive variants among species (Hohenlohe et al. 2013) and the creation of new species (DeMarais et al. 1992; Gerber et al. 2001). The creative potential of natural hybridization (Meier et al. 2017) must be recognized as conservation strategies are developed.

Technical and analytical issues aside, identification of a distinct taxonomic unit (e.g., species) is complicated by the dynamic nature of the evolutionary process itself, especially where patterns of molecular and morphological variation do not fit neatly into an established "species concept." Species have traditionally been identified by phenotypic (typically morphological) traits. The continuous nature of many of these characters, mosaic patterns of variation, and the myriad concepts used to identify species have led to considerable confusion over the taxonomic status of many forms. This situation has been further confounded by the problems with new molecular technologies identified above, which have led to confusion and controversy over the taxonomic status of many endangered taxa (e.g., Theimer et al. 2016; Marsh et al. 2017).

Conservation biologists have recognized the complexities associated with species concepts, and they focus on the importance of protecting evolutionary distinctiveness (i.e., lineages) and processes that generate it (e.g., adaptive potential; Harrisson et al. 2014), rather than explicitly focusing on taxonomic status and presumed genetic purity. Such thinking has led to the development and designation of "process-oriented" conservation units. For example, evolutionary significant units (ESUs) and management units (MUs) are useful for policy setting and for broader-scale conservation efforts (Waples 1995). ESUs are genetically unique groups of populations that are identified by diagnosable features (e.g., reciprocal monophyly of mtDNA in gene trees) and significant divergence among nuclear genes (Moritz 1994). Sanger sequencing of nuclear and mtDNA genes is useful for identifying ESUs (e.g., Hopken et al. 2013; but see Parker et al. 1999, who also used them to identify MUs). Within ESUs, distinct MUs may also be detected and identified based on either divergent mtDNA or nuclear DNA (Moritz 1994). MUs are typically demographically independent, such that population growth is determined by local birth and death rates rather than by immigration (Palsbøll et al. 2007). Functional genomics could facilitate the discovery of ESUs by allowing the underlying genetic causes of adaptive divergence to be identified.

The current emphasis on "defining" units of conservation (typically species) as the endpoint of conservation action can lead to unsuccessful outcomes because of inability to reach consensus on the definitions of these units (Hey et al. 2003). Like others before us, we suggest that resolution of this problem requires a shift in perspective away from defining and managing species and toward protection of the ecological and evolutionary processes that generate and maintain biodiversity. This approach necessitates improved communication between scientists and managers concerning the explicit goals of conservation actions, and it may also require modification of legislative mechanisms for conservation to account for advances in scientific knowledge since inception of the ESA.

The Spatial Dimension: Genetic Diversity within and among Populations

When *Battle against Extinction* was published, the conservation community recognized that species had different geographic genetic patterns related to variation in their habitat needs and evolutionary history. Based on their work with desert fishes, Meffe and Vrijenhoek (1988) devised two zoogeographic models of isolation and gene flow for desert fishes to aid recovery programs. The Death Valley model (DVM), named for the extreme degree of post-Pleistocene isolation in that region, characterizes aquatic organisms with no contemporary gene flow, in which genetic drift and local adaptation result in low diversity within and high divergence among local populations (e.g., Pecos gambusia *Gambusia nobilis*, Echelle et al. 1989; Baja California killifish *Fundulus lima*, Bernardi et al. 2007; and Gila topminnow *Poeciliopsis occidentalis*, Parker et al. 1999). The stream hierarchy model (SHM) describes the geographic variation of aquatic species occupying interconnected, dendritic stream networks, in which local genetic diversity reflects local population size and rates of gene flow among populations. Accordingly, genetic differentiation is expected to reflect geographic connectedness, such that populations in tributaries within the same catchment are more genetically similar to one another (e.g., loach minnow *Tiaroga cobitis*, spikedace *Meda fulgida*, and longfin dace *Agosia chrysogaster*, Tibbets and Dowling 1996; bluehead sucker *Pantosteus discobolus*, Hopken et al. 2013; and arroyo chub *Gila orcuttii*, Benjamin et al. 2016).

From a conservation perspective, species fitting the DVM tend to be at more immediate risk from stochastic or anthropogenic activities that reduce local habitat stability and negatively affect local abundance. Species characterized by the SHM occupy more dynamic stream habitats (and thus experience natural fluctuations in local abundance) and are therefore more at risk from habitat alterations that obstruct dispersal corridors and disrupt connectivity among local patches. In this case, restriction of gene flow among populations leads to increased genetic drift, which reduces within-population diversity and increases among-population divergence for the entire "metapopulation" (Turner et al. 2015).

High-throughput methods for genotyping tens of microsatellite loci to hundreds of SNPs for many (thousands of) individuals provide increased power for hypothesis testing. At the within-population level, microsatellite loci have made (and still make) excellent markers for studies of desert fishes, providing information on population genetics (Pilger et al. 2015), demographics (Alò and Turner 2005; Dowling et al. 2014), functional variation (Krabbenhoft and Turner 2014; Osborne et al. 2017), and population structure (Dowling et al. 2015). Such information is vital for informed management efforts.

Useful conceptual models and new analytical approaches have emerged through the juxtaposition of genetic variation and landscape information. The combination of spatially explicit genetic data and GIS analyses spurred the burgeoning field of landscape genetics. This framework allows researchers to test hypotheses regarding the roles of geographic properties such as distance and elevation, species ecology, and natural and anthropogenic factors that relate to disturbance (Pilger et al. 2017). Landscape genetics often accounts for interactions between evolutionary processes (e.g., gene flow or adaptation) and the environmental template that a species inhabits, providing a framework for understanding how global change may affect patterns of neutral or adaptive genetic variation, as well as a mechanistic understanding of how species are likely to adapt on ecological time scales (Manel and Holderegger 2013). A key finding is that habitat quality is essential for maintaining species and community diversity and persistence (Fitzpatrick et al. 2014; Osborne et al. 2014).

Functional (Adaptive) Genomic Variation

Prior to the widespread use of genomic approaches, most conservation genetic studies were focused on selectively neutral markers—that is, parts of the genome that are presumed not to directly affect fitness. Neutral markers were relatively easy to characterize, and they conform well to the assumptions of models that balance genetic drift and migration (e.g., hierarchical *F*-statistics, Echelle 1991), but they do not offer insight into genetic changes underlying phenotypic (i.e., physiological, morphological, and ecological) variation. NextGen sequencing has opened the field of *functional genomics*, which aims to scan genomes for markers associated with, and potentially causing, phenotypic variation. An advantage of using genomic approaches (e.g., RNAseq and sequence capture) is that neutral and functional variation can be characterized in a single dataset. Identification of functional genetic markers responsible for phenotypic variation can inform us which geographic areas harbor unique, locally adapted populations and can complement ecological or morphology-based studies, particularly when plasticity can be ruled out as the sole cause of phenotypic variation.

Several important themes arise from recent functional genomic studies: (1) local adaptation can occur in the presence of significant gene flow (Hess et al. 2013; Fitzpatrick et al. 2015), even while neutral loci remain homogeneous across populations; (2) genetic markers under divergent natural selection differentiate more rapidly than neutral markers; (3) small parts of the genome can be responsible for significant ecological or morphological divergence among populations (e.g., a single gene, Prince et al. 2016) and are likely to be missed by genetic studies of neutral markers (e.g., microsatellites, mtDNA, and RADseq; Lowry et al. 2016); and (4) such adaptive markers can be especially informative for the identification of ESUs. Each theme has important implications for conservation of desert fishes because studies of neutral genetic markers may miss key genetic changes that underlie important phenotypic variation (Krabbenhoft and Turner 2017). When populations are morphologically or ecologically distinct, lack of genetic differentiation at neutral genetic markers (even large numbers of them, such as in RADseq) should not be taken as evidence of a lack of genetic basis for phenotypic variation, because significant phenotypic variation can arise from very few, but important, genetic changes (i.e., a single gene).

It is possible that genetic variation that is adaptive in a current environment may not be adaptive in the future as environmental change occurs. Environmental changes, such as introductions of nonnative species or novel pathogens, can be rapid and unpredictable. Populations typically respond to rapid, human-induced environmental shifts through selection operating on standing genetic variation (e.g., allele frequency shifts), rather than through the accumulation of new, beneficial mutations (Barrett and Schluter 2008). Thus, all genetic variation, not just currently adaptive polymorphism, is potentially valuable for the long-term persistence of populations.

There is still some debate about the minimum population size needed for maintaining genetic diversity in the face of genetic drift and natural selection (e.g., Lynch and Lande 1998). Genetic variation that is maladaptive in the current environment (often maintained at very low frequencies in large populations) might become selectively advantageous in the future owing to changes in environmental conditions. For example, a key allele of the *ectodysplasin* gene in threespine stickleback *Gasterosteus aculeatus* is responsible for a large portion of its recurrent morphological adaptation to freshwater (Colosimo et al. 2005). This allele is present at a low frequency, and presumably maladaptive, in the "ancestral" marine population, but is adaptive in derived freshwater populations. In a conservation context, populations should be sufficiently large to maintain rare genetic variants, which may require populations even larger than those suggested by Lynch and Lande (1998). Thus, a balance must be struck between maintenance of local adaptive distinctiveness and of species-wide genetic diversity. Such considerations come into play when translocating fishes on the landscape, employing "genetic rescue" approaches, or supplementing natural populations with hatchery-reared stock.

Genetic Monitoring

Over the years, there have been calls to better unite demographic and genetic theory and processes in conservation and recovery planning (Avise 1995; Moran 2002; Lowe et al. 2017). There is an obvious connection between them, but several factors have slowed progress until relatively recently. A misunderstanding of Lande's (1988) seminal paper led to debate on the importance of demographic versus genetic factors in predicting extinction risk, rather than recognition that both act synergistically in determining persistence or extinction. More practical problems, such as the low inferential power of genetic data to estimate demographic processes (e.g., Taylor and Dizon 1996; Palstra and Fraser 2012) and the lack of clarity in definitions of key parameters that have semantic similarities (e.g., migration vs. genetically effective migration), have also slowed progress (Lowe and Allendorf 2010). However, new analytical approaches (e.g., the Bayesian statistics "revolution"; Beaumont and Rannala 2004), an increased appreciation for obtaining demographic and genetic data simultaneously in monitoring programs (e.g., Osborne et al. 2012; Marsh et al. 2015), and a few decades of theoretical, simulation, and empirical study have mostly reconciled demographic and genetic approaches to conservation.

Accordingly, conservation geneticists are focused on time series analysis and its power to reveal connections between demographic and genetic processes in contemporary populations over ecologically relevant time scales. Schwartz et al. (2007) defined genetic monitoring as the taking of at least two temporally spaced genetic samples from the same population. Such temporal genetic sampling offers the advantage of measuring changes in commonly used metrics of genetic diversity such as allelic richness, heterozygosity, and genetically effective population size (N_e) in contemporary focal populations. If demographic information such as census size (N), migration rates, vital rates, and other variables can be simultaneously estimated, it is possible to relate genetic data and metrics to recovery benchmarks like the minimum number of individuals required to stem loss of diversity (e.g., Osborne et al. 2012; Dowling et al. 2014). For example, a positive relationship was found between N_e and density for Pecos bluntnose shiner *Notropis simus pecosensis*, but not for Rio Grande silvery minnow *Hybognathus amarus* (Osborne et al. 2010). The lack of a relationship was due to decoupling of demographic and genetic trajectories in the Rio Grande silvery minnow, presumably because of overwhelming input of hatchery-reared individuals to the Rio Grande (Osborne et al. 2012).

Despite its promise, the adoption of genetic monitoring has been slow in desert fish conservation programs, and it remains somewhat controversial in planning for endangered species (Palstra and Ruzzante 2008). The controversy stems in part from practical questions: Can meaningful and attainable benchmarks for species recovery based on genetic criteria be established and met? Is intensive genetic monitoring justified given the expense? And what should the time interval between temporal samples be (Hoban et al. 2014)? Several papers have commented on a potential downside to estimation of summary statistics over time (termed category II monitoring in Schwartz et al. 2007), where inferential power is sometimes low (and confidence intervals of estimates are broad). In this case, only genetic events of relatively large magnitude (e.g., an order of magnitude decrease in census size) can be detected with high confidence over short time series (Palstra and Ruzzante 2008; Dowling et al. 2014). However, because inferential power is limited by the number of individuals that can be genotyped and the number of gene loci examined, high-throughput targeted SNP approaches have the potential to revolutionize this field because large numbers of loci and individuals can be sampled at each time step (years or generations).

For many endangered fish species, it may be soon possible to genotype nearly all individuals in a population in a cost-effective manner. This will permit genetic monitoring at the individual level (category I monitoring; Schwartz et al. 2007), so that each fish can be uniquely identified and the pedigrees of offspring determined. A natural extension of this technique is parentage-based genetic tagging (PBT; Steele et al. 2013), where pedigree information can be used to directly estimate variation in reproductive success among mating pairs and relate vital rates to environmental variation in the wild. If SNP panels include genes that are under selection, then PBT studies could reveal the rapidity with which evolutionary forces like genetic drift and selection can act on populations. A particularly important application of PBT is to determine the demographic and genetic interactions of hatchery-reared and wild fish to more directly understand the costs and benefits of hatchery programs from an ecological and evolutionary perspective (e.g., Propst et al., chap. 20, this volume).

Hatcheries, Augmentation, and Translocations

For many desert fishes, hatcheries and captive propagation have become integral to management efforts to stave off extinction (Lema et al., chap. 22, this volume). For example, 22 species are held at the Southwestern Native Aquatic Resources and Recovery Center in Dexter, New Mexico (https://www.fws.gov/Southwest/fisheries/dexter/species.html). Fish produced in captivity are typically used as a source for stocking to a species' historical range or to augment extant populations. They are also used as refuge populations in the event of the species' extinction in the wild. In principle, hatchery-reared fish should share the morphological, behavioral, and genetic characteristics of their wild counterparts. Hence, genetic considerations are paramount when planning, establishing, and monitoring breeding programs. These genetic considerations suggest the following strategies: (1) founding of captive stocks that reflect the existing genetic diversity (including adaptive variation), geographic structure, and evolutionary legacy of the wild population; (2) preserving genetic diversity in the captive stock by minimizing genetic drift and inbreeding; and (3) avoiding domestication selection (adaptation to the captive environment). When extinction appears imminent, preserving as much remaining diversity as possible is the primary goal, rather than focusing on spatial or other patterns of diversity (Echelle 1991). Where use of artificial refuges or hatchery propagation is necessary, it is important to reflect natural environmental conditions as much as possible to prevent adaptation to captivity (e.g., Araki et al. 2007; Frankham 2008). Failure to do so can lead to reduced performance and low survival or reproduction in repatriated fish. Adaptation to hatchery conditions is common and can occur rapidly (Christie et al. 2012; Milot et al. 2013), but simple modifications to the hatchery environment, such as the addition of rocks to increase spatial complexity, can prevent some performance impairment (Näslund and Johnsson 2016).

Genetic data have informed captive propagation programs (e.g., Leon Springs pupfish *Cyprinodon bovinus* and Comanche Springs pupfish *Cyprinodon elegans*; Edds and Echelle 1989), genetic management plans (e.g., Rio Grande silvery minnow; USFWS 2009) and recovery plans (Gila trout *Oncorhynchus gilae;* USFWS 2003). In many instances, however, genetic data were not available when captive populations were founded. In other cases, wild populations and the genetic diversity they represented had been greatly depleted (e.g., bonytail *Gila elegans*, Colorado pikeminnow *Ptychocheilus lucius*, woundfin *Plagopterus argentissimus*) by the time of founding, so the captive populations do not reflect historical diversity (Hedrick et al. 2000; Borley and White 2006; Chen et al. 2009). It is prudent, therefore, to collect genetic data from at-risk species prior to their becoming highly imperiled so that relevant data are available if establishment of captive populations becomes necessary. Some captive populations are now the only pure sources of their species (e.g., Leon Springs pupfish, but see Black et al. 2017) and may represent the only relatively secure populations (e.g., bonytail).

Depending on the design of captive rearing and breeding programs, periodic supplementation of a wild population with hatchery-produced individuals can maintain or increase its diversity, but also carries the risk of erosion of the genetic variability, and thus the adaptability, of the wild population (e.g., Ford 2002; McLean et al. 2004). For this reason, hatchery breeding for conservation efforts should be conducted to maximize the genetically effective size and diversity of the captive stock. To accomplish this, candidate breeders should be genotyped to minimize relatedness between crossed individuals (i.e., marker-assisted crosses). This approach has been adopted for delta smelt *Hypomesus transpacificus* (Fisch et al. 2011). However, the use of marker-assisted breeding has been criticized because it is not based on natural mating behaviors (Pitcher and Neff 2007) and because natural mate selection is essential for maintenance of viable wild populations (Quinn 2005).

In some cases, managers use culture practices that accommodate variation in life history of desert fishes. These programs permit wild mating, preserving the benefits of sexual and natural selection (i.e., diminishing domestication selection). For Rio Grande silvery minnow, a short-lived cyprinid native to the Rio Grande basin, some 100,000 eggs from wild-spawning events established the founding captive stock in 2003. Wild-caught eggs or young-of-year fish have been used to supplement the captive brood stock annually to maintain genetic cohesiveness with the wild population. A group-spawning design with equalized sex ratios, to mimic natural aggregate-breeding behavior, produces individuals released to augment the wild population (USFWS 2009). This approach has preserved genetic diversity (Osborne et al. 2012).

For razorback sucker *Xyrauchen texanus*, a long-lived catostomid native to the Colorado River basin, hatcheries rear larvae collected from natural spawning events in Lake Mohave (Marsh et al. 2015) to the juvenile stage in captivity, then release them to protected off-channel locations, where they grow to a size (>450 mm standard length) less susceptible to predation. Subsequently, they are released in the lake. Genetic monitoring (via mtDNA and microsatellite loci) of wild adults, larvae, and repatriates demonstrated that this strategy maintained genetic diversity (Dowling et al. 2005, 2014; Carson et al. 2016). Additionally, parentage analysis via microsatellites has been used to evaluate the potential for using off-channel habitats as management tools for big-river fishes including bonytail and razorback sucker (Minckley et al. 2003; Dowling et al. 2017). The results indicate that in bonytail, a high proportion of males and females contribute offspring, thereby preserving genetic diversity between generations. However, there can also be considerable variation in reproductive success between off-channel habitats and between years (Osborne and Turner 2016).

Incorporation of NextGen approaches means that decisions regarding establishment of captive populations, including identification of source populations, design of breeding programs, and monitoring of captive stocks, can be made based on hundreds to thousands of genetic loci. For example, a panel of SNPs has been developed for delta smelt that is used for brood stock pedigree reconstruction and to improve genetic management (Lew et al. 2015). Technological developments provide a means to assess genomic and fitness consequences of alternative captive breeding and stocking or augmentation protocols to improve management of captive populations (Waters et al. 2016; Willoughby et al. 2017). But no matter what steps are taken to reduce negative genetic effects of hatcheries, these effects cannot be eliminated (Waples 1999). In our view, conservation hatcheries should be used primarily as temporary safe harbors while measures are taken to protect species in their natural habitat or in a suitable substitute. Moreover, population augmentation with hatchery-reared fish should be considered only as an emergency measure because of its negative ecological and evolutionary effects on wild fish (Ford 2002). In other words, hatcheries, no matter how well designed and managed, are not substitutes for adequately restored natural habitats that support self-sustaining populations of desert fishes.

Translocation of individuals from a donor population to newly restored, but unoccupied habitats is often a viable alternative to stocking hatchery-reared fish. Genetic information, including knowledge of local adaptation, is a crucial component for determining which donor populations to use for repatriation efforts. Identification of genetically "pure" individuals or populations (e.g., Kelsch and Baca 1991) and monitoring of translocated populations (e.g., Chen et al. 2013) is also required. Gila trout have been successfully translocated to fishless streams to increase the number of replicated genetic lineages on the landscape in response to threats imposed by wildfire and hybridization with nonnative rainbow trout (see Propst et al., chap. 20, this volume). Similarly, Mohave tui chub *Siphateles bicolor mohavensis*, the only fish endemic to the Mojave River, California, was translocated from its native range to localities at the Mojave National Preserve and Camp Cady Wildlife Area to prevent hybridization and competition with introduced arroyo chub *Gila orcuttii* (Chen et al. 2013).

Comparative Methods in Conservation Genetics

Increased anthropogenic water demand and climate change are impacting whole communities of fishes throughout dryland watersheds of North America and elsewhere. In response, "multi-species conservation plans" are now being developed that attempt to identify common environmental requirements for the persistence of multiple threatened and endangered species (Pikitch et al. 2004; Andersen and Brooks, chap. 12, this volume). Although subject to some criticism (Clark and Harvey 2002; Rahn et al. 2006), multi-species plans provide impetus for comparative conservation genetic studies on co-distributed focal species. The goal is to reveal common and idiosyncratic patterns of genetic diversity in critical habitat, and more importantly, to identify and avoid management actions that might positively affect one species at the expense of another. For example, placement of barriers to preclude introgressive hybridization of one species can impede natural movement in another (e.g., Novinger and Rahel 2003).

In the absence of extensive life history data and genomic characterization, comparative genetic studies offer a practical method for evaluating how ecology and species-specific traits interact with landscapes to shape species' genetic patterns. The distribution of genetic diversity is typically determined by local abundance and effective movement (dispersal) between localities. Because direct dispersal data are often lacking, comparative studies use surrogates such as habitat specificity (Tibbets and Dowling 1996), body size (Pilger et al. 2017), or traits related to vagility (Phillipsen et al. 2015) to infer differences in dispersal across species.

Common findings of such multi-species studies are that traits correlated with dispersal ability are typically inversely related to the degree of genetic differentiation. Importantly, species with different life history strategies respond differently to anthropogenic and natural stream fragmentation (Perkin et al. 2015; Gido et al. 2016) in ways that could lead to management tradeoffs. Thus, comparative conservation genetic studies facilitate generalizations across species and systems (see Keller et al. 2015).

Multiple-species comparisons could be important where natural hybridization and introgression are part of the evolutionary history, as they are for "big-river" fishes of the Colorado River basin. Suckers have historically exhibited varying levels of introgression (Dowling et al. 2016), and here, razorback, flannelmouth *Catostomus latipinnis*, and bluehead suckers co-occur and naturally exchange genes (McDonald et al. 2008). It may be possible to identify environmental features associated with varying levels of introgressive hybridization among these species to benefit each.

eDNA, Archival DNA, and Conservation

Most vertebrates leave DNA in the environment (eDNA) in the form of urine, feces (Poinar 1998; Valiere and Taberlet 2000), scales, mucus, skin, hair (Bunce et al. 2005), and carcasses. Once released, the DNA begins to degrade through exposure to UV radiation and microbial activity. Accordingly, eDNA concentration in a sample is a function of rate of release of DNA and rate of decomposition. This is a complex interaction between the metabolism and ecology of the target species and ambient environmental conditions (Barnes et al. 2014; Strickler et al. 2015). Species detection using eDNA is accomplished by using PCR with taxon-specific primers (typically mtDNA genes characterized by Sanger sequencing) or by using a multi-species approach in which generic PCR primers for the focal group are employed. This technology has been used to detect rare and nonnative species (Jerde et al. 2011), estimate biomass or abundance (Doi et al. 2015), and identify species in gut contents to verify predation (e.g., Ehlo et al. 2017).

The relative ease of sampling and straightforward laboratory and computational procedures have contributed to widespread use of eDNA in a variety of aquatic settings, including the habitats of desert fishes. For example, eDNA methods were developed for detection of two southwestern fishes, loach minnow *Tiaroga cobitis* and spikedace *Meda fulgida* (Dysthe et al. 2016). Although eDNA methods can be powerful, inhibition of PCR can lead to false negatives (i.e., non-detection) even with high concentrations of target-species DNA (Jane et al. 2015). Other limitations of eDNA include its inability to distinguish live from dead animals, to distinguish life stages, or to provide information about habitat use. Use of eDNA for quantifying abundance or biomass can be complicated by unknown effects of environmental conditions (e.g., water temperature) and collection methods (Lacoursière-Roussel et al. 2016), reducing consistency among studies (Iversen et al. 2015). Thus, eDNA should not be considered a replacement for traditional surveys or genetic monitoring and should be used with caution. However, this area is ripe for new developments in remote sample collection and DNA sequencing technologies that could ultimately enhance its applicability for many conservation genetics questions.

Archived samples are also an invaluable source for reconstructing pre-disturbance or historical patterns of genetic diversity. Natural history museums and other curated repositories house materials that can be analyzed by DNA sequencing. A complicating factor is that whole fishes are usually fixed in formalin prior to transfer to ethanol (or isopropanol) for long-term preservation. Sanger sequencing based on PCR is arguably impossible for DNA samples isolated from formalin-fixed tissue. Some NextGen approaches target short fragments as a template for sequencing, and it may be possible to determine nucleotide sequence for formalin-fixed material in the near future. Nonetheless, fishes are sometimes preserved directly in alcohol, or as scales, dried fins, or dried skeletons, all of which can be used for genetic studies. Archived samples collected about 150 years ago were crucial in identifying the historical genetic diversity of greenback cutthroat trout *Oncorhynchus clarkii stomias* and the influx of genetic material from introduced Colorado cutthroat trout *Oncorhynchus clarkii pleuriticus* (Metcalf et al. 2012), which precipitated an important review of previous conservation policy and management actions and necessitated new actions to benefit greenback cutthroat trout.

Computational Advances in Conservation Genetics

Computational and statistical methods (Approximate Bayesian Computation [ABC], Coalescent Theory) have advanced nearly as quickly as technology in molecular genetics. New methods offer fresh insight into conservation genetics and its interaction with demography and evolutionary history. Analytical applications include estimation of historical and contemporary gene flow, which has provided evidence that some desert fishes that are currently isolated had historical patterns consistent with broadly connected populations (e.g., Dunham and Minckley 1998; Blakney et al. 2014), whereas other isolated species show evidence of historical isolation (e.g., Bernardi et al. 2007).

Computer simulation platforms (e.g., simuPOP, EASYPOP) allow us to better understand how violations of modeling assumptions affect estimation of genetic summary statistics and estimates of genetically effective population size, gene flow, and adaptive divergence among populations. For example, Wright's island model (Wright 1931) has been the theoretical backbone relating evolutionary processes to population dynamics. Although computationally tractable, it does not accommodate the network organization of streams (among other limitations; see Whitlock and McCauley 1999). Simulation analysis is one way to validate and check assumptions that underlie estimates of key genetic metrics that describe population structure and gene flow (Meffe and Vrijenhoek 1988). For example, simulation studies improved understanding of demographic responses to fragmentation within dendritic networks (Fagan 2002) and of the role of dendritic geometry in shaping geographic patterns of genetic diversity (e.g.,

Paz-Vinas and Blanchet 2015; Thomaz et al. 2016).

There is still much need for published simulation studies that investigate the interactions of ecology, landscape, and population genetics. Questions pertinent to the conservation genetics of desert fishes that are amenable to simulation modeling include the following: (1) How much do demographic processes affect neutral and adaptive patterns of genetic variation? (2) Can management actions be genetically beneficial for species? (3) Are there threshold levels of fragmentation that begin to affect species' genetic diversity and levels of gene flow?

WHAT'S NEXT?

Technological advances are expected to continue to revolutionize conservation genetics in years to come. Most research conducted to date has focused on "hard-wired" changes to genomes—that is, DNA sequence variation—but epigenetic profiling is an exciting frontier that should provide insight into the early drivers of lineage divergence, the mechanisms underlying phenotypic plasticity, and gene expression variation via so-called soft inheritance (Pigliucci and Müller 2010; Laland et al. 2014). How important epigenetic variation is to desert fish conservation remains to be seen, particularly with respect to providing resilience in rapidly changing environments. Technological and knowledge advances in other areas, such as metabolomics, proteomics, and genome annotation, should further bridge the gap between genotypic and phenotypic variation, leading us toward an integrative understanding of how desert fish interact with their environment and how that information applies to conservation and management.

Remote sampling and characterization of DNA sequences will probably be widely available in the next decade. Combining this with PIT-tagging/scanning and other remote-sensing approaches, along with satellite uplinks, would allow real-time genotype data to be gathered simultaneously with scanning, video capture, or other remotely sensed data. Together, approaches like high-throughput genotyping, genetic monitoring, simulation, and remote sensing could become a cornerstone for proactive management for preservation of genetic diversity for desert fishes.

Yet no matter how sophisticated they become, DNA sequencing technology, informatics, and genomics are only tools that assist comprehensive planning and implementation of management and conservation actions. To be maximally effective, genomic approaches require proper quality control and appropriate interpretation in the context of evolutionary genetic theory and empirical data (Allendorf 2017). Findings from conservation genetic studies must also be accurately and clearly translated into policy and action. Most importantly, genomic technology cannot substitute for on-the-ground conservation activities. Acquisition and restoration of aquatic habitats, restoration of natural flow regimes, reconnection of fragmented habitats, and protection of instream water rights will remain at the forefront of conservation efforts for desert fishes.

ACKNOWLEDGMENTS

We thank Evan Carson and Michael Schwemm for advice about the content of this chapter. Scott Clark and Anthony Echelle provided editorial comments and suggestions on previous drafts of the manuscript.

REFERENCES

Alex Buerkle, C., and Z. Gompert. 2013. Population genomics based on low coverage sequencing: How low should we go? *Molecular Ecology* 22: 3028–35.

Ali, O. A., S. M. O'Rourke, S. J. Amish, M. H. Meek, G. Luikart, C. Jeffres, and M. R. Miller. 2016. RAD capture (Rapture): Flexible and efficient sequence-based genotyping. *Genetics* 202: 389–400.

Allendorf, F. W. 2017. Genetics and the conservation of natural populations: Allozymes to genomes. *Molecular Ecology* 26: 420–30.

Allendorf, F. W., P. A. Hohenlohe, and G. Luikart. 2010. Genomics and the future of conservation genetics. *Nature Reviews Genetics* 11: 697–709.

Alò, D., and T. F. Turner. 2005. Effects of habitat fragmentation on effective population size in the endangered Rio Grande silvery minnow. *Conservation Biology* 19: 1138–48.

Amores, A., J. Catchen, I. Nanda, W. Warren, R. Walter, M. Schartl, and J. H. Postlethwait. 2014. A RAD-tag genetic map for the platyfish (*Xiphophorus maculatus*) reveals mechanisms of karyotype evolution among teleost fish. *Genetics* 197: 625–41.

Arnold, B., R. B. Corbett-Detig, D. Hartl, and K. Bomblies. 2013. RADseq underestimates diversity and introduces genealogical biases due to nonrandom haplotype sampling. *Molecular Ecology* 22: 3179–90.

Arnold, M. L. 2016. *Divergence with Genetic Exchange*. Oxford: Oxford University Press.

Araki, H., B. Cooper, and M. S. Blouin. 2007. Genetic effects of captive breeding cause a rapid, cumulative fitness decline in the wild. *Science* 318: 100–103.

Avise, J. C. 1995. Mitochondrial DNA polymorphism and a connection between genetics and demography of relevance to conservation. *Conservation Biology* 9: 686–90.

Barnes, M. A., C. R. Turner, C. L. Jerde, M. A. Renshaw, W. L. Chadderton, and D. M. Lodge. 2014. Environmental conditions influence eDNA persistence in aquatic systems. *Environmental Science and Technology* 48: 1819–27.

Barrett, R. D., and D. Schluter. 2008. Adaptation from standing genetic variation. *Trends in Ecology and Evolution* 23: 38–44.

Beaumont, M. A., and B. Rannala. 2004. The Bayesian revolution in genetics. *Nature Reviews Genetics* 5: 251–61.

Benestan, L. M., A. L. Ferchaud, P. A. Hohenlohe, B. A. Garner, G. J. P. Naylor, I. B. Baums, M. K. Schwartz, J. L. Kelley, and G. Luikart. 2016. Conservation genomics of natural and managed populations: Building a conceptual and practical framework. *Molecular Ecology* 25: 2967–77.

Benjamin, A., B. May, J. O'Brien, and A. J. Finger. 2016. Conservation genetics of an urban desert fish, the arroyo chub. *Transactions of the American Fisheries Society* 145: 277–86.

Bernardi, G., G. Ruiz-Campos, and F. Camarena-Rosales. 2007. Genetic isolation and evolutionary history of oases populations of the Baja California killifish, *Fundulus lima. Conservation Genetics* 8: 547–54.

Black, A. N., H. A. Seears, C. M. Hollenbeck, and P. B. Samollow. 2017. Rapid genetic and morphologic divergence between captive and wild populations of the endangered Leon Springs pupfish, *Cyprinodon bovinus. Molecular Ecology* 26: 2237–56.

Blakney, J. R., J. L. Loxterman, and E. R. Keeley. 2014. Range-wide comparisons of northern leatherside chub populations reveal historical and contemporary patterns of genetic variation. *Conservation Genetics* 15: 757–70.

Borley, K., and M. M. White. 2006. Mitochondrial DNA variation in the endangered Colorado pikeminnow: A comparison among hatchery stocks and historic specimens. *North American Journal of Fisheries Management* 26: 916–20.

Brito, P. H., and S. V. Edwards. 2009. Multilocus phylogeography and phylogenetics using sequence-based markers. *Genetica* 135: 439–55.

Bunce, M., M. Szulkin, H. R. Lerner, I. Barnes, B. Shapiro, A. Cooper, and R. N. Holdaway. 2005. Ancient DNA provides new insights into the evolutionary history of New Zealand's extinct giant eagle. *PLoS Biology* 31: e9.

Campbell, N. R., S. A. Harmon, and S. R. Narum. 2015. Genotyping-in-thousands by sequencing (GT-seq): A cost effective SNP genotyping method based on custom amplicon sequencing. *Molecular Ecology Resources* 15: 855–67.

Carson, E. W., M. Tobler, W. L. Minckley, R. J. Ainsworth, and T. E. Dowling. 2012. Relationships between spatio-temporal environmental and genetic variation reveal an important influence of exogenous selection in a pupfish hybrid zone. *Molecular Ecology* 21: 1209–22.

Carson, E. W., T. F. Turner, M. J. Saltzgiver, D. Adams, B. R. Kesner, P. C. Marsh, T. J. Pilger, and T. E. Dowling. 2016. Retention of ancestral genetic variation across life-stages of an endangered, long-lived iteroparous fish. *Journal of Heredity* 107: 567–72.

Chen, Y., C. Conway, C. Keeler-Foster, R. Hamman, and S. Meismer. 2009. Genetic characterization of variation in captive and wild woundfin. *North American Journal of Fisheries Management* 29: 843–49.

Chen, Y., S. Parmenter, and B. May. 2013. Genetic characterization and management of the endangered Mohave tui chub. *Conservation Genetics* 14: 11–20.

Chifman, J., and L. Kubatko. 2014. Quartet inference from SNP data under the coalescent model. *Bioinformatics* 30: 3317–24.

Christie, M. R., M. L. Marine, R. A. French, and M. S. Blouin. 2012. Genetic adaptation to captivity can occur in a single generation. *Proceedings of the National Academy of Sciences* 109: 238–42.

Clark, J. A., and E. Harvey. 2002. Assessing multi-species recovery plans under the Endangered Species Act. *Ecological Applications* 12: 655–62.

Colosimo, P. F., K. E. Hosemann, S. Balabhadra, G. Villarreal, M. Dickson, J. Grimwood, J. Schmutz, R. M. Myers, D. Schluter, and D. M. Kingsley. 2005. Widespread parallel evolution in sticklebacks by repeated fixation of ectodysplasin alleles. *Science* 307: 1928–33.

DeMarais, B. D., T. E. Dowling, M. E. Douglas, W. L. Minckley, and P. C. Marsh. 1992. Origin of *Gila seminuda* (Teleostei: Cyprinidae) through introgressive hybridization: Implications for evolution and conservation. *Proceedings of the National Academy of Sciences* 89: 2747–51.

Doi, H., K. Uchii, T. Takahara, S. Matsuhashi, H. Yamanaka, and T. Minamoto. 2015. Use of droplet digital PCR for estimation of fish abundance and biomass in environmental DNA surveys. *PLoS ONE* 10: e0122763.

Dowling, T. E., C. D. Anderson, P. C. Marsh, and M. S. Rosenberg. 2015. Population structure in the roundtail chub (*Gila robusta* complex) of the Gila River basin as determined by microsatellites: Evolutionary and conservation implications. *PLoS ONE* 10: e0139832.

Dowling, T. E., B. D. DeMarais, W. L. Minckley, M. E. Douglas, and P. C. Marsh. 1992a. Use of genetic characters in conservation biology. *Conservation Biology* 6: 7–8.

Dowling, T. E., D. F. Markle, G. J. Tranah, E. W. Carson, D. W. Wagman, and B. P. May. 2016. The role of hybridization in animal evolution: A case study from catostomid fishes. *PLoS ONE* 11: e0149884.

Dowling, T. E., P. C. Marsh, A. T. Kelsen, and C. A. Tibbets. 2005. Genetic monitoring of wild and repatriated populations of endangered razorback sucker (*Xyrauchen texanus*, Catostomidae, Teleostei) in Lake Mohave, Arizona-Nevada. *Molecular Ecology* 14: 123–35.

Dowling, T. E., P. C. Marsh, and T. F. Turner. 2017. *Genetic and Demographic Studies to Guide Conservation Management of Razorback Sucker in Off-Channel Habitats: 2016 Annual Report.* Boulder City, NV: US Bureau of Reclamation.

Dowling, T. E., W. L. Minckley, M. E. Douglas, P. C. Marsh, and B. D. DeMarais. 1992b. Use of molecular characters in conservation biology: Implications for management of the red wolf. *Conservation Biology* 6: 600–603.

Dowling, T. E., T. F. Turner, E. W. Carson, M. J. Saltzgiver, D. Adams, B. Kesner, and P. C. Marsh. 2014. Genetic and demographic responses to intensive management of razorback sucker in Lake Mohave, Arizona-Nevada. *Evolutionary Applications* 7: 339–54.

Dunham, J. B., and W. L. Minckley. 1998. Allozymic variation in desert pupfish from natural and artificial habitats: Genetic conservation in fluctuating populations. *Biological Conservation* 84: 7–15.

Dysthe, J. C., K. J. Carim, Y. M. Paroz, K. S. McKelvey, M. K. Young, and M. K. Schwartz. 2016. Quantitative PCR assays for detecting loach minnow (*Rhinichthys cobitis*) and spikedace (*Meda fulgida*) in the southwestern United States. *PLoS ONE* 11: e0162200.

Echelle, A. 1991. Conservation genetics and genic diversity in freshwater fishes of western North America. In W. L. Minckley and J. E. Deacon, eds., *Battle against Extinction: Native Fish Management in the American West*, 141–53. Tucson: University of Arizona Press.

Echelle, A., A. Echelle, and D. Edds. 1989. Conservation genetics of a spring-dwelling desert fish, the Pecos gambusia (*Gambusia nobilis*, Poeciliidae). *Conservation Biology* 3: 159–69.

Edds, D. R., and A. A. Echelle. 1989. Genetic comparisons of hatchery and natural stocks of small endangered fishes: Leon Springs pupfish, Comanche Springs pupfish, and Pecos gambusia. *Transactions of the American Fisheries Society* 118: 441–46.

Ehlo, C. A., M. J. Saltzgiver, T. E. Dowling, P. C. Marsh, and B. R. Kesner. 2017. Use of molecular techniques to confirm nonnative fish predation on razorback sucker *Xyrauchen texanus* larvae in Lake Mohave, Arizona and Nevada. *Transactions of the American Fisheries Society* 146: 201–5.

Ellegren, H. 2014. Genome sequencing and population genomics in non-model organisms. *Trends in Ecology and Evolution* 29: 51–63.

Fagan, W. F. 2002. Connectivity, fragmentation, and extinction risk in dendritic metapopulations. *Ecology* 83: 3243–49.

Fisch, K. M., J. M. Henderson, R. S. Burton, and B. May. 2011. Population genetics and conservation implications for the endangered delta smelt in the San Francisco Bay-Delta. *Conservation Genetics* 12: 1421–34.

Fitzpatrick, S. W., H. Crockett, and W. C. Funk. 2014. Water availability strongly impacts population genetic patterns of an imperiled Great Plains endemic fish. *Conservation Genetics* 15: 771–88.

Fitzpatrick, S. W., J. C. Gerberich, J. A. Kronenberger, L. M. Angeloni, and W. C. Funk. 2015. Locally adapted traits maintained in the face of high gene flow. *Ecology Letters* 18: 37–47.

Ford, M. J. 2002. Selection in captivity during supportive breeding may reduce fitness in the wild. *Conservation Biology* 16: 815–25.

Fourcade, Y., A. Chaput-Bardy, J. Secondi, C. Fleurant, and C. Lemaire. 2013. Is local selection so widespread in river organisms? Fractal geometry of river networks leads to high bias in outlier detection. *Molecular Ecology* 22: 2065–73.

Frankham, R. 2008. Genetic adaptation to captivity in species conservation programs. *Molecular Ecology* 17: 325–33.

Funk, D. J., and K. E. Omland. 2003. Species-level paraphyly and polyphyly: Frequency, causes, and consequences, with insights from animal mitochondrial DNA. *Annual Review of Ecology, Evolution, and Systematics* 34: 397–423.

Gautier, M., K. Gharbi, T. Cezard, J. Foucaud, C. Kerdelhué, P. Pudlo, J.-M. Cornuet, and A. Estoup. 2013. The effect of RAD allele dropout on the estimation of genetic variation within and between populations. *Molecular Ecology* 22: 3165–78.

Gerber, A. S., C. A. Tibbets, and T. E. Dowling. 2001. The role of introgressive hybridization in the evolution of the *Gila robusta* complex (Teleostei: Cyprinidae). *Evolution* 55: 2028–39.

Gido, K. B., J. E. Whitney, J. S. Perkin, and T. F. Turner. 2016. Fragmentation, connectivity and fish species persistence in freshwater ecosystems. In G. P. Closs, M. Krkosek, and J. D. Olden, eds., *Conservation of Freshwater Fishes*, 292–323. Cambridge: Cambridge University Press.

Harrison, R. G. 2012. The language of speciation. *Evolution* 66: 3643–57.

Harrison, R. G., and E. L. Larson. 2016. Heterogeneous genome divergence, differential introgression, and the origin and structure of hybrid zones. *Molecular Ecology* 25: 2454–66.

Harrisson, K. A., A. Pavlova, M. Telonis-Scott, and P. Sunnucks. 2014. Using genomics to characterize evolutionary potential for conservation of wild populations. *Evolutionary Applications* 7: 1008–25.

Hedrick, P. W., T. E. Dowling, W. L. Minckley, C. A. Tibbets, B. D. Demarais, and P. C. Marsh. 2000. Establishing a captive broodstock for the endangered bonytail chub (*Gila elegans*). *Journal of Heredity* 91: 35–39.

Hess, J. E., N. R. Campbell, D. A. Close, M. F. Docker, and S. R. Narum. 2013. Population genomics of Pacific lamprey: Adaptive variation in a highly dispersive species. *Molecular Ecology* 22: 2898–916.

Hey, J., R. S. Waples, M. L. Arnold, R. K. Butlin, and R. G. Harrison. 2003. Understanding and confronting species uncertainty in biology and conservation. *Trends in Ecology and Evolution* 18: 597–603.

Hoban, S., J. A. Arntzen, M. W. Bruford, J. A. Godoy, A. R. Hoelzel, G. Segelbacher, C. Vilà, and G. Bertorelle. 2014. Comparative evaluation of potential indicators and temporal sampling protocols for monitoring genetic erosion. *Evolutionary Applications* 7: 984–98.

Hoffberg, S. L., T. J. Kieran, J. M. Catchen, A. Devault, B. C. Faircloth, R. Mauricio, and T. C. Glenn. 2016. RADcap: Sequence capture of dual-digest RADseq libraries with identifiable duplicates and reduced missing data. *Molecular Ecology Resources* 16: 1264–78.

Hohenlohe, P. A., M. D. Day, S. J. Amish, M. R. Miller, N. Kamps-Hughes, M. C. Boyer, C. C. Muhlfeld, F. W. Allendorf, E. A., Johnson, and G. Luikart. 2013. Genomic patterns of introgression in rainbow and westslope cutthroat trout illuminated by overlapping paired-end RAD sequencing. *Molecular Ecology* 22: 3002–13.

Hopken, M. W., M. R. Douglas, and M. E. Douglas. 2013. Stream hierarchy defines riverscape genetics of a North American desert fish. *Molecular Ecology* 22: 956–71.

Houston, D. D., D. K. Shiozawa, B. T. Smith, and B. R. Riddle. 2014. Investigating the effects of Pleistocene events on genetic divergence within *Richardsonius balteatus*, a widely distributed western North American minnow. *BMC Evolutionary Biology* 14: 111.

Iversen, L. L., J. Kielgast, and K. Sand-Jensen. 2015. Monitoring of animal abundance by environmental DNA—An increasingly obscure perspective: A reply to Klymus et al. *Biological Conservation* 192: 479–80.

Jane, S. F., T. M. Wilcox, K. S. McKelvey, M. K. Young, M. K. Schwartz, W. H. Lowe, B. H. Letcher, and A. R. Whiteley. 2015. Distance, flow and PCR inhibition: eDNA dynamics in two headwater streams. *Molecular Ecology Resources* 15: 216–27.

Jerde, C. L., A. R. Mahon, W. L. Chadderton, and D. M. Lodge. 2011. "Sight-unseen" detection of rare aquatic species using environmental DNA. *Conservation Letters* 4: 150–57.

Keller, D., R. Holderegger, M. J. van Strien, and J. Bolliger. 2015. How to make landscape genetics beneficial for conservation management? *Conservation Genetics 16:* 503–12.

Kelsch, S. W., and M. J. Baca. 1991. *Biochemical Analysis of Gene Products in the Yaqui Catfish,* Ictalurus pricei. Report prepared for the Dexter National Fish Hatchery, US Fish and Wildlife Service.

Kovach, R. P., B. K. Hand, P. A. Hohenlohe, T. F. Cosart, M. C. Boyer, H. H. Neville, C. C. Muhlfeld, S. J. Amish, K. Carim, S. R. Narum, W. H. Lowe, F. W. Allendorf, and G. Luikart. 2016. Vive la résistance: Genome-wide selection against introduced alleles in invasive hybrid zones. *Proceedings of the Royal Society B* 283: 20161380.

Krabbenhoft, T. J., and T. F. Turner. 2014. *Clock* gene evolution: Seasonal timing, phylogenetic signal, or functional variation. *Journal of Heredity* 105: 407–15.

Krabbenhoft, T. J., and T. F. Turner. 2017. Comparative transcriptomics of cyprinid minnows and carp in a common wild setting: A resource for ecological genomics in freshwater communities. *DNA Research*. https://doi.org/10.1093/dnares/dsx034.

Lacoursière-Roussel, A., M. Rosabal, and L. Bernatchez. 2016. Estimating fish abundance and biomass from eDNA concentrations: Variability among capture methods and environmental conditions. *Molecular Ecology Resources* 16: 1401–14.

Laland, K., G. A. Wray, and H. E. Hoekstra. 2014. Does evolutionary theory need a rethink? *Nature* 514: 161–64.

Lande, R. 1988. Genetics and demography in biological conservation. *Science* 241: 1455–60.

Lew, R. M., A. J. Finger, M. R. Baerwald, A. Goodbla, B. May, and M. H. Meek. 2015. Using next-generation sequencing to assist a conservation hatchery: A single-nucleotide polymorphism panel for the genetic management of endangered delta smelt. *Transactions of the American Fisheries Society* 144: 767–79.

Lowe, W. H., and F. W. Allendorf. 2010. What can genetics tell us about population connectivity? *Molecular Ecology* 19: 3038–51.

Lowe, W. H., R. P. Kovach, and F. W. Allendorf. 2017. Population genetics and demography unite. *Ecology and Evolution*. https://doi.org/10.1016/j.tree.2016.12.002.

Lowry, D. B., S. Hoban, J. L. Kelley, K. E. Lotterhos, L. K. Reed, M. F. Antolin, and A. Storfer. 2016. Breaking RAD: An evaluation of the utility of restriction site–associated DNA sequencing for genome scans of adaptation. *Molecular Ecology Resources* 17: 142–52.

Lynch, M. 1996. A quantitative-genetic perspective on conservation issues. In J. Avise and J. Hamrick, eds., *Conservation Genetics: Case Histories from Nature,* 471–501. New York: Chapman and Hall.

Lynch, M., and R. Lande. 1998. The critical effective size for a genetically secure population. *Animal Conservation* 1: 70–72.

Manel, S., and R. Holderegger. 2013. Ten years of landscape genetics. *Trends in Ecology and Evolution* 28: 614–21.

Marsh, P. C., R. W. Clarkson, and T. E. Dowling. 2017. Molecular genetics informs spatial segregation of two desert stream *Gila* (Cyprinidae). *Transactions of the American Fisheries Society* 146: 47–59.

Marsh, P. C., T. E. Dowling, B. R. Kesner, T. F. Turner, and W. L. Minckley. 2015. Conservation to stem imminent extinction: The fight to save razorback sucker *Xyrauchen texanus* in Lake Mohave and its implications for species recovery. *Copeia* 103: 141–56.

Martin, C. H., J. E. Crawford, B. J. Turner, and L. H. Simons. 2016. Diabolical survival in Death Valley: Recent pupfish colonization, gene flow and genetic assimilation in the smallest species range on Earth. *Proceedings of the Royal Society B* 283: 20152334.

Mastretta-Yanes, A., N. Arrigo, N. Alvarez, T. H. Jorgensen, D. Piñero, and B. C. Emerson. 2015. Restriction site–associated DNA sequencing, genotyping error estimation and de novo assembly optimization for population genetic inference. *Molecular Ecology Resources* 15: 28–41.

McDonald, D. B., T. L. Parchman, M. R. Bower, W. A. Hubert, and F. J. Rahel. 2008. An introduced and a native vertebrate hybridize to form a genetic bridge to a second native species. *Proceedings of the National Academy of Sciences* 105: 10837–42.

McLean, J. E., P. Bentzen, and T. P. Quinn. 2004. Differential reproductive success of sympatric, naturally spawning hatchery and wild steelhead, *Oncorhynchus mykiss*. *Environmental Biology of Fishes* 69: 359–69.

Meffe, G. K., and R. C. Vrijenhoek. 1988. Conservation genetics in the management of desert fishes. *Conservation Biology* 2: 157–69.

Meier, J. I., D. A. Marques, S. Mwaiko, C. E. Wagner, L. Excoffier, and O. Seehausen. 2017. Ancient hybridization fuels rapid cichlid fish adaptive radiations. *Nature Communications* 8: 14363.

Metcalf, J. L., S. Love Stowell, C. M. Kennedy, K. B. Rogers, D. McDonald, J. Epp, K. Keepers, A. Cooper, J. J. Austin, and A. P. Martin. 2012. Historical stocking data and 19th century DNA reveal human-induced changes to native diversity and distribution of cutthroat trout. *Molecular Ecology* 21: 5194–207.

Milot, E., C. Perrier, L. Papillon, J. J. Dodson, and L. Bernatchez. 2013. Reduced fitness of Atlantic salmon released in the wild after one generation of captive breeding. *Evolutionary Applications* 6: 472–85.

Minckley, W. L., P. C. Marsh, J. E. Deacon, T. E. Dowling, P. W. Hedrick, W. J. Matthews, and G. Mueller. 2003. A conservation plan for lower Colorado River native fishes. *BioScience* 53: 219–34.

Moran, P. 2002. Current conservation genetics: Building an ecological approach to the synthesis of molecular and quantitative genetic methods. *Ecology of Freshwater Fish* 11: 30–55.

Moritz, C. 1994. Defining "Evolutionarily Significant Units" for conservation. *Trends in Ecology and Evolution* 9: 373–75.

Narum, S. R., C. A. Buerkle, J. W. Davey, M. R. Miller, and P. A. Hohenlohe. 2013. Genotyping-by-sequencing in ecological and conservation genomics. *Molecular Ecology* 22: 2841–47.

Näslund, J., and J. I. Johnsson. 2016. Environmental enrichment for fish in captive environments: Effects of physical structures and substrates. *Fish and Fisheries* 17: 1–30.

Nielsen, R., J. S. Paul, A. Albrechtsen, and Y. S. Song. 2011. Genotype and SNP calling from next-generation sequencing data. *Nature Reviews Genetics* 12: 443–51.

Novinger, D. C., and F. J. Rahel. 2003. Isolation management with artificial barriers as a conservation strategy for cutthroat trout in headwater streams. *Conservation Biology* 17: 772–81.

Osborne, M. J., E. W. Carson, and T. F. Turner. 2012. Genetic monitoring and complex population dynamics: Insights from a 12-year study of the Rio Grande silvery minnow. *Evolutionary Applications* 5: 553–74.

Osborne, M. J., S. R. Davenport, C. W. Hoagstrom, and T. F. Turner. 2010. Genetic effective size, N_e, tracks density in a small freshwater cyprinid, Pecos bluntnose shiner (*Notropis simus pecosensis*). *Molecular Ecology* 19: 2832–44.

Osborne, M. J., J. S. Perkin, K. B. Gido, and T. F. Turner. 2014. Comparative riverscape genetics reveals reservoirs of genetic diversity for conservation and restoration of Great Plains fishes. *Molecular Ecology* 23: 5663–79.

Osborne, M. J., T. J. Pilger, J. D. Lusk, and T. F. Turner. 2017. Spatio-temporal variation in parasite communities maintains diversity at the major histocompatibility complex class IIβ in the endangered Rio Grande silvery minnow. *Molecular Ecology* 26: 471–89.

Osborne, M. J., and T. F. Turner. 2016. *Genetic and Demographic Studies to Guide Conservation Management of Bonytail in Off-Channel Habitats*. Boulder City: US Bureau of Reclamation.

Palsbøll, P. J., M. Berube, and F. W. Allendorf. 2007. Identification of management units using population genetic data. *Trends in Ecology and Evolution* 22: 11–16.

Palstra, F. P., and D. J. Fraser. 2012. Effective/census population size ratio estimation: A compendium and appraisal. *Ecology and Evolution* 2: 2357–65.

Palstra, F. P., and D. E. Ruzzante. 2008. Genetic estimates of contemporary effective population size: What can they tell us about the importance of genetic stochasticity for wild population persistence? *Molecular Ecology* 17: 3428–47.

Paris, J. R., J. R. Stevens, and J. M. Catchen. 2017. Lost in parameter space: A road map for Stacks. *Methods in Ecology and Evolution*. https://doi.org/10.1111/2041-210X.12775.

Parker, K. M., R. J. Sheffer, and P. W. Hedrick. 1999. Molecular variation and evolutionarily significant units in the endangered Gila topminnow. *Conservation Biology* 13: 108–16.

Paz-Vinas, I., and S. Blanchet. 2015. Dendritic connectivity shapes spatial patterns of genetic diversity: A simulation-based study. *Journal of Evolutionary Biology* 28: 986–94.

Perkin, J. S., K. B. Gido, A. R. Cooper, T. F. Turner, M. J. Osborne, E. R. Johnson, and K. B. Mayes. 2015. Fragmentation and dewatering transform Great Plains stream fish communities. *Ecological Monographs* 85: 73–92.

Peterson, B. K., J. N. Weber, E. H. Kay, H. S. Fisher, and H. E. Hoekstra. 2012. Double digest RADseq: An inexpensive method for de novo SNP discovery and genotyping in model and non-model species. *PLoS ONE* 7: e37135.

Phillipsen, I. C., E. H. Kirk, M. T. Bogan, M. C. Mims, J. D. Olden, and D. A. Lytle. 2015. Dispersal ability and habitat requirements determine landscape-level genetic patterns in desert aquatic insects. *Molecular Ecology* 24: 54–69.

Pigliucci, M., and G. B. Müller. 2010. *Evolution—The Extended Synthesis*. Cambridge, MA: MIT Press.

Pikitch, E., C. Santora, E. A. Babcock, A. Bakun, R. Bonfil, D. O. Conover, P. A. O. Dayton, P. Doukakis, D. Fluharty, B. Heneman, and E. D. Houde. 2004. Ecosystem-based fishery management. *Science* 305: 346–47.

Pilger, T. J., K. B. Gido, D. L. Propst, J. E. Whitney, and T. F. Turner. 2015. Comparative conservation genetics of protected endemic fishes in an arid-land riverscape. *Conservation Genetics* 16: 875–88.

Pilger, T. J., K. B. Gido, D. L. Propst, J. E. Whitney, and T. F. Turner. 2017. River network architecture, genetic effective size and distributional patterns predict differences in genetic structure across species in a dryland stream fish community. *Molecular Ecology* 26: 2687–97.

Pitcher, T. E., and B. D. Neff. 2007. Genetic quality and offspring performance in Chinook salmon: Implications for supportive breeding. *Conservation Genetics* 8: 607–16.

Poinar, H. N., M. Hofreiter, W. G. Spaulding, P. S. Martin, B. A. Stankiewicz, H. Bland, R. P. Evershed, G. Possnert, and S. Pääbo. 1998. Molecular coproscopy: Dung and diet of the extinct ground sloth *Nothrotheriops shastensis*. *Science* 281: 402–6.

Prince, D. J., S. M. O'Rourke, T. Q. Thompson, O. A. Ali, H. S. Lyman, İ. K. Sağlam, T. J. Hotaling, A. P. Spidle, and M. R. Miller. 2016. The evolutionary basis of premature migration in Pacific salmon highlights the utility of genomics for informing conservation. *Science Advances* 3 (8): e1603198.

Quinn, T. P. 2005. Comment: Sperm competition in salmon hatcheries—the need to institutionalize genetically benign spawning protocols. *Transactions of the American Fisheries Society* 134: 1490–94.

Rahn, M. E., H. Doremus, and J. Diffendorfer. 2006. Species coverage in multispecies habitat conservation plans: Where's the science? *BioScience* 56: 613–19.

Sağlam, İ. K., J. Baumsteiger, M. J. Smith, J. Linares-Casenave, A. L. Nichols, S. M. O'Rourke, and M. R. Miller. 2016. Phylogenetics support an ancient common origin of two scientific icons: Devils Hole and Devils Hole pupfish. *Molecular Ecology* 25: 3962–73.

Schwartz, M. K., G. Luikart, and R. S. Waples. 2007. Genetic monitoring as a promising tool for conservation and management. *Trends in Ecology and Evolution* 22: 25–33.

Smith, G. R., and R. F. Stearley. 2018. Fossil cutthroat: Implications for evolution and conservation. In P. Trotter, P. Bisson, L. Schultz, and B. Roper, eds., *Cutthroat Trout: Evolutionary Biology and Taxonomy*, pp. 77–101. Special Publication 36. Bethesda, MD: American Fisheries Society.

Steele, C. A., E. C. Anderson, M. W. Ackerman, M. A. Hess, N. R. Campbell, S. R. Narum, and M. R. Campbell. 2013. A validation of parentage-based tagging using hatchery steelhead in the Snake River basin. *Canadian Journal of Fisheries and Aquatic Sciences* 70: 1046–54.

Strickler, K. M., A. K. Fremier, and C. S. Goldberg. 2015. Quantifying effects of UV-B, temperature, and pH on eDNA degradation in aquatic microcosms. *Biological Conservation* 183: 85–92.

Taylor, B. L., and A. E. Dizon. 1996. The need to estimate power to link genetics and demography for conservation. *Conservation Biology* 10: 661–64.

Theimer, T. C., A. D. Smith, S. M. Mahoney, and K. E. Ironside. 2016. Available data support protection of the Southwestern Willow Flycatcher under the Endangered Species Act. *Condor* 118: 289–99.

Thomaz, A. T., M. R. Christie, and L. L. Knowles. 2016. The architecture of river networks can drive the evolutionary dynamics of aquatic populations. *Evolution* 70: 731–39.

Tibbets, C. A., and T. E. Dowling. 1996. Effects of intrinsic and extrinsic factors on population fragmentation in three species of North American minnows (Teleostei: Cyprinidae). *Evolution* 50: 1280–92.

Tonini, J., A. Moore, D. Stern, M. Shcheglovitova, and G. Ortí. 2015. Concatenation and species tree methods exhibit statistically indistinguishable accuracy under a range of simulated conditions. *PLoS Currents Tree of Life*. Edition 1. https://doi.org/10.1371/currents.tol.34260cc27551a-527b124ec5f6334b6be.

Turner, T. F., M. J. Osborne, M. V. McPhee, and C. G. Kruse. 2015. High and dry: Intermittent watersheds provide a test case for genetic response of desert fishes to climate change. *Conservation Genetics* 16: 399–410.

Unmack, P. J., T. E. Dowling, N. J. Laitenen, C. L. Secor, R. L. Mayden, D. K. Shiozawa, and G. R. Smith. 2014. Influence of introgression and geological processes on phylogenetic relationships of western North American mountain suckers (*Pantosteus*, Catostomidae). *PLoS ONE* 9: e90061.

USFWS (US Fish and Wildlife Service). 2003. *Revised Recovery Plan for Gila Trout* (Oncorhynchus gilae). Albuquerque: US Fish and Wildlife Service.

USFWS (US Fish and Wildlife Service). 2009. *Rio Grande Silvery Minnow Genetics Management and Propagation Plan*. Washington, DC: USFWS, Department of the Interior.

Valiere, N., and P. Taberlet. 2000. Urine collected in the field as a source of DNA for species and individual identification. *Molecular Ecology* 9: 2150–52.

Waples, R. S. 1995. Evolutionary significant units and the conservation of biological diversity under the Endangered Species Act. *American Fisheries Society Symposium* 17: 8–27.

Waples, R. S. 1999. Dispelling some myths about hatcheries. *Fisheries* 24: 12–21.

Waters, C. D., J. J. Hard, M. S. Brieuc, D. E. Fast, K. I. Warheit, R. S. Waples, C. M. Knudsen, W. J. Bosch, and K. A. Naish. 2016. What can genomics tell us about the success of enhancement programs in anadromous Chinook salmon? A comparative analysis across four generations. *bioRxiv*: 087973.

Whitlock, M. C., and D. E. McCauley. 1999. Indirect measures of gene flow and migration: $F_{ST} \neq 1/(4N_m + 1)$. *Heredity* 82: 117–25.

Willoughby, J. R., J. A. Ivy, R. C. Lacy, J. M. Doyle, and J. A. DeWoody. 2017. Inbreeding and selection shape genomic diversity in captive populations: Implications for the conservation of endangered species. *PLoS ONE* 12: e0175996.

Winograd, I. J., T. B. Coplen, J. M. Landwehr, A. C. Riggs, K. R. Ludwig, B. J. Szabo, P. T. Kolesar, and K. M. Revesz. 1992. Continuous 500,000-year climate record from vein calcite in Devils Hole, Nevada. *Science*, 258, 255–60.

Wit, P., M. H. Pespeni, J. T. Ladner, D. J. Barshis, F. Seneca, H. Jaris, N. O. Therkildsen, M. Morikawa, and S. R. Palumbi. 2012. The simple fool's guide to population genomics via RNA-Seq: An introduction to high-throughput sequencing data analysis. *Molecular Ecology Resources* 12: 1058–67.

Wright, S. 1931. Evolution in Mendelian populations. *Genetics* 16: 97–159.

15

Peter N. Reinthal,
Heidi Blasius,
and Mark Haberstich

Long-Term Monitoring of a Desert Fish Assemblage in Aravaipa Creek, Arizona

To understand the living present, and the promise of the future, it is necessary to remember the past.
—RACHEL CARSON, The Edge of the Sea

The freshwater fish fauna of the desert Southwest, which is under immense threat from a variety of sources, is one of the most imperiled groups of vertebrate organisms in North America (Jelks et al. 2008; Minckley and Deacon 1991). Threats to fish faunas include exploitation and manipulation of extremely limited water resources, introduction of nonnative species into these waters, habitat modification and degradation, and climate change (Minckley and Marsh 2009; Ruhí et al. 2016). A central problem associated with both evolutionary and ecological studies of desert fishes is that many of the processes and detectable changes associated with their populations occur over greater temporal spans, decades or longer, than the brief scales of weeks or months over which many scientific studies are conducted. There is no question that the fish faunas of the Southwest and their associated hydrologic systems have undergone dramatic changes on all time scales and will continue to do so. In the past two centuries, human activities have exacerbated natural changes (Sabo 2014). Long-term monitoring can provide important insights into how human activities affect ecosystems.

Desert communities have long been model systems for terrestrial ecosystem monitoring (e.g., Morgan Ernest et al. 2016), but there are few places in the Southwest where there have been long-term studies of how aquatic systems, with relatively intact native fish faunas, respond to changing hydrologic and climate regimes.

When W. L. Minckley arrived at Arizona State University in 1963 and visited Aravaipa Creek, he immediately recognized its importance as a refuge for native fish species. His judgment has proved prescient, as Aravaipa

Table 15.1. Native fish species collected from Aravaipa Creek, Arizona.

Family	Common name	Scientific name	Federal status
Cyprinidae	Longfin dace	*Agosia chrysogaster*	
	Roundtail chub	*Gila robusta*	
	Spikedace	*Meda fulgida*	Endangered
	Speckled dace	*Rhinichthys osculus*	
	Loach minnow	*Tiaroga cobitis*	Endangered
Catostomidae	Sonora sucker	*Catostomus insignis*	
	Desert sucker	*Pantosteus clarkii*	
Cyprinodontidae	Desert pupfish*	*Cyprinodon macularius*	Endangered
Poeciliidae	Gila topminnow*	*Poeciliopsis occidentalis*	Endangered

*Gila topminnow and desert pupfish were extirpated from Aravaipa Creek. They were restored to headwater springs, but it does not appear that these restorations were successful. Neither species has been subsequently encountered at any Aravaipa Creek monitoring site since restoration.

Creek is currently the only water body in Arizona with naturally occurring populations of seven native fish species (table 15.1). Museum records document five additional species that occurred in or near Aravaipa Creek. Two, Gila topminnow *Poeciliopsis occidentalis* and desert pupfish *Cyprinodon macularius,* were extirpated, and repatriation efforts in upper Aravaipa watershed springs failed. Three other native species, Colorado pikeminnow *Ptychocheilus lucius,* razorback sucker *Xyrauchen texanus,* and flannelmouth sucker *Catostomus latipinnis,* all large-river species, historically occurred in the San Pedro River and may have seasonally occupied Aravaipa Creek. Eight nonnative fish species have been detected in Aravaipa Creek during monitoring, but only three, red shiner *Cyprinella lutrensis,* green sunfish *Lepomis cyanellus,* and yellow bullhead *Ameiurus natalis,* have been found during regular monitoring since 2002 (table 15.2).

Minckley's first systematic monitoring of Aravaipa Creek was in the creek's lower reach in fall 1963, followed by additional monitoring of the reach in spring 1964. He expanded the monitoring program to include the upper and canyon portions of the creek in spring 1965, and he more or less continued monitoring at nine sites (fig. 15.1), with both spring and fall surveys, from 1965 until his passing in 2001. Systematic monitoring has continued at the same nine locations since that time in a coordinated effort by the University of Arizona, The Nature Conservancy, and the US Bureau of Land Management.

People often ask, Why monitor this fish population? If populations and conditions are changing, and our goal is to conserve and manage these populations, why don't we evaluate current conditions and respond accordingly with appropriate management? Trends are difficult to detect, in fish or in any species populations, but in our opinion, the value of long-term datasets greatly outweighs any difficulties associated with analyses of such data. Monitoring programs can be classified into three major categories: baseline monitoring, trend monitoring, and project monitoring (Clarkson et al. 2011). Here, we are primarily interested in trend monitoring, which takes measurements at regular, well-spaced time intervals at fixed geographic locations to determine long-term trends. Specifically, the Aravaipa fish monitoring program enables us to detect (1) long-term trends in abundances of native fish species, (2) the proportional composition and distribution of native fish species, (3) the presence and distribution of nonnative fishes in the stream, and (4) how fish populations respond to abiotic and biotic features of the system, such as discharge volume and pattern, temperature, or nonnative species. Frequent monitoring also provides an early warning system that enables rapid detection of critical changes in populations or the presence of a new nonnative species in the stream or drainage. Rapid detection allows for rapid and effective responses, and these monitoring results have proved invaluable in making scientifically sound and evidence-based management decisions. For example, Aravaipa Creek is an important source of spikedace and loach minnow for restoration efforts throughout the Gila

Table 15.2. Nonnative fish species collected from Aravaipa Creek, Arizona, with year first recorded through monitoring and year of most recent record.

Family	Common name	Scientific name	First documented	Last documented
Cyprinidae	Red shiner	*Cyprinella lutrensis*	1990	2017
	Common carp	*Cyprinus carpio*	1988	1988
	Fathead minnow	*Pimephales promelas*	1983	1998
Ictaluridae	Black bullhead	*Ameiurus melas*	1990	1990
	Yellow bullhead	*Ameiurus natalis*	1963	2017
Poeciliidae	Western mosquitofish	*Gambusia affinis*	1981	2002
Centrarchidae	Green sunfish	*Lepomis cyanellus*	1963	2016
	Largemouth bass	*Micropterus salmoides*	1963	1991

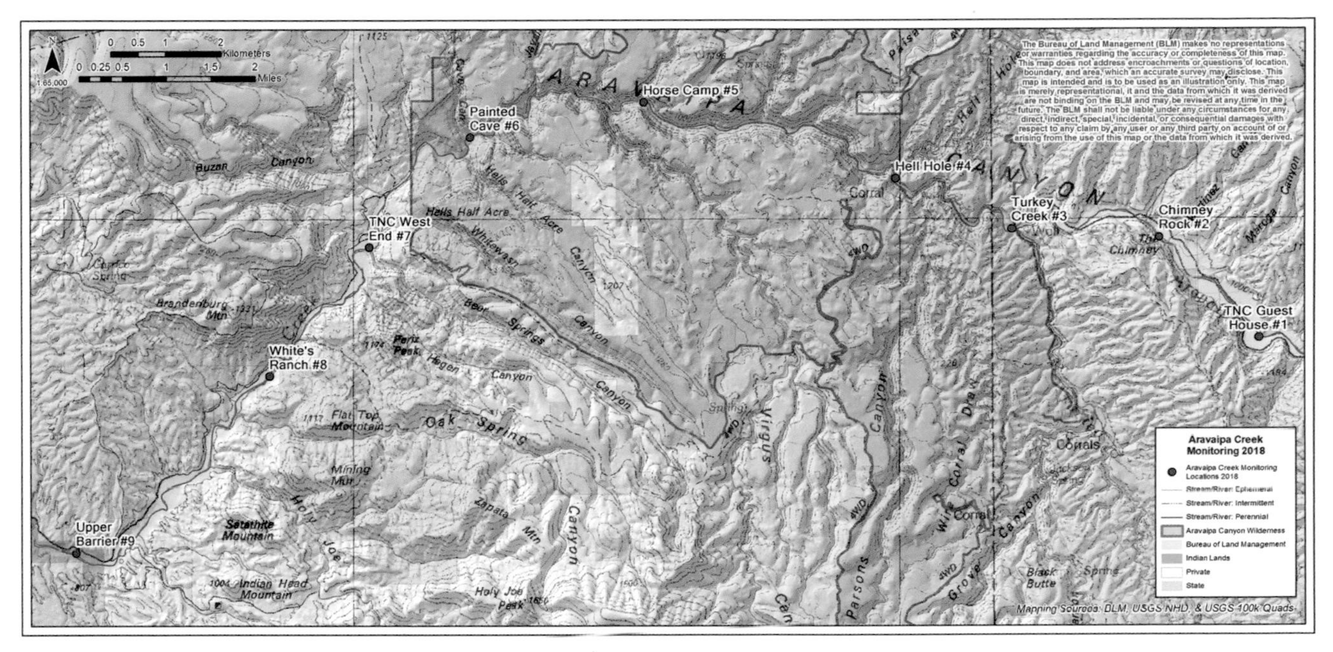

Fig. 15.1 Aravaipa Canyon with fish monitoring sites 1–9. Map by Evan Darrah, US Bureau of Land Management.

River basin. The monitoring data are essential for providing information needed to determine the number of individuals that can be removed with minimal impacts on their populations.

Aravaipa Creek is a complex aquatic ecosystem that has changed dramatically since 1963. Here, we briefly describe fluctuations and trends in both native and nonnative fish abundances and how native fish populations have responded to nonnative fish species and major episodic events, both natural and management-induced. Specifically, we examine how fish populations have responded to major flood events and to the installation of fish-movement barriers at the lower end of Aravaipa Creek. Finally, we emphasize the importance of historical assemblage information for effective evaluation of management actions.

BACKGROUND

The US Bureau of Land Management (BLM) has managed Aravaipa Creek since the 1950s. In recognition of its ecological importance, it was designated Aravaipa Canyon Primitive Area on April 28, 1971, by order of the secretary of the interior. In 1972, Defenders of Wildlife began purchasing and managing portions of the Aravaipa Creek catchment at either end of the

Aravaipa Canyon Primitive Area. In 1988, The Nature Conservancy took over ownership and management of the Defenders of Wildlife lands (Hadley et al. 1991). The Aravaipa Canyon Wilderness Area (ACWA) was proposed (2,701 ha) in 1982 and passed Congress in 1984 (Arizona Wilderness Act of 1984, Public Law no. 98-406). In 1990, the Wilderness Area was expanded (Arizona Wilderness Act of 1990, Public Law no. 101-628) to 7,850 ha. The 1990 act did not affect the 18 stream kilometers of Aravaipa Creek within the ACWA boundaries, but it increased the amount of seasonal tributary drainage area within Wilderness Area boundaries. In 2001, the US Bureau of Reclamation constructed a fish-movement barrier in the lower reaches of Aravaipa Creek, approximately 8 km from the creek's confluence with the San Pedro River (Clarkson 2004).

The Aravaipa Creek catchment has been studied intensively since the early 1960s, beginning with a description of Aravaipa Creek and its native fishes by Barber and Minckley (1966), followed by descriptions of the biology of spikedace *Meda fulgida* (Minckley and Barber 1971) and longfin dace *Agosia chrysogaster* (Barber and Minckley 1983). Subsequent studies reported on the biology of several species (Schreiber and Minckley 1981; Clarkson and Minckley 1988), species-habitat relationships (Deacon and Minckley 1974; Meffe and Minckley 1987), human perturbations (King and Martinez 1998), species taxonomy (Minckley and DeMarais 2000), species genetics (Tibbets and Dowling 1996), assemblage dynamics (Eby et al. 2003; Fagan et al. 2005), and nonnative fish species (Marsh et al. 1989). A summary of this extensive research is presented elsewhere (Minckley 1981; Stefferud and Reinthal 2004).

APPROACH AND STUDY AREA

Since assuming responsibility for monitoring Aravaipa Creek fishes, we have attempted to follow the protocols that Minckley instituted in 1963 and use the same permanent monitoring stations. Unfortunately, Minckley's field notes and data sheets have not been located. To date, we have located only summaries of Minckley's data aggregated for the upper, canyon, and lower reaches, not for the nine individual monitoring sites. Nor have we found a record of his specific sampling methodologies. We believe, however, that we have maintained continuity and comparability of methodologies and data between the monitoring by Minckley (1963–2001) and our own (2002–present).

At each of the nine monitoring stations, a 200-meter section of stream, starting at a fixed point, is sampled by two teams using 3.2 mm mesh, 3.1 × 1.2 m, weighted seines. Fish are identified, counted, and held in buckets before being returned to the stream. Seining is done every year in the spring (March–April) and fall (October–November). Since 2002, young-of-year fish have not been included in counts to minimize mortality. There were, however, often differences in the experience, knowledge, and effort among volunteer teams. We attempted to provide consistency in the 2002–2017 monitoring by having the three principal investigators present during all monitoring efforts.

Since monitoring began in Aravaipa Creek, change in the physical and hydrologic structure of the canyon and associated habitats has been constant, especially that caused by flooding. Changes in geomorphic structure and associated riparian habitat enable us to characterize faunal responses to given events at specific sites. For example, average monthly mean flow varies between about 0.28 and 0.85 m^3/s (10 and 30 cubic feet per second, cfs), whereas winter cyclonic and summer monsoonal storm events produce floods with peak discharges of about 76.5 m^3/s (2,700 cfs) in about 2 of every 3 years and of about 343 m^3/s (12,100 cfs) once every 10 years. A peak discharge of approximately 897 m^3/s (30,000 cfs) occurred in October 1983 and one of 779 m^3/s (27,500 cfs) in August 2006 (Mussetter 2013). These flood events massively altered the physical and hydrologic characteristics of the canyon.

Fish count data were summarized by season and year for graphical presentation. Morisita's index (I_M, Morisita 1959), a density invariant metric of similarity, was used to discern temporal and geographic trends in assemblage structure. Gaps in Minckley's data limited use of his collections in our analyses; consequently, use of Morisita's index was limited mainly to data collected since 2001. The index enables a succession of assemblage comparisons across time or space. Index values range from 0 (no similarity) to 1 (complete similarity). Coefficients of variation were calculated for each fall, spring, and the two seasons combined for each species to assess differences in species numbers. Linear regression was used to evaluate the effects of fall and spring discharge on total native and nonnative fish abundance (2002–2017).

FLUCTUATIONS IN NATIVE FISH SPECIES ABUNDANCES

A summary of 54 years of sampling of native fish species at Aravaipa Creek is presented in figure 15.2. There were two periods of peak native species abundance that stand out: From 1974 through 1976, total counts, from all sites and both seasons combined, ranged from 15,735 to 33,462. Only in 1991, when 22,494 individuals were counted, were any subsequent survey numbers close to those of 1974–1976. This second peak was followed by a count of 12,797 individuals in 1992. These periods represent the five highest counts over the past 54 years, and at no other time have more than 10,000 individuals been counted. Between 2002 and 2017, the peak year was 2009, with a total of 8,333 individuals, less than 25% of the 1975 peak and about 65.1% of the 1992 peak.

These threefold to fourfold differences in population counts raise a cautionary note for those proceeding with analyses of data such as we have done for Aravaipa Creek. The peak years appeared to be ordinary, in terms of both annual runoff volume and annual peak discharge in Aravaipa Creek (USGS Gage no. 09473000). During the 1974–1976 peak, longfin dace *Agosia chrysogaster* accounted for 71.6% of all native fish, and it was the most abundant native species throughout the 54 years of monitoring. A large increase in the number of spikedace *Meda fulgida* was also noted for 1991 and 1992. More spikedace were collected in 1991 ($n = 7{,}481$) and 1992 ($n = 4{,}567$) than in any other year. Again, there was nothing extraordinary about flows during those years. The next greatest total for spikedace was in 2008, when 2,666 individuals were counted; 2008 followed one of the two largest flood events (in July–August 2006) on record for the creek. Flooding conditions may create favorable habitat and conditions for spikedace via sediment mobilization, allochthonous inputs, and nonnative displacement, and the 2008 numbers are therefore thought to be a positive response to the major flood in 2006. The two species discussed above had the greatest fluctuations in individual counts during the monitoring program, but all seven native species demonstrated large interannual fluctuations.

Population variation between successive years was greater than variation between fall and spring surveys in any given year. For all native species, numbers were similar in spring and fall surveys, with slightly more individuals encountered during fall (fig. 15.3). Coefficients of variation (CV) were determined for spring, fall, and combined counts in 2002–2016 (table 15.3). The species having the highest coefficients of variation were speckled dace *Rhinichthys osculus*, spikedace, and loach minnow *Tiaroga cobitis*. Speckled dace was the least abundant native species in Aravaipa Creek, with the most restricted and patchy distribution of any native fish, but spikedace and loach minnow were comparatively common.

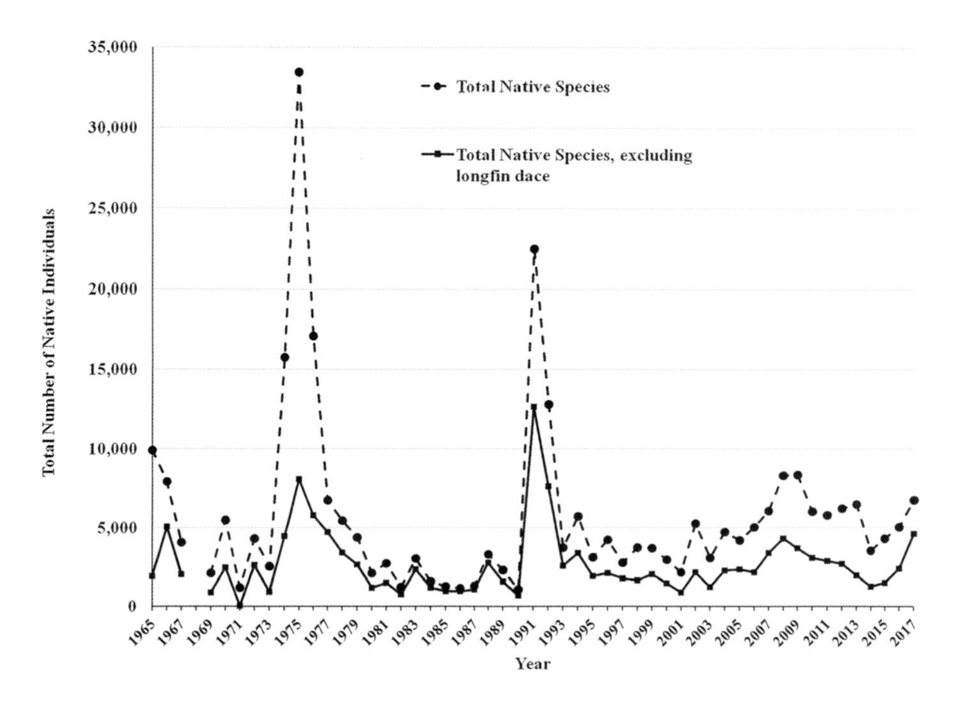

Fig. 15.2 Total numbers of individuals of native species for spring and fall monitoring combined, 1963–2017. Circles = all adults of all seven species present; squares = all adults of six species, excluding longfin dace *Agosia chrysogaster*. Only those years when all nine sites in the upper, canyon, and lower reaches were sampled are included.

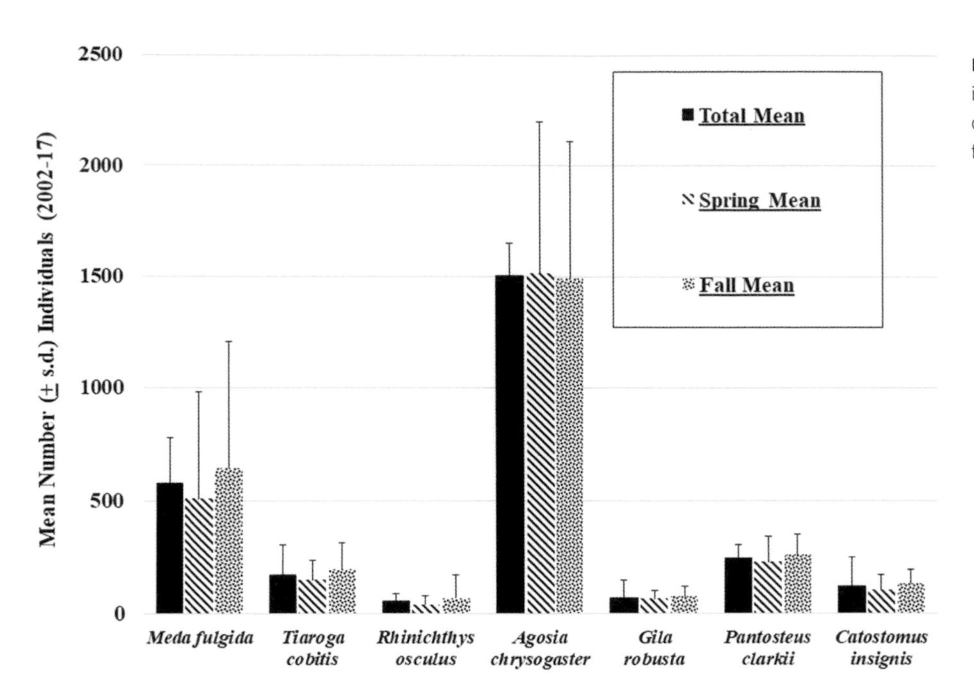

Fig. 15.3 Mean number (±SD) of individuals of seven native species counted during spring, fall, and spring-fall combined sampling, 1963–2017.

Table 15.3. Coefficients of variation (CV) for native fish species numbers in Aravaipa Creek for the spring, fall, and combined spring-fall surveys (2002–2016).

Species	Spring	Fall	Spring & fall
Longfin dace	41.7	44.3	42.5
Roundtail chub	59.7	47.0	52.1
Spikedace	80.0	79.0	78.4
Speckled dace	85.1	153.1	161.3
Loach minnow	62.1	57.5	60.8
Sonora sucker	67.9	42.5	54.1
Desert sucker	52.8	36.7	44.0

NONNATIVE FISH SPECIES

A summary of 54 years of nonnative fish species occurrence in Aravaipa Creek is presented in figure 15.4. A value of monitoring data is the ability to document the appearance, spread, and relative abundance of nonnative species. Minckley's records included sporadic collections by unnamed persons in 1943 (lower reach, fall) and 1950 (lower reach, fall), not included here, with no nonnative fish species reported. However, nonnatives were present in the lower reach when regular monitoring began in 1963. Specifically, nine green sunfish *Lepomis cyanellus*, three yellow bullhead *Ameiurus natalis*, and a single largemouth bass *Micropterus salmoides* were found. The same three species were first recorded in the canyon reach in 1967, 1969, and 1991 and in the upper reach in 1984, 1984, and 1991, respectively. Largemouth bass are no longer found at any Aravaipa monitoring site. Occurrence of green sunfish was restricted, and they may have been eradicated by mechanical removal from Horse Camp (site 5). Yellow bullhead were regularly encountered, but being nocturnal, they may be more common than the monitoring data indicate. They were regularly captured (hundreds of individuals) when electrofishing at night or day and were found throughout the creek. Their impact on the native fishes of Aravaipa Creek is poorly understood. Another nonnative species in Aravaipa Creek with deleterious impacts is red shiner *Cyprinella lutrensis* (Douglas et al. 1994). This species was first recorded during monitoring as a single individual in the lower reach in 1990 and was not found anywhere during subsequent surveys, until it was again found in the lower reach in fall 1997. It spread to the canyon and upper reaches in 1998 and persists in the lower and canyon reaches today. The peak number of red shiners for the entire system during monitoring occurred in fall 2004, when 596 individuals were collected. Aravaipa Creek experienced low flows and much-reduced flooding during a drought from 2001 to 2004, as defined by annual water volume (m^3) less than the lower 95% confidence interval of mean annual volume (Mussetter 2013). The low flows, coupled with the lack of flood events, appear to have allowed the increase in red shiner numbers from

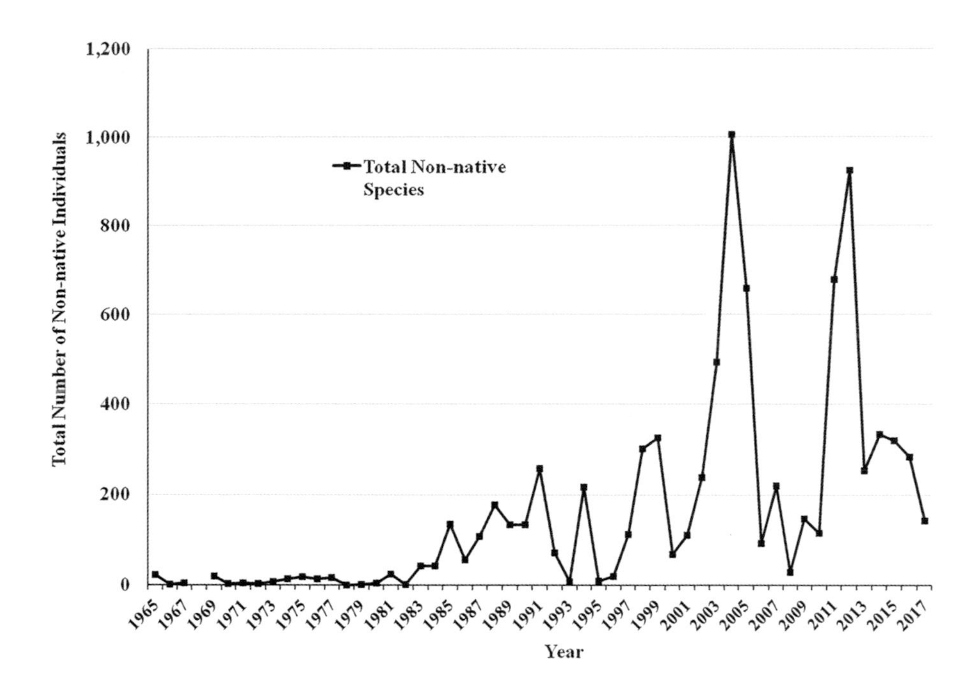

Fig. 15.4 Total number of individuals of nonnative fish species (listed in table 15.2) collected during spring and fall monitoring, 1963–2017. Only those years when all upper, canyon, and lower reaches were surveyed are included.

5 individuals in 2000 to 903 in 2004 and 629 in 2005. Following a major flood in July–August 2006, when flow was estimated to exceed 765 m^3/s (27,000 cfs) by USGS after the gage broke, the number of red shiners encountered during fall monitoring was reduced by 99% compared with previous years (1 red shiner in fall 2006 compared with 596 and 146 in fall 2004 and fall 2005, respectively). Green sunfish had been recorded in all years since 1965, except for 1978, until the 2006 flood; then they were not found until recolonization from a perennial tributary, Horse Camp, occurred in 2008. The 2006 flood, and other monsoonal or cyclonic flooding events, resulted in decreases in red shiner and green sunfish populations at Aravaipa monitoring sites. The importance of flood events for control and displacement of nonnative species had been previously recognized by Meffe and Minckley (1987) and Minckley and Meffe (1987).

The installation of a double fish-movement barrier by the US Bureau of Reclamation (USBR) on the west end of Aravaipa Creek in 2001 has effectively prevented the establishment of additional nonnative species in Aravaipa Creek. The only additional nonnative fish species encountered during monitoring after barrier construction was western mosquitofish *Gambusia affinis*, found in only 2002 and 2004. It is likely that western mosquitofish was present, but undetected, prior to barrier construction. Prior to barrier installation, common carp *Cyprinus carpio*, black bullhead *Ameiurus melas*, and fathead minnow *Pimephales promelas* were regularly recorded by Minckley, but have not been detected upstream of the barrier for the past 15 years.

From 2002 through 2017, monitoring found green sunfish at only one of the nine monitoring sites, Horse Camp (site 5). Green sunfish were encountered in 20 of the 31 surveys during that time, and always in low numbers (<13 individuals). Investigation of the Horse Camp Canyon tributary found a population of green sunfish in its perennial sections. Green sunfish had not spread in Aravaipa Creek beyond Horse Camp (site 5), most likely due to displacement by floods. The detection of green sunfish triggered a concerted effort by the US Bureau of Land Management to remove the source population in Horse Camp Canyon using galvanized fish traps and collapsible live bait traps. These removal efforts were successful, and green sunfish has not been detected above the barrier in Aravaipa Creek since 2014. Without monitoring, it would have been difficult to determine the distribution of green sunfish, initiate its removal, and monitor the effectiveness of the removal effort.

IMPACTS OF NONNATIVE FISHES ON NATIVE FISHES

Nonnative fish species have deleterious impacts on native fish populations and are a major cause of loss of native fish populations in the Colorado River basin (Tyus and Saunders 2000; Olden et al. 2006). Their major impacts include predation on or competition with native species at different life stages, disease or parasite transmission, and disruption of food webs and ecosystem processes (Moyle, chap. 4, this volume). Red shiner is believed to compete directly with native fish for both space and food resources (Schreiber and Minckley 1981; Marsh et al. 1989). Since 2002, red shiner was the most abundant of the three nonnative fish species persisting above the fish-movement barrier. In the past 32 monitoring events, it was detected every spring and fall, but never recorded at sites 1 and 2, rarely at sites 3 (1 survey), 4 (8 surveys) and 5 (3 surveys), and commonly at sites 6 (27 surveys), 7 (15 surveys), 8 (27 surveys), and 9 (28 surveys). Since 2001, 3 of the 5,939 red shiners collected were found in the upper reach and 80.7% were found in the lower reach. Red shiners accounted for 94.8% of all nonnative individuals. The native spikedace was present in the lower reach in the early 1960s and was regularly found there from 1975 until the 1990s, when red shiner was first detected. Currently, the two species have different distributions in Aravaipa Creek, with spikedace occurring mainly in the upper reaches and red shiner in the lower reaches (fig. 15.5). Monitoring provides a systematic approach to following changes in the abundance and distribution of red shiner. Any changes, especially the spread of red shiner to the upper reach, would trigger a rapid and targeted management response to alleviate the threat, such as attempts to physically remove red shiners. There is not a targeted removal program currently in place, but such a program should be considered.

TEMPORAL AND GEOGRAPHIC ASSEMBLAGE STRUCTURE PATTERNS

From 1963 to 2017, there was typically a high degree of similarity in the overall Aravaipa fish community from year to year (fig. 15.6). The lowest values were for the 1975–1976 and 1981–1982 comparisons. From the 1986–1987 through the 2016–2017 comparisons, I_M was always greater than 0.90. In comparison, Gido et al. (1997) reported considerably greater variation in San Juan River fish assemblages.

From 2002 through 2017, there were typically minor differences in mean (±SD) Morisita's indices for spring surveys (all sites, 0.93 ± .07), fall surveys (all sites, 0.93 ± .05) and combined seasons (all sites, 0.97 ± .02) across years (fig. 15.7). Low spring similarity and low fall similarity never occurred in the same year. Combined spring and fall similarity was comparatively constant over time.

The fish assemblage of the upper reach was more similar to that of the canyon reach than either the upper or canyon reach was to the lower reach (fig. 15.8). Over time, the similarity of the upper and canyon reaches has diminished, while the canyon and lower reaches have become more similar in recent years.

FISH POPULATIONS AND FLOW

Monitoring data can also be used in conjunction with Aravaipa Creek hydrologic data to characterize the effects of flow regime on species abundances. Specifically, Reinthal (2013) examined the relationship between

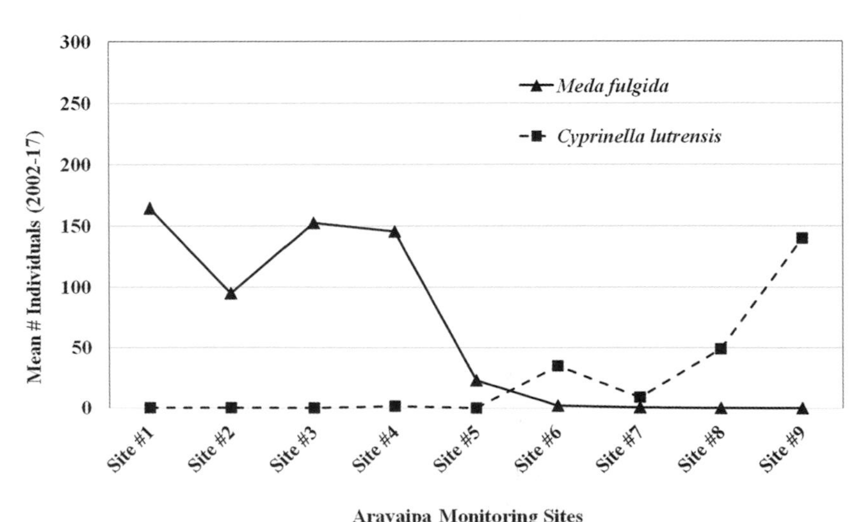

Fig. 15.5 Geographic distribution of spikedace *Meda fulgida* and red shiner *Cyprinella lutrensis* across nine monitoring sites in Aravaipa Creek, 2002–2017.

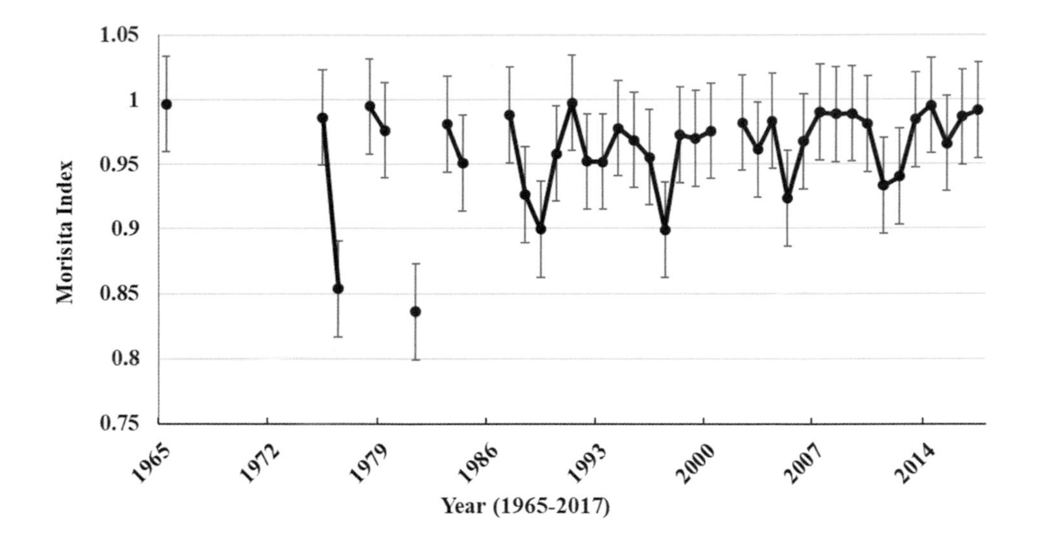

Fig. 15.6 Mean annual Morisita's index values for the fish community of Aravaipa Creek, 1965–2017. Years in which all three study reaches were not sampled are not included.

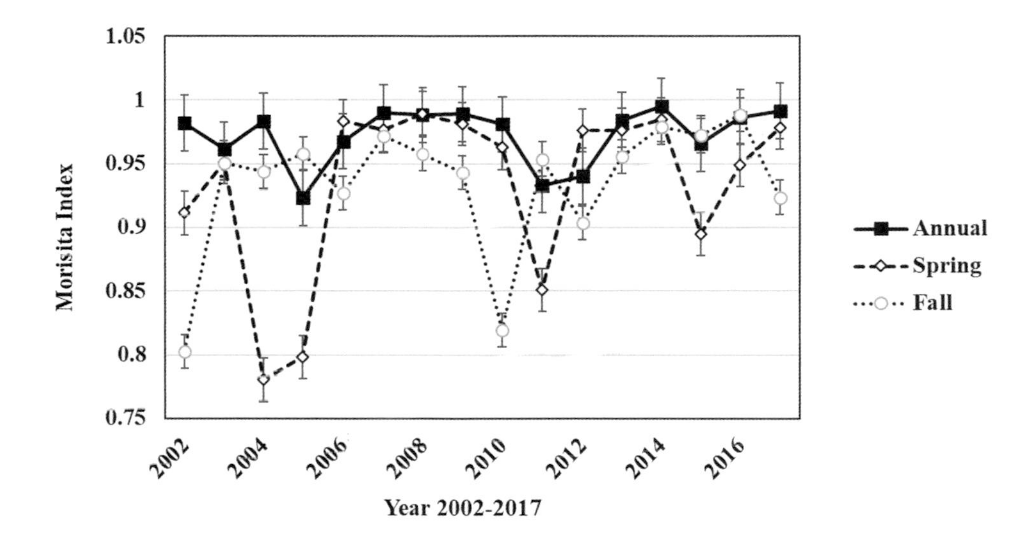

Fig. 15.7 Mean spring, fall, and annual Morisita's index values for the fish community of Aravaipa Creek, 2002–2017.

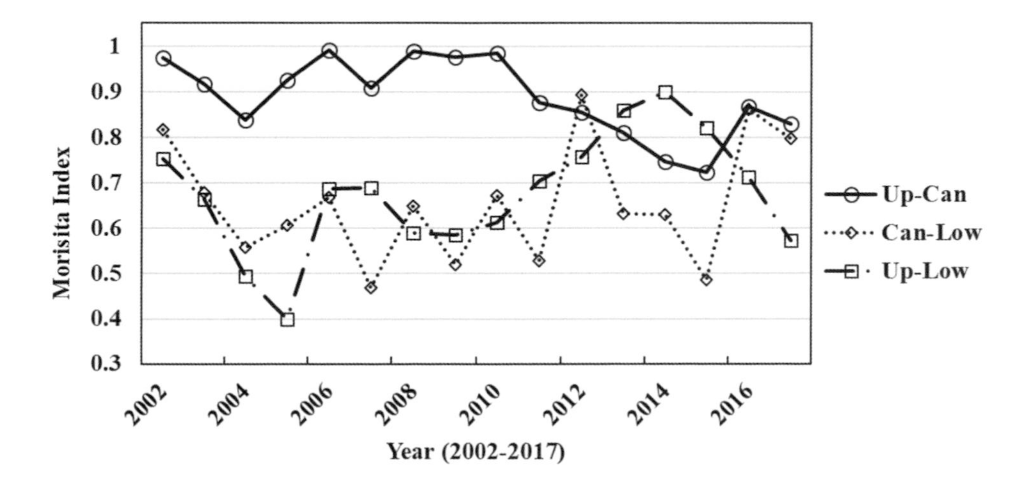

Fig. 15.8 Annual Morisita's index values for the upper reach and canyon reach assemblages, the canyon reach and lower reach assemblages, and the upper reach and lower reach assemblages of Aravaipa Creek, 2002–2017.

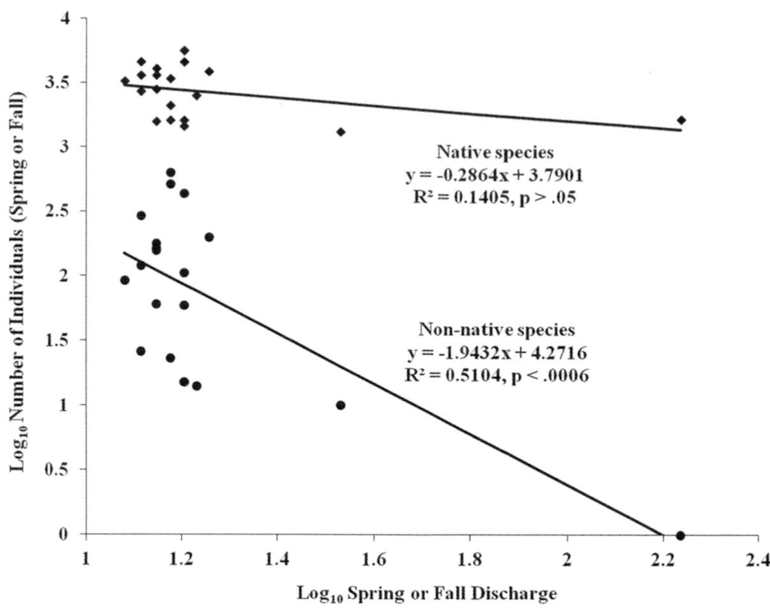

Fig. 15.9 Linear regression of $\log_{10}$ mean spring (January–June) or fall (July–December) monthly discharges for Aravaipa Creek versus $\log_{10}$ spring or fall abundance of native or nonnative fish species collected the same year, 2002–2012. (From Reinthal 2013.)

average monthly flow in six-month increments and fish abundance data for both native and nonnative species from 2002 to 2012. Spring fish abundance data were regressed against mean monthly flows for January–June, and fall fish abundance data were regressed against mean monthly flows for July–December. Mean flows for these six-month periods were used for an overall estimate of the cyclonic rain season (January–April ebbing, with the lowest mean and median flows in June) and the monsoonal rain season (July–October, with the lowest mean and median flows in November). Fish data were calculated as summation of spring (April) and fall (October) collections, and all fish survey sites were combined to yield a total number of individuals of native or nonnative species such that there were no zero counts. Specimen counts and flow data were $\log_{10}$ transformed and linear regressions were conducted for year x versus the number of individuals in year x and in year $x + 1$ for (a) all native specimens and (b) all nonnative specimens. Flow data were unavailable for spring and fall 2004 and spring 2005 due to the failure of the USGS gage, so these collections were excluded from the analyses.

The flow–fish abundance regressions indicated that nonnative species were negatively impacted by elevated flows and native species were not (fig. 15.9). The resistance of native fishes to displacement by elevated flows is probably due to differential selection because native fishes evolved in aquatic systems like Aravaipa Creek that regularly experience flood events (Minckley and Meffe 1987). Alternatively, nonnative fish species have a significant negative response to elevated flows ($p <$ .0006), indicating a naïveté of elevated flow events such as those often experienced by arid-land streams. The results for this limited time period for the nonnatives indicated that two data points pull the regression negatively. Without these points, there was no relationship between flow and nonnatives, but this analysis does illustrate that exceptionally high flows impact nonnatives more than natives and supports the contention that nonnative species are negatively impacted by flood flows because they do not have the behavioral responses needed to avoid displacement, whereas natives are more likely to move to stream margins to avoid displacement (Minckley and Meffe 1987; Gido et al. 2013).

Average spring and fall flows and year $x + 1$ total numbers of natives and nonnatives were regressed to determine if there were delayed responses of either to elevated flows (fig. 15.10). There was no response by total numbers of individuals, but short-lived (i.e., spikedace) and long-lived species (i.e., Sonora sucker) may have different responses.

SUMMARY

When moving forward with a monitoring program, we strongly advocate very clear, well-defined sites, goals, and protocols, and sharing of all original data, including copies of original data sheets. Lack of original data sheets from 1963 to 2001 make comparative analyses

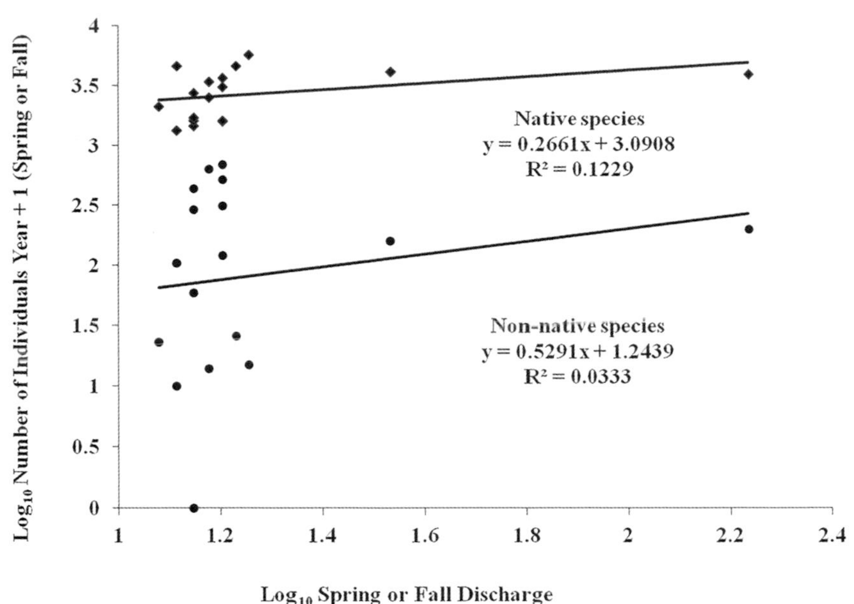

Fig. 15.10 Linear regression of $\log_{10}$ mean spring (January–June) or fall (July–December) monthly discharges for Aravaipa Creek versus $\log_{10}$ spring or fall abundance of native or nonnative fish species collected the following year, 2002–2012. (From Reinthal 2013.)

of fish data at various geographic and temporal scales problematic. Here and elsewhere, many scientists argue for the necessity and importance of long-term datasets, yet no regional commonly used repositories are currently available.

The Aravaipa Creek fish monitoring effort required a substantial voluntary effort through the years. From 2002 to the present, we estimate that over 200 individuals have participated. Without the participation of students, interns, volunteers, and interested parties, this monitoring would not have been possible. Both Minckley and Reinthal, through their affiliations with Arizona State University and the University of Arizona, respectively, have been able to use monitoring to provide an invaluable educational experience otherwise unavailable for hundreds of students.

The 54 years of monitoring Aravaipa Creek fishes has provided a record of both short-term and long-term variation in fish populations. Native species show considerable variability in abundance. There was a high degree of temporal stability and similarity in the fish assemblage from 1963 through 2017. Spatial variation appears to be related to nonnative fish species, habitat quality, and flow regimes. The upper reach, with no nonnatives, consistent flow and flooding, and high-quality riparian habitat, had the most stable fish assemblage, with all seven native species regularly found. From 2002 to 2010, the upper and canyon reaches had high similarity, but since then the canyon has become increasingly like the lower reach.

There was no relationship between flow and total native fish abundance, but nonnative fishes responded negatively to elevated flows. After floods, nonnative numbers were reduced. Native fishes appeared to respond positively to flow variability, and this variability is important for the persistence of native fishes. In contrast, lack of flow variability results in an increase in nonnative fish abundance.

Our initial question was whether fish assemblage monitoring could provide important information on fish assemblage dynamics, species ecology, and responses of native and nonnative fishes to various aspects of the flow regime. The Aravaipa Creek monitoring data, even with their limitations, provide a valuable resource for documenting species presence/absence, within-creek distribution, and relative abundances of native and nonnative fish species. We have been able to document the appearance, loss, and expansion or contraction of nonnative species within the system. The data also provide a means to better understand how fish respond to both biotic and abiotic events. In addition, the monitoring project has been the catalyst for numerous research projects and has greatly enhanced our knowledge of the structure and function of desert aquatic ecosystems. For management, regular monitoring permits rapid and data-driven responses to real and proposed changes in the ecosystem. But obtaining information such as that collected during biannual Aravaipa Creek monitoring requires a tremendous amount of work and coordination. The support of universities, federal and state agencies, and nongovernmental organizations was essential to the effort.

The persistence and stability of the Aravaipa Creek native fish assemblage should not be taken as insurance of its future survival. There are several lower Colorado River drainages that, at one time, had native fish assemblages similar to that of Aravaipa Creek. Many of these, however, no longer survive or have been greatly diminished. The potential for nonnative fish species to colonize Aravaipa Creek, and for other harmful anthropogenic activities, remains. Nonnative fish species can be easily carried above the fish-movement barrier. Without monitoring, such events could not be detected and responded to in a timely manner. Other threats to Aravaipa Creek include proposals for water extraction during flood flows. Models to evaluate effects of water withdrawal contend that "ample scientific evidence exists that there is sufficient surplus in the annual hydrograph to allow for targeted water extraction primarily on the descending hydrograph of destructive flood flows, but also, on prescribed occasions, on the ascending hydrograph of high flows during winter and summer floods" (Carothers et al. 2013, 1). Maintenance of the natural hydrograph is critical to the survival of an intact Aravaipa Creek native fish fauna (Poff et al. 1997; Lytle and Poff 2004), and alterations to the natural hydrograph have deleterious impacts on native faunas (Gido et al. 2013). Both base flows and flood flows are essential and necessary to maintain the native fish fauna by sustaining necessary habitat and, critically, displacing nonnative fishes (Propst et al. 2008; Stefferud et al. 2011). Most importantly, we must ensure that human activities that threaten this fish fauna are prevented. Otherwise, monitoring activities will be nothing more than a documentation of changes in the fish fauna, its gradual demise, and its ultimate extinction, thereby relegating monitoring to a self-gratifying experiential exercise.

REFERENCES

Barber, W. E., and W. L. Minckley. 1966. Fishes of Aravaipa Creek, Graham and Pinal Counties, Arizona. *Southwestern Naturalist* 11: 313–24.

Barber, W. E., and W. L. Minckley. 1983. Feeding ecology of a southwestern cyprinid fish, the spikedace, *Meda fulgida* Girard. *Southwestern Naturalist* 28: 33–40.

Carothers, S. W., W. C. Leibfried, and C. Finch. 2013. *Expert Opinion Report, re Aravaipa Canyon Wilderness Area (W1-11-3342), in the General Adjudication of All Rights to Use Water in the Gila River System and Source, Arizona Supreme Court, Case Nos. W1–W4.* Flagstaff, AZ: SWCA Environmental Consultants.

Clarkson, R. W. 2004. Effectiveness of electrical fish barriers associated with the Central Arizona Project. *North American Journal of Fisheries Management.* 24: 94–105.

Clarkson, R. W., B. R. Kesner, and P. C. Marsh. 2011. *Long-Term Monitoring Plan for Fish Populations in Selected Waters of the Gila River Basin, Arizona* (Revision no. 3). Phoenix: US Fish and Wildlife Service, Arizona Ecological Services.

Clarkson, R. W., and W. L. Minckley. 1988. Morphology and foods of Arizona catostomid fishes: *Catostomus insignis, Pantosteus clarkii,* and their putative hybrids. *Copeia* 1988: 422–33.

Deacon, J. E., and W. L. Minckley. 1974. Desert fishes. In G. W. Brown, Jr., ed., *Desert Biology,* vol. 2, 385–488. New York: Academic Press.

Douglas, M. E., P. C. Marsh, and W. L. Minckley. 1994. Indigenous fishes of western North America and the hypothesis of competitive displacement: *Meda fulgida* (Cyprinidae) as a case study. *Copeia* 1994: 9–19.

Eby, L. A., W. F. Fagan, and W. L. Minckley. 2003. Variability and dynamics of a desert stream community. *Ecological Applications* 13: 1566–79.

Fagan, W. F., C. Kennedy, C. Aumann, and P. J. Unmack. 2005. Rarity, fragmentation and the scale-dependence of extinction risk in desert fishes. *Ecology.* 86: 34–41.

Gido, K. B., D. L. Propst, and M. Molles, Jr. 1997. Spatial and temporal variation of fish communities in the secondary channels of the San Juan River, New Mexico and Utah. *Environmental Biology of Fishes* 49: 417–34.

Gido, K. B., D. L. Propst, J. D. Olden, and K. Bestgen. 2013. Multidecadal responses of native and introduced fishes to natural and altered flow regimes in the American Southwest. *Canadian Journal of Fisheries and Aquatic Sciences* 70: 554–64.

Hadley, D., P. Warshall, and D. Bufkin. 1991. *Environmental Change in Aravaipa, 1870–1970, An Ethnoecological Survey.* Cultural Resource Series Monograph no. 7. Phoenix: US Bureau of Land Management.

Jelks, H. L., S. J. Walsh, N. M. Burkhead, S. Contreras-Balderas, E. Diaz-Pardo, D. A. Hendrickson, J. Lyons, N. E. Mandrak, F. McCormick, J. S. Nelson, S. P. Platania, B. A. Porter, C. B. Renaud, J. J. Schmitter-Soto, E. B. Taylor, and M. L. Warren, Jr. 2008. Conservation status of imperiled North American freshwater and diadromous fishes. *Fisheries* 33: 372–407.

King, K. A., and M. Martinez. 1998. Metals in fish collected from Aravaipa Creek, Arizona, October 1997. Phoenix: US Fish and Wildlife Service.

Lytle, D., and N. L. Poff. 2004. Adaptation to natural flow regimes. *Trends in Ecology and Evolution* 19: 94–100.

Marsh, P. C., F. J. Abarca, M. E. Douglas, and W. L. Minckley. 1989. *Spikedace* (Meda fulgida) *and loach minnow* (Tiaroga cobitis) *relative to the introduced red shiner* (Cyprinella lutrensis). Final Report. Phoenix: Arizona Game and Fish Department.

Meffe, G. K., and W. L. Minckley. 1987. Persistence and stability of fish and invertebrate assemblages in a repeatedly disturbed Sonoran Desert stream. *American Midland Naturalist* 117: 177–91.

Minckley, W. L. 1981. *Ecological Studies of Aravaipa Creek, Central Arizona, Relative to Past, Present, and Future Uses*. Final Report, Contract no. YA-512-CT6-98. Phoenix: US Bureau of Land Management.

Minckley, W. L., and W. E. Barber. 1971. Some aspects of the biology of the longfin dace, a cyprinid fish characteristic of streams in the Sonoran Desert. *Southwestern Naturalist* 15: 459–64.

Minckley, W. L., and J. E. Deacon, eds. 1991. *Battle against Extinction: Native Fish Management in the American West*. Tucson: University of Arizona Press.

Minckley, W. L., and B. D. DeMarais. 2000. Taxonomy of chubs (Teleostei, Cyprinidae, genus *Gila*) in the American Southwest with comments on conservation. *Copeia* 2000: 251–56.

Minckley, W. L., and G. K. Meffe. 1987. Differential selection by flooding in stream-fish communities of the arid American Southwest. In W. J. Matthews and D. C. Heins, eds., *Community and Evolutionary Ecology of North American Stream Fishes*, 93–104. Norman: University of Oklahoma Press.

Minckley, W. L., and P. C. Marsh. 2009. *Inland Fishes of the Greater Southwest: Chronicle of a Vanishing Biota*. Tucson: University of Arizona Press.

Morgan Ernest, S. K., G. M. Yenni, G. Allington, E. M. Christensen, K. Geluso, J. R. Goheen, M. R. Schutzenhofer, S. R. Supp, K. M. Thibault, J. H. Brown, and T. J. Valone. 2016. Long-term monitoring and experimental manipulation of a Chihuahuan desert ecosystem near Portal, Arizona (1977–2013). *Ecology* 97: 1082.

Morisita, M. 1959. Measuring of interspecific association and similarity between communities. *Memoirs Faculty Science Kyushu University, Series E Biology* 3: 65–80.

Mussetter, R. A. 2013. *Hydrologic and Geomorphic Characteristics of Aravaipa Creek within the Aravaipa Wilderness Area: Implications for Instream Flow Water Rights*. Fort Collins, CO: TetraTech.

Olden, J. D., N. L. Poff, and K. R. Bestgen. 2006. Life-history strategies predict fish invasions and extirpations in the Colorado River basin. *Ecological Monographs* 76: 25–40.

Poff, L. N., J. D. Allan, M. B. Bains, J. R. Karr, K. L. Prestegaard, B. D. Richter, R. E. Sparks, and J. C. Stromberg. 1997. The natural flow regime: A paradigm for river conservation and restoration. *BioScience* 47: 769–84.

Propst, D. L., K. B. Gido, and J. A. Stefferud. 2008. Natural flow regimes, nonnative fishes, and native fish persistence in arid-land river systems. *Ecological Applications* 18: 1236–52.

Reinthal, P. N. 2013. Native and exotic fish communities of Aravaipa Creek, Arizona: Impacts of alterations of the current flow regime. Phoenix: Salt River Project.

Ruhí, A., J. D. Olden, and J. L. Sabo. 2016. Declining streamflow induces collapse and replacement of native fish in the American Southwest. *Frontiers in Ecology and the Environment* 14: 465–72.

Sabo, J. L. 2014. Predicting the river's blue line for fish conservation. *Proceedings of the National Academy of Sciences* 111: 13686–87.

Schreiber, D. C., and W. L. Minckley. 1981. Feeding interrelations of native fishes in a Sonoran Desert stream. *Great Basin Naturalist* 41: 409–26.

Stefferud, J. A., K. B. Gido, and D. L. Propst. 2011. Spatially variable response of native fish assemblages to discharge, predators and habitat characteristics in an arid-land river. *Freshwater Biology* 56: 1403–16.

Stefferud, S. E., and P. N. Reinthal. 2004. *Fishes of Aravaipa Creek, Graham and Pinal Counties, Arizona Literature Review and History of Research and Monitoring*. Task Order AAF030025. Phoenix: US Bureau of Land Management.

Tibbets, C. A., and T. E. Dowling. 1996. Effects of intrinsic and extrinsic factors on population fragmentation in three species of North American minnow (Teleostei: Cyprinidae). *Evolution* 59: 1280–92.

Tyus, H. M., and J. F. Saunders III. 2000. Nonnative fish control and endangered fish recovery: Lessons from the Colorado River. *Fisheries* 25: 9, 17–24.

16 Human Impacts on the Hydrology, Geomorphology, and Restoration Potential of Southwestern Rivers

Mark C. Stone and Ryan R. Morrison

Southwestern rivers and streams are dramatically altered from pre–European settlement conditions. In most cases, the priority for managing these systems is to reduce their natural variability to support economic activities by providing a reliable water supply during low-flow periods and by reducing flood risks. Enormous water projects have been constructed to advance this vision (Reisner 1993). This focus has enabled the rise of the large desert cities of the American Southwest, including Phoenix, Las Vegas, and Los Angeles—all of which depend heavily on the taming and exploitation of the Colorado River to support their growth (Udall, chap. 7, this volume). The desert cities of Albuquerque and Reno provide two examples of sprawling communities that occupy historical floodplains, and which depend almost entirely on surface-water diversions or groundwater pumping from hydrologically connected aquifers. The resilience of these desert communities—large and small—is inherently tied to their rivers.

This narrow focus on water as a resource to be developed and consumed for human uses has, however, generated severe stressors on the region's rivers, which have interrupted hydrologic, geomorphic, and ecological processes and contributed to biotic population declines, endangerment, extirpation, and in many cases extinctions of native fish, flora, and other fauna (Williams and Sada, chap. 6, this volume). For example, of the 27 fishes that were historically native to the Rio Grande in New Mexico, only 14 remain, including the federally endangered Rio Grande silvery minnow *Hybognathus amarus* (Cowley 2006). The stark population declines of many southwestern fishes, including Rio Grande silvery minnow, Colorado pikeminnow *Ptychocheilus lucius*, Pecos bluntnose shiner *Notropis simus pecosensis*, loach minnow

Rhinichthys cobitis, spikedace *Meda fulgida*, and razorback sucker *Xyrauchen texanus*, can be tied to a range of structural and nonstructural stressors related to water development and urbanization (Propst and Bestgen 1991; Olden et al. 2006). Structural stressors include the construction of dams, levees, and drains and the straightening of river channels. Nonstructural stressors include flow extraction, watershed development, and floodplain encroachment. In most cases, these stressors interact to modify the natural processes that support river form and function, to which native fishes are adapted. Halting these negative trends and restoring these systems requires an improved understanding of connections between management activities, impacts on natural processes, and ultimately, implications for species of concern.

A growing body of knowledge, increased public awareness, and legal requirements for protecting aquatic species and their habitat have led to a more holistic perspective on water resources and river management (Poff et al. 2009). Restoration projects have been implemented to improve or create critical habitat, and environmental flow programs have restored aspects of the natural flow regime (Poff et al. 1997; Propst and Gido 2004). These programs demonstrate some level of success and offer insights that can guide future efforts (Sabo et al. 2012; Bestgen et al., chap. 21, this volume).

An emerging principle is the need to focus on hydrologic, geomorphic, and ecological processes when engaging in both direct and indirect restoration activities (Bunn and Arthington 2002). Stream restoration projects are increasingly focused on specific ecological process outcomes, such as improved floodplain connectivity (Acreman et al. 2014; Stone et al. 2017). Further, environmental flow prescriptions are growing in complexity to emphasize connections between critical environmental processes and the flows that drive them (Yarnell et al. 2015). Here, we aim to describe the nature of the channel and streamflow modifications that have contributed to the decline of ecological integrity and system resilience, and to highlight efforts to reverse these trends through restoration and environmental flow programs.

MODIFICATIONS OF RIVERS, FLOODPLAINS, AND WATERSHEDS

Humans have significantly altered most river reaches in the American Southwest. The most profound impacts are typically associated with dams, which modify the flow of water, sediments, nutrients, and movement of organisms. Most river reaches have been also subjected to some level of river engineering including flood control projects, adjacent floodplain development, and channel straightening aimed at enhancing water conveyance (Iorns et al. 1965). Furthermore, the contributing watersheds for most river systems are highly altered from baseline conditions (Kenny et al. 2000) and contribute to alterations in the natural flow regime of rivers. Below, we describe the hydrologic, geomorphic, and ecological implications of these modifications.

Dams and Diversions

The construction and management of dams, both large and small, in the American Southwest has profoundly affected the region's rivers and has also enabled development projects to move billions of cubic meters of water away from rivers and toward thirsty cities and agricultural fields. According to the National Inventory of Dams database (https://catalog.data.gov/dataset/national-inventory-of-dams), there are 2,256 major dams in New Mexico, Arizona, Nevada, and Utah alone (USACE 2016). The flagship engineering projects in the lower Colorado River include Hoover Dam and Lake Mead, Glen Canyon Dam and Lake Powell, the Central Arizona Project, and the All-American Canal. Lakes Mead and Powell have a combined water storage capacity of over 6.2^{10} m^3 (50 million acre-feet), and the associated Hoover and Glen Canyon Dams have 3,000 megawatts of installed power generating capacity (Collier et al. 2000). The last dam on the Colorado River, Morelos Dam, diverts the entire remaining flow in most years. While supporting over 202,500 ha (500,000 acres) of farmland in the Colorado River delta, it effectively dewaters 136 km (85 miles) of river channel and thus has eliminated fish habitat and the connection for several fish species that historically moved between the river and the sea (Wichelns and Oster 2006).

The Rio Grande Project includes six diversion dams, two large storage dams (Elephant Butte and Caballo) with a combined capacity of 4.3 billion cubic meters (3.5 million acre-feet), nearly 960 km (600 miles) of irrigation canals, and over 720 km (450 miles) of drainage canals (Stokes 2007). The Middle Rio Grande Project supplies water to the city of Albuquerque and to irrigators in the Middle Rio Grande Valley between Cochiti Dam and Elephant Butte Reservoir. The project's irrigation and flood control facilities were built by the Middle Rio Grande Conservancy District, US Bureau of Reclamation (USBR), and US Army Corps of Engineers (USACE). The Middle Rio Grande Conservancy District built El Vado Dam, with a capacity of 247 million cubic meters (200,000 acre-feet), to provide a reliable water supply for irrigation of the Middle Rio Grande Valley. The project is also closely tied to the San Juan–Chama Project, which provides an additional 118.5 million cubic meters (96,000 acre-feet) of water annually to the Rio Grande basin via diversions from headwater tributaries of the San Juan River in Colorado (Flanigan and Haas 2008).

River Engineering and Floodplain Development

River engineering occurred at a remarkable rate and scale in the Southwest over the course of the twentieth century, with the aim of providing flood control and efficient water conveyance and supporting water diversion activities. Flood control projects included extensive construction of levee networks that disconnected rivers from their natural floodplains and constrained them to significantly smaller areas (Byrne 2017). Channel straightening (channelization) and river training were common river engineering techniques in southwestern rivers, with a goal of reducing channel complexity and ensuring a more efficient and predictable active channel. River training involved an array of bank stabilization techniques, including rip-rap, jetty jacks, and introduced vegetation (Graf et al. 2002). In many cases, nonnative vegetation was used for stabilization projects, which allowed several invasive riparian species to become established, including salt cedar *Tamarix* spp. and Russian olive *Elaeagnus angustifolia* (Shafroth 2010).

Flood risk reduction projects, including levee construction, resulted in increased floodplain development for agricultural, urban, and suburban uses in places like the Colorado River downstream of Lake Mead and the Rio Grande downstream of Cochiti Dam. The extent of floodplain modification can be seen by comparing 1930s-era aerial photographs of historical conditions with aerial photographs showing modern channel and floodplain features (figs. 16.1 and 16.2). For example, Lake Mohave, upstream of Davis Dam, has entirely inundated the Colorado River and its floodplain for 80 km (50 miles) upstream into Black Canyon (fig. 16.1*B*). Figure 16.2 shows the isolation of the Rio Grande from its floodplain just below Cochiti Dam, in Albuquerque, and in the San Acacia reach upstream from Elephant Butte Reservoir.

Watershed Modifications

In the late nineteenth and early twentieth centuries, the Southwest experienced widespread and highly synchronized channel entrenchment (e.g., Wohl et al. 2017), including arroyo development and incision. Although shifting climate regimes probably played a role, widespread channel incision was closely synchronized with the arrival of the railroad and the accompanying activities of deforestation and intense livestock grazing. For example, hundreds of thousands of hectares of forest were cut between 1914 and 1918 in northern New Mexico, and stocking rates for sheep and cattle increased to a maximum of 1.6 million and 150,000, respectively, by 1900 (Phillips et al. 2011). A wet period combined with these modifications increased runoff rates and sediment loads (Scurlock 1998).

The second half of the twentieth century brought a new form of landscape change—urbanization. Large cities emerged in the desert as a result of growing economies and shifting US demographics (table 16.1; US Census Bureau 2014). This rapid urbanization resulted in two profound impacts on watersheds and water resources. First, increased impervious land cover in large cities generated increased runoff and storm flows with higher discharge peaks, which exacerbated channel in-

Table 16.1. Growth of major metropolitan areas between 1950 and 2010.

City	Metropolitan population	
	1950	2010
Las Vegas	48,300	1,951,300
Phoenix	329,266	4,574,351
Los Angeles	4,250,000	12,828,000
Albuquerque	96,815	887,077
Tucson	45,454	980,263
Reno	40,840	420,000

Source: US Census Bureau, 2014.

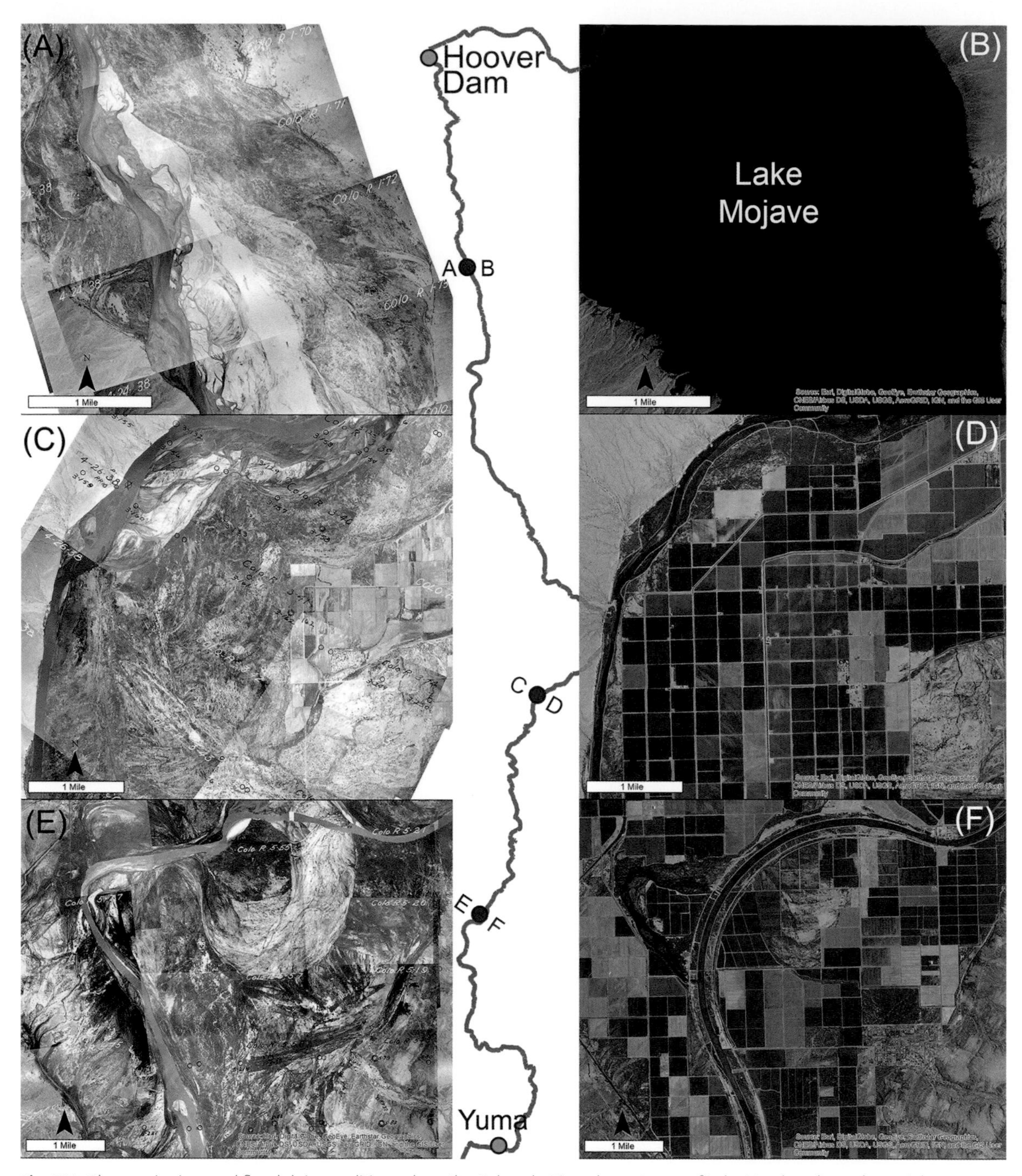

Fig. 16.1 Changes in river and floodplain conditions along the Colorado River downstream of Lake Mead as shown by aerial photographs from 1938 (*A*, *C*, *E*) and the present day (*B*, *D*, *F*). The blue line in the center of the figure represents the Colorado River between Hoover Dam and Yuma, AZ. The purple dots indicate the locations of the aerial photographs along the channel. (Sources: *A*, *C*, *E*, Norman et al. 2006; *B*, *D*, *F*, ESRI 2016.)

cision (Bahner et al. 2006) and required installation of various engineered structures to halt that process. Second, the need of the growing populations for municipal water supplies resulted in increased surface-water withdrawals and groundwater pumping, as well as transfer of water rights from agriculture to urban areas.

Increased wildfire frequency, severity, and size also drove watershed disturbance and change in the late twentieth and early twenty-first centuries as a result of unnaturally high fuel loads, warmer and drier conditions associated with climate change, and development within the wildlife-urban interface (Westerling et al.

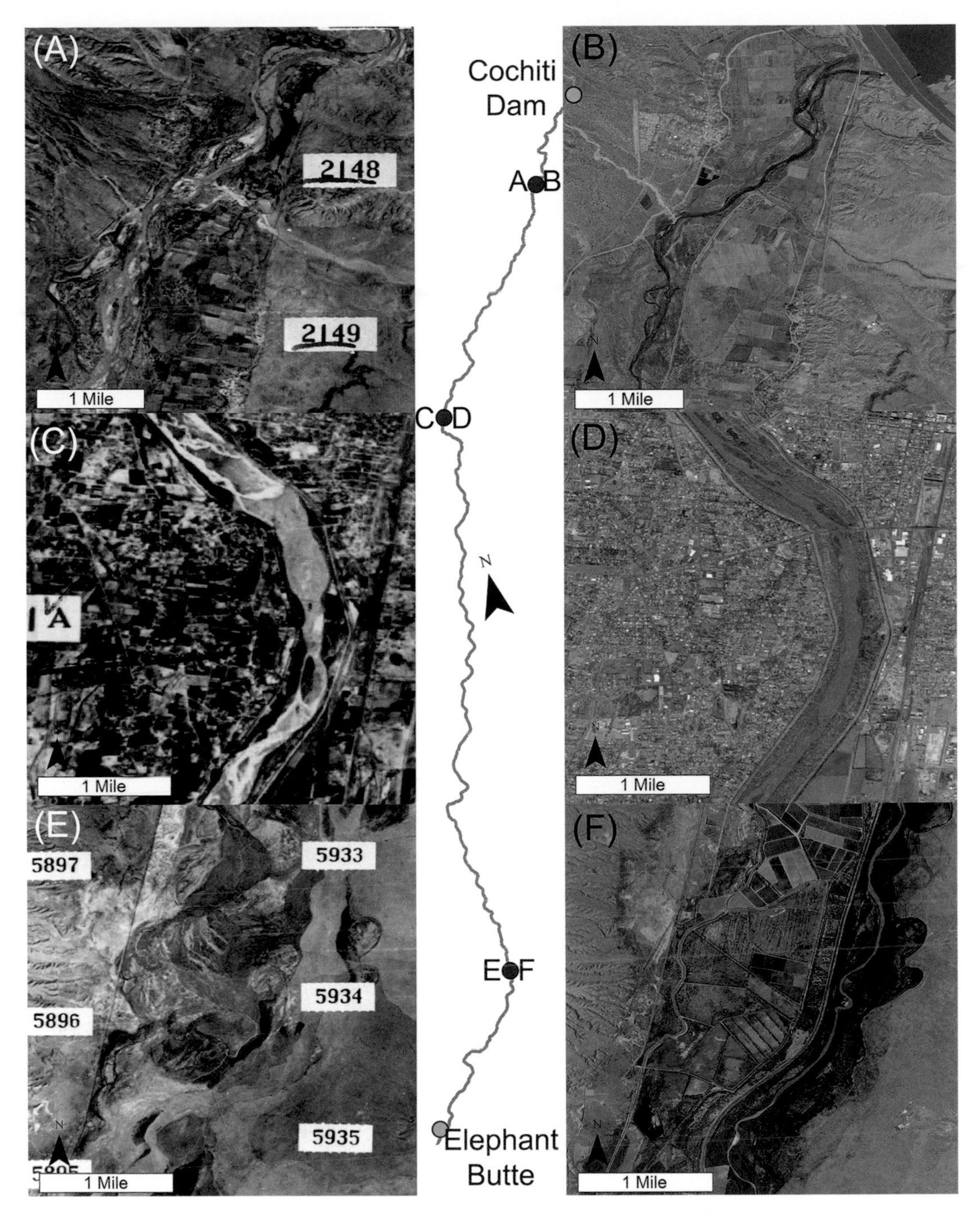

Fig. 16.2 Changes in river and floodplain conditions along the Rio Grande downstream of Cochiti Dam as shown by aerial photographs from 1935 (*A*, *C*, *E*) and the present day (*B*, *D*, *F*). The blue line in the center of the figure represents the Rio Grande between Cochiti Dam and Elephant Butte Reservoir. The purple dots indicate the locations of the aerial photographs along the channel. (Sources: *A*, *C*, *E*, RGIS 2017; *B*, *D*, *F*, ESRI 2016.)

2006). The effects of wildfires on hydrologic processes are highly variable and complex, but typically include increased peak runoff and higher sediment and debris loads (Gresswell 1999). Severe wildfires destroy organic material on the soil surface, reduce infiltration through formation of hydrophobic soils, and reduce soil stability due to the loss of vegetated cover and root reinforcement (Moody et al. 2013).

Hydrologic Alterations

River and watershed modifications have produced

Table 16.2. Components of the natural flow regime and their ecological and geomorphic importance.

Flow regime component	Description	Ecological importance	Geomorphic importance
Magnitude	The volume of water flowing in a river at any given time	Floodplain-river connectivity; maintenance of biodiversity and productivity	Removal of fine sediments; import of large woody debris; inundation of floodplain surfaces
Timing	The regularity with which and time of year when a specific discharge magnitude occurs	Aquatic species life cycle cues; riparian vegetation recruitment	Flushing of seasonal sediment inputs; scouring of floodplain surfaces
Duration	The period over which a specific discharge occurs	Species persistence	Sediment transport flux; channel maintenance
Frequency	How often a specific discharge recurs	Dynamic energy transport in aquatic ecosystems; prevention of channel encroachment	Channel maintenance; sediment flux dynamics
Rate of change	How quickly flow changes from one magnitude to another	Support of riparian vegetation recruitment; aquatic species response (stranding)	Channel maintenance

large-scale hydrologic changes throughout the Southwest. Dams, levees, water withdrawals, and even trans-basin water transfers have altered flow regimes. In many systems, current hydrographs are vastly different than they were historically. For example, peak flows that historically occurred during the spring snowmelt are now captured in reservoirs and slowly released throughout the year, which reduces spring peaks and increases summer base flows. In addition, hydropower production can cause large daily discharge fluctuations.

Ecological processes within river ecosystems evolved in response to the unique hydrologic characteristics of the natural flow regime (e.g., Poff et al. 1997). The natural flow regime is typically separated into five components: timing, magnitude, duration, frequency, and rate of change of flow. Each component is defined, and its general ecological/geomorphic importance summarized, in table 16.2.

Specific flow alterations typically depend on the types of river engineering and watershed modifications that have taken place, which we illustrate with examples from the Rio Chama and Gila River, both located in New Mexico. The Rio Chama, located in northern New Mexico, is a vital water source for farmers and cities in the Rio Grande basin. Three reservoirs hold approximately 118 million cubic meters (96,000 acre-feet) of water that is imported into the basin each year from the Colorado River basin. Due to the storage and release schedules of the dams, as well as the large volume of water imported annually, flows in the Rio Chama differ from natural conditions (fig. 16.3). For example, average minimum discharge has more than doubled due to downstream water demands, but peak flows have decreased by over 25% as a result of spring season water storage. Another conspicuous alteration is the sharp fluctuations in summer flows (e.g., 2005), a result of increased weekend releases to accommodate commercial and private rafting and kayaking. These hydrologic alterations have reduced overbank flooding, limited sediment transport, restricted riparian vegetation recruitment, and degraded spawning habitat for trout (Morrison and Stone 2015).

In contrast, the upper Gila River in southwestern New Mexico represents a system with relatively modest flow regime changes. It has been subjected to little river engineering compared with other southwestern rivers. However, a proposed diversion associated with the 2004 Arizona Water Settlements Act would alter the natural flow regime of the Gila River (Gori et al. 2013), including a 14% decrease in March discharge, because of the unique rules governing withdrawals from the river. Under the proposed actions, minimum Gila River flows over 1-day and 90-day periods would decline by 5% and 14%, respectively (Gori et al. 2013). These changes would restrict recruitment of riparian vegetation on the floodplain, reduce groundwater recharge, and harm fish populations (Morrison and Stone 2014).

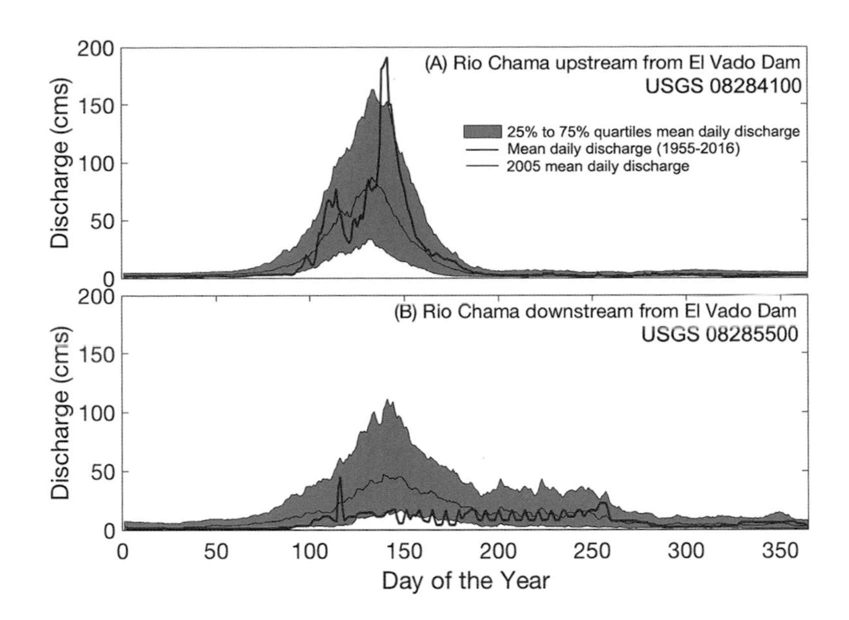

Fig. 16.3 Annual hydrographs for the Rio Chama upstream (*A*) and downstream (*B*) of El Vado Dam. The gray region represents daily mean discharge over the period of record (1955–2016) between the 25% and 75% quartiles. The black line within the gray region represents the median daily discharge over the period of record. The red line indicates daily discharge in 2005 to provide the reader with an example of annual operations. Discharges are given in cubic meters per second (cms).

The flood pulse concept, proposed by Junk et al. (1989), is also useful for understanding river-floodplain interactions and their responses to human impacts. The flood pulse concept proposes that the predictability and duration of floods strongly influence the traits and interactions of river biota. Specifically, geomorphic habitat features, lateral exchanges of organic matter and nutrients, and hydrologic residence times, which are all driven by flooding dynamics, influence species composition and productivity. Therefore, water management activities that alter flood dynamics often result in major changes in river communities. For example, predictable spring snowmelt and sporadic monsoon storms are important for distributing nutrients, providing nursery habitat, and triggering spawning of native fishes (e.g., Pease et al. 2006). Dams and water withdrawals disrupt these flooding patterns, alter native fish productivity, and encourage the spread of nonnative fishes (e.g., Fausch et al. 2001; Gido et al. 2013).

GEOMORPHIC AND ECOLOGICAL IMPLICATIONS OF RIVER AND FLOW REGIME MODIFICATIONS

Geomorphic Implications

The impacts of river engineering projects and streamflow modifications on stream corridors can be described using the interrelated concepts of hydrologic connectivity and sediment connectivity. These concepts connect the structural components (e.g., morphology) and the process components (e.g., transport of water and sediments) of a stream corridor as it is perturbed by outside forces (e.g., dam construction and watershed urbanization). Watershed connectivity can be broadly defined as transfer of energy and matter within a single landscape feature or between two landscape features (Chorley and Kennedy 1971). Hydrologic connectivity, more specifically, describes the water-mediated transfer of matter, energy, and organisms within and between elements of the hydrologic cycle (Pringle 2001). These processes are profoundly impacted by human modifications of rainfall-runoff processes, alterations in the timing and volumes of water discharge through impoundments and diversions, and reductions in lateral and vertical stream connectivity via channelization and groundwater mining.

Sediment connectivity is the transfer of sediment from a source to a sink via sediment detachment, transport, and deposition (Poeppl et al. 2017). Human interactions with a watershed or river corridor can enhance or reduce the detachment, transport, and/or deposition processes and provide an indication of landscape sensitivity to geomorphic change (Harvey 2002). These processes are a function of sediment size distribution, density, and armoring and are influenced by watershed features such as slope, area, and land cover. The various routes, pathways, and scales of sediment transport are sensitive to the frequency and magnitude distributions of disturbances (e.g., floods, wildfires, construction activity, and tilling), which influence landscape processes at broad spatial and temporal scales (Bracken et al. 2015). For example, sediment detachment is enhanced following watershed disturbances such as grazing, wildfire, and logging, but can also be reduced following urbanization. Sediment transport is reduced upstream of impoundments and downstream of water extractions, but is increased following channelization or in watersheds with increased impervious areas.

Channel response to disturbance can be conceptually segmented into discrete stages, each characterized by the dominance of particular adjustment processes (Simon and Rinaldi 2006). These temporally and spatially organized adjustments are collectively termed channel evolution (Simon 1989). Two interrelated relationships helpful for describing channel evolution in response to interrupted hydrologic and sediment connectivity are (1) Lane's relationship (Lane 1955) and (2) Exner's equation (Parker et al. 2000).

$$QS \propto Q_s d_{50} \quad \text{Lane's Relationship}$$

$$(1 - \lambda_p)\frac{\partial z}{\partial t} = -\frac{\partial q_s}{\partial x} \quad \text{Exner's Equation}$$

Within Lane's relationship, Q is the channel-forming discharge, S is the channel gradient, Q_s is the sediment transport capacity, and d_{50} is the median grain size of the bed material. Within Exner's equation, q_s is the sediment transport rate per unit stream width, z is the local bed elevation, and λ_p is the bed porosity. Channel evolution is often triggered when an excess or deficit of sediment transport capacity occurs relative to the sediment load delivered from upstream. Shifts in stages of channel evolution represent the crossing of specific geomorphic thresholds and the dominance of processes associated with those thresholds (Simon and Rinaldi 2006) (table 16.3).

Changes in the supply of sediments and stream discharge downstream of dams, which trap sediments and change flow regimes, may produce adjustments in the river (Graf 1999; O'Connor et al. 2015). Upstream of the dam, the slope (S), and hence the stream power, is reduced, and sediments are deposited in the reservoir. Downstream of the dam, the clear water released from the dam maintains transport capacity that exceeds sediment delivery (Graf et al. 2002). This typically results in erosion of the channel bed and banks and extensive channel incision (Schumm 1977). Changes in sediment supply and flow regime can also induce dramatic changes in channel patterns, such as a transition from a braided river to a single-thread channel. Numerous examples of channel adjustments downstream from dams exist throughout the Southwest. The Rio Grande has experienced 2.4–3.0 m (8–10 feet) of incision downstream from Cochiti Dam, and the channel has transitioned from a complex braided system to a single-thread channel (Richard et al. 2005) with significant narrowing (Swanson et al. 2011).

Watershed modifications also interrupt sediment connectivity and channel dynamic equilibrium. Urbanization results in increased peak discharge as impervious areas are increased and sediment delivery is reduced (Field and Lichvar 2007), resulting in channel incision, headcutting that migrates upstream through

Table 16.3. Examples of morphological response to channel and streamflow modifications. The arrows above and below each equation represent the directional response of the associated variable at upstream (Point A, arrows above the equations) and downstream (Point B, arrows below the equation) portions of the channel reach.

Modification	Lane's Relationship	Exner's Equation	Response
Channelization	*Pt A* ↑ ↑ $QS \propto Q_s d_{50}$ *Pt B* ↓ ↓	*Pt A* ↓ ↑ $(1 - \lambda_p)\frac{\partial \eta}{\partial t} = -\frac{\partial q_s}{\partial x}$ *Pt B* ↑ ↓	*Pt A:* Channel incision *Pt B:* Channel aggradation
Withdrawal & Return of Flow	*Pt A* ↓ ↓ $QS \propto Q_s d_{50}$ *Pt B* ↑ ↑	*Pt A* ↑ ↓ $(1 - \lambda_p)\frac{\partial \eta}{\partial t} = -\frac{\partial q_s}{\partial x}$ *Pt B* ↓ ↑	*Pt A:* Channel aggradation *Pt B:* Channel incision
Up and Downstream of a Dam	*Pt A* ↓ ↓ $QS \propto Q_s d_{50}$ *Pt B* ↓↓ ↓↑	*Pt A* ↑ ↓ $(1 - \lambda_p)\frac{\partial \eta}{\partial t} = -\frac{\partial q_s}{\partial x}$ *Pt B* ↓ ↑	*Pt A:* Delta Formation *Pt B:* Channel incision

the watershed, and temporary liberation of streambed and bank sediments as the channel seeks a new grade. The Las Vegas Wash in Nevada provides an example of channel incision as the result of increased discharge and reduced sediment delivery. Historically, the wash was spring-fed, with lush riparian vegetation, and was inhabited by the endemic, but now extinct, Las Vegas dace *Rhinichthys deaconi* (Jelks et al. 2008). Now, a substantial base flow of about 10 m^3/s is provided by wastewater effluent from Las Vegas; at the same time, peak flows have increased, and sediment delivery has decreased, as the result of increased impervious surfaces throughout the Las Vegas Valley. As a result, the channel experienced incision up to 10.7 m (35 feet) and widening up to 150 m (500 feet), which required over $100 million in engineering interventions to protect infrastructure (Bahner et al. 2006).

In contrast, watershed disturbances such as overgrazing and clear-cutting can increase sediment delivery. If the receiving system lacks the capacity to transport the increased sediment load, the system will experience channel aggradation (Field and Lichvar 2007). Channel aggradation and associated waterlogging occurred in the middle Rio Grande in the early twentieth century, serving as motivation for the Middle Rio Grande Project (Phillips et al. 2011).

A common consequence of flow alterations is loss of floodplain connectivity (Stone et al. 2017). Water exchange between the river and floodplains during overbank flooding has numerous ecological and geomorphic benefits, including shallow groundwater recharge, transport of sediment, nutrient cycling, riparian vegetation recruitment, and provision of temporary aquatic habitat. Studies on the Gila River and the Rio Grande at fine spatial scales have shown decreases in the volume of water entering the floodplain and changes in the timing of flood connectivity (Byrne 2017). Consistent with the flood-pulse concept, the interrupted connectivity has decreased native riparian vegetation recruitment and reduced the availability of spawning and rearing habitat for the endangered Rio Grande silvery minnow.

In addition to surface-water alterations, groundwater pumping can also lead to significant changes in river discharge. Because groundwater and surface-water systems are hydrologically connected, large groundwater withdrawals over a wide area can result in decreased discharge or complete dewatering of rivers or springbrooks (Morrison et al. 2013). The impact of groundwater extraction on arid-land aquatic ecology is described more fully by Garrett et al. (chap. 8, this volume).

Ecological Implications

Variability in streamflow is a natural characteristic of desert rivers, where seasonal and event-based discharge varies widely over space and time. Native species have evolved with these conditions, and efforts by humans to reduce natural variability tend to favor nonnative over native species (e.g., in the Rio Chama, as described above).

A reduction in streamflows across the hydrograph is the most prevalent modification of southwestern rivers. The most obvious ecological impact of reduced streamflows is the direct loss of physical habitat. These effects can include net losses of wetted area or volume along with shifts in habitat suitability. Low or zero flows force organisms to find refuges and prevent free movement throughout the system. They can cause the isolation of certain populations of a species and slow or prevent the recolonization of a river reach following a disturbance. In severe cases, complete desiccation of the river bed can occur, as is regularly observed in portions of the Rio Grande, the lower Colorado River in México, the lower Gila River, and countless other low-order streams and springbrooks.

Reduced streamflow can also degrade water quality, resulting in elevated water temperatures and reduced dissolved oxygen. A reduction in water volume and mass allows water to be heated more quickly by solar radiation and heat convection from the atmosphere. Elevated temperatures can directly affect aquatic organisms, including fish and benthic macroinvertebrates. Low streamflows have led to dramatic drops in dissolved oxygen and fish kills in many situations (e.g., Small et al. 2014). Saturated dissolved oxygen concentrations drop from 11.3 mg/L to 9.1 mg/L as water temperatures rise from 10°C to 20°C.

The reduction of peak flows during the spring season (also referred to as pulse flows) downstream of dams affects many ecological processes, including channel-floodplain interactions and riparian conditions (e.g., Junk et al. 1989). Pulse flow events over a range of magnitudes allow for inundation of floodplain, channel bar, and island surfaces, which in turn allows for the exchange of materials, including nutrients and organic materials (dissolved and particulate). Figure 16.4 shows the change in inundation frequency projected for the Gila River in association with a proposed diversion project and for the Rio Grande resulting from the operation of Cochiti Dam (Stone et al. 2017). Although the diversion proposed for the Gila River is relatively modest, its floodplain is still expected to inundate up to 25% less in areas that are currently important for riparian vegetation recruitment. For the Rio Grande, the combination of reduced peak flows and channel incision associated with Cochiti Dam has reduced floodplain inundation by as much as 100% downstream from the dam (Stone et al. 2017).

Pulse flows play an important role in shaping riparian conditions, including vegetation recruitment and succession processes. The box recruitment model (Mahoney and Rood 1998) provides a conceptual description of the disturbance, establishment, and survival of willow and cottonwood seedlings driven by pulse flows. Large pulse flows have been virtually eliminated from all major rivers in the Southwest as a result of dam operations. An example of the complex inundation patterns associated with overbank flows for an undammed reach of the Gila River is shown in figure 16.5. The loss of such floodplain inundation has resulted in a drastic reduction in recruitment of native willow and cottonwood species for many river systems (Shafroth et al. 2002). In many cases, the resulting physical environment is better suited for nonnative plant species that do not rely upon overbanking disturbance and a shallow groundwater table for recruitment and survival. The salt cedar and Russian olive that now line the banks of many southwestern rivers have had a strong impact on geomorphic processes as described above, including the narrowing of river channels and stabilization of channel islands.

Pulse flows also serve as a seasonal indicator for ecological processes such as a spawning cue for fishes

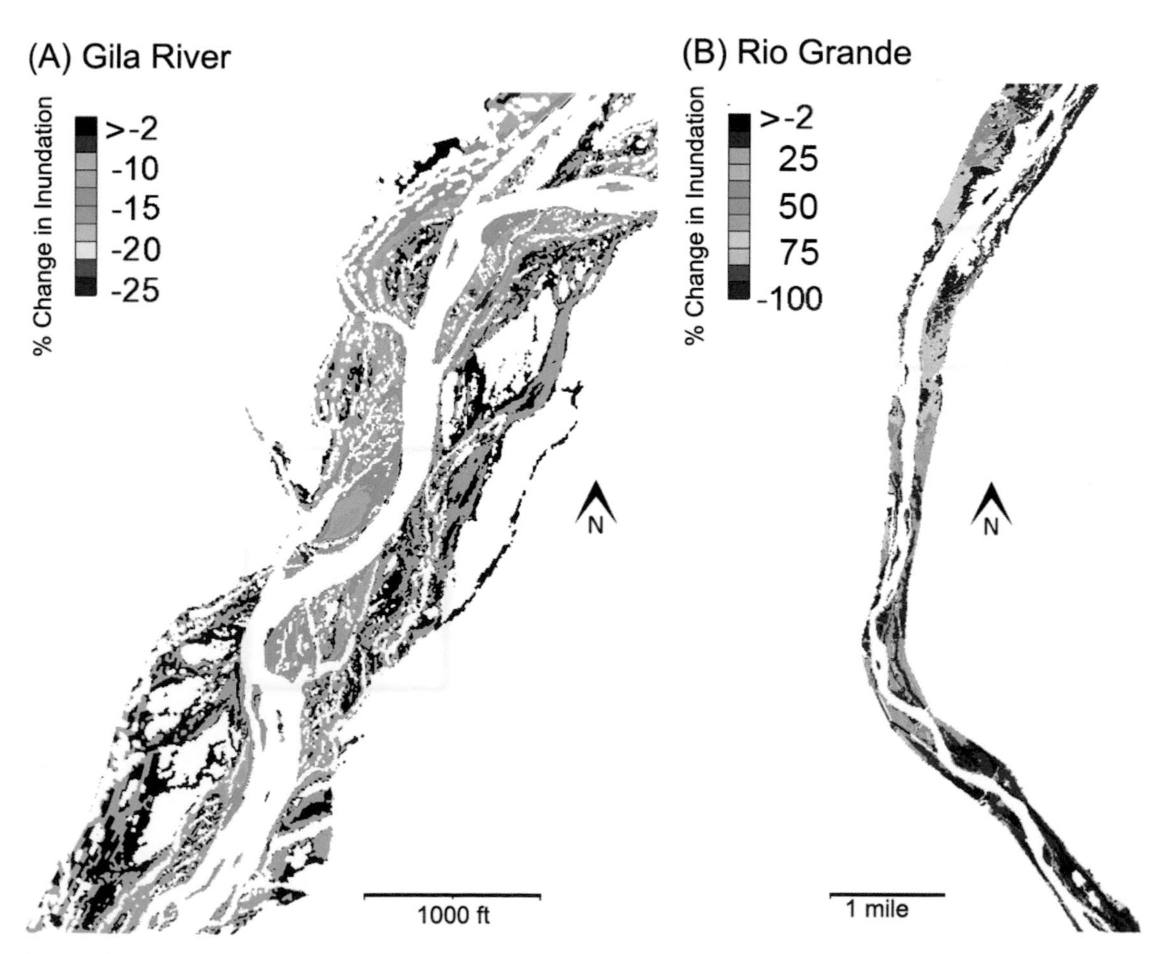

Fig. 16.4 Change in floodplain inundation frequency for (*A*) the Gila River associated with a proposed flow diversion and (*B*) the Rio Grande as a result of Cochiti Dam operations.

(Bunn and Arthington 2002), including Rio Grande silvery minnow (Dudley and Platania 2013) and Colorado pikeminnow (Nesler et al. 1988). Increased streamflow in the spring is an indicator of the impending snowmelt season, which brings steady discharge and increased interactions with floodplain features that increase the availability of allochthonous energy and nutrient inputs.

CONSERVATION AND RESTORATION

Most river restoration efforts are aimed at recovering specific species or groups of species and are motivated by the Endangered Species Act. These efforts include physical restoration techniques (e.g., channel realignments, planting of vegetation, and removal of nonnative species) and implementation of environmental flows. Here, we briefly summarize two projects employing physical stream restoration (on the lower Colorado River and the middle Rio Grande) and two projects emphasizing environmental flows (on the Rio Chama and the Bill Williams River).

The Lower Colorado River Multi-species Conservation Program aims to improve conditions for several threatened, endangered, and sensitive fish and wildlife species, including razorback sucker, humpback chub *Gila cypha*, and flannelmouth sucker *Catostomus latipinnis*. The program aims to benefit at least 26 fish and wildlife species over a period of 50 years. The project, which includes participation from the USBR and state and local entities, sets benchmarks for habitat creation and species recovery, including 3,280 ha (8,100 acres) of habitat for fish and wildlife and production of 1.2 million native fish. Major constraints on the project include commitments to ensure existing water delivery and power operations, which effectively prevents considerations of the natural flow regime, the loss of which is arguably the largest stressor on the Lower Colorado

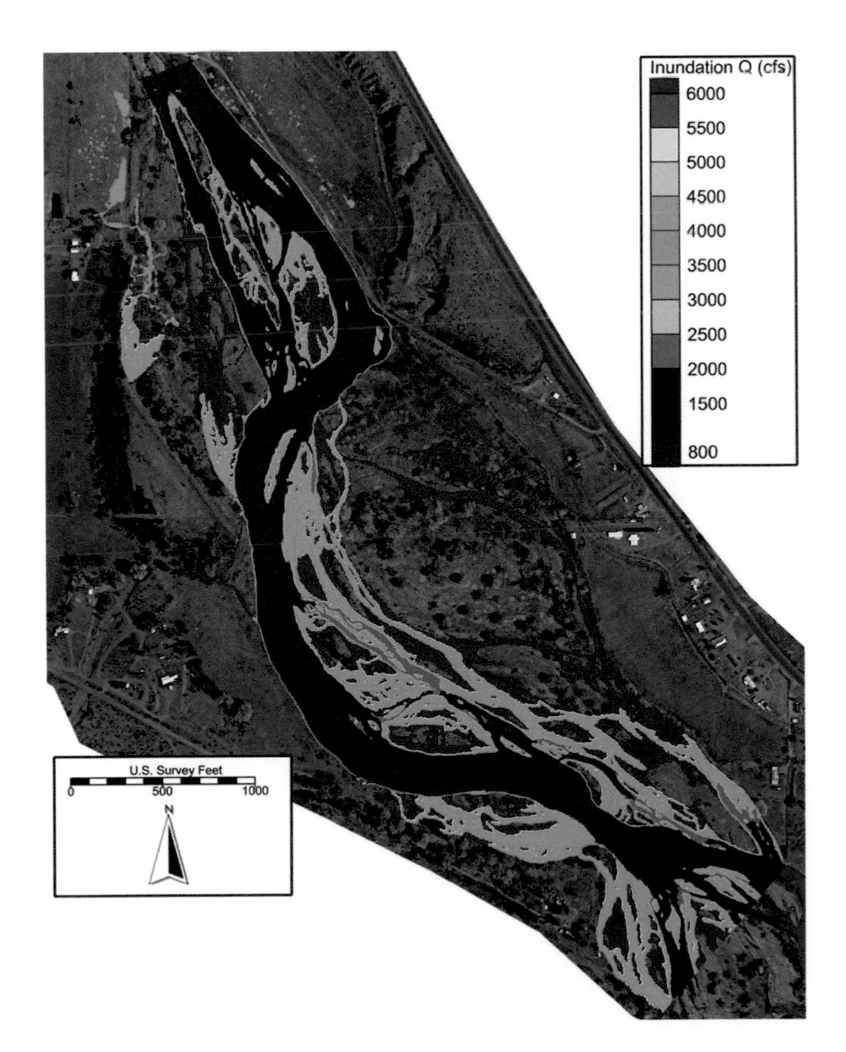

Fig. 16.5 Inundation thresholds for an undammed reach of the Gila River near Cliff, New Mexico. The contours represent the discharge required to inundate various floodplain features and drive riparian recruitment and other geomorphic and ecological processes.

River ecosystem (Marsh et al. 2015). Although the project has succeeded in meeting many of the target habitat benchmarks, recovery of native species, such as the razorback sucker, remains a challenge (Bestgen et al., chap. 21, this volume).

The Middle Rio Grande Endangered Species Collaborative Program is a partnership of over 20 federal, state, and local agencies with the goal of protecting and improving the status of endangered species in the Middle Rio Grande (Cochiti Dam to Elephant Butte Reservoir), such as Rio Grande silvery minnow and southwestern willow flycatcher *Empidonax traillii extimus*. The program involves physical habitat restoration, fish passages, monitoring, fish propagation, and water management. Active restoration efforts have included the construction of side channels and lowered floodplains, clearing of nonnative riparian species, and planting of native species. However, in spite of major investments in habitat restoration projects, the recovery status of the minnow has not improved, and the population has been steady or declining in recent years (USFWS 2016). This status can be attributed to the high degree of correlation between Rio Grande silvery minnow populations and hydrologic conditions, particularly the magnitude, timing, and duration of the spring runoff (Dudley et al. 2016), and the inherent challenges in improving the flow regime to meet species needs in an extremely constrained system.

"Environmental flows" refers to controlled discharges that provide the quantity, magnitude, or timing of water necessary to sustain riverine ecological processes. Numerous guidelines have been presented to assist restoration practitioners select appropriate environmental flow criteria (e.g., Konrad et al. 2011), including the "sustainable boundaries approach" (Richter 2010) and the "designer flows approach" (Acreman et al. 2014). Many rivers in arid and semiarid environments are over-allocated, highly stressed, and significantly altered, and these conditions create unique challenges for implementing environmental flows. In addition, approaches for selecting appropriate environmental flows for intermittent streams have been poorly studied, even though intermittent streams are important ecological systems in the Southwest (Colvin et al. 2019). In spite of these challenges, the incorporation of environmental flow criteria into dam operations has shown promise, as illustrated here by programs on the Rio Chama and Bill Williams River.

The Rio Chama Flows Project is a grassroots effort to investigate alternative water operations for dams on the Rio Chama in New Mexico to improve river health while fulfilling institutional water storage and delivery requirements (Morrison and Stone 2015a). The project team, which includes nonprofit organizations, federal and local government agencies, academics, and environmental consultants, has investigated ecological conditions and operational constraints and developed models to consider alternative water management scenarios (Morrison and Stone 2015a, b). In collaboration with stakeholders, environmental flow guidance has been developed, and three experimental spring-pulse flows were conducted in 2014, 2016, and 2017. Monitoring of the 2014 environmental flow release showed that brown trout spawning habitat was improved in sections of the Rio Chama through the removal of fine sediment from gravel bars (Gregory et al. 2018).

Alamo Dam, on the Bill Williams River in Arizona, is one of 36 dams that are part of the Sustainable Rivers Project (SRP), a collaboration between the USACE and The Nature Conservancy with the aim of managing rivers and dams for the protection and restoration of river health (Shafroth and Beauchamp 2006; Shafroth et al. 2010). Multiple environmental flow releases from Alamo Dam under the SRP have provided opportunities to improve downstream ecological and geomorphic conditions in the river and to study links between vegetation and hydrogeomorphic conditions (Wilcox and Shafroth 2013). Recent work has shown that these environmental flow releases have promoted the growth of native cottonwood and willow while eliminating young nonnative salt cedar populations (Wilcox and Shafroth 2013). Further, Pool and Olden (2015) demonstrated the importance of the natural flow regime in shaping the balance of native and nonnative species at the basin scale. Unfortunately, much of the lower river riparian area has been destroyed as of spring 2018, presumably due to reduced flows and groundwater levels that fell below the root zones of most trees (K. Bestgen, pers. comm.).

CONCLUSION

River engineering, floodplain development, dams and diversions, watershed modifications, and other human activities have dramatically altered stream corridors and their surroundings in the American Southwest (Graf et al. 2002; Schmidt 2010). River systems have been stressed to an extreme degree to support human needs. The use of these systems has enabled the region to experience remarkable population and economic growth over the past 100 years; however, this development has come at a severe price for the aquatic ecosystems on which these communities rely, as it has contributed to population declines, endangerment, extirpation, and in several cases, extinctions of native species (Olden and Poff 2005).

This chapter summarized the history and implications of river modifications in the American Southwest along with efforts to reverse these negative trends using direct and indirect restoration approaches. The examples described are indicative of the challenges we face as a society when attempting to conserve or recover aquatic species in these dramatically altered systems. The over-allocated nature of limited water resources further complicates efforts to restore these systems. Although the effectiveness of ongoing restoration efforts is difficult to quantify at large scales, there is an emerging understanding that an increased focus on hydrologic, geomorphic, and ecological processes is needed. By advancing research and adaptive management in ways that improve our understanding of the connections between underlying processes and ecosystem conditions, we can improve opportunities for restoration success. Such advances will become even more critical in the presence of continued population growth and the reduced water supply associated with drought and climate change (Udall and Overpeck 2017; Udall, chap. 7, this volume).

REFERENCES

Acreman, M., A. H. Arthington, M. J. Colloff, C. Couch, N. D. Crossman, F. Dyer, I. Overton, C. A. Pollino, M. J. Stewardson, and W. Young. 2014. Environmental flows for natural, hybrid, and novel riverine ecosystems in a changing world. *Frontiers in Ecology and the Environment* 12: 466–73.

Bahner, C., G. A. Hester, and L. Berger. 2006. Restoration of the Lower Las Vegas Wash. *Proceedings of the Eighth Federal Interagency Sedimentation Conference*, Reno, NV.

Bracken, L. J., L. Turnbull, J. Wainwright, and P. Bogaart. 2015. Sediment connectivity: A framework for understanding sediment transfer at multiple scales. *Earth Surface Processes and Landforms* 40: 177–88.

Bunn, S. E., and A. H. Arthington. 2002. Basic principles and ecological consequences of altered flow regimes for aquatic biodiversity. *Environmental Management* 30: 492–507.

Byrne, C. F. 2017. Evaluating dynamic flood wave processes as indicators for process-based river science. PhD diss., University of New Mexico.

Chorley, R. J., and B. A. Kennedy. 1971. *Physical Geography: A Systems Approach*. Upper Saddle River, NJ: Prentice Hall.

Collier, M., R. H. Webb, and J. C. Schmidt. 2000. *Dams and Rivers: A Primer on the Downstream Effects of Dams*. US Geological Survey Circular 1126.

Colvin, S. A. R., S. M. P. Sullivan, P. D. Shirey, R. W. Colvin, K. O. Winemiller, R. M. Hughes, K. D. Fausch, D. M. Infante, J. D. Olden, K. R. Bestgen, R. J. Danehy, and L. Eby. 2019. Headwater streams and wetlands are critical for sustaining fish, fisheries, and ecosystem services. *Fisheries* 44: 73–91.

Cowley, D. E. 2006. Strategies for ecological restoration of the Middle Rio Grande in New Mexico and recovery of the endangered Rio Grande Silvery Minnow. *Reviews in Fisheries Science* 14: 169–86.

Dudley, R. K., and S. P. Platania. 2013. *Spatial Spawning Periodicity of Rio Grande Silvery Minnow during 2013*. Albuquerque: US Fish and Wildlife Service, Middle Rio Grande Endangered Species Collaborative Program.

Dudley, R. K., S. P. Platania, and G. C. White. 2016. *Rio Grande Silvery Minnow Population Monitoring Program Results from February to December 2015*. Albuquerque: US Fish and Wildlife Service, Middle Rio Grande Endangered Species Collaborative Program.

ESRI. 2016. World imagery basemap. *ESRI, DeLorme, NAVTEQ, TomTom, Intermap, Increment P Corp, GEBCO, USGS, FAO, NPS, NRCAN, GoBase, Kadaster NL, Ordinance Survey, ESRI Japan, METI, and the GIS User Community*. Accessed via ArcGIS, December 1, 2016.

Fausch, K. D., Y. Taniguchi, S. Nakano, G. D. Grossman, and C. R. Townsend. 2001. Flood disturbance regimes influence rainbow trout invasion success among five holarctic regions. *Ecological Applications* 11: 1438–55.

Field, J. J., and R. W. Lichvar. 2007. *Review and Synopsis of Natural and Human Controls on Fluvial Channel Processes in the Arid West*. Hanover, NH: US Army Corps of Engineers Engineering Research and Development Center.

Flanigan, K. G., and A. I. Haas. 2008. The impact of full beneficial use of San Juan–Chama Project water by the City of Albuquerque on New Mexico's Rio Grande Compact obligations. *Natural Resources Journal* 371–405.

Gido, K. B., D. L. Propst, J. D. Olden, and K. R. Bestgen. 2013. Multidecadal responses of native and introduced fishes to natural and altered flow regimes in the American Southwest. *Canadian Journal of Fisheries and Aquatic Sciences* 70: 554–64.

Gori, D., M. S. Cooper, E. S. Soles, M. C. Stone, R. R. Morrison, T. F. Turner, D. L. Propst, G. Grafin, and K. Kindscher. 2013. *Gila River Flow Needs Assessment*. Santa Fe: The Nature Conservancy. https://kuscholarworks.ku.edu/handle/1808/21029. Accessed December 8, 2017.

Graf, W. L. 1999. Dam nation: A geographic census of American dams and their large-scale hydrologic impacts. *Water Resources Research* 35: 1305–11.

Graf, W. L., J. Stromberg, and B. Valentine. 2002. Rivers, dams, and willow flycatchers: A summary of their science and policy connections. *Geomorphology* 47: 169–88.

Gregory, A., R. R. Morrison, and M. C. Stone. 2018. Assessing the hydrogeomorphic effects of environmental flows using hydrodynamic modeling. *Environmental Management* 62: 352–64.

Gresswell, R. E. 1999. Fire and aquatic ecosystems in forested biomes of North America. *Transactions of the American Fisheries Society* 128: 193–221.

Harvey, A. M. 2002. Effective timescales of coupling within fluvial systems. *Geomorphology* 44: 175–201.

Iorns, W. V., C. H. Hembree, and G. L. Oakland. 1965. *Water Resources of the Upper Colorado River Basin*. US Geological Survey Professional Paper 441. Washington, DC: US Government Printing Office.

Jelks, H. L., S. J. Walsh, N. M. Burkhead, S. Contreras-Balderas, E. Diaz-Pardo, D. A. Hendrickson, J. Lyons, N. E. Mandrak, F. McCormick, J. S. Nelson, S. P. Platania, B. A. Porter, C. B. Renaud, J. J. Schmitter-Soto, E. B. Taylor, and M. L. Warren, Jr. 2008. Conservation status of imperiled North American freshwater and diadromous fishes. *Fisheries* 33: 372–407.

Junk, W., P. Bayley, and R. Sparks. 1989. The flood pulse concept in river-floodplain systems. In D. P. Dodge, ed., *Proceedings of the International Large River Symposium (LARS)*, 110–27. Canadian Special Publication of Fisheries and Aquatic Sciences 106.

Kenny, D. S., S. T. McAllister, W. H. Caile, and J. S. Peckham. 2000. *The New Watershed Source Book: A Directory and Review of Watershed Initiatives in the Western United States*. Boulder: Natural Resources Law Center, University of Colorado School of Law.

Konrad, C. P., J. D. Olden, D. A. Lytle, T. S. Melis, J. C. Schmidt, E. N. Bray, M. C. Freeman, K. B. Gido, N. P. Hemphill, M. J. Kennard, L. E. McMullen, M. C. Mims, M. Pyron, C. T. Robinson, and J. G. Williams. 2011. Large-scale flow experiments for managing river systems. *BioScience* 61: 948–59.

Lane, E. W. 1955. *Importance of Fluvial Morphology in Hydraulic Engineering*. Proceedings of the American Society of Civil Engineers 81, paper no. 745.

Mahoney, J. M., and S. B. Rood. 1998. Streamflow requirements for cottonwood seedling recruitment—An integrative model. *Wetlands* 18: 634–45.

Marsh, P. C., T. E. Dowling, B. R. Kesner, T. F. Turner, and W. L. Minckley. 2015. Conservation to stem imminent extinction: The fight to save razorback sucker *Xyrauchen texanus* in Lake Mohave and its implications for species recovery. *Copeia* 2015: 141–56.

Moody, J. A., R. A. Shakesby, P. R. Robichaud, S. H. Cannon, and D. A. Martin. 2013. Current research issues related to post-wildfire runoff and erosion processes. *Earth-Science Reviews* 122: 10–37.

Morrison, R. R., and M. C. Stone. 2014. Spatially implemented Bayesian network model to assess environmental impacts of water management. *Water Resources Research* 50: 8107–24.

Morrison, R. R., and M. C. Stone. 2015a. Evaluating the impacts of environmental flow alternatives on reservoir and recreational operations using system dynamics modeling. *Journal of the American Water Resources Association* 51 (1): 33–46.

Morrison, R. R., and M. C. Stone. 2015b. Investigating environmental flows for riparian vegetation recruitment using system dynamics modelling. *River Research and Applications* 31: 485–96.

Morrison, R. R., M. C. Stone, and D. Sada. 2013. Response of a desert springbrook to incremental discharge reductions, with tipping points of non-linear environmental change. *Journal of Arid Environments* 99: 5–13.

Nesler, T. P., R. T. Muth, and A. F. Wasowicz. 1988. Evidence for baseline flow spike as spawning cues for Colorado squawfish in the Yampa River, Colorado. *American Fisheries Society Symposium* 5: 68–79.

Norman, L. M., M. Gishey, L. Gass, B. Yanites, E. Pfeifer, R. Simms, and R. Ahlbrandt. 2006. *Processes Aerial Photography for Selected Areas of the Lower Colorado River, Southwestern United States*. US Geological Survey Open-File Report 2006-1141.

O'Connor, J. E., J. J. Duda, and G. E. Grant. 2015. 1000 dams down and counting. *Science* 348: 496–97.

Olden, J. D., and N. L. Poff. 2005. Long-term trends in native and non-native fish faunas in the American Southwest. *Animal Biodiversity and Conservation* 28: 75–89.

Olden, J. D., N. L. Poff, and K. R. Bestgen. 2006. Life-history strategies predict invasions and extirpations in the Colorado River basin. *Ecological Monographs* 76: 25–40.

Parker, G., C. Paola, and S. Leclair. 2000. Probabilistic Exner sediment continuity equation for mixtures with no active layer. *Journal of Hydraulic Engineering* 126: 818–26.

Pease, A. A., J. J. Davis, M. S. Edwards, and T. F. Turner. 2006. Habitat and resource use by larval and juvenile fishes in an arid-land river (Rio Grande, New Mexico). *Freshwater Biology* 51: 475–86.

Phillips, F. M., G. E. Hall, and M. E. Black. 2011. *Reining In the Rio Grande: People, Land, and Water*. Albuquerque: University of New Mexico Press.

Poeppl, R. E., S. D. Keesstra, and J. A. Maroulis. 2017. Conceptual connectivity framework for understanding geomorphic change in human-impacted fluvial systems. *Geomorphology* 277: 237–50.

Poff, N. L. 2009. Managing for variation to sustain freshwater ecosystems. *Journal of Water Resources Planning and Management* 135: 1–4.

Poff, N. L., J. D. Allan, M. B. Bain, J. R. Karr, K. L. Prestegaard, B. D. Richter, R. E. Sparks, and J. C. Stromberg. 1997. The natural flow regime. *BioScience* 47: 769–84.

Pool, T. K., and J. D. Olden. 2015. Assessing long-term fish responses and short-term solutions to flow regulation in a dryland river basin. *Ecology of Freshwater Fish* 24: 56–66.

Pringle, C. M. 2001. Hydrologic connectivity and the management of biological reserves: A global perspective. *Ecological Applications* 11: 981–98.

Propst, D. L., and K. R. Bestgen. 1991. Biology and habitat of the loach minnow, *Tiaroga cobitis*, in the Gila River drainage, New Mexico. *Copeia* 1991: 29–38.

Propst, D. L., and K. B. Gido. 2004. Responses of native and nonnative fishes to natural flow regime mimicry in the San Juan River. *Transactions of the American Fisheries Society* 133: 922–31.

Reisner, M. 1993. *Cadillac Desert: The American West and Its Disappearing Water*. Rev. ed. New York: Penguin Books.

RGIS (Resource Geographic Information System). 2017. Online data portal. Earth Data Analysis Center, University of New Mexico. https://rgis-data. Accessed December 8, 2017.

Richard, G. A., P. Y. Julien, and D. C. Baird. 2005. Statistical analysis of lateral migration of the Rio Grande, New Mexico. *Geomorphology* 71: 139–55.

Richter, B. D. 2010. Re-thinking environmental flows: From allocations and reserves to sustainability boundaries. *River Research and Applications* 26: 1052–63.

Sabo, J. L., K. R. Bestgen, W. Graf, T. Sinha, and E. E. Wohl. 2012. Dams in the Cadillac Desert: Downstream effects in a geomorphic context. *Annals of the New York Academy of Science* 1249: 227–46.

Schmidt, J. C. 2010. A watershed perspective of changes in streamflow, sediment supply, and geomorphology of the Colorado River. In T. S., Melis, J. F. Hamill, L. G. Coggins, Jr., G. E. Bennett, P. E. Grams, T. A. Kennedy, D. M. Kubly, and B. E. Ralston, eds., *Proceedings of the Colorado River Basin Science and Resource Management Symposium*, November 18–20, 2008, Scottsdale, AZ. US Geological Survey Scientific Investigations Report 2010-5135.

Schumm, S. A. 1977. *The Fluvial System*. New York: Wiley.

Scurlock, D. 1998. *From the Rio to the Sierra: An Environmental History of the Middle Rio Grande Basin*. USDA Forest Service General Technical Report RMRS-GTR-5. Fort Collins, CO: USDA Forest Service, Rocky Mountain Research Station.

Shafroth, P. 2010. *Saltcedar* (Tamarix *spp.*) *and Russian olive* (Elaeagnus angustifolia) *in the Western United States—A Report on the State of the Science*. US Geological Survey Fact Sheet 2009-3110.

Shafroth, P. B., and V. B. Beauchamp. 2006. Defining ecosystem flow requirements for the Bill Williams River, Arizona. *U.S. Geological Survey Open-File Report* 2006–1314, 135 p.

Shafroth, P. B., J. C. Stromberg, and D. T. Patten. 2002. Riparian vegetation response to altered disturbance and stress regimes. *Ecological Applications*12: 107–23.

Shafroth, P. B., A. C. Wilcox, D. A. Lytle, J. T. Hickey, D. C. Andersen, V. B. Beauchamp, A. Hautzinger, L. E. McMullen, and A. Warner. 2010. Ecosystem effects of environmental flows: Modelling and experimental floods in a dryland river. *Freshwater Biology* 55: 68–85.

Simon, A. 1989. A model of channel response in disturbed alluvial channels. *Earth Surface Processes and Landforms* 14: 11–26.

Simon, A., and M. Rinaldi. 2006. Disturbance, stream incision, and channel evolution: The roles of excess transport capacity and boundary materials in controlling channel response. *Geomorphology* 79: 361–83.

Small, K., R. K. Kopf, R. J. Watts, and J. Howitt. 2014. Hypoxia, blackwater and fish kills: Experimental lethal oxygen thresholds in juvenile predatory lowland river fishes. *PLoS ONE* 9 (4): e94524.

Stokes, C. 2007. *Managing Water Resources in New Mexico: Climate Trends and Cropping Patterns in the Lower Rio Grande*. Water Resources Professional Project Reports. http://digitalrepository.unm.edu/wr_sp/69. Accessed December 8, 2017.

Stone, M. C., C. F. Byrne, and R. R. Morrison. 2017. Evaluating the impacts of hydrologic and geomorphic alterations on floodplain connectivity. *Ecohydrology* 10: e1833.

Swanson, B. J., G. A. Meyer, and J. E. Coonrod. 2011. Historical channel narrowing along the Rio Grande near Albuquerque, New Mexico in response to peak discharge reductions and engineering: Magnitude and uncertainty of change from air photo measurements. *Earth Surface Processes and Landforms* 36: 885–900.

Udall, B., and J. Overpeck. 2017. The twenty-first century Colorado River hot drought and implications for the future. *Water Resources Research* 53: 2404–18.

USACE (US Army Corps of Engineers). 2016. *The National Inventory of Dams*. http://nid.usace.army.mil/cm_apex-/f?p=838:12. Accessed December 12, 2017.

US Census Bureau. 2014. *Profile of General Population and Housing Characteristics: 2010—2010 Demographic Profile Data*. http://census.gov. Accessed December 12, 2017.

USFWS (US Fish and Wildlife Service). 2016. *Final Biological and Conference Opinion for Bureau of Reclamation, Bureau of Indian Affairs, and Non-Federal Water Management and Maintenance Activities on the Middle Rio Grande, New Mexico*. https://www.fws.gov/southwest/es/NewMexico/BO_MRG.cfm. Accessed December 20, 2017.

Westerling, A. L., H. G. Hidalgo, D. R. Cayan, and T. W. Swetnam. 2006. Warming and earlier spring increase western US forest wildfire activity. *Science* 313: 940–43.

Wichelns, D., and J. D. Oster. 2006. Sustainable irrigation is necessary and achievable, but direct costs and environmental impacts can be substantial. *Agricultural Water Management* 86: 114–27.

Wilcox, A. C., and P. B. Shafroth. 2013. Coupled hydrogeomorphic and woody-seedling responses to controlled flood releases in a dryland river. *Water Resources Research* 49: 2843–60.

Wohl, E. 2017. Connectivity in rivers. *Progress in Physical Geography* 41(3): 345–62.

Yarnell, S. M., G. E. Petts, J. C. Schmidt, A. A. Whipple, E. E. Beller, C. N. Dahm, P. Goodwin, and J. H. Viers. 2015. Functional flows in modified riverscapes: Hydrographs, habitats and opportunities. *BioScience* 65: 963–72.

17

Donald W. Sada
and Lawrence E. Stevens

Conservation and Ecological Rehabilitation of North American Desert Spring Ecosystems

Springs are functionally zero-order, surface-linked aquatic and terrestrial ecosystems, occurring where groundwater reaches, and usually flows from, the Earth's surface, in subaqueous as well as subaerial settings (Meinzer 1923; Springer et al. 2008). Aridity, the impacts of burgeoning human populations on water resources, and the dire conservation status of a plethora of rare and endemic aquatic, wetland, and stream-riparian organisms all challenge the stewardship of desert spring ecosystems throughout the world. Interest in their conservation in southwestern North America was initiated by concern for fishes (Miller 1961; Minckley and Deacon 1968; Williams 1991), but more recently has expanded to include other rare, endemic, and sometimes federally listed wetland and riparian biota, including aquatic macroinvertebrate (AMI) and vertebrate crenobiontic (spring-dependent) species (SDS). In addition, spring ecosystems are increasingly recognized for their internal complexity and individuality, their roles as highly valued habitat patches within landscapes, and their biological, sociocultural, and economic importance (Stevens and Meretsky 2008; Hershler et al. 2014). Ecosystem rehabilitation is a widely appreciated and implemented form of ecological reconciliation, but rehabilitation of springs, the subject of this chapter, has received relatively little scientific attention.

In relation to their area, springs are arguably the most productive and bioculturally diverse ecosystems on Earth. However, springs are globally threatened by groundwater depletion and pollution, flow regulation, overgrazing, and other underinformed stewardship practices, including the introduction of nonnative species. Rehabilitation of spring ecosystem geomorphology, habitat structure, and SDS are often readily accomplished if the aquifer is intact and degrading disturbances are remedied. For example, Davis et al. (2011) reported that most spring rehabilitation projects were successful; however, the majority of such projects they reviewed were focused on the conservation of a single taxon, rather than on improving overall ecosystem integrity, and monitoring and feedback were typically insufficient to demonstrate long-term success. Rehabilitation actions are often as simple as improving fencing, constructing trails to reduce erosion, and reconsidering livestock watering practices, the costs of which are trivial but can result in substantial conservation success. Thus, spring ecosystems can often be sustainably managed for native species and ecological processes, and with compatible levels of human uses (Nabhan 2008).

This chapter provides guidance to those involved in rehabilitation of the common spring types in the arid Southwest. We present (1) the basic aspects of crenecology; (2) a description of successional responses of aquatic and riparian assemblages to rehabilitation; (3) stewardship guidelines for restoration of different types of springs; (4) the information available from reference sites; and (5) information gaps and rehabilitation lessons learned. Such information can help improve scientific understanding and stewardship of these remarkable, but often overlooked and underappreciated, ecosystems.

DESERT SPRING DISTRIBUTION AND ECOHYDROLOGY

Geography

In the American Southwest, springs occur at elevations ranging from 100 m below to 4,000 m above sea level and are particularly numerous in regions with high topographic relief (Stevens and Meretsky 2008; fig. 17.1). Nearly 50,000 springs of all types have been reported in the desert Southwest (Springer and Stevens 2009; SSI 2016), of which nearly two-thirds oc-

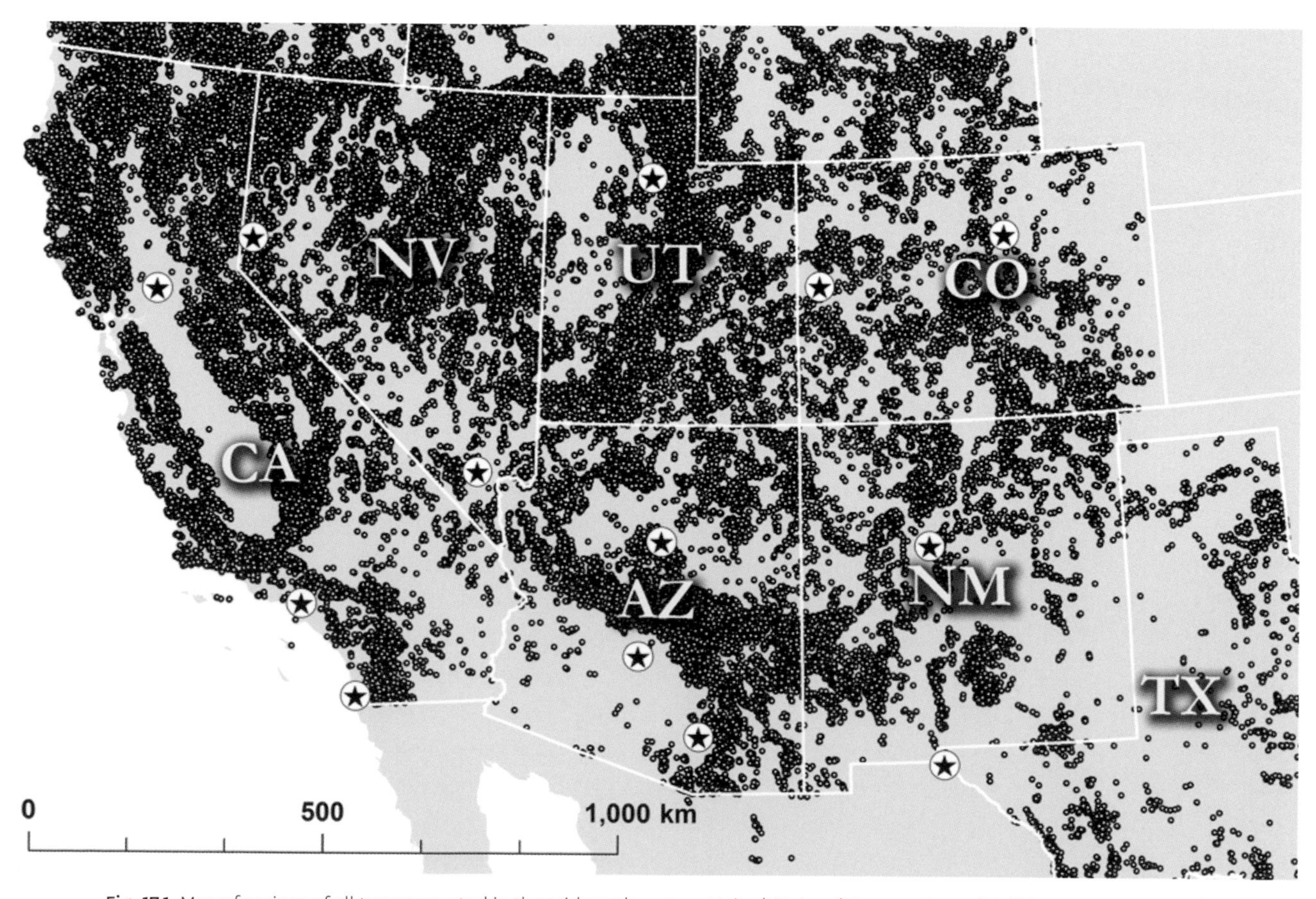

Fig. 17.1 Map of springs of all types reported in the arid southwestern United States. (Map courtesy of Jeff Jenness, Museum of Northern Arizona Springs Stewardship Institute, Flagstaff.)

cur in Nevada and Arizona. Springs often provide the only reliable source of water in large areas, and they support rural economies by providing water for drinking, pastures, crops, livestock, and recreation (Mueller et al. 2017). Additionally, they provide the base flow of many, if not most, streams and rivers in the nation (Arnold et al. 1993). However, springs are generally poorly mapped. For example, Junghans et al. (2016) used repeated spring inventories over time to develop an accumulation curve–based estimate of total spring density in Death Valley National Park and concluded that additional unreported springs existed in that landscape.

Geomorphology

Overview

Springs vary enormously in physical form, environmental setting, biotic assemblage diversity and structure, and sociocultural significance, and multiple spring types can emerge from an individual aquifer. Springs have been variously classified on the basis of aquifer-landscape setting, geochemistry or water temperature, vegetation (Bryan 1919; Springer et al. 2008), and spheres of discharge (geomorphic types; Meinzer 1923; Springer and Stevens 2009), of which the last provides the most resolution with regard to stewardship and rehabilitation. The ecological functions of springs vary among geologic provinces, geomorphic settings, aquifer characteristics, spheres of discharge, microhabitat array, and the extent of natural and anthropogenic disturbance (Keleher and Sada 2012; fig. 17.2). In this section, we describe springs emerging from aquifers in mountain, bajada, valley floor, and plateau canyon landscape settings, and indicate which support Springer and Stevens's (2009) 12 spheres of discharge. We note that hillslope, rheocrene, helocrene, and on the Colorado Plateau, hanging gardens are the most common spring types in the Southwest (table 17.1), each of which poses unique rehabilitation challenges.

Aquifer–spring type distribution

Mountain, valley floor, and plateau canyon aquifers can support carbonate mound springs that emerge as mineral precipitate mound, hillslope, rheocrene, limnocrene, or hypocrene spring types, often with much

Fig. 17.2 Thunder River, a large gushet spring ecosystem in Grand Canyon National Park, Arizona. This site supports many different microhabitats and a high diversity of aquatic and wetland-riparian plants and invertebrate species, including several crenobiontic species. (L. E. Stevens photo.)

evidence of paleocrene deposits. Karstic terrain occupies 16% of the Earth's surface and provides about 25% of the world's drinking water. Many southwestern aquifers flow through carbonate rock, and some of those springs have deposited enormous quantities of travertine over time (e.g., Blue Springs, Fossil Springs, Havasu Springs, and Salado Springs in Arizona). Carbonate mound complexes often include hillslope, rheocrene, and limnocrene springs, along with much evidence of paleosprings. Elevated mineral content may restrict AMI diversity to geochemically stenotolerant taxa, and travertine deposition often armors springbrook channels, reducing hyporheic flow and restricting riparian groundwater availability. Due to the dominant impacts of travertine, carbonate mound spring rehabilitation requires planning for mineral precipitation along a gradient from the source to downstream reaches (Dreybrodt et al. 1992; Carter and Marks 2007).

Table 17.1. Classification and characteristics of spring ecosystems in the American Southwest.

Discharge sphere	Landscape hydrogeology	Flow path length	Probability of endemic biota occurrence	Maximum number of endemic SDS	Example	Comments	References
Carbonate	Valley floors to cliff faces	Moderate–long	High	High	Montezuma Well, Arizona	Travertine deposits	Blinn 2008; Sada and Cooper 2012
Cave	Karstic or volcanic terrains	Variable	High	Moderate	Roaring Springs, Grand Canyon National Park, AZ	Emergence in cave	Drost and Blinn 1997
Exposure	Bedrock-dominated valley floors	Long	High	Low	Devils Hole, Ash Meadows–Death Valley National Park, CA-NV	Low gradient, open exposure of groundwater; no outflow or springbrook	Wilson and Blinn 2007
Fountain	Valley floors, floodplains	Moderate	Moderate	None?	Vulcans Bidet, AZ	Artesian upwellings	SSI 2015
Geyser	Geothermal fields, usually flow-travertine-dominated	Variable	Moderate	Low	Crystal Geyser, UT	Geothermal or gas-driven	Allis et al. 2005
Geothermal	Valley floors, bajadas	Moderate	Moderate	Moderate	Railroad Valley, NV	Mostly rheocrenes and limnocrenes	Sada and Rosamond 2013
Gushet	Karst; cliff faces	Short	Moderate	Low	Thunder River, AZ	Waterfall springs	SSI 2015
Hanging garden	Small sandstone aquifers	Short	High	Low–moderate	Weeping Rock, Zion National Park, UT	Sandstone seeping springs	Welsh 1984; Malanson 1993
Helocrene	Valley floors, floodplain margins, meadows, playas	Variable	High	Moderate	Balmorhea State Park, Reeves County, TX	Marshy, wet meadow, ciénega (low elev) or GDE fen (high elev)	
Hillslope	Low- to high-gradient colluvial slopes or alluvial terraces	Variable	Low–moderate	Moderate	Deer Creek Spring, Grand Canyon National Park, AZ	No upslope channel or regular surface flooding	SSI 2015
Hypocrene	Low-gradient colluvial slopes	Variable	Variable	None?	Mile 71L Spring, Grand Canyon, AZ	Transitional type through dewatering of hillslope and other spring types	SSI 2015
Limnocrene	Valley floors	Long	High	High	King Spring, Ash Meadows, NV	Subaqueous emergence in a pool; rare on CP	
Paleocrene	Extinct springs of several types	None	Low	None	Salado Paleosprings, Salado, AZ	Indicated by paleo-peat deposits or travertine mounds	SSI 2015
Rheocrene		Variable	Low	None–Low	Parsons Spring, Prescott National Forest, AZ	Emergence in a stream channel; sometimes subaqueous	SSI 2015

Source: Modified from Springer and Stevens 2009.

Mountain and bajada aquifers can also produce low-mineral-content hillslope, rheocrene, gushet, helocrene, and hanging garden springs. These springs emerge either on mountain blocks or bajadas (figs. 17.3, 17.4, and 17.5). Mountain springs are often supported by small to medium-sized aquifers in a single surface watershed, while bajada springs are fed by larger recharge areas along mountain blocks. Both spring groups are characterized by cool to cold, stenothermic, low-conductance waters, reflecting relatively short flow paths. Many mountain springs have highly variable flow, and some dry frequently due to the small size and capacity of their aquifers, but most bajada springs are persistent. Fish are uncommon in these springs, but the contrast between their cool water and ambient desert heat can support complex, taxonomically rich aquatic, wetland, and riparian invertebrate assemblages. Mountain rheocrene riparian habitats are similar to higher-order lotic systems and are typically dominated by herbaceous and woody plant assemblages that reach maximum diversity at intermediate elevations and slope angles (Sinclair 2018; see fig. 17.4). Hanging gardens are a distinctive and abundant spring type on the Colorado Plateau, where they are fed by relatively small sandstone aquifers with impervious clay beds (Welsh 1989; fig.17.6).

Valley floor springs can include hillslope, helocrene, hypocrene, limnocrene, and rheocrene springs. Due to the extensive use of water for agriculture in these landscapes, undisturbed examples of valley floor springs are rare. Aquifers feeding these springs are often larger and fed by montane groundwater. Valley floor springs can be perennial and cool; their mineral content is low, but higher than that of mountain and bajada springs. In protected areas, valley floor springbrooks are often strongly dominated by *Salix exigua* or other willows, particularly in the northern Great Basin Desert, but woody vegetation is usually absent, and grasses and

Fig. 17.3 *(upper left)* Maple Grove Springs, Utah, an undisturbed mountain rheocrene spring occupied by the bifid duct pyrg *Pyrgulopsis peculiarius*. (D. W. Sada photo.)

Fig. 17.4 *(upper right)* Hannapah Spring, Inyo County, California, an undisturbed mountain/bajada spring occupied by the southeast Nevada pyrg *Pyrgulopsis turbatrix*. (D. W. Sada photo.)

Fig. 17.5 *(lower left)* Upper Royal Arch Alcove, a large hanging garden spring at Colorado River Mile 41 in Grand Canyon National Park, Arizona. (L. E. Stevens photo.)

facultative wetland vegetation dominate where soils are hypersaturated and/or alkaline. Sada and Thomas (unpublished data) found that valley floor spring invertebrate assemblages had fewer taxa than those of mountain and bajada springs, and that amphipods and ostracods often numerically dominated the communities. These invertebrate assemblages included more pollution-tolerant taxa than did those of mountain and bajada springs. Fish and crenobiontic invertebrates are common in these springs, and riparian vegetation is similar to that around bajada springs.

Fig. 17.6 Geothermal springs occupied by desert dace (*Eremichthys acros*) and two springsnails (*Pyrgulopsis noticola* and *P. limaria*) endemic to Soldier Meadow, Humboldt County, Nevada. (D. W. Sada photo.)

Fig. 17.7 Montezuma Well, a regional aquifer limnocrene carbonate mound spring ecosystem in Montezuma Castle National Monument in central Arizona, which has a 13,300-year flow path from recharge on the southern Colorado Plateau and is home to at least six endemic taxa. (L. E. Stevens photo.)

The water quality of geothermal springs is distinctive, but their geomorphology generally conforms to that of carbonate mound, rheocrene, limnocrene, and helocrene springs. Geothermal springs can be fed by local or regional aquifers (Crossey et al. 2009), but while bajada and valley floor springs are cool, the water emerging at geothermal springs has circulated deeply, where it is heated by magmatic forces and becomes laden with dissolved solids. These harsh environments support depauperate assemblages of extremophilic organisms, such as bacterial and archaeal assemblages (Costa et al. 2009), but as the water cools, downstream reaches can also support crenobiontic invertebrates and some fish. Woody vegetation is typically sparse or absent due to mineralized soils, but salt grass (*Distichlis* sp.), sedges, rushes, and cattails (*Typha* sp.) may proliferate (see fig. 17.6).

Regional aquifers support springs of most types across vast areas, and biologically, these aquifers support locally endemic crenobiontic invertebrates and fishes (Blinn 2008; EARIP 2012; Hershler and Liu 2017). Regional aquifers are typically rich in carbonate waters derived from long flow paths (some exceeding several hundred kilometers), which results in travertine deposition, and they are distinguished by nearly constant slightly alkaline pH, elevated conductivity, stenothermal water temperature (22°C–34°C), and nearly constant discharge. Examples of regional aquifer spring provinces include Ash Meadows in Nevada, Furnace Creek Springs in California, Montezuma Well in Arizona, the Edwards Aquifer in Texas, and Cuatro Ciénegas in Coahuila, México (Blinn 2008; Hendrickson et al. 2008; Hershey et al. 2010; EARIP 2012). Springs that emerge from regional aquifers can be persistent, evolutionarily stable environments that can support endemic biota, including crenobiontic springsnails, crustaceans, aquatic insects, and sometimes fish. AMI richness is generally less than in mountain and bajada springs of similar size, but greater than in valley floor and geothermal springs; chironomids and other midges are less common, and stoneflies are absent. Riparian vegetation is dominated by wetland herbs and graminoids, and subarborescent mesquite, ash, and nonnative woody species, as well as a dozen endemic plant species, occur in the harsh, alkaline wetlands surrounding spring sources in Ash Meadows. When undisturbed by human activity, springbrook channels can be armored with travertine, which naturally reduces hyporheic fauna and limits water for riparian vegetation (fig. 17.7).

Flow

The size, structure, and water quality of arid-land aquifers vary in relation to tectonic setting, parent rock geology, groundwater depth, and the distance and duration of the flow path. For example, the extensional Basin and Range geologic province produces many small aquifers on mountain slopes as well as larger graben valley and regional karstic aquifers. In contrast, the Colorado Plateau contains stacked and perched aquifers, which may drain downward. In both provinces, infiltration is predominantly derived from montane snowmelt. Flow path duration for small local aquifers may be brief (even just a few days), while regional aquifer flow duration may exceed 13,300 years (Monroe et al. 2005; Adams et al. 2006; Johnson et al. 2011).

Spring flow is rarely monitored, and flow variation among spring types has not generally been synthesized. Surveys of more than 1,500 springs in 10 southwestern US national parks and monuments revealed that average discharge ranged from 1 to 10 L/min (Sada 2013). Large springs were relatively uncommon, occurring mostly in spring provinces such as the Blue Springs complex in Arizona and Ash Meadows, Nevada (SSI 2016). On the basis of flow data from 1,342 of Arizona's 10,400 reported springs, Stevens et al. (2016) estimated that springs contributed 17,176 L/s of flow, for a total discharge of 542 billion liters per year, or roughly 6% of the state's 2012 annual water budget. Most of that flow is used to support livestock or wildlife.

Water Chemistry

Spring water temperature and geochemistry vary with the geology of the supporting aquifer as well as with flow path length and duration. In arid North America, spring water temperature varies from below 5°C in boreal and montane environments to above 100°C in geothermal settings. For local springs, the temperature of the water is similar to the mean annual air temperature of the local area, with some modest seasonal variation. Annual variation in groundwater geochemistry is generally meager in comparison with that in surface waters, but it varies widely across the Southwest. Often, high-elevation springs are derived from crystalline bedrock aquifers with short flow paths and cool water, while geothermal springs and springs with long, slow flow paths emerge with elevated solute concentrations. Clear isotope geochemistry differences exist between local and regional groundwater flow systems in the Southwest (Springer et al. 2017).

SPRING ECOSYSTEM ECOLOGY

Biota

Springs are biologically distinctive, supporting a substantial portion of southwestern aquatic, wetland, and riparian biodiversity. Most spring assemblages include widespread aquatic, wetland, riparian, and upland taxa (Keleher and Sada 2012; Sinclair 2018), and many also support locally endemic or rare crenobiontic biota, including plants, annelids, molluscs, aquatic arthropods (especially Acarini, Arachnida, crustaceans such as *Hyalella*, and many Hexapoda), fish (e.g., many pupfish [Cyprinodontidae] and dace and chubs [Cyprinidae]), amphibians (e.g., Amargosa toad *Anaxyrus nelsoni*), and some reptiles (e.g., several southwestern *Thamnophis* garter snakes) and mammals (e.g., *Microtus* voles) (Hershler and Sada 1987; Schmude 1999; Baldinger et al. 2000; Sada and Vinyard 2002; Hershler et al. 2014).

AMIs and fish occupy spring sources as well as long lengths of springbrooks, but many hydrobiid snails and other crenobiontic AMIs are most abundant near spring sources (McCabe 1998). The structure of AMI assemblages also changes from source areas to downstream reaches (Dumnicka et al. 2013; Reiss et al. 2016), with sources dominated by pollution-intolerant taxa and downstream reaches increasingly dominated by tolerant taxa. Many crenobiontic AMI are specifically adapted to the physicochemical characteristics of their environment, which makes them highly sensitive to natural and anthropogenic environmental alterations (e.g., O'Brien and Blinn 1999; Sada 2001, 2007).

Remarkably high levels of species packing have been found in springs (Springer et al. 2015; Sinclair 2018). Vegetation inventories conducted on springs in southern Nevada and northern Arizona reported 20%–25% of the regional flora in the small amount of surveyed spring habitat (Ledbetter et al. 2015). In addition, the inventories found high concentrations of aquatic, wetland, and riparian invertebrates that occurred nowhere else, along with much evidence of native wildlife use. Species density at springs was often an order of magnitude greater than in surrounding uplands (Grand Canyon Wildlands Council 2002).

Ecosystem Ecology

The generally small size of springs might suggest that they are "simple" ecological systems with low environmental and biological diversity. However, springs are highly productive, with estimates of net annual primary productivity (NAPP) exceeding 6 kg/m/yr^{-1}, three times greater than the highest known terrestrial rates (Odum 1957; Blinn 2008), and several orders of magnitude greater than those in desert landscapes. Arid-land springs are often isolated from other aquatic ecosystems, which increases their importance as water sources and supports genetic differentiation of crenobiontic taxa. Perla and Stevens (2008) emphasized that arid-land springs often function as keystone ecosystems: ecologically influential habitat patches that organize surrounding upland ecosystems. For example, spring surveys throughout the Southwest commonly reveal evidence of top predators, indicating that springs function as focal points for prey ambush. From a conservation perspective, their isolation also often protects springs from invasion by nonnative plant, crustacean, fish, and amphibian species.

SPRING REHABILITATION: BACKGROUND

Arid-land stream-riparian ecosystem rehabilitation has a lengthy history in the United States, with an initial and ongoing emphasis on the conservation of fishes (Echelle and Echelle, chap. 23, this volume), particularly short-lived desert species (Minckley et al. 1991). In contrast, rehabilitation of the physical dynamics of springs has received relatively little attention in comparison with that undertaken in stream and lake management. Several agency-based efforts have attempted to apply stream-riparian management concepts to springs; however, the hydraulic and geomorphic processes that regulate spring ecosystems differ substantially from those in streams, making methods used to rehabilitate those systems inappropriate for springs (Stevens et al. 2005; SSI 2016).

Recent attention to the great diversity of crenobiontic and crenophilic plant, invertebrate, and non-fish vertebrate species has somewhat expanded the focus of spring conservation actions from single species to the entire ecosystem. Davis et al. (2011) reviewed available information on several dozen spring rehabilitation projects and concluded that many of the projects achieved their specific goals, but that whole-ecosystem rehabilitation was rarely their focus. Rehabilitation planning should consider the unique, linked nature of the aquifer-aquatic-riparian ecosystem dynamics at springs and remain cognizant of the influences of natural, historical, and contemporary anthropogenic disturbance on the springs' ecological integrity.

Rehabilitation planning that fails to consider a spring's ecosystem integrity runs the risk of extirpating rare taxa by not taking into account the effects of rehabilitation on its specific water chemistry, temperature, and microhabitat availability and by accidentally enhancing habitat for nontarget and nonnative species (Blinn 2008). Restoration of several springs in Ash Meadows, Nevada, focused on endangered fish and endemic snails, but neglected a rarely detected belostomatid giant water bug and a *Catinella* succineid land snail, neither of which have been detected since rehabilitation. Ponds excavated during Duckwater Little Warm Spring rehabilitation for Railroad Valley springfish *Crenichthys nevadae* decreased springsnail habitat, but created optimal habitat for nonnative bullfrogs *Lithobates (Rana) catesbeianus*. Similar results occurred when springbrooks were ponded to rehabilitate springs for the Amargosa toad *Anaxyrus nelsoni* in Nevada. Several other rehabilitation and management challenges are unique to springs. High levels of natural ecological disturbance (e.g., episodic drying or rheocrene flooding) can overwhelm management actions (Keleher and Sada 2102), reducing the priority for rehabilitation of such sites.

The concept of managing for a "pristine ecosystem," one without a human imprint, is inappropriate for springs because most large springs, and many others, have been disturbed by human activities since the Pleistocene (Haynes 2008; Unmack and Minckley 2008). While quantification of the effects of early human disturbance on spring ecology is limited, Freemont (1845) observed that Native Americans often dug ponds to increase the availability of water in Great Basin and Mojave Desert springs. However, such impacts may have been less severe in larger, more resilient springs. Native Americans used spring water to irrigate crops (Courtenay et al. 1985), and they burned spring-fed riparian vegetation (Mehringer and Warren 1976), but the presence of many recently extinct Great Basin crenobiontics in fossil middens indicates that such activities apparently did not lead to extinction.

ECOSYSTEM STEWARDSHIP

A General Stewardship Formula

Springs are diverse across the landscape, and each spring has unique, site-specific environmental characteristics (e.g., temperature, aspect, elevation, water chemistry, substrate composition, flow variability, and history of use; Kreamer and Springer 2008; Springer and Stevens 2009). Successful spring restoration projects must recognize both the individuality of springs and the extent of differences among spring types. A spring's rehabilitation requires applying treatments that are appropriate for its environment and aquatic life. As an example, restoring a rheocrene spring by impounding the springbrook to create a limnocrene pool creates ecological instability, the cost of which is the long-term commitment to maintain an unstable, but naturalized, ecosystem.

Stewardship of spring ecosystems is most effective when it is based on explicitly defined project objectives and a clear understanding of site conditions, history, aquifer integrity, geomorphic context (sphere of discharge or spring type), the bio-cultural and economic context of the site (Paffett et al. 2018), and environmental-biotic relationships (e.g., Sada et al. 2001; Sada and Cooper 2012). Stevens et al. (2016) described spring ecosystem stewardship as a progressive management formula:

Administrative context → Inventory → Assessment →
Planning → Implementation → Monitoring →
Feedback to improve project success

Each of these elements requires a basic understanding of spring ecology and appropriate reference conditions as well as high-quality information management. These challenges are difficult, but the information and guidance provided by sources such as Springs Online (SpringsData.org), a spring ecosystem management database provided by the Museum of Northern Arizona Springs Stewardship Institute (SSI) (www.SpringStewardshipInstitute.org) is helpful. This user-friendly database is designed for any spring ecosystem steward to securely archive and readily document, analyze, and report on a wide array of spring ecosystem information. The SSI website provides spring ecosystem ecology background, database tutorials, notice of workshops, inventory and assessment protocols, and information on rehabilitation projects throughout western North America.

Spring Rehabilitation Inventory, Assessment, and Planning

Several comprehensive spring ecological inventory and assessment protocols have been developed, including Sada and Pohlmann (2006) for the Mojave, Sonoran, and Chihuahuan National Park Networks; Stevens et al. (2006) for Colorado Plateau national parks; Keleher and Rader (2008) for Bonneville Basin artesian springs in Utah; US Forest Service (2012); and the broad-based SSI spring inventory protocol (SSI 2016). While all of these protocols involve measurement of similar variables, they differ in the specific inventory and assessment metrics employed. Resource managers often wish to inventory and assess different characteristics for different types of springs under their jurisdiction. This issue is resolved in the SSI inventory and assessment protocols, which allow the user to select variables of interest and preferred protocols. Protocols that use qualitative, "yes-no," presence-absence, or check-box answers typically suffer from limited utility in prioritization and planning, while those that emphasize ordinal or quantitative data and estimates are more directly applicable to those tasks (Paffett et al. 2018; L. E. Stevens, unpublished data).

RESTORING ECOSYSTEM ATTRIBUTES

Overview

Spring ecosystem rehabilitation planning and implementation is best and most realistically informed by comparison with reference sites to improve ecological integrity and to plan and test monitoring metrics of success (Stevens and Meretsky 2008). However, no regional suite of reference springs has been proposed for the Southwest, nor have more than a few springs been subjected to long-term monitoring of basic ecosystem characteristics. Springs Online (SSI 2015) provides much available information, and as that database is populated, it can be queried to identify a suitable suite of springs with relatively high ecological integrity. Nonetheless, those planning spring rehabilitation should visit one or more regionally comparable reference springs.

Successful rehabilitation programs have restored functional characteristics of spring environments and their assemblages (Davis et al. 2011; Stevens et al. 2016). Concomitantly, unsuccessful spring rehabilitation programs typically have not considered whole-ecosystem rehabilitation, too often focusing on creating or re-creating habitat for a single target species. Such projects may unwittingly advantage competitive and predatory native and nonnative species (e.g., creating ponds that eliminate flowing springbrooks or enhancing conditions that allow expansion of riparian vegetation). Several of the most important, interrelated, and salient factors to consider when rehabilitating springs are discussed below.

Geomorphic Rehabilitation

Restoring a spring to its pre-disturbance geomorphology is a primary consideration due to the influence of geomorphology on the composition, structure, and function of biotic assemblages. As a guide, Springer and Stevens (2009) and Stevens et al. (2016) have identified spring types and the factors degrading springs in the Great Basin, Mojave, Sonoran, and Chihuahuan Deserts as well as on the Colorado Plateau. Possible rehabilitation actions range from minor activities, such as planting native species or removing small man-made structures (e.g., springboxes, tanks, leaking piping) to large geomorphic reconfiguration efforts. Regardless of the project, strategies to alleviate stressors should focus on restoring functional aquatic and riparian characteristics appropriate to the spring type and the specific habitat requirements of target species.

The ecological characteristics of each spring are strongly influenced by water chemistry, thermal gradients, substrate composition, and a myriad of other physicochemical factors. For example, travertine formation is important for springs with high carbonate concentrations; harsh conditions and fine sediments typify geothermal springs; and mountain springs are cold, with large channel substrate particle sizes (Thomas et al. 1996; Kreamer and Springer 2008). Restoring these conditions within the appropriate geomorphic context creates appropriate aquatic and riparian habitat that is consistent with restoring the ecological health of the spring ecosystem.

Vegetation

Early seral stage native herbaceous plants, such as cattails, reeds, and rushes, and nonnative white top *Lepidium draba*, *Bromus* spp., and other nonnative grasses are likely to quickly colonize the spring riparian zone and springbrook channel habitats, as evidenced during the rehabilitation of Kings Spring in Ash Meadows, Nevada. Management may be needed to control woody nonnative vegetation, but active management of early seral stage native herbaceous vegetation is generally discouraged. Herbaceous plant species are adapted to disturbed conditions, and reduced disturbance following site treatment is likely to have little long-term impact. Depending on the spring type and the desired conditions to be achieved, the density of weedy herbs can be reduced by planting native woody species that shade springbrook channels. The density of early seral stage herbaceous colonizers is also likely to decrease in springbrooks where travertine is deposited, which reduces water availability for emergent and riparian vegetation.

Ecosystem Responses

Many studies have documented extirpations, extinctions, and effects of human disturbance on spring biota (e.g., Sada and Vinyard 2002; Williams and Sada, chap. 6, this volume); however, few studies have examined the response of spring ecosystems to rehabilitation. Sada and Friese (2017) reported the response of AMIs to fire, mechanical removal of nonnative palm trees, and subsequent rewatering of a spring in Death Valley National Park, with the following results: (1) species richness increased, and pollution-tolerant taxa dominated the assemblage, immediately following fire and palm tree removal; (2) species richness declined as the assemblage recovered over 4.5 years; (3) no discernible long-term impact of this disturbance on crenobiontic AMI abundance was detected; (4) crenobiontic amphipods from the aquifer quickly recolonized the spring (which had been dry for more than 80 years), and springsnails colonized from a population occupying a tributary spring. Two other crenobiontic species did not recolonize after rewatering, indicating that some species are not vagile and can be reestablished only through translocation. Qualitative observations at other southwestern springs also revealed a dramatic increase in AMI abundance (including crenobiontic molluscs) immediately after rehabilitation, followed by a dramatic decrease in subsequent years (e.g., Corn Creek Springs, Nevada).

Riparian vegetation may respond differently to rehabilitation and disturbance. Early seral stage emergent taxa (e.g., cattails and reeds) rapidly and densely colonize springbrook channel margins, and invasive weed species (e.g., *Bromus*, *Lepidium*, *Salsola*, *Tamarix*) and native grasses colonize the springbrook banks (Burke et al. 2015). The presence of cattails and bulrushes is problematic because these clonal species quickly choke rehabilitated channels, but removing them increases post-rehabilitation disturbance. Management intervention may be necessary to accelerate the successional process by planting woody vegetation to shade the springbrook, thereby reducing the abundance of early seral stage and problematic wetland taxa.

Invasive Species Management

Habitat restoration often includes eliminating and preventing colonization by nonnative species. Large earth-moving equipment may be needed to remove undesired trees, such as salt cedar (*Tamarisk* sp.), palms (*Washingtonia* sp.), Russian olive (*Elaeagnus angustifolia*), and elms (*Elmus* sp.). Barriers, weirs, or other structures may be constructed to exclude nonnative fish. Rehabilitation should consider the ecological opportunism of nonnative species, and consideration should be given to creating habitat that discourages colonization by nontarget and nonnative species.

SUMMARY AND LESSONS LEARNED

The wide environmental and biological diversity of springs requires that rehabilitation design and implementation be customized for each spring ecosystem. The following considerations generally apply to all spring rehabilitation programs.

- Develop a resource stewardship formula to guide the rehabilitation process:
  - Inventory, assess, plan, implement, monitor, and adjust as necessary.
  - Prioritize spring types and crenobiontic species that are most at risk.
  - Manage information to ensure long-term project success (e.g., via Springs Online).
- Develop a rehabilitation plan:
  - Prioritize among or within springs in the management area to efficiently accomplish rehabilitation projects.
  - Integrate both scientific and stewardship concerns into the plan.
  - Determine water rights associated with each restoration site and integrate this information into the restoration plan.
  - Initially focus on moderately altered sites with moderate levels of restoration potential—sites at which a modest investment of funding is likely to improve ecological integrity and reduce risk (Paffett et al. 2018).
  - Search for and evaluate regional reference springs as analogs for desired site geomorphology (Griffiths et al. 2008), hydrogeology, and landscape setting for the spring being restored and at which to test metrics of success.
  - If no reference springs exist, thoroughly discuss and clearly articulate the rehabilitation characteristics that are required, realistic, and achievable.
  - Consider beginning with relatively small programs, which are likely to have greater and more rapid success. Larger, highly disturbed springs may be difficult, expensive, and time-consuming to rehabilitate.
  - In contrast, rehabilitation of larger springs may allow conservation of more species and microhabitats and may have greater appeal to the public.
- Rehabilitate or re-create natural, functioning characteristics:
  - Determine the aquifer provenance and landscape setting (e.g., mountain, bajada, regional).
  - Clearly identify the spring type or morphology (see Springer and Stevens 2009).
- Remove or mitigate deleterious human disturbances:
  - Avoid overbuilding habitats (e.g., excessive channel reconfiguration).

- Determine the actual and desired post-treatment potential disturbance regime (e.g., site rehabilitation sometimes increases recreational visitation and site attractiveness to grazing wildlife or livestock).
- Some springs are more seriously degraded than others. If rehabilitation is possible for only a portion of the spring system, restoring and protecting eucrenal habitat (at least 50 m to 100 m of springbrook) is a priority due to its importance to crenobiontic taxa.
- In springs where water is diverted into a pipe from a springbox, return the flow, or as much of it as possible, to the springbrook. Leaving the springbox in place is optional.
- If diversion from the spring is required, attempt to limit it to the amount required for the designated use (e.g., if used to supply a trough to water livestock, use a float valve to stop diversion when the trough is full).
- Consider alternatives to diverting water, such as collecting water from a downstream gallery rather than the source or developing a rain-harvesting guzzler system. Leave the source and at least 50 m of the springbrook unaffected by diversion. Protect longer reaches of the source springbrook in larger springs.
- Consider contingencies: what unanticipated physical, biological, and social (e.g., policy) changes might occur?

• If rehabilitating springs to conserve crenobiontic taxa, determine the target species' habitat requirements and ensure maintenance of suitable habitat diversity (e.g., both lentic and lotic habitats with appropriate substrates and depths).
• Carefully consider which other crenobiontic taxa may be translocated into the restored site.
• Develop and implement a monitoring plan to track environmental and biological changes and regularly assess rehabilitation progress (Sada and Cooper 2012; Stevens et al. 2016).
• Maintain the information management system in a long-term database like Springs Online (Springs-Data.org) to document ecosystem changes, inform management for subsequent rehabilitation projects, and report to the public, managerial, and scientific communities on project success.

Additional recommendations on spring rehabilitation for common desert spring types can be found in Stevens et al. (2016) and online at http://docs.springstewardship.org/PDF/SIA-Handbook_010916.pdf.

REFERENCES

Adams, E. A., S. A. Monroe, A. E. Springer, K. W. Blasch, and D. J. Bills. 2006. Flow timing of South Rim springs of Grand Canyon, Arizona using electrical resistance sensors. *Ground Water* 44: 630–41.

Allis, R., D. Bergfeld, J. Moore, K. McClure, C. Morgan, T. C. Chidsey, J. Hearth, and B. J. McPherson. 2005. Implications of results from CO_2 flux surveys over known CO_2 systems for long-term monitoring. *Proceedings of the Fourth Annual Conference on Carbon Capture and Sequestration, May 2–5, Alexandria, Virginia.*

Arnold, J. G., P. M. Allen, and G. Bernhardt. 1993. A comprehensive surface-groundwater flow model. *Journal of Hydrology* 142: 47–69.

Baldinger, A. J., W. D. Shepard, and D. L. Threloff. 2000. Two new species of *Hyalella* (Crustacea: Amphipoda: Hyalellidae) from Death Valley National Park, California, U.S.A. *Proceedings of the Biological Society of Washington* 113: 443–57.

Blinn, D. W. 2008. The extreme environment, trophic structure, and ecosystem dynamics of a large, fishless desert spring. In L. E. Stevens and V. J. Meretsky, eds., *Aridland Springs of North America: Ecology and Conservation*, 98–126. Tucson: University of Arizona Press.

Bryan, K. 1919. Classification of springs. *Journal of Geology* 27: 552–61.

Burke, K. J., K. A. Harcksen, L. E. Stevens, R. J. Andress, and R. J. Johnson. 2015. Collaborative rehabilitation of Pakoon Springs in Grand Canyon–Parashant National Monument, Arizona. In L. F. Huenneke, C. van Riper III, and K. A. Hayes-Gilpin, eds., *The Colorado Plateau VI: Science and Management at the Landscape Scale*, 312–30. Tucson: University of Arizona Press.

Carter, C. D., and J. C. Marks. 2007. Influences of travertine dam formation on leaf litter decomposition and algal accrual. *Hydrobiologia* 575: 329–41.

Costa, K. C., J. B. Navarro, E. L. Shock, C. L. Zhang, D. Soukup, and B. P. Hedlund. 2009. Microbiology and geochemistry of great boiling and mud host springs in the United States Great Basin. *Extremophiles* 13: 447–59.

Courtenay, W. R., J. E. Deacon, D. W. Sada, R. C. Allan, and G. L. Vinyard. 1985. Comparative status of fishes along the course of the pluvial White River, Nevada. *Southwestern Naturalist* 30: 503–24.

Crossey, L. J., K. E. Karlstrom, A. E. Springer, D. Newell, D. R. Hilton, and T. Fischer. 2009. Degassing of mantle-derived CO_2 and He from springs in the southern Colorado Plateau region—Neotectonic connections and implications for groundwater system. *Geological Society of America Bulletin* 121: 1034–53.

Davis, C. J., A. E. Springer, and L. E. Stevens. 2011. *Have Arid Land Springs Rehabilitation Projects Been Effective in Restoring*

Hydrology, Geomorphology, and Invertebrate and Plant Species Composition Comparable to Natural Springs with Minimal Anthropogenic Disturbance? CEE Systematic Review 10-002. Collaboration for Environmental Evidence.

Dreybrodt, W., D. Buhamn, D. Michaelis, and E. Usdowski. 1992. Geochemically controlled calcite precipitation by CO_2 outgassing: Field measurements of precipitation rates in comparison to theoretical predictions. *Chemical Geology* 97: 285–94.

Drost, C. A., and D. W. Blinn. 1997. Invertebrate community of Roaring Springs Cave, Grand Canyon National Park, Arizona. *Southwestern Naturalist* 42: 497–500.

Dumnicka, E., J. Galas, I. Jatulewicz, J. Karlikowska, and B. Rzonca. 2013. From spring sources to springbrook: Changes in environment and benthic fauna. *Biologia* 68: 142–49.

EARIP (Edwards Aquifer Recovery and Implementation Program). 2012. *Edwards Aquifer Recovery Implementation Program: Habitat Conservation Plan.* San Diego: RECON Environmental, Inc.

Freemont, J. C. 1845. *Report of the Exploring Expedition to the Rocky Mountains in the Year 1842, and to Oregon and North California in the Years 1843 and 1844.* Printed by Order of the Senate of the United States. Washington, DC: Gales and Seaton, Printers.

Grand Canyon Wildlands Council. 2002. *Arizona Strip Springs, Seeps, and Natural Ponds: Inventory, Assessment, and Development of Recovery Priorities.* Final Project Report. Flagstaff, AZ. http://docs.springstewardship.org/PDF/ASSP_Final_Report020430.pdf.

Griffiths, R. E., D. E. Anderson, and A. E. Springer. 2008. The morphology and hydrology of small spring-dominated channels. *Geomorphology* 102: 511–21.

Haynes, C. V. 2008. Quaternary calcron springs as paleoecological archives. In L. E. Stevens and V. J. Meretsky, eds., *Aridland Springs of North America: Ecology and Conservation*, 76–97. Tucson: University of Arizona Press.

Hendrickson, D. A., J. C. Marks, A. B. Moline, E. Dinger, and A. E. Cohen. 2008. Combining ecological research and conservation: A case study in Cuatro Ciénegas, Coahuila, Mexico. In L. E. Stevens and V. J. Meretsky, eds., *Aridland Springs of North America: Ecology and Conservation*, 127–57. Tucson: University of Arizona Press.

Hershey, R. L., S. A. Mizell, and S. Earman. 2010. Chemical and physical characteristics of springs discharging from regional flow systems of the carbonate-rock province of the Great Basin, western United States. *Hydrogeology Journal.* https://doi.org/10.1007/s10040-009-0571-7.

Hershler, R., and H.-P. Liu. 2017. *Annotated Checklist of Freshwater Truncatelloidean Gastropods of the Western United States, with an Illustrated Key to the Genera.* US Department of the Interior Bureau of Land Management Technical Note 449. Denver: National Operations Center.

Hershler, R., H.-P. Liu, and J. Howard. 2014. Springsnails: A new conservation focus in western North America. *BioScience* 64: 693–700.

Hershler, R., and D. W. Sada. 1987. Springsnails (Gastropoda: Hydrobiidae) of Ash Meadows, Amargosa basin, California-Nevada. *Proceedings of the Biological Society of Washington* 100: 776–843.

Johnson, R. H., E. DeWitt, L. Wirt, L. R. Arnold, and J. D. Horton. 2011. *Water and Rock Geochemistry, Geologic Cross Sections, Geochemical Modeling, and Groundwater Flow Modeling for Identifying the Source of Groundwater to Montezuma Well, a Natural Spring in Central Arizona.* US Geological Survey Open-File Report 2011-1063.

Junghans, K. M., A. E. Springer, L. E. Stevens, and J. D. Ledbetter. 2016. Springs ecosystem distribution and density for improving stewardship. *Freshwater Science* 35: 1330–39.

Keleher, M. J., and R. Rader. 2008. Bioassessment of artesian springs in the Bonneville Basin, Utah, USA. *Wetlands* 28: 1048–59.

Keleher, M. J., and D. W. Sada. 2012. Desert spring wetlands of the Great Basin. In D. P. Batzer and A. H. Baldwin, eds., *Wetland Habitats of North America*, 329–41. Berkeley: University of California Press.

Kreamer, D. K., and A. E. Springer. 2008. The hydrology of desert springs in North America. In L. E. Stevens and V. J. Meretsky, eds., *Aridland Springs of North America: Ecology and Conservation*, 35–48. Tucson: University of Arizona Press.

Ledbetter, J. D., L. E. Stevens, M. Hendrie, and A. Leonard. 2015. Ecological inventory and assessment of springs ecosystems in Kaibab National Forest, northern Arizona. In B. E. Ralston, ed., *Proceedings of the 12th Biennial Conference of Research on the Colorado Plateau*, 25–40. US Geological Survey Scientific Investigations Report 2015-5180.

Malanson. G. P. 1993. *Riparian Landscapes.* New York: Cambridge University Press.

McCabe, D. J. 1998. Biological communities in springbrooks. In L. Botosaneau, ed., *Studies in Crenobiology: The Biology of Springs and Springbrooks*, 221–28. Leiden, The Netherlands: Backhuys Publishers.

Mehringer, P. J., Jr., and C. N. Warren. 1976. Marsh, dune, and archaeological chronology, Ash Meadows, Amargosa Desert, Nevada. In R. Elston, ed., *Holocene Environmental Change in the Great Basin*, 120–50. Nevada Archaeological Survey Research Paper 6.

Meinzer, O. E. 1923. *Outline of Ground Water Hydrology, with Definitions.* Water Supply Paper 494. US Geological Survey.

Miller, R. R. 1961. Man and the changing fish fauna of the American Southwest. *Papers of the Michigan Academy of Science, Arts, and Letters* 46: 365–404.

Minckley, W. L., and J. E. Deacon. 1968. Southwestern fishes and the enigma of "endangered species." *Science* 159: 1424–32.

Minckley, W. L., G. K. Meffe, and D. L. Soltz. 1991. Conservation and management of short-lived fishes: The cyprinodonts. In W. L. Minckley and J. E. Deacon, eds., *Battle against Extinction: Native Fish Management in the American West*, 247–82. Tucson: University of Arizona Press.

Monroe, S. A., R. C. Antweiler, J. R. Hart, H. E. Taylor, M. Truini, J. R. Rihs, and T. J. Felger. 2005. *Chemical Characteristics of Ground-Water Discharge along the South Rim of Grand Canyon in Grand Canyon National Park, Arizona, 2000–2001.* US Geological Survey Scientific Investigations Report 2004-5146.

Mueller, J. M., R. E. Lima, and A. E. Springer. 2017. Can environmental attributes influence protected area designation? A case study valuing preferences for springs in Grand Canyon

National Park. *Land Use Policy* 63. https://doi.org/10.1016/j.landusepol.2017.01.029.

Nabhan, G. P. 2008. Plant diversity influenced by indigenous management of freshwater springs. In, L. E. Stevens and V. J. Meretsky, eds., *Aridland Springs in North America: Ecology and Conservation*, 244–67. Tucson: University of Arizona Press.

O'Brien, C., and D. W. Blinn. 1999. The endemic springsnail *Pyrgulopsis montezumensis* in a high CO_2 environment: Importance of extreme chemical habitats as refugia. *Freshwater Biology* 42: 225–34.

Odum, H. T. 1957. Trophic structure and productivity of Silver Springs, Florida. *Ecological Monographs* 27: 55–112.

Paffett, K., L. E. Stevens, and A. E. Springer. 2018. Ecological assessment and rehabilitation prioritization for improving springs ecosystem stewardship. In J. Dorney, R. Savage, R. Tiner, and P. Adamus, eds., *Wetland and Stream Rapid Assessments: Development, Validation, and Application.* Atlanta: Elsevier.

Perla, B. S., and L. E. Stevens. 2008. Biodiversity and productivity at an undisturbed spring in comparison with adjacent grazed riparian and upland habitats. In L. E. Stevens and V. J. Meretsky, eds., *Arid Land Springs in North America: Ecology and Conservation*, 230–43. Tucson: University of Arizona Press.

Reiss, M., O. Martin, R. Gereke, and S. von Fumetti. 2016. Limno-ecological characteristics and distribution patterns of spring habitats and invertebrates from the lowlands to the Alps. *Environmental Earth Sciences* 75: 1033. https://doi.org/10.1007/s12665-016-5818-8.

Sada, D. W. 2001. Demography and habitat use of the Badwater snail (*Assiminea infima*), with observations on its conservation status. *Hydrobiologia* 466: 255–65.

Sada, D. W. 2007. Synecology of a springsnail (Prosobranchia: Family Hydrobiidae) assemblage in a western US thermal spring province. *Veliger* 50: 59–71.

Sada, D. W. 2013. *Environmental and Biological Characteristics of Springs in the Chihuahuan Desert Network of National Parks, with a Prioritized Assessment of Suitability to Monitor for Effects of Climate Change.* Las Cruces, NM: US National Park Service, Chihuahuan Desert Inventory and Monitoring Network.

Sada, D. W., and D. J. Cooper. 2012. *Furnace Creek Springs Rehabilitation and Adaptive Management Plan, Death Valley National Park, California.* Death Valley National Park, CA: US National Park Service.

Sada, D. W., and R. Friese. 2017. *Resistance and Resilience of Benthic Macroinvertebrate Communities to the Effects of Drying, Fire, and Invasive Plant Removal in Travertine Springs, Death Valley National Park, USA.* Ft. Collins, CO: US National Park Service.

Sada, D. W., and K. F. Pohlmann. 2006. *US National Park Service Mojave Inventory and Monitoring Network Spring Survey Protocols*: Level I. Reno: Desert Research Institute.

Sada, D. W., and C. Rosamond. 2013. *Abundance, Distribution, and Habitat Use of the Elongate Mud Meadow Springsnail* (Pyrgulopsis notidicola), *Soldier Meadow, NV.* Reno: US Fish and Wildlife Service.

Sada, D. W., and G. L. Vinyard. 2002. Anthropogenic changes in historical biogeography of Great Basin aquatic biota. In R. Hershler, D. B. Madsen, and D. R. Currey, eds., *Great Basin Aquatic Systems History*, 277–92. Smithsonian Contributions to Earth Sciences, no. 33.

Sada, D. W., J. E. Williams, J. C. Silvey, A. Halford, J. Ramakka, P. Summers, and L. Lewis. 2001. *Riparian Area Management. A Guide to Managing, Restoring, and Conserving Springs in the Western United States.* US Bureau of Land Management Technical Reference 1735-17.

Schmude, K. L. 1999. Riffle beetles in the genus *Stenelmis* (Coleoptera: Elmidae) from warm springs in southern Nevada: New species, new status, and a key. *Entomological News* 110: 1–12.

Sinclair, D. A. 2018. Geomorphology influences on springs of the Grand Canyon ecoregion, Arizona, USA. MS thesis, Northern Arizona University.

Springer, A. E., E. M. Boldt, and K. M. Junghans. 2017. Local vs. regional groundwater flow from stable isotopes at Western North America springs. *Groundwater* 55: 100–109.

Springer, A. E., and L. E. Stevens. 2009. Spheres of discharge of springs. *Hydrogeology Journal* 17: 83–93.

Springer, A. E., L. E. Stevens, D. E. Anderson, R. A. Parnell, D. K. Kreamer, L. A. Levin, and S. P. Flora. 2008. A comprehensive springs classification system: Integrating geomorphic, hydrochemical, and ecological criteria. In L. E. Stevens and V. J. Meretsky, eds., *Arid Land Springs in North America: Ecology and Conservation*, 49–76. Tucson: University of Arizona Press.

Springer, A. E., L. E. Stevens, J. D. Ledbetter, E. M. Schaller, K. Gill, and S. B. Rood. 2015. Ecohydrology and stewardship of Alberta springs ecosystems. *Ecohydrology* 8: 896–910.

SSI (Springs Stewardship Institute). 2015. *Developing a Geodatabase and Geocollaborative Tools to Support Springs and Springs-Dependent Species Management for the Desert Landscape Conservation Cooperative.* Flagstaff: Museum of Northern Arizona. http://springstewardshipinstitute.org/desert-lcc.

Stevens, L. E., J. D. Ledbetter, A. E. Springer, C. Campbell, L. Misztal, M. Joyce, and G. Hardwick. 2016. *Arizona Springs Rehabilitation Handbook.* Flagstaff: Museum of Northern Arizona; Tucson: Sky Island Alliance.

Stevens, L. E., and V. J. Meretsky, eds. 2008. *Aridland Springs in North America: Ecology and Conservation.* Tucson: University of Arizona Press.

Stevens, L. E., A. E. Springer, and R. Harms. 2006. *Springs Ecosystem Inventory and Assessment Protocols in National Parks on the Colorado Plateau, Southwestern USA.* Flagstaff: US National Park Service.

Stevens, L. E., P. B. Stacey, A. Jones, D. Duff, C. Gourley, and J. C. Caitlin. 2005. Protocol for rapid assessment of southwestern stream-riparian ecosystems. In C. van Riper III and D. J. Mattson, eds., *Fifth Conference on Research on the Colorado Plateau*, 397–420. Tucson: University of Arizona Press.

Thomas, J. M., A. H. Welch, and M. D. Dettinger. 1996. *Geochemistry and Isotope Hydrology of Representative Aquifers in the Great Basin Region of Nevada, Utah, and Adjacent States.* US Geological Survey Professional Paper 1409-C.

US Forest Service. 2012. *Groundwater-Dependent Ecosystems: Level II Inventory Field Guide.* US Forest Service General Technical Report WO-86b.

Unmack, P. J., and W. L. Minckley. 2008. The demise of desert springs. In L. E. Stevens and V. J. Meretsky, eds., *Arid Land*

Springs in North America: Ecology and Conservation, 11–34. Tucson: University of Arizona Press.

Welsh, S. L. 1984. *Flora of Glen Canyon National Recreation Area.* Unpublished report to Glen Canyon National Recreation Area, Page, Arizona.

Welsh, S. L. 1989. On the distribution of Utah's hanging gardens. *Great Basin Naturalist* 49: 1–30.

Williams, J. E. 1991. Preserves and refuges for native western fishes: History and management. In W. L. Minckley and J. E. Deacon, eds., *Battle against Extinction: Native Fish Management in the American West,* 171–89. Tucson: University of Arizona Press.

Wilson, K. P., and D. W. Blinn. 2007. Food web structure, energetics, and importance of allochthonous carbon in a desert cavernous limnocrene: Devils Hole, Nevada. *Western North American Naturalist* 67: 185–98.

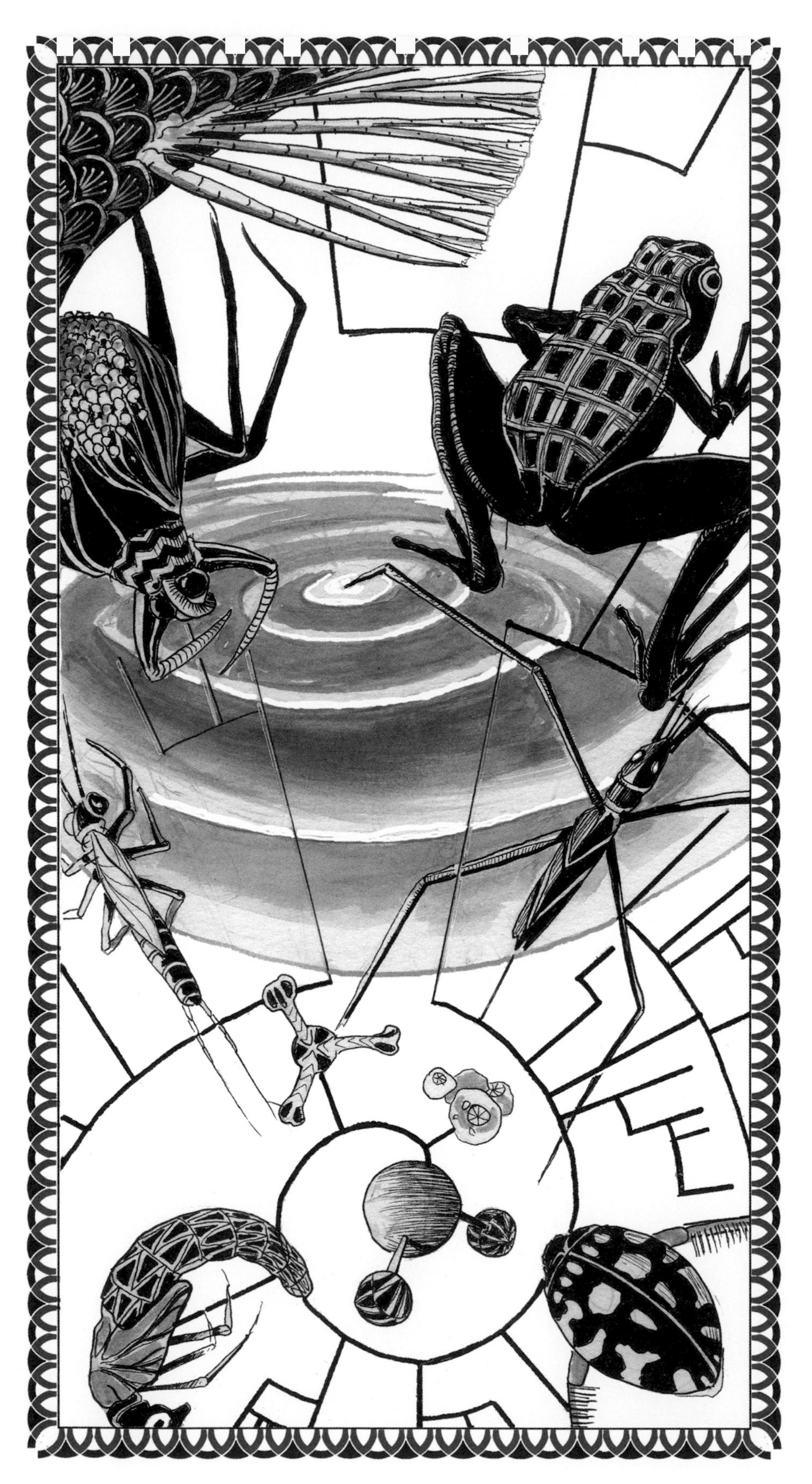

::Dependencies

SECTION 4

Searching for Recovery

Fig. 18.1 The Sierra El Aguaje (*A*, with Gulf of California in background) is home to numerous steep-walled canyons with perennial bedrock pools known as tinajas (*B*, *C*) and spring-fed oases ringed by native palms (*D*) or fig trees with large buttressed roots (*E*).

18 Oases

Finding Hidden Biodiversity Gems in the Southern Sonoran Desert

Michael T. Bogan, Carlos Alonso Ballesteros-Córdova, Scott E. K. Bennett, Michael H. Darin, Lloyd T. Findley, and Alejandro Varela-Romero

The Sierra El Aguaje dominates the skyline of the southern Sonoran Desert, with its western flank rising directly out of the Gulf of California. This jagged volcanic mountain range is characterized by towering rhyolite cliffs and deep, shaded canyons (fig. 18.1). Summers are hot (mean July temperature ~32°C) and annual rainfall is less than 230 mm, most of which falls between July and September. In some years, either summer or winter rains, or both, can fail (Felger 1999). Despite this aridity, many canyons support lush pockets of tropical vegetation, including dense groves of native fig trees and palms (e.g., *Ficus* spp., *Brahea brandeegii*, and *Washingtonia robusta*; Felger et al. 2017). Furthermore, there is a surprising amount of freshwater hidden deep within these rugged canyons, providing essential habitat for aquatic biota and drinking water for terrestrial animals. In fact, the word *aguaje* signifies "water source" in northern México, so the mountain range is named for these freshwater habitats. While some canyons only harbor small pools for brief periods after torrential rains, others are spring-fed and support perennial freshwater oases (Bogan et al. 2014). The most isolated oases are hidden far up rocky canyons, the mouths of which can be reached only by boat from the Gulf of California.

The incredible landscape of the Sierra El Aguaje, however, is not an unpopulated wilderness or national park, but has been used by humans for millennia. The Comcáac (Seri) people have occupied the central coast of Sonora since at least AD 220, and probably much longer (Bowen 1976; Felger and Moser 1985). They developed a rich taxonomic understanding of the region's biota, including marine, terrestrial, and riparian plants and animals (Felger and Moser 1985; Nabhan 2003). One band of Comcáac, the *Guaíma* or *Xiica xnaai iicp coii,* lived along the coast of the Sierra El Aguaje and relied on the water resources of its canyons (Moser 1963; Dixon 1990). By 1628, though, this band had been forcibly relocated by Spanish colonists to distant villages along the Río Yaqui, where they amalgamated with the *Yoeme* (Yaqui), a separate and distinct indigenous culture (Spicer 1980). Since the 1800s, Mexican ranchers have used the more accessible portions of the Sierra El Aguaje for grazing cattle, and many canyons and surrounding areas are part of *ejidos,*

Fig. 18.2 Faults and freshwater habitats of the Sierra El Aguaje. (*A*) Shaded-relief map illustrating known and hypothesized faults (red and black lines) and locations of documented springs (blue diamonds). (*B*) M. Darin and S. Bennett measuring the orientation of a fault surface in La Navaja Canyon. (*C*, *D*) Waterfalls and seeps where groundwater emerges along nearly vertical fault zones in La Navaja Canyon.

large parcels of land that are communally owned and managed for livestock and agriculture. In recent times, the most accessible canyons of the Sierra El Aguaje have become popular recreational areas for residents and tourists from the nearby towns of San Carlos and Guaymas (Bogan et al. 2014).

Modern scientists first took an abiding notice of the Sierra El Aguaje, and its dramatic canyons, roughly 50 years ago. The preeminent botanist for the region, Richard Felger, began studying the remarkable flora of the range's desert-bound tropical canyons in the 1960s (Felger 1999). But it has been only in the last 15 years or so that the remote freshwater habitats in the range have received scientific attention (e.g., Varela-Romero 2001; Pfeiler and Markow 2008). The main goal in this chapter is to summarize what is known about the hydrology, deep history, and ecology of the range's freshwater ecosystems. In doing so, we hope to call attention to their notable biodiversity and promote their conservation. In the following sections, we (1) describe the variety of freshwater habitats found in the Sierra El Aguaje, (2) explore their hydrology, geology, and deep history, (3) relate what is known about their biodiversity, and (4) discuss pressing conservation concerns in the region. Finally, new research avenues to pursue in these beautiful, often remote, and underexplored habitats are proposed.

FRESHWATER HABITATS IN THE SIERRA EL AGUAJE

Perennial freshwater habitats in the canyons of the Sierra El Aguaje range from tiny seeps that barely moisten the surface of rock faces to rock-bound pools and palm-fringed oases (see fig. 18.1). At least 18 locations support freshwater habitat (fig. 18.2*A*). Previous research identified four habitat types: (1) seeps (with laminar flow <1 L/min), (2) riffles (with turbulent flow >1 L/min) (3) tinajas (bedrock-bound pools), and (4) oases (deep pools lined with riparian vegetation) (Bogan et al. 2014). Seeps and tinajas occur throughout, but oases are less common. In general, freshwater inputs via springs (fig. 18.2*C, D*) barely offset evaporative losses, meaning that oases and tinajas are kept full, but do not spill into flowing riffles. Spring discharge is large enough to produce perennial riffle habitat in only a few locations, making riffles the rarest freshwater habitat type. To our knowledge, only four canyons in the Sierra El Aguaje support perennially flowing riffle habitats, the total surface area of which may be less than 10 m^2.

In addition to perennial spring-fed habitats, seasonal freshwater habitats can be found after summer monsoon or hurricane rains. Seasonal tinajas may hold water for weeks to months after rain events. Some canyons support such habitat fairly reliably following monsoon seasons having average precipitation, whereas others have seasonal water only following relatively rare, intense hurricane or hurricane-edge storm events (Bogan 2012). Such an event occurred in September 2009, during Hurricane Jimena, when 720 mm of rain fell in 36 hours (Pérez-Tello et al. 2016). This impressive deluge resulted in massive flooding through the canyons of the Sierra El Aguaje, producing seasonal riffles and pools that persisted for several months (Findley et al. 2010; Pérez-Tello et al. 2016). It also redistributed much sediment through the canyons and their arroyos farther downstream, filling oases in some canyons and scouring out new pools in others (Felger et al. 2017). However, strong hurricanes strike the Sonoran Desert only once or twice each decade (Antinao et al. 2016), so only perennial habitats serve as refugia for aquatic taxa and as permanent water sources for terrestrial animals.

GEOLOGY, HYDROLOGY, AND DEEP HISTORY OF SPRING-FED HABITATS IN THE SIERRA EL AGUAJE

How can there be so many freshwater spring-fed habitats in the middle of a desert mountain range? What controls their locations (see fig. 18.2)? How long have they been there? Answers to these questions lie at the intersection of geology, hydrology, and climate.

Desert spring ecosystems commonly occur along geologic faults. Most fault systems are composed of a complex network of interconnected faults and fractures that collectively facilitate the movement of adjacent rock units (Davis and Reynolds 1996; Caine et al. 1996). In the Sonoran Desert, a long history of crustal extension and transtension has engraved a complex network of faults across the landscape (Bennett et al. 2013; Darin et al. 2016). Although basin-and-range-style faulting began about 30 million years ago, the most significant faulting in this region occurred over the last 12 million years as the boundary between the Pacific and

North American plates gradually reorganized from a subduction zone to a rift basin—the Gulf of California (Atwater and Stock 1998; Bennett and Oskin 2014). In addition to landscape-scale alterations, geologic faults also provide conduits through which groundwater stored in subsurface aquifers discharges to the land surface. In the Sierra El Aguaje, known spring locations correspond remarkably well with the locations of faults known from geologic mapping studies. Here, we present a map (see fig. 18.2*A*) of previously published faults, supplemented by our own fault mapping from field reconnaissance (fig. 18.2*B*) and our analysis of satellite imagery and digital topography.

No studies have identified the sources of groundwater that feed springs in the Sierra El Aguaje, but we can speculate about potential sources on the basis of the region's known geology and climate and studies from similar areas. Although rainfall in most years is not sufficient to explain the volume of water present on the surface, occasional hurricane-induced deluges could recharge regional aquifers, as they do in hyper-arid central Oman, where rainfall from infrequent cyclones percolates through fractured bedrock into deep freshwater aquifers (Young et al. 2004). Although the Sierra El Aguaje consists primarily of rhyolite, granite, and other typically low-porosity igneous rocks, faulting can greatly increase the porosity and permeability of bedrock, suggesting that such aquifer recharge processes could occur in the Sierra El Aguaje. Relatively low conductivity values documented from the range's aquatic habitats (e.g., 100–350 μS/cm; Bogan et al. 2014) also suggest that spring discharge is due to neither highly mineralized water emerging from deep aquifers (e.g., Smith et al. 2002) nor saltwater intrusion from the Gulf of California.

HISTORICAL ECOLOGY OF FAUNA IN SPRING-FED HABITATS IN THE SIERRA EL AGUAJE

Whatever their water sources, the reliability of freshwater oases in the Sierra El Aguaje is of primary importance to the aquatic and terrestrial faunas. Have individual freshwater habitats persisted as discrete, recognizable oases over millennia? Or do they come and go through time, as effects of faulting and flooding alter the volume and locations of spring discharge? How long has it been since aquatic habitats in the Sierra El Aguaje have had surface-water connections to other freshwater habitats?

Although these important biogeographic questions remain largely unanswered, we can gain insight into the history and connectivity of these isolated habitats from the landscape genetics of freshwater-obligate species such as fishes and frogs. For instance, two populations of a fish closely related to the desert chub *Gila* cf. *eremica* occur in the Arroyo El Tigre basin of the Sierra El Aguaje (Varela-Romero 2001). Their isolation from other desert chub populations may have begun as long as 10–15 million years ago, caused by volcanic and tectonic events that built the nearby Sierra Santa Úrsula (Mora-Álvarez and McDowell 2000). These events probably caused hydrologic disconnection of the Arroyo El Tigre basin from the Río Mátape to the east. Estimates of the timing of genetic divergence between chub populations in the Sierra El Aguaje and the Rio Mátape (Ballesteros-Córdova 2017) are compatible with this hypothesis. Within the Sierra El Aguaje, the two chub populations also appear to be reproductively isolated from each other (Ballesteros-Córdova 2017), despite sharing some genetic haplotypes and being morphologically similar (Ballesteros-Córdova et al. 2016).

While fish cannot disperse overland between oases or mountain ranges, frogs may be able to do so during periods of rain or high humidity (Mims et al. 2015; Rorabough and Lemos-Espinal 2016). Genetic studies of Northwest México leopard frogs (*Lithobates magnaocularis*) suggest that populations in the Sierra El Aguaje are distinct from those in other ranges, but that individuals may be able to disperse among oases within the Sierra (Pfeiler and Markow 2008). Together, these findings suggest that (1) freshwater habitats in the Sierra El Aguaje have continuously supported fish populations for at least the last 5–10 million years; (2) there has been little or no hydrologic connectivity with other river basins during that time; and (3) dispersal among oases within the range is common for frogs, but not for fish.

BIODIVERSITY OF FRESHWATER HABITATS IN THE SIERRA EL AGUAJE

The long history of perennial oases in the Sierra El Aguaje and their hydrologic isolation from neighboring river basins set the stage for development of diverse and unique biotic communities. In the following sections, recent discoveries about the biodiversity of these communities, including aquatic and wetland plants, aquatic invertebrates, and aquatic and riparian vertebrates, are described.

Aquatic and Wetland Plants

Although aquatic and wetland plants of some freshwater oases in Baja California Sur have been well studied (e.g., León de la Luz et al. 1997; León de la Luz and Domínguez-Cadena 2006), those of Sonora have received less attention. In the Sierra El Aguaje, Nacapule Canyon is the only place that has been the subject of extensive botanical surveys. Felger et al. (2017) identified at least 22 species of wetland plants growing in and along the canyon's seeps and pools. Species found in deeper pools included cattail *Typha dominguezi*, common reed *Phragmites australis*, sedges *Cyperus* spp., water-primrose *Ludwigia peploides*, umbrella grass *Fuirena simplex*, and water plantain *Echinodorus berteroi*. Previously, whisk-fern *Psilotum nudum* grew from a seep there, the only place it had been documented from the entire Sonoran Desert. However, the species was extirpated from Nacapule due to drought and declining flows sometime after its collection in 1970 (Felger 1999). Fortunately, another population of whisk-fern was recently discovered at La Balandrona Canyon, several kilometers north of Nacapule, so the species persists in the range (Felger et al. 2017). Although Nacapule Canyon has received significant botanical attention, much remains to be done to document aquatic and wetland plants in other canyons across the range.

Aquatic Invertebrates

The freshwater habitats of the Sierra El Aguaje support a rich community of aquatic invertebrates (fig. 18.3). At least 210 taxa are documented, and total richness is certainly much higher because many of those taxa were identified only to the level of genus or family, and only about 70% of freshwater habitats in the range have been sampled (Bogan et al. 2014). Some species represent extreme northern outlier populations for otherwise tropical taxa, while others are either temperate species at their southern limit or are endemic to the Sonoran Desert (Bailowitz et al. 2015). Over 70% of the aquatic invertebrate taxa recorded are beetles (Coleoptera), true flies (Diptera), and true bugs (Hemiptera), but the fauna also includes numerous species of mayflies (Ephemeroptera), caddisflies (Trichoptera), dragonflies (Odonata), aquatic moths (Lepidoptera), and non-insects, including sponges, snails, mites, and flatworms (Bogan et al. 2014). This diverse fauna contrasts sharply with reports from colder desert springs. For example, Sada et al. (2005) reported a total of only 48 invertebrate taxa from 45 springs in the Mojave Desert.

High taxonomic richness in the Sierra El Aguaje allows for a trophically complex and diverse aquatic invertebrate fauna. The top predator in spring-fed oases and tinajas is the giant water bug *Lethocerus medius* (Belostomatidae). This impressive insect uses its raptorial forelimbs, piercing and sucking mouthparts, and digestive enzymes to capture and consume amphibians, reptiles, and fish that may be twice as large as the bug itself (fig. 18.3*F*). Other true bugs and aquatic beetles readily scavenge what is left of such vertebrate prey and also feed upon smaller invertebrates. Perhaps the most surprising aquatic invertebrates found in the Sierra El Aguaje are spongilla-flies (Neuroptera: Sisyridae) and their obligate hosts and prey items, freshwater sponges (Spongillidae) (fig. 18.3*A*). Sponges and spongilla-flies occur in tinajas and oases in at least six canyons across the desert mountain range (Bogan et al. 2014).

Broadly, the composition of aquatic invertebrate communities tends to be strongly shaped by local habitat characteristics (Bogan 2012), and this is certainly the case in the Sierra El Aguaje. Each of the four major perennial habitat types (seeps, riffles, tinajas, and oases) supports distinct assemblages. For example, riffles at Nacapule Canyon support 10 taxa not found elsewhere in the Sierra El Aguaje, including three species of riffle beetles (Bogan et al. 2014). Oasis habitats are home to species that prefer deeper water and complex habitat structure (e.g., submerged roots, overhanging vegetation), such as spongilla-flies and several species of diving beetles. Seeps are the smallest habitat type and support the fewest species, but are home to several unique true flies (Diptera) that do not occur in other habitats.

Seasonal freshwater habitats can also be rich with aquatic invertebrates. For example, over 60 taxa were found in temporary pools of Arroyo El Palmar (Bogan et al. 2014), despite these ephemeral pools being dry

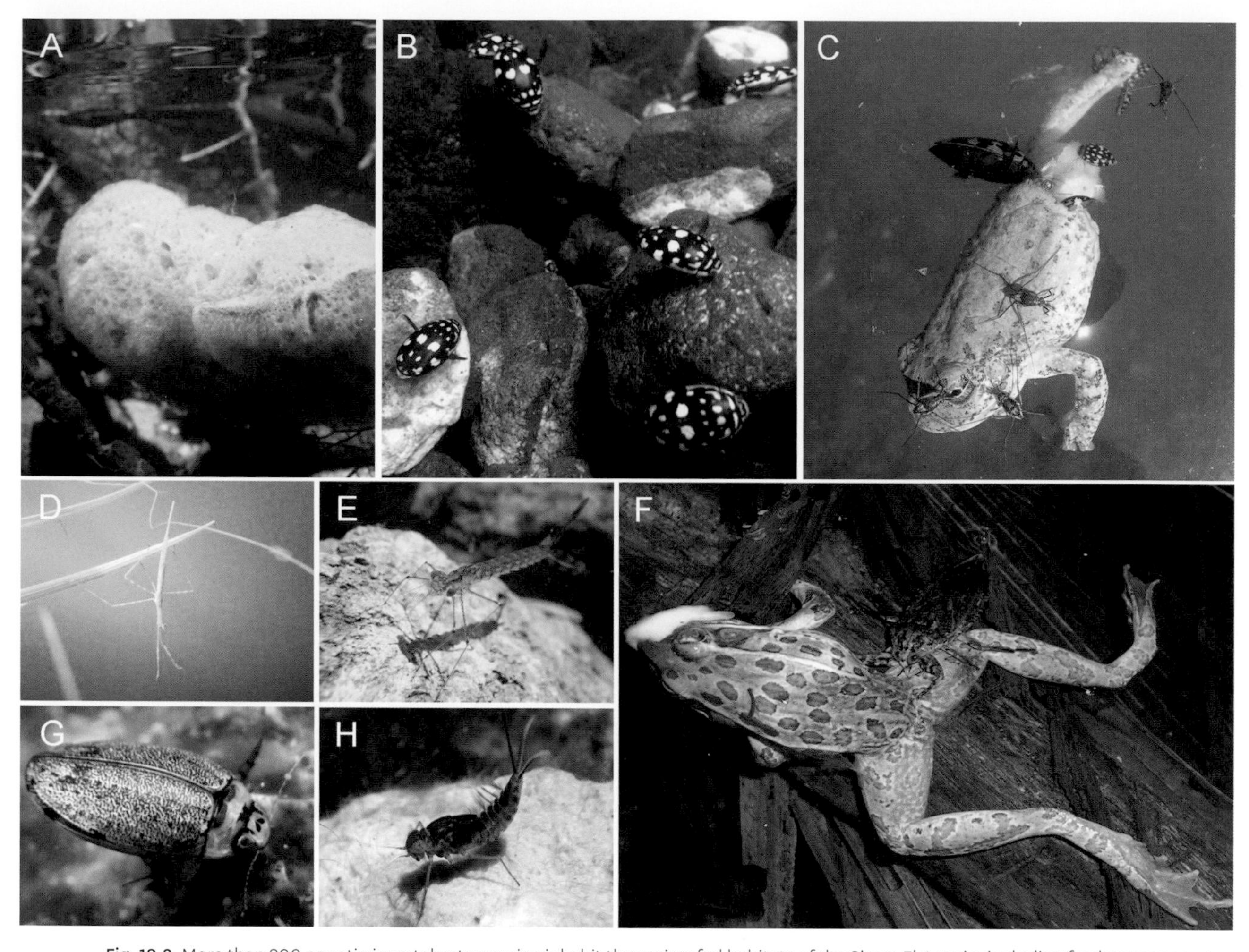

Fig. 18.3 More than 200 aquatic invertebrate species inhabit the spring-fed habitats of the Sierra El Aguaje, including freshwater sponges (Spongillidae) (*A*), sunburst beetles (*Thermonectus marmoratus*) (*B*, *C*), water striders (*Trepobates* sp.) (*C*, feeding on dead treefrog), water scorpions (*Ranatra quadridentata*) (*D*), damselflies (*Archilestes grandis*) (*E*), giant water bugs (*Lethocerus medius*) (*F*, feeding on dead leopard frog), diving beetles (*Thermonectus succinctus*) (*G*), and mayflies (*Callibaetis* sp.) (*H*).

for more than eight months each year (fig. 18.4). Some taxa appear to be more common in temporary habitats. For example, blackflies (Diptera: Simuliidae) are early successional species in many stream systems around the world (Hammock and Bogan 2014). In the Sierra El Aguaje, blackflies occur mainly in temporary riffles that flow for brief periods (i.e., weeks to months) after summer rains (Bogan 2012).

Surveys of the terrestrial adult life stages of aquatic insects have documented several new species that may be endemic to the Sierra El Aguaje. At least six new species of caddisflies and diving beetles have been found in two canyons, Nacapule and La Navaja (Bogan et al. 2014). Though taxonomists are still working to describe several of these species, the caddisfly *Neotrichia ruiteri* (Trichoptera: Hydroptilidae) was recently described (Keth et al. 2015). Currently, the population of this caddisfly at the oasis in La Navaja Canyon (see fig. 18.1*D*) is the only one known. There is still much to learn about aquatic invertebrates across the Sierra El Aguaje and other nearby ranges, and higher taxonomic resolution is needed for many difficult-to-identify groups (e.g., oligochaete worms, chironomid midges, snails).

Aquatic Vertebrates

At least six native species of aquatic and semiaquatic vertebrates occur in the Sierra El Aguaje. The most conspicuous and widespread is the Northwest México leopard frog (fig. 18.5*A*). This relatively large frog (~65 mm snout-vent length) occupies tinaja and oasis

Fig. 18.4 Seasonal aquatic habitat in Arroyo El Palmar, southern Sierra El Aguaje, as seen in matching photographs taken in April (*A*) and November (*B*), 2008.

habitats across at least five canyons (Pfeiler and Markow 2008; Bogan et al. 2014). Leopard frogs consume large invertebrates, such as adult dragonflies and scarab beetles, and in turn are often consumed by giant water bugs (see fig. 18.3*F*). Canyon treefrogs *Hyla arenicolor* and red spotted toads *Anaxyrus punctatus* occur in low abundances in at least four canyons. Although leopard frogs and treefrogs appear to be restricted to perennial water, red-spotted toads also breed in temporary habitats (like those in fig. 18.4). A distinct form of desert chub *Gila* cf. *eremica* is restricted to perennial pools in Las Pirinolas and La Balandrona canyons. As previously mentioned, these populations have probably been isolated from other freshwater fish populations for millions of years.

We have scattered observations for two species of semiaquatic vertebrates, but very little information on their distributions, feeding habits, or breeding habitats in the Sierra El Aguaje. To date, we have found only a single black-necked garter snake *Thamnophis cyrtopsis* from Las Pirinolas Canyon (fig. 18.5*B*). However, given the ubiquity of the species across the Sonoran Desert (Rorabough and Lemos-Espinal 2016), it probably inhabits other canyons. We also found a dried carapace of a mud turtle *Kinosternon* sp. in one canyon and observed live turtles at a second canyon, but failed to capture them to confirm their species identity. The Mexican mud turtle *K. integrum* is the only species currently documented from the region (Rorabough and Lemos-Espinal 2016), so it is probably the species we observed.

In addition to the chub, which lives only in freshwater, euryhaline marine fishes occasionally colonize the Sierra El Aguaje from the Gulf of California. These fishes include two sleepers (Eleotridae), a goby (Gobiidae), a mojarra (Gerreidae), and a mullet (Mugilidae) (table 18.1; fig. 18.5*D–F*). These fishes colonize freshwater habitats following intense monsoon or hurricane rainfall events, when ephemeral flow briefly connects inland freshwater habitats to the sea. These flow events usually last just a few hours, but in extreme cases, such as following Hurricane Jimena in 2009, they can last for weeks or even months (Findley et al. 2010). Unfortunately for the fishes, however, most of the freshwater habitats they colonize dry after a few weeks or months (see fig. 18.4).

In contrast to populations in more mesic regions (Miller et al. 2006), euryhaline colonists generally do not survive long enough to reproduce in freshwater or return to the ocean. Most will travel upstream only a kilometer or two before stopping or reaching impassible waterfall barriers in the Sierra El Aguaje. However, juveniles (total length ~40–50 mm) of the catadromous mountain mullet *Agnostomus monticola* swim upstream as far as 8 km, through incredible torrents, in search of permanent freshwater habitats (Bogan et al.,

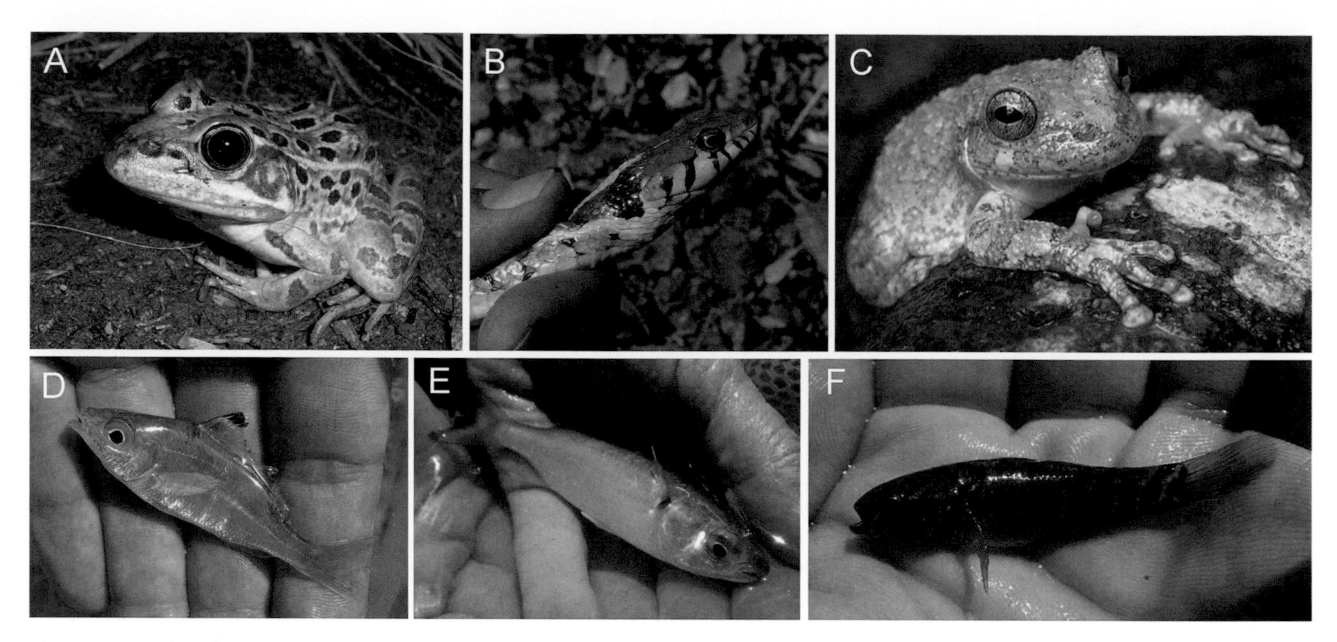

Fig. 18.5 Examples of aquatic vertebrates found in oases and tinajas of the Sierra El Aguaje, including Northwest México leopard frog (*Lithobates magnaocularis*) (*A*), black-necked garter snake (*Thamnophis cyrtopsis*) (*B*), canyon treefrog (*Hyla arenicolor*) (*C*), flag mojarra (*Eucinostomus currani*) (*D*), mountain mullet (*Agnostomus monticola*) (*E*), and Pacific fat sleeper (*Dormitator latifrons*) (*F*).

Table 18.1. Euryhaline fishes that have temporarily colonized inland freshwater habitats in the Sierra El Aguaje, Sonora, México.

Family	Species	Common name	No. of observations	No. of localities	Habitat type
Eleotridae	*Dormitator latifrons*	Pacific fat sleeper	3	2	Temporary
Eleotridae	*Eleotris picta*	Spotted sleeper	1	1	Temporary
Gerreidae	*Eucinostomus currani*	Pacific flagfin mojarra	2	2	Temporary
Gobiidae	*Awaous banana*	River goby	1	1	Temporary
Mugilidae	*Agnostomus monticola*	Mountain mullet	7	3	Perennial, temporary

Sources: Findley et al. 2010; Bogan 2012; M. T. Bogan, unpublished data.

unpublished data). With this impressive innate drive, small mountain mullet reached permanent freshwater habitat in Las Barajitas Canyon at least twice in the past 10 years. However, given that more than 6 years can pass between rainfall events that reconnect the oasis with the Gulf of California, it is unlikely they lived long enough to return to the ocean. Why these fish so readily colonize habitats that are likely to be reproductive "dead ends" is unknown, but mountain mullet remain a fascinating component of the Sierra El Aguaje's aquatic fauna.

Riparian Vertebrates

Several terrestrial vertebrates that occur in riparian areas of the Sierra El Aguaje highlight the canyons' roles as tropical outposts in the Sonoran Desert. For example, desert-outlier populations of tropical snake species such as Mexican boa *Boa constrictor imperator* and Yaqui black-headed snake *Tantilla yaquia* occur in several canyons (Felger 1999; Bogan et al. 2009). White-nosed coati *Nasua narica* are frequently seen at the Sierra El Aguaje's oases, feeding on ripe fruits in dense riparian groves of native fig trees (*Ficus* spp.). Sonoran cliff chipmunk *Tamias dorsalis sonorensis*, or *chichimoco* as it is locally known, normally occurs at much higher elevations (~2,000 m above sea level) in eastern Sonora. However, the steep canyon walls and riparian areas of the Sierra El Aguaje support highly isolated desert populations (~100 m above sea level).

Because few biological surveys have been done in the Sierra El Aguaje, much remains to be learned about its vertebrate fauna. Intriguing place names hint at some potential additions to the known fauna. For example, although no current substantiated records of jaguar *Panthera onca* exist, local ranchers have claimed sightings, and large, jaguar-like scratch marks on the trunk of a jito tree have been photographed (A. Gardea, pers. comm.). Further, *Arroyo El Tigre* (Jaguar Wash) presumably refers to the historical presence of this species. Additionally, a jaguar population was reported just 40 km to the east, in the Sierra Bacatete (Brown and López Gonzales 2001). It is thus easy to imagine that oases in the Sierra El Aguaje have been, or could one day be, home to jaguars.

CONSERVATION ISSUES

The freshwater oases of the Sierra El Aguaje are biodiversity gems that face numerous conservation concerns. Of particular concern, human water use and groundwater pumping are rapidly increasing for the growing town of San Carlos, which draws its municipal supply from wells in the Arroyo El Tigre aquifer, and groundwater withdrawal for agricultural and municipal use in the nearby cities of Guaymas and Empalme has resulted in plummeting water tables and decreased surface flows in the Río Mátape basin to the east (Custodio 2002). Research is needed to determine connectivity between montane aquifers in the Sierra El Aguaje and those in the adjacent coastal plain, where most pumping is occurring. Reduced groundwater levels have the potential to dry up seeps and springs and reduce tropical oases to arid canyons. Flowing-water habitat at Nacapule Canyon has already decreased significantly since the 1960s (Felger 1999), and the highly diverse riffle community present today probably arose in a much larger historical habitat. This unique community may be exhibiting an "extinction debt" (Jackson and Sax 2010; Halley et al. 2016), where the reduction in water levels has already occurred, but there is a temporal lag before species start disappearing. Thus, careful hydrologic and biodiversity monitoring is warranted at Nacapule.

The beauty of spring-fed oases in the Sierra El Aguaje also draws human visitors, and visitation rates have increased dramatically in recent years (Felger 1999; Carrizales 2007). While it is positive that people are learning more about these oases, some visitors destroy habitat by cutting vegetation, leaving behind trash and graffiti, and starting unnatural wildfires (fig. 18.6*A*, *B*). Fortunately, some local private ranchers and communal landowners (*ejiditarios*) are collaborating with state and federal agencies to guide visitors' experiences, control unrestrained access to some of the canyons, and develop a more sustainable model of ecotourism (fig. 18.6*C*, *D*).

Introductions of nonnative species also pose a grave threat to the aquatic biodiversity of the Sierra El Aguaje. In the neighboring Santa Úrsula and Bacatete ranges, nonnative tilapia *Tilapia* spp. have been introduced to most oases, causing extirpations of native amphibians and dramatic reductions in the diversity and density of aquatic invertebrates (Bogan et al. 2014). Additionally, strong negative effects of nonnative tilapia on native desert fishes have been observed in the oases of Baja California Sur (Andreu-Soler and Ruiz-Campos 2013). Tilapia had not been documented in the Sierra El Aguaje until they were introduced into a previously fishless habitat in February 2011, without permission from the local *ejiditario* (D. Magallanes-Molina, pers. comm.). Fortunately, flash floods appear to have destroyed the population. We did not find tilapia there in 2015, 2016, or 2017. Other exotic species, such as the American bullfrog *Lithobates catesbeianus*, have yet to invade the Sierra El Aguaje, but have had detrimental impacts on native species in the oases of Baja California Sur (Luja and Rodríguez-Estrella 2010).

FUTURE EXPLORATION

In recent years, we have learned much about the aquatic and riparian habitats in the Sierra El Aguaje. However, there remain numerous subjects to be investigated. A better understanding of regional hydrology is particularly needed. Currently, we have only hypotheses about the sources and modes of recharge for groundwater that feeds oases. Further geologic and hydrologic studies would help reveal how vulnerable these oases are to regional groundwater pumping and climate change. Stable isotope and other geochemical studies would help determine the source and age of life-giving groundwater.

Over 200 species of aquatic animals occupy the Sierra El Aguaje, but we are just beginning to understand the connectivity of their populations among isolated aquatic habitats. Animals with a terrestrial or semi-

Fig. 18.6 Impacts of human recreation and evidence of new land management strategies in the Sierra El Aguaje. In 2004, La Navaja Canyon oasis was rarely visited, and vegetation was lush (*A*). Visitation increased dramatically when the oasis was featured on a local television program in 2009. In February 2011, visitors set individual palm front skirts on fire lower in the canyon and started a palm-forest fire (*B*). At Nacapule Canyon, the most popular canyon in the range, local landowners (the Dávila family) and the state ecology agency of Sonora (CEDES) work together to promote sustainable ecotourism (*C, D*). They control time of access (daytime use only), offer guided recreational activities, inform visitors about the canyon's biodiversity, and collect trash after periods of heavy visitation.

terrestrial adult stage (e.g., aquatic insects, frogs) can disperse across the harsh desert at opportune times. In contrast, wholly aquatic animals (e.g., fishes) must move during wet periods or hurricane-induced floods, and sessile animals (e.g., sponges) and plants have no obvious ability to actively disperse. Understanding population connectivity is essential to predicting which species will be resilient to natural and human-caused disturbances (e.g., hurricane flooding, groundwater withdrawal, wildfire). Much work is needed to document how aquatic and riparian plants, invertebrates, and vertebrates move, or fail to move, across the Sierra El Aguaje. Such information is essential to effective conservation planning.

Finally, additional studies will probably reveal new species endemic to the Sierra El Aguaje. Without knowledge of endemic species, efforts to protect them may not be implemented in time to ensure their conservation. For example, in the neighboring Chihuahuan Desert, some species of springsnails (Gastropoda: Hydrobiidae) were described from museum specimens only after their extinction because the springs to which they were endemic had already dried (Hershler et al. 2011). If scientists and local land managers had known about the endemic species prior to their extinction, dedicated efforts could have been made to better protect the water sources or salvage the species from the drying springs. Efforts should also be made to survey the freshwater microbial communities in the Sierra El Aguaje. Aquatic microbial communities represent a new frontier for the fields of biodiversity and biogeography. Recent genomic and environmental DNA studies from other deserts have revealed astounding biodiversity and endemism among aquatic microbes in desert springs (Souza et al. 2008; Espinosa-Asuar et al. 2015), suggesting that the age of biodiversity exploration is still young in the Sierra El Aguaje.

ACKNOWLEDGMENTS

The Sierra El Aguaje is the ancestral home of the *Xiica xnaai iicp coii* band of the Comcáac people. We are grateful to Delfino Magallanes-Molina, Ramón Villafraña, and the Dávila family for access to several canyons in the Sierra El Aguaje. Dave Lytle, Kate Boersma, Rick Ziegler, Lacey Greene, Dean Wilde, Norm Swider, Oscar Gutiérrez-Ruacho, Andrés Alvarado-Castro, Nohemí Noriega-Felix, and Sylvette Vidal-Aguilar assisted us on visits to several canyons. We thank Ben Wilder and the Next-Generation Sonoran Desert Researchers network for encouraging interdisciplinary studies of the Sierra El Aguaje. Reviews by David Miller, David Rogowski, and one anonymous reviewer greatly improved our chapter. This chapter is dedicated to the memory of Enrique Castillo-Grijalva, who facilitated access to several study sites and fearlessly clambered up canyons and cliffs with us in search of hidden water.

REFERENCES

Andreu-Soler A., and G. Ruiz-Campos. 2013. Effects of exotic fishes on the somatic condition of the endangered killifish *Fundulus lima* (Teleostei: Fundulidae) in oases of Baja California Sur, Mexico. *Southwestern Naturalist* 58: 192–201.

Antinao, J., E. McDonald, E. Rhodes, N. Brown, W. Barrera, J. Gosse, and S. Zimmerman. 2016. Late Pleistocene-Holocene alluvial stratigraphy of southern Baja California, Mexico. *Quaternary Science Reviews* 146: 161–81.

Atwater, T., and J. Stock. 1998. Pacific North America plate tectonics of the Neogene southwestern United States: An update. *International Geology Review* 40: 375–402.

Bailowitz, R., D. Danforth, and S. Upson. 2015. *A Field Guide to the Damselflies and Dragonflies of Arizona and Sonora*. Tucson: Nova Granada Publications.

Ballesteros-Córdova, C. A. 2017. Identidad específica de la carpa del desierto (*Gila eremica* DeMarais, 1991) en las cuencas de los Ríos Sonora y Mátape, Sonora. PhD diss., Universidad de Sonora.

Ballesteros-Córdova, C. A., G. Ruiz-Campos, L. T. Findley, J. M. Grijalva-Chon, L. E. Gutiérrez-Millán, and A. Varela-Romero. 2016. Morphometric and meristic characterization of the endemic desert chub *Gila eremica* (Teleostei: Cyprinidae), and its related congeners in Sonora, México. *Revista Mexicana de Biodiversidad* 87: 390–98.

Bennett, S. E. K., and M. E. Oskin. 2014. Oblique rifting ruptures continents: Example from the Gulf of California shear zone. *Geology* 42: 215–18.

Bennett, S. E. K., M. E. Oskin, and A. Iriondo. 2013. Transtensional rifting in the proto-Gulf of California, near Bahía Kino, Sonora, México. *Geological Society of America Bulletin* 125: 1752–82.

Bogan, M. T. 2012. Drought, dispersal, and community dynamics in arid-land streams. PhD diss., Oregon State University.

Bogan, M. T., L. T. Findley, and E. F. Endersen. 2009. Geographic distribution: *Tantilla yaquia* (Yaqui black-headed snake). *Herpetological Review* 40: 458.

Bogan, M. T., N. Noriega-Felix, S. L. Vidal-Aguilar, L. T. Findley, D. A. Lytle, O. G. Gutiérrez-Ruacho, J. A. Alvarado-Castro, and A. Varela-Romero. 2014. Biogeography and conservation of aquatic fauna in spring-fed tropical canyons of the southern Sonoran Desert, Mexico. *Biodiversity and Conservation* 23: 2705–48.

Bowen, T. 1976. Seri prehistory: The archaeology of the central coast of Sonora, Mexico. *University of Arizona Anthropological Papers* 27: 1–120.

Brown, D. E., and C. A. López Gonzales. 2001. *Borderland Jaguars*. Salt Lake City: University of Utah Press.

Caine, J. S., J. P. Evans, and C. B. Forster. 1996. Fault zone architecture and permeability structure. *Geology* 24: 1025–28.

Carrizales, J. 2007. Cañón del Nacapule: Peligra paraíso ecológico. *El Imparcial* (newspaper; Hermosillo, Sonora). http://expresionguaymas.wordpress.com/2007/01/22/canon-del-nacapule-peligra-paraiso-ecologico. Accessed February 8, 2014.

Custodio, E. 2002. Aquifer overexploitation: What does it mean? *Hydrogeology Journal* 10: 245–77.

Darin, M. H., S. E. K. Bennett, R. J. Dorsey, M. E. Oskin, and A. Iriondo. 2016. Late Miocene extension in coastal Sonora, México: Implications for the evolution of dextral shear in the proto–Gulf of California oblique rift. *Tectonophysics* 693: 378–408.

Davis, G. H., and S. J. Reynolds. 1996. *Structural Geology of Rocks and Regions*. 2nd ed. Hoboken, NJ: John Wiley & Sons.

Dixon, K. A. 1990. La Cueva de la Pala Chica: A burial cave in the Guaymas region of coastal Sonora, Mexico. *Vanderbilt University Publications in Anthropology* 38: 1–97.

Espinosa-Asuar, L., A. E. Escalante, J. Gasca-Pineda, J. Blaz, L. Peña, L. E. Eguiarte, and V. Souza. 2015. Aquatic bacterial assemblage structure in Pozas Azules, Cuatro Ciénegas Basin, Mexico: Deterministic vs. stochastic processes. *International Microbiology* 18: 105–15.

Felger, R. S. 1999. The flora of Cañón de Nacapule: A desert-bounded tropical canyon near Guaymas, Sonora, Mexico. *Proceedings of the San Diego Society of Natural History* 35: 1–42.

Felger, R. S., S. D. Carnahan, and J. J. Sánchez-Escalante. 2017. Oasis at the Desert Edge: Flora of Cañón del Nacapule, Sonora, Mexico. *Proceedings of the Desert Laboratory* 1: 1–220.

Felger, R. S., and M. B. Moser. 1985. *People of the Desert and Sea: Ethnobotany of the Seri Indians*. Tucson: University of Arizona Press.

Findley, L. T., M. Pérez-Tello, and W. L. Montgomery. 2010. A "thousand-year" flood provides long-term ephemeral habitat for "peripheral" fishes in the southern Sonoran Desert, Mexico. *Program and Abstracts, Desert Fishes Council 42nd Annual Meeting*, 11–12.

Halley, J. M., N. Monokrousos, A. D. Mazaris, W. D. Newmark, and D. Vokou. 2016. Dynamics of extinction debt across five taxonomic groups. *Nature Communications* 7: 12283.

Hammock, B. G., and M. T. Bogan. 2014. Black fly larvae facilitate community recovery in a mountain stream. *Freshwater Biology* 59: 2162–71.

Hershler, R., H.-P. Liu, and J. J. Landye. 2011. New species and records of springsnails (Caenogastropoda: Cochliopidae: *Tryonia*) from the Chihuahuan Desert (Mexico and United States), an imperiled biodiversity hotspot. *Zootaxa* 3001: 1–32.

Jackson, S. T., and D. F. Sax. 2010. Balancing biodiversity in a changing environment: Extinction debt, immigration credit and species turnover. *Trends in Ecology and Evolution* 25: 153–60.

Keth, A. C., S. C. Harris, and B. J. Armitage. 2015. *The Genus* Neotrichia *Morton (Trichoptera: Hydroptilidae) in North America, Mexico, and the Caribbean Islands*. Columbus, OH: Caddis Press.

León de la Luz, J. L., and R. Domínguez-Cadena. 2006. Hydrophytes of the oases in the Sierra de la Giganta of Central Baja California Sur, Mexico: Floristic composition and conservation status. *Journal of Arid Environments* 67: 553–65.

León de la Luz, J. L., R. Domínguez-Cadena, M. Domínguez-León, and J. J. Perez-Navarro. 1997. Floristic composition of the San Jose del Cabo Oasis, Baja California Sur, México. *SIDA Contributions to Botany* 17: 599–614.

Luja, V. H., and R. Rodríguez-Estrella. 2010. The invasive bullfrog *Lithobates catesbeianus* in oases of Baja California Sur, Mexico: Potential effects in a fragile ecosystem. *Biological Invasions* 12: 2979–83.

Miller, R. R., W. L. Minckley, and S. M. Norris. 2006. *Freshwater Fishes of México*. Chicago: University of Chicago Press.

Mims, M. C., I. C. Phillipsen, D. A. Lytle, E. Hartfield Kirk, and J. D. Olden. 2015. Ecological strategies predict associations between aquatic and genetic connectivity for dryland amphibians. *Ecology* 96: 1371–82.

Mora-Álvarez, G., and F. W. McDowell. 2000. Miocene volcanism during late subduction and early rifting in the Sierra Santa Ursula of western Sonora, Mexico. *Geological Society of America Special Papers* 334: 123–41.

Moser, E. 1963. Seri bands. *Kiva* 28: 14–27.

Nabhan, G. P. 2003. *Singing the Turtles to Sea: The Comcaac (Seri) Art and Science of Reptiles*. Berkeley: University of California Press.

Pérez-Tello, M. G., J. P. Gallo-Reynoso, J. L. Villalobos-Hiriart, and J. G. Mondragón-Mota. 2016. Registro de *Macrobrachium olfersii* en un arroyo temporal de San Carlos, Guaymas, Sonora, México. *Revista Mexicana de Biodiversidad* 87: 1379–82.

Pfeiler, E., and T. A. Markow. 2008. Phylogenetic relationships of leopard frogs (*Rana pipiens* complex) from an isolated coastal mountain range in southern Sonora, Mexico. *Molecular Phylogenetics and Evolution* 49: 343–48.

Rorabough, J. C., and J. A. Lemos-Espinal. 2016. *A Field Guide to the Amphibians and Reptiles of Sonora, Mexico*. Tucson: ECO Wear & Publishing.

Sada, D. W., E. Fleishman, and D. D. Murphy. 2005. Associations among spring-dependent aquatic assemblages and environmental and land use gradients in a Mojave Desert mountain range. *Diversity and Distributions* 11: 91–99.

Smith, G. I., I. Friedman, G. Veronda, and C. A. Johnson. 2002. Stable isotope composition of waters in the Great Basin, United States. 3. Comparison of groundwaters with modern precipitation. *Journal of Geophysical Research* 107: ACL 16-1–16-15.

Sousa, V., L. Espinosa-Asuar, A. E. Escalante, L. E. Eguiarte, J. Farmer, L. Forney, L. Lloret, J. M. Rodríguez-Martínez, X. Soberón, R. Dirzo, and J. J. Elser. 2008. An endangered oasis of aquatic microbial biodiversity in the Chihuahuan Desert. *Proceedings of the National Academy of Sciences* 103: 6565–70.

Spicer, E. H. 1980. *The Yaquis: A Cultural History*. Tucson: University of Arizona Press.

Varela-Romero, A. 2001. Newly documented localities for desert chub, *Gila eremica*, in tropical canyons, Río Mátape Basin, Sonora, México. *Proceedings of the Desert Fishes Council* 32: 33.

Young, M. E., P. G. Macumber, M. D. Watts, and N. Al-Toqy. 2004. Electromagnetic detection of deep freshwater lenses in a hyper-arid limestone terrain. *Journal of Applied Geophysics* 57: 43–61.

19

Alejandro Varela-Romero,
Carlos Alonso Ballesteros-Córdova,
Gorgonio Ruiz-Campos,
Sergio Sánchez-González,
and James E. Brooks

Recent Discoveries and Conservation of Catfishes, Genus Ictalurus, in México

North American catfishes of the genus *Ictalurus* (family Ictaluridae) inhabit freshwaters from southern Canada to Belize and northern Guatemala. On the Pacific Slope of México, the genus is distributed from the basins of the Ríos Sonora and Yaqui in the north to the Río Balsas basin in the south (Miller 1959; Berra 2007; Miller et al. 2009), and on the Gulf Coastal side from the Río Bravo in the north to the Río Usumacinta in the south (table 19.1).

Currently, there are eight described species of *Ictalurus* in México (Miller et al. 2009; Ruiz-Campos et al. 2009; and Rodiles-Hernández et al. 2010). A phylogenetic analysis of the family based on morphology (Lundberg 1992) places these species in two sister clades, the *I. punctatus* and *I. furcatus* species groups (fig. 19.1). Forms described as *I. australis, I. ochoterenai,* and *I. meeki,* once considered Mexican endemics, now are treated as synonyms of *I. punctatus, I. dugesii,* and *I. pricei,* respectively (Lundberg 1992; Miller et al. 2009).

Miller et al. (2009) noted that the number of Mexican species of *Ictalurus* and their systematic relationships are poorly known, despite México being the center for speciation and adaptive radiation of the genus. This paucity of knowledge is due to inadequate sets of specimens and tissues for morphological and molecular analyses, the physical difficulty of accessing their habitats, and dangers presented by drug trafficking for scientists collecting specimens. Nonnative catfishes continue to expand, and hybridization with *I. punctatus* is well documented and widespread in several basins in México, which hampers understanding and conservation efforts (Varela-Romero et al. 2011). The purpose of this chapter is to synthesize scattered information about undescribed species of native catfish, describe the limited knowledge of this important fauna in México, and promote studies for better management and conservation.

Table 19.1. Occurrence, conservation status, and assignment to *I. punctatus* and *I. furcatus* clades for the described catfish species in México.

Species	Common name		Occurrence	Clade	Conservation status
	Spanish	English			
Ictalurus punctatus	Bagre de canal	Channel catfish	Atlantic Slope, Río Bravo del Norte in north, south into Río Cazones, eastern México	*punctatus*	México (SEMARNAT, 2010): Unlisted IUCN: Least concern
Ictalurus lupus	Bagre lobo	Headwater catfish	Atlantic Slope, Ríos Salado and San Juan in north southward into Río Soto La Marina basin, eastern México	*punctatus*	México (SEMARNAT, 2010): Special Protection IUCN: Data deficient Texas, New Mexico: Species of Special Concern
Ictalurus dugesii	Bagre de Lerma	Lerma catfish	Pacific Slope, basins of Ríos Lerma and Ameca	*punctatus*	México (SEMARNAT, 2010): Threatened IUCN: Unlisted
Ictalurus mexicanus	Bagre del Verde	Río Verde catfish	Pacific Slope, Río Pánuco basin, eastern México	*punctatus*	México (SEMARNAT, 2010): Threatened IUCN: Vulnerable
Ictalurus pricei	Bagre Yaqui	Yaqui catfish	Pacific Slope, Ríos Yaqui, Fuerte, and Mayo (Mayo population may be extirpated), northwestern México	*punctatus*	México (SEMARNAT, 2010): Threatened IUCN: Endangered ESA of USFWS: Threatened
Ictalurus furcatus	Bagre azul	Blue catfish	Atlantic Slope, Ríos Bravo and Conchos in north southward into the Río Soto La Marina, eastern México	*furcatus*	México (SEMARNAT, 2010): Unlisted IUCN: Least concern
Ictalurus meridionalis	Bagre del Usumacinta	Southern blue catfish	Atlantic Slope, eastern México from the Río Pánuco basin, Veracruz, south into Ríos Usumacinta and Honda of México, Guatemala, and Belize	*furcatus*	México (SEMARNAT, 2010): Unlisted IUCN: Unlisted
Ictalurus balsanus	Bagre del Balsas	Balsas catfish	Pacific Slope, Río Balsas basin, southwestern México	*furcatus*	México (SEMARNAT, 2010): Unlisted IUCN: Unlisted

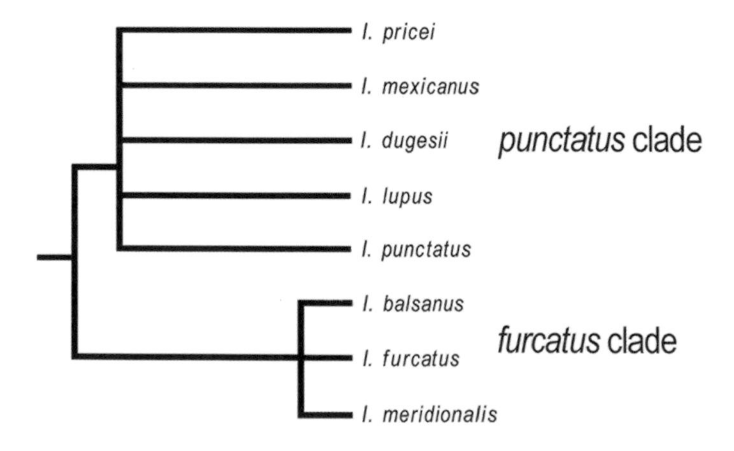

Fig. 19.1 Phylogeny based on morphological classification of catfish species of the genus *Ictalurus*. (Modified from Lundberg 1992.)

DISCOVERIES

Literature on Mexican catfishes has long alluded to the presence of several undescribed species of *Ictalurus* in multiple drainages of México. Miller et al. (2009) summarized information regarding several undescribed taxa, but in spite of their recognition, no new species have been described (e.g., Varela-Romero 2007; Minckley and Marsh 2009; Varela-Romero et al. 2011; van der Heiden et al. 2016; Ruiz-Campos et al. 2016). These forms are scattered among major drainages of México and are discussed below according to areas of suspected occurrence.

Pacific Coast, Sierra Madre Occidental

Undescribed catfishes of Pacific drainages include members of the *Ictalurus pricei* complex that have taxonomically messy descriptions (Varela-Romero et al. 2011). *Ictalurus pricei* is historically native to the highlands of northwestern México and the southwestern United States in the basins of the Ríos Yaqui, Mayo, and Fuerte (Hendrickson 1983; Minckley et al. 1986; Minckley and Marsh 2009; Varela-Romero et al. 2011). Besides *I. pricei*, the complex includes populations in drainages on the Mexican Pacific versant south of the Río Fuerte, in the basins of the Ríos Sinaloa, Culiacán, San Lorenzo, Baluarte, and San Pedro–Mezquital. These populations apparently represent at least one, and possibly two, undescribed species, differing from *I. pricei* in cephalic morphology and in having lower anal-fin ray counts (fig. 19.2) (Varela-Romero et al. 2011).

Phylogenetic analysis of mitochondrial cytochrome *b* (*cytb*) sequences identified the populations south of the Río Fuerte as sister to, but distinct from, *I. pricei* (fig. 19.3; Varela-Romero 2007), which was also supported by analyses of the cytochrome oxidase subunit I (*COI*) (Castañeda-Rivera et al. 2014) and the entire mitochondrial DNA sequence (Ballesteros-Córdova et al. 2016). Data for the Bayesian inference were obtained

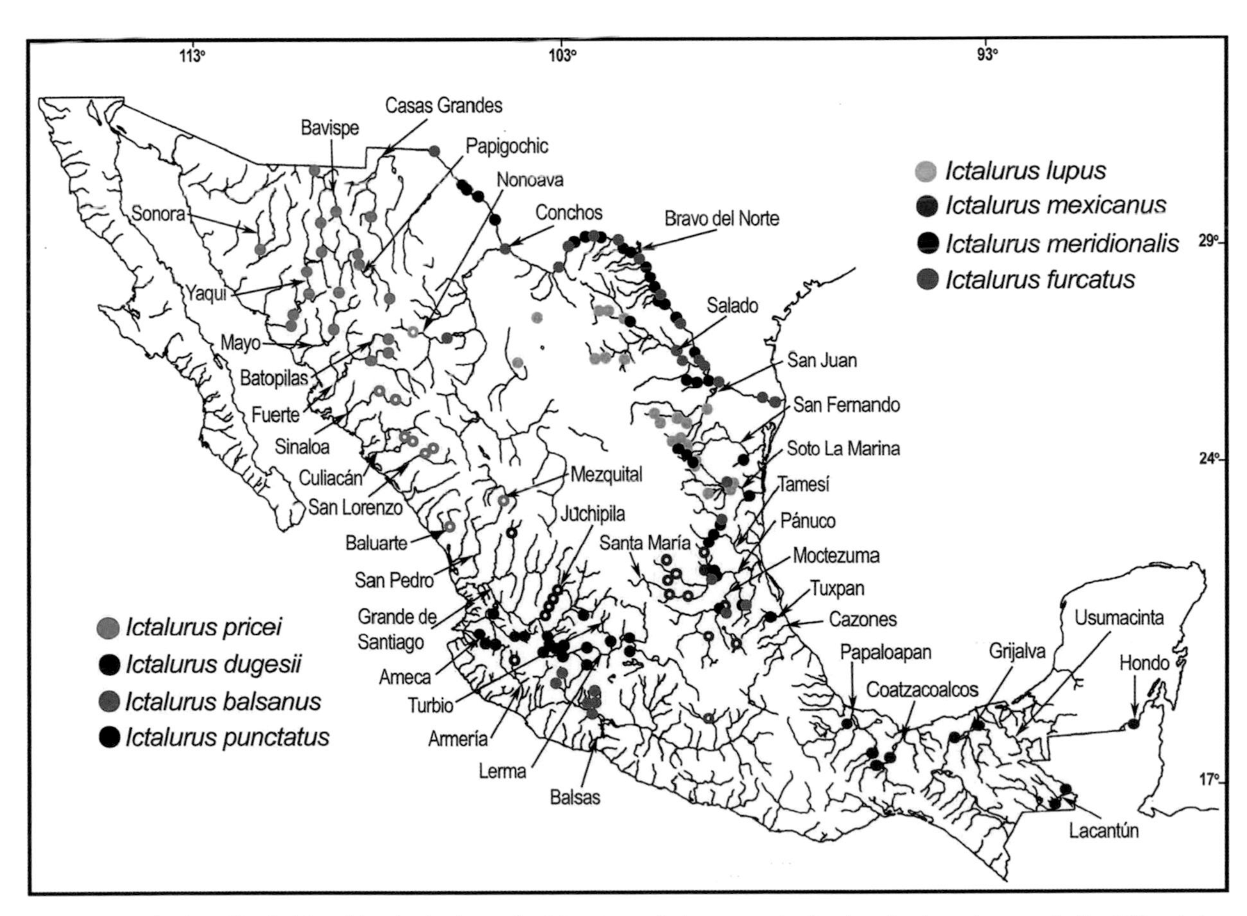

Fig. 19.2 Distribution of ictalurid catfishes in the rivers of México. Open circles = records of undescribed species reported by Miller et al. (2009), Varela-Romero (2007), and van der Heiden et al. (2016). (Map modified from Miller et al. 2009.)

using a modified universal PCR set of primers (Kocher et al. 1989) to obtain the *cytb* gene and run a maximum likelihood analysis with tree bisection–reconnection (TBR) branch-swapping heuristic searches. Using the GTR+I+G model of nucleotide substitution, we calculated tree topology and parameter values in a maximum likelihood (ML) search using PAUP. This tree was used as the starting point for a non-parametric bootstrap analysis of the data in PHYML (Guindon and Gascuel 2003; 1,000 pseudo-replicates, settings as for initial tree search) to estimate support for the nodes of the ML tree. Bayesian inference was performed in the software MrBayes 3.1 (Ronquist and Huelsenbeck 2003). We performed two independent runs starting from random trees in which the Metropolis-coupled Markov chain Monte Carlo (MC3) process included four chains (three heated and one cold) running simultaneously for 4×10^6 generations and in which parameter values and trees were sampled every 1,000 generations. Default settings were used, and we determined the approximate number of generations required for convergence of the two runs onto the distribution of posterior probabilities by examining the change in average standard deviation of split frequencies and a plot of cold chain log likelihoods versus generations. We excluded trees and model parameter values recorded in the preconvergence "burn-in" phases of the two runs from the calculation of parameter values and trees (see fig. 19.3).

The southernmost population of the *I. pricei* complex (Río San Pedro–Mezquital basin) may represent an undescribed species. It had a single haplotype that was 0.02900 *p* (table 19.2) divergent from its sister group, and this undescribed form (informally called the Sinaloa catfish) is restricted to the Ríos Culiacán and San Lorenzo. Molecular differentiation of the Río Culiacán/San Lorenzo catfish from the Yaqui catfish was confirmed with a meristic and morphometric comparison of the two forms, and the differences are now supported by both molecular and morphological analyses (Varela-Romero 2007; Ruiz-Campos et al.

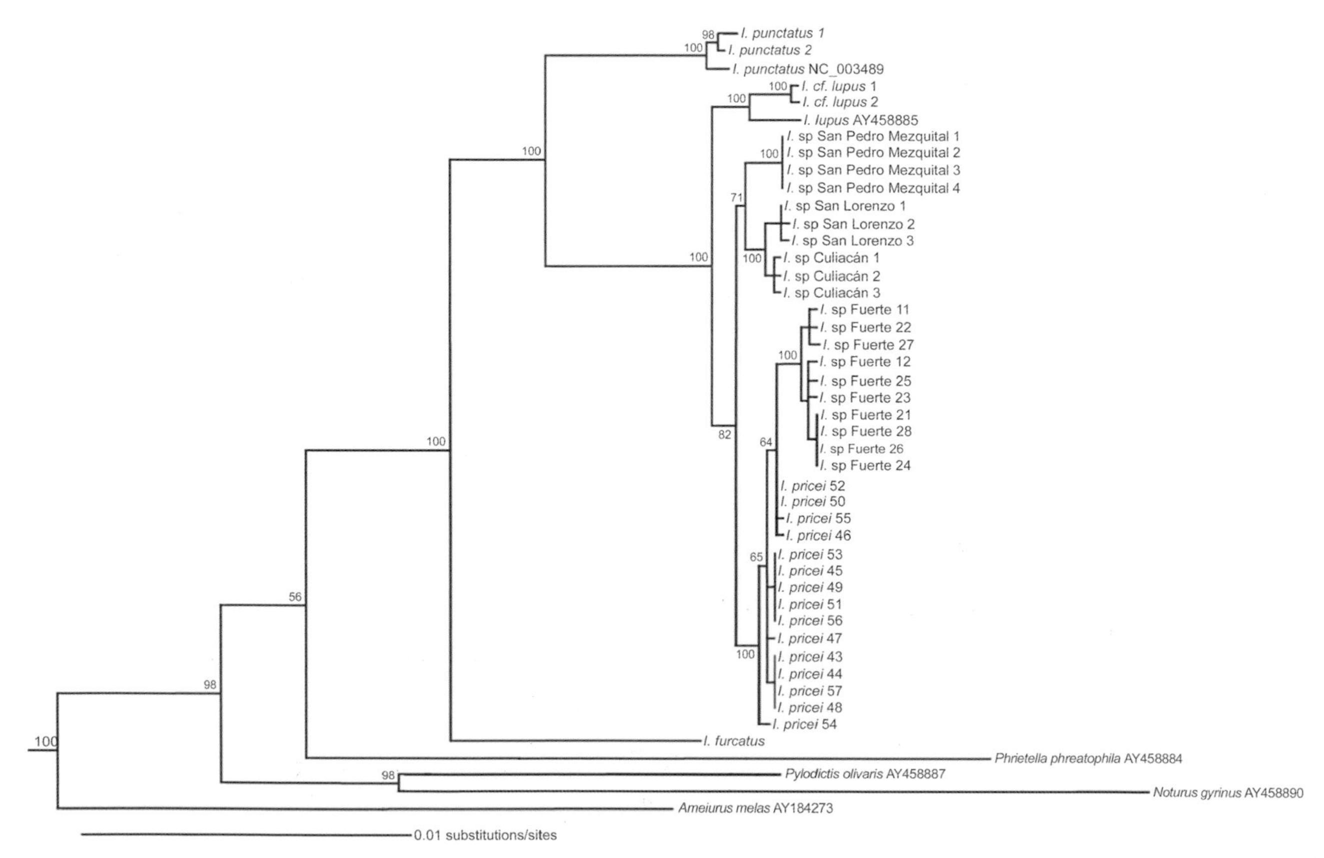

Fig. 19.3 Bayesian analysis of *cytb* sequences for native and nonnative catfishes of northwestern México. Branching pattern = the 50% majority-rule consensus of combined post-burn-in trees from two independent runs in MrBayes 3.1. The *cytb* sequences were partitioned by codon position, each with an unlinked GTR+I+G model of substitution. Posterior probabilities at nodes are expressed as percentages. Numbers on the terminal node labels are specimen identification numbers. The numbers with AY prefix are GenBank accession numbers.

2016). Other indications of distinct species in Pacific drainages of west-central México include mention of an undescribed species of the *I. pricei* complex in the Arroyo Colorado of the Río Baluarte basin near Pánuco, Sinaloa (Van der Heiden et al. 2016), and Miller's (1986) report of an undescribed species in the Ríos Tunal and Santiaguillo of the Río San Pedro–Mezquital basin in Durango.

Mesa Central

The account by Miller et al. (2009) on *I. dugesii* mentions that populations from three basins draining the Mesa Central—the Ríos Balsas, Grande de Santiago, and Armería—represent one or more undescribed species (see fig. 19.2).

Gulf Coast, Río Bravo del Norte/Rio Grande

Sampling in the range of *I. lupus* in the Río Nonoava of the Río Conchos system (see fig. 19.2) in Chihuahua, northwestern México (Varela-Romero 2007), verified presence of an undescribed catfish species. This catfish potentially represents the undescribed *I. lupus*-like form previously reported from the Río Conchos and associated sections of the Pecos River and Río Bravo del Norte by Williams et al. (1989) and Miller et al. (2009). A phylogenetic analysis combining data from two mtDNA *cytb* studies (Varela-Romero 2007; McClure-Baker et al. 2010) groups this form (*I.* cf. *lupus* 1 and 2 of fig. 19.3) in the same clade as *I. lupus* from San Felipe Creek of the Devils River in Texas (*I. lupus* AY458885, fig.19.2), and the percentage sequence divergence between *I. lupus* and this undescribed Conchos River form was 0.07293 based on uncorrected *p* distances (see table 19.2).

Gulf Coast, Eastern México

An undescribed species occurs in eastern México in tributaries of the upper Río Pánuco basin of Hidalgo, Querétaro, and San Luis Potosí states (Miller et al. 2009). This species was previously referred to *I. mexicanus*, an endemic of the basin. It is unclear whether the work published on the parasites of *I. mexicanus* (Salgado-Maldonado et al. 2004; Gutiérrez-Cabrera et al. 2005; Miranda et al. 2012) actually refers to *I. mexicanus* or to one or more undescribed forms of the Río Pánuco basin.

In addition, Ruiz-Campos et al. (2009) and Rodiles-Hernández et al. (2010) resurrect the name of *I. meridionalis* for the blue catfish from Río Pánuco to Río Usumacinta in Belize based on morphological and molecular characters, recognizing it as a distinct species from *I. furcatus*.

CONSERVATION STATUS OF DESCRIBED SPECIES

Only four *Ictalurus* species are assigned protected status by the Mexican government. Undescribed forms tend to occur in relatively limited geographic areas and are under the same stressors faced by the listed species. Although the current conservation status of these forms is unresolved, the most conservative approach would be to consider all undescribed species imperiled until

Table 19.2. Uncorrected *p* sequence divergence distances (obtained using PAUP version 4.0b10) from *cytb* gene dataset from various described species of catfish and undescribed species of the *pricei* complex in the Pacific coast, Sierra Madre Occidental.

		1	2	3	4	5	6	7	8
1	*I. punctatus*	—							
2	*I.* sp. (Conchos)	0.07821	—						
3	*I. lupus*	0.07206	0.07293	—					
4	*I.* sp. (San Pedro–Mezquital)	0.06767	0.06855	0.02900	—				
5	*I.* sp. (San Lorenzo)	0.07118	0.07206	0.02636	0.02900	—			
6	*I.* sp. (Culiacán)	0.07030	0.07118	0.02373	0.02636	0.01406	—		
7	*I.* sp. (Fuerte)	0.06942	0.07030	0.02812	0.02724	0.01933	0.01757	—	
8	*I. pricei* 50	0.07293	0.07381	0.02460	0.02636	0.01670	0.01582	0.01318	—

proven otherwise. In reality, until they are officially described and better understood, they will receive no protection. Research on their phylogenetics, distribution, and ecology should be given the highest priority, or some of these species might disappear before they are described.

Ictalurus dugesii

This species is categorized as threatened in México (SEMARNAT 2010). Main causes for population declines are habitat loss due to water depletion and pollution, introduction of nonnative common carp *Cyprinus carpio* and Mozambique tilapia *Oreochromis mossambicus*, and overfishing (Soto-Galera and Alcántara-Soria 2007).

Ictalurus mexicanus

This catfish is listed as threatened in México (SEMARNAT 2010) and vulnerable by the International Union for Conservation of Nature (IUCN) (Contreras-Balderas and Almada-Villela 1996). Primary risk factors are urban, industrial, and agricultural pollution, as well as habitat modification due to hydraulic works and deforestation (Soto-Galera 2003).

Ictalurus lupus

The official conservation status of *I. lupus* in México is "under special protection," and IUCN (NatureServe 2013a) lists it as "data deficient." The current distribution and abundance of *I. lupus* is probably declining, but the rate of decline is unknown. Miller et al. (2009, 163) remarked that it is "one of the least known or studied North American fishes," and Hudson et al. (2005) noted that it was surprising that a basin as important as the Conchos was so poorly sampled for fish. Although *I. lupus* formerly had a relatively wide distribution in the Rio Grande basin of Texas and New Mexico, it is now listed as a Species of Special Concern in those states (Williams et al. 1989).

Hybridization and competition with introduced channel catfish *I. punctatus* pose the most imminent threats for *I. lupus*, especially where anthropogenic habitat changes may favor the former species (McClure-Baker et al. 2010). Koster (1957) and Sublette et al. (1990) noted the nonnative status of *I. punctatus* in the Rio Grande basin in New Mexico. Additionally, McClure-Baker et al. (2010) found no verifiable specimens of channel catfish from the middle or lower Rio Grande of Texas or adjacent México prior to 1951, and suggested that *I. punctatus* was most likely introduced into the Rio Grande after 1930. Conservation efforts for *I. lupus* should focus on detecting and conserving populations that are not genetically introgressed with channel catfish. According to Sublette et al. (1990, 240), "Competition and/or hybridization with *I. punctatus* in the greatly perturbed streams of New Mexico has eliminated *I. lupus* from most of its original range."

In Río Bravo del Norte tributaries in México, *I. lupus* faces the same threats as in New Mexico (Soto-Galera and Rincón-Sandoval 2007). In addition, dramatic flow reductions and increases in agricultural, chemical, and petrochemical pollution over the last 50 years have strongly impacted the lower Rio Grande region, with dramatic negative effects on the ichthyofauna of the region (Edwards and Contreras-Balderas 1991). Museum collections of *I. lupus* from México date as early as 1930, with a total of 171 specimens from 71 collections, 35 of them reporting only a single specimen (Soto-Galera and Rincón-Sandoval 2007). These records reflect the scarcity of this species in particular and are representative of an increasingly evident generalization: native catfishes are unable to thrive in altered habitats of the American Southwest.

Ictalurus pricei

Wild populations of *Ictalurus pricei* are listed as critically endangered by IUCN. In México between 2002 and 2010, the species was granted only "special protection" status (SEMARNAT 2002), but was upgraded to threatened (SEMARNAT 2010) after it was proposed as endangered to the Mexican federal government by Varela-Romero et al. (2011). Status changes were recommended following a survey of the entire Mexican range of *I. pricei*. Only three sites in the Río Yaqui basin and two in the Río Fuerte basin yielded Yaqui catfish; none were found in the Río Mayo basin, and it is extirpated from the Sonora and Casas Grandes basins (Lundberg 1992; Miller et al. 2009; Minckley and Marsh 2009; Varela-Romero and Hendrickson 2009). In the United States, Yaqui catfish have been listed as a threatened species since 1984 under the Endangered Species Act (USFWS 1984), and recently the IUCN increased its classification level to endangered (NatureServe 2013b).

Conservation status for undescribed forms of this complex is basically unknown due to lack of survey information. Surveys should be accompanied by assessment of species status using thorough morphological and molecular studies of additional sequence information including nuclear loci (Castañeda-Rivera et al. 2014; Ballesteros-Córdova et al. 2016). The upper San Pedro–Mezquital basin deserves special conservation attention because it harbors at least one undescribed species of the *I. pricei* complex and is the last watercourse of the Pacific versant to remain undammed (WWF 2010). It is the seventh largest (by volume) river in México and flows through a canyon 800–1,000 m deep in the Sierra Madre Occidental. This river corridor effectively links the Chihuahuan Desert with the Gulf of California, Nearctic with Neotropical regions, and the arid Mexican Plateau with tropical deciduous forests of the western slope of the Sierra Madre Occidental (López-González et al. 2014; González-Díaz et al. 2015). Anthropogenic threats include the large hydroelectric Las Cruces Dam proposed for the area, unrestricted cattle grazing, and pollution.

SUMMARY AND RECOMMENDATIONS

Many of the conservation concerns for native catfish diversity in México are related to inappropriate conservation status for most of the listed species and the difficulty of assigning an official conservation status to species without formal descriptions. Although only four of the catfish species in México are listed by the Mexican government, others are probably imperiled, including the several undescribed species discussed here and perhaps even the large and specialized Balsas catfish, *I. balsanus*, given that *I. punctatus* has recently been introduced into the Río Balsas basin (Mejía-Mojica et al. 2013). Major threats to most of these forms include nonnative fishes, habitat alteration, and water depletion. The effects of these threats are magnified by the restricted distributions of most *Ictalurus* taxa.

Natural ecosystems across México are increasingly stressed as the human population increases. New reservoirs have been constructed and stocked with nonnative blue and channel catfishes as well as the highly predatory flathead catfish *Pylodictis olivaris* (González et al. 2014). Widespread introductions of nonnative catfishes into Mexican rivers have historically threatened the genetic integrity of native catfishes via hybridization (Baker et al. 2008; Contreras-Balderas et al. 2003; Hendrickson and Varela-Romero 2002; Kelsch and Hendricks 1986; McClure-Baker et al. 2010; Miller et al. 2009; Minckley and Marsh 2009; Varela-Romero and Hendrickson 2009; Varela-Romero et al. 2011; Yates et al. 1984). Unfortunately, these nonnative catfish species support commercial fisheries that are promoted by stocking programs of federal and state fisheries agencies (SAGARPA 2003). Preventing continued introduction of nonnative catfish species into drainages with native catfishes and promoting the aquaculture of native catfishes within their historical ranges would help protect this unique and threatened fauna. The burgeoning global industry of channel catfish aquaculture (Jolly et al. 2001) emphasizes the potentially substantial commercial value that could derive from conservation and cultivation of the genetic diversity of native catfishes in México (Varela-Romero et al. 2011). Public outreach and altered management programs are desperately needed.

Insufficient field sampling and representation of voucher specimens in museums have contributed to confusion about the taxonomic and conservation status of native catfishes of northwestern México and elsewhere (Varela-Romero et al. 2011). This paucity of knowledge could be resolved by systematic and comprehensive surveys, including collection of specimens for molecular, morphometric, and meristic analysis. Each species should be individually evaluated for the Mexican Endangered Species list (NOM) for protection of wild populations and establishment of captive populations. Binational projects should be promoted to collaborate on studies of reproduction, management, and conservation of the native catfishes that are shared with the United States. The native catfishes of México, which represents the center of speciation for the genus *Ictalurus*, deserve devoted study, attention, and conservation.

ACKNOWLEDGMENTS

We thank Dean A. Hendrickson, John G. Lundberg, Gloria Yepiz-Velázquez, Jae Ahn, Jessica Rosales, Anthony A. Echelle, and Peter Reinthal for support in the development of the study. Richard L. Mayden, David A. Neely, and John P. Sullivan provided software and advice for the phylogenetic analyses. Howard Brandenburg, Nick Smith, José G. Soñanez-Organis, and Iván Anduro-Corona helped with field sampling. We thank Alice F. Echelle and Norman Mercado-Silva for comments and edits of the manuscript. This work was partially funded by CIAD, DICTUS Universidad de Sonora, and All Catfish Species Inventory of the United States National Science Foundation (DEB-0315963).

REFERENCES

Baker, S., C. Keeler-Foster, and W. Radke. 2008. Assessing hybridization between Yaqui catfish, *Ictalurus pricei*, channel catfish, *I. punctatus*, and blue catfish, *I. furcatus* using microsatellite markers. *Proceedings of the Desert Fishes Council* 40: 3.

Ballesteros-Córdova, C. A., J. M. Grijalva-Chon, R. A. Castillo-Gámez, F. Camarena-Rosales, and A. Varela-Romero. 2016. The complete mitochondrial genome of the Yaqui catfish *Ictalurus pricei* (Teleostei: Ictaluridae) and evidence of a cryptic *Ictalurus* species in northwest Mexico. *Mitochondrial DNA Part A* 27 (6): 4439–41.

Berra, T. M. 2007. *Freshwater Fish Distribution*. Chicago: University of Chicago Press.

Castañeda-Rivera, M., J. M. Grijalva-Chon, L. E. Gutiérrez-Millán, G. Ruiz-Campos, and A. Varela-Romero. 2014. Analysis of the *Ictalurus pricei* complex (Teleostei: Ictaluridae) in northwest Mexico based on mitochondrial DNA. *Southwestern Naturalist* 59: 434–38.

Contreras-Balderas, S., and P. Almada-Villela. 1996. *Ictalurus mexicanus*. *IUCN Red List of Threatened Species*. http://dx.doi.org/10.2305/IUCN.UK.1996.RLTS.T10769A3215184. Accessed June 8, 2017.

Contreras-Balderas, S., P. Almada-Villela, M. L. Lozano-Vilano, and M. E. García-Ramírez. 2003. Freshwater fish at risk or extinct in México A checklist and review. *Reviews in Fish Biology and Fisheries* 12: 241–51.

Edwards, R. J., and S. Contreras-Balderas. 1991. Historical changes in the ichthyofauna of the Lower Rio Grande (Río Bravo del Norte) Texas and Mexico. *Southwestern Naturalist* 36: 201–12.

González, A. I., Y. Barrios, G. Born-Schmidt, and P. Koleff. 2014. El sistema de información sobre especies invasoras. In R. Mendoza and P. Koleff, eds., *Especies Acuáticas Invasoras en México*, 95–112. Ciudad de México: Comisión Nacional para el Conocimiento y Uso de la Biodiversidad.

González-Díaz, A. Á., M. Soria-Barreto, L. Martínez-Cardenas, and M. Blanco. 2015. Fishes in the lower San Pedro Mezquital River, Nayarit, Mexico. *Check List* 11(6): 1–7.

Guindon, S., and O. Gascuel. 2003. A simple, fast, and accurate algorithm to estimate large phylogenies by maximum likelihood. *Systematic Biology* 52: 696–704.

Gutiérrez-Cabrera, A. E., G. Pulido-Flores, S. Monks, and J. C. Gaytán-Oyarzún. 2005. Presencia de *Bothriocephalus acheilognathi* Yamaguti, 1934 (Cestoidea: Bothriocephalidae) en peces de Metztitlán, Hidalgo, México. *Hidrobiológica* 15: 283–88.

Hendrickson, D. A. 1983. Distribution records of native and exotic fishes in the Pacific drainages of northern México. *Journal of the Arizona-Nevada Academy of Sciences* 18: 33–38.

Hendrickson, D. A., and A. Varela-Romero. 2002. Fishes of the Río Fuerte Drainage. In M. L. Lozano-Vilano, ed., *Libro Jubilar en Honor al Dr. Salvador Contreras Balderas*, 172–95. Monterrey: Universidad Autónoma de Nuevo León.

Hudson, P. F., D. A. Hendrickson, A. C. Benke, A. Varela-Romero, R. Rodiles-Hernández, and W. Minckley. 2005. Rivers of Mexico. In A. C. Benke, and C. E. Cushing, eds., *Rivers of North America*, 1030–84. Burlington: Elsevier Academic Press.

Jolly, C., C. Ligeon, J. Crews, Z. Morley, and R. Dunham. 2001. Present and future trends in the US catfish industry: Strategies and concerns for millennium years. *Reviews in Fisheries Science* 9: 271–95.

Kelsch, S. W., and F. S. Hendricks. 1986. An electrophoretic and multivariate morphometric comparison of the American catfishes *Ictalurus lupus* and *I. punctatus*. *Copeia* 1986: 646–52.

Kocher, T. D., W. K. Thomas, A. Meyer, S. V. Edwards, S. Paabo, F. X. Villablanca, and A. C. Wilson. 1989. Dynamics of mitochondrial DNA evolution in animals—amplification and sequencing with conserved primers. *Proceedings of the National Academy of Sciences* 86: 6196–200.

Koster, W. J. 1957. *Guide to the Fishes of New Mexico*. Albuquerque: University of New Mexico Press.

López-González, C., A. Lozano, D. F. García-Mendoza, and A. I. Villanueva-Hernández. 2014. Mammals of the San Pedro–Mezquital River Basin, Durango-Nayarit, Mexico. *Check List 10*: 1277–89.

Lundberg, J. G. 1992. The phylogeny of ictalurid catfishes: A synthesis of recent work. In R. Mayden, ed., *Systematics, Historical Ecology and North American Freshwater Fishes*, 392–420. Stanford: Stanford University Press.

McClure-Baker, S. A., A. A. Echelle, R. A. Van den Bussche, A. F. Echelle, D. A. Hendrickson, and G. P. Garrett. 2010. Genetic status of headwater catfish in Texas and New Mexico: A perspective from mtDNA and morphology. *Transactions of the American Fisheries Society* 139: 1780–91.

Mejía-Mojica, H., M. E. Paredes-Lira, and R. G. Beltrán-López. 2013. Primer registro y establecimiento del bagre de canal *Ictalurus punctatus* (Siluriformes: Ictaluridae) en un tributario del Río Balsas, México. *Hidrobiológica* 23: 456–59.

Miller, R. R. 1959. Origin and affinities of the freshwater fish fauna of western North America. In C. L. Hubbs, ed., *Zoogeography*. Publications of the American Association for the Advancement of Science 51: 187–222.

Miller, R. R. 1986. Composition and derivation of the freshwater fish fauna of México. *Anales de la Escuela Nacional de Ciencias Biológicas IPN* 30: 121–53.

Miller, R. R., W. L. Minckley, and S. Norris. 2009. *Peces dulceacuícolas de México*. CONABIO, Sociedad Ictiológica Mexicana, El Colegio de la Frontera Sur, Consejo de los Peces del Desierto.

Minckley, W. L., D. A. Hendrickson, and C. E. Bond. 1986. Geography of western North American freshwater fishes: Description and relationships to intracontinental tectonism. In C. H. Hocutt and E. O. Wiley, eds., *The Zoogeography of*

Freshwater Fishes of North America, 519–613. New York: Wiley Inter-Flores.

Minckley, W. L., and P. C. Marsh. 2009. *Inland Fishes of the Greater Southwest: Chronicle of a Vanishing Biota*. Tucson: University of Arizona Press.

Miranda, R., D. Galicia, S. Monks, and G. Pulido-Flores. 2012. Diversity of freshwater fishes in Reserva de la Biosfera Barranca de Metztitlán, Hidalgo, Mexico, and recommendations for conservation. *Southwestern Naturalist* 57: 285–91.

NatureServe. 2013a. *Ictalurus lupus. IUCN Red List of Threatened Species*. e.T10768A19035323. http://dx.doi.org/10.2305/IUCN.UK.2013-1.RLTS.T10768A19035323. Accessed June 15, 2017.

NatureServe. 2013b. *Ictalurus pricei. IUCN Red List of Threatened Species*. e.T10770A19034686. http://dx.doi.org/10.2305/IUCN.UK.2013-1.RLTS.T10770A19034686. Accessed June 15, 2017.

Rodiles-Hernández, R., J. G. Lundberg, and J. P. Sullivan. 2010. Taxonomic discrimination and identification of extant blue catfishes (Siluriformes: Ictaluridae: *Ictalurus furcatus* Group). *Proceedings of the Academy of Natural Sciences of Philadelphia* 159: 67–82.

Ronquist, F., and J. P. Huelsenbeck. 2003. MrBayes 3: Bayesian phylogenetic inference under mixed models. *Bioinformatics* 19: 1572–74.

Ruiz-Campos, G., D. Ceseña-Gallegos, S. Sánchez-González, and A. Varela-Romero. 2016. Caracterización merística y morfométrica del bagre del Río Culiacán (*Ictalurus* sp.) y su comparación con el bagre del "yaqui" (*Ictalurus pricei* Rutter, 1896) de la Sierra Madre Occidental, México. *XV Congreso Nacional de Ictiología, V Simposio Latinoamericano de Ictiología, I Simposio Internacional de Genómica de Peces*. Universidad Autónoma de Aguascalientes, México.

Ruiz-Campos, G., M. L. Lozano-Vilano, and M. E. García-Ramírez. 2009. Morphometric comparison of blue catfish *Ictalurus furcatus* (Lesueur, 1840) from northern and southern Atlantic drainages of Mexico. *Bulletin of the Southern California Academy of Sciences* 108: 36–44.

SAGARPA (Secretaría de Agricultura, Ganadería, Desarrollo Rural, Pesca y Alimentación). 2003. Anuario estadístico de pesca 2003. Comisión Nacional de Pesca y Acuacultura, México, Distrito Federal, México.

Salgado-Maldonado, G., G. Cabañas-Carranza, E. Soto-Galera, R. F. Pineda-López, J. M. Caspeta-Mandujano, E. Aguilar-Castellanos, and N. Mercado-Silva. 2004. Helminth parasites of freshwater fishes of the Pánuco River basin, east central Mexico. *Comparative Parasitology* 71: 190–202.

SEMARNAT (Secretaría de Medio Ambiente y Recursos Naturales). 2002. *Norma Oficial Mexicana NOM-059-ECOL-2001, Protección ambiental-Especies nativas de México de flora y fauna silvestres—Categorías de riesgo y especificaciones para su inclusión, exclusión o cambio—Lista de especies en riesgo*. Diario Oficial de la Federación (Segunda Sección): 1–81.

SEMARNAT (Secretaría de Medio Ambiente y Recursos Naturales). 2010. *Norma Oficial Mexicana NOM-059-ECOL-2010, Protección ambiental—Especies nativas de México de flora y fauna silvestres—Categorías de riesgo y especificaciones para su inclusión, exclusión o cambio—Lista de especies en riesgo*. Diario Oficial de la Federación (Segunda Sección): 1–78.

Soto-Galera, E. 2003. Ficha técnica de *Ictalurus mexicanus*. Elaboración de las fichas técnicas para la evaluación del riesgo de extinción de 18 especies de peces dulceacuícolas mexicanos. Laboratorio de Ictiología y Limnología. Departamento de Zoología, Escuela Nacional de Ciencias Biológicas, Instituto Politécnico Nacional. Bases de datos SNIB-CONABIO. Proyecto no. W040. México, D.F.

Soto-Galera, E., and L. Alcántara-Soria. 2007. Ficha técnica de *Ictalurus dugesii*. In E. Soto-Galera, ed., *Conocimiento biológico de 32 especies de peces dulceacuícolas mexicanos incluidos en la Norma Oficial Mexicana 059-SEMARNAT-2001*. Escuela Nacional de Ciencias Biológicas, IPN. Bases de datos SNIB-CONABIO. Proyecto no. CK011. México, D.F.

Soto-Galera, E., and L. A. Rincón-Sandoval. 2007. Ficha técnica de *Ictalurus lupus*. In E. Soto-Galera, ed., *Conocimiento biológico de 32 especies de peces dulceacuícolas mexicanos incluidos en la Norma Oficial Mexicana 059-SEMARNAT-2001*. Escuela Nacional de Ciencias Biológicas, IPN. Bases de datos SNIB-CONABIO. Proyecto no. CK011. México, D.F.

Sublette, J. E., M. D. Hatch, and M. Sublette. 1990. *The Fishes of New Mexico*. Albuquerque: University of New Mexico Press.

USFWS (United States Fish and Wildlife Service). 1984. Endangered and threatened wildlife and plants; final rule to determine the Yaqui chub to be an endangered species with critical habitat, and to determine the beautiful shiner and Yaqui catfish to be threatened species with critical habitat. *Federal Register* 49: 34490–97.

Van der Heiden, A. M., H. G. Plascencia-González, J. L. Villalobos-Hiriarti, and H. S. Espinoza-Pérez. 2016. Los peces y macrocrustáceos decápodos de la cuenca media del río Pánuco en la Sierra Madre Occidental del sur de Sinaloa, México. In M. S. González-Elizondo, M. M. González-Elizondo, C. Cortés-Montaño, eds., *Biodiversidad y paisaje de la Sierra Madre Occidental*. Ciudad de México: Instituto Politécnico Nacional-CONABIO.

Varela-Romero, A. 2007. *Variación genética mitocondrial en bagres del género* Ictalurus *(Pisces: Ictaluridae) en el noroeste de México*. Tesis de Doctorado, Centro de Investigación en Alimentación y Desarrollo, Hermosillo.

Varela-Romero, A., and D. A. Hendrickson. 2009. Peces dulceacuícolas. In F. E. Molina-Fraener and T. R. Van Devender, eds., *Diversidad biológica de Sonora*, 339–56. Distrito Federal: Universidad Nacional Autónoma de México, México.

Varela-Romero, A., D. A. Hendrickson, G. Yepiz-Plascencia, J. E. Brooks, and D. A. Neely. 2011. Status of the Yaqui catfish (*Ictalurus pricei*) in the United States and northwestern Mexico. *Southwestern Naturalist* 56: 277–85.

Williams, J. E., J. E. Johnson, D. A. Hendrickson, S. Contreras-Balderas, J. D. Williams, M. Navarro-Mendoza, D. E. McAllister, and J. E. Deacon. 1989. Fishes of North America endangered, threatened, or of special concern: 1989. *Fisheries* 14: 2–20.

WWF (World Wildlife Fund). 2010. La Cuenca Alta del Río San Pedro Mezquital, Caudal de Vida y Cultura. WWF/Nokia/Coca Cola. http://awsassets.panda.org/downloads/cuenca_alta_del_Río_spm_2010_1.pdf.

Yates, T. L., M. A. Lewis, and M. D. Hatch. 1984. Biochemical systematics of three species of catfish (Genus *Ictalurus*) in New Mexico. *Copeia* 1984: 97–101.

20

Ecology, Politics, and Conservation of Gila Trout

David L. Propst, Thomas F. Turner, Jerry A. Monzingo, James E. Brooks, and Dustin J. Myers

Viewed from a figurative distance, the conservation of Gila trout *Oncorhynchus gilae* would seem a rather straightforward, relatively simple proposition. The reasons for its decline and imperilment are known, actions to rectify them have been identified, many of those actions have been implemented, and efforts to diminish, mitigate, and eliminate the remaining threats are under way. But the reality of Gila trout conservation has proved more complicated, contentious, and challenging than anyone who has participated in this effort might have imagined. Concepts and approaches taken initially were found wanting or insufficient to the task. New information, especially genetic information, necessitated changes in recovery strategies. New and improved technologies and equipment made many activities easier and more efficient, but actual recovery was still largely carried out on remote streams in designated wilderness. And major changes on the landscape, wrought by climate change, extended drought, and mega-wildfires, necessitated rethinking and expanding upon what is necessary to ensure the persistence of Gila trout. Our objectives in this chapter are to provide an overview of Gila trout biology and environment and a narrative history of the challenges and accomplishments of this effort to conserve an iconic species of the American Southwest.

ORIGINS AND RELATIONSHIPS

Genetic and geologic data suggest that about 1 to 2 million years ago, an ancestral steelhead *Oncorhynchus* sp. made its way from the Gulf of California upstream to the headwaters of the Gila River through the newly connected lower Colorado River corridor (Behnke 1992; Abadía-Cardoso et al. 2015).

Fig. 20.1 Gila trout *Oncorhynchus gilae*. This ca. 290 mm TL fish of the South Diamond Creek lineage was captured in Willow Creek in August 2017. (Photograph by D. J. Myers.)

Long before human settlement, this ancestral form diversified into landlocked trout species and populations. Yet it was not until 1950 that a descendant of the ancestral form, Gila trout, was recognized by the scientific community (Miller 1950) (fig. 20.1). In 1972, a second species, Apache trout *O. apache*, which occupied upland streams of the Salt River portion of the Gila drainage and Little Colorado River in Arizona, was described (Miller 1972). The identification of these unique forms, and their distinctiveness compared with other trout species, was based on differences in physical characteristics and numbers of chromosomes (Beamish and Miller 1976), but how Gila and Apache trout fit into the broader salmonid evolutionary tree was still an open question.

Needham and Gard (1959) proposed an affinity with rainbow trout *O. mykiss* for Gila trout. Miller (1972) speculated that Apache trout was derived from cutthroat trout *O. clarkii*, but he was less certain about the origin of Gila trout, proposing that it represented a fifth phyletic line of *Oncorhynchus* (in addition to Mexican golden trout *O. chrysogaster*, cutthroat trout, rainbow trout, and at the time undescribed redband trout). By applying principles of systematic biology to morphological and karyological data, other researchers proposed a more robust hypothesis of close evolutionary ancestry with rainbow trout (e.g., Gold 1977; Thorgaard 1983), which was later confirmed using DNA sequence data (Wilson and Turner 2009; Crête-Lafrenière et al. 2012).

HISTORICAL DISTRIBUTION

Prior to European settlement, trout occupied suitable habitat throughout the Gila River drainage of Arizona and New Mexico, including streams in the Agua Fria, Verde, Tonto, Salt, San Francisco, and Gila drainages (fig. 20.2). Determining the taxonomic status of each drainage's native trout, however, has proved problematic. The paucity of historical collections, losses of collected specimens, extensive population extinctions, the introduction of hybridizing congeners, semantics, and conflicting interpretations of taxonomic data contributed to the uncertainty of the relatedness of trout among these drainages. By the time systematic cataloguing began in the mid-twentieth century, native trout were limited to a few isolated headwater populations.

Presumably, the first official note of trout in the region was by personnel associated with the US and Mexican border surveys (Sublette et al. 1990), but the Gila trout reported and illustrated by Emory (1848) was instead a species of chub (Cyprinidae), probably roundtail chub *Gila robusta*, from the Gila River. Based

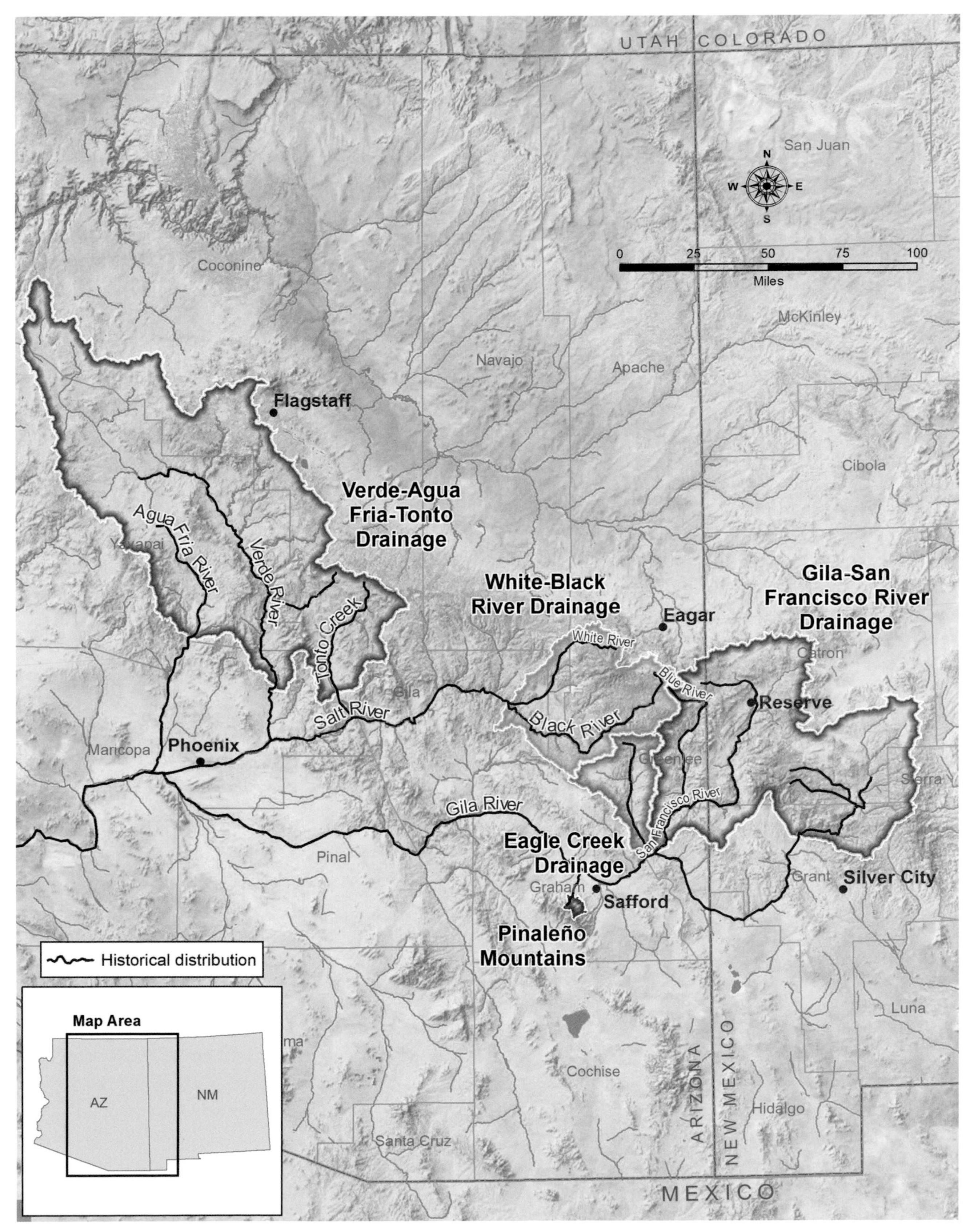

Fig. 20.2 Historical distribution of trout *Oncorhynchus* sp. in the Gila River drainage, Arizona and New Mexico.

upon the illustration and his comments, Emory termed chubs (*Gila* sp.) "trout" and referred to trout (*Oncorhynchus* sp.) as "mountain trout." No illustration was provided by Bartlett (1854) of a putative trout, but his description of capture location in the lower Mimbres River near Faywood Hot Springs makes it almost certain that his trout was instead Chihuahua chub *G. nigrescens*. Although adjacent to the Gila River basin, the Mimbres River is in the endorheic Guzmán Basin. An early record of what was almost certainly Gila trout was from French (1965). While a rancher in southwestern New Mexico in the late nineteenth century, French and a colleague rode across what is now the Gila Wilderness and caught "mountain trout" in the upper West Fork Gila River. By the 1920s, however, the elimination of mountain trout from many streams had been noted (Dinsmore 1924; Miller 1950).

Elsewhere within the Gila River drainage, the historical record for trout is equally sketchy. Leopold (1921) reported observing trout in 1885 in the Blue River drainage, a tributary to the San Francisco River, before stocking of nonnative trout (Miller 1950). However, he did not retain specimens. Although Miller (1950) doubted the natural occurrence of Gila trout in Spruce Creek, a San Francisco River tributary, David (1976) and Behnke and Zarn (1976) provided evidence to support its native occurrence. Later, Riddle et al. (1998) presented additional information supporting natural presence of Gila trout in Spruce Creek. Initially, Miller (1950) determined that trout collected by F. M. Chamberlain in 1904 from KP Creek in the Blue River drainage were "almost surely" Gila trout, based on Chamberlain's description (although the specimens were lost). Subsequently, Miller (1972) decided that the KP Creek trout were instead Apache trout. Miller made this change because KP Creek trout, although hybrid, had a spotting pattern more like that of Apache than Gila trout. Behnke (2002), however, considered Gila trout the native trout of the entire San Francisco River drainage.

The taxonomic designation of trout native to Eagle Creek, a tributary to the mainstem Gila River in Arizona, remains unresolved. Mulch and Gamble (1956) and Miller and Lowe (1964) reported Gila trout, but that was before Miller (1972) described Apache trout. Examinations of hybrid specimens from Chitty Creek (an Eagle Creek tributary) yielded mixed results, with one study considering this population a variant of Apache trout (Kynard 1976) and another identifying specimens as Gila × rainbow trout (Papoulias et al. 1988).

Miller (1950) examined trout from Oak Creek, a Verde River tributary, and decided these fish were hybrid rainbow × Gila trout. Since then, Gila trout has been considered the native trout of the Verde River drainage (e.g., Minckley 1973; Behnke 1992), but Behnke (2002) stated that preserved specimens had traits of both Gila and Apache trout and illustrated it as "Verde River population" in his treatise.

Behnke and Zarn (1976) proposed that Gila trout was native to the Agua Fria River drainage based on specimens from Sycamore Creek. Although these specimens were hybridized with rainbow trout, their spotting pattern was typical of Gila trout. Behnke (1992) stated that Gila trout was "probably" native to the Agua Fria River, but subsequently (Behnke 2002) did not include, without explanation, the Agua Fria River in the historical range of Gila trout.

The currently recognized historical range of Gila trout encompasses at least the higher-elevation tributaries of the San Francisco River and upper Gila River (Behnke 2002). Minckley and Marsh (2009) included the Verde River within the native range of Gila trout. Apache trout is restricted to higher-elevation tributaries of the Salt River and several tributaries of the Little Colorado River (Behnke 2002; Minckley and Marsh 2009).

PRE-ENDANGERED SPECIES ACT CONSERVATION (1900–1973)

Long before the US Forest Service (USFS) and the first foresters, livestock grazing, timber harvest, and mining were well established in the Southwest. Shortly after the turn of the twentieth century, there were nearly 370,000 cattle and over 1 million sheep and goats on the forest reserves in New Mexico and Arizona. There were so many livestock grazing on public lands, in times of both adequate and inadequate forage, that signs of range deterioration began to appear even in "good" years (Roberts 1963; Baker et al. 1988). By 1901, overstocking of sheep on southwestern forest ranges caused natural forest regeneration to come to a standstill, and the forest floor in some places was "as bare and compact as a road bed" (Toumey 1901, 16). In 1912, 19,500 sheep were counted in the high country of the USFS McKenna District (Woodrow 1943; Baker et al. 1988), which included many streams occupied by Gila trout. Cattle and sheep numbers on forest ranges in Arizona and New Mexico peaked during or shortly after World War I, with approximately 550,000 cattle and horses and 875,000 sheep and goats permitted (Dahms and Geils 1997).

Timber production in Arizona and New Mexico was estimated at 8 million board feet in 1879, which rose to 22 million in 1889 and 67 million in 1900 (Baker et al. 1988). By 1904, Leiberg et al. (1904) reported that 60,280 ha (148,845 acres) on the San Francisco Mountains Forest Reserve in Arizona had been cut, and more than 40,500 ha (100,000 acres) had had 60% or more of the stand removed in building the Atlantic and Pacific Railroad. Similarly, on the Gila River Forest Reserve, cutting had removed all timber of "marketable value" by the early twentieth century, and Rixon (1905, 15–16) reported that "logging operations have been carried on in a desultory manner for some years in different parts of the reserve." He also reported that most of the damage from timber operations had occurred in and near the creek bottoms. By 1907, timber managers were worrying that the timber harvest was exceeding the sustained yield capacity of the national forests (Woolsey 1907).

Although the mineral resources of the Gila River basin were mined by Native Americans long before the arrival of Spanish conquistadors searching for the Seven Cities of Gold, it was not until the mid-nineteenth century that mining began in earnest, especially in Arizona. For the next 100 years, prospectors and miners swarmed over the landscape in search of riches. While more activity and larger efforts occurred at lower elevations, higher-elevation streams occupied by Gila trout were not spared. Most such mining activity was probably on the westward-flowing Mogollon Mountain streams, such as Big Dry, Spruce, and Mineral Creeks (Ratté et al. 1979). Elsewhere, there were few exploitable mineral resources (e.g., the Blue River drainage in Arizona) in the vicinity of streams occupied by Gila trout (Ratté et al. 1969).

Thus, over the course of a few decades, resource exploitation, including unregulated fish harvest, so depleted Gila trout populations that a captive propagation effort was initiated in 1923 at Jenks Cabin Hatchery in the Gila Primitive Area (now Wilderness). The limited success and remoteness of the hatchery forced its closure in 1935. A second attempt to rear Gila trout, at Glenwood Hatchery on Whitewater Creek, was likewise unsuccessful, and efforts at captive rearing ceased in 1947 (Miller 1950).

At the request of Elliot S. Barker, state game warden of New Mexico, R. R. Miller collected specimens of a trout that "Mr. Barker and others felt sure was new to science" in 1939 (Miller 1950, 10). At that time, Gila trout was known to exist in only Main Diamond (type locality) and Spruce Creeks. The persistence of Gila trout was possible, in part, because the New Mexico Department of Game and Fish (NMDGF) did not stock nonnative trout in "certain tributaries" (Miller 1950, 11), and because Main Diamond was closed to angling in 1933 (Regan 1964). During the 1930s, the Civilian Conservation Corps installed log stream improvement structures in numerous streams, which contributed to survival of the species during the severe drought of the 1950s.

By the mid-twentieth century, Gila trout was known to persist in fewer than 10 km of stream. Miller (1950,18) cited overfishing and "changing weather conditions," coupled with failure to produce Gila trout in a hatchery, as reasons for widespread stocking of nonnative rainbow trout in Gila River drainage streams. Interestingly, he made no mention of the intensive livestock grazing that occurred in the uplands and the associated habitat degradation (Bryan 1925). During 1952–1955, Huntington (1955) conducted an extensive survey of the Gila River drainage and reported Gila trout in 17 streams, including Main Diamond and Spruce Creeks.

No formal investigation of Gila trout biology was conducted until the early 1960s, when Regan (1964, 1966) studied the Main Diamond population. He reported that Gila trout lived six years, attained about 210 mm total length (TL), consumed aquatic insects, and had low fecundity (<150 eggs per female). Regan (1964) included McKenna Creek among those supporting Gila trout, but did not attribute its discovery. Hanson (1971) collected specimens from several putative Gila trout populations that were examined by R. J. Behnke, and those from South Diamond and McKenna Creeks—both streams Huntington (1955) had identified as supporting Gila trout—were identified as Gila trout.

Hanson (1971) also surveyed several streams, including McKnight Creek (a Mimbres River tributary), to assess their potential for sustaining Gila trout populations. Following construction of a waterfall and a piscicide (rotenone) treatment to reduce native Rio Grande sucker *Pantosteus plebeius* numbers, 307 Gila trout from Main Diamond Creek were released in McKnight Creek in autumn 1970. In 1972, Sheep Corral Canyon, a fishless tributary of Sapillo Creek protected by a natural waterfall, was stocked with Gila trout from Main Diamond Creek. Based on information acquired in these studies, a Gila Trout Management Plan was finalized in early 1973. Its primary objectives were to establish additional Gila trout populations and "provide limited use of this fishery resource by sports fishermen" (Bickle 1973, 10).

POST-ENDANGERED SPECIES ACT CONSERVATION (1973–2000)

On the eve of passage of the Endangered Species Act (ESA) in 1973, Gila trout was known to occur in six streams (Main Diamond, South Diamond, McKenna, Spruce, McKnight, and Sheep Corral), which collectively provided about 24 km of habitat. The largest population, estimated at 4,000 individuals (excluding age = 0 fish; Hanson 1971), was undoubtedly in Main Diamond Creek. By its inclusion on the federal Endangered Species Preservation Act of 1966 (USFWS 1967) list of endangered species, Gila trout was listed as endangered under the authority of the US Endangered Species Act of 1973. In 1975, NMDGF listed it as a Category 1 (endangered) species under provisions of the New Mexico Wildlife Conservation Act of 1975.

Passage of the ESA provided additional impetus for Gila trout conservation. In 1974, Main Diamond fish were stocked in Gap Creek, a Verde River tributary, in Arizona (Minckley and Brooks 1985). Under auspices of Section 4 of the ESA, a Gila Trout Recovery Team was formed in 1974, and in 1975 it was directed to draft a Gila Trout Recovery Plan. Section 6 of the ESA provided a mechanism for USFWS to transfer funds to state agencies to support work to benefit federally protected species. Two such projects were a taxonomic analysis of documented and putative Gila trout populations (David 1976) and a status survey of known populations (Mello and Turner 1980). Detailed morphological analysis suggested that genetic differences existed among remnant populations (David 1976). To the six known populations of Gila trout, David (1976) added Iron Creek, but also determined that six streams Huntington (1955) had identified as having Gila trout instead had Gila × rainbow trout.

The primary objective of the Gila Trout Recovery Plan (USFWS 1979) was to ensure the security of each of the presumed genetically pure, but morphologically different, Gila trout lineages in the wild. The recovery plan laid out a series of issues that should be considered in selecting streams for Gila trout restoration. Adherence to these conditions effectively limited the streams selected for restoration to small headwaters. The recovery plan revealed a tension between conservation imperatives and political concerns. Much was made of the need for Gila trout to be a worthy sport fish to gain public and agency acceptance for its conservation. This tension between providing for a species' security and political pressures to treat it primarily, if not solely, as a sport fish persists and occasionally sidetracks and stalls recovery activities.

The 1979 Gila Trout Recovery Plan was revised in 1984 (USFWS 1984). A key difference between the two versions was the explicit statement regarding criteria for reclassifying the species from endangered to threatened. The 1984 Recovery Plan (p. 21) stated that the species would be "considered for downlisting . . . to a threatened status when survival of the five original ancestral populations is secured and when all morphotypes are successfully replicated or their status is otherwise appreciably improved." Importantly, an allozyme study (Loudenslager et al. 1986) confirmed the genetic distinctiveness of each relict population, except McKenna Creek, which was not included in the study.

Activities more tangible to Gila trout recovery were implemented in comparatively rapid succession. The first was extension of Gila trout habitat in Iron Creek using piscicide (antimycin A) to remove nonnative brown trout and construction of a waterfall barrier (Coman 1981) to prevent their recolonization. That effort was followed by successful removal of brown trout, construction of a waterfall barrier, and stocking of McKenna Gila trout in Little Creek during 1982–1983. Renovation of upper Big Dry Creek was completed in 1985, and Spruce Gila trout were stocked there that autumn. Renovation of Mogollon Creek and subsequent stocking of its tributaries with South Diamond fish was a major effort from 1986 through 1988, and was accompanied by scattered, but vocal, public opposition to Gila trout conservation. Opposition centered mainly on the closure of renovated streams to angling, but concerns about the safety of piscicides and their effects on non-target organisms were also voiced. Public meetings did little to assuage opposition.

With the stocking of South Diamond Gila trout in Trail Canyon (a Mogollon Creek tributary) in 1987, each relict population, except Iron Creek, had at least one replicate. Extension of the species' habitat by about 3 km, however, was deemed to have "appreciably improved" the security of its population, and the USFWS proposed to reclassify the species as threatened (USFWS 1987). The proposal was ultimately not acted upon because the 1989 Divide Fire eliminated Gila trout from Main Diamond Creek and greatly reduced the population in South Diamond Creek (Propst et al. 1992). Thus began a decade of challenges on several fronts and a rethinking of strategy and approaches to Gila trout conservation. It was apparent that recovery would not be accomplished in small, isolated headwaters; repatriation to larger, hydrologically complex drainages would be necessary, and captive propagation would be needed to supply the numbers of fish required to establish populations in larger habitats in a timely manner. The elimination of trout from Main Diamond Creek clearly demonstrated that if wildfire threatened a stream supporting Gila trout, evacuation would be necessary, if possible, especially for relict populations.

During the 1980s, several studies enhanced understanding of the biology of Gila trout. Spawning occurred between 1300 and 1600 hours from mid-April through May to early June, when water temperature was 8°C–10°C (Rinne 1980). In the small streams it occupied, Rinne (1982) reported that Gila trout were very sedentary; few moved more than several meters over the course of his eight-month study. Females matured at age 3 or 4 (ca. 130 mm TL; Nankervis 1988). Van Eimeren (1988) found that Gila trout fed mainly between 0900 and 1300 hours and that it was piscivorous, consuming both its own species and speckled dace *Rhinichthys osculus*.

The Divide Fire dramatically exposed the fallacy of "appreciably improving" the species' security by merely extending the length of occupied habitat in a stream, especially a small headwater, and thereby made urgent the need for a replicate Iron lineage population. Although fishless, Sacaton Creek provided suitable habitat, and Iron fish were translocated there in 1990. During surveys of the upper West Fork Gila River in 1990, the peripatetic Nick Smith observed trout in a small unnamed tributary (subsequently called Whiskey Creek; fig. 20.3). These fish were determined to be pure Gila trout (Leary and Allendorf 1999). Following piscicide application to eliminate rainbow trout, White Creek received

Fig. 20.3 Whiskey Creek, Gila Wilderness, Gila National Forest, June 2003. (Photograph by J. E. Brooks.)

Iron lineage Gila trout in 1993. In 1994, the Spruce Fire prompted evacuation of Gila trout from Spruce Creek. The Bonner Fire eliminated Gila trout in South Diamond Creek in 1995, and the Sprite (1995) and LL Complex (1996) fires necessitated evacuation of Gila trout from Mogollon Creek. The effort to establish Gila trout in Gap Creek failed by the early 1990s, probably as a consequence of the stream providing marginal habitat (it was fishless prior to receiving Gila trout).

Despite substantial differences among streams in disturbance history and physiographic attributes, Propst and Stefferud (1997) reported that demographic attributes (e.g., length-frequency, length-mass, and density) of Gila trout populations were strikingly similar, especially among intermediate and large streams. A mtDNA study that included all presumed genetically pure Gila trout relict populations (n = 6) found that the McKenna Creek population was introgressed with rainbow trout (Riddle et al. 1998). Consequently, the McKenna replicate in Little Creek was likewise contaminated. A study to characterize genetically all extant Gila trout populations (using protein electrophoresis) revealed that the recently established Mogollon Creek population contained F_1 Gila × rainbow trout, that the Iron population was apparently introgressed with rainbow trout (albeit at low presumed rainbow allele frequencies), and confirmed that the McKenna population was a hybrid swarm (Leary and Allendorf 1999). Paired matings of fish evacuated from Mogollon Creek that phenotypically appeared to be Gila trout, separate rearing of offspring of each paired mating, and post-spawning genetic characterization of parents enabled resurrection of the South Diamond lineage (Leary et al. 1999). Following piscicide treatments of Mogollon Creek to eliminate fish not evacuated in 1996, offspring of parents deemed genetically pure (based on absence of presumed "diagnostic" alleles) were stocked there in 1997. Although the habitat was deemed marginal (as a consequence of the 1995 Bonner Fire), pure offspring were also stocked in South Diamond Creek in 1997. Elimination of McKenna lineage fish from Little Creek and Iron lineage fish from upper White Creek was completed, and Main Diamond trout were translocated to both in 2000.

Negative public opinion of Gila trout recovery in rural western New Mexico was long standing. The presence of species protected under the ESA was constantly contested by various local groups, including county commissions, hunter/angler groups, and Sagebrush Rebellion adherents. Numerous public meetings, positive media coverage, and dissemination of information on Gila trout conservation in pamphlets and popular articles (e.g., Propst 1994) did little to quell criticism of recovery activities. There were increasing questions about the safety of piscicides, but stream closures to angling remained a key issue. Despite an official opinion (Opinion no. 94-01) and a subsequent advisory letter (November 1994) from the attorney general of New Mexico stating that county ordinances that "seek to restrict the public land regulatory authority of the United States and the State of New Mexico . . . have no legal effect," Grant County passed the Grant County Water Pollution Nuisance Ordinance (#95-05-08) in May 1995, which stipulated that Grant County approval was required before introduction of "any object or substance into a body of public water." To appease Grant County, both federal and state agencies agreed that henceforth county concurrence would be sought for restoration activities, a county representative would attend recovery team meetings, and a county representative would participate in piscicide applications. Although the Grant County ordinance did not stop recovery activities, it delayed several projects, including treatment of Mogollon Creek. For several years, Grant County sent an observer to recovery team meetings, but otherwise showed little inclination to be involved.

At the same time, Gila trout was swept into the controversies over grazing on public lands. Environmental groups charged that uncontrolled grazing destroyed Gila trout habitat, while grazing interests insisted there was no damage. Agencies were accused both of coddling livestock growers and of bowing to the whims of radical environmentalists. In retrospect, and better appreciated now than at the onset of ESA actions, poor articulation of natural resource conservation needs to policy makers as well as non-experts (e.g., Sarewitz 2004; Sullivan et al. 2006; Dahlstrom 2013) hampered the implementation of recovery actions.

Prior to 1999, the primary environmental regulatory requirements for piscicide application were compliance with the National Environmental Policy Act (NEPA) and the ESA, and a USFS Pesticide Use Permit. The New Mexico Surface Water Quality Bureau (NMWQCC), which was responsible for Clean Water Act (CWA) compliance within the state, allowed piscicide applications because they were temporary water quality impairments and were to achieve an objective of the CWA (to "restore and maintain the . . . biological integrity of the Nation's waters"), but in 1999, NMWQCC determined that its formal approval was needed. Consequently, a formal petition for temporary impairment, public hearing(s) in the project area, and the hearing officer's report on findings with recommendations were required. Then NMWQCC decided to accept, or not accept, the hearing officer's recommendations.

For much of the 1980s and 1990s, Arizona was marginally involved in Gila trout recovery, but population losses in New Mexico clearly demonstrated the need to expand conservation to Arizona. Unfortunately, the timing (late 1990s) of planned expansion to Arizona was not fortuitous; increasing opposition to native fish conservation from politically powerful special interests, and the Arizona Game Commission's requirement that there be no net loss of angling opportunities when restoring streams to native fishes, greatly diminished options. These factors effectively limited opportunities to streams that were historically fishless or where natural events had eliminated fishes. The Dude Fire eliminated nonnative trout from Dude Creek (an East Verde River tributary) in 1989; in 1999, it was deemed sufficiently recovered to support trout and was stocked in 2000 with Gila trout from Spruce Creek. The fishless Raspberry Creek (a Blue River tributary) was stocked with Spruce lineage fish in 2000.

Through the early 1990s, all restored Gila trout populations were established with fish from a wild donor population. Although this method was successful, it was evident that small donor populations could not be regularly depleted to establish additional populations. Consequently, hatchery propagation of Gila trout was initiated in 1988 at Mescalero National Fish Hatchery (NFH). With evacuation of Gila trout from Main Diamond Creek during the 1989 Divide Fire, an additional role for hatcheries—refuge—became an increasingly important, and challenging, conservation tool. Maintenance of Gila trout at Mescalero NFH proved difficult, however. Because its primary mission was to produce rainbow trout, addition of Gila trout stretched its limited facilities. During its tenure at Mescalero NFH, Gila trout was plagued by deteriorating facilities, water quality issues, poor survival, and potential contamination by rainbow trout. In 1999, the recently constructed Mora NFH began acquiring Gila trout, and in 2001, it assumed all responsibilities for captive rearing of Gila trout.

While wildfire posed a major threat to extant Gila trout populations, it also created opportunities. In addition to eliminating Gila trout from South Diamond Creek, the 1995 Bonner Fire eliminated nonnative brown and rainbow trout from upper Black Canyon, while native longfin dace *Agosia chrysogaster*, speckled dace, Sonora sucker *Catostomus insignis*, and desert sucker *Pantosteus clarkii* survived. With surprising speed, funds were obtained, a cooperative agreement among agencies signed, and NEPA compliance completed to enable construction, with assistance of volunteers, of a gabion basket waterfall in 1998 to prevent upstream fish movement. A final check of the stream in June found 1 rainbow (225 mm TL) and 4 brown (70–74 mm TL) trout. Because the presence of four native species precluded piscicide use, multiple electrofishing passes over three months were made to remove nonnative trout that summer (Brooks and Propst 1999). These efforts yielded an additional 23 rainbow (171–240 mm TL), 341 brown (all age 1), and heretofore undocumented cutthroat trout *Oncorhynchus clarkii* (7 individuals, all about 300 mm TL). Several factors pointed to sabotage. First, radii of brown trout scales were uniformly distributed, indicative of captive rearing, rather than being irregularly spaced, as in wild fish (Seelbach and Whelan 1988). Second, no adult brown trout were captured. Third, subadult and adult rainbow and adult cutthroat trout were found, but no young of either species were captured. And, finally, the appearance of cutthroat trout where none had been previously collected additionally substantiated sabotage. Gila trout were repatriated that autumn, but the persistence of brown trout meant repeated removal efforts.

GILA TROUT CONSERVATION IN THE NEW MILLENNIUM (2001–2018)

The 1993 revision of the Gila Trout Recovery Plan (USFWS 1993, 35) stated: "Downlisting to threatened status will be considered when all known indigenous lineages are replicated in the wild. In addition to replications, Gila trout must be established in a sufficient number of drainages such that no natural or human-caused event may eliminate a lineage." By 2000, each lineage deemed genetically pure ($n = 4$) was replicated in at least one stream. Whereas the Main and South

Diamond lineages met standards for downlisting, it was doubtful that the Whiskey and Spruce lineages met those standards. Main Diamond was replicated in five streams, and South Diamond was established in the dendritically complex Mogollon Creek drainage. Upper Little Creek provided, at best, limited and marginal habitat for the Whiskey replicate. Persistence of Spruce replicates in Dude and Raspberry Creeks was not certain, and only a narrow ridge separated Spruce Creek from a replicate population in Big Dry Creek. Nonetheless, the recovery team recommended that Gila trout be downlisted to threatened in December 2000. The recovery team also recommended that neither Iron Creek nor McKenna Creek be treated to remove putative hybrid trout. But before the downlisting recommendation could be acted upon, a third revision of the Gila Trout Recovery Plan was completed (USFWS 2003), with criteria for downlisting that were more demanding and specific than those in the 1993 recovery plan. In addition to each lineage having at least one replicate in the wild, Gila trout had to occupy a minimum of 85 km of habitat, with each lineage "replicated in a stream geographically separate . . . such that no natural or human-caused event may eliminate . . . a lineage" (p. 41). The revised recovery plan also required adoption of an emergency evacuation plan to be implemented if a population was threatened by natural or human-caused events.

Recognition of the risks wildfire posed for Gila trout (Brown et al. 2001) prompted development of a protocol for fish rescue. While wildfire impacts were the initial impetus for development of the protocol, other risks (drought, flood, barrier failure, and nonnative trout invasion) were also addressed (Brooks 2004). All wildfire situations where the protocol might be implemented were on USFS lands and required close coordination with the USFS unit managing the fire. The protocol clearly defined USFS as the lead agency in the event of a wildfire and required agency coordination and communication.

With all environmental compliance requirements met, the largest renovation effort to date for Gila trout began in autumn 2003 in the upper West Fork Gila River drainage. Wildfires in 2002 (Cub) and 2003 (Dry Lakes Complex) had burned much of the watershed and had nearly eliminated brown and rainbow trout, giving hope that the project could be finished in three to five years. Restoration of the drainage provided an opportunity to naturally mix lineages. Each lineage (except Spruce) was to be stocked in a different tributary. In time, individuals from each tributary would move to the main stem West Fork Gila River, where spawning by individuals from different lineages would yield progeny more representative of historical Gila trout than occurred in small, isolated populations.

Despite three antimycin applications in 2004, brown and rainbow trout were found in post-treatment assessments. But the greater setback was a decision by the New Mexico State Game Commission in August 2004 disallowing NMDGF use of piscicides to eliminate nonnative trout. Piscicides had become increasingly controversial tools, and opposition to their use stymied native trout restoration projects across the American West, from Montana (Cherry Creek) to California (Silver King Creek). Misinformation and scaremongering about EPA-registered piscicides (rotenone and antimycin) characterized public meetings and commonly appeared in the popular press. Noted conservation writer Ted Williams skewered "chemophobes" in his column in *Fly Rod & Reel* (July 2005), exposing and rebutting their illogical tirades with scientific data. The West Fork Gila River restoration endured a 12-month hiatus before lobbying of New Mexico Governor Bill Richardson by Trout Unlimited prompted the New Mexico State Game Commission to reverse itself in June 2005 and allow piscicide use.

Promptly after the commission rescinded its order, renovation of the West Fork Gila River resumed, but efforts were hampered by monsoon rains. An all-out effort to complete the renovation was mounted in 2006, but before it could be completed, the Bear Fire forced cessation of treatment. Before leaving the treatment area, however, personnel captured Gila trout from fire-threatened Whiskey Creek and released them in Langstroth Canyon (a West Fork Gila drainage stream). Surveys later that summer found nonnative trout persisting in lower reaches of the West Fork Gila drainage. In Arizona, nonnative trout survival after treatment was documented and prompted assays to determine if the antimycin A used for the treatment was compromised. It was: tested batches were below 20% label strength. Because environmental compliance considered only antimycin A, only it could be used, and the 2007 West Fork Gila efforts using compromised chemical failed to eliminate nonnative trout. Consequently, all of 2008 was consumed in obtaining a new NMWQCC order allowing use of rotenone, another environmental assessment, and another ESA Section 7 consultation. Another two years were required to remove all nonnative trout, and Gila trout from South and Main Diamond Creeks were stocked in Cub and White Creeks, respectively, in October 2010.

Drought in the early twenty-first century was at least partially responsible for losses of Gila trout populations in streams providing marginal habitat (Dude, Raspberry, Sheep Corral, and Sacaton). Wildfires prompted evacuation of Gila trout from Whiskey Creek in 2002 and 2006 (Cub and Bear fires, respectively), Mogollon Creek in 2003 (Dry Lakes Complex Fire), and Raspberry Creek in 2004 (Raspberry Fire).

Despite the facts that only three of the four relict populations (Main Diamond, South Diamond, Spruce, and Whiskey) had been replicated, and that the two Spruce lineage replicates were tenuous, USFWS proceeded with a proposal to reclassify Gila trout as threatened in 2006 (USFWS 2006). In addition to downlisting the species, USFWS also proposed a special rule under Section 4(d) of the ESA that would enable NMDGF and the Arizona Game and Fish Department (AZGFD) "to promulgate special regulations in collaboration with the Service, allowing recreational fishing of Gila trout." At this time, Gila trout was in 12 streams, which collectively provided about 100 km of habitat. Both NMDGF and Trout Unlimited opposed the downlisting. In the time between the proposal and the formal downlisting of Gila trout in May 2006 (USFWS 2006), Langstroth Canyon received Whiskey fish, and failure of the Dude Creek effort was documented. In anticipation of the downlisting and of angling "take" in selected streams, Arizona initiated environmental compliance for renovation of several streams in the Verde and Blue drainages. Following the downlisting, failure of the Raspberry stocking was confirmed.

Under propagation protocols for Gila trout at Mora NFH (Kincaid and Reisenbichler 2002), a specific number of paired matings was required to satisfy each request for Gila trout to establish or augment populations. Typically, more young were produced than needed. Special regulations (artificial fly or lure, barbless hook, catch and release) were promulgated for angling of two recovery populations (Black Canyon and Mogollon Creek). To estimate angling pressure and promote angler satisfaction, a free "Gila Trout Permit" was required to fish these streams. Retired brood fish were released into easily accessed waters. In Arizona, angling for Gila trout was allowed in Frye Mesa Reservoir, Goldwater Lake, and West Fork of Oak Creek.

Through the last decade of the twentieth century and the first decade of the twenty-first, substantial advances were made in characterizing genetic diversity and relationships among putative Gila trout populations. Following Loudenslager et al. (1986), additional allozyme work (Leary and Allendorf 1999; Leary et al. 1999), mitochondrial (mt)DNA-based assays (Riddle et al. 1998), and microsatellite (Wares et al. 2004), and single-nucleotide polymorphism (SNP; Peters and Turner 2008; Turner and Camak 2017) studies characterized genetic variation within and among lineages of Gila trout. The lineages differed in their evolutionary relationships with one another, with the Spruce lineage the most ancestral and most divergent from the other lineages (Riddle et al. 1998). The Main Diamond and South Diamond lineages were the most closely related, and Whiskey was more distantly related to these two lineages. The fifth lineage, Iron, was purportedly hybridized with rainbow trout (Leary and Allendorf 1999), but more recent analyses with microsatellite DNA (Turner et al. 2014) and single-nucleotide polymorphism data (Turner and Camak 2017) confirmed that it was as genetically intact as any other relict Gila trout lineage (fig. 20.4). The five relict lineages had low (e.g., Spruce) to moderate (e.g., Iron) genetic diversity compared with other salmonid fishes (Wares et al. 2004; Turner et al. 2014), and each lineage represented a unique part of the genetic legacy of the species.

Hybridization and introgression with rainbow trout has been perceived as the most urgent and persistent threat to the genetic integrity of Gila trout. Securing that integrity through movement barriers and removal of hybridizing individuals remains at the forefront of restoration efforts. These measures, however, often directly oppose the goal of maintaining genetic diversity within Gila trout by precluding movement and gene flow among populations. When gene flow does not occur, genetic diversity is lost through founder effects and genetic drift (Peters and Turner 2008). Genetically depauperate populations are more susceptible to total loss from inbreeding and disease, and thus overall persistence probability decreases dramatically relative to more diverse populations. The restoration strategy in the West Fork Gila River was designed to provide for mixing of lineages to yield a natural metapopulation.

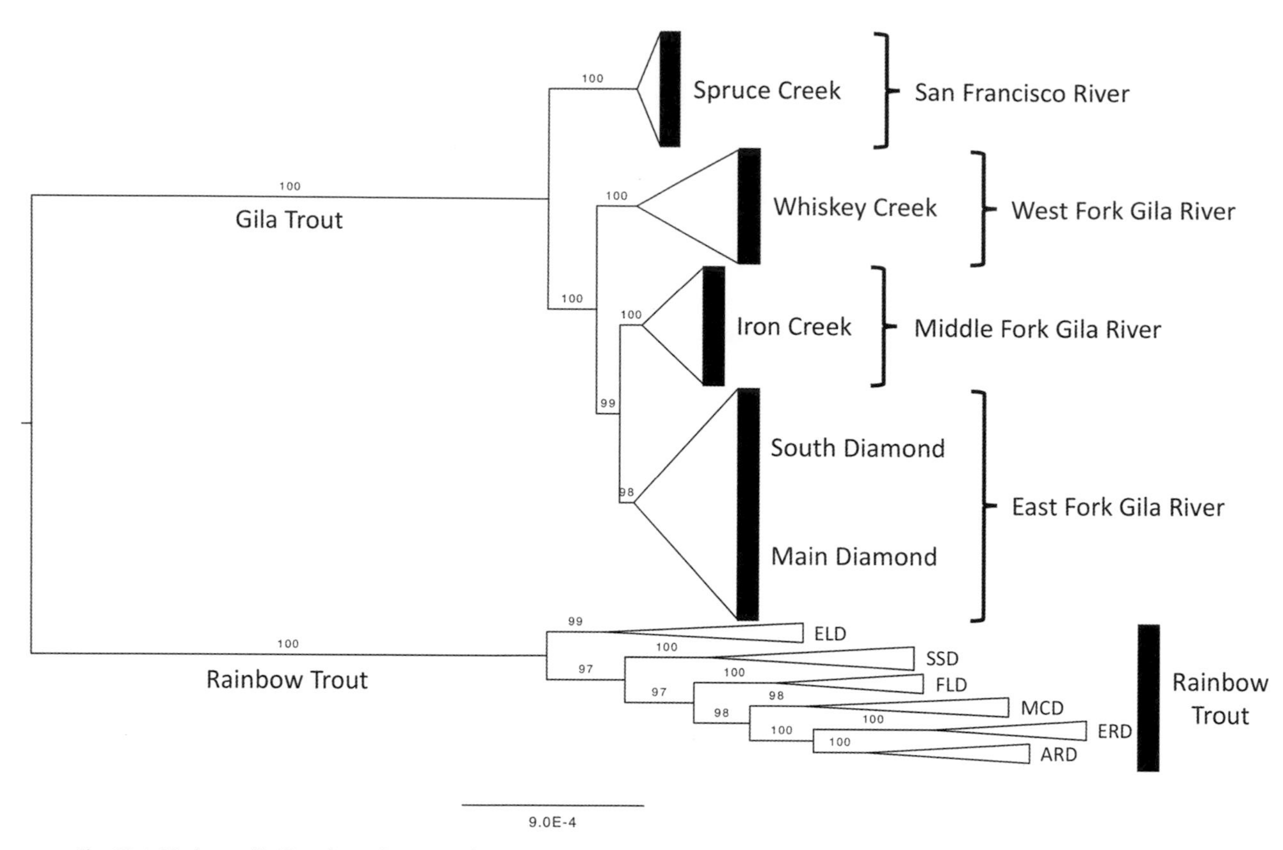

Fig. 20.4 Maximum likelihood tree (generated via RAxML; Stamatakis 2014) depicting evolutionary relationships of Gila trout and rainbow trout. Phylogenetic analysis was based on DNA sequence variation across roughly 23,000 single-nucleotide polymorphic (SNP) loci assayed for 190 individuals (154 Gila trout [*n* = 30 or 31 per lineage] and 36 rainbow trout from six strains [*n* = 6 per strain]). Bootstrap support for relationships is indicated on branches. All branches received over 95% support except those joining Main and South Diamond lineages. These lineages may have experienced relatively recent gene flow and are most closely related. Rainbow trout samples were provided by W. D. Wilson, US Fish and Wildlife Service.

Large-scale and intense wildfires in 2012 and 2013 negatively affected abundance in nearly all extant Gila trout populations. In addition to their demographic toll, drought and wildfire depleted genetic resources (Turner et al. 2014). Comparisons of genetic diversity and heterozygosity from samples obtained in 2003 with those from samples obtained in 2013 showed dramatic losses in both measures for the Spruce and Whiskey lineages. The Main and South Diamond lineages did not show appreciable changes in heterozygosity, but Main Diamond lost allelic diversity at the MHC class II locus. Future environmental disturbances are likely to have strong effects on neutral and adaptive genetic diversity in Gila trout, making the conservation of remaining diversity in all lineages paramount to its long-term persistence. Recovery planning must anticipate future environmental changes and trends, but must also maintain existing diversity to ensure resistance and resilience to change. In the case of Gila trout, losses of total species diversity from wildfire and drought have been offset somewhat by bringing the relatively diverse Iron lineage back into recovery implementation.

By the end of 2010, 16 populations of Gila trout occupied 160 km of habitat in New Mexico and Arizona, and there was some confidence that Gila trout recovery was achievable. The gabion waterfall on Black Canyon was replaced with a permanent concrete barrier in 2011. Besides having four of five relict lineages replicated, widely distributed populations, with several exceeding 5,000 individuals, and a presence in two hydrologically complex drainages, Gila trout had gained general support for continued expansion of its range. Certainly, angling opportunities contributed to the newly found public support.

But over the next three years, that confidence was severely tested. In 2011, the Arizona State Game Commission ordered a one-year moratorium on all Arizona stream renovation activities while AZGFD drafted and obtained approval of its *Piscicide Treatment Planning and Procedures Manual* (AZGFD 2014). The 2011 Wallow Fire eliminated the Raspberry Gila trout population. In spring 2012, brown and rainbow trout were found above the presumed waterfall barrier on the West Fork Gila River. During high flows, a large boulder lodged in a slot below the waterfall had caused water to pool and thus enabled upstream movement of nonnative trout into renovated portions of the stream, thereby compromising over 20 km of Gila trout habitat. The Whitewater-Baldy Fire (May–June 2012) eliminated nonnative trout above the West Fork waterfall, but it also eliminated the relict Whiskey Creek population (fig. 20.5), its replicate in Langstroth Canyon, the Main Diamond population in upper White Creek, and the South Diamond population in Cub Creek. Outside the West Fork drainage, the 120,500 ha fire eliminated Gila trout from Spruce Creek as well as nonnative trout from Willow and Mineral Creeks, and it substantially reduced nonnative trout in Whitewater, West Fork Mogollon, and Rain Creeks. The likelihood that trout had survived in Iron Creek, which was in the center of the Whitewater-Baldy Fire and flowed through some of the most intensely burned portions, was deemed low, but Gila trout did survive. Iron Creek now supports the only relict population that has not been eliminated from its natal stream by wildfire. The following year, the Silver Fire eliminated Main Diamond fish from Black Canyon and McKnight Creek. In the span of three years, seven populations in 60 km of habitat were lost (table 20.1). Mora NFH provided refuge for fish evacuated during the fires, but maintaining and rearing fish from Spruce Creek remained challenging.

Although hatcheries are widely used to respond to demographic losses, they remain a controversial tool for conservation and management of genetic diversity in general, and for Gila trout specifically. Hatchery rearing is controversial for at least three reasons: (1) domestication and relaxed selection can reduce hatchery fish performance in the wild; (2) hatchery fish compete with wild fish for limited resources in a density-dependent fashion; and (3) not all relict lineages can be propagated reliably in the hatchery. It is the third reason that offers the most pressing challenge to conservation of Gila trout because Main Diamond and South Diamond lineages have been extensively propagated over the years, whereas Spruce, Whiskey, and Iron Creek fish are difficult to culture reliably. This difference could lead to long-term bias in lineages repatriated to the wild, and to ultimate losses of genetic diversity of the species while individual lineages lose diversity in response to environmental change.

A brood stock management plan that explicitly considered genetic factors was developed in 2002 (Kincaid and Reisenbichler 2002), but its implementation was not fully realized due to the biological complications of rearing recalcitrant lineages as well as limitations on hatchery space and resources at Mora NFH. More success has been achieved through naturalized culture and rearing, and these approaches have the added benefit of potentially reducing hatchery selection effects on the performance and genetic background of the cultured fish (N. Wiese, pers. comm.). Although microsatellite genotyping has been successfully used to quantify relatedness among parents to determine mate pairings that minimize risks of inbreeding (W. D. Wilson, pers. comm.), these approaches do not overcome the negative effects of lineage bias in repatriation activities. Because of lineage bias in stocking and uncertainties about the effects of hatchery selection, it is preferable to replicate lineages from wild donor populations. Wild populations can have higher genetic diversity than those that have had a history of hatchery origin or support, and use of wild donor populations reduces the risk of loss of fitness (sensu Neff et al. 2011). Turner et al. (2014) identified the highest diversity at all loci examined, including genes involved in immune competence (e.g., MHC class II genes), in the Iron Creek lineage, which has no history of hatchery influence. Maintenance of this critical variation in Iron Creek supports the idea that "wild" donor populations hold ecologically relevant genetic diversity that can potentially benefit replicate populations in recipient streams (Peters and Turner 2008).

Following the mega-wildfires of 2011–2013, considerable effort was expended to restore Gila trout to formerly occupied and newly available streams. Timely installation of a gabion waterfall in 2013, followed by construction of a concrete waterfall in 2016, secured 17 km of Willow Creek for Gila trout. In McKenna Creek, mechanical removal of trout that survived the Miller Fire in 2011 was completed in summer 2012, and the creek was stocked with Whiskey fish in 2012. Gila trout were also repatriated to Black Canyon (2014), upper White (2014), Dude (2015), Langstroth (2015), Ash (2015) and Mineral (2016) Creeks (table 20.2).

Fig. 20.5 Whiskey Creek drainage following the 2012 Whitewater-Baldy Fire (*top*). Collecting Gila trout from Whiskey Creek for evacuation to Mora NFH (*middle*). A Whiskey Creek Gila trout (*bottom*). (Photographs by J. E. Brooks.)

Table 20.1 History of Gila trout lineages. **Boldface** indicates streams supporting Gila trout as of September 2019. Use population number to locate streams on fig. 20.6. Streams of the Mogollon Creek drainage are considered a single population (no. 10). Willow Creek (no. 11) includes its tributary, Little Turkey Creek.

Lineage	Stream	Population number	Date established	Habitat (km)	Source	Notes
MAIN DIAMOND	**Main Diamond**	1	Relict/1994	6.3	Main Diamond/McKnight & Mescalero NFH	Eliminated 1989 (Divide Fire)
	McKnight		1970	8.5	Main Diamond	Eliminated 2013 (Silver Fire)
	Sheep Corral	6	1972/2009	1.0	Main Diamond & Mora NFH	Eliminated 2002 (drought)
	Gap		1974	2.0	Main Diamond	Eliminated early 1990s (poor habitat)
	Black	7	1989/2014	17.4	Mescalero NFH/Mora NFH	Nonnatives eliminated 1995 (Bonner Fire), eliminated 2013 (Silver Fire)
	Little	9	2000	4.6	Mora NFH	Introgressed, eliminated 1998–1999 (piscicide)
	Upper White		2000	11.0	Mora NFH	Eliminated 2012 (Whitewater-Baldy Fire)
	Upper Langstroth	8	2015	5.9	Mora NFH	
	Lower White		2010/2013	2.6	Main Diamond/Mora NFH	Eliminated 2012 (Whitewater-Baldy Fire)/contaminated
	Lower Langstroth		2010	0.9	Main Diamond	Eliminated 2012 (Whitewater-Baldy Fire)
	Rawmeat		2010	0.4	Main Diamond	Eliminated 2012 (Whitewater-Baldy Fire)
SOUTH DIAMOND	**South Diamond**	2	Relict/1997	5.2	South Diamond/ Mescalero NFH	Eliminated 1995 (Bonner Fire)
	Trail	10	1987/1997	1.0	South Diamond	Eliminated 1997 (piscicide)
	Woodrow	10	1988/1997	0.8	South Diamond	Eliminated 1997 (piscicide)
	Mogollon	10	1989/1997	19.1	Trail & Mescalero NFH	Eliminated 1997 (piscicide)
	South Fork Mogollon	10	1998	2.6	Mescalero NFH	
	Cub		2010/2013	7.4	South Diamond/Mora NFH	Eliminated 2012 (Whitewater-Baldy Fire), contaminated 2014
	Grapevine	17	2009	3.2	Mora NFH	Likely eliminated 2017 (fire)
	Frye	18	2009	8.0	Mora NFH	Likely eliminated 2017 (fire)
	Willow	11	2013	19.0	Mora NFH	
WHISKEY	Whiskey	3	Relict	2.3	Whiskey	Eliminated 2012 (Whitewater-Baldy Fire)
	Upper Little		2000	0.5	Whiskey	Eliminated mid-2000s (drought)
	Upper Langstroth		2006	4.3	Whiskey & Mora NFH	Eliminated 2012 (Whitewater-Baldy Fire)
	McKenna	12	2012	1.6	Langstroth & Mora NFH	
	Upper White	13	2014	11.0	Mora NFH	
	Mineral	14	2016	18.8	Mora NFH	
	Sacaton	19	2018	3.8	Mora NFH	
	Raspberry	20	2018	4.1	Mora NFH	
IRON	**Iron**	4	relict	4.4	Iron	
	Sacaton		1990	2.2	Iron	Eliminated mid-2000s (drought)
	Upper White		1993	8.8	Iron	Eliminated 1998 & 1999 (piscicide)

Lineage	Stream	Population number	Date established	Habitat (km)	Source	Notes
SPRUCE	**Spruce**	5	Relict/2018	5.7	Spruce/Dry	Eliminated 2012 (Whitewater-Baldy Fire)
	Big Dry	15	1985	2.9	Spruce	
	Dude		2000	3.1	Spruce	Eliminated 2005 (degraded habitat)
	Raspberry		2000	4.1	Spruce	Eliminated 2011 (Wallow Fire)
MIXED LINEAGES	Ash		2011	5.5	Spruce/Mora NFH	South Diamond & Whiskey lineages Likely eliminated 2017 (fire)
	Dude	16	2015	3.1	Mora NFH	Main Diamond, South Diamond, Whiskey, & Spruce-Whiskey lineages

Table 20.2. Extant Gila trout populations as of September 2019. Use population number to locate streams on figure 20.6.

Lineage	State	Drainage	Stream	Population number	Habitat (km)	Origin
Main Diamond	NM	Upper Gila	Main Diamond	1	6.3	Mescalero NFH
	NM	Upper Gila	Sheep Corral	6	1.0	Mora NFH
	NM	Upper Gila	Black	7	17.4	Mora NFH
	NM	Upper Gila	Upper Langstroth	8	5.9	Mora NFH
	NM	Upper Gila	Little	9	4.6	Mora NFH
South Diamond	NM	Upper Gila	South Diamond	2	5.2	Mescalero NFH
	NM	Upper Gila	Mogollon[a]	10	23.5	Mescalero NFH
	NM	Upper Gila	Willow[b]	11	19.0	Mora NFH
	AZ	Gila	Frye	18	8.0	Mora NFH
Whiskey	AZ	Agua Fria	Grapevine	17	3.2	Mora NFH
	NM	Upper Gila	Upper White	13	11.0	Mora NFH
	NM	San Francisco	Mineral	14	18.8	Mora NFH
	NM	Gila	Sacaton	19	3.8	Mora NFH
	AZ	Blue	Raspberry	20	4.1	Mora NFH
Iron	NM	Upper Gila	Iron	4	4.4	Iron
Spruce	NM	San Francisco	Spruce	5	5.7	Big Dry
	NM	San Francisco	Big Dry	15	2.9	Spruce
Main-South Diamond	AZ	Verde	Dude	16	3.1	Mora NFH

[a]In addition to Mogollon Creek, includes its tributaries Trail, Woodrow, and South Fork Mogollon creeks.
[b]In addition to Willow Creek, includes Little Turkey Creek.

Collectively, these efforts regained much of what the Wallow, Whitewater-Baldy, and Silver Fires eliminated. Unfortunately, wildfires in Arizona during 2017 eliminated repatriated populations in Grapevine, Frye, and Ash Creeks, thereby reducing Gila trout to 14 wild populations in 116 km of habitat. Because McKnight Creek was outside the historical range of Gila trout, NMDGF opposed restoration of Gila trout to that stream.

AN UNCERTAIN FUTURE (2018 AND BEYOND)

After almost 50 years of effort, Gila trout is arguably more secure than it was in 1970, when the first effort

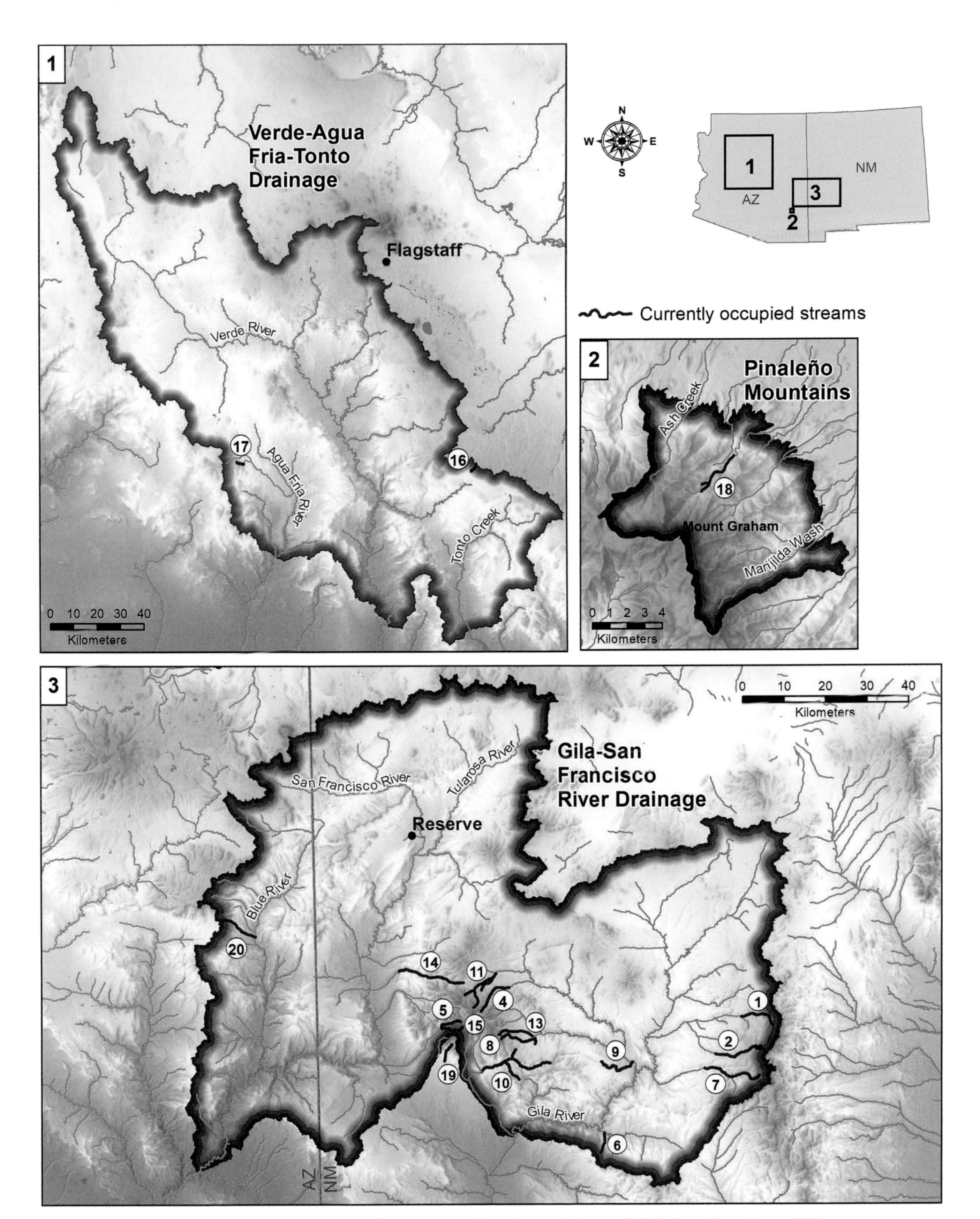

Fig. 20.6 Distribution of Gila trout populations in 2019 (*1*) Agua Fria, Verde, and Tonto drainages, (*2*) Pinaleno Mountains, and (*3*) upper Gila and San Francisco drainages. Refer to table 20.1 for numbered stream names and fig. 20.2 for the general location of each drainage.

to expand its much-diminished range was made by establishing a population in McKnight Creek. From 5 populations in about 23 km of habitat, the species now is represented by 19 populations in about 150 km of habitat (fig. 20.6). In 1970, extant populations were concentrated in the Black Range (Main and South Diamond) and the Mogollon Mountains (Spruce, Iron, and Whiskey). Currently, populations remain concentrated in the Mogollon Mountains and Black Range; only 5 extant populations occur elsewhere. In 1970, each population existed in comparatively short, single-thread streams. Now, Gila trout occupy two comparatively large, hydrologically complex drainages, those of Mogollon and Willow Creeks. The remaining populations, however, are in single-thread streams ranging in length from less than 2 km to about 18 km. While the demographic status of Gila trout has improved, its genetic health has eroded; from 2003 to 2013, the Spruce and Whiskey lineages showed losses in both genetic diversity and heterozygosity, and the Main Diamond lineage lost allelic diversity at the MHC class II locus (Turner et al. 2014).

Repatriation of Gila trout to additional streams within its historical range will continue to be essential to securing its future. Establishment of a metapopulation composed of naturally mixing lineages in West Fork Gila River remains a goal (USFWS 2018). While recent mega-wildfires eliminated Gila trout from many streams, to which it must be restored, they also eliminated or greatly reduced nonnative trout populations in others. Expansion of Gila trout into these habitats will provide additional demographic and, hopefully, genetic security.

When efforts to conserve Gila trout began in earnest in the 1970s, climate change was not an identified threat to the species (e.g., USFWS 1979). By the early twenty-first century, however, altered precipitation patterns, higher ambient temperatures, and increasing aridity because of global warming were projected for the arid regions of North America (Christensen et al. 2004; Milly et al. 2005). Associated with a warming climate was a projected increase in the frequency of wildfires in the American Southwest (Brown et al. 2004), a decrease in surface water (Seager et al. 2013), and an increase in drought risk (Cook et al. 2015). Beginning in 2000, the American Southwest experienced an extended drought that persisted through 2014 (Udall and Overpeck 2017). During that time, Gila trout was plagued by a series of wildfires that dramatically compromised its persistence. Kennedy et al. (2008) estimated that with climate change, 70% of the habitat Propst and Stefferud (1997) reported occupied by Gila trout would be thermally unsuitable by the mid-twenty-first century. The extent to which these projected changes will limit or diminish survival of Gila trout remains to be determined. It is likely, however, that mixed stream assemblages of native cyprinids, catostomids, and Gila trout will become more common and present new challenges to managers.

The conservation of Gila trout is at a critical juncture. Through considerable human intervention, it has survived multiple major natural and human-caused catastrophes and mishaps. The tensions between conservation and exploitation, agency mandates and constituencies, and perceived or real political pressures will persist. The challenge has been, and continues to be, retention of remaining genetic diversity and quality, restoring and maintaining viable populations across the historical range, balancing competing resource demands and expectations, and accomplishing this in the context of changing environmental conditions wrought by a warming climate.

ACKNOWLEDGMENTS

Numerous individuals have contributed to the conservation of Gila trout. Among them, B. Anderson, M. Anderson, H. Bishop, W. Britton, G. Burton, D. Camak, J. Carter, S. Coleman, J. Conway, S. Davenport, R. David, A. Dean, J. Fowler-Propst, W. Furr, G. Jimenez, A. Kingsbury, J. Krammer, Z. Law, V. Lente, R. Maes, A. Monié, P. Mullane, K. Patten, Y. Paroz, C. Pease, D. Peters, J. Pittenger, N. Smith, J. Stefferud, A. Telles, P. Turner, A. Unthank, R. Ward, J. Whitney, M. Whitney, J. Wick, N. Wiese, B. Wiley, and J. Zapata have made significant and lasting contributions. The comments and suggestions of D. Dauwalter improved this chapter. Maps were provided by M. Mayfield. Tissue, specimen, and data archives are held at the Museum of Southwestern Biology, University of New Mexico.

REFERENCES

Abadía-Cardoso, A., J. C. Garza, R. L. Mayden, and F. J. García de León. 2015. Genetic structure of Pacific trout at the extreme southern end of their native range. *PLoS ONE* 10 (10): e0141775.

AZGFD (Arizona Game and Fish Department). 2014. *Piscicide Treatment Planning and Procedures Manual.* Phoenix: Arizona Game and Fish Department.

Baker, R. D., R. S. Maxwell, V. H. Treat, and H. C. Dethloff. 1988. *Timeless Heritage: A History of the Forest Service in the Southwest.* FS409. Washington, DC: USDA Forest Service.

Bartlett, J. R. 1854. *Personal Narrative of Explorations and Incidents in Texas, New Mexico, California, Sonora, and Chihuahua: Connected with the US and Mexican Boundary Commission during the Years 1850, '51, '52, and '53.* New York: D. Appleton & Company.

Beamish, R. J., and R. R. Miller. 1976. Cytotaxonomic study of Gila trout, *Salmo gilae. Journal of the Fisheries Research Board of Canada* 34: 1041–45.

Behnke, R. J. 1992. *Native Trout of Western North America.* Monograph 6. Bethesda, MD: American Fisheries Society.

Behnke, R. J. 2002. *Trout and Salmon of North America.* New York: Free Press.

Behnke, R. J., and M. Zarn. 1976. *Biology and Management of Threatened and Endangered Western Trouts.* USDA Forest Service General Technical Report RM-28. Fort Collins, CO: USDA Forest Service, Rocky Mountain Forest and Range Experimental Station.

Bickle, T. S. 1973. *Gila Trout Management Plan.* Silver City, NM: Gila National Forest.

Brooks, J. E. 2004. *Emergency Evacuation Procedures for Gila Trout.* Albuquerque: US Fish and Wildlife Service, NM Fisheries Resource Office.

Brooks, J. E., and D. L. Propst. 1999. *Nonnative Salmonid Removal from Black Canyon Drainage, East Fork Gila River, June–October 1998.* Albuquerque: US Fish and Wildlife Service, NM Fisheries Resource Office.

Brown, D. K., A. A. Echelle, D. L. Propst, J. E. Brooks, and W. L. Fisher. 2001. Catastrophic wildfire and number of populations as factors influencing risk of extinction for Gila trout (*Oncorhynchus gilae*). *Western North American Naturalist* 61: 139–48.

Brown, T. J., B. L. Hall, and A. L. Westerling. 2004. The impact of twenty-first century climate change on wildland fire danger in the western United States: An application perspective. *Climatic Change* 62: 365–88.

Bryan, K. 1925. Date of channel trenching (arroyo cutting) in the arid Southwest. *Science* 62: 338–44.

Christensen, N. S., A. W. Wood, N. Voisin, D. P. Lettenmaier, and R. N. Palmer. 2004. The effects of climate change on the hydrology and water resources of the Colorado River basin. *Climatic Change* 62: 337–63.

Coman, C. H. 1981. *Gila Trout Management and Recovery Activities with Emphasis on Iron Creek Recovery Efforts.* Report for the United States Forest Service. Silver City, NM: Gila National Forest.

Cook, B. I., T. R. Ault, and J. E. Smerdon. 2015. Unprecedented 21st century drought risk in the American Southwest and Central Plains. *Science Advances* 1: e1400082.

Crête-Lafrenière, A., L. K. Weir, and L. Bernatchez. 2012. Framing the Salmonidae family phylogenetic portrait: A more complete picture from increased taxon sampling. *PLoS ONE* 7 (10): e46662.

Dahlstrom, M. F. 2013. Using narratives and storytelling to communicate science with nonexpert audiences. *Proceedings of National Academy of Sciences* 111: 13614–20.

Dahms, C. W., and B. W. Geils, tech. eds. 1997. *An Assessment of Forest Ecosystem Health in the Southwest.* USDA Forest Service General Technical Report RM-GTR-295. Fort Collins, CO: USDA Forest Service, Rocky Mountain Forest and Range Experiment Station.

David, R. E. 1976. Taxonomic analysis of Gila and Gila × Rainbow trout in southwestern New Mexico. MS thesis, New Mexico State University.

Dinsmore, A. E. 1924. Extract from A. E. Dinsmore's report of January 23, 1924, to the Commissioner of Fisheries covering his investigations during July 1923 of the trout streams in the Gila National Forest and other New Mexico trout waters. Aldo Leopold Papers, Series 9/25/10-4 (Species and Subjects), Box 005 (Fish: Trout), Folder 008, pp. 796–807. https://uwdc.library.wisc.edu/collections/AldoLeopold/. Accessed April 1, 2016.

Emory, W. H. 1848. *Notes of a Military Reconnaissance from Ft. Leavenworth, in Missouri, to San Diego, in California, Including Part of the Arkansas, del Norte, and Gila Rivers. By Lt. Emory, Made in 1846–47, with the Advanced Guard of the Army of the West.* Washington, DC: Wendell van Benthuysen, Printers.

French, W. 1965. *Further Recollections of a Western Ranchman: New Mexico, 1883–1899.* New York: Argosy-Antiquarian Ltd.

Gold, J. R. 1977. Systematics of western North American trout (*Salmo*), with notes on the reband trout of Sheepheaven Creek, California. *Canadian Journal of Zoology* 55: 1858–73.

Hanson, J. N. 1971. Investigations on Gila trout, *Salmo gilae* Miller, in southwestern New Mexico. MS thesis, New Mexico State University.

Huntington, E. H. 1955. *Fisheries Survey of the Gila and Mimbres rivers drainages, 1952–1955.* Project Completion Report, Federal Aid Project F-1-R. Santa Fe: New Mexico Department of Game and Fish.

Kennedy, T. L., D. S. Gutzler, and R. L. Leung. 2008. Predicting future threats to the long-term survival of Gila trout using a high-resolution simulation of climate change. *Climatic Change* 94 (3–4): 503–15.

Kincaid, H. L., and R. Reisenbichler. 2002. *Gila Trout: Genetic Broodstock Management Plan.* Albuquerque: U.S. Fish and Wildlife Service, Region 2, New Mexico Fisheries Resource Office.

Kynard, B. E. 1976. *A Study of the Pollution Sources and Their Effect on the Aquatic Habitat of Eagle Creek Watershed, Apache-Sitgreaves National Forest, Arizona.* Final Report, Cooperative Agreement no. 16-514-CA, USDA Forest Service.

Leary, R. F., and F. W. Allendorf. 1999. *Genetic Issues in the Conservation and Restoration of the Endangered Gila Trout: Update.* Wild Trout and Salmon Genetics Laboratory Report 99/2. Missoula: University of Montana.

Leary, R. F., F. W. Allendorf, and N. Kanda. 1999. *Recovery of Gila Trout Descended from South Diamond Creek from Recently Hybridized Populations in the Mogollon Creek Drainage.* Wild Trout and Salmon Genetics Laboratory Report 99/1. Missoula: University of Montana.

Leiberg, J. B., T. F. Rixon, and A. Dodwell. 1904. *Forest Conditions in the San Francisco Mountains Forest Reserve, Arizona*. Professional Paper no. 22, Series H, Forestry, 7. Washington, DC: USDI, US Geological Survey.

Leopold, A. 1921. A plea for recognition of artificial works in forest erosion control policy. *Journal of Forestry* 19: 269–70.

Loudenslager, E. J., J. N. Rinne, G. A. E. Gall, and R. E. David. 1986. Biochemical genetic studies of native Arizona and New Mexico trout. *Southwestern Naturalist* 31: 221–34.

Mello, K., and P. R. Turner. 1980. *Population Status and Distribution of Gila Trout in New Mexico*. Endangered Species Report no. 6. Albuquerque: US Fish and Wildlife Service.

Miller, R. R. 1950. Notes on the cutthroat and rainbow trouts with the description of a new species from the Gila River, New Mexico. *Occasional Papers of the Museum of Zoology, University of Michigan*, no. 529. Ann Arbor: University of Michigan Press.

Miller, R. R. 1972. Classification of the native trouts of Arizona, with the description of a new species, *Salmo apache*. *Copeia* 1972: 401–22.

Miller, R. R., and C. H. Lowe. 1964. An annotated check list of the fishes of Arizona. In C. H. Lowe, ed., *The Vertebrates of Arizona*, 133–51. Tucson: University of Arizona Press.

Milly, P. C. D., K. A. Dunne, and A. V. Vecchia. 2005. Global patterns of trends in streamflow and water availability in a changing climate. *Nature* 438: 347–50.

Minckley, W. L. 1973. *Fishes of Arizona*. Phoenix: Arizona Game and Fish Department.

Minckley, W. L., and J. E. Brooks. 1985. Transplantations of native Arizona fishes: Records through 1980. *Journal of the Arizona-Nevada Academy of Science* 20: 73–89.

Minckley, W. L., and P. C. Marsh. 2009. *Inland Fishes of the Greater Southwest: Chronicle of a Vanishing Biota*. Tucson: University of Arizona Press.

Mulch, E. E., and W. C. Gamble. 1956. *Game Fishes of Arizona*. Phoenix: Arizona Game and Fish Department.

Nankervis, J. M. 1988. Age, growth, and reproduction of Gila trout in a small headwater stream in the Gila National Forest. MS thesis, New Mexico State University.

Needham, P. R., and R. Gard. 1959. Rainbow trout in Mexico and California, with notes on the cutthroat series. *University of California Publications in Zoology* 67: 1–124.

Neff, B. D., S. R. Garner, and T. E. Pitcher. 2011. Conservation and enhancement of wild fish populations: Preserving genetic quality versus genetic diversity. *Canadian Journal of Fisheries and Aquatic Sciences* 68: 1139–54.

Papoulias, D., D. Valenciano, and D. Hendrickson. 1988. *A Fish and Riparian Survey of the Clifton Ranger District.* Draft Final Report to USDA Forest Service, Apache-Sitgreaves National Forest. Phoenix: Arizona Game and Fish Department.

Peters, M. B., and T. F. Turner. 2008. Genetic variation of the major histocompatibility complex (MHC class II β gene) in the threatened Gila trout, *Oncorhynchus gilae*. *Conservation Genetics* 9: 257–70.

Propst, D. L. 1994. The status of Gila trout. *New Mexico Wildlife*, November/December 1994.

Propst, D. L., and J. A. Stefferud. 1997. Population dynamics of Gila trout in the Gila River drainage of the south-western United States. *Journal of Fish Biology* 51: 1137–54.

Propst, D. L., J. A. Stefferud, and P. R. Turner. 1992. Conservation and status of Gila trout, *Oncorhynchus gilae*. *Southwestern Naturalist* 37: 117–25.

Ratté, J. C., E. R. Landis, D. L. Gaskill, and R. G. Raabe. 1969. *Mineral Resources of the Blue Range Primitive Area, Greenlee County, Arizona, and Catron County, New Mexico*. Geological Survey Bulletin 1261-E. Washington, DC: US Government Printing Office.

Ratté, J. C., D. L. Gaskill, G. P. Eaton, D. L. Peterson, R. B. Stotelmeyer, and H. C. Meeves. 1979. *Mineral Resources of the Gila Primitive Area and Gila Wilderness, New Mexico*. Geological Survey Bulletin 1451. Washington, DC: US Government Printing Office.

Regan, D. M. 1964. Ecology of Gila trout, *Salmo gilae*, in Main Diamond Creek, New Mexico. MS thesis, Colorado State University.

Regan, D. M. 1966. *Ecology of Gila Trout in Main Diamond Creek in New Mexico*. Technical Paper 5. Washington, DC: Bureau of Sport Fisheries and Wildlife.

Riddle, B. R., D. L. Propst, and T. L. Yates. 1998. Mitochondrial DNA variation in Gila trout, *Oncorhynchus gilae*: Implications for management of an endangered species. *Copeia* 1998: 31–39.

Rinne, J. N. 1980. Spawning habitat and behavior of Gila trout, a rare salmonid of the southwestern United States. *Transactions of the American Fisheries Society* 109: 83–91.

Rinne, J. N. 1982. Movement, home range, and growth of a rare southwestern trout in improved and unimproved habitats. *North American Journal of Fisheries Management* 2: 150–57.

Rixon, T. F. 1905. *Forest Conditions in the Gila River Forest Reserve, New Mexico*. Professional Paper no. 39, Series H, Forestry, 13. Washington, DC: USDI, US Geological Survey.

Roberts, P. H. 1963. *Hoof Prints on the Forest Ranges: The Early Years of National Forest Range Administration*. San Antonio: The Naylor Company.

Sarewitz, D. 2004. How science makes environmental controversies worse. *Environmental and Science Policy* 7: 385–403.

Seager, R., M. Ting, C. Li, N. Naik, B. Cook, and J. Nakamura. 2013. Projections of the declining surface-water availability for the southwestern United States. *Nature Climate Change* 3: 482–86.

Seelbach, P. W., and G. E. Whelan. 1988. Identification and contribution of wild and hatchery steelhead stocks in Lake Michigan tributaries. *Transactions of the American Fisheries Society* 117: 444–51.

Stamatakis A. 2014. RAxML version 8: A tool for phylogenetic analysis and post-analysis of large phylogenies. *Bioinformatics* 30: 1312–13.

Sublette, J. E., M. D. Hatch, and M. Sublette. 1990. *The Fishes of New Mexico*. Albuquerque: University of New Mexico Press.

Sullivan, P. J., J. M. Acheson, P. L. Angermeier, T. Faast, J. Flemma, C. M. Jones, E. E. Knudsen, T. J. Minello, D. H. Secor, R. Wunderlich, and B. A. Zanetell. 2006. Defining and implementing best available science for fisheries and environmental science, policy, and management. *Fisheries* 31: 460–65.

Thorgaard, G. H. 1983. Chromosomal differences among rainbow trout populations. *Copeia* 1983: 650–62.

Toumey, J. W. 1901. Our Forest Reservations. *Popular Science Monthly* 59: 115–28.

Turner, T. F., and D. Camak. 2017. *Next-Generation Genetic Resource Management in Gila Trout, Phase I*. Arlington: Trout Unlimited.

Turner, T. F., M. J. Osborne, T. J. Pilger, W. D. Wilson, and D. L. Propst. 2014. *Genetic Monitoring of Gila Trout Lineages Restored to the Upper West Fork Gila River and Implications for Future Conservation Strategies.* Annual Report for Project Work Order CSD 120731-B. Santa Fe: New Mexico Department of Game and Fish.

Udall, B., and J. Overpeck. 2017. The twenty-first century Colorado River hot drought and implications for the future. *Water Resources Research* 53: 2404–18.

USFWS (US Fish and Wildlife Service). 1967. Native fish and wildlife, endangered species. *Federal Register* 32: 4001.

USFWS (US Fish and Wildlife Service). 1979. *Gila Trout Recovery Plan*. Albuquerque: US Fish and Wildlife Service.

USFWS (US Fish and Wildlife Service). 1984. *Gila Trout Recovery Plan* (first revision). Albuquerque: US Fish and Wildlife Service.

USFWS (US Fish and Wildlife Service). 1987. Proposed reclassification of the Gila trout (*Salmo gilae*) from endangered to threatened. *Federal Register* 52: 37424–27.

USFWS (US Fish and Wildlife Service). 1993. *Gila Trout Recovery Plan* (second revision). Albuquerque: US Fish and Wildlife Service.

USFWS (US Fish and Wildlife Service). 2003. *Gila Trout* (Oncorhynchus gilae) *Recovery Plan* (third revision). Albuquerque: US Fish and Wildlife Service.

USFWS (US Fish and Wildlife Service). 2006. Reclassification of the Gila trout (*Oncorhynchus gilae*) from endangered to threatened; Special rule for Gila trout in New Mexico and Arizona. *Federal Register* 71: 40657–74.

USFWS (US Fish and Wildlife Service). 2018. *Revised Recovery Plan for Gila Trout* (Oncorhynchus gilae). Technical review draft. Albuquerque: US Fish and Wildlife Service.

Van Eimeren, P. A. 1988. Comparative food habits of Gila trout and speckled dace in a southwestern headwater stream. MS thesis, New Mexico State University.

Wares, J. P., D. Alò, and T. F. Turner. 2004. A genetic perspective on management and recovery of federally endangered trout (*Oncorhynchus gilae*) in the American Southwest. *Canadian Journal of Fisheries and Aquatic Sciences* 61: 1890–99.

Wilson, W. D., and T. F. Turner. 2009. Phylogenetic analysis of the Pacific cutthroat trout (*Oncorhynchus clarki* ssp.: Salmonidae) based on partial mtDNA ND4 sequences: A closer look at the highly fragmented inland species. *Molecular Phylogenetics and Evolution* 52: 406–15.

Woodrow, H. 1943. *History of the McKenna Park District*. Silver City: Gila National Forest.

Woolsey, T. S., Jr. 1907. Some government timber sales in the Southwest from the practical and technical standpoint. *Proceedings of the Society of American Foresters* 2: 115–29.

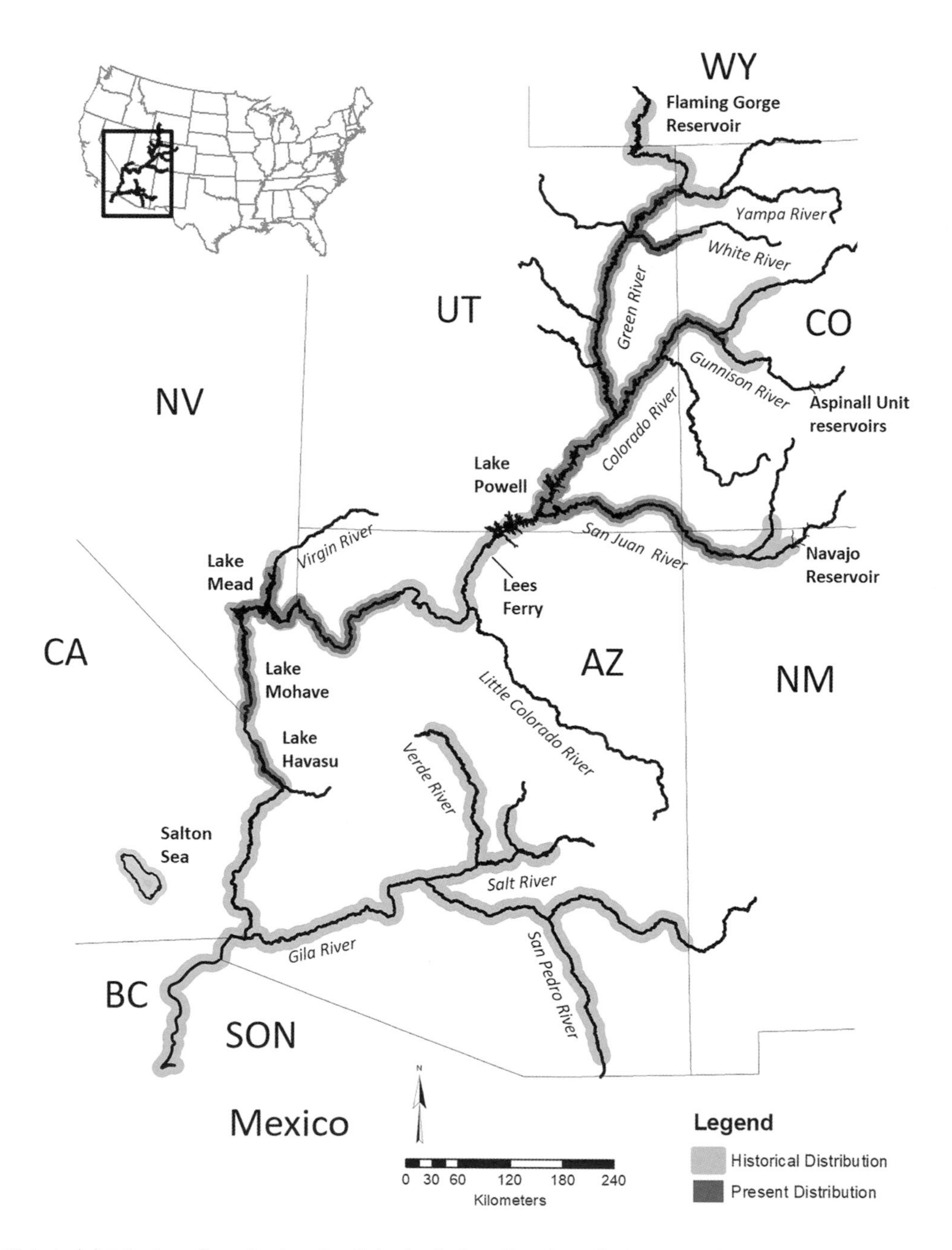

Fig. 21.1 Historical distribution of razorback sucker (light shading) and locations of existing population concentrations (dark shading) in the Colorado River basin. The three basin divisions described in the chapter text are Lake Mohave and downstream (including the Gila River basin); Lake Mead to Lake Powell, including the intervening Grand Canyon; and the upper Colorado River basin upstream of Lake Powell (including the Green, Colorado, and San Juan River basins).

21

Kevin R. Bestgen, Thomas E. Dowling, Brandon Albrecht, and Koreen A. Zelasko

Large-River Fish Conservation in the Colorado River Basin

Progress and Challenges with Razorback Sucker

In the Colorado River of the arid American Southwest, large-river fishes such as endemic and endangered razorback sucker *Xyrauchen texanus* are emblematic of the global plight of river-dwelling organisms (Dudgeon et al. 2006; Jelks et al. 2008). Historically, razorback sucker (Williams and Propst, chap. 1, this volume, fig. 1.4) was one of the most widespread and abundant fishes in warmwater streams of the Colorado River basin, occurring in more than 5,000 km of mainstem river and large tributaries. Razorback suckers were abundant enough to serve as crop fertilizer and supported small commercial fisheries (Miller 1961; Minckley et al. 1991). The distribution and abundance of razorback suckers has declined dramatically over the past 100 years, a period when human population growth increased greatly in the American Southwest. By the 1980s, the species was extirpated from 90% or more of its historical range. Here, we summarize research and management conducted on behalf of the razorback sucker to illustrate the challenges associated with conservation of large-river fishes globally and to highlight efforts to restore self-perpetuating populations of the species in portions of its native range. Over time, these efforts helped us to identify needed course corrections, and the knowledge gained will guide future conservation.

BACKGROUND

The Colorado River drains portions of seven states in the southwestern United States and two in northwestern México (fig. 21.1). It supports more than 40 million people, but that number is growing rapidly (Udall and Overpeck 2017; Stone and Morrison, chap. 16, this volume). To serve that expanding

Table 21.1. Native fishes of the Colorado River basin, their status, and their historical distribution.

Species	Common name	Federal statusa	River association	Upper basin	Lower basin
SALMONIDAE					
Oncorhynchus clarkii pleuriticus	Colorado River cutthroat trout		Headwaters	X	
O. apache	Apache trout	Threatened	Headwaters		X
O. gilae	Gila trout	Threatened	Headwaters		X
Prosopium williamsoni	Mountain whitefish		Headwaters	X	
CYPRINIDAE					
Agosia chrysogaster	Longfin dace		Small		X
Gila cypha	Humpback chub	Endangered	Large, canyons	X	X
G. elegans	Bonytail	Endangered	Large	X	X
G. robusta	Roundtail chub		Small to large	X	X
G. r. jordani	Pahranagat roundtail chub	Endangered	Small, springs		X
G. intermedia	Gila chub	Endangered	Small, springs		X
G. nigra	Headwater chub		Small		X
G. seminuda	Virgin chub	Endangered	Small		X
Lepidomeda albivallis	White River spinedace	Endangered	Springs		X
L. altivelis	Pahranagat spinedace	Extinct	Springs		X
L. mollispinis mollispinis	Virgin spinedace		Small		X
L. mollispinis pratensis	Big Spring spinedace	Threatened	Small		X
L. vittata	Little Colorado spinedace	Threatened	Small		X
Meda fulgida	Spikedace	Endangered	Small		X
Moapa coriacea	Moapa dace	Endangered	Small, springs		X
Plagopterus argentissimus	Woundfin	Endangered	Small to large		X
Ptychocheilus lucius	Colorado pikeminnow	Endangered	Large	X	X
Rhinichthys osculus	Speckled dace		Small to large	X	X
R. o. thermalis	Kendall Warm Springs dace	Endangered	Springs	X	
R. cobitis	Loach minnow	Endangered	Small		X
R. deaconi	Las Vegas dace	Extinct	Springs		X
CATOSTOMIDAE					
Catostomus platyrhynchus	Mountain sucker		Headwaters	X	
C. discobolus	Bluehead sucker		Small to large	X	X
C. d. yarrowi	Zuni bluehead sucker	Endangered	Headwaters		X
C. latipinnis	Flannelmouth sucker		Large	X	X
C. clarkii	Desert sucker		Small		X
C. insignis	Sonora sucker		Small		X
Xyrauchen texanus	Razorback sucker	Endangered	Large	X	X
CYPRINODONTIDAE					
Cyprinodon arcuatus	Santa Cruz pupfish	Extinct	Springs		X
C. macularius	Desert pupfish	Endangered	Springs		X
GOODEIDAE					
Crenichthys baileyi	White River springfish	Endangered	Springs		X

FAMILY					
Species	Common name	Federal statusa	River association	Upper basin	Lower basin
C. b. grandis	Hiko White River springfish	Endangered	Springs		X
C. nevadae	Railroad Valley springfish	Threatened	Springs		X
POECILIIDAE					
Poeciliopsis occidentalis	Gila topminnow	Endangered	Small, springs		X
COTTIDAE					
Cottus bairdii	Mottled sculpin		Headwaters	X	
Cottus beldingii	Paiute sculpin		Headwaters	X	
ELEOTRIDAE					
Eleotris picta	Spotted sleeper		Large, estuarine		X
ELOPIDAE					
Elops affinis	Machete		Large, estuarine		X
MUGILIDAE					
Mugil cephalus	Striped mullet		Large, estuarine		X

Sources: Names follow Page et al. (2013). Other sources include Olden et al. (2006) and Minckley and Marsh (2009).
[a] US Fish and Wildlife Service federal listing.

population, extensive infrastructure was built to store limited, seasonally available water for agricultural, domestic, and recreational use, making the basin's water supplies some of the most tightly controlled in the world (Iorns et al. 1965). Widespread and continuing water development is a major reason for the decline of large-river fishes everywhere (Pikitch et al. 2005), and in the Colorado River basin it has resulted in depleted or altered streamflow and sediment patterns, locally reduced water temperatures, blocked fish passage, and river channel narrowing due to establishment of nonnative vegetation (Grams and Schmidt 2002). Long-term habitat degradation due to dams and flow alteration began early in the twentieth century and will continue well into the future. Predation and competition from over 65 nonnative fishes also limit native fish survival (Carlson and Muth 1989; Olden et al. 2006). Unabated species introductions began over 100 years ago, in part to support sport fisheries. All of these environmental changes have created a challenging riverscape for native fishes and those who would conserve them.

Modifications to the physical and biological environments in the Colorado River basin have resulted in one of the highest proportions of imperiled native fishes in the world (table 21.1). Of 35 native freshwater fishes, 3 are extinct, 23 species or subspecies are federally protected as threatened or endangered, and many remaining forms receive some level of state protection in the United States and México (Jelks et al. 2008). Large-river fishes have suffered some of the greatest declines (Bezzerides and Bestgen 2002; Bestgen et al. 2007; Marsh et al. 2015), and razorback sucker was the last of those listed for federal protection in 1991 (56 FR 54957).

RAZORBACK SUCKER CONSERVATION

In upper reaches of the Colorado River, razorback suckers were widespread in mainstems and larger tributaries of the Green, Colorado, and San Juan Rivers in reaches below 1,800 m elevation (Bestgen 1990; Platania et al. 1991). Downstream, razorback suckers were historically present in the Colorado River mainstem, including in Grand Canyon, to the Gulf of California, and were widespread in the Gila River (Minckley et al. 1991; Marsh et al. 2015). Razorback sucker occurrence in reservoirs indicates adaptable habitat use (Minckley et al. 1991). Although for the purposes of water allocation, the Colorado River has historically been divided into upper and lower basins at Lees Ferry, just downstream of Glen Canyon Dam, we divide it for conservation discussions into three areas: (1) Lake Mohave and downstream, (2) Lake Mead, Grand Canyon, and Lake Powell, and (3) the upper Colorado River basin upstream of Lake Powell, including the San Juan, Colorado, and Green River basins.

Lake Mohave and Downstream

Negative effects of habitat change and nonnative fishes on razorback sucker are most pronounced in the lower Colorado River, where numerous dams on the mainstem Colorado and tributary Gila, Salt, and Verde Rivers have affected habitat and diverse nonnative fish assemblages have become established (Olden et al. 2006; Minckley and Marsh 2009). For example, the lower 500 km of the Colorado River downstream of Lake Mead has eight large dams (Mueller and Marsh 2002). Razorback sucker and other large-river fishes (see table 21.1) were mostly eliminated from this area by 1970 (Minckley et al. 1991). Remaining razorback sucker populations were thought to be limited to Lake Mohave, Lake Havasu, and adjacent riverine sections.

Recognizing the plight of razorback sucker, agency and academic biologists established a brood stock of Lake Mohave fish at Willow Beach National Fish Hatchery, Arizona, in 1974 (Hamman 1985). In 1981, those fish were transferred to Dexter National Fish Hatchery (now Southwestern Native Aquatic Resources and Recovery Center [SNARRC]), New Mexico, where a brood stock remains. That action, taken when thousands of adult razorback suckers remained in Lake Mohave, but recruitment was nil, was among the first of many human interventions that would benefit future recovery efforts. Although all hatchery stocks are valuable, Lake Mohave fish were especially so, given their high genetic diversity (Dowling et al. 1996a), and their hatchery-reared progeny were used for most early conservation stocking actions, especially in central Arizona (Marsh et al. 2015). Widespread experimental stocking of early-life-stage razorback sucker was generally unsuccessful because little or no effort was made to enhance habitat or mitigate effects of nonnative fishes (Marsh and Brooks 1989). This result was a first indication that any progress toward recovery of razorback sucker was unlikely to be successful without direct and various human interventions.

Research and monitoring in Lake Mohave and elsewhere increased knowledge of the ecology, genetic structure, and specific mechanisms for population declines of the razorback sucker (McAda and Wydoski 1980; Minckley et al. 1991). Early-life-stage research initiated in the early 1980s (Bozek et al. 1990) documented high fecundity but low larval survival, mainly due to predation by nonnative fishes (Minckley et al. 1991). Late 1980s research documented a large adult population and annual reproduction in Lake Mohave, but lack of recruitment was evident in the aging population (McCarthy and Minckley 1987). Once estimated at over 100,000 individuals, the abundance of wild razorback sucker in Lake Mohave plummeted to about 2,000 individuals by 2000, and the original wild fish were thought extirpated by 2011 (Marsh et al. 2003, 2015).

Lack of recruitment in Lake Mohave prompted intensive culture efforts beginning in 1990 to replace the aging wild population with younger individuals. To maintain high genetic diversity, the strategy involved annual harvest of larvae produced from wild and, later, repatriated adults in Lake Mohave, rearing to 250–300 mm total length (TL) in protected predator-free conservation habitats such as aquaria, ponds, or backwaters, and release back into the reservoir. The protected conservation habitats, the key element in this "protective custody" approach (Minckley et al. 2003; Marsh et al. 2015), were variable in size and other physical characteristics, but had in common the requirement that they be predator-free.

Because of their large size, razorback sucker repatriated to lower Colorado River reservoirs such as Lake Mohave were thought to be immune to predation, and the population goal of 50,000 adults in Lake Mohave was thought to be attainable by the early twenty-first century. However, through 2013, the population in Lake Mohave numbered less than 3,000 repatriated individuals, in spite of the over 165,000 subadult or older fish stocked through 2012 (Dowling et al. 2014; Marsh et al. 2015). A main reason for low survival was predation, as Karam et al. (2008) found that acoustically monitored repatriated razorback suckers 355–455 mm TL were quickly eliminated by striped bass *Morone saxatilis*. Efforts to increase the size of stocked fish to 500 mm TL were initially hampered by slow growth, but the hatchery program reduced densities in rearing ponds and increased holding times up to several years to more often attain that length (Kesner et al. 2016). Razorback sucker stocked in Lake Havasu, the next reservoir downstream from Lake Mohave, also had low survival due to predation by nonnative fish and piscivorous birds (Humphrey et al. 2016; Kesner et al. 2017), and at this time all populations downstream of Lake Mead, with the exception of those in conservation habitats, are maintained only by stocking.

Despite issues with predation by nonnative fishes, stocking has increased the razorback sucker population in the lower basin, albeit segregated in reservoirs, to about 10,000 individuals while maintaining genetic diversity (Dowling et al. 2005, 2014), and abundant larvae are produced in both Lake Mohave and Lake Havasu each year (Kesner et al. 2017). Predator presence essentially precludes main-channel reservoirs or riverine habitats from making a substantial contribution toward long-term recovery of the species, although surviving adults produce larvae that can contribute to annual propagation efforts (Marsh et al. 2015).

The protected conservation habitat concept was demonstrated successfully in a Colorado River oxbow (Mueller 2006), Cibola High Levee Pond, where razorback sucker reproduced and young recruited annually for more than a decade before the population was destroyed by illicitly stocked largemouth bass *Micropterus salmoides*. Bass have since been removed, and endangered fish, including razorback sucker, were reestablished and have persisted since 2008 (J. Lantow, pers. comm.). The conservation habitat program continues at Cibola High Levee Pond and at Imperial National Wildlife Refuge, where populations have been established in one permanent and two ephemeral backwaters adjacent to Lake Mohave (Dowling et al. 2017a).

Lake Mead, Grand Canyon, and Lake Powell

Razorback sucker was relatively common throughout Lake Mead during the 1950s and 1960s following closure of Hoover Dam in 1935 (Holden et al. 1997). As it did elsewhere, razorback sucker abundance declined in Lake Mead by the 1970s, and few were collected there during the 1980s (Minckley et al. 1991; Holden et al. 1997). In the early 1990s, angler observations of razorback sucker in the Las Vegas Bay and Echo Bay arms of Lake Mead were confirmed, which led to investigations of the ecology of the species there (Holden et al. 1997; Albrecht et al. 2010). The origin of Lake Mead razorback suckers is unknown, but they may be from individuals that persisted there or in Grand Canyon after reservoir filling, or they may be from intermittent stocking (Sjoberg 1995; Schooley and Marsh 2007). Escape of juvenile razorback suckers from local culture facilities was considered a possible source, but to date no evidence supports this hypothesis (Albrecht et al. 2017).

The Lake Mead population is biologically important because it sustains reproduction, juvenile survival, and recruitment to the adult life stage. Whether the population is self-sustaining is still uncertain, but if recruitment does in fact equal or exceed mortality, it would be the only self-sustaining population in the basin. In contrast to remnant populations with only old individuals, the Lake Mead razorback sucker population has a mix of ages and has exhibited nearly annual recruitment since the 1970s (Albrecht et al. 2017). Annual survival rates of adults are estimated between 70% and 80%, and the population size is roughly 500–1,500 individuals that are 2 to 36+ years old (Shattuck et al. 2011; Mohn et al. 2016). Razorback sucker use Las Vegas Bay, Echo Bay, the Virgin River/Muddy River inflow area, and the Colorado River inflow area as spawning areas, and individuals move among those areas between years (Albrecht et al. 2010; Mohn et al. 2016). Annual recruitment of razorback sucker in Lake Mead may occur because vegetation and turbidity in inflow areas provide cover and enhance survival of young, despite the presence of nonnative predators (Albrecht et al. 2010).

Upstream in Grand Canyon, few razorback suckers were historically documented. From 1944 through 1990, only 10 adults were captured between Lees Ferry (river kilometer [RK] 0) and Shinumo Creek (RK 175) (Kegerries et al. 2017). Razorback sucker was thought functionally extirpated from this reach (Clarkson and Childs 2000), but individuals, including those moving upriver from Lake Mead, were subsequently documented in several Grand Canyon locations upstream nearly to Phantom Ranch (RK 142) by 2016 (Kegerries et al. 2016, 2017). Juveniles were also captured just downstream of Grand Canyon during 2013 and 2014 (Kegerries et al. 2016).

Captures of larvae in 2014, 2015, and 2016 (n = 462, 81, and 46, respectively) indicated that razorback sucker reproduction occurred in lower Grand

Canyon, and larvae were present in samples down to and in Lake Mead (Kegerries et al. 2017; S. P. Platania, pers. comm.). Tracking sonic- and radio-tagged individuals indicated use of multiple spawning locations, but to date, no juveniles have been captured in Grand Canyon (Kegerries et al. 2016). The discovery of reproducing razorback suckers and their larvae, as well as other abundant native fishes, including endangered humpback chub (Van Haverbeke et al. 2017), was coincident with increasing water temperatures in Grand Canyon due to lowered water levels in Lake Powell and warmer releases from Glen Canyon Dam (Ross and Vernieu 2013; Van Haverbeke et al. 2017). A return to colder thermal regimes in that reach may have negative consequences for Grand Canyon razorback suckers and other native fishes.

The San Juan and Colorado River inflow areas of Lake Powell may likewise provide important habitats for razorback sucker (Karp and Mueller 2002; Albrecht et al. 2017). A few adult razorback suckers were found there in the 1980s, but hundreds have been found since 2011, including many tagged hatchery-reared fish released upstream in the Green, Colorado, and San Juan Rivers (Durst and Francis 2016; Francis et al. 2017). Reproduction has been detected in or near reservoir inflows each year from 2011 through 2016 (Albrecht et al. 2017). Whether these populations would persist in the absence of upstream stocking is unknown, and no information is available regarding in situ recruitment, but if additional investigations reveal local recruitment, the conservation importance of these populations may be substantial.

Upper Colorado River Basin

In contrast to the fragmented and sometimes cool Colorado River downstream of Lake Powell, there are long, connected warmwater reaches in the upper Colorado River basin where large-river fishes persist (Muth et al. 2000; Bestgen 2015). For example, the Green River flows essentially unimpeded for nearly 600 km from Flaming Gorge Reservoir downstream to its confluence with the Colorado River, with additional connected habitat in the White and Yampa rivers (169 km and 225 km, respectively; see fig. 21.1), and seven large-bodied, large-river fishes occur there (see table 21.1). Upper Colorado River streams support most of the large-river fishes found in the basin (Bezzerides and Bestgen 2002; Franssen et al. 2014) and are considered conservation strongholds for those taxa (Bestgen et al. 2007; Zelasko et al. 2010).

Despite the streams in the upper Colorado River basin being connected and having minimally modified flow regimes (Yampa and White Rivers), limited sampling indicated that wild razorback sucker populations were declining by the 1960s (Vanicek et al. 1970; Holden and Stalnaker 1975), which was verified with more intensive sampling in the 1980s (Tyus 1987; Platania et al. 1991). As it did in the lower Colorado River, lack of recruitment in the upper basin doomed all wild populations by the early twenty-first century or before (Platania et al. 1991; Bestgen et al. 2002). For example, the Green River population numbered 300 to 600 individuals from 1985 to 1992, but the few remaining old fish had disappeared by about 2000. Although captures were irregular prior to 1990, no razorback sucker has been captured in the Colorado River near the confluence of the Colorado and Gunnison Rivers, Colorado, since then (Bestgen 1990). Wild razorback sucker was last collected in the San Juan River in 1988 (Platania et al. 1991).

Limited stocking of razorback sucker in the upper basin began in the mid-1990s, but by the following decade had expanded greatly in the San Juan, Colorado, and Green Rivers (table 21.2). A total of over 545,000 razorback suckers, most 250 mm TL or longer, have been stocked since 1994, with comparable numbers in each river system. Although the numbers stocked declined slightly after 2013 due to intermittent survival issues in hatcheries and a shift to stocking larger and fewer individuals, several thousand individuals are released into each system every year. Separate Colorado River and Green River hatchery stocks, derived from remnant wild populations, were gathered before those populations were extirpated. The individuals stocked in the San Juan River are from a variety of sources, mainly from the genetically diverse SNARRC stock, along with a few individuals from the San Juan arm of Lake Powell (Bestgen et al. 2009).

Survival of razorback sucker in the first year after stocking was generally low throughout the upper basin. For example, first-year survival rates of razorback sucker stocked in the San Juan, Colorado, and Green Rivers were only 5%–10%, but increased with fish length (Bestgen et al. 2009; Zelasko et al. 2010), and subsequent survival rose to 50%–80%. Those rates were similar to those found in Lake Mohave and other wild fish populations (Bestgen et al. 2002; Kesner et al. 2017).

In the Green River basin, wild razorback sucker regularly moved 100 km or more in spring to documented spawning areas in the Green and Yampa Rivers (Tyus 1987; Bestgen et al. 2011). In contrast, spawning movements of fish in the Colorado and San Juan River sub-basins are poorly understood. Individuals stocked in the mainstem Green and Colorado Rivers moved up to 55 km per day after stocking in a variety of seasons and in a mostly downstream direction (Zelasko et al. 2010). Stocked fish also dispersed from the mainstem to Green River tributaries, including the San Rafael, White, and Yampa Rivers (Bottcher et al. 2013; Webber et al. 2013), indicating wide dispersal in upstream and downstream directions in spawning and non-spawning seasons. Several stocked fish moved over 550 km down the San Juan River, through Lake Powell, and upstream into the Colorado and Green Rivers (Durst and Francis 2016). Fixed and mobile PIT-tag antennas yield thousands of detections annually (Webber and Beers 2014; Cathcart et al. 2015), which, when analyzed, will improve understanding of large-scale movements and survival rates. Importantly, movement patterns of razorback sucker verified that stocked fish found and successfully used documented and newly discovered spawning sites, resolving an uncertainty that existed before stocking began (Bestgen et al. 2002; Webber et al. 2013).

Table 21.2. Number of razorback suckers stocked in upper Colorado River basin drainages. Most stocked razorbacks exceeded 250 mm TL, and since 2010, most were 350 mm TL or greater.

Year	Upper Colorado River basin drainage			Total
	Green River	Colorado River	San Juan River	
1994	0	0	687	687
1995	905	316	146	1,367
1996	1,067	287	237	1,591
1997	0	3,718	2,883	6,601
1998	389	606	1,275	2,270
1999	1,357	6,185	0	7,542
2000	224	31,730	1,044	32,998
2001	0	6,253	686	6,939
2002	274	11,378	139	11,791
2003	10,868	5,658	883	17,409
2004	15,576	6,703	2,977	25,256
2005	9,095	11,555	1,992	22,642
2006	20,212	11,557	13,748	45,517
2007	16,286	10,111	16,884	43,281
2008	18,548	13,200	4,419	36,167
2009	19,282	17,740	8,312	45,334
2010	21,428	9,956	28,404	59,788
2011	21,117	12,025	18,778	51,920
2012	21,106	10,511	15,810	47,427
2013	8,811	10,074	15,338	34,223
2014	3,696	6,148	6,134	15,978
2015	5,055	3,331	5,206	13,592
2016	1,850	5,664	7,630	15,144
Total	197,146	194,706	153,612	545,464

Sources: Data were obtained from a centralized database of the Upper Colorado River Endangered Fish Recovery Program and annual reports.

Reproduction by stocked razorback suckers has occurred in all upper basin rivers since about 1999 (Farrington et al. 2016; Zelasko et al. 2018), the time when most or all of the wild fish disappeared. Unfortunately, the presence of adults and larvae is not sufficient for recovery, as few juvenile or subadult fish have been captured in recent years. Thus, the main goal of all basin recovery programs—survival of sufficient numbers of young razorback suckers so that adults are replaced as they die—has not yet been achieved.

Recent studies aimed at increasing early-life-stage survival, assisted by otolith micro-increment analyses, identified flow-dependent mechanisms that may increase recruitment of razorback sucker in the Green River (Bestgen et al. 2011; LaGory et al. 2012). Those studies prescribed later and higher-magnitude water releases from Flaming Gorge Reservoir, precisely timed with the springtime emergence of larvae, to reconnect floodplain wetlands with the Green River (fig. 21.2; Bestgen et al. 2011). Larvae hatch at 7–9 mm TL, and those entrained into wetlands grow quickly over the summer to 100–200 mm TL in these warm, food-rich environments (Modde 1996; Bestgen 2008). Green River floodplain studies also indicated that the wetlands most useful for increasing razorback sucker survival were highly managed and outfitted with substantial infrastructure (Speas et al. 2017). These wetlands have gates that allow exclusion of water at lower flow levels until larvae are available for entrainment, and also retain higher water levels through summer and early autumn. Gates fitted with screens limit the passage of large-bodied nonnative fish into these wetlands during springtime filling.

The combination of higher and later reservoir releases triggered by larval presence and gated wetland operation has facilitated increased survival of larvae. For example, from 2013 through 2019, more than 4,000 juveniles were produced in the Stewart Lake floodplain wetland (Schelly et al. 2016), even in the presence of substantial populations of nonnative fishes, including green sunfish *Lepomis cyanellus*, black bullhead *Ameiurus melas*, and fathead minnow *Pimephales promelas* (Bestgen et al. 2017). Juvenile razorback suckers produced there were released to the Green River in autumn, and those large enough (most, but not all) were PIT-tagged to enable later detection. Reproduction by stocked bonytail *Gila elegans* was also documented for the first time in over 50 years in the Green River system, and larvae survived to autumn in two wetlands (Stewart Lake and Johnson Bottom; Bestgen et al. 2017), in-

Fig. 21.2 A floodplain wetland (Johnson Bottom) connecting with the Green River in Spring 2017 (river flow is right to left). *Inset:* Biologists seining a fish capture kettle in the gate infrastructure (located just right of picture) that is used to fill the wetland in spring and retain water through summer. Water also fills and drains the wetland through the inlet indicated at Green River flows greater than about 453 m^3/s. (Photograph credits USFWS/T. Jones.)

dicating that floodplain and other management actions for razorback sucker may also assist with conservation of other large-river fishes. The success of floodplain-wetland management for razorback sucker recruitment will be more confidently known if individuals produced in these habitats grow to adulthood, spawn, and produce their own recruits.

MOLECULAR INVESTIGATIONS THROUGH TIME IN THE COLORADO RIVER BASIN

Variation among Populations

Initial studies of genetic variation in razorback sucker focused on quantifying differences among populations. Buth et al. (1987, 1995) characterized variation at allozyme loci among upper- and lower-basin stocks of razorback sucker as well as flannelmouth sucker, finding limited differentiation among razorback sucker and low levels of introgression with flannelmouth sucker (<5%). Dowling et al. (1996a) used mtDNA restriction site variation to characterize structure among the same populations that Buth et al. (1995) studied, and found most variation within, not among, populations. The majority of variation was found in downstream populations (e.g., Lake Mohave), and they concluded that razorback sucker should be managed as a single population.

Newer molecular techniques (Turner et al., chap. 14, this volume) characterized nuclear-gene variation within and among populations. The USFWS (2011) and Dowling et al. (2012b, using the same specimens as Dowling et al. 1996a) each examined 13 different microsatellite loci and again found lower genetic variation in upper-basin fish than in Lake Mohave fish. Differences among populations from upstream and downstream of Grand Canyon identified it as a potential impediment to dispersal in the pre-dam era, leading to the new recommendation that the populations be managed separately. Dowling et al. (2012b) further noted that small numbers of individuals from the diverse Lake Mohave population could be used to augment all other populations to increase genetic variation while maintaining local genetic variants.

Variation within Populations

Genetic management practices differ between upstream and downstream portions of the basin. Local populations in the upper basin are augmented through stocking of hatchery-reared fish generated by paired matings of brood stock (USFWS 2012). Management in Lake Mohave is based on the rearing of field-captured larvae in predator-free environments, followed by repatriation (Marsh et al. 2015). The latter approach reduces the effect of adaptation to hatchery conditions (Lande 1988; McClure et al. 2008), an important consideration in the captive rearing of endangered species, and avoids the reduced fitness previously identified in hatchery populations of razorback sucker (Dowling et al. 1996b). This approach also provides a better representation of the genetic diversity of the entire population. Razorback sucker larvae stocked in other lower Colorado River locations (e.g., Lake Havasu) derive from a variety of hatchery sources.

Genetic monitoring effort has been variable among populations, and only limited work has occurred in the upper basin. For example, Dowling et al. (2012a) had samples only from the late 1980s and early 1990s, but none since then. Although the USFWS (2011) provided updated baseline information for genetic monitoring, we are unaware of recent genetic studies of repatriated razorback sucker from the upper basin, so patterns of genetic diversity post-stocking are unknown.

In the lower basin, genetic monitoring has been ongoing since the mid-1990s, especially for the Lake Mohave population. Assays of mtDNA and microsatellite variation (Turner et al. 2007; Dowling et al. 2014) have demonstrated an increased number of breeders relative to population size since implementation of the repatriation program in Lake Mohave. Further analysis of mtDNA and microsatellite variation of adult individuals (Carson et al. 2016) has indicated that repatriates contain levels of diversity similar to those of the original wild Lake Mohave population, suggesting that this approach has maintained neutral genetic variation in razorback sucker while limiting the concerns associated with standard hatchery programs discussed by Marsh et al. (2015).

The protected conservation habitat program (Minckley et al. 2003) includes genetic monitoring, initiated in 2010, to examine patterns of reproductive success in one permanent and two ephemeral off-channel ponds (Dowling et al. 2017a). Results indicate considerable variation in reproductive success among all three ponds and among years, and adult male mortality is inexplicably higher than that for females. Additional investigations may sort out the significance of those patterns for management.

Recent studies in the lower Colorado River have identified two additional populations that may contribute to the conservation of razorback sucker. A substantial number of adults is present in the Colorado River just upstream of Lake Mohave, adding to the store of genetic material that persists there (Wisenall et al. 2015; Kesner et al. 2016). Larvae from that reach are being incorporated into the genetic monitoring and captive rearing program. In addition, the riverine portion of the Colorado River downstream of Lake Mohave, but upstream of Lake Havasu, supports a substantial number of adult razorback suckers, most derived from Lake Mohave via SNARRC (Wydoski and Mueller 2006; Kesner et al. 2017). Adults from this reach produce thousands of larvae annually, which may eventually contribute to the Lake Mohave larval capture program. Genetic monitoring was recently initiated for both of these riverine populations, following the Lake Mohave protocol (Dowling et al. 2005). Too few annual samples are available to contrast temporal patterns of variation with the large dataset from the main basin of Lake Mohave. However, both new locations exhibit levels of variation similar to those of the once large, wild main basin population (Dowling et al. 2017b). Therefore, while the riverine reach downstream of Lake Mead contributes a substantial number of individuals to the Lake Mohave population, the riverine stretch of the Lake Havasu region potentially represents a much-needed replicated population in the lower basin.

Genetic variation in razorback sucker from Lake Mead has been regularly monitored since 2011. Dowling et al. (2012a) compared earlier samples, collected in the late 1980s, 1997, and 2002, with those from 2011 and noted reduced genetic variation and increased relatedness relative to those from Lake Mohave. Samples through 2016 supported the conclusion of reduced variation and increased relatedness relative to the original wild population in Lake Mohave, but also identified divergence of the post-2011 samples from the original population (Dowling et al. 2017b). Differences among temporal samples could result from genetic drift associated with the small population size in Lake Mead or from influx of Grand Canyon fish. Therefore, plans have been developed to introduce a small number of Lake Mohave fish into Lake Mead, which may enhance genetic variation.

Synthesis of Molecular Studies

Attempts to differentiate genetic variation among local razorback sucker populations are ongoing, using newer tools as they become available. Unfortunately, these efforts have focused on older samples (pre-1990s) and do not provide an updated picture of the patterns of genetic variation after considerable management. A comprehensive study of extant populations basin-wide with next-generation sequencing techniques (Turner et al., chap. 14, this volume) would aid in conservation efforts for the species.

There has been considerable progress in maintaining genetic variation in relatively small populations of razorback sucker (as compared with historical levels), especially in the lower basin. As future management options are formulated, establishment of genetic baselines for upper-basin populations will be critical. Once quantified, continued genetic monitoring will allow for development of appropriate management actions to ensure that genetic variation is maintained where razorback sucker is extant.

SUCCESSES, FAILURES, AND WAYS FORWARD

Despite the long lifespan and high reproductive potential of razorback sucker, the outcomes of oftentimes intensive management to secure self-sustaining populations in the Colorado River basin have been mixed (Bestgen et al. 2012; Marsh et al. 2015). On one hand, all wild stocks (except possibly the Lake Mead population) were at one time or another extirpated, including populations from Lake Mohave and stream-resident populations from the upper Colorado River basin. On the other hand, most populations present in the 1980s and then extirpated have been reestablished with hatchery stocks and are reproducing.

Early efforts at stocking hatchery-reared razorback suckers met with failure (Schooley and Marsh 2007), primarily because of predation on those mostly small fish by nonnative predators (Minckley 1983; Marsh et al. 2015). However, the hatchery stocks in several locations were essential to reestablishment of razorback sucker populations throughout the Colorado River basin as new strategies, such as harvest of wild-produced larvae and stocking of larger fish, were employed. Without those captive stocks, recovery action options would have been limited, and the species would be closer to extinction. Although success of stocked razorback sucker in most locations continues to be hampered by avian and piscine predators (Karam and Marsh 2010; Ehlo et al. 2017), and the efficacy of floodplain wetlands to boost recruitment is yet unproven, hatchery stocks have provided managers with valuable time and options to experiment with management techniques. The full array of options for razorback sucker management will depend on the maintenance of high-quality wild and hatchery stocks, as outlined by Marsh et al. (2015). Monitoring will be required to determine if captive fish stocked in the wild represent the genetic diversity of original wild populations without developing adaptations to hatchery conditions, an implicit goal of any hatchery program aimed at native fish conservation.

Despite recent successes, continued shifts in the landscape of native fish conservation have thwarted long-term goals. The introduction and establishment of striped bass, northern pike *Esox lucius*, and the suite of other nonnative piscivores throughout the basin place even the largest adult razorback sucker at risk (Karam et al. 2008; Zelasko et al. 2016). Recent expansions of walleye *Sander vitreus*, smallmouth bass *Micropterus dolomieu*, and most recently, burbot *Lota lota* have increased the predator load in upper-basin streams such that nonnatives sometimes greatly exceed native taxa in abundance or biomass (Johnson et al. 2008). Yet managers have few tools to combat these invaders (Franssen et al. 2016; Zelasko et al. 2016). The short-term positive effects of mechanical-removal programs are often offset by increases in nonnative fish abundance due to favorable environmental conditions for recruitment and new or emerging sources of invaders, usually off-channel locations or reservoirs (Breton et al. 2014; Zelasko et al. 2016). Thus, introduction and establishment of new taxa and leakage from existing sources are substantial threats to native fish recovery. Implementing flow management actions, which benefit razorback sucker in wetlands and may be useful to disadvantage smallmouth bass reproduction (Bestgen and Hill 2016; Bestgen 2018), will require flexibility by managers and acceptance by the public because those actions may be controversial or reduce water supplies for other uses. The demonstration that dams and water stored in reservoirs can benefit fish conservation is a step forward, and additional cooperative efforts among agencies and better public relations should be fostered.

The successes in razorback sucker recovery rely on human intervention, which may promote debate regarding the "naturalness" of recovery actions and whether such actions meet the intent of conservation planning and the Endangered Species Act (Andersen and Brooks, chap. 12, this volume). A status assessment was recently completed for razorback sucker (USFWS 2018a), and the USFWS subsequently recommended downlisting the species from endangered to threatened status (USFWS 2018b), based on continued management and its finding that "the current risk of extinction is low, such that the species is not in danger of extinction throughout all of its range" (USFWS 2018b). Regardless of listing status and to the extent possible, recovery programs should strive for the most natural settings and processes, but the reality is that few locations in the Colorado River basin, or in other large rivers globally, are "natural" with respect to undisturbed flow regimes and intact native fish communities. The limited restoration success for razorback sucker throughout the basin to date makes it clear that recovery without human intervention is impossible until threats and limiting factors are eliminated. Acceptance from resource managers and the public that direct interventions are needed for razorback sucker and other severely imperiled large-river fishes (e.g., bonytail) to survive in the wild will save time and resources and allow progress to be made.

CONCLUSIONS AND IMPLICATIONS FOR CONSERVATION OF LARGE-RIVER FISHES

Freshwater fishes and associated ecosystems are among the most endangered globally, and large-river fishes are especially at risk and in need of conservation (Dudgeon et al. 2006; Jelks et al. 2008). The life histories and ecological requirements of most long-lived and endangered riverine fishes are poorly understood (Phelps et al. 2010; Pracheil et al. 2013), and such information is difficult to acquire when populations are depleted and entire life stages are absent from natural environments. Such was the case for razorback sucker in the Colorado River basin, where the few adults occupied limited habitat, and younger life stages were rare and escaped detection for years. For example, razorback sucker juveniles in rivers of the upper Colorado River basin were essentially unknown between about 1965 and the early 1990s (Taba et al. 1965; Gutermuth et al. 1994), even with extensive sampling in that interval (Platania et al. 1991; Bestgen et al. 2002). Likewise, riverine razorback sucker larvae were commonly documented in collections only after 1991 (Muth et al. 2000), both because they were rare and because basic morphological descriptions to distinguish them from other catostomids were available only after 1981 (Snyder and Muth 2004). Pallid sturgeon *Scaphirhynchus albus* recovery progress has been similarly limited by lack of early life history information and difficulties differentiating them from morphologically similar congeners (Hrabik et al. 2007; Phelps et al. 2010). These and other issues among recovery programs for large and long-lived riverine fishes (e.g., Acipenseriformes; Pikitch et al. 2005) indicate that lessons learned about razorback sucker conservation in the Colorado River basin have broad implications for large-river fish conservation efforts elsewhere.

The relatively new discoveries of razorback sucker reproducing in Grand Canyon and recruiting in Lake Mead provide hope that some atypical environmental settings may provide conservation habitat for the species (Albrecht et al., chap. 11, this volume). Assessment of flow regimes that produce viable riverine-floodplain connections for razorback sucker recruitment, extension of valuable long-term monitoring that yields insights into population responses to stocking and habitat management, and discovery of additional tools to control nonnative fishes, including flow management, are needed.

Recovery programs for Colorado River basin endangered fishes are sometimes criticized because they are expensive and progress is slow (Mueller 2005). We submit that a more realistic view of time to recovery should be considered when conservation programs are initiated because managers often understand the scope and magnitude of the issues facing endangered organisms, and their ecological requirements, only after years or decades of intensive study. This is true for several long-lived taxa (e.g., razorback sucker, pallid and other sturgeons), such that increases in reproduction or recruitment may not manifest for 5–10 years or more (Pikitch et al. 2005; Bestgen et al. 2011). In a longer view, stakeholders must realize that certain threat abatement or recovery activities need to continue in perpetuity to avoid re-listing of the same species. At this writing, the cooperative agreement that implements the Upper Colorado River Endangered Fish Recovery Program expires in 2023, after which razorback sucker and other endangered fishes were presumably expected to meet established recovery criteria. Even the most optimistic supporters would struggle to say that goal is achievable, and discussions are under way regarding required recovery and conservation actions post-2023.

Perceptions of the slow progress of recovery programs should also be tempered with knowledge that the scope and magnitude of problems causing endangerment (e.g., flow alterations, nonnative organisms) were created over many decades and exist over broad geographic scales (e.g., Moyle, chap. 4, this volume). Witness the negative effects of channel changes over thousands of kilometers of large and small southwestern rivers associated with altered flows and invasive woody plants (Stone and Morrison, chap. 16, this volume), some of which began over 100 years ago (Grams and Schmidt 2002; Stromberg et al. 2007). All the while, time disfavors conservation of native fishes because the fast-growing human population in the Southwest will impose further alterations to flows and habitat and enable proliferation of nonnative fishes. While barely discussed herein, the effects of climate change loom large (Dettinger et al. 2015; Udall and Overpeck 2017) and will further reduce water supplies, shift runoff timing, magnitude, and duration (Udall, chap. 7, this volume), and render habitat more suitable for invasive than native fishes (Olden et al. 2006; Gido et al. 2013). This chapter was prepared as the president of the United States announced that the nation would withdraw from the landmark Paris Agreement (Norment, chap. 25, this volume), the global effort to reduce effects of climate change, signaling even more challenges to come.

Another obstacle is the potential for conflict between native fish conservation and other natural resource values and the need for management to accommodate both. For example, the dual mandate of state agencies to provide recreational anglers with fish to catch—usually nonnatives—and to perpetuate native fish at the same time creates a difficult intersection of opposing management actions (Clarkson et al. 2005). Any proposals to stock nonnative species in Colorado River basin aquatic systems should be viewed skeptically by management agencies (e.g., Brandenburg et al. 2019), given their well-documented history of negative effects on native biota. The negative effects of ubiquitous nonnative fishes in the Colorado River basin demonstrate that native fishes rarely prosper in their presence (Clarkson et al. 2005). Thus, management agencies must have the political will, and the financial and other means, to provide suitable native fish habitat. Unfortunately, acquisition of sanctuary areas with secure water resources for native fishes, ranging in size from the protected backwater conservation habitats used in the lower Colorado River to entire drainage basins, is an increasingly difficult prospect given water scarcity, drought, widespread habitat alteration, and climate change. Furthermore, opportunities to fulfill sanctuary requirements grow increasingly rare as more nonnative fishes occupy new locations, often aided by introductions, illicit or otherwise (e.g., Brandenburg et al. 2019).

Thus, the outlook for recovery of large-river fishes, in the Colorado River basin and elsewhere, is mixed in both the short and long term as environmental changes not yet realized or imagined unfold. A public that is increasingly distanced from the natural world and less sympathetic to conservation issues also clouds the future (Louv 2005), and this situation is not aided by governing bodies willing to reduce, rather than increase, support for conservation. Now more than 25 years removed from the publication of *Battle against Extinction*, we have learned much about how to manage and conserve native fishes such as razorback sucker, but we have not been able to successfully implement the recovery actions that were known even then to secure their future. Despite our mixed results and slow progress, remnants of wild native fish populations and their habitat remain in the Colorado River basin. New observations of juvenile razorback sucker in some areas of the basin indicate recovery progress and offer evidence that successful management is possible, but many challenges remain. Our generation must not squander the opportunity to ensure the existence of iconic native species such as razorback sucker in the Colorado River basin, because we may not be afforded another.

REFERENCES

Albrecht, B., P. B. Holden, R. Kegerries, and M. E. Golden. 2010. Razorback sucker recruitment in Lake Mead, Nevada-Arizona, why here? *Lake and Reservoir Management* 26: 336–44.

Albrecht, B., H. E. Mohn, R. Kegerries, M. C. McKinstry, R. Rogers, T. Francis, B. Hines, J. Stolberg, D. Ryden, D. Elverud, B. Schleicher, K. Creighton, B. Healy, and B. Senger. 2017. Use of inflow areas in two Colorado River basin reservoirs by endangered razorback sucker *Xyrauchen texanus*. *Western North American Naturalist* 77: 500–514.

Bestgen, K. R. 1990. *Status Review of the Razorback Sucker,* Xyrauchen texanus. Larval Fish Laboratory Contribution 44. Salt Lake City: US Bureau of Reclamation.

Bestgen, K. R. 2008. Effects of water temperature on growth of razorback sucker larvae. *Western North American Naturalist* 68: 15–20.

Bestgen, K. R. 2015. *Aspects of the Yampa River Flow Regime Essential for Maintenance of Native Fishes.* Larval Fish Laboratory Contribution 181. Natural Resource Report NPS/NRSS/WRD/NRR-2015/962. Fort Collins, CO: National Park Service. DOI: 10.13140/RG.2.2.12857.98402.

Bestgen, K. R. 2018. *Evaluate Effects of Flow Spikes to Disrupt Reproduction of Smallmouth Bass in the Green River Downstream of Flaming Gorge Dam.* Final report to the Upper Colorado River Endangered Fish Recovery Program. Larval Fish Laboratory Contribution 214. Denver: Department of Fish, Wildlife, and Conservation Biology, Colorado State University, Fort Collins. DOI: 10.13140/RG.2.2.33277.61926.

Bestgen, K. R., G. B. Haines, R. Brunson, T. Chart, M. Trammell, R. T. Muth, G. Birchell, K. Christopherson, and J. Bundy. 2002. *Status of Wild Razorback Sucker in the Green River Basin, Utah and Colorado, Determined from Basinwide Monitoring and Other Sampling Programs.* Larval Fish Laboratory Contribution 126. Denver: US Fish and Wildlife Service, Upper Colorado River Endangered Fish Recovery Program.

Bestgen, K. R., G. B. Haines, and A. A. Hill. 2011. *Synthesis of Flood Plain Wetland Information: Timing of Razorback Sucker Reproduction in the Green River, Utah, Related to Stream Flow, Water Temperature, and Flood Plain Wetland Availability.* Larval Fish Laboratory Contribution 163. Denver: US Fish and Wildlife Service, Upper Colorado River Endangered Fish Recovery Program.

Bestgen, K. R., J. A. Hawkins, G. C. White, K. Christopherson, M. Hudson, M. Fuller, D. C. Kitcheyan, R. Brunson, P. Badame, G. B. Haines, J. Jackson, C. D. Walford, and T. A. Sorensen. 2007. Population status of Colorado pikeminnow in the Green River basin, Utah and Colorado. *Transactions of the American Fisheries Society* 136: 1356–80.

Bestgen, K. R., and A. A. Hill. 2016. *River Regulation Affects Reproduction, Early Growth, and Suppression Strategies for Invasive Smallmouth Bass in the Upper Colorado River Basin.* Larval Fish Laboratory Contribution 187. Denver: US Fish and Wildlife Service, Upper Colorado River Endangered Fish Recovery Program.

Bestgen, K. R., R. C. Schelly, R. R. Staffeldt, M. J. Breen, D. E. Snyder, and M. T. Jones. 2017. First reproduction by stocked bonytail *Gila elegans* in the upper Colorado River basin. *North American Journal of Fisheries Management* 37: 445–55.

Bestgen, K. R., K. A. Zelasko, and G. C. White. 2009. *Survival of Hatchery-Reared Razorback Suckers* Xyrauchen texanus *Stocked in the San Juan River Basin, New Mexico, Colorado, and Utah.* Larval Fish Laboratory Contribution 160. Albuquerque: US Fish and Wildlife Service, San Juan River Basin Recovery Implementation Program.

Bestgen, K. R., K. A. Zelasko, and G. C. White. 2012. *Monitoring Reproduction, Recruitment, and Population Status of Razorback Suckers in the Upper Colorado River Basin*. Larval Fish Laboratory Contribution 170. Denver: US Fish and Wildlife Service, Upper Colorado River Endangered Fish Recovery Program.

Bezzerides, N., and K. R. Bestgen. 2002. *Status of Roundtail Chub* Gila robusta, *Flannelmouth Sucker* Catostomus latipinnis, *and Bluehead Sucker* Catostomus discobolus *in the Colorado River Basin*. Larval Fish Laboratory Contribution 118. Salt Lake City: US Bureau of Reclamation.

Bottcher, J. L., T. E. Walsworth, G. P. Thiede, P. Budy, and D. W. Speas. 2013. Frequent usage of tributaries by the endangered fishes of the upper Colorado River basin: Observations from the San Rafael River, Utah. *North American Journal of Fisheries Management* 33: 585–94.

Bozek, M. A., L. J. Paulson, and G. R. Wilde. 1990. Effects of ambient Lake Mohave temperatures on development, oxygen consumption, and hatching success of the razorback sucker. *Environmental Biology of Fishes* 27: 255–63.

Brandenburg, W. H., T. A. Francis, D. E. Snyder, K. R. Bestgen, B. A. Hines, W. D. Wilson, S. Bohn, A. S. Harrison, and S. L. Clark Barkalow. 2019. Discovery of grass carp *Ctenopharyngodon idella* larvae in the Colorado River arm of Lake Powell. *North American Journal of Fisheries Management* 39: 166–71.

Breton, A. R., D. L. Winkelman, J. A. Hawkins, and K. R. Bestgen. 2014. *Population Trends of Smallmouth Bass in the Upper Colorado River Basin with an Evaluation of Removal Effects*. Larval Fish Laboratory Contribution 169. Denver: US Fish and Wildlife Service, Upper Colorado River Endangered Fish Recovery Program.

Buth, D. G., T. R. Haglund, and S. L. Drill. 1995. Allozyme variation and population structure in the razorback sucker, *Xyrauchen texanus*. *Proceedings of the Desert Fishes Council* 26: 30.

Buth, D. G., R. W. Murphy, and L. Ulmer. 1987. Population differentiation and introgressive hybridization of the flannelmouth sucker and of hatchery and native stocks of razorback sucker. *Transactions of the American Fisheries Society* 116: 103–10.

Carlson, C. A., and R. T. Muth. 1989. The Colorado River: Lifeline of the American Southwest. In D. P. Dodge, ed., *Proceedings of the International Large River Symposium (LARS)*, 220–39. Canadian Special Publication of Fisheries and Aquatic Sciences 106.

Carson, E. W., T. F. Turner, M. J. Saltzgiver, D. Adams, B. R. Kesner, P. C. Marsh, T. J. Pilger, and T. E. Dowling. 2016. Retention of ancestral genetic variation across life-stages of an endangered, long-lived iteroparous fish. *Journal of Heredity* 107: 567–72.

Cathcart, C. N., K. B. Gido, and M. C. McKinstry. 2015. Fish community distributions and movements in two tributaries of the San Juan River, USA. *Transactions of the American Fisheries Society* 144: 1013–28.

Clarkson, R. W., and M. R. Childs. 2000. Temperature effects of hypolimnial-release dams on early life stages of Colorado River basin big-river fishes. *Copeia* 2000: 402–12.

Clarkson, R. W., P. C. Marsh, S. E. Stefferud, and J. A. Stefferud. 2005. Conflicts between native fish and nonnative sport fish management in the southwestern United States. *Fisheries* 30: 20–27.

Dettinger, M., B. Udall, and A. Georgakakos. 2015. Western water and climate change. *Ecological Applications* 25: 2069–93.

Dowling, T. E., P. C. Marsh, A. T. Kelsen, and C. A. Tibbets. 2005. Genetic monitoring of wild and repatriated populations of endangered razorback sucker (*Xyrauchen texanus*, Catostomidae, Teleostei) in Lake Mohave, Arizona-Nevada. *Molecular Ecology* 14: 123–35.

Dowling, T. E., P. C. Marsh, and T. F. Turner. 2017a. *Genetic and Demographic Studies to Guide Conservation Management of Razorback Sucker in Off-Channel Habitats*. 2016 annual report. Boulder City, NV: US Bureau of Reclamation.

Dowling, T. E., P. C. Marsh, and T. F. Turner. 2017b. *Razorback Sucker Genetic Diversity Assessment*. 2016 annual report. Boulder City, NV: US Bureau of Reclamation.

Dowling, T. E., W. L. Minckley, and P. C. Marsh. 1996a. Mitochondrial DNA diversity within and among populations of razorback sucker (*Xyrauchen texanus*) as determined by restriction endonuclease analysis. *Copeia* 1996: 542–50.

Dowling, T. E., W. L. Minckley, P. C. Marsh, and E. Goldstein. 1996b. Mitochondrial DNA diversity in the endangered razorback sucker (*Xyrauchen texanus*): Analysis of hatchery stocks and implications for captive propagation. *Conservation Biology* 10: 120–27.

Dowling, T. E., M. J. Saltzgiver, D. Adams, and P. C. Marsh. 2012a. Assessment of genetic variability in a recruiting population of endangered fish, the razorback sucker (*Xyrauchen texanus*, Family Catostomidae), from Lake Mead, AZ-NV. *Transactions of the American Fisheries Society* 141: 990–99.

Dowling, T. E., M. J. Saltzgiver, and P. C. Marsh. 2012b. Genetic structure within and among populations of the endangered razorback sucker (*Xyrauchen texanus*) as determined by analysis of microsatellites. *Conservation Genetics* 13: 1073–83.

Dowling, T. E., T. F. Turner, E. W. Carson, M. J. Saltzgiver, D. Adams, B. Kesner, and P. C. Marsh. 2014. Genetic and demographic responses to intensive management of razorback sucker in Lake Mohave, Arizona-Nevada. *Evolutionary Applications* 7: 339–54.

Dudgeon, D., A. H. Arthington, M. O. Gessner, Z.-I. Kawabata, D. J. Knowler, C. Leveque, R. J. Naiman, A.-H. Prier-Richard, D. Soto, M. L. J. Stiassny, and C. A. Sullivan. 2006. Freshwater biodiversity: Importance, threats, status and conservation challenges. *Biological Reviews* 81: 163–82.

Durst, S. L., and T. A. Francis. 2016. Razorback sucker transbasin movement through Lake Powell, Utah. *Southwestern Naturalist* 61: 60–63.

Ehlo, C. A., M. J. Saltzgiver, T. E. Dowling, P. C. Marsh, and B. R. Kesner. 2017. Use of molecular techniques to confirm nonnative fish predation on razorback sucker larvae in Lake Mohave, Arizona and Nevada. *Transactions of the American Fisheries Society* 146: 201–5.

Farrington, M. A., R. K. Dudley, J. L. Kennedy, S. P. Platania, and G. C. White. 2016. *Colorado Pikeminnow and Razorback Sucker Larval Fish Survey in the San Juan River during 2015*. Albuquerque: US Fish and Wildlife Service, San Juan River Basin Recovery Implementation Program.

Francis, T. A., D. S. Elverud, B. J. Schleicher, D. W. Ryden, and B. Gerig. 2017. *San Juan River Arm of Lake Powell Razorback*

Sucker (Xyrauchen texanus) *Survey: 2012.* Albuquerque: US Fish and Wildlife Service, San Juan River Basin Recovery Implementation Program.
Franssen, N. R., J. E. Davis, D. Ryden, and K. B. Gido. 2014. Fish community responses to mechanical removal of nonnative fishes in a large southwestern river. *Fisheries* 39: 352–63.
Franssen, N. R., S. L. Durst, K. B. Gido, D. W. Ryden, V. Lamarra, and D. L. Propst. 2016. Long-tern dynamics of large-bodied fishes assessed from spatially intensive monitoring of a managed desert river. *River Research and Applications* 32: 348–61.
Gido, K. B., D. L. Propst, J. D. Olden, and K. R. Bestgen. 2013. Multi-decadal responses of native and introduced fishes to natural and altered flows in streams of the American Southwest. *Canadian Journal of Fisheries and Aquatic Science* 70: 554–64.
Grams, P. E., and J. C. Schmidt. 2002. Streamflow regulation and multi-level flood plain formation: Channel narrowing on the aggrading Green River in the eastern Uinta Mountains, Colorado and Utah. *Geomorphology* 44: 337–60.
Gutermuth, F. B., L. D. Lentsch, and K. R. Bestgen. 1994. Collection of age-0 razorback suckers (*Xyrauchen texanus*) in the Green River, Utah. *Southwestern Naturalist* 39: 389–91.
Hamman, R. L. 1985. Induced spawning of hatchery-reared razorback sucker. *Progressive Fish Culturist* 47: 187–89.
Holden, P. B., P. D. Abate, and J. B. Ruppert. 1997. *Razorback Sucker Studies on Lake Mead, Nevada. 1996–1997 Annual Report.* PR-578-1. Logan, UT: BIO-WEST, Inc.
Holden, P. B., and C. B. Stalnaker. 1975. Distribution and abundance of fishes in the middle and upper Colorado River basins, 1967–1973. *Transactions of the American Fisheries Society* 104: 217–31.
Hrabik, R. A., D. P. Herzog, D. E. Ostendorf, and M. D. Petersen. 2007. Larvae provide first evidence of successful reproduction by pallid sturgeon, *Scaphirhynchus albus*, in the Mississippi River. *Journal of Applied Ichthyology* 23: 436–43.
Humphrey, K. G., B. R. Kesner, and P. C. Marsh. 2016. *Distribution and Post-stocking Survival of Bonytail in Lake Havasu, 2013–2016.* Boulder City, NV: US Bureau of Reclamation.
Iorns, W. V., C. H. Hembree, and G. L. Oakland. 1965. *Water Resources of the Upper Colorado River Basin.* US Geological Survey Professional Paper 441. Washington, DC: US Government Printing Office.
Jelks, H. L., S. J. Walsh, N. M. Burkhead, S. Contreras-Balderas, E. Diaz-Pardo, D. A. Hendrickson, J. Lyons, N. E. Mandrak, F. McCormick, J. S. Nelson, S. P. Platania, B. A. Porter, C. B. Porter, C. B. Renaud, J. J. Schmitter-Soto, E. B. Taylor, and M. L. Warren, Jr. 2008. Conservation status of imperiled North American freshwater and diadromous fishes. *Fisheries* 33: 372–407.
Johnson, B. M., P. J. Martinez, J. A. Hawkins, and K. R. Bestgen. 2008. Ranking relative predatory threats by nonnative fishes in the Yampa River, Colorado, via bioenergetics modeling. *North American Journal Fisheries Management* 28: 1941–53.
Karam, A. P., B. R. Kesner, and P. C. Marsh. 2008. Acoustic telemetry to assess post-stocking dispersal and mortality of razorback sucker *Xyrauchen texanus. Journal of Fish Biology* 73: 719–27.
Karam, A. P., and P. C. Marsh. 2010. Predation of adult razorback sucker and bonytail by striped bass in Lake Mohave, Arizona-Nevada. *Western North American Naturalist* 70: 117–20.
Karp, C. A., and G. Mueller. 2002. Razorback sucker movements and habitat use in the San Juan River inflow, Lake Powell, Utah, 1995–1997. *Western North American Naturalist* 62: 106–11.
Kegerries, R. B., B. C. Albrecht, E. I. Gilbert, W. H. Brandenburg, A. L. Barkalow, M. C. McKinstry, H. E. Mohn, B. D. Healy, J. R. Stolberg, E. C. Omana Smith, C. B. Nelson, and R. J. Rogers. 2017. Occurrence and reproduction by razorback sucker *Xyrauchen texanus* in the Grand Canyon, Arizona. *Southwestern Naturalist* 62: 227–32.
Kegerries, R. B., B. Albrecht, E. I. Gilbert, W. H. Brandenburg, A. L. Barkalow, H. Mohn, R. Rogers, M. McKinstry, B. Healy, J. Stolberg, E. Omana Smith, and M. Edwards. 2016. *Razorback Sucker* Xyrauchen texanus *Research and Monitoring in the Colorado River Inflow Area of Lake Mead and the Lower Grand Canyon, Arizona and Nevada.* Boulder City, NV: US Bureau of Reclamation.
Kesner, B. R., C. A. Ehlo, J. B. Wisenall, and P. C. Marsh. 2017. *Comparative Survival of Repatriated Razorback Sucker in Lower Colorado River Reach 3, 2014–2016.* Boulder City, NV: US Bureau of Reclamation, Lower Colorado River Multi-species Conservation Program.
Kesner, B. R., J. B. Wisenall, and P. C. Marsh. 2016. Adaptive management of an imperiled catostomid in Lake Mohave, lower Colorado River, USA. In M. Nageeb Rashed, ed., *Lake Sciences and Climate Change,* 119–33. InTech, DOI: 10.5772/63808.
LaGory, K., T. Chart, K. R. Bestgen, J. Wilhite, S. Capron, D. Speas, H. Hermansen, K. McAbee, J. Mohrman, M. Trammell, and B. Albrecht. 2012. *Study Plan to Examine the Effects of Using Larval Razorback Sucker Occurrence in the Green River as a Trigger for Flaming Gorge Dam Peak Releases.* Denver: US Fish and Wildlife Service, Upper Colorado River Endangered Fish Recovery Program.
Lande, R. 1988. Genetics and demography in biological conservation. *Science* 241: 1455–60.
Louv, R. 2005. *Last Child in the Woods: Saving Our Children from Nature Deficit Disorder.* Chapel Hill, NC: Algonquin Books.
Marsh, P. C., and J. E. Brooks. 1989. Predation by ictalurid catfishes as a deterrent to re-establishment of hatchery-reared razorback sucker. *Southwestern Naturalist* 34: 188–95.
Marsh, P. C., T. E. Dowling, B. R. Kesner, T. F. Turner, and W. L. Minckley. 2015. Conservation to stem imminent extinction: The fight to save razorback sucker *Xyrauchen texanus* in Lake Mohave and its implications for species recovery. *Copeia* 103: 141–56.
Marsh, P. C., C. A. Pacey, and B. R. Kesner. 2003. Decline of the razorback sucker in Lake Mohave, Colorado River, Arizona and Nevada. *Transactions of the American Fisheries Society* 132: 1251–56.
McAda, C. W., and R. S. Wydoski. 1980. *The Razorback Sucker,* Xyrauchen texanus, *in the Upper Colorado River Basin, 1974–76.* US Fish and Wildlife Service Technical Papers, no. 99.
McCarthy, M. S., and W. L. Minckley. 1987. Age estimation for razorback sucker (Pisces: Catostomidae) from Lake Mohave, Arizona and Nevada. *Journal of the Arizona-Nevada Academy of Sciences* 21: 87–97.

McClure, M. M., F. M. Utter, C. Baldwin, R. W. Carmichael, P. F. Hassemer, P. J. Howell, P. Spruell, T. D. Cooney, H. A. Schaller, and C. E. Petrosky. 2008. Evolutionary effects of alternative artificial propagation programs: Implications for viability of endangered anadromous salmonids. *Evolutionary Applications* 1: 356–75.

Miller, R. R. 1961. Man and the changing fish fauna of the American Southwest. *Papers of the Michigan Academy of Science, Arts, and Letters* 46: 365–404.

Minckley, W. L. 1983. Status of the razorback sucker, *Xyrauchen texanus* (Abbott), in the lower Colorado River basin. *Southwestern Naturalist* 28: 165–87.

Minckley, W. L., and P. C. Marsh. 2009. *Inland Fishes of the Greater Southwest: Chronicle of a Vanishing Biota*. Tucson: University of Arizona Press.

Minckley, W. L., P. C. Marsh, J. E. Brooks, J. E. Johnson, and B. L. Jensen. 1991. Management toward recovery of the razorback sucker. In W. L. Minckley and J. E. Deacon, eds., *Battle against Extinction: Native Fish Management in the American West*, 303–57. Tucson: University of Arizona Press.

Minckley, W. L., P. C. Marsh, J. E. Deacon, T. E. Dowling, P. W. Hedrick, W. J. Matthews, and G. Mueller. 2003. A conservation plan for native fishes of the lower Colorado River. *BioScience* 53: 219–34.

Modde, T. 1996. Juvenile razorback sucker (*Xyrauchen texanus*) in a managed wetland adjacent to the Green River. *Great Basin Naturalist* 56: 375–76.

Mohn, H. E., B. Albrecht, and R. Rogers. 2016. *Razorback Sucker* Xyrauchen texanus *Studies on Lake Mead, Nevada and Arizona*. 2015–2016 Final Annual Report. Boulder City, NV: US Bureau of Reclamation.

Mueller, G. A. 2005. Predatory fish removal and native fish recovery in the Colorado River main stem: What have we learned? *Fisheries* 30: 10–19.

Mueller, G. A. 2006. *Ecology of Bonytail and Razorback Sucker and the Role of Off-Channel Habitats in Their Recovery*. US Geological Survey Scientific Investigations Report 2006-5056.

Mueller, G. A., and P. C. Marsh. 2002. *Lost: A Desert River and Its Native Fishes: A Historical Perspective of the Lower Colorado River.* US Geological Survey Information and Technology Report USGS/BRD/ITR-2002-0010.

Muth, R. T., L. W. Crist, K. E. LaGory, J. W. Hayse, K. R. Bestgen, T. P. Ryan, J. K. Lyons, and R. A. Valdez. 2000. *Flow and Temperature Recommendations for Endangered Fishes in the Green River Downstream of Flaming Gorge Dam*. Larval Fish Laboratory Contribution 120. Denver: US Fish and Wildlife Service, Upper Colorado Endangered Fish Recovery Program.

Olden, J. D., N. L. Poff, and K. R. Bestgen. 2006. Life-history strategies predict invasions and extirpations in the Colorado River basin. *Ecological Monographs* 76: 25–40.

Page, L. M., H. Espinosa-Pérez, L. T. Findley, C. R. Gilbert, R. N. Lea, N. E. Mandrak, R. L. Mayden, and J. S. Nelson. 2013. *Common and Scientific Names of Fishes from the United States, Canada, and Mexico*. 7th ed. Special Publication 34. Bethesda, MD: American Fisheries Society.

Phelps, Q. E., S. J. Tripp, J. E. Garvey, D. P. Herzog, D. E. Ostendorf, J. W. Ridings, J. W. Crites, and R. A. Hrabik. 2010. Habitat use during early life history infers recovery needs for shovelnose sturgeon and pallid sturgeon in the middle Mississippi River. *Transactions of the American Fisheries Society* 139: 1060–68.

Pikitch, E. K., P. Doukakis, L. Lauck, P. Chakrabarty, and D. L. Erickson. 2005. Status, trends and management of sturgeon and paddlefish fisheries. *Fish and Fisheries* 6: 233–65.

Platania, S. P., K. R. Bestgen, M. A. Moretti, D. L. Propst, and J. E. Brooks. 1991. Status of Colorado squawfish and razorback sucker in the San Juan River, Colorado, New Mexico, and Utah. *Southwestern Naturalist* 36: 147–50.

Pracheil, B. M., P. B. McIntyre, and J. D. Lyons. 2013. Enhancing conservation of large-river biodiversity by accounting for tributaries. *Frontiers in Ecology and the Environment* 11: 124–28.

Ross, R. P., and W. S. Vernieu. 2013. *Nearshore Temperature Findings for the Colorado River in Grand Canyon, Arizona: Possible Implications for Native Fish*. US Geological Survey Fact Sheet 2013-3104.

Schelly, R. C., R. R. Staffeldt, and M. J. Breen. 2016. *Use of Stewart Lake Floodplain by Larval and Adult Endangered Fishes*. Denver: US Fish and Wildlife Service, Upper Colorado River Endangered Fish Recovery Program.

Schooley, J. D., and P. C. Marsh. 2007. Stocking of endangered razorback sucker in the lower Colorado River basin over three decades: 1974–2004. *North American Journal of Fisheries Management* 27: 43–51.

Shattuck, Z., B. Albrecht, and R. J. Rogers. 2011. Razorback sucker studies on Lake Mead, Nevada and Arizona: 2010–2011. Boulder City, NV: Lower Colorado River Multi-species Conservation Program.

Sjoberg, J. C. 1995. Historic distribution and current status of the razorback sucker in Lake Mead, Nevada-Arizona. *Proceedings of the Desert Fishes Council* 26: 24–27.

Snyder, D. E., and R. T. Muth. 2004. *Catostomid Fish Larvae and Early Juveniles of the Upper Colorado River Basin—Morphological Descriptions, Comparisons, and Computer-Interactive Key*. Larval Fish Laboratory Contribution 139. Technical Publication 42. Fort Collins, CO: Colorado Division of Wildlife.

Speas, D., M. Breen, T. Jones, and B. Schelly. 2017. Updated floodplain wetland priorities for recovery of endangered fish in the Middle Green River. White paper submitted to the Biology Committee, Upper Colorado River Endangered Fish Recovery Program, Denver, Colorado.

Stromberg, J. C., V. B. Beauchamp, M. D. Dixon, S. J. Lite, and C. Paradzick. 2007. Importance of low-flow and high-flow characteristics to restoration of riparian vegetation along rivers in arid south-western United States. *Freshwater Biology* 52: 651–79.

Taba, S. S., J. R. Murphy, and H. H. Frost. 1965. Notes on the fishes of the Colorado River near Moab, Utah. *Proceedings of the Utah Academy of Science, Arts, and Letters* 42: 280–83.

Turner, T. F., T. E. Dowling, P. C. Marsh, B. R. Kesner, and A. T. Kelsen. 2007. Effective size, census size, and genetic monitoring of the endangered razorback sucker, *Xyrauchen texanus*. *Conservation Genetics* 8: 417–25.

Tyus, H. M. 1987. Distribution, reproduction, and habitat use of the razorback sucker in the Green River, Utah, 1979–1986. *Transactions of the American Fisheries Society* 116: 111–16.

Udall, B., and J. Overpeck. 2017. The twenty-first century Colorado River hot drought and implications for the future. *Water Resources Research* 53: 1–15.

USFWS (US Fish and Wildlife Service). 2011. *Razorback Sucker Broodstock Evaluation and Genetic Monitoring*. Dexter NFHTC 2011 Annual Report. Salt Lake City: US Bureau of Reclamation.

USFWS (US Fish and Wildlife Service). 2012. *Razorback Sucker* (Xyrauchen texanus) *Captive Population and Genetics Management Plan for Southwestern Native Aquatic Resources and Recovery Center*. Inter-agency Acquisition R11PG30006.

USFWS (U.S. Fish and Wildlife Service). 2018a. *Species Status Assessment Report for the Razorback Sucker* (Xyrauchen texanus). Denver: U.S. Fish and Wildlife Service, Mountain-Prairie Region (6). https://ecos.fws.gov/ServCat/DownloadFile/166375.

USFWS (U.S. Fish and Wildlife Service). 2018b. *Razorback Sucker* (Xyrauchen texanus) *5-Year Review: Summary and Evaluation*. Denver: US Fish and Wildlife Service, Mountain-Prairie Region (6). http://www.coloradoriverrecovery.org/documents-publications/foundational-documents/recovery-goals.html.

Van Haverbeke, D. R., D. M. Stone, M. J. Dodrill, K. L. Young, and M. J. Pillow. 2017. Population expansion of humpback chub in western Grand Canyon and hypothesized mechanisms. *Southwestern Naturalist* 62: 285–92.

Vanicek, C. D., R. H. Kramer, and D. R. Franklin. 1970. Distribution of Green River fishes in Utah and Colorado following closure of Flaming Gorge Dam. *Southwestern Naturalist* 14: 297–315.

Webber, P. A., and D. Beers. 2014. Detecting razorback suckers using passive integrated transponder tag antennas in the Green River, Utah. *Journal of Fish and Wildlife Management* 5(1): 191–96.

Webber, P. A., K. R. Bestgen, and G. B. Haines. 2013. Tributary spawning by endangered Colorado River basin fishes in the White River. *North American Journal of Fisheries Management* 33: 1166–71.

Wisenall, J. B., B. R. Kesner, C. A. Pacey, and P. C. Marsh. 2015. *Demographics and Monitoring of Repatriated Razorback Sucker in Lake Mohave, 2011–2014*. Boulder City, NV: US Bureau of Reclamation.

Wydoski, R., and G. Mueller. 2006. *The Status of Razorback Sucker in the Colorado River between Davis and Parker Dams (Lake Havasu), 2003–2005*. US Geographical Survey Technical Memorandum 86-68220-06-19, Denver, CO.

Zelasko, K. A., K. R. Bestgen, J. A. Hawkins, and G. C. White. 2016. Evaluation of a long-term predator removal program: Abundance and population dynamics of invasive northern pike *Esox lucius*, Yampa River, Colorado. *Transactions of the American Fisheries Society* 145: 1153–70.

Zelasko, K. A., K. R. Bestgen, and G. C. White. 2010. Survival rate estimation and movement of hatchery-reared razorback suckers *Xyrauchen texanus* in the upper Colorado River basin, Utah and Colorado. *Transactions of the American Fisheries Society* 139: 1478–99.

Zelasko, K. A., K. R. Bestgen, and G. C. White. 2018. *Abundance and Survival Rates of Razorback Suckers* Xyrauchen texanus, *Green River Basin, Utah, 2011–2013*. Larval Fish Laboratory Contribution 203. Denver: US Fish and Wildlife Service, Upper Colorado River Endangered Fish Recovery Program.

22 Assisting Recovery

Intensive Interventions to Conserve Native Fishes of Desert Springs and Wetlands

Sean C. Lema,
Jennifer M. Gumm,
Olin G. Feuerbacher, and
Michael R. Schwemm

Conflicts between maintaining surface water as habitat for aquatic species and using water for municipal and agricultural purposes occur worldwide. Nowhere in North America are these conflicts more evident than in the arid Southwest, where surface water is scarce, most rivers and streams are dammed, and the majority of contemporary surface waters (reservoirs, ponds) are artificially maintained. Despite these pressures, small groundwater-fed springs, seeps, and wetlands throughout this region historically provided habitat for an endemic ichthyofauna with remarkable physiological tolerances, morphologies, and behaviors. Many of these fishes occurred in groundwater springs or wetlands as one or a few populations. The small size and geographic isolation of these habitats, however, make them vulnerable to the loss of reliable water and introductions of nonnative taxa, which have caused most of the region's endemic fishes to become imperiled (Williams and Sada, chap. 6, this volume). One of these fishes—the Tecopa pupfish, *Cyprinodon nevadensis calidae*—even holds the ill-fated distinction of being the first species listed under the US Endangered Species Act (ESA) declared extinct, because its spring-marsh habitat was converted into bathhouses (Miller 1948; Miller et al. 1989).

Because suitable aquatic habitat for native desert fishes is so limited, conservation efforts have often adopted intensive interventional approaches, such as establishing populations of imperiled species in geographically isolated *refuges* (box 22.1). Such refuges may consist of natural habitats, habitats partially modified from their natural state, or entirely artificial or captive environments that are selected or engineered with the aim of mitigating threats (Williams 1991). In practice, however, establishing populations of threatened

[Box 22.1] Terminology Associated with the Use of Refuges as an Assisted Recovery Strategy for Fishes

Assisted colonization: Human-assisted relocation of individuals to a different location, inside or outside their native range.

Assisted gene flow: Managed movement or transfer of individuals or gametes between populations to mitigate local maladaptation or genetic drift.

Assisted recovery: Managed restoration of populations that is informed by historical information on the past range or population size of the species.

Captive propagation: Breeding and rearing of fish in captivity for subsequent release into the wild or for maintaining captive populations.

Captive rearing: Raising fish in captive environments for at least a portion of their life cycle to help support wild populations.

Domestication: Genetic and phenotypic divergence resulting from evolutionary selection for a captive environment.

Environmental enrichment: Improving biological functioning or fitness of captive-reared animals via modifications to their captive environment.

Genetic load: The fitness difference between the average genotype in a population and the theoretically fittest genotype possible. This difference may be caused by deleterious alleles, detrimental epistatic interactions, genetic drift, or introgression of less fit alleles.

Introgressive hybridization: Spread of alleles from one species to another wherein hybrids backcross within one of the parental populations, ultimately diluting its genetic distinctiveness.

Inbreeding depression: A reduction in fitness of offspring resulting from reproduction by individuals of closely related genetic composition.

Outbreeding depression: A reduction in fitness of offspring resulting from reproduction by individuals with divergent genetic composition.

Refuge: A natural or artificial habitat managed specifically for one or several species, rather than for an entire biological community.

Translocation: Capturing and transporting species and releasing them in another location.

or endangered species in refuges—whether located within or outside the species' native range (Minckley 1995)—amounts to a major intervention and begets new challenges.

In this chapter, the roles that refuges have played in the protection and recovery of imperiled desert fishes are examined. Some of these efforts were initiated recently, while others have been ongoing for over 50 years. The successes and setbacks of these efforts offer lessons on how future recovery programs can be implemented effectively. Here, we briefly review selected species as examples that highlight the benefits and challenges associated with maintaining native spring or wetland fishes in refuges. These case studies illustrate how the success of these approaches relies on several interacting genetic, phenotypic, and ecological factors, including the inability to remedy the environmental changes that led to the wild population's decline.

REFUGES FOR DESERT SPRING AND WETLAND FISHES

Use of refuges for desert fish conservation began in the 1960s and 1970s as emergency efforts to preserve species under imminent threat (Johnson and Jensen 1991; Williams 1991). Today, many conservation programs rely on refuges as primary strategies to protect and recover threatened and endangered desert fishes. Contemporary use of the term *refuge* applies to protected areas used with varying degrees of intervention, including (1) *captive rearing* that maintains fish in an artificial environment for a portion of the species' life cycle when individuals are unlikely or unable to survive in the wild, (2) *captive propagation* to produce individuals for supplementing extirpated or declining populations, and (3) preserving the species or population in a managed environment—whether natural or artificial—in perpetuity. For a given species, the degree of intervention depends on the severity and type of threats involved, the recovery goals, and the likelihood of remedying ecological causes for the species' decline (fig. 22.1).

The biology and ecology of desert fishes native to springs and wetlands contribute to the extensive use of refuges. Spring and wetland endemics often share small body sizes, short generation times, generalist feeding strategies, and broad requirements for reproduction that result in rapid and spontaneous breeding in captivity (e.g., Minckley et al. 1991). These characteristics generally make desert spring and wetland fishes amenable to *translocations* to new environments, including refuges.

Even so, the use of refuges as a conservation strategy is not without challenges. Refuges established in natural habitats can remain vulnerable to local or regional groundwater pumping, vandalism, or invasion by nonnative species. When refuges are established in artificially constructed environments, they are typically smaller, have lower habitat complexity, and may require costly monitoring and maintenance. Refuges may also have negative ecological and evolutionary consequences for an imperiled taxon (Stockwell and Leberg 2002). For fishes naturally occurring in only one or a few small, isolated populations, establishing multiple refuges may reduce extinction risk, but can also reduce genetic diversity via founder effects (Stockwell et al. 1996; Dunham and Minckley 1998). Additionally, refuges with novel conditions may alter phenotypes or fitness via local evolutionary adaptation and phenotypic plasticity (Lema and Nevitt 2006; Christie et al. 2012). Therefore, establishing populations of imperiled taxa in refuges requires efforts to maintain their genetic and phenotypic composition.

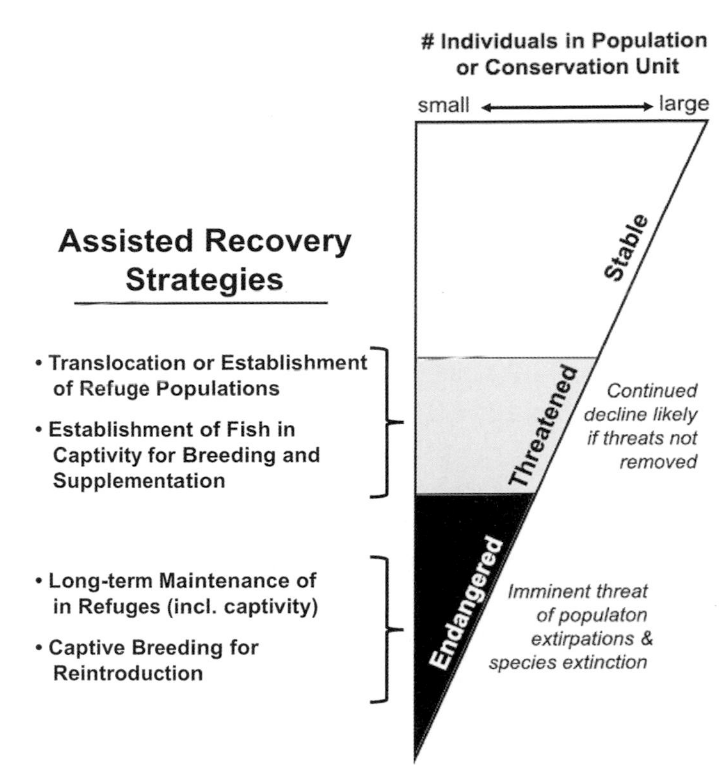

Fig. 22.1 The selection of an assisted recovery strategy for a native fish species is highly dependent on the severity of its decline. Other factors, including the ecological cause(s) underlying the decline or loss of the species in the wild, will influence the specific recovery goals. (Adapted with permission from Fraser 2008.)

Challenges in Maintaining Genetic Diversity in Refuge Populations

Genetic diversity contributes to phenotypic diversity, and genetic variation is considered critical if populations are to cope with future environmental change (Crandall et al. 2000; Frankham 2005). Threatened and endangered desert fishes typically already occur as small populations when efforts toward *assisted recovery* are initiated. Thus, it is often unavoidable that refuge populations are founded with few individuals, and there is a high likelihood that founder effects will reduce genetic variation further (Stockwell et al. 1996; Dunham and Minckley 1998). Because many desert fishes have declined rapidly from large populations, remaining populations are also unlikely to have purged recessive deleterious alleles. This combination can depress effective population sizes (N_e), increasing vulnerability to *inbreeding depression* (Turner et al., chap. 14, this volume). In the absence of gene flow, populations with a small N_e generally lose genetic diversity at relatively high rates due to genetic drift (Keller and Waller 2002). Drift is further accelerated by logistical constraints that may be difficult to avoid in some refuge populations, including highly unequal family sizes across limited breeding events, unbalanced sex ratios, and fluctuations in population size from stochastic events (Frankham 2005).

Populations maintained in multiple, disjunct refuges can diverge in genetic composition as they experience differential genetic drift or local selective pressures. Unlike wild populations, where individuals may disperse between habitats, if only rarely, gene flow does not naturally occur among isolated refuge populations.

Populations maintained in artificial structures may also experience evolutionary *domestication* selection, which has occurred repeatedly in freshwater and marine fishes (e.g., salmonids) propagated in captivity (e.g., Frankham 2008; Christie et al. 2012). Domesticated fish often have lower fitness in the wild, which makes *outbreeding depression* a concern if they are released to supplement wild populations (Tymchuk et al. 2007; Le Cam et al. 2015).

Phenotypic Integrity and the Refuge Environment

Ecological differences between refuges and a species' historical habitats can also alter development to affect behavioral, morphological, life history, and physiological phenotypes. Rearing studies of salmon in hatcheries show that phenotypic attributes including (1) behavior (Berejikian et al. 1999; Weber and Fausch 2003), (2) physiology (Tymchuk et al. 2009; Madison et al. 2015), (3) morphology (Hard et al. 2000; Kihslinger et al. 2006), and (4) life history (Unwin and Glova 1997; Kallio-Nyberg et al. 2013) can be altered in artificial captive environments. Such changes can arise from domestication or developmental plasticity (Brown and Day 2002; Salvanes and Braithwaite 2006). Importantly, while developmental plasticity and genetic change are often viewed as separate processes, they are part of a bidirectional interaction between environmental and genetic change during phenotypic evolution (West-Eberhard 2005; Moczek et al. 2011). Developmental shifts in phenotype that emerge from altered patterns of gene expression expose hidden components of genetic variation to evolutionary selection, so that even short periods in a refuge can begin domestication (Christie et al. 2016).

Most desert fishes in refuges, however, have not been examined for changes in phenotype, in part because many species have insufficient—or in some cases, no—conspecifics remaining in the wild to allow for such comparisons. Even so, artificial environments probably have phenotypic consequences for desert fishes. Belk et al. (2008, 2016), for instance, found that June sucker *Chasmistes liorus liorus* reared in a hatchery differed in head and mouth shape compared with wild relatives. Similarly, Williams (1977) identified differences in size and body shape only a few years after a population of endangered Devils Hole pupfish *Cyprinodon diabolis* was established in an artificial refuge. In practice, even seemingly minor differences in environmental conditions, if experienced during a critical period of development, could lead to phenotypic change.

The importance of phenotypic change, however, depends both on the extent of differentiation and on the goals of the conservation program. Phenotypic divergence or domestication in traits important to survival and reproductive success (e.g., foraging and antipredator behaviors, thermal tolerance, morphology) may decrease fitness if fish from a refuge are released into the wild. The challenge, however, is determining whether phenotypic changes have occurred and whether they are significant to long-term conservation goals. While there is no simple solution to preventing phenotypic changes that lower fitness, refuges could be designed to incorporate *environmental enrichment* by creating environmental heterogeneity within a single refuge, or environmental variation across refuges, both of which promote phenotypic variation (e.g., Johnsson et al. 2014).

SINGLE-SPECIES REFUGES: SUCCESSES AND ONGOING CHALLENGES

One of the most common strategies in desert fish conservation is the use of single-species refuges, which aligns well with the aims of the US Endangered Species Act because many recovery plans are traditionally oriented toward protecting individual species. In many instances, species had declined to such a critical state that the establishment of refuge populations was viewed as imperative to reduce extinction risk, making such single-species refuges a reactive measure rather than a proactive strategy. Yet these refuges vary widely in complexity, size, and success. Here, the history and status of refuges for four imperiled desert fishes are reviewed. Each case study highlights a crucial "take-away" recommendation to be considered when adopting single-species refuges for conservation.

Case Study: Owens Pupfish *Cyprinodon radiosus*

Multiple refuge populations distribute risk and reduce the likelihood of extinction

The small-bodied Owens pupfish is endemic to the Owens Valley, California, where it was once widespread in shallow marshes bordering the Owens River (Snyder 1917). The Owens pupfish disappeared from nearly its entire range due to nonnative species introductions and habitat loss, and was thought extinct until a population of some 200 fish was rediscovered in 1964 (Miller and Pister 1971). On August 18, 1969—in one of the most unassuming, but biologically momentous, events in species conservation—Phil Pister of the California Department of Fish and Game, alone, walked the world's population of Owens pupfish in two buckets across the desert north of Bishop, California, preventing the species' extinction due to a drying habitat (Pister 1993). Since that fortuitous event, the Owens pupfish has relied entirely on refuges.

The first managed population of Owens pupfish was established in 1969, when about 400 fish were translocated to BLM Spring (fig. 22.2*A*). That population served as the source for all other refuge populations. Pupfish from BLM Spring were translocated to two man-made ponds at Warm Springs in 1970, to an overflow channel from a spring-fed pond (Well 368) in 1986 and 1988, and to the lower of two man-made ponds at a refuge called Mule Springs in 1995 (fig. 22.2*B*) (USFWS 2009). In total, over 83 translocations of Owens pupfish to 27 different refuges, many outside of the species' historical range, have occurred (Finger et al. 2013). However, nearly all the translocations failed, and several populations that did become established initially later succumbed to invasions by nonnative species. Owens pupfish at BLM Spring, for instance, were lost following introduction of largemouth bass *Micropterus salmoides* circa 1997. This refuge population was reestablished in 2003, only to decline again following another introduction of largemouth bass in 2007 (USFWS 2009). This pattern of repeated extirpations of refuge populations illustrates why it is critical to establish multiple refuge populations to ensure long-term persistence.

Existing refuge populations of Owens pupfish continue to be vulnerable to threats, including nonnative species and habitat loss from the encroachment of bulrush *Scirpus americanus* (Moyle 2002). Some refuges require regular monitoring to remove vegetation to keep substrate open for pupfish (Davis and Parmenter 2012).

Genetic studies using neutral microsatellite markers also indicate that refuge populations exhibit small N_e values and reduced genetic diversity, which appear to have resulted from genetic bottlenecks linked to the repeated founding and loss of refuge populations over the past 50 years (Finger et al. 2013). To address that loss of genetic variation, Finger et al. (2013) recommended a management program of *assisted gene flow*, which involves transferring 1 to 10 pupfish from each refuge to be distributed among other refuges each generation to dampen the rate of genetic heterozygosity loss in each refuge. Assisted gene flow may enhance population sizes in the short term while providing genetic diversity to facilitate adaptation over the long term (Aitken and Whitlock 2013). Transferring only a small number of individuals regularly could therefore minimize the potential for negative fitness consequences from outbreeding depression while still helping prevent the further erosion of genetic variation in Owens pupfish.

Case Study: Mohave Tui Chub *Siphateles bicolor mohavensis*

Refuges may need to be established outside a species' historical range to reduce extinction risk

The Mohave tui chub is the only fish endemic solely to the Mojave River, California (Hubbs and Miller 1943), and conservation efforts for this species illustrate the challenges that arise when habitat is severely limited. Historically, Mohave tui chub occupied deep pools and sloughs in the Mojave River (Snyder 1918; Miller 1938). It declined in the early twentieth century due to damming of the Mojave River's headwaters, water withdrawal, and possibly hybridization (Hubbs and Miller 1943; Chen et al. 2013) and asymmetric competition (Castleberry and Cech 1986) with arroyo chub *Gila orcuttii* (presumably introduced from coastal streams).

The Mohave tui chub survived as a single, relict population in Mohave Chub Spring (MC Spring), a spring-fed pool about 3 m in diameter and 1.5 m deep near the terminus of the Mojave River in Soda Dry Lake (Miller 1938). Around 1955, Curtis Howe Springer, owner of the Zzyzx Mineral Springs and Health Resort, excavated a large (~38 × 150 m), shallow (maximum depth ~3.3 m) spring-fed pond located about 260 m from MC Spring—named Lake Tuendae—where a second population was established (fig. 22.2*C*). During the 1960s and 1970s, Mohave tui chub were translocated to 13 additional habitats (Hoover and St. Amant 1983; USFWS 1984), but only one population—an introduction of some 400 chub in 1971 into Lark Seeps on the China Lake Naval Air Weapons Station—became established successfully (St. Amant 1983). Ultimately, three additional populations were established: (1) in 1986, at the Camp Cady Wildlife Area along the Mojave River; (2) in 2008, when 548 chub from China Lake were released into two artificial ponds at the Lewis Center for Educational Research in Apple Valley, California; and (3) in

Fig. 22.2 Select examples of refuge and captive-rearing facilities for native desert fishes. Both (*A*) BLM Spring and (*B*) the lower pond at Mule Spring, in Owens Valley, California, are refuges for Owens pupfish (*C. radiosus*). (*C*) Lake Tuendae, an artificial pond at Zzyzx Springs within Mojave National Preserve, California, supports Mohave tui chub (*S. b. mohavensis*). Artificial refuges for Pahrump poolfish (*E. l. latos*) are located at (*D*) Corn Creek Springs, Desert National Wildlife Range, Nevada, and (*E*) Lake Mead Fish Hatchery, Nevada. (*F*) Lake Harriet, Nevada, is under renovation (removal of nonnative species) for future poolfish reestablishment. (Photographs *A–B* by Steve Parmenter, *C–F* by Sean Lema.)

2011, when about 1,000 tui chub from China Lake and Lake Tuendae were introduced into the defunct Morning Star Mine site in Mojave National Preserve, California. With five refuge populations and the MC Spring population, the number of Mohave tui chub is now estimated at 12,000–16,000 fish (S. Henkanaththegedara, pers. comm.). However, the refuges differ ecologically from natural conditions in the Mojave River. Four are man-made, and only two (Camp Cady and Lewis Center) are within the species' native range.

In the absence of assisted gene flow, local divergence and loss of genetic diversity are likely via selection or drift. Recent genetic analyses revealed that Mohave tui chub at Camp Cady had a small N_e and low neutral genetic variation compared with those at the other refuges (Chen et al. 2013). The Camp Cady refuge was founded with only 65 fish from Lake Tuendae, so low genetic diversity probably stems from a founder effect. The MC Spring population also exhibited significant genetic divergence from Lake Tuendae, Camp Cady, and China Lake, despite those populations being founded with MC Spring fish.

Chen et al. (2013) proposed slowing this local divergence by translocating 10 fish from the large population in Lake Tuendae to each refuge annually. The extent of genetic divergence observed for the MC Spring population, however, indicated significant local differentiation, possibly resulting from selection for the distinctive conditions of MC Spring, which has low dissolved oxygen (<5 mg/L) and a stable temperature (15°C). Translocating Mohave tui chub into MC Spring therefore runs a heightened risk of outbreeding depression. For that reason, translocations into MC Spring were recommended only if the MC Spring population declined.

Even if assisted gene flow is implemented, the Mohave tui chub is likely to face continuing threats. Under a shifting climate regime, refuge populations may also be vulnerable to increasing water temperature and declining dissolved oxygen (e.g., McClanahan et al. 1986). The Mojave River region is also expected to experience altered evapotranspiration dynamics with climate change, and aquatic habitats in the region may experience reduced primary productivity (Archer and Predick 2008). Yet, with as many as 19 nonnative fishes now occupying the Mojave River (Moyle 2002), successful restoration of Mohave tui chub to the river is improbable without nonnative species removal. Establishing more refuge populations may therefore provide the greatest likelihood for long-term persistence of Mohave tui chub.

Case Study: Pahrump Poolfish *Empetrichthys latos*

Refuges supporting species with no remaining natural habitat may require engineering and regular monitoring to address persistent threats

The Pahrump poolfish is endemic to the Pahrump Valley in southern Nevada. It has been the only extant taxon within the genus *Empetrichthys* since its sole congener, the Ash Meadows poolfish *E. merriami*, went extinct in the 1950s (Miller et al. 1989). Three subspecies of *E. latos* were described based on morphology and geographic isolation (Miller 1948), but *E. l. pahrump* and *E. l. concavus* became extinct due to water withdrawal and other human activities (Minckley and Deacon 1968). The third subspecies, *E. l. latos*, occurred in Manse Spring, which also suffered ecological changes. Increased groundwater extraction in the Pahrump Valley reduced water discharge from Manse Spring (Minckley and Deacon 1968; Deacon and Williams 2010), and nonnative goldfish *Carassius auratus* were introduced in 1961 (Deacon et al. 1964). Ultimately, water discharge from Manse Spring failed in 1975, eliminating the last native habitat for Pahrump poolfish (Deacon 1979).

Conservation efforts for *E. latos* required the use of refuges, as the species had no remaining natural habitat. Auspiciously, increasing awareness of the tenuous state of poolfish in Manse Spring, spurred in large part by W. L. Minckley and Jim Deacon, led to translocations of *E. l. latos* into three fishless habitats in Nevada: Los Latos Pools, near Lake Mohave in the Lake Mead National Recreation Area, in 1970; Corn Creek Springs, on the Desert National Wildlife Refuge, in 1971; and Shoshone Ponds Natural Area, in 1972 (Goodchild 2016; Jimenez et al. 2017). The Los Latos Pools population was lost during the late 1970s and never reestablished. The Shoshone Ponds population failed in 1974 when vandals turned off the habitat's water source, but it was reestablished in 1976 with poolfish from Corn Creek and has since remained stable (LaVoie 2004). In 1983, a new population was founded, also with fish from Corn Creek, in an irrigation and storage reservoir—

Lake Harriet—in Spring Mountain Ranch State Park, Nevada.

Each of the refuge populations has faced continuing challenges. In the late 1990s, the Corn Creek population declined following introduction of nonnative red swamp crayfish *Procambarus clarkii*. In response, an artificial tank was constructed at Corn Creek in 2002 to provide a refuge secure from nonnative species (fig. 22.2*D*). That artificial refuge was populated with poolfish from Lake Harriet in 2003. After habitat restoration and nonnative species removal via rotenone and pond desiccation, 175 poolfish from Lake Harriet and 500 fish from Shoshone Ponds were released back into Corn Creek in 2014, followed by another 775 poolfish from Lake Harriet in 2015 (B. Senger and K. Guadalupe, pers. comm.). Smaller translocations from Lake Harriet to Corn Creek have also occurred to slow genetic and phenotypic differentiation of the refuge populations (B. Senger and K. Guadalupe, pers. comm.).

Despite supporting a secure population of poolfish for some 30 years, Lake Harriet was found in 2013 to contain nonnative red swamp crayfish and western mosquitofish *Gambusia affinis*. The population of poolfish crashed from more than 10,000 in 2015 to fewer than 1,000 in 2016, coincident with the population explosions of these invasive species. To salvage what was left of this refuge population, all poolfish remaining in Lake Harriet (~700 fish) were trapped in fall 2016 and spring 2017. Forty-four fish were translocated to Corn Creek, and the rest were moved to the Lake Mead Fish Hatchery, where they are currently maintained (fig. 22.2*E*). Lake Harriet is being rehabilitated to remove nonnatives (fig. 22.2*F*), and reestablishment of poolfish there is planned. As of 2017, the US Fish and Wildlife Service was working to establish a fourth refuge population of Pahrump poolfish in engineered ponds at Springs Preserve in the city of Las Vegas (*Federal Register* DOC #: 2016-08344).

The survival of Pahrump poolfish continues to rely entirely on refuge habitats. Notably, all refuge populations of Pahrump poolfish are located outside of the species' native range, and each refuge has required some degree of engineering to prevent nonnative invasions (see fig. 22.2*D*). With no native habitat remaining, the Pahrump poolfish is likely to rely on refuges in perpetuity. To date, however, almost nothing is known about patterns of genetic diversity across the refuge populations, or whether any refuge population has diverged phenotypically from the others. It will therefore be important for some refuge fish to be allocated toward research efforts that provide the genetic and phenotypic data necessary to inform management procedures supporting recovery of the Pahrump poolfish.

Case Study: Devils Hole Pupfish *Cyprinodon diabolis*

Difficult decisions may arise about whether refuge conditions should optimize fish reproduction and survival or precisely mimic natural habitat

One of the most enigmatic species in the desert Southwest, the Devils Hole pupfish occupies what may be the most restricted habitat for any vertebrate—Devils Hole. Devils Hole is a collapsed cave with a small pool (~3.5 m wide by 22 m long) characterized by extreme conditions: low dissolved oxygen, low food availability, and high temperature (~33.5°C) (Riggs and Deacon 2002; Bernot and Wilson 2012). Establishing replicate "lifeboat" populations of *C. diabolis* as insurance against extinction has been a conservation priority for decades (Deacon and Baugh 1985; Riggs and Deacon 2002). Several attempts were made to propagate *C. diabolis* in aquaria (Castro 1971). While eggs were produced in some cases, all captive breeding efforts ultimately failed (Baugh and Deacon 1983; Deacon et al. 1995).

Efforts to establish refuge populations of Devils Hole pupfish fared somewhat better, and three populations were established in similarly designed concrete refuge tanks between 1972 and 1990: one near Hoover Dam and two (School Springs and Point of Rocks) in Ash Meadows National Wildlife Refuge (Baugh and Deacon 1988; Karam et al. 2012). However, all three refuge populations were eventually lost. A water-pump failure extirpated the School Springs refuge population. In the Point of Rocks refuge, the closely related Ash Meadows pupfish, *Cyprinodon nevadensis mionectes*, entered the tank and hybridized with *C. diabolis* (Martin et al. 2012). In response to the loss of those two refuge populations, the population in the Hoover Dam refuge was moved to Willow Beach National Fish Hatchery, Arizona, where it also failed. While ultimately unsuccessful, these refuge efforts exposed one of the major challenges of using artificial refuges. In all three refuges, water temperature, algal composition, and other ecological parameters differed from those in Devils Hole (Karam et al. 2012). This ecological variation appears to have contributed to changes in the body size and behavior of pupfish in the refuges (Lema and Nevitt 2006; Wilcox and Martin 2006). Such phenotypic changes may have undermined the usefulness of the refuges for reestablishing *C. diabolis* in Devils Hole, should the need have arisen.

With *C. diabolis* still under threat in Devils Hole, a new state-of-the-art refuge facility—Ash Meadows Fish Conservation Facility (AMFCF)—was constructed in 2013. The heart of AMFCF is the "refuge tank" (fig. 22.3), a 6.7 m deep concrete pool with a volume of 378,541 L, roughly 20 times larger than any of the three previous refuges. One end of the tank has a fiberglass-encased spawning shelf constructed with the precise dimensions and contours of the shallow "shelf" in Devils Hole—an area characterized by higher primary productivity, where *C. diabolis* typically gathers. The refuge is also stocked with algae and invertebrates from Devils Hole to simulate the natural biota.

Following completion of the AMFCF, a major concern became how best to bring *C. diabolis* into captivity without adversely affecting the already small wild population in Devils Hole. Spawning occurs at low levels during winter and late summer in Devils Hole, although eggs produced during those times do not survive. Risk analysis confirmed that collecting eggs posed the lowest risk to the Devils Hole population (Beissinger 2014). Following this rationale, eggs were collected using "recovery mats" constructed of nylon carpet glued to travertine tiles. The first collection of eggs from Devils Hole in winter 2013 yielded some 60 eggs, and 29 pupfish hatched in captivity from those eggs were introduced into the AMFCF refuge tank in June 2014.

Previous work with hybrid *C. diabolis* × *C. n. mionectes* indicated that pupfish kept in environmental conditions true to those in Devils Hole have diminished reproductive output, and that maintaining pupfish at lower temperatures increased egg production and hatching rates (Feuerbacher et al. 2015). Given that the utmost concern for *C. diabolis* was to establish a self-supporting refuge population, the decision was made to have the environmental parameters of the refuge differ from those in Devils Hole. Despite the possibility that less extreme conditions might influence the phenotypic development of refuge pupfish (e.g., Williams 1977; Lema and Nevitt 2006), the temperature is 2°C–3°C cooler in the refuge than in Devils Hole, and dissolved oxygen is about 1 ppm higher. While the effects of these less extreme conditions are not yet fully known, reproduction by *C. diabolis* has been confirmed in the refuge, and the refuge population has increased, albeit slowly. However, this new refuge population has not grown as rapidly as anticipated, which implies that extreme conditions may not be the only factor limiting the growth of the Devils Hole population. Rather, *genetic load* or other unidentified factors may be contributing to the impaired growth of *C. diabolis* populations (e.g., Martin et al. 2012).

REFUGES THAT SUPPORT THE CONSERVATION OF MULTIPLE SPECIES

Some imperiled fishes are the sole fish in their habitats, but many are not, and interactions with non-threatened taxa, or even between co-occurring endangered species, may complicate recovery actions (Gumm et al. 2008; Robinson and Ward 2011). Fortuitously, refuges that need to support multiple species can be structured with attention to the spatial or temporal allocation of resources (i.e., food, breeding territories) to alter the dynamics of competition, predation, or reproduction in ways that affect population size and evolutionary potential (Watters et al. 2003, 2017).

Case Study: Diamond Y Draw

Conflicts between imperiled fishes maintained in the same refuge require creative approaches to remedy negative interactions

Diamond Y Draw, near Fort Stockton, Texas, illustrates the challenges that can arise with multi-species refuges. A flood tributary of the Pecos River, Diamond Y Draw includes two disjunct watercourses separated by dry streambed: (1) an upper watercourse with a headwater spring (Diamond Y Spring) and outflow (fig. 22.4*A*), and (2) a lower watercourse with a spring (Monsanto Pool), outflow, and multiple isolated pools. Historically, Diamond Y Draw was managed for the endangered Leon Springs pupfish *Cyprinodon bovinus*, which was rediscovered there by W. L. Minckley and W. E. Barber in 1965, after its extinction from its type locality in Leon Springs (Echelle and Miller 1974).

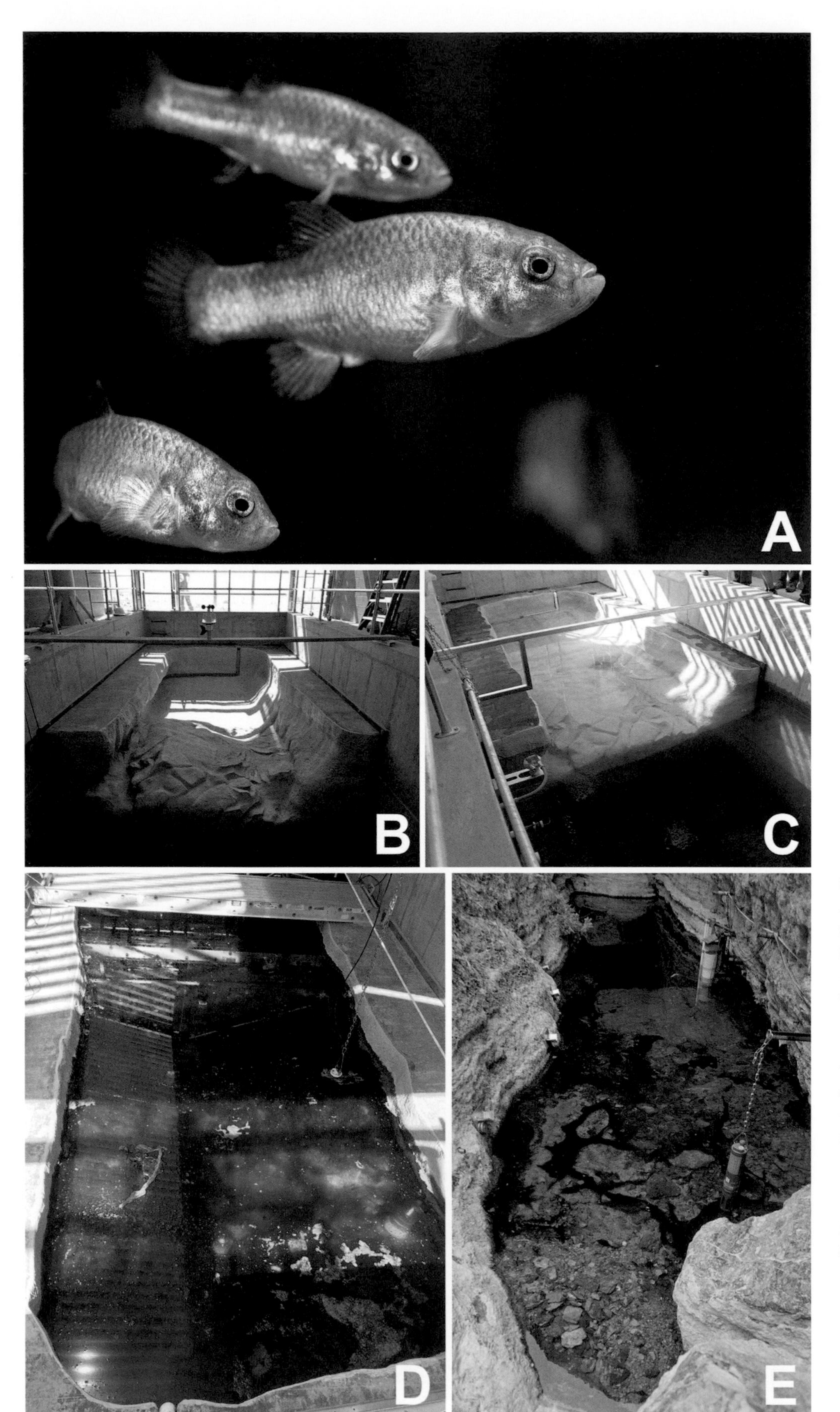

Fig. 22.3 Development of the Ash Meadows Fish Conservation Facility (AMFCF) for Devils Hole pupfish (*C. diabolis*). (*A*) Focal species. (*B*) Upper fiberglass shelf in the AMFCF. (*C*) AMFCF refuge tank in November 2014, after being filled with water. (*D*) AMFCF showing growth of algae. (*E*) Devils Hole—the sole native habitat for *C. diabolis*, Death Valley National Park, Nevada—for comparison. (Photographs *A*, *B*, *D*, and *E* by Olin Feuerbacher; *D* by Sean Lema.)

In 1974, the nonnative sheepshead minnow *Cyprinodon variegatus* invaded the lower watercourse of Diamond Y Draw. In response, 80 *C. bovinus* from the upper watercourse were captured and moved to the Southwestern Native Aquatic Resources and Recovery Center (SNARRC, formerly Dexter National Fish Hatchery) near Dexter, New Mexico, to establish a captive population, and Diamond Y Draw was treated with chemical piscicides to remove sheepshead minnow (Hubbs et al. 1978; Hubbs 1980). However, a second introduction of sheepshead minnow had resulted in *introgressive hybridization* by the early 1990s (Echelle and Echelle 1997). Eradication efforts were again conducted, and *C. bovinus* from SNARRC were restored to both watercourses in 1998–2001 (Echelle et al. 2004). Subsequent genetic analyses of specimens collected in 2013–2014 found no evidence of *C. variegatus* introgression or hybrids in either watercourse (Black et al. 2017).

Compared with the previous case studies, conservation of Leon Springs pupfish in Diamond Y Draw is complicated because this habitat also supports the endangered Pecos gambusia *Gambusia nobilis*. As is typical for pupfishes (e.g., Kodric-Brown 1977), larger *C. bovinus* males in these springs defend breeding territories, intermediate-sized males use a "satellite" reproductive strategy, and small males behave as "sneakers" that mimic females. By 2006, however, the number of territorial males in Diamond Y Spring had plummeted, and by 2007, only a lone territorial male was observed. At that time, a new behavior was observed, in which the co-occurring Pecos gambusia swarmed pairs of spawning pupfish (fig. 22.4*B*). On average, 20 gambusia crowded each pair, suggesting that a challenge existed wherein the endangered Pecos gambusia appeared to decrease the reproductive success of the endangered Leon Springs pupfish by consuming newly spawned pupfish eggs (Gumm et al. 2008). Because both species are endangered, an innovative, multi-species solution was required.

In partnership with The Nature Conservancy, the US Fish and Wildlife Service, and the Texas Parks and Wildlife Department, Murray Itzkowitz and associates applied the theory of "behavioral management" (Watters et al. 2003, 2017) to this refuge (e.g., Gumm et al. 2011; Black et al. 2016). This theory is well suited to species for which resource availability can determine the reproductive tactics expressed by males. In pupfishes, the availability of high-quality breeding habitat—typically areas with heterogeneous substrates—limits the number of territorial males, and therefore may constrain population-level reproductive output, as territorial males have higher reproductive success than satellite or sneaker males (Kodric-Brown 1977). Watters et al. (2003) predicted that altering habitat could change social competition, so that constructing habitats intentionally with specific structural features could optimize a population's reproductive output.

Thus, the goals of habitat modification in the Diamond Y system were twofold (Gumm et al. 2011; Black et al. 2016): (1) increase the number of territorial males in the pupfish population, and (2) decrease predation pressure by Pecos gambusia on pupfish eggs. In January 2007, about 15 m^2 of bulrush were removed from Di-

Fig. 22.4 Interactions between sympatric endangered species in a refuge habitat. (*A*) Upper spring pool of Diamond Y Draw, Texas, occupied by endangered Leon Springs pupfish (*C. bovinus*) and Pecos gambusia (*G. nobilis*). Habitat modifications (cement tiles) are visible adjacent to the bank (*middle right*). (*B*) Endangered Pecos gambusia swarming a spawning pair of endangered Leon Springs pupfish in the spring pool. (Photographs by Jennifer Gumm.)

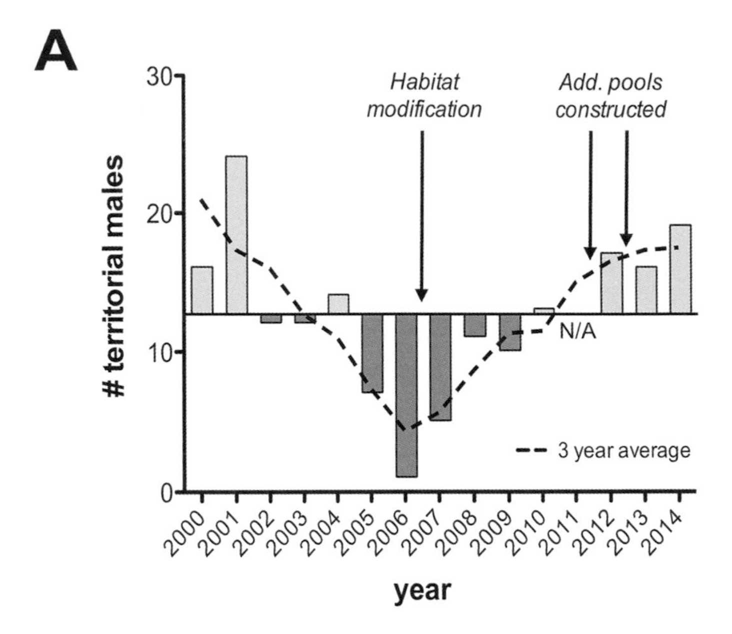

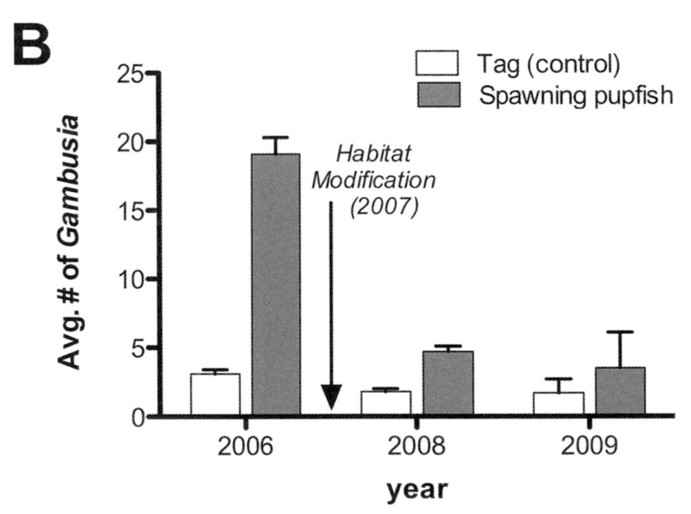

Fig. 22.5 Successful behavioral management. (*A*) The number of territorial male *C. bovinus* pupfish observed in Diamond Y Spring near the turn of the century declined steadily to a low in 2006. The upper spring pool was modified in 2007 by the addition of cement tiles to provide additional breeding substrate for *C. bovinus* (see fig. 22.4*A*). Habitat was again modified in 2012 and 2013 by the construction of new pools, also provided with cement tiles, downstream of the spring pool. N/A indicates data are unavailable for 2011. (Data adapted with permission from Black et al. 2016.) (*B*) Mean number (±SE) of Pecos gambusia (*G. nobilis*) observed within one body length (~5 cm) of spawning pairs of *C. bovinus* pupfish. Gambusia preferentially swarmed spawning pairs of pupfish compared with stationary plastic tags placed in the habitat that served as a comparison (control) for overall gambusia density. The number of gambusia swarming spawning pupfish declined following the habitat modification in 2007. (Data adapted with permission from Gumm et al. 2011.)

amond Y Spring, and cement tiles were placed in the cleared area to provide additional breeding substrate for pupfish. From 2008 to 2010, territorial pupfish increased to numbers similar to those in 2001–2005 (fig. 22.5*A*). Pecos gambusia became more widely distributed so that they were no longer associated with territorial pupfish in the newly expanded breeding habitat, and gambusia density around spawning pupfish declined sixfold (fig. 22.5*B*). Additional pools for pupfish spawning were later added downstream, further elevating territorial males from an average of 4 observed per year (2005–2007) to 17 per year (2012–2014) in the upper water course. In 2013, bulrush was also removed from the spring pool of the lower watercourse, cement tiles were added, and 400 Leon Springs pupfish from SNARRC were introduced; these actions likewise reduced breeding interference by Pecos gambusia, despite their continued presence in pupfish breeding territories (Paciorek et al. 2014).

The history of *C. bovinus* in Diamond Y Draw illustrates how habitat structure in a refuge can be used to shape the behavior of endangered desert fishes. Ecological interactions between habitat structure, genetics, and other factors generate complex patterns of behavioral variation, and—if those interactions are understood sufficiently—designing refuges with the intention of managing those interactions may be a viable approach to promote behaviors that improve conservation outcomes (Watters et al. 2003, 2017).

THE FUTURE OF REFUGES FOR ASSISTING RECOVERY OF DESERT FISHES

Taken together, these case studies illustrate the successes and ongoing challenges of using refuges to recover imperiled desert fishes. Populations in even the best-designed refuges are vulnerable to genetic drift or local adaptation, which could lead to unintended changes in phenotype or the loss of genetic diversity. Under ideal circumstances, species would need to be maintained in a natural or artificial refuge only for the portion of their life cycle when individuals are unlikely or unable to survive in the wild. Unfortunately, for many native desert fishes, the severity of decline and extent of habitat loss instead requires long-term maintenance in refuges, possibly in perpetuity. The history of refuge use for native desert fishes does, however, offer several guidelines for future conservation efforts.

1. **Multiple refuge populations reduce risk but require regular monitoring and genetic management.** Multiple refuges hedge against stochasticity and vandalism. Even though many desert fishes occurred naturally in only one or a few habitats, they may need to be in several refuges—which then need to be monitored regularly—to ensure long-term persistence. Because multiple refuge populations are subject to divergent selection pressures and genetic drift (e.g., Koike et al. 2008), the challenge is to establish multiple refuges while using other interventions (i.e., assisted gene flow) to protect against potentially divergent evolutionary trajectories. The balance between these two strategies should be expected to vary depending on the historical geographic distribution, extent of decline, and natural history of each imperiled taxon.
2. **Assisted recovery efforts may need to balance trade-offs among multiple imperiled species.** While some imperiled species are the sole fish in their habitats, many are not, and in some springs or wetlands, almost the entire community of native fishes is imperiled (e.g., USFWS 1990, 1995). Even movement toward recovery could lead to new conflicts for some fishes if there is resource competition or trophic interaction among imperiled species. Aquatic habitats that are suitable as refuges are limited, and the rarity of potential refuges for multiple species can restrict management options, thus requiring creative approaches to remedy interactions between imperiled species.
3. **Environmental conditions in refuges can be selected to achieve specific conservation goals.** Phenotypes can change rapidly due to developmental plasticity or domestication selection when conditions in a refuge or captive environment differ from those in natural habitat. Minimizing ecological dissimilarities between refuges and natural habitats might lessen this risk (e.g., Lema and Nevitt 2006; Christie et al. 2012). Constructed refuges could be designed and engineered to promote phenotypes that foster species recovery. Even seemingly modest structural changes, such as the placement of rocks, could alter reproductive dynamics and help achieve recovery goals (Watters et al. 2003).
4. **Both neutral genetic variation *and* quantitative genetic traits should be monitored in refuge populations.** Even with informed rearing and genetic management approaches, captive environments may unavoidably lead to genetic changes. There is evidence that neutral genetic variation (i.e., microsatellite markers) can be maintained in captive populations for several generations by maintaining a high N_e (Fraser 2008), and many assisted recovery programs now include periodic genotyping of wild and captive stocks to monitor changes in genetic composition. However, studies of genetic composition in refuge populations of desert fishes have focused almost exclusively on neutral genetic markers (e.g., Chen et al. 2013; Finger et al. 2013), and quantitative genetic traits are rarely evaluated, even though research suggests that changes in quantitative genetic traits linked to phenotypes may occur more rapidly than the loss of neutral genetic variation in captive populations (Fraser 2008). Efforts should therefore be made to monitor both neutral markers and quantitative genetic traits in refuge populations.
5. **Recovery strategies should consider whether long-term persistence requires more than just reestablishing populations in historical locations of record.** Conservation programs that need to identify potential refuge habitats and conduct translocations can consider *assisted colonization*. Assisted colonization strategies aim to translocate threatened and endangered species to new habitats in regions where the taxon did not occur, but where habitat may be more suitable or protected for its long-term survival (Shirley and Lamberti 2010; Collyer et al. 2011). Refuge and reintroduction sites within a species' native range may be limited. The condition and availability of habitat may also shift under climate change (Moyle et al. 2013; Hausner et al. 2014). In such cases, recovery efforts could consider habitats outside the species' historical range if those sites better secure long-term survival in the face of a changing regional climate and do not threaten other native species.
6. **Research into novel management procedures that promote long-term persistence of refuge populations is needed.** More research is needed on imperiled desert fishes to better evaluate the phenotypic, genetic, and fitness effects of refuges. The best methods for maintaining refuge populations of taxonomically and ecologically diverse desert fishes should be expected to vary widely among species,

and any fitness reductions in refuge fish are likely to emerge from changes in multiple traits (i.e., foraging behavior, swimming performance, stress responses). Yet, managers still lack even basic data on the phenotypic and genetic composition of refuge populations for many imperiled fishes. Assisted recovery programs should allocate some proportion of refuge or captive-reared individuals toward research aimed at enhancing management procedures (e.g., environmental enrichment of captive or refuge habitats, minimizing stress during translocations; see Johnsson et al. 2014; Näslund and Johnsson 2016).

Refuges will continue to play an important role in the conservation and management of desert fishes, but when and how refuges are used must be considered carefully. Should populations be kept in refuges in perpetuity if doing so will prevent species extinction, but lead to evolutionary changes? How are phenotypic shifts in refuge populations to be evaluated when a species has no wild populations remaining to compare them against? Should artificial refuges be constructed to re-create a species' natural habitat, or is it better to select conditions that best promote reproduction and survival, therefore increasing population size even at costs to the species' phenotypic or genetic composition?

Decisions to adopt or reject controversial strategies for conservation will require a critical evaluation of what is truly valued about desert fishes. Biodiversity is not static in time or space, and the natural processes that generated the current diversity of desert fishes will continue to act in a changing world. Assessing what is imperative to preserve—the phenotypic and genetic integrity of populations, evolutionary potential, or other biotic processes—is necessary to guide assisted recovery efforts for desert fishes facing an uncertain future.

Ultimately, however, the recovery of native fishes requires remedying the ecological causes for their decline. As stated by Johnson and Jensen (1991, 200): "Captive propagation is not an end in itself, but rather a means of perpetuating species until suitable habitat can be found, rebuilt, or created." Refuge programs should be adopted in combination with other efforts, including habitat restoration, flow management, and removal of nonnative species. Without amelioration of the environmental changes that reduced wild populations in the first place, refuges and other assisted recovery strategies are unlikely to conserve native desert fishes over the long term.

ACKNOWLEDGMENTS

The authors thank Brandon Senger and Kevin Guadalupe for providing information about the conservation status of Pahrump poolfish. The authors also thank Steve Parmenter for providing photographs, and three anonymous reviewers for thoughtful comments and suggestions that improved this chapter.

REFERENCES

Aitken, S. N., and M. C. Whitlock. 2013. Assisted gene flow to facilitate local adaptation to climate change. *Annual Review of Ecology, Evolution, and Systematics* 44: 367–88.

Archer, S. R., and K. I. Predick. 2008. Climate change and ecosystems of the Southwestern United States. *Rangelands* 30: 23–28.

Baugh, T. M., and J. E. Deacon. 1983. Maintaining the Devils Hole pupfish, *Cyprinodon diabolis* Wales, in aquaria. *Journal of Aquaculture and Aquatic Science* 3: 73–75.

Baugh, T. M., and J. E. Deacon. 1988. Evaluation of the role of refugia in conservation efforts for the Devils Hole pupfish, *Cyprinodon diabolis* Wales. *Zoo Biology* 7: 351–58.

Beissinger, S. R. 2014. Digging the pupfish out of its hole: Risk analyses to guide harvest of Devils Hole pupfish for captive breeding. *PeerJ* 2: e549.

Belk, M. C., L. K. Benson, J. Rasmussen, and S. L. Peck. 2008. Hatchery-induced morphological variation in an endangered fish: A challenge for hatchery-based recovery efforts. *Canadian Journal of Fisheries and Aquatic Sciences* 65: 401–8.

Belk, M. C., M. Maxwell, C. Laidlaw, and J. Wesner. 2016. Building a better June sucker: Characterization of mouth shape in the captive brood stock. *Open Fish Science Journal* 9: 29–36.

Berejikian, B. A., E. P. Tezak, S. L. Schroder, T. A. Flagg, and C. M. Knudsen. 1999. Competitive differences between newly emerged offspring of captive-reared and wild coho salmon. *Transactions of the American Fisheries Society* 128: 832–39.

Bernot, M. J., and K. P. Wilson. 2012. Spatial and temporal variation of dissolved oxygen and ecosystem energetics in Devils Hole, Nevada. *Western North American Naturalist* 72: 265–75.

Black, A. N., H. A. Seears, C. M. Hollenbeck, and P. B. Samollow. 2017. Rapid genetic and morphologic divergence between captive and wild populations of the endangered Leon Springs pupfish, *Cyprinodon bovinus*. *Molecular Ecology* 26: 2237–56.

Black, A. N., J. L. Snekser, L. Al-Shaer, T. Paciorek, A. Bloch, K. Little, and M. Itzkowitz. 2016. A review of the Leon springs pupfish (*Cyprinodon bovinus*) long-term conservation strategy and response to habitat restoration. *Aquatic Conservation* 26: 410–16.

Brown, C., and R. L. Day. 2002. The future of stock enhancements: Lessons for hatchery practice from conservation biology. *Fish and Fisheries* 3: 79–94.

Castleberry, D. T., and J. J. Cech, Jr. 1986. Physiological responses of a native and an introduced desert fish to environmental stressors. *Ecology* 67: 912–18.

Castro, A. 1971. Steinhart Aquarium log for Devil's Hole pupfish colony, *Cyprinodon diabolis*. *Proceedings of the Desert Fishes Council* 3: 30–31.

Chen, Y., S. Parmenter, and B. May. 2013. Genetic characterization and management of the endangered Mohave tui chub. *Conservation Genetics* 14: 11–20.

Christie, M. R., M. L. Marine, S. E. Fox, R. A. French, and M. S. Blouin. 2016. A single generation of domestication heritably alters the expression of hundreds of genes. *Nature Communications* 7: 10676.

Christie, M. R., M. L. Marine, R. A. French, and M. S. Blouin. 2012. Genetic adaptation to captivity can occur in a single generation. *Proceedings of the National Academy of Sciences* 109: 238–42.

Collyer, M. L., J. S. Heilveil, and C. A. Stockwell. 2011. Contemporary evolutionary divergence for a protected species following assisted colonization. *PLoS ONE* 6: e22310.

Crandall, K. A., O. R. P. Bininda-Emonds, G. M. Mace, and R. K. Wayne. 2000. Considering evolutionary processes in conservation biology. *Trends in Ecology and Evolution* 15: 290–95.

Davis, M., and S. Parmenter. 2012. Winning the cattail battle. *Proceedings of the Desert Fishes Council* 44: Abstract ID #116.

Deacon, J. E. 1979. Endangered and threatened fishes of the West. *Great Basin Naturalist Memoirs* 3: 41–64.

Deacon, J. E., and T. M. Baugh. 1985. *Population Fluctuations of the Devils Hole Pupfish—1972–1984*. Death Valley National Monument: National Park Service.

Deacon, J. E., C. Hubbs, and B. J. Zahuranec. 1964. Some effects of introduced fishes on the native fish fauna of southern Nevada. *Copeia* 1964: 384–88.

Deacon, J. E., F. R. Taylor, and J. W. Pedretti. 1995. Egg viability and ecology of Devils Hole pupfish: Insights from captive propagation. *Southwestern Naturalist* 40: 216–23.

Deacon, J. E., and J. E. Williams. 2010. Retrospective evaluation of the effects of human disturbance and goldfish introduction on endangered Pahrump poolfish. *Western North American Naturalist* 70: 425–36.

Dunham, J. B., and W. L. Minckley. 1998. Allozyme variation in desert pupfish from natural and artificial habitats: Genetic conservation in fluctuating populations. *Biological Conservation* 84: 7–15.

Echelle, A. A., and A. F. Echelle. 1997. Genetic introgression of endemic taxa by nonnatives: A case study with Leon Spring pupfish and sheepshead minnow. *Conservation Biology* 11: 153–61.

Echelle, A. A., and R. R. Miller. 1974. Rediscovery and redescription of the Leon Springs pupfish, *Cyprinodon bovinus*, from Pecos County, Texas. *Southwestern Naturalist* 19: 179–90.

Echelle, A. F., A. A. Echelle, L. K. Bonnel, N. L. Allen, J. E. Brooks, and J. Karges. 2004. Effects of a restoration effort on an endangered pupfish (*Cyprinodon bovinus*) after genetic introgression by a non-native species. In M. L. Lozano-Vilano and A. J. Contreras-Balderas, eds., *Homenaje al Doctor Andrés Reséndez Medina*, 129–39. Monterrey: Universidad Autónoma de Nuevo León.

Feuerbacher, O. G., J. A. Mapula, and S. A. Bonar. 2015. Propagation of hybrid Devils Hole pupfish × Ash Meadows Amargosa pupfish. *North American Journal of Aquaculture* 77: 513–23.

Finger, A. J., S. Parmenter, and B. P. May. 2013. Conservation of the Owens pupfish: Genetic effects of multiple translocations and extirpations. *Transactions of the American Fisheries Society* 142: 1430–43.

Frankham, R. 2005. Genetics and extinction. *Biological Conservation* 126: 131–40.

Frankham, R. 2008. Genetic adaptation to captivity in species conservation programs. *Molecular Ecology* 17: 325–33.

Fraser, D. J. 2008. How well can captive breeding programs conserve biodiversity? A review of salmonids. *Evolutionary Applications* 1: 535–86.

Goodchild, S. C. 2016. Life History and Interspecific Co-Persistence of Pahrump Poolfish and Amargosa Pupfish in Ex Situ Refuges. PhD diss., North Dakota State University.

Gumm, J. M., J. L. Snekser, and M. Itzkowitz. 2008. Conservation and conflict between endangered desert fishes. *Biology Letters* 4: 655–58.

Gumm, J. M., J. L. Snekser, K. Little, J. Leese, V. Imhoff, B. Westrick, and M. Itzkowitz. 2011. Management of interactions between endangered species using habitat restoration. *Biological Conservation* 144: 2171–76.

Hard, J. J., B. A. Berejikian, E. P. Tezak, S. L. Schroder, C. M. Knudsen, and L. T. Parker. 2000. Evidence for morphometric differentiation of wild and captively reared adult coho salmon: A geometric analysis. *Environmental Biology of Fishes* 58: 61–73.

Hausner, M. B., K. P. Wilson, D. B. Gaines, F. Suárez, G. G. Scoppettone, and S. W. Tyler. 2014. Life in a fishbowl: Prospects for the endangered Devils Hole pupfish (*Cyprinodon diabolis*) in a changing climate. *Water Resources Research* 50: 7020–34.

Hoover, F., and J. A. St. Amant. 1983. Results of Mohave chub, *Gila bicolor mohavensis*, relocations in California and Nevada. *California Fish and Game* 69: 54–56.

Hubbs, C. 1980. The solution to the *Cyprinodon bovinus* problem: Eradication of a pupfish genome. *Proceedings of the Desert Fishes Council* 10: 9–18.

Hubbs, C., T. Lucier, E. Marsh, G. P. Garrett, R. J. Edwards, and E. Milstead. 1978. Results of an eradication program on the ecological relationships of fishes in Leon Creek, Texas. *Southwestern Naturalist* 23: 487–96.

Hubbs, C. L., and R. R. Miller. 1943. Mass hybridization between two genera of cyprinid fishes in the Mohave Desert, California. *Papers of the Michigan Academy of Science, Arts, and Letters* 28: 343–78.

Jimenez, M., S. C. Goodchild, C. A. Stockwell, and S. C. Lema. 2017. Characterization and phylogenetic analysis of complete mitochondrial genomes for two desert cyprinodontoid fishes, *Empetrichthys latos* and *Crenichthys baileyi*. *Gene* 626: 163–72.

Johnson, J. E., and B. L. Jensen. 1991. Hatcheries for endangered freshwater fishes. In W. L. Minckley and J. E. Deacon, eds., *Battle against Extinction: Native Fish Management in the American West*, 199–217. Tucson: University of Arizona Press.

Johnsson, J. I., S. Bockmark, and J. Näslund. 2014. Environmental effects on behavioural development consequences for fitness of captive-reared fishes in the wild. *Journal of Fish Biology* 85: 1946–71.

Kallio-Nyberg, I., E. Jutila, I. Saloniemi, and E. Jokikokko. 2013. Effects of hatchery rearing and sea ranching of parents on the life history traits of released salmon offspring. *Aquaculture* 402–403: 76–83.

Karam, A. P., M. S. Parker, and L. T. Lyons. 2012. Ecological comparison between three artificial refuges and the natural habitat for Devils Hole pupfish. *North American Journal of Fisheries Management* 32: 224–38.

Keller, L. F., and D. M. Waller. 2002. Inbreeding effects in wild populations. *Trends in Ecology and Evolution* 17: 230–41.

Kihslinger, R. L., S. C. Lema, and G. A. Nevitt. 2006. Environmental rearing conditions produce forebrain differences in wild Chinook salmon *Oncorhynchus tshawytscha*. *Comparative Biochemistry and Physiology A* 145: 145–51.

Kodric-Brown, A. 1977. Reproductive success and the evolution of breeding territories in pupfish (*Cyprinodon*). *Evolution* 31: 750–66.

Koike, H., A. A. Echelle, D. Loftis, and R. A. Van Den Bussche. 2008. Microsatellite DNA analysis of success in conserving genetic diversity after 33 years of refuge management for the desert pupfish complex. *Animal Conservation* 11: 321–29.

LaVoie, A. 2004. Endangered and threatened wildlife and plants: Withdrawal of proposed rule to reclassify the Pahrump poolfish (*Empetrichthys latos*) from endangered to threatened status. *Federal Register* 69.64: 17383–86.

Le Cam, S., C. Perrier, A.-L. Besnard, L. Bernatchez, and G. Evanno. 2015. Genetic and phenotypic changes in an Atlantic salmon population supplemented with non-local individuals: A longitudinal study over 21 years. *Proceedings of the Royal Society B* 282: 20142765.

Lema, S. C., and G. A. Nevitt. 2006. Testing an ecophysiological mechanism of morphological plasticity in pupfish and its relevance to conservation efforts for endangered Devils Hole pupfish. *Journal of Experimental Biology* 209: 3499–509.

Madison, B. N., J. W. Heath, D. D. Heath, and N. J. Bernier. 2015. Effects of early rearing environment and breeding strategy on social interactions and the hormonal response to stressors in juvenile Chinook salmon. *Canadian Journal of Fisheries and Aquatic Sciences* 72: 673–83.

Martin, A. P., A. A. Echelle, G. Zegers, S. Baker, and C. L. Keeler-Foster. 2012. Dramatic shifts in the gene pool of a managed population of an endangered species may be exacerbated by high genetic load. *Conservation Genetics* 13: 349–58.

McClanahan, L. L., C. R. Feldmeth, J. Jones, and D. L. Soltz. 1986. Energetics, salinity and temperature tolerances in the Mohave tui chub, *Gila bicolor mohavensis*. *Copeia* 1986: 45–52.

Miller, R. R. 1938. Description of an isolated population of the freshwater minnow *Siphateles mohavensis* from the Mohave River basin, California. *Pomona College Journal of Entomology and Zoology* 30: 65–67.

Miller, R. R. 1948. *The Cyprinodont Fishes of the Death Valley System of Eastern California and Southwestern Nevada*. Miscellaneous Publications of the Museum of Zoology, University of Michigan, 68. Ann Arbor: University of Michigan Press.

Miller, R. R., and E. P. Pister. 1971. Management of the Owens pupfish, *Cyprinodon radiosus*, in Mono County, California. *Transactions of the American Fisheries Society* 100: 502–9.

Miller, R. R., J. D. Williams, and J. E. Williams. 1989. Extinctions of North American fishes during the past century. *Fisheries* 14: 22–38.

Minckley, W. L. 1995. Translocation as a tool for conserving imperiled fishes: Experiences in western United States. *Biological Conservation* 72: 297–309.

Minckley, W. L., and J. E. Deacon. 1968. Southwestern fishes and the enigma of "endangered species." *Science* 159: 1424–32.

Minckley, W. L., G. K. Meffe, and D. L. Soltz. 1991. Conservation and management of short-lived fishes: The cyprinodontoids. In W. L. Minckley and J. E. Deacon, eds., *Battle against Extinction: Native Fish Management in the American West*, 247–82. Tucson: University of Arizona Press.

Moczek, A. P., S. Sultan, S. Foster, C. Ledon-Rettig, I. Dworkin, H. F. Nijhout, E. Abouheif, and D. W. Pfennig. 2011. The role of developmental plasticity in evolutionary innovation. *Proceedings of the Royal Society B* 278: 2705–13.

Moyle, P. B. 2002. *Inland Fishes of California*. Berkeley: University of California Press.

Moyle, P. B., J. D. Kiernan, P. K. Crain, and R. M. Quiñones. 2013. Climate change vulnerability of native and alien freshwater fishes of California: A systematic assessment approach. *PLoS ONE* 8: e63883.

Näslund, J., and J. I. Johnsson. 2016. Environmental enrichment for fish in captive environments: Effects of physical structures and substrates. *Fish and Fisheries* 17: 1–30.

Paciorek, T., L. Al-Shaer, and M. Itzkowitz. 2014. How territoriality affects the density of an egg predator: Habitat renovation and reintroduction as a method of conserving two endangered desert spring fish. *Current Zoology* 60: 527–33.

Pister, E. P. 1993. Species in a bucket. *Natural History* 102: 14–19.

Riggs, A. C., and J. E. Deacon. 2002. Connectivity in desert aquatic ecosystems: The Devils Hole story. In D. W. Sada, and S. E. Sharpe, eds., *Conference Proceedings. Spring-Fed Wetlands: Important Scientific and Cultural Resources of the Intermountain Region May 7–9,2002*. Paradise, NV: Desert Research Institute. http://www.dri.edu/images/stories/conferences_and_workshops/spring-fed-wetlands/spring-fed-wetlands-riggs-deacon.pdf.

Robinson, A. T., and D. L. Ward. 2011. Interactions between desert pupfish and Gila topminnow can affect reintroduction success. *North American Journal of Fisheries Management* 31: 1093–99.

Salvanes, A. G. V., and V. Braithwaite. 2006. The need to understand the behaviour of fish reared for mariculture or restocking. *ICES Journal of Marine Sciences* 63: 346–54.

Shirley, P. D., and G. A. Lamberti. 2010. Assisted colonization under the US Endangered Species Act. *Conservation Letters* 3: 45–52.

Snyder, J. O. 1917. An account of some fishes from Owens River, California. *Proceedings of the US National Museum* 54: 201–5.

Snyder, J. O. 1918. Fishes of the Mohave River, California. *Proceedings of the US National History Museum* 54: 297–99.

St. Amant, J. 1983. Report on reestablishment of the Mohave chub *Gila mohavensis* (Snyder), an endangered species. *Proceedings of the Desert Fishes Council* 15: 18–19.

Stockwell, C. A., and P. L. Leberg. 2002. Ecological genetics and the translocation of native fishes: Emerging experimental approaches. *Western North American Naturalist* 62: 32–38.

Stockwell, C. A., M. Mulvey, and G. L. Vinyard. 1996. Translocations and the preservation of allelic diversity. *Conservation Biology* 10: 1133–41.

Tymchuk, W. E., B. Beckman, and R. H. Devlin. 2009. Altered expression of growth hormone/insulin-like growth factor I axis hormones in domesticated fish. *Endocrinology* 150: 1809–16.

Tymchuk, W. E., L. F. Sundström, and R. H. Devlin. 2007. Growth and survival trade-offs and outbreeding depression in rainbow trout (*Oncorhynchus mykiss*). *Evolution* 61: 1225–37.

Unwin, M. J., and G. J. Glova. 1997. Changes in life history parameters in a naturally spawning population of chinook salmon (*Oncorhynchus tshawytscha*) associated with releases of hatchery-reared fish. *Canadian Journal of Fisheries and Aquatic Sciences* 54: 1235–45.

USFWS (US Fish and Wildlife Service). 1984. *Recovery Plan for the Mohave Tui Chub,* Gila bicolor mohavensis. Portland: US Fish and Wildlife Service.

USFWS (US Fish and Wildlife Service). 1990. *Recovery Plan for the Endangered and Threatened Species of Ash Meadows, Nevada.* Portland: US Fish and Wildlife Service.

USFWS (US Fish and Wildlife Service). 1995. *San Bernardino and Leslie Canyon National Wildlife Refuges: Comprehensive Management Plan 1995–2015.* Albuquerque: US Fish and Wildlife Service.

USFWS (US Fish and Wildlife Service). 2009. *Owens Pupfish* (Cyprinodon radiosus)*: 5-Year Review: Summary and Evaluation.* Ventura, CA: US Fish and Wildlife Service.

Watters, J. V., S. Bremner-Harrison, and D. M. Powell. 2017. Phenotypic management: An inclusive framework for supporting individuals' contributions to conservation populations. In J. Vonk, J. Weiss, and S. A. Kuczaj, eds., *Personality in Nonhuman Animals,* 277–94. Switzerland: Springer Nature.

Watters, J. V., S. C. Lema, and G. A. Nevitt. 2003. Phenotype management: A new approach to habitat restoration. *Biological Conservation* 112: 435–45.

Weber, E. D., and K. D. Fausch. 2003. Interactions between hatchery and wild salmonids in streams: Differences in biology and evidence for competition. *Canadian Journal of Fisheries and Aquatic Sciences* 60: 1018–36.

West-Eberhard, M. J. 2005. Developmental plasticity and the origin of species differences. *Proceedings of the National Academy of Sciences* 102 (Suppl. 1): 6543–49.

Wilcox, J. L., and A. P. Martin. 2006. The devil's in the details: Genetic and phenotypic divergence between artificial and native populations of the endangered pupfish (*Cyprinodon diabolis*). *Animal Conservation* 9: 316–21.

Williams, J. E. 1977. Observations on the status of the Devil's Hole pupfish in the Hoover Dam refugium. Denver: US Bureau of Reclamation.

Williams, J. E. 1991. Preserves and refuges for native western fishes: History and management. In W. L. Minckley and J. E. Deacon, eds., *Battle against Extinction: Native Fish Management in the American West,* 171–89. Tucson: University of Arizona Press.

23 Restoration of Aquatic Habitats and Native Fishes in the Desert

Some Successes in Western North America

Anthony A. Echelle and Alice F. Echelle

The US Environmental Protection Agency defines restoration as the return of a degraded ecosystem to a close approximation of its *remaining* (our emphasis) natural potential (USEPA 2000). This definition acknowledges the reality that restoration to the native state is generally precluded because systems of concern have passed some threshold of anthropogenic degradation that is effectively irreversible, given political, socioeconomic, and environmental realities. The results are compromise solutions for recovery (Winemiller and Anderson 1997). Most restorations are compromise solutions because managing agencies must contend with unrelenting threats that cannot be eliminated. Predominant among these threats are anthropogenic alterations and losses of habitats to which native species are adapted and the associated ubiquity of nonnative species. Nonnative species are generally favored over natives in the artificial environments created by human activities (Moyle et al. 1986; Minckley 1991).

An example of the relationship between altered habitat and prevalence of nonnative species is the lower Colorado River from Grand Canyon downstream to the Gulf of California (Mueller and Marsh 2002; Minckley and Marsh 2009). The river, effectively remade for human needs, supports mostly nonnative fishes (Mueller and Marsh 2002). Of 13 native species (Minckley et al. 2003), 6 are gone (e.g., the large Colorado pikeminnow *Ptychocheilus lucius*), and the remainder are rare and restricted to remnants of their former occurrence (e.g., humpback chub *Gila cypha*). The nonnatives include large predators (e.g., catfishes *Ictalurus* spp., black basses *Micropterus* spp., and striped bass *Morone saxatilis*), small predators (e.g., red shiner *Cyprinella lutrensis*, western mosquitofish *Gambusia affinis*, and sunfishes *Lepomis* spp.), a planktivore (threadfin shad *Dorosoma petenense*), and an omnivore (common carp *Cyprinus carpio*).

Activities to mitigate effects of nonnative fishes on the status of native fishes and the potential for their recovery have consumed a great deal of state and federal agency resources (Mueller 2005; Albrecht et al., chap. 11, this volume). An example illustrating compromise restoration is the Gila River Basin Native Fishes Conservation Program (GRBNFCP). The program is funded by the US Bureau of Reclamation (USBR) to mitigate negative effects of nonnative fishes escaping the Central Arizona Project (CAP), an aqueduct transporting water from the Colorado River to southern and central Arizona (Duncan and Clarkson 2013). In exchange, USBR is permitted under the Endangered Species Act (ESA) to continue CAP operation, despite the risks associated with nonnatives in the canal.

Some have argued that if not for the presence of nonnatives, many native species would thrive in the highly altered aquatic environments of the American Southwest (Minckley 1991; Minckley and Marsh 2009), but this depends heavily on the species or situation. There are many examples where the predominant factor in the decline of native fishes is an anthropogenically altered abiotic environment. One such example is the Yaqui catfish *Ictalurus pricei* at San Bernardino National Wildlife Refuge, in southeastern Arizona. Nonnatives apparently were not an issue in the various spring-fed impoundments supporting reestablished populations (W. Radke, pers. comm.), yet the population has performed poorly and is predicted to be extinct by sometime in 2018, probably because of the absence of critical, but poorly understood, habitat features (Stewart et al. 2017). Another example is the struggle to maintain the last population of Rio Grande silvery minnow *Hybognathus amarus* in the face of an altered hydrology that is directly interfering with completion of its life cycle (Osborne et al. 2012). Especially stark examples of direct effects of anthropogenic hydrologic alterations are geographically restricted spring-dwelling species driven to extinction or threatened with such by over-mining of groundwater and resultant loss of spring flows (Contreras-Balderas and Lozano-Vilano 1996; Garrett et al., chap. 8, this volume).

The original ESA imposed rigorous restrictions on land use activities, by both private landowners and government agencies, that were potentially harmful to local populations of species listed under the ESA as in danger of extinction (endangered) or likely to become so in the foreseeable future (threatened). Subsequent amendments and reinterpretations of the ESA, however, provide a variety of mechanisms promoting compromise recovery solutions involving relaxation of ESA restrictions on land use (Andersen and Brooks, chap. 12, this volume). Examples include designation of reestablished populations of endangered species as experimental or non-essential and the development of Habitat Conservation Plans, Candidate Conservation Agreements with Assurances, and Safe Harbor Agreements. On the positive side, these mechanisms often reduce the time lag and procedural burden required for implementation of conservation actions, and relaxed ESA restrictions encourage approval and participation from the public, private landowners, tribes, and government agencies. On the negative side, they reduce legal protection for the populations involved. The overall goal of a net benefit for conservation can be hampered by inadequate biological information relating to management options and the potential irreversibility of habitat loss resulting from land use decisions (Langpap and Wu 2004).

Restorations are done under constraints limiting the scale and effectiveness of the projects, usually falling far short of creating self-sustaining populations. As Minckley (1995, 298) noted, "A formidable gap exists between saving an organism from extinction and ensuring it to be capable of programming its own future." Indeed, most federally listed species are conservation-reliant to the point that no amount of restoration will restore them to self-sufficiency (Doremus and Pagel 2001; Goble 2010). Given these limitations, success in restoration projects must be defined relative to what might have been, or would be, without the effort. In this sense, success refers to positive results for the target species, even if they are likely to be short-lived without continued effort.

The case studies described herein are a select few of those worthy of inclusion. For example, without emergency rescue efforts, Big Bend gambusia *Gambusia gaigei* (Hubbs and Brodrick 1963), Owens pupfish *Cyprinodon radiosus* (Miller and Pister 1971), and Pahrump killifish *Empetrichthys latos* (Minckley et al. 1991) would now be extinct. Without translocations to unoccupied sites, the least chub *Iotichthys phlegethontis* probably would not have been removed from the list of candidates for federal listing (Thompson et al. 2015). Without habitat renovation with a piscicide and restocking with captive-reared fish (Echelle et al. 2004), the Leon Springs pupfish *Cyprinodon bovinus* would have been effectively eliminated by hybridization with an introduced congener (Echelle and Echelle 1997), and habitat manipulation shows promise of heightening the size of the post-renovation breeding population (Black et al. 2016). Without prescribed burning and excavation to create open habitat, the population size of Foskett speckled dace *Rhinichthys osculus* ssp. would be precariously small compared with the post-restoration size (about 2,000 versus 25,000; Scheerer et al. 2015). Without a multitude of efforts aimed at restoring habitats and controlling nonnatives on the Ash Meadows National Wildlife Refuge (e.g., Weissenfluh 2007; Scoppettone et al. 2005), the status of a variety of endemic aquatic taxa would be far less secure than it is. The restoration projects discussed herein encapsulate the range of conservation efforts for native arid-land fishes in North America. We end by considering the commonalities of success and by commenting on some sources of uncertainty for these efforts and their implications for native fish restoration.

CASE STUDIES

Floodplain Species

Oregon chub *Oregonichthys crameri*

Oregon chub is a small (<9 cm total length [TL]) floodplain minnow (Cyprinidae) (fig. 23.1) endemic to the Willamette River system, a basin containing two-thirds of Oregon's human population. The species evolved in an unconstrained and dynamically changing riverscape, where it inhabited oxbow lakes and other slack-water, off-channel (floodplain) habitats. Historically, the river underwent frequent floods (Benner and Sedell 1997), permitting dispersal and genetic exchange among populations (Scheerer 2002).

The species had experienced severe decline by the mid-1900s after the Willamette River underwent extensive modifications (Markle et al. 1991), which included flood control projects (dams and levees), channelization of the river and its tributaries for navigation, and introductions of nonnative species (USFWS 1993a; Benner and Sedell 1997). The result was reduced

Fig. 23.1 Off-channel habitat for Oregon chub: Berry Slough, Willamette River, Lane County, Oregon. *Inset:* Oregon chub. (Photograph by and published with permission of B. Bangs [ODFW]. Oregon chub drawing copyright J. Tomelleri.)

availability and quality of chub floodplain habitats and reduced potential for dispersal and genetic exchange among such habitats. In the late 1980s, the species was detected at only 8 of 29 historically documented locations (Pearsons 1989), prompting its listing as federally endangered (USFWS 1993a). Listed threats were habitat degradation, competition and predation by nonnative species, and inadequacy of regulatory mechanisms. The Oregon Chub Recovery Plan (USFWS 1998) deemed the species to have low recovery potential. Thus, it is remarkable that on February 17, 2015, it became the first fish to be removed, because of recovery, from the US endangered species list (USFWS 2015a).

Implementation of the recovery plan was central to delisting Oregon chub. The plan included guidelines for research, a detailed inventory of sites for previously undetected populations or potential for reestablishing populations, and methods for public outreach and education. It required protection of existing wild populations and reestablishment of populations at sites within the historical range but unoccupied by nonnatives. Importantly, starting in 2002, the US Army Corps of Engineers (USACE) implemented minimum outflows from dams on the Willamette River to sustain downstream floodplain habitat (USFWS 2015a).

The recovery plan included three criteria for downlisting and delisting [in brackets]: (1) establish and manage 10 [20] populations of at least 500 adults each, all showing a stable or increasing trend for 5 [7] years; (2) at least 3 [4] such populations had to be in each of the three sub-basins supporting the species; and (3) for delisting, management of the 20 populations had to be guaranteed in perpetuity. A population was defined as a group of chubs in a single defined water body; those in adjacent sloughs or ponds with an open connection were treated as one population.

In 2007, there were 38 known populations, and all downlisting criteria had been met. By 2010, with 50 populations and an upward trend in stability, the chub was reclassified as threatened (USFWS 2010a), and in 2015 it was delisted (USFWS 2015a). At the time of delisting, there were 77 populations, 41 with 500 or more adult fish. Of the latter, 23 showed stable or increasing abundance (USFWS 2015a). The 77 populations included the 8 known at the time of listing, 48 discovered after sampling of more than 1,000 locations over 20 years, and 21 representing translocations to new locations (B. Bangs, pers. comm.). Most of the newly discovered populations probably reflected sampling effort, rather than range expansion, because many of them were in isolated habitats (Scheerer 2007). However, several previously unknown populations apparently became established through natural colonization of locations with frequent or continuous connections between the river and the floodplain (B. Bangs, pers. comm.).

Sites chosen for reestablished populations were hydrologically isolated habitats devoid of nonnative fishes. Each population in a given sub-basin was founded with large numbers of wild-caught individuals (hundreds) from the same sub-basin. Founding procedures were partially based on microsatellite DNA variation among populations (DeHaan et al. 2012). A follow-up study indicated that genetic variation in the natural populations had been successfully transferred to reestablished populations (DeHaan et al. 2016). The overall result was the presence of protected and abundant populations across several Willamette River basin tributaries (USFWS 2015a).

Partnerships were the cornerstone of success. Even before the species was listed, the Oregon Chub Working Group was created to improve the status of the fish. This diverse group included the Oregon Department of Fish and Wildlife (ODFW), the US Forest Service (USFS), USACE, McKenzie River Trust, the Oregon Parks and Recreation Department, the Confederated Tribes of the Grand Ronde, local watershed councils, and private landowners. Most, along with the National Resources Conservation Service, were directly involved in restoring the species. Safe Harbor Agreements facilitated the participation of private landowners. A nine-year monitoring plan (USFWS 2013a) became effective immediately following delisting. Monitoring can be terminated in 2022 if there is no evidence of a population decline and no basis for expecting one in the foreseeable future.

Gila topminnow *Poeciliopsis occidentalis*

Gila topminnow is a small (females ≤60, males ≤35 mm TL) livebearer (Poeciliidae). Its range encompassed the Gila River basin of Arizona and New Mexico and the small Río Concepción basin in northern Sonora. Until the 1940s and 1950s, Gila topminnow was widely distributed and abundant in its native range (Hubbs and

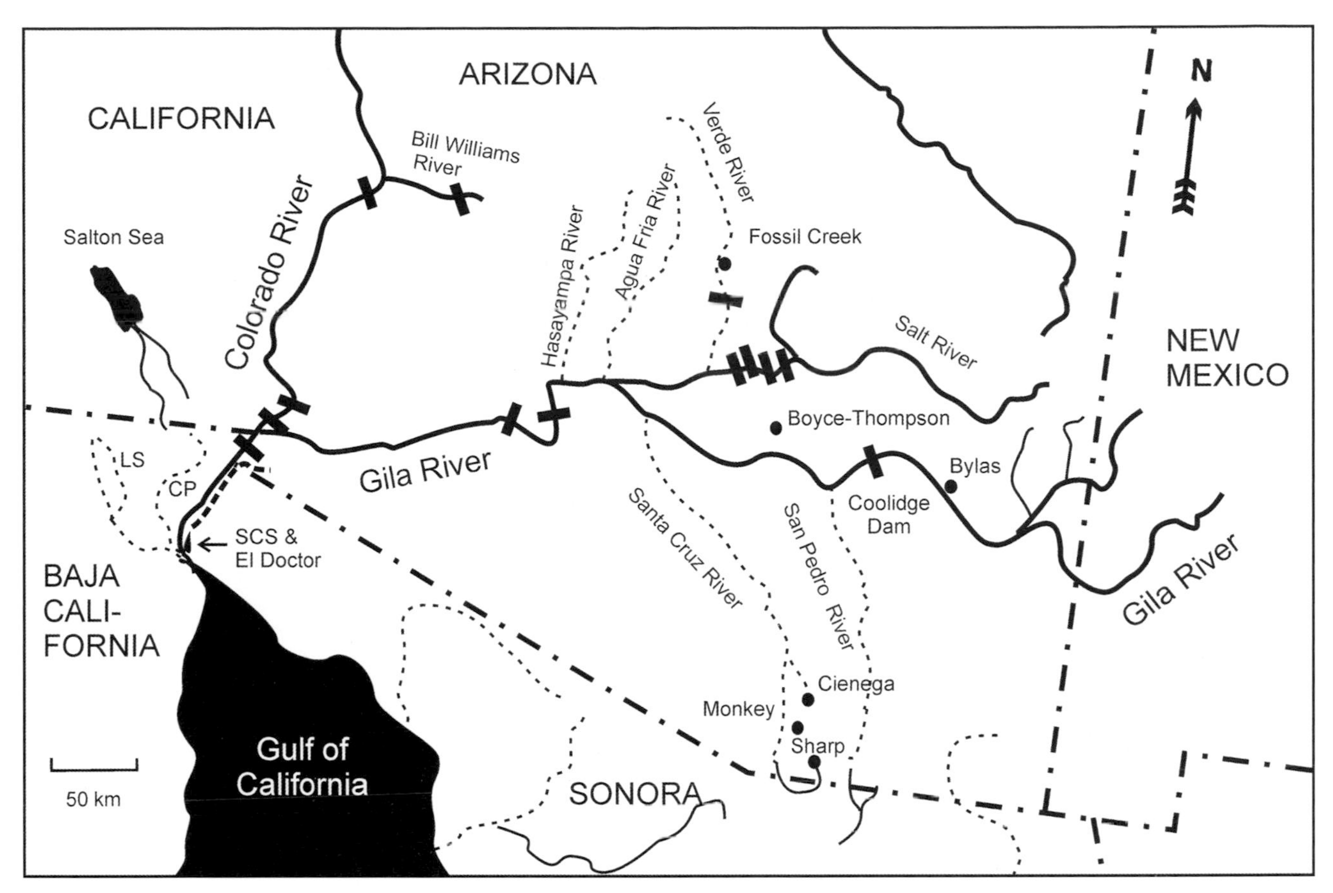

Fig. 23.2 Gila River and Lower Colorado River, with some localities and streams mentioned in the text. Bars = major dams. LS = Laguna Salado, CP = Cerro Prieto, and SCS = Santa Clara Slough, created by water diverted from the Colorado River into the Wellton-Mohawk Irrigation District (southern Arizona). Water from irrigated fields drains into the MODE (Main Outlet Drain Extension) Canal (heavier dashed line) and is carried to SCS.

Miller 1941; Miller 1961). Although mainly a floodplain species, it also occupied small streams and spring systems. However, by the 1960s, natural populations in the United States were confined to 10–12 isolated localities (Minckley et al. 1991), primarily in the upper Santa Cruz River watershed (fig. 23.2). This range contraction mirrored the spread of nonnative fishes, the most problematic being western mosquitofish *Gambusia affinis*, a small, ecologically similar poeciliid (Minckley and Marsh 2009). Gila topminnow repeatedly disappeared from habitats within months of western mosquitofish detection (Meffe et al. 1983).

Gila topminnow recovery efforts were (Minckley et al. 1991), and still are, focused on establishing populations, maintaining captive (non-aquarium) stocks, and monitoring. Populations are established in natural waters as well as semi-natural habitats such as small stock ponds and metal tanks (fig. 23.3). Preferred sites are below 1,600 m elevation in small watersheds and in vegetated spring runs and spring-fed ponds (Meffe et al. 1983; Sheller et al. 2006).

Until the 1990s, stocks used for transplantation of Gila topminnow were chosen with little attention to the possibility of geographic variation in life histories or genetics of natural populations (Hendrickson and Brooks 1991). To preserve genetic diversity, the revised recovery plan (Weedman 1998) recommended establishing and protecting wild replicates of each of the known remnant populations and giving equal protection to wild populations derived from a mixed stock held at Boyce-Thompson Arboretum State Park. Genetic studies and other considerations led to recognition of two evolutionarily significant units (ESUs; Hedrick et al. 2006; Hedrick and Hurt 2012) among populations of the species, enabling management with attention to the natural pattern of diversity.

Recovery has benefited from an umbrella Safe Harbor Agreement granted to the Arizona Game and

Fig. 23.3 Three sites of reestablishment or attempted reestablishment of Gila topminnow and desert pupfish. *Upper left:* Heron Spring, wild habitat, Santa Cruz basin; Gila topminnow (*inset*) reestablished in 1981, blinked out by 2007. *Upper right:* Larry Creek, Agua Fria River basin; wild habitat stocked in 2005–2006 with Gila topminnow (extant) and in 2006 with desert pupfish (*inset*, not extant). *Bottom:* Antelope Tank being stocked with pupfish in 2013; water is supplied to this cement tank by a windmill. BLM property, Apple-Whittell Audubon Research Ranch. (Photographs and notes provided by D. Duncan, USFWS. Gila topminnow and desert pupfish drawings copyright J. Tomelleri.)

Fish Department (AZGFD) (Duncan and Voeltz 2005). Under this agreement, AZGFD has enlisted the Bureau of Land Management, the Arizona Parks and Recreation Department, municipalities, The Nature Conservancy, and private landowners to allow recovery activities on their lands. Between 2006 and 2017, there were 45 translocations and 119 augmentations, with 25 populations surviving three or more years, 11 surviving five or more years, and only 2 failures (AZGFD files). Four populations were established in renovated habitat, the largest of which were 4.2 km of Bonita Creek (Robinson 2008) and 14 km of Fossil Creek (see below). Estimates indicate that most populations contain several hundred (15 populations) or several thousand (20) individuals, and four populations have more than 10,000; only two have fewer than 100. This success reflects heightened monitoring and augmentation activities as well as a focus on quality in choosing sites for transplantation.

The revised Gila topminnow recovery plan considers delisting of the species infeasible for the foreseeable future because of the extremely degraded condition of the Gila River watershed (Weedman 1998). The recovery plan criterion for downlisting from endangered to threatened requires, in addition to persistence of the remnant natural populations, 20 contemporaneous, reestablished wild populations persisting for three or more years (USFWS 1984). This criterion was satisfied from 1985 through 1990, and the species was considered for downlisting (Simons et al. 1989). However, the number of populations fell below the requirement in 1991. Currently, there are 27 populations, mostly on lands administered by the federal government, the state of Arizona, or The Nature Conservancy, and downlisting is a possibility (A. Robinson, pers. comm.).

Desert pupfish *Cyprinodon macularius*

Desert pupfish (see fig. 23.3) is a small (≤65 mm TL), hardy species (Cyprinodontidae). Spotty records show historical occurrences in the Gila River basin from central and southern Arizona and northern Sonora west to the Colorado River, and south and west into the Colorado River delta and the Salton Sea (see fig. 23.2) (Minckley and Marsh 2009). Surveys of extant populations on the delta (Hendrickson and Varela-Romero 1989) suggest that the species was especially abundant in harsh habitats relatively intolerable to other species. Such habitats (e.g., shallow backwaters and drying marsh pools) were probably common on the Gila River floodplain and the Colorado River delta before upstream overexploitation of the Colorado River. By the 1970s, the desert pupfish was extirpated from Arizona, had declined markedly in the Salton Sea area, and occurred only at scattered localities on the delta (Minckley et al. 1991). The decline coincided with extensive losses of surface waters from delta and floodplain habitats in the Gila River watershed. Additionally, nonnative fishes probably contributed to decline of desert pupfish in the Colorado River delta (Hendrickson and Varela-Romero 1989; Varela-Romero et al. 2002) and the Salton Sea area (Schoenherr 1988).

Remnants of natural desert pupfish populations persist in about 10 local areas, with different degrees of isolation from neighboring populations. Five are in the Salton Sea area and five on the Colorado River delta in México (USFWS 2010b). In the Salton Sea, the species occurs at low densities in shallow, nearshore waters, and sometimes abundantly in shoreline pools and irrigation drains emptying to the lake (S. Keeney, pers. comm.); it also occurs in short (<0.5 km) segments of Salt Creek and Hot Mineral Spa Wash, and in about a 10 km reach of San Felipe Creek, including San Sebastian Marsh. In México, it occurs in a small spring (Pozo del Tules) at the edge of the usually dry lakebed of Laguna Salada (C. O. Minckley, pers. comm.), a wetland near the Cerro Prieto Geothermal Power Station, a stream downstream of the power station, two or three localities in Santa Clara Slough, which receives its water via the MODE Canal (see fig. 23.2), and El Doctor, where a series of small springs emerges from a terrace at the edge of the delta.

Conservation actions for desert pupfish focus on establishing refuge populations (wild or semi-wild, non-aquarium) within its historical range. The founding stocks for such activities are obtained directly, or indirectly (via captive stocks), from the Colorado River Delta and the Salton Sea area. Since 2007, refuges have been established with a combination of fish from the Southwestern Native Aquatic Resources and Recovery Center (SNARRC) and one or another of the captive populations in Arizona (A. Robinson, pers. comm.). The use of mixed stock is based on the finding that individual refuge stocks, including the SNARRC stock, had markedly lower genetic diversity than the wild delta stocks, but that the global refuge stock was only slightly less diverse than the wild stocks (Koike et al. 2008).

Today in Arizona, there are 43 desert pupfish populations in refuges. The populations average 9 years in age and range from 1 to 32 years. In 2010, 16 such populations had persisted 10 or more years in Arizona (USFWS 2010b), but only 3 of those populations exist today, along with 10 others, demonstrating a high degree of turnover. In California, 1–4 refuge ponds or artificial streams support populations (including a number that have been lost and restocked) in each of seven localities in the Salton Sea Basin (USFWS 1993b; S. Keeney, pers. comm.). Downlisting of the species is not being considered because of the precarious existence of remnant natural populations on the delta and the uncertain future of the Salton Sea (D. Duncan, pers. comm.).

Stream-Dwelling Species

Modoc sucker *Catostomus microps*

Modoc sucker, a relatively diminutive fish (≤34 cm TL; Catostomidae), inhabits small perennial and intermittent streams at elevations of about 1,280 to 1,768 m in the Ash Creek and Turner Creek watersheds of the Pit River, in northeastern California, and the Thomas Creek watershed of Goose Lake Basin, in southern Oregon and northern California (figs. 23.4 and 23.5; Reid 2007, 2008). Historically, Goose Lake has overflowed into the North Fork of the Pit River, and on several occasions it has completely dried. The lake reportedly last overflowed in 1881, so it has been somewhat disconnected from the Pit River since that time (Laird 1971). The Thomas Creek population was unknown when the species was listed as endangered (USFWS 1985). Its apparent extirpation from much of its historical range was attributed to hybridization with a native Pit River species, Sacramento sucker *Catostomus occidentalis*, and to habitat loss and degradation caused by overgrazing, siltation, and channelization. The original objectives for species recovery were to restore and maintain habitat for Modoc sucker in Turner and Ash Creeks and, in select stream reaches, to control potential hybridization with Sacramento sucker (USFWS 2015b, c).

The listing of Modoc sucker (USFWS 1985) assumed that human activities had eliminated natural barriers and thus exposed it to predatory nonnatives and hybridization with Sacramento sucker. Subsequent studies suggested that such barriers never existed (USFWS 2015b, c). Separation was primarily ecological, usually with Modoc suckers in smaller, sometimes intermittent, headwater streams and Sacramento suckers in larger, warmer, perennial downstream reaches of tributaries and river mainstems (Moyle and Marciochi 1975; Moyle et al. 1982). Reproductive isolation was probably reinforced by other natural history differences (Reid 2008; Smith et al. 2011). Post-listing morphological and genetic studies indicated that genetic introgression via hybridization is part of the evolutionary legacy (of both species) and apparently poses no genetic threat (Kettratad 2001; Dowling 2005). The Modoc sucker maintains morphological and ecological distinctiveness, even in the face of genetic introgression (Topinka 2006; Smith et al. 2011).

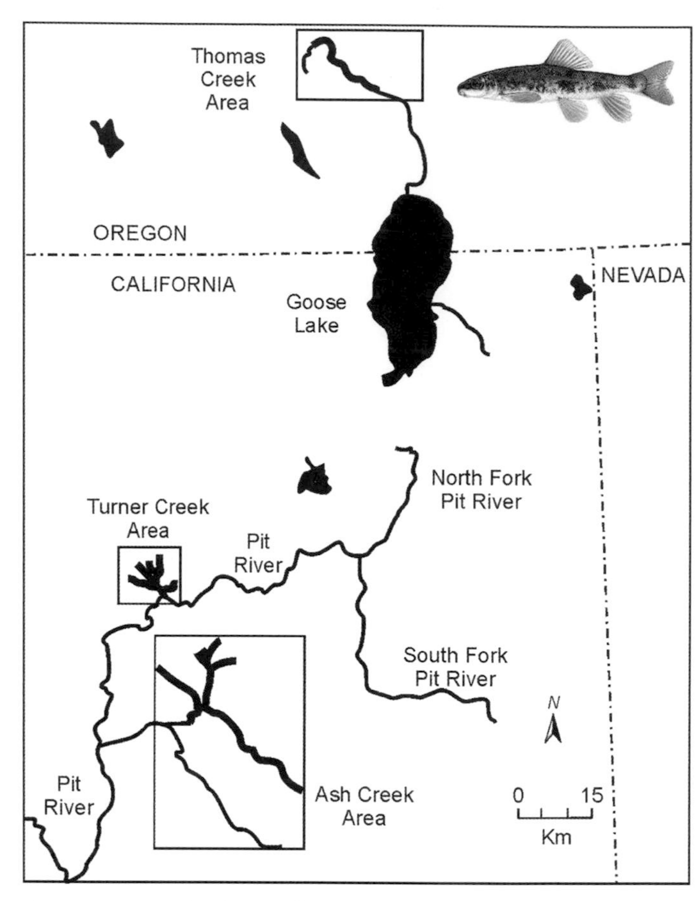

Fig. 23.4 Present distribution of Modoc sucker. Occupied stream reaches are those drawn with thick lines. *Inset:* Modoc sucker. (Redrawn from USFWS 2009. Modoc Sucker drawing copyright of J. Tomelleri.)

Fig. 23.5 Modoc sucker habitat, Turner Creek, California. (Photograph provided by N. Banish, USFWS.)

Predatory nonnative fishes appear not to pose a threat to the Modoc sucker, despite the presence, in two of the three drainages supporting Modoc sucker, of two primary predators; brown trout *Salmo trutta* in Ash Creek and largemouth bass *Micropterus salmoides* in Turner Creek (USFWS 2009, 2015c). A natural waterfall in the lowest reach of Thomas Creek precludes nonnative invasion from downstream. Since 2005, the USFWS has supported removal efforts (by angling and by hand) that have eliminated largemouth bass from upper Turner Creek, illustrating the feasibility of control in small streams once the species is detected. Although suppressed by nonnatives, Modoc sucker has coexisted with largemouth bass and/or brown trout for decades—more than 80 years in Ash Creek (USFWS 2015d). Importantly, the California Department of Fish and Wildlife has discontinued stocking of brown trout in the Pit River, and the ODFW does not stock nonnatives in the Goose Lake Basin (USFWS 2015c).

Improved livestock management was also instrumental in recovery of the species. Fencing projects excluded cattle and sheep from substantial reaches of habitat, reduced erosion, improved channel morphology (narrower and deeper, vegetation-bordered channels) to favor the sucker, and reestablished riparian vegetation and instream cover. Other restoration work included use of off channel stock tanks, stabilizing streambanks with cut-juniper revetments, creating and expanding pools, placement of boulders in streambeds for instream cover, and restoration of head-cut areas to prevent downcutting and channel incision (Reid 2008).

When Modoc sucker was listed, known populations were confined to 20.8 km in 7 streams of the Turner Creek and Ash Creek sub-basins. After listing, populations were discovered in Thomas Creek (Reid 2007), and successful transplants into unoccupied streams led to the species being present and well established in each sub-basin of its historical occurrence (USFWS 2015b). When delisted in 2015, the species occupied 68.4 km of habitat in 12 streams.

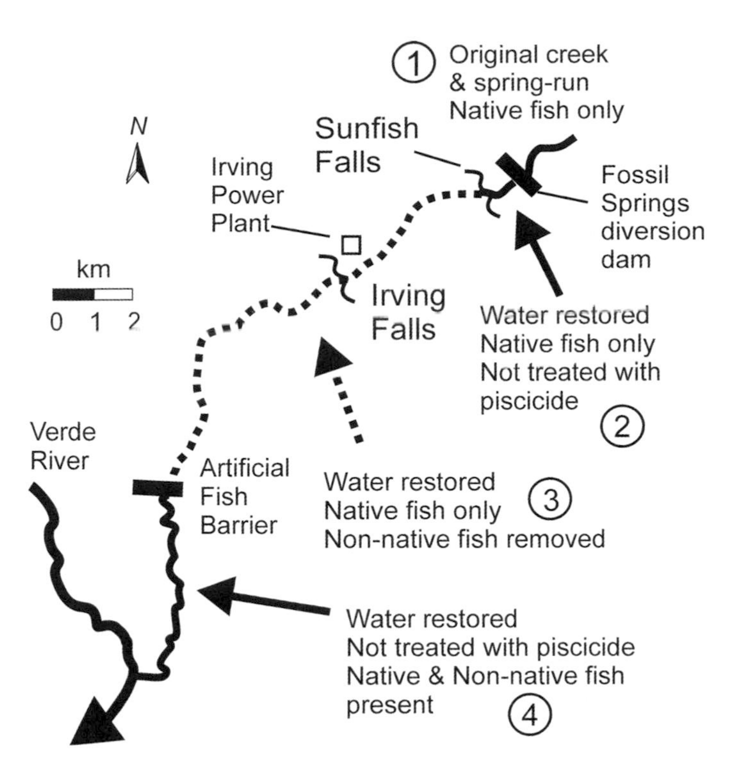

Fig. 23.6 Map of Fossil Creek. Segment 1 received Fossil Springs flow before restoration; flows were restored to segments 2, 3, and 4. Segments 1 and 2 were free of nonnative fish and were not treated with piscicide; segment 3 was treated with piscicide; segment 4 was untreated. (Modified from Marks et al. 2010.)

Fossil Creek fishes

Fossil Creek, a spring-fed tributary of the Verde River (fig. 23.6; see also fig. 23.2) in the Mazatzal Mountains of central Arizona, underwent severe water depletion for nearly a century following the construction of Fossil Springs Diversion Dam (fig. 23.7; Weedman et al. 2005; Marks 2007). Dewatering, together with invasions by nonnative fishes downstream of the dam, caused a severe decline in the native Fossil Creek fauna, which included two endemic invertebrates (Fossil springsnail *Pyrgulopsis simplex* and a microcaddisfly *Metrichia nigritta*) and six native fishes (Sonora sucker *Catostomus insignis*, desert sucker *Pantosteus clarkii*, headwater chub *Gila nigra*, roundtail chub *Gila robusta*, longfin dace *Agosia chrysogaster*, and speckled dace *Rhinichthys osculus*) (Marks et al. 2005).

Perennial flow in Fossil Creek originates with Fossil Springs, a complex of several outlets in a 200 m stream reach about 22 km upstream of the creek's confluence with the Verde River (Malusa et al. 2003). The diversion dam, about 0.5 km downstream from the spring source, diverted nearly the entire base flow into a system of flumes and pipes serving two hydropower plants (Weedman et al. 2005). Historically, the creek had a step-pool stream morphology, with each pool bounded by travertine dams, over a reach of about 10 km (Fuller et al. 2011). The dams were formed of stream sediments and woody debris embedded in a matrix of $CaCO_3$ precipitate, deposited when CO_2-charged subterranean water emerges as spring water. Most of the dams were destroyed by floods after water diversion (Malusa et al. 2003; Fuller et al. 2011), and travertine deposition was restricted to less than 1 stream kilometer (5–6 km downstream of the dam) that was receiving 57–150 L/s of water, primarily as return flow from one of the power plants.

Prior to restoration, only natives occurred upstream of the diversion dam and in a short segment between the diversion dam and a natural dam of boulders at Sunfish Falls. Downstream of Sunfish Falls, native fishes were relatively uncommon, but they coexisted with eight nonnative taxa, six fish and two crayfish (Marks et al. 2005). Above the dam, the springs supported headwater chub, speckled dace, longfin dace, desert sucker, and the two endemic invertebrates. Only desert sucker occupied isolated pools upstream of the springs.

Fig. 23.7 Fossil Creek. *Top*, diversion dam, March 24, 2009, after full flow was restored and the dam was lowered by 36 cm. *Bottom*, Fossil Creek downstream of the diversion dam, August 19, 2008, after full flow over the dam was restored. (Photos by and published with permission from K. Ashcraft, Arizona Public Services.)

In 1999, following a long public campaign to restore the natural Fossil Creek system, Arizona Public Services (the power plant operator) signed an agreement in principle to return full flows by December 31, 2004 (Weedman et al. 2005). In preparation for this event, a variety of public, state, federal, and tribal stakeholders developed a plan to eradicate with piscicide all nonnative fishes from a section of Fossil Creek after salvaging natives for subsequent return to the treated section (Weedman et al. 2005). Prior to restoration of full flow, and after construction by USBR of a fish-movement barrier, a piscicide, Fintrol (active ingredient antimycin-A), was used to eliminate nonnative fishes from a 14 km stream reach during October and November 2004. Stock tanks in the watershed upstream of the fish barrier were surveyed; 5 of 48 contained nonnatives and were treated with Fintrol. In June 2005, Fossil Springs flow was directed around the diversion dam, restoring full flow to the creek. Base flow rose from less than 1 to 1,220 L/s (Marks et al. 2005). In 2010, the dam height was lowered by 36 cm, directing flow over the dam.

Between 2007 and 2015, AZGFD and USBR repeatedly introduced four native Gila River fishes into Fossil Creek (Mosher 2017). As of autumn 2017, Gila topminnow and spikedace *Meda fulgida* had self-sustaining populations, but razorback sucker *Xyrauchen texanus* and loach minnow *Tiaroga* (*Rhinichthys*) *cobitis* did not (A. Robinson, pers. comm.).

Events of 2009–2011 highlighted the importance of vigilance in maintaining the hard-won gains of efforts like the Fossil Creek project. An unusually large flood in winter 2009–2010 deposited boulders and other sediments on the downstream side of the fish-movement barrier, reducing its effectiveness. This apparently explains the detection, in July 2011, of nonnative smallmouth bass *M. dolomieu* as far as 200 m upstream of the barrier (USFS 2016). This discovery led to repair of the barrier and successful removal of smallmouth bass above it. Such activities were effectively mandated at the highest federal level when, on March 30, 2009, President Obama signed the Omnibus Public Land Management Act, a detail of which protected 27 km of Fossil Creek and required federal managing agencies to "protect and enhance" the river's free-flowing condition, water quality, and "outstandingly remarkable values" (USFS 2016, 1).

Spring Endemics

Borax Lake chub *Gila boraxobius*

Borax Lake chub is, at a maximum of 110 mm TL, a dwarf member of its genus (Williams and Bond 1980). The species is endemic to Borax Lake and locally adjacent wetlands (fig. 23.8) in the Alvord Desert of southeastern Oregon. Water emerges from several thermal springs at the bottom of the lake, and precipitation of sodium borate over the millennia has raised the lake shoreline approximately 10 m above the adjacent desert playa. The lake is small (about 4.1 ha; average depth <1 m) with surface-water temperatures ranging from 16°C to 38°C. Lake effluent feeds surrounding marshes, small pools, and Lower Borax Lake, an artificial reservoir. The chub is inherently at high risk of extinction because it is short-lived (rarely >2 years) and restricted to a single water source in a perched topographic position (Williams and Bond 1980; Scoppettone et al.1995).

Borax Lake chub was listed as endangered under emergency provisions of the ESA (USFWS 1980) in the year it was described as a species (Williams and Bond 1980). This emergency listing was followed by a final listing rule designating 259 ha (private and public) surrounding the lake and adjacent wetlands as critical habitat (USFWS 1982). The emergency listing was prompted by threats centered on vulnerability of its limited habitat. In 1980, the perimeter of the lake was modified for irrigation, and the lake level dropped about 0.3 m, decreasing available habitat and increasing water temperatures. Another threat was the prospect of leasing of Borax Lake and surrounding land for geothermal development. Additionally, the potential for vehicular destruction of the fragile salt-encrusted shoreline was a concern.

The Borax Lake Chub Recovery Plan (USFWS 1987) set four criteria for federal downlisting of the chub from endangered to threatened (Scheerer and

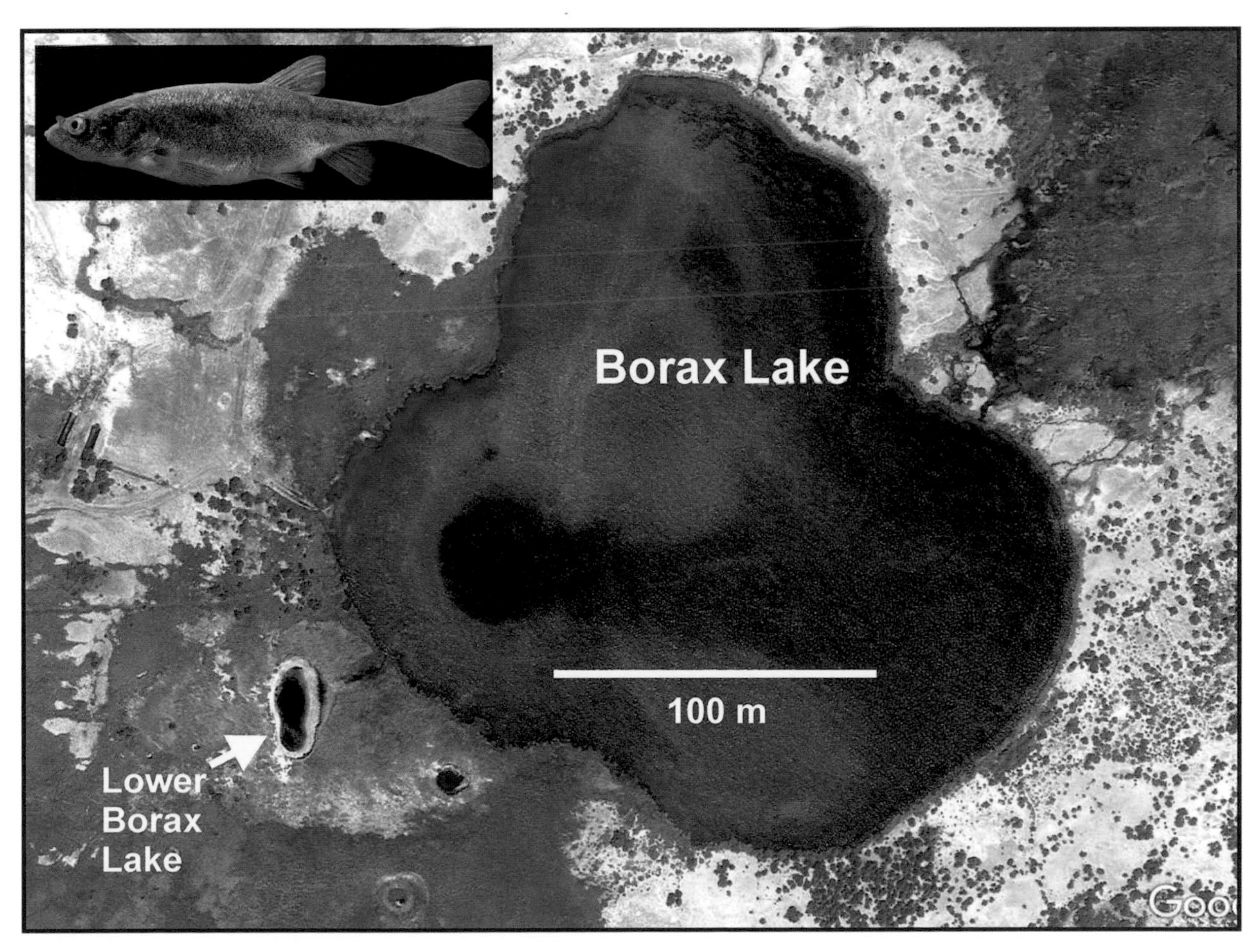

Fig. 23.8 Borax Lake, Harney County, Oregon. (Image from Google Earth, April 25, 2017; Map data ©2017 Google.) *Inset:* Borax Lake chub. (Fish image by and published with permission from D. Markle, Oregon State University.)

Clements 2015). Criterion 1, natural reproduction in Borax Lake, with an absence of nonnatives, has been met. Between 2005 and 2012, there was good recruitment and a stable trend in population size, with estimates of numbers varying from 8,246 to 26,571. Criterion 2, permanent protection for 65 ha surrounding and including Borax Lake, has also been met (Scheerer and Clements 2015). In 1983, The BLM designated the public land in the Borax Lake vicinity as an Area of Critical Environmental Concern, and 130 ha of private land surrounding and including Borax Lake was leased and eventually purchased (1993) by The Nature Conservancy, bringing all critical habitat into public or conservation ownership and terminating water diversion for livestock and irrigation. Criterion 3, removal of threats to subsurface waters, was considered to have been achieved within practical bounds (Scheerer and Clements 2015). Finally, criterion 4, reestablishment of ponds and natural marshes, was met, except for Lower Borax Lake, which has been blocked to chub entry (USFWS 2012) because it dries in the summer and is a potential population sink (Williams and Macdonald 2003).

By the early twenty-first century, Borax Lake chub was considered on the brink of recovery (Motivans and Balis-Larsen 2003), and with subsequent recovery activities, USFWS (2012) recommended its downlisting from endangered to threatened. By then, its recovery had surpassed the criteria for downlisting and had satisfied criteria for delisting (Scheerer and Clements 2015). Meeting the downlisting requirements effectively satisfied the first three (of six) delisting criteria. The fourth delisting criterion was met with the completion, in 2011, of a fence to exclude vehicles from critical habitat. The last two criteria, failure to identify new threats to the ecosystem for five consecutive years and establishment of a monitoring program for both the fish and the habitat, have also been met. Importantly, the Steens Mountain Cooperative Management and Protection Act, approved by President Clinton on October 24, 2000, heightened protection for most BLM land within the Alvord Basin by banning mineral and geothermal development and by designating it as a wilderness area. In 2019, the USFWS declared the species to be recovered and proposed delisting (USFWS 2019). Although criteria for delisting were met, Borax Lake chub will probably remain vulnerable and of management concern for the foreseeable future because of its brief lifespan and restriction to a single, geologically fragile site (Williams et al. 2005).

Rancho Nuevo springs

Perennial water in Ejido Rancho Nuevo, Guzmán Ba-

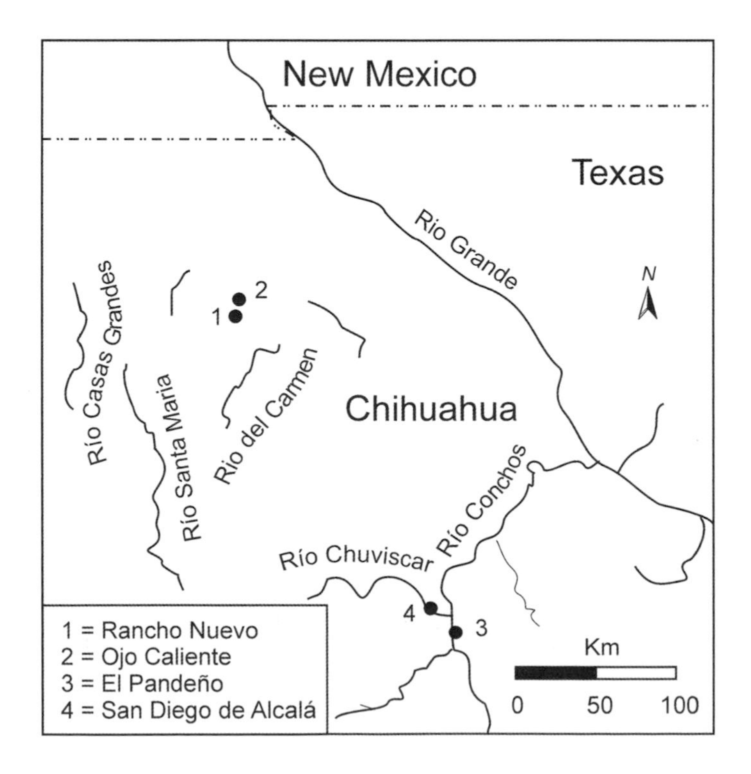

Fig. 23.9 Mexican localities associated with conservation efforts mentioned in the accounts for Rancho Nuevo springs and El Pandeño. Localities 1 and 2 are in the closed (endorheic) Guzmán Basin.

sin, Chihuahua (fig. 23.9), once emerged from five primary spring systems: Ojo del Apache, Ojo de Carbonera, Ojo de en Medio, Ojo de las Varas, and Ojo Solo. The known endemic fauna included Carbonera pupfish *Cyprinodon fontinalis* (fig. 23.10), largemouth shiner *Cyprinella bocagrande*, a springsnail *Tryonia contrerasi*, an isopod *Thermosphaeroma milleri*, and the Chihuahuan dwarf crayfish *Cambarellus chihuahuae*. The crayfish had been considered extinct, but was rediscovered in Ojo Solo (Carson et al. 2015).

The five Rancho Nuevo springs had a long history of anthropogenic disturbance (Smith and Miller 1980). All except Ojo de Carbonera were impounded, with the spring pools emptying directly into irrigation ditches. Ojo de Carbonera ran from a series of small headsprings for about 100 m before entering the irrigation system. By 2009, four of the five springs were dry, apparently because of groundwater mining (Alvarez et al. 2010). By 2012, the remaining spring (Ojo Solo) was declining (see fig. 23.10; Carson et al. 2015), and it was dry in 2016 (M. de la Maza-Benignos, pers. comm.).

In 2013, prior to the loss of Ojo Solo, a refuge was constructed in a marsh fed by Ojo Caliente (see fig. 23.10), a hot spring system about 10 km to the north (see fig. 23.9) in Ejido Villa Ahumada y Anexos. The suitability of Ojo Caliente as a refuge, and the absence of native fishes and crayfish there, had been confirmed by a biotic survey and hydrology and topology studies (Carson and de la Maza-Benignos 2014). Translocation of Carbonera pupfish, largemouth shiner, and Chihuahuan dwarf crayfish from Ojo Solo to Ojo Caliente began in February 2014, and all appear established in the refuge (Carson et al. 2015). Long-term plans include expansion of the Ojo Caliente refuge, continued outreach programs, and population and genetic monitoring of the translocated species (Carson and de la Maza-Benignos 2014; M. de la Maza-Benignos, pers. comm.).

El Pandeño

El Pandeño is the largest of a complex of small hot springs (38°C–45°C) originally draining to the Río

Fig. 23.10 The rapidly drying Ojo Solo, September 8, 2012 (*top*) and the refuge constructed at Ojo Caliente, June 25, 2014 (*bottom*). The mound in the background is spoil from the construction of a pool (<1 m deep, 300 m²) in the marsh fed by Ojo Caliente. *Inset:* Carbonera pupfish. (Photographs by and published with permission of E. Carson, USFWS.)

Conchos, Chihuahua, in the municipality of Julimes, a small farming community (fig. 23.11; see also fig. 23.9). The known endemic biota includes Julimes pupfish *Cyprinodon julimes* and an undescribed (presumably endemic) species of *Gambusia*, an isopod *Thermosphaeroma macrura*, a springsnail *Tryonia julimesensis*, and a sedge *Eleocharis aresenifera* (de la Maza-Benignos et al. 2014). The endemic Julimes springsnail was once more widespread in the local area, but was considered extinct until its rediscovery in El Pandeño (Hershler et al. 2014). Another springsnail, *T. minckleyi*, is known only from the Julimes springs and another hot spring system at San Diego de Alcalá, 24 km to the north (Hershler et al. 2011).

Until the 1950s, the El Pandeño outflow supported a wetland about 1.5 km long and up to 200 m wide (de la Maza-Benignos et al. 2012, 2014). The original streambed has been obliterated by agricultural fields. In recent decades, El Pandeño emerged at the eastern edge of the Río Conchos floodplain and emptied directly into an earthen canal, from which a system of pipes and canals provided water for recreation and irrigation (de la Maza-Benignos et al. 2014). The canal, less than 1 m deep and 300 m^2 in area, provided the main habitat for the fish and invertebrates (Carson et al. 2014; Lozano-Vilano and de la Maza-Benignos 2016).

The El Pandeño restoration project began in 2007 with community outreach by a nongovernmental organization (NGO), Pronatura Noreste, and included workshops where local stakeholders and municipal authorities eventually agreed to participate in developing an integrated management plan (de la Maza-Benignos et al. 2012, 2014). The plan included the establishment of a community-controlled, natural protected area cen-

Fig. 23.11 Main spring (*top*) and restored marsh (*bottom*) at the site of the El Pandeño restoration project. (Photos by and published with permission of M. De la Maza-Benignos, Pronatura.)

tered on the spring and a monitoring program for the Julimes pupfish as a bioindicator of change. The importance of El Pandeño Spring to the local economy was acknowledged, as was the fact that mismanagement and abuse of the spring system posed a risk to municipal stability. The restoration process included the chartering of Amigos del Pandeño, A.C., an NGO founded by the local irrigation society. The entire flow from the spring was legally allocated to the 36 farmers of the irrigation society, and the NGO recognized "explicitly that stewardship of El Pandeño Spring . . . [should be] focused on sustainable development of the water resource" (de la Maza-Benignos et al. 2014, 16). As part of the management plan, a marsh just downstream of the El Pandeño outlet canal was restored in 2012. The marsh, which was quickly colonized by the pupfish and *Gambusia* sp., covers 300 m^2, about doubling the area occupied by the fish (M. de la Maza-Benignos, pers. comm.).

The El Pandeño project presented legal challenges requiring a review of Mexican water laws and documentation of the environmental and social benefits of the spring (de la Maza-Benignos et al. 2012, 2014). Most waters of México are owned by the federal government and controlled by the federal executive, directly or indirectly through the National Water Commission (CONAGUA). In September 2010, CONAGUA granted Amigos del Pandeño control over federal lands in the El Pandeño water district (12,731 m^2) for a 10-year period. The same response denied granting water rights specifically for preservation of the endemic fauna in the waters fed by the spring. On the positive side, CONAGUA's argument for the denial revived old federal regulations prohibiting concessions of water rights in the Río Conchos Basin, potentially setting a legal precedent with "far reaching consequences in favor of water protection in the long term" for the Río Conchos Basin, and for El Pandeño in particular (de la Maza-Benignos et al. 2014, 18).

Balmorhea Springs Complex

The Balmorhea area, in Reeves and Jeff Davis Counties, Texas, once supported seven springs within a radius of 13.4 km. They included three large artesian springs: Phantom Lake, San Solomon, and Giffin, and four smaller, gravity-fed springs: West Sandia, East Sandia, Saragosa, and Toyah Creek (White et al. 1941). The native aquatic fauna included six species federally listed as threatened or endangered (USFWS 1967, 1970, 2013b): two fishes, Comanche Springs pupfish *Cyprinodon elegans* and Pecos gambusia *Gambusia nobilis*; three snails, Phantom Cave snail *Pyrgulopsis texana*, Phantom springsnail *Tryonia cheatumi*, and Pecos assiminea *Assiminea pecos*; and the diminutive amphipod *Gammarus hyalleloides*. The springs have been declining for at least the past 100 years (Sharp et al. 2003). Saragosa, West Sandia, and Toyah Creek springs are now dry or not producing consistent flow, Phantom Lake Spring flows only with the aid of a pump, and San Solomon Spring is slowly declining (USGS 2005).

San Solomon Spring, in Balmorhea State Park, is the largest remaining spring in the Pecos River basin (Bumgarner et al. 2012). The outflow (708–850 L/s; USFWS 2013c) emerges at the bottom of the largest spring-fed swimming pool (0.71 ha) in the world (fig. 23.12), constructed in 1936–1941 by the Civilian Conservation Corps. The cement-sided pool is nonchlori-

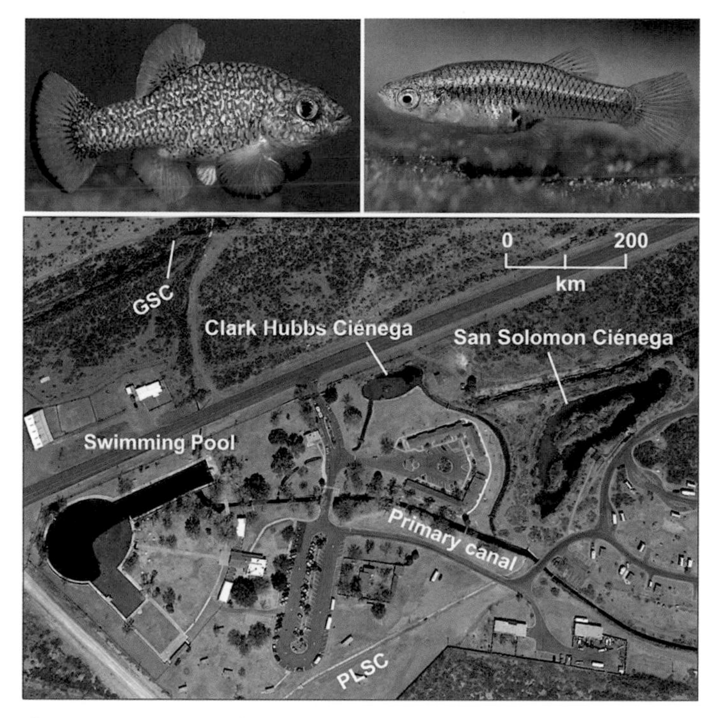

Fig. 23.12 Swimming pool over San Solomon Spring and downstream ciénega-like refuges for Pecos gambusia and Comanche Springs pupfish at Balmorhea State Park, Reeves County, Texas. PLSC = dry irrigation canal from Phantom Lake Spring, 6 km W of the park; GSC = irrigation canal from Giffin Spring (just off the map to the left). (Image from Google Earth, April 25, 2017; Map data ©2017 Google.) *Insets:* Photographs of a male Comanche Springs pupfish (*left*) and a female Pecos gambusia (*right*). (Copyright Engbretson Underwater Photography.)

nated and has a bottom of natural substrate. Outflow from the pool, and from the smaller, dammed and channelized Giffin Spring, supplies the Reeves County Water Improvement District's (RCWID) system of irrigation canals (USFWS 2013c).

Starting in the mid-1870s (Brune 1981), the spring flows of the Balmorhea area were intensively modified to irrigate about 4,300 ha of farmland (Simonds 1996). In 1917, RCWID built Lake Balmorhea (230 ha) about 8 km SE of Balmorhea to impound water. The lake is regularly stocked with sport fish (Scott and Farooqi 2010), and since the 1960s, it has supported a nonnative pupfish, sheepshead minnow *Cyprinodon variegatus* (Stevenson and Buchanan 1973). This pupfish, also reported in nearby East Sandia Spring, threatens Comanche Springs pupfish through competitive replacement and hybridization (Stevenson and Buchanan 1973; Echelle and Echelle 1994). Poor habitat in the concrete canal system and a sharp drop (10–40 cm) at the terminus of the flume emptying into Lake Balmorhea seem to have precluded upstream invasion of the springs by sheepshead minnow and hybrids.

Federal listing of the Comanche Springs pupfish in 1967 and the Pecos gambusia in 1970 precipitated decades of effort to mitigate habitat loss in the Balmorhea area. In 1974, the Texas Parks and Wildlife Department (TPWD) and the Texas Organization for Endangered Species built a shallow (≤80 cm), concrete-lined refuge canal (275 m long) that receives water from the swimming pool (Cokendolpher 1978). Pupfish habitat was further augmented during the cooler months when water not needed for irrigation was shunted from the canal into a marshy section of an old irrigation ditch. By 2009, the canal had deteriorated and was damaging nearby motel units, and a new version, the Clark Hubbs Ciénega (see fig. 23.12), named for the pioneer of fish conservation in Texas, was built in the general vicinity. A larger refuge, the San Solomon Ciénega (see fig. 23.12), completed in 1996, is a 1 ha wetland supporting a diversity of taxa, including the two federally listed fishes (Garrett 2003). Its special features include an outside glass viewing wall, an observation deck, and educational signage.

Maintenance of the federally listed species in Phantom Lake Spring (6 km W of San Solomon Spring) has been challenging. The spring once flowed from Phantom Cave at the base of a limestone bluff into what was a large ciénega (White et al. 1941). In 1945, 7.1 ha including the cave opening and surrounding land were purchased by USBR, and a concrete-lined irrigation canal was built to deliver spring flow to RCWID (USBR 2011). In 1900 and 1904, the spring produced 1,300 L/s (Brune 1981), but declined to less than half that in the 1930s, and by 1999 had failed completely (USFWS 2013c). In 1990, spring flow was critically low, and a captive stock of the morphologically and genetically divergent (Echelle 1975; Echelle et al. 1987) Phantom Lake Spring population of Comanche Springs pupfish was established at the Uvalde National Fish Hatchery (UNFH) in Texas (USFWS 2002). Later, 400 UNFH fish were used to establish a stock at SNARRC. The wild Phantom Lake Spring population subsequently lost genetic diversity (Robinson and Wilson 2012), affirming the value of captive stocks in managing genetic diversity of critically declining forms.

In 1993, in response to continued decline in Phantom Lake Spring, a refuge channel was built parallel to the old irrigation canal (Young et al. 1993; Winemiller and Anderson 1997). Efforts to stock the refuge with fish from canals fed by Phantom Cave produced 135 Pecos gambusia, but only two Comanche Springs pupfish, which were supplemented with 109 pupfish from UNFH (Young et al. 1993; USFWS 2013c). By 1999, the refuge and canals were dry, leaving only a small pool at the cave mouth. In 2001, USBR installed a pump 23 m inside the cavern to circulate water between the cave, the pool at the cave mouth, and the refuge (USFWS 2005). Since then, the refuge area has undergone periodic and intensive upgrading to secure the system against desiccation and loss of the five resident endangered species. Modifications have included filling in the original refuge channel, building a larger, more natural ciénega (USBR 2011; USFWS 2013c), installing a more reliable pumping system, and removing salt cedar (*Tamarisk* spp.).

Monitoring shows thriving populations of Pecos gambusia and Comanche Springs pupfish in each of the three artificial ciénegas (Hargrave 2014, 2016). The greatest densities for both species are in the more recently constructed Clark Hubbs and Phantom Lake Spring ciénegas, but the three ciénegas combined provide added living space for potentially thousands of fish of both species. This success involved cooperation and efforts by diverse state and federal agencies and the local farming community (Garrett 2002). The farmers forfeited a small amount of irrigation water for the ciénegas. In return, the Texas Department of Agriculture granted them permission to use pesticides in fields adjacent to downstream irrigation canals. The project was notably successful due to a sense of mutual benefit among the parties involved (Sharp 2001).

DISCUSSION

These case studies have indicated some of what can be done, given adequate resources, to restore fishes and aquatic habitats in western North America. Without conservation efforts, extinction would have been the fate of the Carbonera pupfish, largemouth shiner, Phantom Lake Spring forms of Comanche Springs pupfish and Pecos gambusia, and potentially Borax Lake chub. The remaining species and populations would be less secure than they are today. Gila topminnow is being considered for downlisting from federally endangered to threatened; desert pupfish, Comanche Springs pupfish, and Pecos gambusia have expanded living space in semi-natural habitats; and Oregon chub and Modoc sucker have been removed from the federal list of endangered species.

Common threads among these restoration successes from the western United States include governmental interagency cooperation, the participation of conservation-oriented NGOs, and the approval and participation of private landowners and other stakeholders. Behind all this is the ESA, "the strongest piece of environmental legislation in history" (Evans et al. 2016, 2). Despite the successes, however, there are problems in how species and habitat restorations are implemented. These problems are exemplified by the many cases in which inadequate funding for restoration and the nonscientific influence of social and political factors on how funding is allocated have introduced a degree of uncertainty about the reliability of support for monitoring and restoration into the future (Evans et al. 2016).

The Rancho Nuevo and El Pandeño projects in México are models of restoration in local situations where conservation of natural resources has not been a high government priority. The projects are largely driven by NGOs in cooperation with local farmers, and with funding primarily from donations and grants from private foundations. The El Pandeño project grew in part out of a water crisis in the Conchos River basin. Water availability became a critical issue during the severe 1993–2005 drought, triggering adaptive responses by water users, government agencies, and NGOs (Barrios et al. 2009; de la Maza-Benignos et al. 2014). Some of these activities reflect a potential shift in water management in the face of a somewhat unpredictable, but generally dire, water future for the region (Barrios et al. 2009). This shift would involve reduced emphasis on delivering more water and heightened emphasis on local and regional water conservation measures, simultaneously addressing present human needs and striving for sustainability.

The long-term success of most projects described in this chapter depends heavily on continued human intervention. For example, the need for management programs for native fishes of the Gila River basin is not likely to abate by 2027, when the GRBNFCP expires. Water use in Arizona is projected to increase from 7.1 million to 8.1 million acre-feet per year by 2035 (Arizona Water Resources Development Commission 2011), largely due to growth in Phoenix and Tucson, where the human population is expected to increase by more than a million people from 2017 to 2027 (Arizona Office of Economic Opportunity, https://population.az.gov/population-projections). Although water conservation is playing a role in preparing for Arizona's future, closing the projected gap between water demand and supply will dictate intensified exploitation of surface and groundwater sources (Eden et al. 2015).

Other sources of uncertainty for restoration projects in western North America include shifts in political ideology and powerful socioeconomic forces negatively affecting conservation goals. For example, repealing bans on mineral and geothermal development on public lands could adversely affect conservation-sensitive areas, such as the BLM land surrounding critical habitat for Borax Lake chub. Another example is the shrinking of the Salton Sea and its associated increasing salinity, which began in 2003 with an agriculture-to-urban transfer of Colorado River water from the Imperial Irrigation District to the San Diego area in Southern California. This transfer accelerated in 2018, and within 5 years or so, the salinity of the lake will exceed the tolerance (about 80 ppt) of desert pupfish (Cohen and Hyun 2006). The state of California has a 10-year plan, starting in 2018, to mitigate the environmental impacts of the transfer, but it remains to be seen what this will mean for desert pupfish and other native species of the region. Still another example is the recent discovery of massive oil reserves (Alpine High Play) in the Delaware Basin of West Texas and what it might mean for the already declining springs in the Toyah Creek drainage (meeting notes, Trans-Pecos Oil and Gas Workgroup, Austin, Texas, 2016).

Most imperiled fishes of the American West clearly exist as remnants of larger metapopulations with varying degrees of connectedness among dynamically fluctuating subpopulations. Historically, subpopulations extirpated by chance and catastrophic events (e.g., drought and destructive floods) had greater potential to be reestablished by dispersal from elsewhere. In the present degraded environment, the rate of extirpation is exacerbated and the potential for replacement by natural dispersal is greatly diminished (Fagan et al. 2005). Thus, modern-day managers must be committed to Sisyphean programs of monitoring, augmentation, reestablishment, and restoration of populations and habitats just to maintain the status quo. This commitment, in turn, requires the political and societal will to protect and restore natural systems and their native biota.

ACKNOWLEDGMENTS

An obvious factor in efforts to restore and protect natural aquatic systems in the American West is the multi-year, often multi-decadal, commitment of one or a few dedicated individuals who are the prime movers behind a given project. Some such people are included among the following whom we thank for assistance with this chapter: B. Bangs, S. Reid, and P. Scheerer (Oregon chub and Modoc sucker); J. Williams (Borax Lake chub); R. Clarkson, J. Marks, P. Marsh, J. Stefferud, and D. Weedman (Fossil Creek); D. Duncan, A. Robinson, and J. Voeltz (Gila topminnow and desert pupfish); S. Keeney (desert pupfish); E. Carson and M. de la Maza-Benignos (Mexican springs); and M. Bean, P. Conner, G. Garrett, and W. Wilson (Balmorhea area). We thank J. Tomelleri for fish drawings, and K. Ashcraft, B. Bangs, N. Banish, E. Carson, M. de la Maza-Benignos, D. Duncan, J. Marks, Doug Markle, P. Scheerer, L. Sepúlveda-Hernández, P. Sponholtz, and D. Weedman for assistance with locating or providing photographs. We thank the editors for the invitation to participate in this volume and for their suggestions, and those from H. Blasius, R. Edwards, D. Hendrickson, W. Radke, and D. Stewart for improvement of the manuscript.

REFERENCES

Alvarez, F., M. López-Mejía, and C. Pedraza-Lara. 2010. *Cambarellus chihuahuae. IUCN Red List of Threatened Species.* http://www.iucnredlist.org/details/153621/0. Accessed February 21, 2017.

Arizona Water Resources Development Commission. 2011. *Final Report.* Phoenix: Department of Water Resources.

Barrios, E., J. A. Rodríguez-Pineda, and M. de la Maza-Benignos. 2009. Integrated river basin management in the Conchos River basin, Mexico: A case study of freshwater climate change adaptation. *Climate and Development* 1: 249–60.

Benner, P. A., and J. R. Sedell. 1997. Upper Willamette River landscape: A historic perspective. In A. Laenen and D. A. Dunnette, eds., *River Quality: Dynamics and Restoration,* 23–47. Boca Raton, FL: CRC Press.

Black, A. N., J. L. Snekser, L. Al-Shaer, T. Paciorek, A. Bloch, K. Little, and M. Itzkowitz. 2016. A review of the Leon Springs pupfish (*Cyprinodon bovinus*) long-term conservation strategy and response to habitat restoration. *Aquatic Conservation: Marine and Freshwater Ecosystems* 26: 410–16.

Brune, G. 1981. *Springs of Texas.* Vol. 1. Fort Worth: Branch-Smith.

Bumgarner, J. R., G. P. Stanton, A. P. Teeple, J. V. Thomas, N. A. Houston, J. D. Payne, and M. Musgrove. 2012. *A Conceptual Model of the Hydrogeologic Framework, Geochemistry, and Groundwater-Flow System of the Edwards-Trinity and Related Aquifers in the Pecos County Region, Texas.* US Geological Survey Scientific Investigations Report 2012-5124.

Carson, E. W., and M. de la Maza-Benignos. 2014. *Feasibility, Design, and Establishment of a Refuge Population for the Critically Endangered Carbonera Pupfish,* Cyprinodon fontinalis. Final Report to the Desert Fishes Council. Bishop, CA: Desert Fishes Council.

Carson, E. W., M. de la Maza-Benignos, M. L. Lozano-Vilano, L. Vela-Valladares, I. Banda-Villanueva, and T. F. Turner. 2014. Conservation genetic assessment of the critically endangered Julimes pupfish, *Cyprinodon julimes. Conservation Genetics* 15: 483–88.

Carson, E. W., C. Pedraza-Lara, M. L. Lozano-Vilano, G. A. Rodríguez-Almaráz, I. Banda-Villanueva, L. A. Sepúlveda-Hernández, L. Vela-Valladares, A. Cantú-Garza, and M. de la Maza-Benignos. 2015. The rediscovery and precarious status of Chihuahuan dwarf crayfish *Cambarellus chihuahuae. Occasional Papers of the Museum of Southwestern Biology* 12: 1–7.

Cohen, M. J., and K. H. Hyun. 2006. *Hazard: The Future of the Salton Sea with No Restoration Project.* Oakland: Pacific Institute.

Cokendolpher, J. C. 1978. *Cyprinodon elegans* (Cyprinodontidae). *American Currents* January–March 1978: 11.

Contreras-Balderas, S., and M. L. Lozano-Vilano. 1996. Extinction of most Sandia and Potosí valleys (Nuevo León, México) endemic pupfishes, crayfishes and snails. *Ichthyological Exploration of Freshwaters* 7: 334–40.

DeHaan, P. W., B. A. Adams, P. D. Scheerer, and B. L. Bangs. 2016. Influence of introduction history on genetic variation in introduced populations: A case study of Oregon chub. *North American Journal of Fisheries Management* 36: 1278–91.

DeHaan, P. W., P. D. Scheerer, R. Rhew, and W. R. Ardren. 2012.

Analyses of genetic variation in populations of Oregon chub, a threatened floodplain minnow in a highly altered environment. *Transactions of the American Fisheries Society* 141: 533–49.

De la Maza-Benignos, M., J. A. Rodriguez-Pineda, A. de la Mora-Covarrubias, E. W. Carson, M. Quiñones-Martínez, P. Lavín-Murcio, L. Vela-Valladares, M. L. Lozano-Vilano, H. Parra-Gallo, A. Macías-Duarte, T. Lebgue-Keleng, E. Pando-Pando, M. Pando-Pando, M. Andazola-González, A. Anchondo-Najera, G. Quintana-Martínez, J. Zapata-López, I. A. Banda-Villanueva, and H. J. Ibarrola-Reyes. 2012. *Planes de Manejo y Programa de Monitoreo de Signos Vitales para las Áreas de Manantiales de la UMA El Pandeño; y San Diego de Alcalá en el Desierto Chihuahuense.* Vol 1. Pronatura Noreste, A. C., Amigos del Pandeño, A. C.

De la Maza-Benignos, M., L. Vela-Valladares, M. L. Lozano-Vilano, M. E. García-Ramírez, J. Zapata-López, A. J. Contreras-Balderas, and E. W. Carson. 2014. The potential of holistic approaches to conservation of desert springs: A case study of El Pandeño Spring and its microendemic pupfish *Cyprinodon julimes* in the Chihuahuan Desert at Julimes, Chihuahua, Mexico. In M. de la Maza-Benignos, M. L. Lozano-Vilano, and E. W. Carson, eds., *Conservation of Desert Wetlands and Their Biotas*, 1–45. Special Publications, vol. 1. Albuquerque: Museum of Southwestern Biology, Pronatura Noreste A. C., and Universidad Autónoma de Nuevo León.

Doremus, H., and J. E. Pagel. 2001. Why listing may be forever: Perspectives on delisting under the US Endangered Species Act. *Conservation Biology* 15: 1258–68.

Dowling, T. E. 2005. *Conservation Genetics of Modoc Sucker.* Klamath Falls, OR: US Fish and Wildlife Service.

Duncan, D., and R. W. Clarkson. 2013. Gila River basin native fishes conservation program. In G. J. Gottfried, P. F. Folliott, B. S. Gebow, L. G. Eskew, and L. C. Collins, compilers, *Merging Science and Management in a Rapidly Changing World: Biodiversity and Management of the Madrean Archipelago III*, 376–80. US Forest Service Proceedings. RMRS-P-67. Fort Collins, CO: USDA Forest Service.

Duncan, D., and J. Voeltz. 2005. Safe Harbor: A tool to help recover topminnow and pupfish in Arizona. *USDA Forest Service Proceedings* RMRS-P-36: 392–94.

Echelle, A. A. 1975. A multivariate analysis of variation in an endangered fish, *Cyprinodon elegans*, with an assessment of populational status. *Texas Journal of Science* 26: 529–38.

Echelle, A. A., and A. F. Echelle. 1997. Genetic introgression of endemic taxa by non-natives: A case study with Leon Springs pupfish and sheepshead minnow. *Conservation Biology* 11: 153–61.

Echelle, A. F., and A. A. Echelle. 1994. Assessment of genetic introgression between two pupfish species, *Cyprinodon elegans* and *C. variegatus* (Cyprinodontidae), after more than 20 years of secondary contact. *Copeia* 1994: 590–97.

Echelle, A. F., A. A. Echelle, L. K. Bonnel, N. L. Allan, J. E. Brooks, and J. Karges. 2004. Effects of a restoration effort on an endangered pupfish (*Cyprinodon bovinus*) after genetic introgression by a non-native species. In M. L. Lozano-Vilano and A. J. Contreras-Balderas, eds., *Homenaje al Doctor Andrés Reséndez Medina*, 129–39. Monterrey: Universidad Autónoma de Nuevo León.

Echelle, A. F., A. A. Echelle, and D. R. Edds. 1987. Conservation genetics of a spring-dwelling desert fish, the Pecos gambusia (*Gambusia nobilis*, Poeciliidae). *Conservation Biology* 3: 159–69.

Eden, S., M. Ryder, and M. A. Capehart. 2015. Closing the water demand-supply gap in Arizona. *Arroyo.* Tucson: University of Arizona Water Resources Research Center.

Evans, D. M., J. P. Che-Castaldo, D. Crouse, F. W. Davis, R. Epanchin-Niell, C. H. Flather, R. K. Frohlich, D. D. Goble, Y.-W. Li, T. D. Male, L. L. Master, M. P. Moskwik, M. C. Neel, B. R. Noon, C. Parmesan, M. W. Schwartz, J. M. Scott, and B. K. Williams. 2016. Species recovery in the United States: Increasing the effectiveness of the Endangered Species Act. *Issues in Ecology* 20: 1–28.

Fagan, W. F., C. Aumann, C. M. Kennedy, and P. J. Unmack. 2005. Rarity, fragmentation, and the scale dependence of extinction risk in desert fishes. *Ecology* 86: 34–41.

Fuller, B. M., L. S. Sklar, G. Z. Compson, K. J. Adams, J. C. Marks, and A. C. Wilcox. 2011. Ecogeomorphic feedbacks in regrowth of travertine step-pool morphology after dam decommissioning, Fossil Creek, Arizona. *Geomorphology* 126: 314–32.

Garrett, G. P. 2002. Community involvement—a more comprehensive approach to recovering endangered species. In D. W. Sada and S. E. Sharpe, eds., *Conference Proceedings, Spring-Fed Wetlands: Important Scientific and Cultural Resources of the Intermountain Region*, 1–10. Las Vegas: DHS Publication no. 41210.

Garrett, G. P. 2003. Innovative approaches to recover endangered species. In G. P. Garrett and N. L. Allan, eds., *Aquatic Fauna of the Northern Chihuahuan Desert*, 151–60. Museum of Texas Tech University, Special Publication no. 46.

Goble, D. D. 2010. A fish tale: A small fish, the ESA, and our shared future. *Environmental Law* 40: 339–62.

Hargrave, C. 2014. *Conservation of Fish Species in Phantom Lake Spring.* Final Report. Agreement no. R10AP40038 US Bureau of Reclamation.

Hargrave, C. 2016. *Developing a Predictive Habitat Model for the Comanche Springs Pupfish* (Cyprinodon elegans) *to Be Used in Species Recovery.* Interim Report, Grant TX ET-159-R. Austin: Texas Parks and Wildlife Department.

Hedrick, P. W., and C. R. Hurt. 2012. Conservation genetics and evolution in an endangered species: Research in Sonoran topminnows. *Evolutionary Applications* 5: 806–19.

Hedrick, P. W., R. Lee, and C. Hurt. 2006. The endangered Sonoran topminnow: Examinations of species and ESUs using three mtDNA genes. *Conservation Genetics* 7: 483–93.

Hendrickson, D. A., and J. E. Brooks. 1991. Transplanting short-lived fishes in North American deserts: Review, assessment, and recommendations. In W. L. Minckley and J. E. Deacon, eds., *Battle against Extinction: Native Fish Management in the American West*, 283–98. Tucson: University of Arizona Press.

Hendrickson, D. A., and A. Varela-Romero. 1989. Conservation status of Desert pupfish, *Cyprinodon macularius*, in México and Arizona. *Copeia* 1989: 478–85.

Hershler, R., J. J. Landye, H.-P. Liu, M. de la Maza-Benignos, P. Ornelas, and E. W. Carson. 2014. New species and records of Chihuahuan Desert springsnails with a new combination for *Tryonia brunei. Western North American Naturalist* 74: 47–65.

Hershler, R., H.-P. Liu, and J. Landye. 2011. New species and records of springsnails (Caenogastropoda: Cochliopidae: *Tryonia*) from the Chihuahuan Desert (Mexico and United States), an imperiled biodiversity hotspot. *Zootaxa* 3001: 1–32.

Hubbs, C., and H. J. Brodrick. 1963. Current abundance of *Gambusia gaigei*, an endangered fish species. *Southwestern Naturalist* 8: 46–48.

Hubbs, C. L., and R. R. Miller. 1941. Studies of the fishes of the order Cyprinodontes. XVII. Genera and species of the Colorado River system. *Occasional Papers Museum of Zoology, University of Michigan* 433: 1–9.

Kettratad, J. 2001. Systematic study of Modoc suckers (*Catostomus microps*) and Sacramento suckers (*Catostomus occidentalis*) in the upper Pit River system, CA. MS thesis, Humboldt State University.

Koike, H., A. A. Echelle, D. Loftis, and R. A. Van Den Bussche. 2008. Microsatellite DNA analysis of success in conserving genetic diversity after 33 years of refuge management of the desert pupfish complex. *Animal Conservation* 11: 321–29.

Laird, I. W. 1971. *The Modoc Country*. Alturas, CA: Lawton and Kennedy Printers.

Langpap, C., and J. Wu. 2004. Voluntary conservation of endangered species: When does regulatory assurance mean no conservation? *Journal of Environmental Economics and Management* 47: 435–57.

Lozano-Vilano, M. L., and M. de la Maza-Benignos. 2016. Diversity and status of Mexican killifishes. *Journal of Fish Biology* 90: 3–38.

Malusa, J., S. T. Overby, and R. A. Parnell. 2003. Potential for travertine formation: Fossil Creek, Arizona. *Applied Geochemistry* 18: 1081–93.

Markle, D. F., T. N. Pearsons, and D. T. Bills. 1991. Natural history of *Oregonichthys* (Pisces: Cyprinidae), with a description of a new species from the Umpqua River of Oregon. *Copeia* 1991: 277–93.

Marks, J. C. 2007. Down go the dams. *Scientific American* 296: 66–71.

Marks, J. C., G. A. Haden, E. Dinger, and K. Adams. 2005. *A Survey of the Aquatic Community at Fossil Creek, AZ*. Phoenix: Arizona Game and Fish Department, Arizona Game and Fish Heritage Fund.

Marks, J. C., G. A. Haden, M. O'Neill, and C. Pace. 2010. Effects of flow restoration and exotic species removal on recovery of native fish: Lessons from a dam decommissioning. *Restoration Ecology* 18: 934–43.

Meffe, G. K., D. A. Hendrickson, and W. L. Minckley. 1983. Factors resulting in decline of the endangered Sonoran topminnow *Poeciliopsis occidentalis* (Atheriniformes: Poeciliidae) in the United States. *Biological Conservation* 25: 135–59.

Miller, R. R. 1961. Man and the changing fish fauna of the American Southwest. *Papers of the Michigan Academy of Science, Arts, and Letters* 46: 365–404.

Miller, R. R., and E. P. Pister. 1971. Management of the Owens pupfish, *Cyprinodon radiosus*, in Mono County, California. *Transactions of the American Fisheries Society* 100: 502–9.

Minckley, W. L. 1991. Native fishes of the Grand Canyon region: An obituary? In *Colorado River Ecology and Dam Management: Proceedings of a Symposium, May 24–25, 1990, Santa Fe, New Mexico*, 124–77. Washington, DC: National Academy of Sciences Press.

Minckley, W. L. 1995. Translocation as a tool for conserving imperiled fishes: Experiences in western United States. *Biological Conservation* 72: 297–309.

Minckley, W. L., and P. C. Marsh. 2009. *Inland Fishes of the Greater Southwest: Chronicle of a Vanishing Biota*. Tucson: University of Arizona Press.

Minckley, W. L., P. C. Marsh, J. E. Deacon, T. E. Dowling, P. W. Hedrick, W. J. Matthews, and G. A. Mueller. 2003. A conservation plan for native fishes of the lower Colorado River. *BioScience* 53: 219–33.

Minckley, W. L., G. K. Meffe, and D. L. Soltz. 1991. Conservation and management of short-lived fishes: The cyprinodontoids. In W. L. Minckley and J. E. Deacon, eds., *Battle against Extinction: Native Fish Management in the American West*, 247–82. Tucson: University of Arizona Press.

Mosher, K. 2017. Featured story: Update on the Fossil Creek Native Fish Restoration Project. *In the Current*. https://inthecurrent.org/uncategorized/featured-story-update-on-the-fossil-creek-native-fish-restoration-project/. Accessed February 6, 2017.

Motivans, K., and M. Balis-Larsen. 2003. Species on the brink of recovery. *Endangered Species Bulletin* 28: 10–11.

Moyle, P. B., H. W. Li, and B. A. Barton. 1986. The Frankenstein effect: Impact of introduced fishes on native fishes in North America. In R. H. Stroud, ed., *Fish Culture in Fisheries Management*, 415–26. Bethesda, MD: American Fisheries Society.

Moyle, P. B., and A. Marciochi. 1975. Biology of the Modoc sucker, *Catostomus microps*, in northeastern California. *Copeia* 3: 556–60.

Moyle, P. B., J. J. Smith, R. A. Daniels, T. L. Taylor, D. G. Price, and D. M. Baltz. 1982. *Distribution and Ecology of Stream Fishes of the Sacramento–San Joaquin Drainage System, California*. Berkeley: University of California Press.

Mueller, G. A. 2005. Predatory fish removal and native fish recovery in the Colorado River mainstem: What have we learned? *Fisheries* 30: 10–19.

Mueller, G. A., and P. C. Marsh. 2002. *Lost: A Desert River and Its Native Fishes: A Historical Perspective of the Lower Colorado River*. US Geological Survey Information and Technology Report USGS/BRD/ITR-2002-0010.

Osborne, M. J., E. W. Carson, and T. F. Turner. 2012. Genetic monitoring and complex population dynamics: Insights from a 12-year study of the Rio Grande silvery minnow. *Evolutionary Applications* 5: 553–74.

Pearsons, T. N. 1989. Ecology and decline of a rare western minnow: The Oregon chub (*Oregonichthys crameri*). MS thesis, Oregon State University.

Reid, S. B. 2007. *Surveys for the Modoc Sucker,* Catostomus microps, *in the Goose Lake Basin, Oregon*. Klamath Falls, OR: US Fish and Wildlife Service.

Reid, S. B. 2008. *Conservation Review: Modoc Sucker* Catostomus microps. Klamath Falls, OR: US Fish and Wildlife Service.

Robinson, A. 2008. *Arizona Native Fish Recovery and Non-native Fish Control*. Agreement no. 201816J808. Phoenix: Arizona Game and Fish Department.

Robinson, M. L., and W. D. Wilson. 2012. *Comanche Springs Pupfish Genetic Monitoring*. Dexter, NM: US Fish and Wildlife

Service, Southwestern Native Aquatic Resources and Recovery Center.

Scheerer, P. D. 2002. Implications of floodplain isolation and connectivity on the conservation of an endangered minnow, Oregon chub, in the Willamette River, Oregon. *Transactions of the American Fisheries Society* 131: 1070–80.

Scheerer, P. D. 2007. Improved status of the endangered Oregon chub in the Willamette River, Oregon. In M. J. Brouder and J. A. Scheurer, eds., *Status, Distribution, and Conservation of Native Freshwater Fishes of Western North America: Symposium Proceedings*, 91–102. Symposium 53. Bethesda, MD: American Fisheries Society.

Scheerer, P. D., and S. Clements. 2015. *Borax Lake Chub Investigations: An Analysis of Recovery Actions and Current Threats, with Recommendations for Future Management*. Bureau of Land Management Cooperative Agreement L10AC20301. Corvallis: Oregon Department of Fish and Wildlife.

Scheerer, P. D., J. T. Peterson, and S. Clements. 2015. *Foskett Creek Speckled Dace Investigations*. Corvallis: Oregon Department of Fish and Wildlife.

Schoenherr, A. 1988. A review of the life history and status of the desert pupfish, *Cyprinodon macularius*. *Bulletin of the Southern California Academy of Science* 87: 104–34.

Scoppettone, G. G., P. H. Rissler, C. Gourley, and C. Martinez. 2005. Habitat restoration as a means of controlling non-native fish in a Mojave Desert oasis. *Restoration Ecology* 13: 247–56.

Scoppettone, G. G., P. H. Rissler, B. Nielsen, and M. Grader. 1995. *Life History and Habitat Use of Borax Lake Chub* (Gila boraxobius *Williams and Bond*) *with Some Information on the Borax Lake Ecosystem*. Reno: US Geological Survey, Northwest Biological Science Center.

Scott, M., and M. Farooqi. 2010. *Statewide Freshwater Fisheries Monitoring and Management Program, 2009 Survey Report: Balmorhea Reservoir*. Federal Aid Project F-30-R-35. Austin: Texas Parks and Wildlife Department.

Sharp, J. M., Jr. 2001. Regional groundwater flow systems in Trans-Pecos, Texas. In R. E. Mace, W. F. Mullican III and E. S. Angle, eds., *Aquifers of West Texas*, 41–55. Texas Water Development Board Report no. 356.

Sharp, J. M., Jr., R. Boghici, and M. M. Uliana. 2003. Groundwater systems feeding the springs of West Texas. In G. P. Garrett and N. L. Allan, eds., *Aquatic Fauna of the Northern Chihuahuan Desert*, 1–11. Museum of Texas Tech University, Special Publication no. 46.

Sheller, F. J., W. F. Fagan, and P. J. Unmack. 2006. Using survival analysis to study translocations success in the Gila topminnow (*Poeciliopsis occidentalis*). *Ecological Applications* 16: 1771–84.

Simonds, W. J. 1996. *Balmorhea Project*. Historic Reclamation Projects. Washington, DC: US Bureau of Reclamation.

Simons, L. H., D. A. Hendrickson, and D. Papoulias. 1989. Recovery of the Gila topminnow: A success story? *Conservation Biology* 3: 11–15.

Smith, C. T., S. B. Reid, L. Godfrey, and W. R. Ardren. 2011. Gene flow among Modoc sucker and Sacramento sucker populations in the upper Pit River, California and Oregon. *Journal of Fish and Wildlife Management* 2: 72–84.

Smith, M. L., and R. R. Miller. 1980. Systematics and variation of a new cyprinodontid fish, *Cyprinodon fontinalis*, from Chihuahua, Mexico. *Proceedings of the Biological Society of Washington* 93: 405–16.

Stevenson, M. M., and T. M. Buchanan. 1973. An analysis of hybridization between the cyprinodont fishes *Cyprinodon variegatus* and *C. elegans*. *Copeia* 1973: 682–92.

Stewart, D. R., M. J. Butler, G. Harris, and W. R. Radke. 2017. Mark-recapture models identify imminent extinction of Yaqui catfish *Ictalurus pricei* in the United States. *Biological Conservation* 209: 45–53.

Thompson, P. D., P. A. Webber, and C. D. Mellon. 2015. The role of introduced populations in the management and conservation of Least Chub. *Fisheries* 40: 546–56.

Topinka, J. R. 2006. Conservation genetics of two narrow endemics: The Modoc sucker (*Catostomus microps*) and Kearney's bluestar (*Amsonia kearneyana*). PhD diss., University of California, Davis.

USBR (US Bureau of Reclamation). 2011. *Final Environmental Assessment: Ciénega Wetland Rehabilitation at Phantom Lake Spring, Toyavale, Texas*. Albuquerque: US Bureau of Reclamation.

USEPA (US Environmental Protection Agency). 2000. *Principles for the Ecological Restoration of Aquatic Resources*. Report no. EPA841-F-00-003. Office of Water. Washington, DC: US Environmental Protection Agency,

USFS (US Forest Service). 2016. *Fossil Creek Wild and Scenic River Resource Assessment*. Sedona, AZ: US Forest Service, Coconino National Forest.

USFWS (US Fish and Wildlife Service). 1967. Native fish and wildlife: Endangered species. *Federal Register* 32: 4001.

USFWS (US Fish and Wildlife Service). 1970. Appendix-D. United States list of endangered native fish and wildlife. *Federal Register* 35: 16047–48.

USFWS (US Fish and Wildlife Service). 1980. Endangered and threatened wildlife and plants: Emergency determination of endangered status and critical habitat for Borax Lake chub. *Federal Register* 45: 35821–23.

USFWS (US Fish and Wildlife Service). 1982. Endangered and threatened wildlife and plants: Endangered status and critical habitat for Borax Lake chub (*Gila boraxobius*). *Federal Register* 47: 43957–63.

USFWS (US Fish and Wildlife Service). 1984. *Sonoran Topminnow (Gila and Yaqui) Recovery Plan*. Albuquerque: US Fish and Wildlife Service.

USFWS (US Fish and Wildlife Service). 1985. Endangered and threatened wildlife and plants: Determination of endangered status and critical habitat for the Modoc sucker. *Federal Register* 50: 24526–30.

USFWS (US Fish and Wildlife Service). 1987. *Recovery Plan for the Borax Lake Chub,* Gila boraxobius. Portland: US Fish and Wildlife Service.

USFWS (US Fish and Wildlife Service). 1993a. Determination of endangered status for the Oregon chub. *Federal Register* 58: 53800–4.

USFWS (US Fish and Wildlife Service). 1993b. *Desert Pupfish Recovery Plan*. Phoenix: US Fish and Wildlife Service.

USFWS (US Fish and Wildlife Service). 1998. *Oregon Chub* (Oregonichthys crameri) *Recovery Plan*. Portland: US Fish and Wildlife Service.

USFWS (US Fish and Wildlife Service). 2002. *A Genetic Reserve Population and Stock Management Process*. Dexter, NM: Dexter National Fish Hatchery and Technology Center.

USFWS (US Fish and Wildlife Service). 2005. Endangered and threatened wildlife and plants: listing Roswell springsnail, Koster's springsnail, Noel's amphipod, and Pecos assiminea as endangered with critical habitat. *Federal Register* 70: 46304–33.

USFWS (US Fish and Wildlife Service). 2009. *Modoc Sucker* (Catostomus microps) *5-Year Review: Summary and Evaluation*. Klamath Falls, OR: US Fish and Wildlife Service.

USFWS (US Fish and Wildlife Service). 2010a. Endangered and threatened wildlife and plants: Reclassification of the Oregon chub from endangered to threatened. *Federal Register* 75: 21179–89.

USFWS (US Fish and Wildlife Service). 2010b. *Desert Pupfish* (Cyprinodon macularius) *5-Year Review: Summary and Evaluation*. Phoenix: US Fish and Wildlife Service.

USFWS (US Fish and Wildlife Service). 2012. *Borax Lake Chub* (Gila boraxobius) *5-Year Review: Summary and Evaluation*. Portland: US Fish and Wildlife Service.

USFWS (US Fish and Wildlife Service). 2013a. *Draft Post-delisting Monitoring Plan for the Oregon Chub* (Oregonichthys crameri). Portland: US Fish and Wildlife Service.

USFWS (US Fish and Wildlife Service). 2013b. Endangered and threatened wildlife and plants: Determination of endangered species status for six West Texas aquatic invertebrates. *Federal Register* 78: 41228–58.

USFWS. (US Fish and Wildlife Service). 2013c. *Comanche Springs Pupfish* (Cyprinodon elegans) *5-Year Review: Summary and Evaluation*. Austin: US Fish and Wildlife Service.

USFWS (US Fish and Wildlife Service). 2015a. Endangered and threatened wildlife and plants: Removing the Oregon chub from the federal list of endangered and threatened wildlife; Final rule. *Federal Register* 80: 9126–50.

USFWS (US Fish and Wildlife Service). 2015b. Endangered and threatened wildlife and plants: Removal of the Modoc sucker from the federal list of endangered and threatened wildlife. *Federal Register* 80: 76235–49.

USFWS (US Fish and Wildlife Service). 2015c. *Species Report for the Modoc Sucker* (Catostomus microps). Sacramento: US Fish and Wildlife Service.

USFWS (US Fish and Wildlife Service). 2015d. *Modoc Sucker* (Catostomus microps) *Post-delisting Monitoring Plan*. Klamath Falls, OR: US Fish and Wildlife Service.

USFWS (US Fish and Wildlife Service). 2019. Endangered and threatened wildlife and plants: Removing the Borax Lake chub from the list of endangered and threatened wildlife. *Federal Register* 84: 6110–26.

USGS (US Geological Survey). 2005. *Diminished Springflows in the San Solomon Springs System, Trans-Pecos, Texas*. Austin: US Geological Survey.

Varela-Romero, A., G. Ruiz-Campos, L. M. Yípiz-Velázquez, and J. Alaníz-García. 2002. Distribution, habitat and conservation status of desert pupfish (*Cyprinodon macularius*) in the lower Colorado River basin, Mexico. *Reviews in Fish Biology and Fisheries* 12: 157–65.

Weedman, D. A. 1998. *Gila Topminnow,* Poeciliopsis occidentalis occidentalis, *Revised Recovery Plan*. Phoenix: Arizona Game and Fish Department.

Weedman, D. A., P. Sponholtz, and S. Hedwall. 2005. *Fossil Creek Native Fish Restoration Project*. Phoenix: Arizona Game and Fish Department.

Weissenfluh, D. 2007. *The Upper Jackrabbit Restoration (Phase 1) Site Step-by-Step Report*. Ash Meadows National Wildlife Refuge, NV: US Department of the Interior, US Fish and Wildlife Service, Ash Meadows National Wildlife Refuge.

White, W. N., H. S. Gale, and S. S. Nye. 1941. *Geology and Ground-Water Resources of the Balmorhea Area, Western Texas*. Geological Survey Water-Supply Paper 849-C. US Department of the Interior.

Williams, J. E., and C. E. Bond. 1980. *Gila boraxobius*, a new species of cyprinid fish from southeastern Oregon with a comparison to *G. alvordensis* Hubbs and Miller. *Proceedings of the Biological Society of Washington* 93: 293–98.

Williams, J. E., and C. A. Macdonald. 2003. *A Review of the Conservation Status of the Borax Lake Chub, an Endangered Species*. Portland: US Fish and Wildlife Service.

Williams, J. E., C. A. Macdonald, C. Deacon Williams, H. Weeks, G. Lampman, and D. W. Sada. 2005. Prospects for recovering endemic fishes pursuant to the US Endangered Species Act. *Fisheries* 30: 24–29.

Winemiller, K. O., and A. A. Anderson. 1997. Response of endangered desert fish populations to a constructed refuge. *Restoration Ecology* 5: 204–13.

Young, D. A., K. J. Fritz, G. P. Garrett, C. Hubbs, and C. Day. 1993. Status review of construction, native species introductions, and operation of an endangered species refugium channel, Phantom Lake Spring, Texas. *Proceedings of the Desert Fishes Council* 25: 22–25.

::Bernoulli Trials

SECTION 5

Exploring Our Future

Fig. 24.1 Devils Hole. (*Left*) In June 1972, Robert R. Miller (*left, in profile*) started conducting monthly pupfish censuses in April 1972 under NPS contract. Note the people standing on the dry shallow shelf. The deep pool lies below the floating artificial shelf that was installed to promote primary production and a spawning habitat for pupfish. (*Right*) The shallow shelf (dark green and tan) and the deep pool (blue) in March 2016. Notice how much of the shallow shelf is exposed in 1972 compared with 2016, and that most of the infrastructure has been removed. (*Left*, photo by and courtesy of Edwin P. Pister; *Right*, photo courtesy of the National Park Service.)

24 The Devils Hole Pupfish

Science in a Time of Crises

Kevin P. Wilson,
Mark B. Hausner, and
Kevin C. Brown

Tucked away along miles of dirt roads some 50 km east of Death Valley National Park headquarters is a lonely, disjunct 16 ha parcel of park property. Here, in the Amargosa Desert of Nevada, surrounded by Bureau of Land Management lands and a US Fish and Wildlife Service wildlife refuge, lives a critically endangered desert fish, the Devils Hole pupfish *Cyprinodon diabolis*. Genetic isolation in a tiny desert habitat has made this pupfish unique, but the story of the pupfish since the 1950s is not one of solitude. Instead, it is the Devils Hole pupfish's connection to the wider world—through science, endangered species management, and federal water politics—that have made it both iconic and imperiled.

Administrative and ecological crises have been a recurring feature of the pupfish's history for more than half a century. When Devils Hole was added to what was then Death Valley National Monument in 1952, President Truman's proclamation declared Devils Hole and its "peculiar race of desert fish" to be "of such outstanding scientific importance that it should be given special protection." The security afforded by that reservation was tested in the 1960s, however, when groundwater development in the Ash Meadows area led to nearly complete dewatering of the ecologically critical shallow shelf (fig. 24.1*A*) in Devils Hole and a dramatic decline in the pupfish population. The response to this crisis resulted in a dramatic victory for conservation when in 1976, the US Supreme Court affirmed that the federal government's reservation of Devils Hole implicitly included the water necessary to sustain the pupfish (Deacon and Deacon Williams 1991; Williams and Propst, chap. 1, this volume).

[Box. 24.1]

Devils Hole is a fracture in the marine carbonate bedrock that intersects the regional groundwater aquifer, exposed by a series of ceiling collapses that created a water-filled cave open at the land surface (Riggs and Deacon 2002). The rectangular water surface encompasses an area of 3.5 × 22 m. The majority of biological activity in the ecosystem occurs on a 2.6 × 6.1 m submerged "shallow shelf" (see fig. 24.1A) formed by a boulder perched between the fractured walls. The depth of water on the shallow shelf varies due to floods and earthquake activity, but at this writing it ranges from 0.02 to 0.87 m, with a mean depth of 0.29 m. Below the water surface, the cavern descends to a depth of at least 133 m, beyond which it has not been explored.

Physicochemical conditions in the deep pool are remarkably stable, with water temperatures of 33.4°C–34.0°C, dissolved oxygen concentrations of 1.9–2.2 mg/L, and pH values of 7.0–7.8 (Miller 1948; Bernot and Wilson 2012). Although background physicochemical conditions are relatively constant, they may be limiting to some biological processes. For example, dissolved oxygen in Devils Hole is low, and water temperatures are high, in relation to the physiological tolerances of many fishes. Over the shallow shelf, where water and the substrate receive direct solar radiation during some months, physicochemical conditions are more variable. For example, in summer, midday water temperatures can exceed 36°C (Hausner et al. 2013) and dissolved oxygen, driven by seasonally variable photosynthesis, can be 1.9–6.0 mg/L (Bernot and Wilson 2012).

This tiny habitat supports a wide variety of life, including over 80 species of algae (Sheppard et al. 2000) and approximately 15 benthic macroinvertebrates (Herbst and Blinn 2003). The Devils Hole pupfish *Cyprinodon diabolis* (Wales 1930) is the only vertebrate species in Devils Hole. The small (typically <30 mm long) and short-lived (10–14 months) fish was listed as endangered by the US Department of the Interior in 1967 (James 1969; Beissinger 2014).

Despite the US Supreme Court decision, the Devils Hole pupfish saga continues. Since the publication of *Battle against Extinction*, resource managers at Devils Hole have faced three subsequent crises—each one as severe as the struggle of the 1970s—triggered by shifts in pupfish population counts. Since the 1970s, scuba divers have regularly counted the Devils Hole pupfish population, compiling one of the longest-running datasets collected at Devils Hole (fig. 24.1*B*). Because these surveys have relied on consistent methodology over decades, the population counts (termed "relative abundance") are comparable to one another over time. In 1996 (crisis 2), autumn population counts declined, and by 2006 (crisis 3), a new low spring count of 38 fish was recorded. In 2013 (crisis 4), divers tallied the lowest number of pupfish in 40 years, at just 35 individuals. By autumn 2017, numbers had rebounded slightly to more than 100 pupfish, but are still well below historical levels.

In this chapter, we explore the varying management responses to crises in the Devils Hole ecosystem. We show that the ever-changing scope of scientific research and monitoring at Devils Hole has played a central role in defining—and at times limiting—the options available to conserve this iconic species and ecosystem. The Devils Hole experience demonstrates the critical importance of proactive monitoring, research, and planning, especially when endangered populations are not in decline. Devils Hole may be a tiny place, but these hard-earned lessons from the last two and a half decades can apply to endangered species conservation efforts in many arenas (Forbes et al. 2016; Mortenson and Reed 2016).

THE PAST IS STILL PRESENT

Until the 1960s, university researchers and National Park Service (NPS) managers knew very little about Devils Hole beyond the unique morphology of the pupfish that lived there (Wales 1930; Miller 1948) and the pool's water level, which the US Geological Survey (USGS) began monitoring in the mid-1950s. Then, in 1969, a local ranch began pumping groundwater from wells near Devils Hole to irrigate more than 4,050 ha (10,000 acres) of land. Almost immediately, the water level in Devils Hole began to fall, and the agencies tasked with management of Devils Hole faced the first of four management crises.

Crisis 1: 1969

Managers from federal and state agencies, along with university scientists, responded rapidly with efforts to save the pupfish as the Devils Hole habitat shrank. These efforts—which coalesced into the Desert Fishes Council and later forced the federal government to pursue legal action to protect the pupfish—immediately confronted the fact that almost no scientific research existed to guide emergency management measures or support a legal remedy for the water level decline.

The US Department of the Interior Solicitor's Office, for example, refused to pursue an injunction against the nearby agricultural groundwater pumping without more definite proof that those wells were actually responsible for the decline observed in Devils Hole (an alternative theory in 1970 suggested that nuclear detonations at the nearby Nevada Test Site could be partly responsible). It took federal agencies months to allocate funds for USGS to conduct a detailed analysis, ultimately delaying legal action by a full year (Brown 2017).

As the water level continued to decline in the early 1970s, scientists and agency managers—cognizant of the fish's short lifespan—believed they could not wait for a long-term solution and began to take emergency actions to forestall the pupfish's extinction. Beginning in 1970, managers transplanted over 200 Devils Hole pupfish to several natural and artificial refuges around the Mojave Desert. Inside Devils Hole, they installed artificial lighting over the pool to stimulate algal growth as well as a fiberglass shelf suspended over the deep portion of Devils Hole to simulate the rapidly desiccating shallow shelf.

None of these actions included any consistent monitoring of their effects, and some actually may have been detrimental to the pupfish. For example, by spring 1972, the green algae in Devils Hole had turned brown. Although incandescent lights had been in operation for almost two years, no one had identified the algae, or what light would work best to stimulate their growth. After consultation with an algae expert in Utah, fluorescent lights were installed (Brown 2017).

If the initial response to the 1969 crisis demonstrated how little ecological and hydrologic research the agencies had available for incorporating into management decisions, the resolution of the crisis confirmed the power of monitoring and science as the bedrock for pupfish conservation. In 1976, the US Supreme Court decided in *Cappaert v. United States* that the federal government had a water right sufficient to support the pupfish in Devils Hole. The court did not specify exactly how much water the pupfish needed, returning that critical question to the federal district court in Las Vegas.

Using data collected from monthly pupfish counts (started in 1972 using scuba divers), data on solar insolation, and a model of the shallow shelf size at different water levels, James Deacon argued in November 1976 that the pupfish required at least enough water to reach 0.82 m (2.7 feet) below a reference marker installed in Devils Hole, which was enough water to cover the shallow shelf, but still more than 0.46 m (1.5 feet) below the pre-pumping level. Judge Roger Foley agreed and modified the permanent injunction to this minimum water level (*United States v. Cappaert* 1978). In a further nod to the importance of scientific data, the court maintained continuing jurisdiction over the water level, allowing for the possibility of raising or lowering the legally mandated level in response to future research. Ecological and hydrologic research—as much as federal water law—led to the resolution of this crisis. It was a profound lesson in the importance of monitoring, but one that over the following decades proved easily forgotten.

Crisis 2: 1996

In autumn 1996, Doug Threloff, a Death Valley National Park employee, was alarmed by a lower than usual autumn pupfish count (fig. 24.2), and he wrote in a letter to cooperators in the pupfish program that "the number of fish which were counted . . . appeared to be significantly lower" than past counts (Threloff 1996). Threloff's concern was justified: autumn pupfish counts have never been higher since.

In contrast to the decline in pupfish numbers in the early 1970s, the reasons for this decline were not obvious. The water level remained comparatively stable and did not correlate with the observed decline in the autumn population. James Deacon had conducted limited ecological monitoring, along with population counts, in the late 1970s and early 1980s, but this was discontinued in 1985 (Brown 2017). As long as population counts were high and backup refuge populations existed at Hoover Dam and two other sites (Karam 2005;

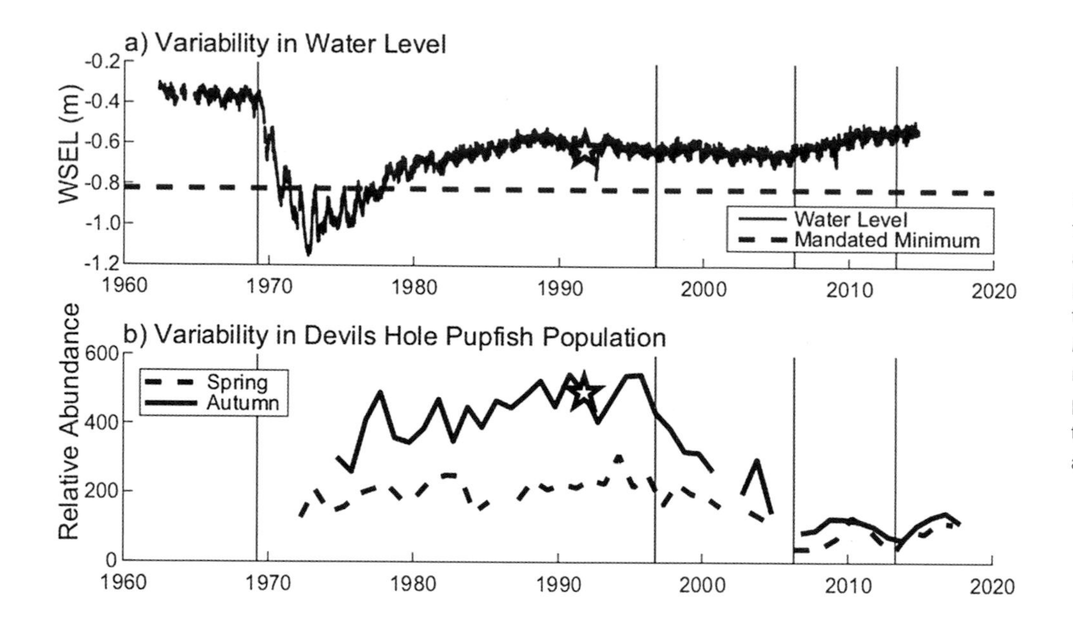

Fig. 24.2 Variability in (*A*) water surface elevation (WSEL) and (*B*) Devils Hole pupfish population numbers over time. The 1991 publication of *Battle Against Extinction* is marked by the star in each plot, and the time of each of the four crises is indicated by a vertical line.

Brown 2017), Devils Hole was not thought to warrant the resources necessary for ecosystem-level monitoring from already thin NPS budgets.

Besides water level data and pupfish population counts, the managing agencies (NPS, US Fish and Wildlife Service [USFWS], and the Nevada Department of Wildlife [NDOW]) had little data to go on. During the late 1990s and early 2000s, the agencies, informed by a 1998 workshop of university scientists and resource managers, entertained a lengthy list of hypotheses to explain the population decline. The explanations considered included the illegal removal of some of the population, errors in the scuba population survey methodology, unidentified natural population cycles, and phthalates and other contaminants reducing recruitment success (Brown 2017). More complex and troubling hypotheses came from ecologists, who wondered whether observed changes in invertebrate and algal communities had made the habitat less welcoming to pupfish, and from geneticists, who wondered whether the chronically low population numbers of the pupfish were resulting in a kind of genetic meltdown (Riggs and Deacon 2002).

With such divergent ideas, the agencies' first concrete action was to undertake more extensive research. This research began in 1999 with funding from the USGS to examine the trophic ecology, food-web dynamics, and energy flow of the Devils Hole ecosystem (Wilson and Blinn 2007) and to perform a more detailed aquatic macroinvertebrate study (Herbst and Blinn 2003). Furthermore, the mid-1990s decline in pupfish coincided with the availability of new techniques in genetic analysis. Martin and Wilcox (2004) examined the evolutionary history of the Ash Meadows pupfishes (including the Devils Hole pupfish), revealing extremely low genetic diversity among Devils Hole pupfish and reinforcing concerns about inbreeding depression.

The community ecology and bioenergetics study dramatically expanded the agencies' understanding of the ecosystem and provided the rationale for specific management actions aimed at reversing the slide in the pupfish population. For example, the agencies removed gravel deposited on the shallow shelf by floods (Brown 2017) after research suggested that such deposits made Devils Hole less hospitable to pupfish (NPS, unpublished data). Unfortunately, these actions were not coupled with a clear monitoring protocol, and their effects remained uncertain.

In 2005, genetic analysis was used to address a specific management issue: Had the Devils Hole pupfish population in the Point of Rocks refuge hybridized with Ash Meadows Amargosa pupfish *Cyprinodon nevadensis mionectes*? Managers suspected hybridization after a pupfish accidentally killed in the course of research at the refuge was found to have pelvic fins (Karam 2005). Unlike other Death Valley pupfishes, Devils Hole pupfish rarely express pelvic fins (Wales 1930). Development of these fins may be thermally regulated (Lema and Nevitt 2006), and it was therefore possible that the lower water temperature in the refuge, and not hybridization, had caused the morphological change in the fish. However, microsatellite DNA analysis from the refuge fish revealed that they were indeed hybridized (Martin 2005). In 2006, the managing agencies removed the hybrid pupfish to two captive propagation sites for use in future research (Brown 2017).

Although this second management crisis spurred a growing research effort and increasingly informed management at Devils Hole, pupfish population numbers continued their downward trajectory through this period. Despite proposals for long-term ecological monitoring, no such plan was adopted. Moreover, researchers and managers continued to lament the fact that ecosystem-level research had not been conducted during the 1980s and early 1990s when the population was stable. The newly implemented studies could explain how Devils Hole worked after 1998, but could shed little light on ecosystem dynamics during that earlier period when pupfish numbers were greater. And an unanswered question remained: What caused the sudden decline in the pupfish population?

Crisis 3: 2006

The continued decline in pupfish numbers after 1995 revealed substantial shortcomings not only in knowledge of the Devils Hole ecosystem, but also in the capacity of NPS, USFWS, and NDOW to effectively cooperate on management decisions. By April 2006, when the managing agencies faced a third crisis after a population survey counted just 38 pupfish in Devils Hole, there had been a near-complete breakdown in communication among the agencies (Brown 2017). With extinction looming, the managing agencies were addressing fast-moving management questions with little regard for coordinated long-term research and planning. The low spring survey numbers indicated that the natural population of Devils Hole pupfish was not sustainable, and therefore the refuge-based backup populations were thought to represent the best—and only—hope for survival of the species. However, the existing refuges had largely failed. The Devils Hole pupfish population at Point of Rocks had hybridized with *C. n. mionectes* (Martin 2005), and the Hoover Dam refuge population had been declining for three years. Pure individuals were harvested from both the Hoover Dam refuge and Devils Hole and transferred to Willow Beach National Fish Hatchery, a USFWS facility on the Colorado River in Arizona (Brown 2017). No backup population survived, leaving the few remaining fish in Devils Hole as the sole extant Devils Hole pupfish population. With the failure of these transplantations, the managing agencies had few other actions available to them.

Through the spring and summer of 2006, management decisions were made on an ad hoc basis, and communication problems and disagreements continued. When J. T. Reynolds, then superintendent of Death Valley National Park, pulled an idea from an unlikely source—wildland firefighting—the managing agencies adopted a new management structure to improve cooperative decision making and planning. This change proved a turning point in the agencies' ability to coordinate research and management. The "incident command system" adopted by the Devils Hole pupfish recovery team in autumn 2006 allows personnel from multiple state and federal agencies to share responsibility for an emergency response, and it integrates staff from agencies with varying management structures into a unified decision-making framework (Cole 2000). The system provides each agency (USFWS, NDOW, and NPS) an equal vote in each of a three-tiered set of decision-making groups. When the lowest-level group (the incident command team, or ICT) cannot come to a consensus, or when the issue at hand involves "take" as defined by the Endangered Species Act, a higher-level supervisory group steps in to resolve the impasse.

One of the first decisions made by the ICT came during the winter of 2006–2007, when biologists noticed that the fish in Devils Hole appeared emaciated. With pupfish numbers critically low, the ICT directed NPS staff to begin supplemental feeding of the pupfish, despite a lack of historical data for comparison. Although the frequency and composition of the feedings have been adjusted over time, the practice is ongoing.

The ICT has been enabled by the creation of the Death Valley National Park Aquatics Program. Initially staffed by a single full-time fish biologist and supported by a second biologist temporarily detailed to the office, the Aquatics Program now includes four full-time NPS staff. The increase in staffing allowed ICT-approved projects, such as the supplemental feeding, to move forward. Moreover, the program collects, and archives, ecological data with the guidance of a long-sought and recently developed Long Term Ecosystem Monitoring Plan (LTEMP) (Wilson et al. 2009), which is a compendium of standard operating procedures for collecting a wide variety of ecological data. Reviewed and updated regularly, the LTEMP calls for the collection and maintenance of data to support the management of Devils Hole, and is not intended to be a vehicle for academic research. However, the collected data can lead to the identification of unanswered questions and point toward knowledge gaps that need to be addressed. These data gaps can be filled through collaborations with academic researchers, who can perform in-depth analyses of data that cannot be done in-house. Prioritized and steered by the ICT, such projects produce peer-reviewed work to inform management decisions. With the LTEMP as a guide for identifying questions, these studies have examined oxygen dynamics and ecosystem energetics (Bernot and Wilson 2012; Madinger et al. 2016), circulation of water and heat in the system (Hausner et al. 2012, 2013), errors and uncertainties in pupfish surveys (Dzul et al. 2012, 2013b), climate change and pupfish recruitment (Hausner et al. 2014, 2016), pupfish population dynamics (Barrett 2009; Dzul et al. 2013a), spawning behavior (Chaudoin et al. 2015), evolutionary history (Reed and Stockwell 2014; Martin et al. 2016; Sağlam et al. 2016), and physiology (Heuton et al. 2015; Jones et al. 2016), as well as artificial refuge populations (Karam et al. 2012) and the propagation of both hybrid and Devils Hole pupfish in captivity (Feuerbacher et al. 2015, 2016, 2017).

Crisis 4: 2013

The importance of this work became immediately apparent in 2013, when the managing agencies were faced with a spring survey that counted just 35 pupfish. In contrast to previous crises, the management response to this survey was informed by the research that had been undertaken since 2006. For example, implementation of the LTEMP had shown recent changes in both the quantity and composition of cover on the shallow shelf, which were exacerbated by floods and seiches in 2012 and early 2013 (NPS, unpublished data). In response, the ICT recommended packets of artificial cover (native vegetation bundled with organic twine) to add spatial complexity to the habitat. The recommendation was based on experimental data showing that such cover reduced cannibalism on early-life-stage (ELS) hybrid pupfish in aquaria (S. Hillyard, pers. comm.; O. G. Feuerbacher, pers. comm.), and it was accompanied by a monitoring plan (including control areas) to assess its effectiveness in Devils Hole.

Nowhere was the value of recent research more apparent than in the opening and stocking of the newest Devils Hole pupfish refuge, the Ash Meadows Fish Conservation Facility (AMFCF). Initially conceived as a response to the low numbers found in the 2006 and 2007 surveys, the $4.5 million AMFCF features a 418,000 L tank and a state-of-the-art automated control room designed to prevent the mechanical and oversight failures that doomed previous refuge populations. The system mimics the water chemistry and environmental conditions of the upper reaches of Devils Hole. Construction of the facility was completed in 2012, and it was stocked with Devils Hole pupfish in 2013.

Drawing on recent research conducted at Devils Hole, Beissinger (2014) constructed a population model analyzing the risk of extinction under a variety of harvesting scenarios. Ultimately, Beissinger recommended harvesting eggs from Devils Hole in autumn, when Devils Hole is entering a period of seasonal food limitation and newly hatched pupfish are less likely to survive to maturity. This recommendation was first implemented in autumn 2013, and the harvested eggs were hatched and reared to adulthood in aquaria in the AMFCF. The adult fish were transferred to the refuge tank in the spring of 2014, and ELS pupfish were first observed in the tank in August 2014. To maintain genetic diversity in the captive pupfish population, eggs are regularly harvested from Devils Hole, hatched in captivity, and added to the captive stock. The stocking of the AMFCF appears to have had little effect on the Devils Hole population, which has steadily increased in abundance since the spring 2013 low (the autumn 2017 survey counted 115 pupfish).

The crises in 2006 and 2013 had the same root (low spring population counts), and the goal of the management response to both crises was the same: to establish a sustainable backup population. However, the specific responses and their differing results show the importance of ecosystem data and scientific research. In 2006, adult pupfish were transferred both from Devils Hole and between refuges on an ad hoc basis. Although this action was considered the best of a suite of bad options, understanding of the feasibility of such transfers was lacking. Removing adult fish from Devils Hole risked

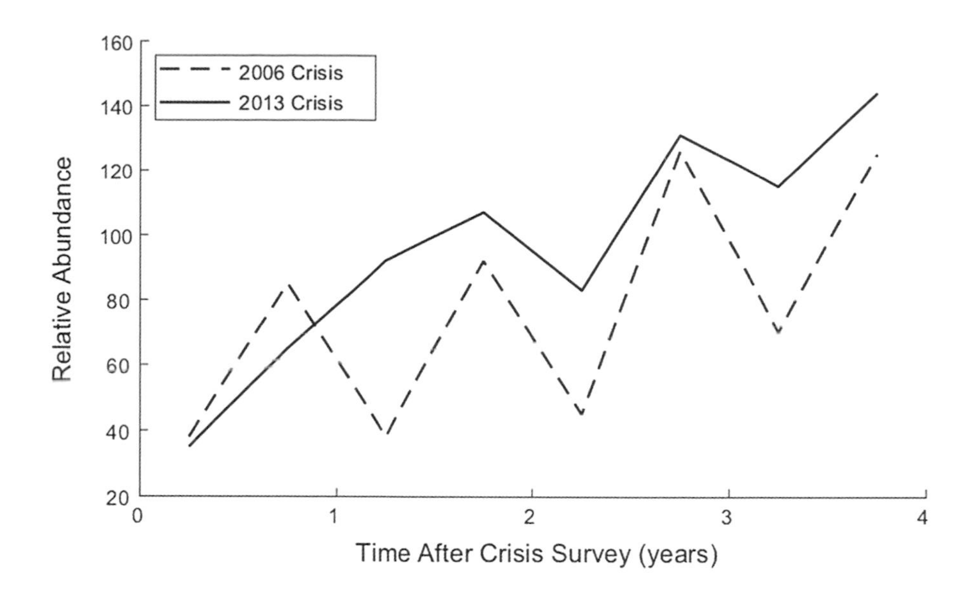

Fig. 24.3 Population surveys in the four years following crises 3 (2006) and 4 (2013) plotted on the same axes. Both crises begin in the spring of year 0; the dashed line represents 2006–2009, and the solid line represents 2013–2016.

both the relocated fish (who seldom survived more than 24 hours) and the extant Devils Hole population (Brown 2017). And while the attempt to establish a captive population as a response to the 2006 crisis was a clear failure, it raised questions about the stocking and management of potential future refuges. By 2013, the ICT could draw on years of research to stock the refuge population with minimal adverse effect on the Devils Hole population (Beissinger 2014). It was not possible to directly attribute results to the use of cover packets and stocking strategies, but the population counts in the three years following the 2013 crisis were consistently larger than those after 2006, especially in the spring (fig. 24.3).

WHERE WE NEED TO GO

Today, the cooperating agencies have built the capacities to ask questions, generate information, and make collective decisions to protect Devils Hole and the Devils Hole pupfish in ways they had not in the past. Still, the legacy of earlier crises at Devils Hole is evident, both in the historical data available and in the silences that surround questions and data that went unasked and uncollected, respectively, in earlier eras. And despite the current upward trajectory of the pupfish population, its numbers are still less than half of what they were before the crash of the mid-1990s, and the extent of the existing population's dependence on supplemental feeding and habitat manipulations is unknown.

The path forward for pupfish management requires integrating existing scientific research into decision making and continually assessing what issues, problems, and questions require further study and analysis. One new endeavor aimed at meeting this goal is the development of a strategic plan. The first stage of this project, which has been completed, reviews past stresses and changes in the Devils Hole ecosystem and management responses to them, and identifies potential future management actions for Devils Hole (Hausner 2017). The project has also identified several areas in which future scientific research and policy decisions can play clear roles in improving the understanding of the ecological dynamics of Devils Hole and the conservation of the pupfish. Below, we consider three broad areas we believe require further research to provide managers with more information: climate change and ecosystem disturbances, pupfish genetics and artificial refuge management, and desert water and land use.

Climate Change and Disturbances

Climate change has already directly affected the Devils Hole pupfish population by shortening the period each spring during which conditions are best for recruitment (Hausner et al. 2014). Water temperatures in Devils Hole are at the extreme for fish-occupied ecosystems (Riggs and Deacon 2002), and the reproductive life stage of *Cyprinodon* spp. is strongly sensitive to water temperature (Shrode 1975; Shrode and Gerking 1977). As the desert around Devils Hole has warmed, maximum daily water temperatures have increased, and egg-killing temperature spikes have already reduced the potential recruitment period in Devils Hole by about a week; by the end of this century, the optimal spring recruitment period is predicted to be shorter by two and a half weeks (Hausner 2014), and potentially more if the warming is accompanied by changes in water depth (Hausner et al. 2016).

The usually placid surface of the Devils Hole ecosystem is occasionally agitated by two kinds of major disturbance events: floods and earthquake-induced seiches. Both disturbances rearrange sediments on the shallow shelf, changing the bathymetry of this crucial habitat. Flooding from intense localized rainstorms can deliver sediment and nutrients that can reduce water depth and bury eggs and algae, driving potentially toxic anaerobic conditions in the sediment (Lyons 2005; Hausner 2017). The frequency of extreme precipitation events in the southwestern United States is projected to fluctuate significantly with climate change (Kunkel et al. 2013). In addition to the effects of floods, earthquake-induced seiches can generate large waves within Devils Hole (Cutillo and Ge 2006; Weingarten and Ge 2014) that tend to sort and remove sediments, as well as removing algae, aquatic invertebrates, and pupfish eggs, from the shallow shelf. Both of these phenomena present unique challenges for management at Devils Hole. Not only do they occur sporadically and unpredictably, but research suggests that they have both positive and negative effects on the Devils Hole ecosystem—floods deliver beneficial allochthonous carbon to an energy-limited ecosystem (Wilson and Blinn 2007), and earthquake-induced seiches remove fine sediments from the shelf and appear to induce spawning activity (Minckley and Deacon 1973; Chaudoin et al. 2015).

Genetic and Refuge Considerations

Genetic research represents another area that will shape future management. A major management failure at the Hoover Dam, School Spring, and Point of Rocks refuges from the 1970s through 2006 was lack of attention to genetic diversity in the refuge stock. Although the Devils Hole Pupfish Recovery Plan called for the annual transfer of 10 individuals to each refuge (Deacon et al. 1980), this practice was rarely followed (Brown 2017). The lack of monitoring led to substantial genetic drift between *C. diabolis* populations in Devils Hole and the refuge populations (Wilcox and Martin 2006). The USFWS is currently developing a new genetic management plan for both the naturally occurring Devils Hole pupfish population and the refuge population in the AMFCF. The genetic management plan will support the two main purposes of the refuge: (1) to serve as a backup in the case of a catastrophe in Devils Hole, and (2) to offer a productive place for research deemed too risky for Devils Hole itself.

Effects of Water and Land Use

The effect of local and regional groundwater pumping on Devils Hole remains an important danger to the pupfish's long-term survival. The local groundwater basin is over-allocated, and future developments using existing water rights—such as a long-planned solar energy development several kilometers west of Devils Hole—may reduce the local water level (BLM 2010). Additionally, existing and future water withdrawals from within the regional groundwater system may alter regional flow patterns and water availability at Devils Hole (Deacon et al. 2007).

Litigation offers a powerful tool for confronting such problems. In 1976, the US Supreme Court (in *Cappaert v. United States*) affirmed a minimum water level in Devils Hole that was set by the district court in 1974 and revised in 1978 (*United States v. Cappaert* 1974, 1978). The Nevada Division of Water Resources, which manages Nevada water rights, has issued multiple orders to protect this minimum level (Newman 1979; Taylor 2008). However, the minimum is not set in stone, and the current water level in Devils Hole is approximately 30 cm above the mandated level. If the federal government could prove in court that the pupfish required a higher water level to survive, it could argue for an increased minimum. An increased water right dating back to 1952 (the original date of reservation) would have priority over many existing users in the basin, but would be extremely controversial in the community.

Water issues are often inextricably tied to land issues. Currently, the US Bureau of Land Management is reviewing its policy on a range of issues, including the sale (disposal) of public lands. In 2014, the agency's preferred option included the disposal of additional federal land on the outskirts of Pahrump (BLM 2014), making it available for residential development. Domestic wells on residential properties are not subject to the same state water permitting required for other uses, and therefore wells drilled on new residential lots pose a potential threat to the water level in Devils Hole.

If impersonal structural forces potentially threaten Devils Hole, so too do individual actions, including those of researchers. In 2004, for example, improperly stored larval fish traps washed into Devils Hole during a flash flood and killed approximately 80 pupfish, perhaps one-third of the total population (Lyons 2005). Malicious individual acts also pose threats to Devils Hole, an issue vividly demonstrated by the April 2016 alcohol-fueled break-in during which a trespasser waded across the shallow shelf; NPS later discovered a dead pupfish (Rocha 2016). This act was just the latest in a long line of incidents dating back to the 1950s (Brown 2017). Although the fish trap incident suggests the need for continued vigilance in research protocols and safety requirements, it is more difficult for NPS to respond to malicious threats. When compared with many earlier events, the unique aspect of the 2016 break-in is simply that that the episode was captured on video—a result of security equipment installed after the recent completion of NPS's Devils Hole Site Plan.

SUMMARY

In *Battle against Extinction*, Deacon and Deacon Williams (1991) described the "legacy" of the Devils Hole pupfish. They argued that the fight to save the species in the 1960s and 1970s showed that society could recognize the right of a species to persist, evolve, and go extinct on its own terms. In the Devils Hole pupfish, they saw a "symbiotic evolution of species and ethics." The last 25 years suggest that another legacy has emerged from the management of the iconic pupfish and its ecosystem: a reminder to scientists and resource managers of the importance of studying critical habitats and ecosystems, regardless of whether or not a species is in crisis. Today, managers at Devils Hole have internalized this lesson. They continue to implement an adaptive management approach that considers the best available science, proactively identifies future scientific needs, and continues ecosystem-level research and monitoring.

What will the legacy of Devils Hole be in another 25 years? Although we cannot predict the future of Devils Hole, whatever challenges the ecosystem faces will be met with decisions made in the context of ecosystem-level research, with monitoring data compiled and analyzed over time to drive specific research hypotheses. This science-based approach to management will be a central facet of that future legacy. With better and more informed decisions, we can hope that another legacy will be the continued persistence of the incredible Devils Hole ecosystem and its devilish pupfish.

REFERENCES

Barrett, P. J. 2009. Estimating Devils Hole pupfish lifestage ratios using the Delphi method. *Fisheries* 34: 73–79.

Beissinger, S. R. 2014. Digging the pupfish out of its hole: Risk analyses to guide harvest of Devils Hole pupfish for captive breeding. *PeerJ* 2: e549. https://doi.org/10.7717/peerj.549.

Bernot, M. J., and K. P. Wilson. 2012. Spatial and temporal variation of dissolved oxygen and ecosystem energetics in Devils Hole, Nevada. *Western North American Naturalist* 72: 265–75.

BLM (US Bureau of Land Management). 2010. Notice of Availability of Final Environmental Impact Statement for the Solar Millennium, Amargosa Farm Road Solar Power Project, Nye County Nevada, 75 FR 63503.

BLM (US Bureau of Land Management). 2014. *Draft Resource Management Plan/Environmental Impact Statement*. Las Vegas: US Bureau of Land Management.

Brown, K. C. 2017. *Recovering the Devils Hole Pupfish: An Environmental History*. Death Valley National Park, CA: National Park Service.

Cappaert v. United States, 426 US 128 (1976).

Chaudoin, A. L., O. G. Feuerbacher, S. A. Bonar, and P. J. Barrett. 2015. Underwater videography outperforms above-water videography and in-person surveys for monitoring the spawning of Devils Hole pupfish. *North American Journal of Fisheries Management* 35: 1252–62.

Cole, D. 2000. The Incident Command System: A 25-Year Evaluation by California Practitioners. National Fire Academy. http://www.alnap.org/pool/files/efo31023.pdf.

Cutillo, P. A., and S. Ge. 2006. Analysis of strain-induced groundwater fluctuations at Devils Hole, Nevada. *Geofluids* 6: 319–33.

Deacon, J. E., and C. Deacon Williams. 1991. Ash Meadows and the legacy of the Devils Hole pupfish. In W. L. Minckley and J. E. Deacon, eds., *Battle against Extinction: Native Fish Management in the American West*, 69–87. Tucson: University of Arizona Press.

Deacon, J. E., D. Lockard, O. Casey, G. Kobetich, J. Radtke, H. Gunther, and D. Soltz. 1980. *Devils Hole Pupfish Recovery Plan*. Portland: US Fish and Wildlife Service.

Deacon, J. E., A. E. Williams, C. Deacon Williams, and J. E. Williams. 2007. Fueling population growth in Las Vegas: How large-scale groundwater withdrawal could burn regional biodiversity. *BioScience* 57: 688–98.

Dzul, M. C., S. J. Dinsmore, M. C. Quist, D. B. Gaines, K. P. Wilson, M. R. Bower, and P. M. Dixon. 2013a. A simulation model of the Devils Hole pupfish population using monthly length-frequency distributions. *Population Ecology* 55: 325–41.

Dzul, M. C., P. M. Dixon, M. C. Quist, S. J. Dinsmore, M. R. Bower, K. P. Wilson, and D. B. Gaines. 2013b. Using variance components to estimate power in a hierarchically nested sampling design. *Environmental Monitoring and Assessment* 185: 405–14.

Dzul, M. C., M. C. Quist, S. J. Dinsmore, P. M. Dixon, M. R. Bower, K. P. Wilson, and D. B. Gaines. 2012. Identifying source of error in surveys of Devils Hole pupfish (*Cyprinodon diabolis*). *Southwestern Naturalist* 57: 44–50.

Feuerbacher, O., S. A. Bonar, and P. J. Barrett. 2016. Design and testing of a mesocosm-scale habitat for culturing the endangered Devils Hole pupfish. *North American Journal of Aquaculture* 78: 259–69.

Feuerbacher, O., S. A. Bonar, and P. J. Barrett. 2017. Enhancing hatch rate and survival in laboratory-reared hybrid Devils Hole pupfish through application of antibiotics to eggs and larvae. *North American Journal of Aquaculture* 79: 106–14.

Feuerbacher, O. G., J. A. Mapula, and S. A. Bonar. 2015. Propagation of hybrid Devils Hole pupfish × Ash Meadows Amargosa pupfish. *North American Journal of Aquaculture* 77: 513–23.

Forbes, V. E., N. Galic, A. Schmolke, J. Vavra, R. Pastorok, and P. Thorbek. 2016. Assessing the risks of pesticides to threatened and endangered species using population modeling: A critical review and recommendations for future work. *Environmental Toxicology and Chemistry* 35: 1904–13.

Hausner, M. B. 2017. *A Chronological Inventory and Review of Management Practices at Devils Hole to Support Development of a Strategic Plan*. Death Valley National Park, CA: National Park Service.

Hausner, M. B., K. P. Wilson, D. B. Gaines, F. Suárez, G. G. Scoppettone, and S. W. Tyler. 2014. Life in a fishbowl: Prospects for the endangered Devils Hole pupfish (*Cyprinodon diabolis*) in a changing climate. *Water Resources Research* 50: 1–15.

Hausner, M. B., K. P. Wilson, D. B. Gaines, F. Suárez, G. G. Scoppettone, and S. W. Tyler. 2016. Projecting the effects of climate change and water management on Devils Hole pupfish (*Cyprinodon diabolis*) survival. *Ecohydrology* 9: 560–73.

Hausner, M. B., K. P. Wilson, D. B. Gaines, F. Suárez, and S. W. Tyler. 2013. The shallow thermal regime of Devils Hole, Death Valley National Park. *Limnology and Oceanography: Fluids and Environments* 3: 119–38.

Hausner, M. B., K. P. Wilson, D. B. Gaines, and S. W. Tyler. 2012. Interpreting seasonal convective mixing in Devils Hole, Death Valley National Park, from temperature profiles observed by fiber-optic distributed temperature sensing. *Water Resources Research* 48: W05513.

Herbst, D. B., and D. W. Blinn. 2003. *Devils Hole Benthic Invertebrate Community Dynamics: Distribution, Seasonality and Production*. Death Valley National Park, CA: National Park Service.

Heuton, M., L. Ayala, C. Burg, K. Dayton, K. McKenna, A. Morante, G. Puentedura, N. Urbina, S. Hillyard, S. Steinberg, and F. van Breukelen. 2015. Paradoxical anaerobism in desert pupfish. *Journal of Experimental Biology* 218: 3739–45.

James, C. J. 1969. Aspects of the ecology of the Devils Hole pupfish, *Cyprinodon diabolis* Wales. MS thesis, University of Nevada, Las Vegas.

Jones, A. C., D. Lim, J. J. Wayne-Thompson, N. Urbina, G. Puentedura, S. Hillyard, and F. van Breukelen. 2016. Oxygen consumption is limited at an ecologically relevant rearing temperature in pupfish eggs. *Journal of Experimental Zoology Part A—Ecological Genetics and Physiology* 325: 539–47.

Karam, A. P. 2005. History and development of refuge management for Devils Hole pupfish (*Cyprinodon diabolis*) and an ecological comparison of three artificial refuges. MS thesis, Southern Oregon University.

Karam, A. P., M. S. Parker, and L. T. Lyons. 2012. Ecological comparison between three artificial refuges and the natural habitat for Devils Hole pupfish. *North American Journal of Fisheries Management* 32: 224–38.

Kunkel, K. E., L. E. Stevens, S. E. Stevens, L. Sun, E. Janssen, D. Wuebbles, K. T. Redmond, and J. G. Dobson. 2013. *Regional Climate Trends and Scenarios for the US National Climate Assessment*: Part 5. *Climate of the Southwest US* Technical Report NESDIS 142-5. Washington, DC: NOAA.

Lema, S. C., and G. A. Nevitt. 2006. Testing an ecophysiological mechanism of morphological plasticity in pupfish and its relevance to conservation efforts for endangered Devils Hole pupfish. *Journal of Experimental Biology* 209: 3499–509.

Lyons, L. T. 2005. Temporal and spatial variation in larval Devils Hole pupfish (*Cyprinodon diabolis*) abundance and associated microhabitat variables in Devils Hole, Nevada. MS thesis, Southern Oregon University.

Madinger, H. L., K. P. Wilson, J. A. Goldstein, and M. J. Bernot. 2016. Biogeochemistry and nutrient limitation of microbial biofilms in Devils Hole, Nevada. *Western North American Naturalist* 76: 53–71.

Martin, A. P. 2005. Genetic analysis of *Cyprinodon diabolis*: Hybridization with *C. nevadensis* in the Point of Rocks Refuge. Department of the Interior, National Park Service, US Fish and Wildlife Service.

Martin, A. P., and J. L. Wilcox. 2004. Evolutionary history of Ash Meadows pupfish (genus *Cyprinodon*) populations inferred using microsatellite markers. *Conservation Genetics* 5: 769–82.

Martin, C. H., J. E. Crawford, B. J. Turner, and L. H. Simons. 2016. Diabolical survival in Death Valley: Recent pupfish colonization, gene flow, and genetic assimilation in the smallest species range on Earth. *Proceedings of the Royal Society B* 283: 20152334.

Miller, R. R. 1948. *The Cyprinodont Fishes of the Death Valley System of Eastern California and Southwestern Nevada*. Miscellaneous Publications of the Museum of Zoology, University of Michigan, 68. Ann Arbor: University of Michigan Press.

Minckley, C. O., and J. E. Deacon. 1973. Observations on the reproductive cycle of *Cyprinodon diabolis*. *Copeia* 1973: 610–13.

Mortenson, J. L., and J. M. Reed. 2016. Population viability and vital rate sensitivity of an endangered avian cooperative breeder, the white-breasted thrasher (*Ramphocinclus brachyurus*). *PLoS ONE* 11: e0148928.

Newman, W. J. 1979. Order 724. Office of the State Engineer, Nevada Division of Water Resources, Carson City, NV.

Reed, J. M., and C. A. Stockwell. 2014. Evaluating an icon of population persistence: The Devils Hole pupfish. *Proceedings of the Royal Society B* 281: 20141648.

Riggs, A. C., and J. E. Deacon. 2002. Connectivity in desert aquatic ecosystems: The Devils Hole story. In D. W. Sada and S. E. Sharpe, eds., *Conference Proceedings, 2002, Spring-Fed Wetlands: Important Scientific and Cultural Resources of the Intermountain Region, May 7–9,2002*. Paradise, NV: Desert Research Institute. http://www.dri.edu/images/stories/conferences_and_workshops/spring-fed-wetlands/spring-fed-wetlands-riggs-deacon.pdf.

Rocha, V. 2016. Guns, beer and vomit: Rampage leaves endangered fish dead in Death Valley. *Los Angeles Times*, May 9, 2016.

Sağlam, İ. K., J. Baumsteiger, M. J. Smith, J. Linares-Casenave, A. L. Nichols, S. M. O'Rourke, and M. R. Miller. 2016. Phylogenetics support an ancient common origin of two scientific icons: Devils Hole and Devils Hole pupfish. *Molecular Ecology* 25: 3962–73.

Shepard, W. D., D. W. Blinn, R. J. Hoffman, and P. T. Kantz. 2000. Algae of Devils Hole, Nevada, Death Valley National Park. *Western North American Naturalist* 60: 410–19.

Shrode, J. B. 1975. Developmental temperature tolerance of a Death Valley pupfish (*Cyprinodon nevadensis*). *Physiological Zoology* 48: 378–89.

Shrode, J. B., and S. D. Gerking. 1977. Effects of constant and fluctuating temperature on reproductive performance of a desert pupfish, *Cyprinodon n. nevadensis*. *Physiological Zoology* 50: 1–10.

Taylor, T. 2008. Order 1197. Office of the State Engineer, Nevada Division of Water Resources, Carson City, NV.

Threloff, D. 1996. Subject: Fall 1996 population counts for *Cyprinodon diabolis*. Death Valley National Park, Death Valley, CA. Memo to Files, November 7.

United States v. Cappaert, 375 F.Supp. 456 (1974).

United States v. Cappaert, 455 F.Supp. 81 (1978).

Wales, J. H. 1930. Biometrical studies of some races of cyprinodont fishes, from the Death Valley region, with description of *Cyprinodon diabolis*, n. sp. *Copeia* 1930: 61–70.

Weingarten, M., and S. Ge. 2014. Insights into water level response to seismic waves: A 24-year high-fidelity record of global seismicity at Devils Hole. *Geophysical Research Letters* 41: 74–80.

Wilcox, J. L., and A. P. Martin. 2006. The devil's in the details: Genetic and phenotypic divergence between artificial and native population of endangered pupfish. *Animal Conservation* 3: 316–21.

Wilson, K. P., and D. W. Blinn. 2007. Food web structure, energetics, and importance of allochthonous carbon in a desert cavernous limnocrene: Devils Hole, Nevada. *Western North American Naturalist* 67: 185–98.

Wilson, K. P., M. R. Bower, C. A. Gable, J. T. Back, G. M. Nelson, S. D. Hillyard, Z. L. Marshall, D. B. Gaines, and D. W. Blinn. 2009. *Devils Hole Ecosystem Monitoring Plan*. Death Valley National Park, NV: National Park Service.

25 Politics, Imagination, Ideology, and the Realms of Our Possible Futures

Christopher Norment

We live in a political world.
—BOB DYLAN

As I consider the fate of southwestern aquatic habitats—the realms of their possible futures and the ways in which these possibilities are affected by our priorities and values—I know that my thoughts are shaped by the context of this particular historical moment. If I had written this chapter in 2015, it would have been far different from the one that I began during the first half of 2017 and finished revising in early 2019: more optimistic, not so political, and less burdened by despair and uncertainty. In a similar way, the chapters in *Battle against Extinction* (Minckley and Deacon 1991a) were written and edited from a particular intellectual, emotional, scientific, political, and historical vantage point: the transition between the Ronald Reagan and George H. W. Bush administrations, 20 years after the founding of the Desert Fishes Council, and 12 years after the US Supreme Court's *Cappaert v. United States* decision upheld the federal government's right to reserve groundwater to protect the Devils Hole pupfish *Cyprinodon diabolis* (Riggs and Deacon 2002; Williams and Propst, chap. 1, this volume). Ash Meadows National Wildlife Refuge was 4 years old, while the US Endangered Species Act (ESA) of 1973, although criticized by some (Clark 1994; Carroll et al. 1996), promised at least partial protection for the most vulnerable taxa and their habitats.

Although *Battle against Extinction* begins with the chapter "Discovery and Extinction of Western Fishes: A Blink of the Eye in Geologic Time" (Minckley and Douglas 1991), and the second section of the book bears the title "Spirals toward Extinction: Actions and Reactions," its overall mood seems cautiously optimistic. *Battle against Extinction* acknowledged the myriad extinctions of desert fishes and the widespread destruction of their habitats, but it also described the positive effects of increased scientific knowledge, improved management techniques, a strengthening environmental movement, and better judicial, regulatory, and legislative protection. Progress toward conserving native fishes and their habitats in the American West was hesitant, and many species remained vulnerable, but there was a clear chronology of success, including formation of the Desert Fishes Council (1968), passage of the ESA (1973), the US Supreme Court's decision (1976), and establishment of Ash Meadows National Wildlife Refuge (1984). Thus, at the end of *Battle against Extinction*, Deacon and Minckley (1991, 405) could write, "A review of the information presented in the book demonstrates a significant and positive transition from attempts to promote awareness of the plight of endangered species to development of comprehensive programs aimed at preventing their further decline." The Battle against Extinction was not won, but it had been joined, and there were many reasons for hope.

I interviewed Jim Deacon several times toward the end of his life, while working on *Relicts of a Beautiful Sea: Survival, Extinction, and Conservation in a Desert World* (Norment 2014)—my argument for the importance of conserving aquatic species and their habitats in the West. Jim remained concerned about threats to the region's aquatic biodiversity posed by urban growth and increased water demand (Deacon et al. 2007) and was disheartened by the Southern Nevada Water Authority's (SNWA) proposal to develop groundwater resources in northern and central Nevada (SNWA 2009). However, he retained some of the cautious optimism that characterized *Battle against Extinction*. He had witnessed the near-extinction of the Devils Hole pupfish, and in the early 1970s thought that "many of the endemic Ash Meadows taxa would go extinct" due to excessive groundwater pumping and destruction of springs that accompanied industrial-scale agricultural development of the area. But the Devils Hole pupfish, and most Ash Meadows taxa, survived the onslaught, in part due to the ESA and the *Cappaert* decision, and he hoped that these powerful tools would continue to protect regional aquatic biodiversity, at least from extinction.

In considering Jim Deacon's concerns and the struggle to protect native desert fishes and their ecosystems, it is seductive to craft a narrative that emphasizes the triumph of science, reason, hard work, determination, and environmental ethics over reactionary forces (those who would dump thousands of gallons of rotenone into the Green River to kill native fish or pump the springs of Ash Meadows dry to grow alfalfa and cotton). When viewed from this particular historical moment, the sequence of events that led from the mid-1960s (prior to the founding of the Desert Fishes Council) to the late 1980s (when the chapters in *Battle against Extinction* were written) to the present (summarized by the chapters in *Standing between Life and Extinction*) might appear to be a series of predictable outcomes arising from distinctive and understandable cause-effect relationships. We then might conclude that these same factors will produce similar results when played forward, 30 years or so into the future. And while such a narrative is in one sense true—science, reason, hard work, determination, and environmental ethics were crucial factors in the fight to protect native desert fishes and their ecosystems—it is important to understand that history is contingent and victories in places such as Devils Hole and Ash Meadows were close things, their outcomes enmeshed in politics and far less than certain (Deacon and Deacon Williams 1991; Pister 1991). Biotic systems and abiotic environments are rarely, if ever, static. Likewise, the challenges of our current moment are not identical to the challenges of 1965 or 1989, and the political and ideological issues faced by conservation biologists and environmentalists in 2050 will differ from those of 2017 (when both houses of Congress were controlled by Republicans) and 2019 (when control of Congress was split between the Republicans and Democrats). However, even though our political and cultural landscapes are constantly changing, it still is critical to ask, "Where are we, where might we go in the years ahead, and why might we go there?"

Thus, this chapter.

THE POLITICAL PRESENT, CIRCA 2019

Jim Deacon died in February 2015, 20 months before Donald J. Trump's election as president of the United States, combined with Republican control of both the US House of Representatives and the US Senate, altered the political landscape. Relative to issues central to *Battle against Extinction* and *Standing between Life and Extinction*, the implications of Trump's election are negative, and I wonder how much optimism Jim would retain if he were alive today. The president and his administration have moved to slash federal discretionary spending on scientific research and roll back federal environmental laws and regulations, actions that have affected the missions of agencies such as the US Fish and Wildlife Service, US National Park Service, US Environmental Protection Agency, and US Geological Survey. He has also selected cabinet secretaries and other administrators with little sympathy for environmental protection, such as Scott Pruitt, former head of the EPA, and former Secretary of the Interior Ryan Zinke, who, as a Montana congressman, had a League of Conservation Voters lifetime score of 3% (League of Conservation Voters 2017). Zinke once served as a member of the House Committee on Natural Resources, which at the time was chaired by Rob Bishop of Utah. Bishop has advocated for "reform" of the ESA (ESA Working Group 2014) in ways that would undermine the statute's ability to protect at-risk species and habitats (Ecological Society of America 2016; Zaffos 2016), and under his leadership the committee's website displayed a "report card" giving the ESA an F (House Committee on Natural Resources 2017a).

Changes to the Endangered Species Act are not the only threat to the federal government's ability to protect rare and endangered species. The US Fish and Wildlife Service has primary responsibility for administering the ESA, but all federal agencies must comply with federal environmental laws and regulations (Stein et al. 2008). Federal agencies manage 29% of the land area in the United States (Stein et al. 2008), although the percentage in western states is much higher. For example, the federal government manages 85% of Nevada (Vincent et al. 2014), including 93% of Nye County (Harris et al. 2001), the site of Devils Hole and Ash Meadows National Wildlife Refuge. Large federal landholdings in the West, combined with the primary role of federal agencies in protecting at-risk species and ecosystems, means that major changes to federal land ownership and regulatory standards may harm society's ability to protect biodiversity. Given the hostility of the Trump administration to federal regulatory authority (Yardley 2016; Zinke 2017), and the desire of many lawmakers in the region to transfer public lands or their management to the states (Frazzini 2015; Shogren 2017), it is difficult to remain optimistic about the federal government's ability to conserve aquatic resources and at-risk species, at least in the immediate future.

Although the other linchpin of protection for aquatic ecosystems and endangered species in the arid West, the unanimous *Cappaert v. United States* decision, appears less threatened than the ESA, the current Supreme Court has a much different ideological composition than the Burger court of the 1970s. Because persistent "doctrinal uncertainty" surrounds the federal reserved water rights protected by *Cappaert* and the 1908 *Winters v. United States* decision (Leonard 2010), further judicial or legislative action could weaken these rulings. Conflicts over water among western states, the federal government, and stakeholders abound, and "the Supreme Court has sought to protect state law water users by narrowly construing the intent of Congress in reserving water and by attempting to integrate federal and state water law" (Rusinek 1990, 408). Because President Trump already—as of 2019—has selected two justices for the US Supreme Court, and has appointed a host of district and circuit court judges, erosion of federal water rights could occur in a changing judicial environment.

CLIMATE CHANGE, DROUGHT, POPULATION GROWTH, AND WATER SCARCITY

Nature, politics, and science interact in complex ways (Wulf 2015)—but if the political implications of Donald Trump's presidency relative to the concerns of *Standing between Life and Extinction* remain somewhat unclear in 2019, the role of nature in this triad appears more certain. The importance of ensuring adequate water supplies for conserving aquatic species in the American West is obvious, especially because the region has entered a prolonged period of increased temperatures and decreased precipitation and runoff (Cayan et al. 2010; USGS 2011). The American West has experienced major, decade-long droughts over the last 1,200 years (Cook et al. 2004; Woodhouse et al. 2010), and climate models predict a substantially greater probability of extended drought and reduced streamflows in the latter half of the twenty-first century than in the twentieth century and earlier paleoclimates (Seager and Vecchi 2010; Cook et al. 2015).

The recent drought has greatly affected runoff in the Colorado River basin, the "critical conduit of water in the Southwest" (MacDonald 2010, 21256)—an effect that may be exacerbated by continued warming and temperature-induced declines in streamflows (Udall and Overpeck 2017; Overpeck and Bonar, chap. 9, this volume). In early 2017, Lake Mead, which supplies roughly 90% of the water used in Clark County, Nevada (Pavelko et al. 1999; SNWA 2009), was at 41% of its storage capacity, and its elevation fell to 1,079 feet (329 m), its third-lowest minimum since it was filled in the late 1930s, while Lake Powell was at 46% of its storage capacity (USBR 2017a, b); it is possible that these reservoirs may never completely refill (Barnett and Pierce 2008). Although surface-water reserves in the Colorado River basin have fallen, groundwater reserves in the Southwest and California have decreased even more dramatically. Satellite data indicate that between 2004 and 2013, the Colorado River basin experienced a substantial depletion of its groundwater reserves (Castle et al. 2014; see also Konikow 2013), an analysis supported by measurements of recent uplift derived from GPS displacement time series in the western United States (Borsa et al. 2014). During roughly the same period (2003 to 2010), satellite data also indicated that groundwater depletion in California's Central Valley was proceeding at unsustainable rates (Famiglietti et al. 2011).

Challenges to freshwater sustainability in the Southwest from increasing dryness and declining water resources will be exacerbated by increases in the region's population (Theobald et al. 2013). By 2030, over 67 million people will live in the Southwest and California, with Nevada, Arizona, and Utah ranking among the top five states in the nation in terms of population increase (MacDonald 2010; US Bureau of the Census 2017). Past and projected increases in the Southwest's population are concentrated in urban and suburban areas. Since the "baby boom" (approximately 1943–1964), the population of western cities has increased by an average of 32% (US Bureau of the Census 2012; City Mayors 2017). And the population of Clark County, Nevada, is projected to climb from about 2 million in 2012 to more than 3.1 million by 2050 (Tra 2015), while Arizona's population is projected to increase by 108% between 2000 and 2030 (US Bureau of the Census 2017). Sabo et al. (2010) estimate that by the time the population of the arid Southwest "superregion" doubles, humans will appropriate 86% of current mean annual streamflow, up from the current 76%.

The capacity of freshwater ecosystems to support native biodiversity is negatively affected by declines in the sustainability of regional freshwater supplies (Deacon et al. 2007; Sabo et al. 2010). Given the cumulative negative impacts of population growth and drought on freshwater supplies predicted for the Southwest, it is difficult to escape the conclusion that native freshwater biodiversity in the region will face increasing threats as the twenty-first century progresses, despite impressive efforts by some metropolitan areas to conserve water supplies and increase water use efficiencies (Deacon et al. 2007; Sabo et al. 2010). Regionally, extinctions of freshwater fauna due to unsustainable surface and groundwater use have occurred many times; examples include the Pahrump Ranch poolfish *Empetrichthys latos pahrump* in Pahrump Valley; the Vegas Valley leopard frog *Lithobates fisheri* in Las Vegas Valley; the Ash Meadows poolfish *Empetrichthys merriami* at Ash Meadows; the Fish Lake pyrg *Pyrgulopsis ruinosa* in Esmerelda County, Nevada; the Pahranagat spinedace *Lepidomeda altivelis* in Pahranagat Valley, Nevada; and the Rio Grande bluntnose shiner *Notropis simus simus* from the upper Rio Grande (Miller et al. 1989; Sada and Vinyard 2002; Norment 2014). These local extinctions are part of a larger, regional loss of aquatic biodiversity due to many anthropogenic factors, including water diversions, nonnative species introductions, livestock grazing, pollution, and habitat alteration (Miller et al. 1989; Sada and Vinyard 2002). These factors have resulted in the listing of at least 13 invertebrate and 51 vertebrate aquatic taxa under the ESA in arid portions of Arizona, Nevada, New Mexico, and Utah (USFWS 2018).

IDEOLOGY AND THE DEATH OF FACTS

Beyond the threats posed to southwestern aquatic biodiversity by climate change and drought, population growth, overuse of scarce water resources, erosion of federal public land management policy and regulatory authority, and weakening of the ESA and potentially the *Cappaert* decision, lies an even more profound and worrisome danger: what commentators have called "the death of facts," or "post-truth" (Huppke 2012; Higgins 2016). Indeed, CNN commentator Scottie Nell Hughes said on the November 30, 2016, edition of *The Diane Rehm Show*, "There is no such thing, unfortunately, anymore of [sic] facts." Hughes went on to add, "People say that facts are facts, they're not really facts. Everybody has a way, it's kind of like looking at a glass half-full of water. Everybody has a way of interpreting them to be the truth or not true" (Wemple 2016). Less than two months later, White House senior advisor

Kellyanne Conway used the term "alternative facts" to describe Trump press secretary Sean Spicer's false description of the size of the inaugural crowd (Rutenberg 2017). The above attitudes were foreshadowed by an unnamed aide to President George W. Bush, who was interviewed by the journalist Ron Suskind (2004):

> The aide said that guys like me were "in what we call the reality-based community," which he defined as people who "believe that solutions emerge from your judicious study of discernible reality." . . . "That's not the way the world works anymore," he continued. "We're [the United States] an empire now, and when we act, we create our own reality. And while you're studying that reality—judiciously, as you will—we'll act again, creating other new realities, which you can study too, and that's how things will sort out. We're history's actors . . . and you, all of you, will be left to just study what we do."

The worldview of Hughes, Conway, President Trump, and the unnamed George W. Bush aide (most likely Karl Rove)—that truth and facts are a matter of one's perspective (Hughes and Conway) or power (Rove, Trump)—have a curious postmodern twist, in that postmodernism holds that what we think of as facts and reality are socially constructed (Ayelsworth 2015; Lukes 2017)—in other words, a matter of opinion. Many people take as "truth" Donald Trump's statements that "the concept of global warming was created by and for the Chinese in order to make US manufacturing non-competitive" (Trump 2012), and that there is no drought in California (Hiltzik 2016). Trump's attitude toward California's water problem—that it is a "man-made," avoidable situation caused by environmentalists "trying to protect a certain kind of 3-inch fish [probably the delta smelt *Hypomesus transpacificus*]"—has also been promoted by the free-market Heartland Institute (Crews 2014) and, until recently, by the House Committee on Natural Resources (2017b), whose web pages featured an article entitled "The Man-Made California Drought." This article argued that recent declines in available groundwater and surface water in California are due to governmental incompetence, rather than any connection between drought and anthropogenic climate change. Such assertions about the nonreality of global warming are supported by many conservatives (Pew Research Center 2016), from Senator James Inhofe (2012) of Oklahoma to the majority of the Utah state legislature, which in 2010 passed H.J.R. 12, a resolution attacking "global warming science" (Utah State Legislature 2010).

Rejection of the reality of climate-driven drought, and related attitudes about unsustainable water use in the American West, run counter to the overwhelming weight of scientific evidence. For example, global atmospheric carbon dioxide concentrations have risen to over 400 ppm, the highest in over 800,000 years (USEPA 2017; NOAA 2017a), and the years 2016, 2015, 2014, 2010, 2013, and 2005 were the six warmest since 1880, with 2016 breaking modern temperature records (NOAA 2017b). Between 2012 and 2014, the California drought was the most severe in the last 1,200 years, with "record high temperatures" and "anomalously low precipitation" (Griffin and Anchukaitis 2014; see also AghaKouchak et al. 2014). As of January 2017, much of California's Central Valley and southern coastline remained in the "Exceptional Drought" category (US Drought Monitor 2017), and all indicators (decreasing groundwater levels and streamflows, soil water deficits, and local ground subsidence) supported the hypothesis of a major California drought (USGS 2017). Although heavy rain and snow fell in California during the winter and spring of 2018–2019, there is a critical difference between one year's weather and a climate trend, and we face a future with a higher probability of sustained drought (Cook et al. 2015; Udall and Overpeck 2017).

It is tempting to view denial of climate change and climate-driven drought as a product of a more general growth in public mistrust of science and willingness to dismiss scientific authority (e.g., Gawande 2016), in part due to the influence of cable TV and social media. However, research suggests a more complex situation. Although "a sizable 'opinion gap' exists between the general public and scientists on a range of science and technology topics" (Pew Research Center 2015), the American public's general trust in science has remained high since the 1970s, and in 2012, 41% of Americans expressed a "great deal" of confidence in scientific leaders (National Science Board 2016; Nisbet and Markowitz 2016). However, between 2008 and 2015, public confidence in scientists as sources of information about climate change has declined (Nisbet and Markowitz 2016), in concert with the increased politicization of science that has occurred during the last four decades—a trend most obvious among those who self-identify as "conservative" (Gauchat 2012).

At its core, the difficulty many people have with accepting facts and scientific consensus is one of ideology. How one chooses to see the world—in this case, the world of global change, drought, and the issues that surround protection of aquatic biodiversity in the arid West—is often filtered through one's ideological lens and belief system (Gauchat 2012; Pew Research Center 2015). The novelist Barbara Kingsolver brilliantly summarized the situation (Supin 2014):

> We believe we collect evidence and then use it to make up our minds, but in fact we make up our minds and then collect evidence to support our beliefs. Almost all of us work this way, more or less, unless we are scientists trained to ask unbiased questions. . . . We make these kind of animal decisions about who's on our team, and then we pretty much believe what they say.

Kingsolver captures the essence of President Barack Obama's (2006, 59) distinction between values and ideology: "Values are faithfully applied to the facts before us, while ideology overrides whatever facts call theory into question." It is understandable that people's views differ on the effectiveness of the ESA, the importance of protecting at-risk species, the role of the federal government in managing western lands, or the wisdom of the *Cappaert* decision; these are values-based attitudes. It is less understandable, and indefensible and dangerous, that many individuals refuse to accept the reality of anthropogenic climate change, a California drought caused by something other than bureaucratic malfeasance, or even the efficacy and safety of vaccines, because their ideological perspective (their "team") prevents them from fairly evaluating scientific evidence. Data, hypotheses, theories, and paradigms should be open to rigorous testing and healthy skepticism—that is the foundation of the scientific method (Popper 1959). But when ideology and intellectual dishonesty affect how scientific information is presented and interpreted—as in the case of the tobacco lobby, the pesticide industry, and many climate change deniers (Oreskes and Conway 2010)—as well as our willingness to consider alternative explanations for how the world works, there is little hope for either a constructive, substantive, and well-informed debate on issues or policy decisions that benefit society.

In the preface to *Battle against Extinction*, Minckley and Deacon (1991b, xiv–xv) wrote,

> The most discouraging thing about editing and contributing to a book of this nature is the realization that one may be working mostly for an audience that already is convinced of the values of conservation and perpetuation of natural systems. . . . Those who embrace other philosophies are often just as firmly convinced of their alternative views, *and only education based on tangible data and logical, documentable results of research and observation can be expected to change their minds* (emphasis added).

The problem with Minckley and Deacon's logic is that "education based on tangible data and logical, documentable results of research" is insufficient to change many people's minds on issues from inconsequential (crowd size at President Trump's inauguration) to important (drought in the arid West). Consider many Americans' resistance to the scientific consensus held by 97% of publishing climate scientists (Cook et al. 2013, 2016) that recent trends in global warming can be explained by anthropogenic factors. Or consider that in April 2017, the Environmental Protection Agency removed much of the information about climate science from its website (Mooney and Eilperin 2017). Thus, we (society as a whole, conservation biologists, wildlife managers, citizens) cannot rely solely on evidence—or even laws and regulations—based on science to protect and enhance aquatic ecosystems and biodiversity in the Southwest. Although scientific understanding is crucial for achieving this outcome, it is insufficient in the context of our hyper-politicized, hyper-ideological, "post-fact," "fake news" environment. Political action and advocacy, along with continual litigation, will be necessary—perhaps even more so than following the founding of the Desert Fishes Council, when battles over the fate of Devils Hole and Ash Meadows were being fought (Deacon and Deacon Williams 1991; Pister 1991). Simply being an active member of what Karl Rove called the "reality-based community" will not be enough—particularly in the context of today's society and its political polarization.

ON THE IMPORTANCE OF IMAGINATION

Protecting Devils Hole and establishing Ash Meadows National Wildlife Refuge were important political and scientific victories, but they were also acts of the imagination. As Guy Davenport (1981, 3) has written,

> The difference between the Parthenon and the World Trade Center, between a French wine glass and a German beer mug, between Bach and John Phillip Sousa, between Sophocles and Shakespeare, between a bicycle and a horse, though explicable by historical moment, necessity, and destiny, is before all a difference of the imagination.

In like manner, one expression of the human imagination brought corporate agriculture to Ash Meadows and populated it with cattle and alfalfa; another vision hoped to create "Calvada Lakes," a planned development for 30,000 (Deacon and Deacon Williams 1991), while another, wiser and more lovely act of the imagination led to the creation of Ash Meadows National Wildlife Refuge and restoration of its springs.

Although Ash Meadows and Devils Hole are protected, they remain threatened—as much by different imaginations as by the vagaries of the Southwestern climate or by the region's population growth and increasing demand for water. Take, for example, the contrasting views of the former manager of Ash Meadows National Wildlife Refuge, Sharon McKelvey, and Gary Hollis, former Nye County commissioner. Both told me that their overriding concerns were ensuring an adequate supply of water for their constituencies. However, in McKelvey's case, her primary constituencies were the ecosystems and native species of Ash Meadows (and by extension, due to her position as a federal employee, the citizens of the United States), while Hollis's primary concern was for the people of Pahrump and Amargosa Valley, Nevada (Norment 2014). Even though the interests of these disparate constituencies (the people of Pahrump and Amargosa Valley, the other citizens of the United States, the endemic species and unique ecosystems of Ash Meadows) are not necessarily different, Hollis believed that Ash Meadows National Wildlife Refuge brought no important benefits to Nye County and took more water than it should (Norment 2014). Or take the following online comments posted in the aftermath of the November 26, 2010, article in the *Pahrump Valley Times*, "Devils Hole Being Upgraded":

> I am just sooooo glad that our government is spending OUR money on this pupfish. The world would surely not end without this particular fish. . . . How many millions will we spend?
>
> I guess it's like Obamacare for fish.
>
> As a friend and co-worker once said to the Las Vegas media in an early 1980s interview about the Devil's Hole pupfish: "If the damn fish aren't big enough to put in a frying pan, they aren't worth keeping."
>
> I really enjoyed the comments on the pupfish. Glad to see not everyone buys into the waste of money spent out there.

The comment about the diminutive pupfish and their lack of value as human food apparently expresses a common sentiment. I encountered this attitude at Tecopa Hot Springs, when discussing the extinct Tecopa pupfish *Cyprinodon nevadensis calidae* with one employee: "I think they were pretty tiny, not good for much of anything. You couldn't eat them. Not like trout" (Norment 2014). And the former manager of the Ash Meadows Fish Conservation Facility heard much the same thing from one of the contractors involved in its construction (D. Weissenfluh, pers. comm.). These sentiments are as much about imagination as they are about utilitarian value: pupfish as either worthless as a human food resource, or important as desirable elements of biodiversity worth protecting for their intrinsic right to exist, importance to ecosystems, or option value.

In a similar way, Marc Reisner's (1986) *Cadillac Desert: The American West and Its Disappearing Water* was as much a product of the imagination as it was a remarkably prescient analysis of political folly, human greed, population growth, and the unsustainability of twentieth-century water management policies in the arid West. Reisner's book followed an earlier visionary work, John Wesley Powell's *Report on the Lands of the Arid Region of the United States* (Powell 1879). Powell's basic premise was that "water scarcity would place limits on the growth of a new civilization in the region"

(Sabo et al. 2010, 21268). Reisner's (and Powell's) argument has been mostly supported by recent history and scientific analyses: over the next 100 years, the combined effects of climate change, increasing population, and higher agricultural demand appear likely to create unprecedented demands on the region's water supply (Sabo et al. 2010). How society responds to this situation—and how it collectively imagines the future—will determine the sustainability of the region's freshwater resources, the fate of its native aquatic biodiversity, and ultimately, the quality of life in the Southwest (Gleick 2010).

UNCERTAINTY AND OUR POSSIBLE FUTURES

In 1993, Ludwig et al. published a short but important paper, "Uncertainty, Resource Exploitation, and Conservation: Lessons from History." Although the authors were concerned primarily with maximum sustained yield and harvested fish populations, their analysis is as germane to southwestern water issues and aquatic biodiversity conservation as it is to managing Pacific salmonid stocks. Ludwig and colleagues suggested four major "drivers" of the consistent historical pattern of overharvesting resources and the requisite need to "confront uncertainty":

> (i) Wealth or the prospect of wealth generates political and social power that is used to promote unlimited exploitation of resources. (ii) Scientific understanding and consensus is hampered by the lack of controls and replicates, so that each new problem involves learning about a new system. (iii) The complexity of the underlying biological and physical systems precludes a reductionist approach to management. . . . (iv) Large levels of natural variability mask the effects of overexploitation.

In terms of sustaining the Southwest's aquatic resources, three of Ludwig et al.'s drivers have had particularly important, negative effects in the past and continue to haunt us today: natural variability, system complexity, and wealth. Natural variability, particularly in terms of precipitation, has masked some of the Southwest's vulnerability to water shortages and made it difficult to manage its water resources sustainably (Milly et al. 2008; Gober and Kirkwood 2010). For example, allocation of Colorado River waters under the 1922 Colorado River Compact was based on data collected in the early twentieth century, when sustained flows were greater than at any other time between 1520 and the present (Woodhouse et al. 2006, 2010). Conversely, water policies facilitating agricultural development and urban growth in the Southwest have never adapted to the reality of the megadroughts that have occurred in the region during the last millennium (Cook et al. 2004; Woodhouse et al. 2010).

System complexity (which includes environmental and social variability) has also negatively affected our ability to craft sustainable water policies in the region. Consider the complicated relationships among climate variability; the monumental infrastructure for storing and moving water; the political, bureaucratic, and economic inertia generated by water management policies created during the twentieth century; massive agricultural development; uncontrolled urban growth; the multiplicity of stakeholders in the region; and even the effort needed to understand the complex ways in which groundwater flows through and between hydrographic basins, as with the Death Valley regional groundwater flow system (Belcher and Sweetkind 2010; Belcher et al. 2016). These factors, and many others, have created an intricate and massive nexus resistant to fundamental change.

And then there is wealth, which has supported efforts to contravene and distort scientific findings on environmental issues ranging from climate change to the harmful effects of pesticides, ozone, and acid rain (Oreskes and Conway 2010). The Heartland Institute (2017), which describes its mission as "to discover, develop, and promote free-market solutions to social and economic problems," has actively supported climate skeptics, often with funds provided by donors such as the billionaire Koch brothers (Oreskes and Conway 2010; Hickman 2012). Although "comprehensive, multisectoral, and transregional policies" (MacDonald 2010, 21262) will be required to achieve sustainable water use in the Southwest (Gleick 2010; Sabo et al. 2010), the Heartland Institute strongly opposes most governmental regulation (Heartland Institute 2017), as does the Pacific Legal Foundation, which also receives funds from the Koch brothers (Carrk 2011) and aggressively pursues lawsuits to prevent regulation of surface and groundwater use in California and elsewhere (Pacific Legal Foundation 2017). And the economic impact of California agriculture (in 2014, the state produced $53.5 billion in farm receipts; California Department of Food and Agriculture 2015) gives that industry substantial economic and political leverage in protecting "business-as-usual" groundwater and surface-water use. Yet California agriculture accounts for approximately 72% of the state's total water use (USGS 2017), which means that sustainable water use will be achieved only through major changes in agricultural practices, including irrigation methods and crop types (Gleick 2010; Sabo et al. 2010).

Wealthy interests may aggressively promote policies and perspectives that are "anti-science" and anti-regulatory, but wealth itself may act as an impediment to policies promoting sustainable water use and the protection of aquatic biodiversity. A primary example would be the economic wealth generated by Las Vegas and the surrounding metropolitan area in Clark County, Nevada. In 2015, Clark County, which contains 72% of Nevada's population, produced an estimated $103.3 billion in GDP (Center for Business and Economic Research 2016). Any substantial decrease in the ability of the SNWA to deliver water to Clark County, due to declining supplies and increasing demand, would have serious negative effects on the local, state, regional, and even national economies. The easiest way to ameliorate future water deficits in Clark County might be to implement SNWA's (2009) plans to mine groundwater reserves in northern and central Nevada, which could seriously impact regional biodiversity (Deacon et al. 2007). Even though these plans are currently on hold (Great Basin Water Network 2017), and *Cappaert* and the present version of the ESA protect the region's biodiversity (Deacon et al. 2007), there is no guarantee that future conflicts will be resolved in favor of sustainable groundwater use and biodiversity.

In considering future conflicts over an increasingly scarce water supply, perhaps we might take some comfort in how a previous encounter between Clark County's growth and the ESA was resolved. In the 1990s, concern that development would negatively affect the desert tortoise *Gopherus agassizii* and other federally listed species led to implementation of the Clark County Multispecies Habitat Conservation Plan (MSHCP), which granted an incidental take permit in exchange for mitigation efforts, including a $550 per acre "disturbance fee" paid by developers (Clark County 2000). However, the impetus for development of the Clark County MSHCP was conflict over space—which southern Nevada has plenty of—rather than water, which is in short supply and absolutely crucial to Clark County's continued growth. Given the current political climate in the United States, Clark County's economic juggernaut could provide a rationale for casting aside legal and regulatory protection for threatened aquatic species and ecosystems. After all, the political constituencies of native pupfish, dace, and springsnails are orders of magnitude less powerful than those of the Las Vegas gaming industry and developers—and attempts to deal with drought-related water shortages affecting business interests by circumventing the ESA are not without precedent. For example, in 2014, House Resolution 3964, which mandated that the California State Water Project and Central Valley Project be managed "without regard to the Endangered Species Act of 1973," passed the House of Representatives, though it stalled in the Senate. One sponsor of H.R. 3964, Devin Nunes, called the federally threatened delta smelt, which is endemic to the Sacramento–San Joaquin Delta, a "stupid little fish" (Norment 2014).

In contrast to natural variability, system complexity, and wealth, one of Ludwig et al.'s (1993) drivers appears less problematic today: there is a strong scientific basis for understanding water scarcity in the Southwest, as well as a reasonable level of scientific (and policy) consensus regarding how best to address the problem, even in the face of uncertainty. It is not my intention here to summarize the extensive literature on the topic, but instead to note that many of the chapters in *Standing between Life and Extinction* directly or indirectly address it, as do the papers collected in the December 14, 2010, special feature in *Proceedings of the National Academy of Sciences*, entitled "Climate Change and Water in Southwestern North America"; Reisner's *Cadillac Desert* (1986); Robert Glennon's *Unquenchable: America's Water Crisis and What to Do About It* (2009); and the myriad publications of the Pacific Institute, including two reports on the Colorado River (Cohen 2011; Cohen et al. 2013). Taken collectively, these publications—and many more—point toward a robust understanding of the issues involved with, and potential solutions to, sustainable water management in the arid Southwest.

System complexity, natural variability, and (particularly) wealth present daunting challenges to developing a comprehensive approach to sustainably managing water in the arid Southwest and protecting the region's aquatic biodiversity. However, Ludwig et al. (1993) do not explicitly discuss an additional driver of resource overexploitation, even if it is implicit in their analysis: the blinding effects of ideology, which seems to have achieved an almost unprecedented level of prominence in contemporary American society. Thus, it is difficult to chart a way forward when the balance of political power at both the federal and state levels is held by many who embrace or fail to criticize "fake news" and "alternative facts," reject compromise, scorn the notion of objective truth, view the world through a distorted ideological filter, and rather than ignoring science, often actively disdain it. The unsustainable exploitation of water in the American Southwest may be yet another example of what the historian Barbara Tuchman termed the "March of Folly" (Keller and Day 2007), and it is unclear if society can muster the necessary political will, foresight, and imagination to deal effectively with the issues of water, climate change, and sustainability in the region.

Given this pessimistic view—while also acknowledging the wealth of information and creative ideas that point us toward sustainable water use in the arid West—we can imagine several possible futures for aquatic biodiversity in the region, just as John Fowles (1969) imagined three different endings for his novel *The French Lieutenant's Woman*. One is encouraging: there is hope (but not certainty) that Charles Smithson, a Victorian gentleman and naturalist who has risked his social standing and financial security for Sarah Woodruff, a mysterious woman tainted by scandal and far beneath his "station," will triumph over convention and fate and win the love of his life. A second ending is decidedly mixed; here, Fowles describes Charles as capitulating to society's expectations, marrying his fiancée (whom he cares for but does not love deeply) and going into "business," which Victorian gentlemen did not do willingly. Although Charles and his wife "did not live happily ever after," by most standards his life was successful; he sired seven children and eventually became a prominent businessman, driven to do so by the loss of his inheritance. The third ending is the bleakest: Charles risks everything for Sarah—all chance for wealth, the prospects of a respectable marriage—but she rejects him, his ultimate reward being nothing more than "a celibacy of the heart" (Fowles 1969, 466).

In a manner similar to Fowles's novel, we might embrace uncertainty and consider several plausible scenarios for the future of water use and aquatic biodiversity in the Southwest. Some outcomes may appear more likely than others, but each is subject to the same uncertainty that haunts all efforts to sustainably manage resources (Ludwig et al. 1993) and prevent biodiversity loss (Minteer and Collins 2005). All, too, are products of how we imagine the world and our place in it: visions shaped by a complex mix of inclination, historical moment, politics, values, and ideology.

In my most pessimistic scenario, a rapidly warming and drying climate, coupled with an expanding population, weakening or elimination of the protection provided by the ESA and *Cappaert*, and a desperate attempt to preserve traditional attitudes about economic growth and water use practices, leads to further, widespread damage to aquatic habitats and impoverishment of biodiversity throughout the region. Too many politicians ignore science and the wisdom of good policy, embrace the mindset of "alternative facts," and lead the arid West away from a sustainable future. To paraphrase Fowles, a "celibacy of the heart and mind" triumphs. Cities like Las Vegas struggle to supply water to their expanding populations, in spite of aggressive conservation measures; the Colorado River's flow continues to decline, and Lake Mead's elevation falls below the operating level for the intake that came online in 2015 (SNWA 2017). The pipeline constructed by the SNWA drains groundwater from aquifers in northern and central Nevada. Springs dry, and instream flows decline across region. Agencies like the US Fish and Wildlife Service, hampered by loss of regulatory and statutory authority and shrinking budgets, struggle to fulfill their missions, and species like the Devils Hole pupfish become extinct in the wild. Other species persevere, through either their preference for out-of-the-way places or tenacity like that displayed by many species in Ash Meadows as they survived the onslaught of Cappaert Enterprises and Spring Meadows Inc. In 2050, a university press publishes *The Battle Lost: Mass Extinction of Aquatic Species in the American Southwest*.

My dystopian view is extreme, although possible. A more likely outcome might resemble Fowles's middle road: nothing by way of a truly happy ending, but an adaptation to circumstance. The climate of the Southwest continues to warm and dry, but not in an extreme way; enough years with good precipitation counter the worst effects of the trend toward increased aridity. The Colorado River delivers sufficient water to Lake Mead to supply Las Vegas with a substantial portion of its needs. Aggressive conservation measures and tiered rate structures drive down per capita water demand, agriculture moderately reforms its irrigation practices and shifts toward drought-tolerant crops, and resource-driven forces produce a net demographic shift away from the arid West and into the water-rich Great Lakes Basin. Ideologues are repudiated by voters in enough places to overcome the worst excesses of the "alternative facts" mindset. The competing interests of myriad stakeholders slow progress, but enough halting, imperfect policy gains are made to move the Southwest toward a partially sustainable future. These gains, plus increased funding for conservation, a healthy Endangered Species Act, and the resilient *Cappaert* and *Winters* decisions, help maintain aquatic biodiversity in the region. Some species and ecosystems prosper while others decline, but the overall story is one of progress. In 2050, a university press publishes *The Battle Continues: Gains and Losses in the Struggle to Conserve Aquatic Species in the American Southwest.*

And finally, at the other end of my imaginary spectrum, is the happiest ending, equivalent to Fowles's "the hero (probably) gets the woman." The climate continues to warm and dry—again, not severely—but there is widespread acceptance of the policy initiatives needed to craft a sustainable future for the Southwest. This acceptance is driven by a general rejection of extreme ideology and the "alternative facts" universe, which resulted in the transformative 2020 and 2024 elections. Science is broadly seen in a more positive light than in 2017, and the preservation of aquatic ecosystem health and the "needs of society" are understood as inextricably linked by a wide spectrum of the public. The regulatory, judicial, and legislative environments are such that substantial progress is made in restoring ecosystem health and creating a sustainable regional economy. Uncertainty, too, is accepted as a fundamental property of all natural and human systems, and is incorporated into economic and resource management policy—but in 2050, the Desert Fishes Council convinces a university press to publish *The Battle Won: Restoration of Aquatic Biodiversity and Ecosystem Health in the American West.*

CONCLUSION

In the epilogue to *Battle against Extinction*, Deacon and Minckley (1991, 413) wrote,

> While progress is being made, both in acceptance of environmental values and in programs to reduce loss of biodiversity, we still lose more battles than we win. The proximate cause for failure is more often attributable to bureaucratic intransigence than to inadequate legislation or knowledge. The ultimate cause is growth of resource use by a growing population that fails to recognize or acknowledge the rights of and its duties toward natural objects.

My current perspective on the above passage is that since 1991, we have made substantial progress toward crafting a sustainable future and conserving aquatic biodiversity in the American West. But many of the gains in the areas of scientific understanding, policy, and management are now threatened less by "bureaucratic intransigence" than by the twin plagues of ideology and ignorance—by what Karl Popper described as an "active aversion to knowledge" (Gifford 2011, 91). We still have the robust knowledge base, and the laws and regulations, to craft the policies and practices necessary for responsible, sustainable water use in the American West and the consequent conservation of its aquatic biodiversity—although during its first two years in office, the Trump administration has acted to eliminate or weaken many existing environmental rules (Popovich et al. 2018). Implementation of such policies, though, is an entirely different matter, one that will require either wise forethought, as we muddle through a frustrating series of painfully slow and halting advances and retreats, or a terrible crisis, such as a megadrought that drags on for decades, rendering "alternative facts" and hyperbole about "man-made droughts" irrelevant—for although "we may live in a post-truth era . . . nature does not" (Barnett 2017).

If we are to avoid a "backs to the wall" scenario, it will demand much from people with good knowledge, strength, and values, and with the courage of their convictions. It will require continued advances in science, management, and public policy, as well as faith in the value of accurate information, tireless action, and an understanding of the role that politics plays in determining public policy outcomes. And it will require reaching out to people and organizations with worldviews that may, at first glance, diverge substantially from those of many conservation biologists and members of organizations such as the Desert Fishes Council. Such an approach means seeking out whatever common ground might exist among people with divergent viewpoints, particularly related to sustainable use of arid-land water resources. Examples of successful efforts along these lines include the Malpai Borderlands Group and the Great Basin Water Network, both of which represent cooperative efforts among conservationists, rural communities, and local landowners to promote sustainable resource use in arid-land systems As Don Sada, an expert on the ecology and biogeography of springsnails, told me about times when his work for the US Fish and Wildlife Service required him to stand in front of hostile audiences, "I didn't talk about springsnails; I talked about springs. If there's enough high-quality water, you can use it for many things, including humans—and if you take too much water, it will hurt people, as much as snails and fish" (Norment 2014). And finally, many scientists will be challenged to behave in ways that might run counter to their inclinations and training, by immersing themselves in the political process and engaging more directly with the public (Lubchenco 2017).

Perhaps my analysis in this chapter is unduly influenced by the political turmoil of late 2016, 2017, and 2018, and the chapter will quickly become dated. People such as Devin Nunes, Ryan Zinke, Scott Pruitt, and Donald Trump (and their collective mindset) have been, or soon will be, relegated to the past, and the intense political conflicts of the present will ease. As the worst excesses of the Trump administration and its fellow travelers are curbed by public outcry, political reversals, and strong judicial oversight, perhaps the federal Endangered Species Act and the *Cappaert* decision will survive, and continue to protect aquatic ecosystems and endangered species throughout the arid West . . . But the world is an uncertain place, and we must embrace the understanding that the future is always unknowable even if it carries some measure of predictability.

Still: "The future is dark, the present a knife's edge. It's the past that's knowable, incandescent, real" (Taylor 2017, 166). At any moment in time, we stand balanced on the knife's edge of the present, more or less certain about the past, but unsure as to what the next few years and decades will bring to the arid West, its waters, ecosystems, and native biota. In spite of this uncertainty, we still must understand that "standing between life and extinction" will require that a critical mass of people transcend blinding ideology, convention, and human short-sightedness to think about the past and present and where they might lead us—whether in 2019, 2025, or 2040. And finally, "standing between life and extinction" will require acts of courage and imagination exemplified by those who fought and won the battles for Devils Hole and Ash Meadows National Wildlife Refuge, but on a larger and even more profound scale—for the threats are immediate, real, and innumerable.

REFERENCES

AghaKouchak, A., L. Cheng, O. Mazdiyasni, and A. Farahmand. 2014. Global warming and changes in risk of concurrent climate extremes: Insights from the 2014 California drought. *Geophysical Research Letters* 41: 8847–52. doi:10.1002/2014GL062308.

Ayelsworth, G. 2015. Postmodernism. In E. Zalta, ed., *Stanford Encyclopedia of Philosophy*, Spring 2015 ed. https://plato.stanford.edu/entries/postmodernism/.

Barnett, C. 2017. We may live in a post-truth era, but nature does not. *Los Angeles Times*, February 10, 2017. http://www.latimes.com/opinion/op-ed/la-oe-barnett-nature-alternative-facts-20170210-story.html. Accessed February 14, 2017.

Barnett, T. P., and D. W. Pierce. 2008. When will Lake Mead go dry? *Water Resources Research* 44: W03301. https://doi.org/10.1029/2007WR006704.

Belcher, W. R., and D. S. Sweetkind. 2010. *Death Valley Regional Groundwater Flow System, Nevada and California: Hydrogeologic Framework and Transient Groundwater Flow Model*. US Geological Survey Professional Paper 1711.

Belcher, W. R., D. S. Sweetkind, C. C. Faunt, M. T. Pavelko, and M. C. Hill. 2016. *An Update of the Death Valley Regional Groundwater Flow System Transient Model, Nevada and California*. US Geological Survey Scientific Investigations Report 2016-5150. https://doi.org/10.3133/sir20165150.

Borsa, A. A., D. C. Agnew, and D. R. Cayan. 2014. Ongoing drought-induced uplift in the western United States. *Science* 345: 1587–90.

California Department of Food and Agriculture. 2015. *California Agricultural Statistics Review, 2014–2015*. https://www.cdfa.ca.gov/statistics/PDFs/2015Report.pdf. Accessed February 11, 2017.

Carrk, T. 2011. The Koch brothers: What you need to know about the financiers of the radical right. Center for American Progress Action Fund. https://www.americanprogressaction.org/wp-content/uploads/issues/2011/04/pdf/koch_brothers.pdf/. Accessed February 11, 2017.

Carroll, R., C. Augspurger, A. Dobson, J. Franklin, G. Orians, W. Reid, R. Tracy, D. Wilcove, and J. Wilson. 1996. Strengthening the use of science in achieving the goals of the Endangered Species Act: An assessment by the Ecological Society of America. *Ecological Applications* 6: 1–11.

Castle, S. L., B. F. Thomas, J. T. Reager, M. Rodell, S. C. Swenson, and J. S. Famiglietti. 2014. Groundwater depletion during drought threatens future water security of the Colorado River basin. *Geophysical Research Letters* 41: 5904–11.

Cayan, D. R., T. Das, D. W. Pierce, T. P. Barnett, M. Tyree, and A. Gershunov. 2010. Future dryness in the Southwest US and

the hydrology of the early 21st century drought. *Proceedings of the National Academy of Sciences* 107: 21271–76.
Center for Business and Economic Research. 2016. Economic data—Las Vegas/Clark County. UNLV Lee Business School. http://cber.unlv.edu/CCEconData.html/. Accessed February 11, 2017.
City Mayors. 2017. Statistics: The fastest growing US cities. http://www.citymayors.com/gratis/uscities_growth.html/. Accessed February 11, 2017.
Clark County. 2000. Final Clark County Multiple Species Habitat Conservation Plan and Environmental Impact Statement. September 2000, Clark County Department of Comprehensive Planning. http://www.clarkcountynv.gov/airquality/dcp/Pages/CurrentHCP.aspx/. Accessed February 11, 2017.
Clark, J. A. 1994. The Endangered Species Act: Its history, provisions, and effectiveness. In T. W. Clark, R. P. Reading, and A. L. Clarke, eds., *Endangered Species Recovery: Finding the Lessons, Improving the Process*, 19–46. Washington, DC: Island Press.
Cohen, M. J. 2011. *Municipal Deliveries of Colorado River Basin Water*. Oakland: Pacific Institute. http://pacinst.org/app/uploads/2013/02/crb_water_8_21_2011.pdf. Accessed February 11, 2017.
Cohen, M., J. Christian-Smith, and J. Berggren. 2013. *Water Supply to the Land: Irrigated Agriculture in the Colorado River Basin*. Oakland: Pacific Institute. http://pacinst.org/app/uploads/2013/05/pacinst-crb-ag.pdf. Accessed February 11, 2017.
Cook, B. I., T. R. Ault, and J. E. Smerdon. 2015. Unprecedented 21st century drought risk in the American Southwest and Central Plains. *Science Advances* 1: e1400082.
Cook, E. R., C. A. Woodhouse, C. M. Eakin, D. M. Meko, and D. W. Stahle. 2004. Long-term aridity changes in the western United States. *Science* 306: 1015–18.
Cook, J., D. Nuccitelli, S. A. Green, M. Richardson, B. Winkler, R. Painting, R. Way, P. Jacobs, and A. Skuce. 2013. Quantifying the consensus on anthropogenic global warming in the scientific literature. *Environmental Research Letters* 8: http://dx.doi.org/10.1088/1748-9326/8/2/024024.
Cook, J., N. Oreskes, P. T. Doran, W. R. L. Anderegg, B. Verheggen, E. W. Maibach, J. S. Carlton, S. Lewandowsky, A. C. Skuce, and S. A. Green. 2016. Consensus on consensus: A synthesis of consensus estimates on human-caused global warming. *Environmental Research Letters* 11: http://dx.doi.org/10.1088/1748-9326/11/4/048002.
Crews, C. W., Jr. 2014. *Fountains of Solutions: Western Water and California Drought*. Chicago: Heartland Institute. https://www.heartland.org/publications-resources/publications/fountains-of-solutions-western-water-and-california-drought. Accessed February 11, 2017.
Davenport, G. 1981. *The Geography of the Imagination*. San Francisco: North Point Press.
Deacon, J. E., and C. Deacon Williams. 1991. Ash Meadows and the legacy of the Devils Hole pupfish. In W. L. Minckley and J. E. Deacon, eds., *Battle against Extinction: Native Fish Management in the American West*, 69–87. Tucson: University of Arizona Press.
Deacon, J. E., and W. L. Minckley. 1991. Western fishes and the real world: The enigma of "endangered species" revisited. In W. L. Minckley and J. E. Deacon, eds., *Battle against Extinction: Native Fish Management in the American West*, 405–13. Tucson: University of Arizona Press.
Deacon, J. E., A. E. Williams, C. Deacon Williams, and J. E. Williams. 2007. Fueling population growth in Las Vegas: How large-scale groundwater withdrawal could burn regional biodiversity. *BioScience* 57: 688–98.
Ecological Society of America. 2016. Policy News. Federal Budget: Endangered Species & Defense, Energy and Interior Authorization Bills. October 11, 2016. http://www.esa.org/esablog/ecology-in-policy/policy-news-october-11-2016/.
ESA Working Group. 2014. *Report, Findings, and Recommendations*. http://naturalresources.house.gov/uploadedfiles/esa_working_group_final_report__and_recommendations_02_04_14.pdf. Accessed February 11, 2017.
Famiglietti, J. S., M. Lo, S. L. Ho, J. Bethune, K. J. Anderson, T. H. Syed, S. C. Swenson, C. R. de Linage, and M. Rodell. 2011. Satellites measure recent rates of groundwater depletion in California's Central Valley. *Geophysical Research Letters* 38. https://doi.org/10.1029/2010GL046442.
Fowles, J. 1969. *The French Lieutenant's Woman*. Boston: Little, Brown.
Frazzini, K. 2015. This land is whose land? *State Legislatures Magazine*, July/August 2015, 30–33. http://www.ncsl.org/research/environment-and-natural-resources/this-land-is-whose-land.aspx.
Gauchat, G. 2012. Politicization of science in the public sphere: A study of public trust in the United States, 1974 to 2010. *American Sociological Review* 77: 167–87.
Gawande, A. 2016. The mistrust of science. June 10, 2017. *New Yorker*. http://www.newyorker.com/news/news-desk/the-mistrust-of-science. Accessed July 6, 2107.
Gifford, D. 2011. *Zones of Re-membering: Time, Memory, and (un) Consciousness*. New York: Rodopi.
Glennon, R. 2009. *Unquenchable: America's Water Crisis and What to Do about It*. Washington, DC: Island Press.
Gleick, P. H. 2010. Roadmap for sustainable water resources in southwestern North America. *Proceedings of the National Academy of Sciences* 107: 21300–21305.
Gober, P., and C. W. Kirkwood. 2010. Vulnerability assessment of climate-induced water shortage in Phoenix. *Proceedings of the National Academy of Sciences* 107: 21295–99.
Great Basin Water Network. 2017. Litigation, information and resources. http://www.greatbasinwater.net/litigation.htm/. Accessed February 11, 2017.
Griffin, D., and K. J. Anchukaitis. 2014. How unusual is the 2012–2014 California drought? *Geophysical Research Letters* 41: 9017–23.
Harris, T. R., W. W. Riggs, and J. Zimmerman. 2001. *Public Lands in the State of Nevada: An Overview*. University Center for Economic Development, University of Nevada, Reno. Fact Sheet-01-32: https://www.unce.unr.edu/publications/files/cd/2001/fs0132.pdf.
Heartland Institute. 2017. *Regulation*. https://www.heartland.org/topics/regulation/index.html. Accessed February 12, 2012.
Hickman, L. 2012. Leaked Heartland Institute documents pull back curtain on climate skepticism. *Guardian*, February 15, 2012. Accessed February 11, 2017. https://www.theguardian.com/environment/blog/2012/feb/15/leaked-heartland-

institute-documents-climate-scepticism.
Higgins, K. 2016. Post-truth: A guide for the perplexed. *Nature* 540: 9.
Hiltzik, M. 2016. California's drought: How Trump's blustering caricatured a genuine crisis. *Los Angeles Times*, June 6, 2016. http://www.latimes.com/business/hiltzik/la-fi-hiltzik-trump-westlands-20160606-snap-story.html. Accessed February 12, 2017.
House Committee on Natural Resources. 2017a. 21st Century Endangered Species Transparency Act (H.R. 4315). http://republicans-naturalresources.house.gov/legislation/hr4315/. Accessed February 11, 2017.
House Committee on Natural Resources. 2017b. *The Man-Made California Drought*. http://republicans-naturalresources.house.gov/drought/. Accessed February 11, 2017.
Huppke, R. 2012. Facts, 360 B.C.–A.D. 2012. *Chicago Tribune*, April 19, 2012. http://articles.chicagotribune.com/2012-04-19/news/ct-talk-huppke-obit-facts-20120419_1_facts-philosopher-opinion.
Inhofe, J. M. 2012. *The Greatest Hoax: How the Global Warming Conspiracy Threatens Your Future*. Washington, DC: WND Books.
Keller, E. A., and J. W. Day, Jr. 2007. Untrammeled growth as an environmental "March of Folly." *Ecological Engineering* 30: 206–14.
Konikow, L. F. 2013. *Groundwater Depletion in the United States (1900–2008)*. US Geological Survey Scientific Investigations Report 2013-5079. https://pubs.usgs.gov/sir/2013/5079/.
League of Conservation Voters. 2017. National Environmental Scorecard: Representative Ryan Zinke (R). http://scorecard.lcv.org/moc/ryan-zinke. Accessed February 11, 2017.
Leonard, D. 2010. Doctrinal uncertainty in the law of federal reserved water rights: The potential impact on renewable energy development. *Natural Resources Journal* 50: 611–43.
Lubchenco, J. 2017. Environmental science in a post-truth era. *Frontiers in Ecology and Environment* 15: 3.
Ludwig, D., R. Hilborn, and C. Walters. 1993. Uncertainty, resource exploitation, and conservation: Lessons from history. *Science* 260: 17, 36.
Lukes, S. 2017. Morals in a post-truth era. *Spiked*, March 2017. https://www.spiked-online.com/2017/03/31/morals-in-a-post-truth-era/. Accessed July 6, 2017.
MacDonald, G. M. 2010. Water, climate change, and sustainability in the Southwest. *Proceedings of the National Academy of Sciences* 107: 21256–62.
Miller, R. R., J. D. Williams, and J. E. Williams. 1989. Extinctions of North American fishes during the past century. *Fisheries* 14: 22–35.
Milly, P. C. D., J. Betancourt, M. Falkenmark, R. M. Hirsch, Z. W. Kundzewicz, D. P. Lettenmaier, and R. J. Stouffer. 2008. Stationarity is dead: Whither water management? *Science* 319: 573–74.
Minckley, W. L., and J. E. Deacon, eds. 1991a. *Battle against Extinction: Native Fish Management in the American West*. Tucson: University of Arizona Press.
Minckley, W. L., and J. E. Deacon. 1991b. Preface. In W. L. Minckley and J. E. Deacon, eds., *Battle against Extinction: Native Fish Management in the American West*, xiii–xv. Tucson: University of Arizona Press.
Minckley, W. L., and M. E. Douglas. 1991. Discovery and extinction of western fishes: A blink of the eye in geologic time. In W. L. Minckley and J. E. Deacon, eds., *Battle against Extinction: Native Fish Management in the American West*, 7–18. Tucson: University of Arizona Press.
Minteer, B. A., and J. P. Collins. 2005. Ecological ethics: Building a new tool kit for ecologists and biodiversity managers. *Conservation Biology* 19: 1803–12.
Mooney, C., and J. Eilperin. 2017. EPA website removes climate science site from public view after two decades. *Washington Post*, April 29, 2017. https://www.washingtonpost.com/news/energy-environment/wp/2017/04/28/epa-website-removes-climate-science-site-from-public-view-after-two-decades/? utm_term=.d5bc86df55ac. Accessed July 6, 2017.
National Science Board. 2016. *Science & Engineering Indicators 2016*. Chapter 7: Science and technology: Public attitudes and understanding. Arlington, VA: National Science Foundation (NSB-2016-1). https://www.nsf.gov/statistics/2016/nsb20161/#/report. Accessed July 6, 2017.
Nisbet, M. C., and E. Markowitz. 2016. *Americans' Attitudes about Science and Technology: The Social Context for Public Communication*. Washington, DC: American Association for the Advancement of Science. https://www.aaas.org/sites/default/files/content_files/NisbetMarkowitz_ScienceAttitudesReview_AAAS_Final_March10.pdf. Accessed July 6, 2017.
NOAA. 2017a. *Recent Monthly Average Mauna Loa CO2*. NOAA Earth Systems Research Laboratory. https://www.esrl.noaa.gov/gmd/ccgg/trends/. Accessed February 11, 2017.
NOAA. 2017b. *Global Analysis—Annual 2016*. NOAA National Centers for Environmental Information. https://www.ncdc.noaa.gov/sotc/global/201613. Accessed February 11, 2017.
Norment, C. J. 2014. *Relicts of a Beautiful Sea: Survival, Extinction, and Conservation in a Desert World*. Chapel Hill: University of North Carolina Press.
Obama, B. H. 2006. *The Audacity of Hope*. New York: Crown/Three Rivers Press.
Oreskes, N., and E. M. Conway. 2010. *Merchants of Doubt*. New York: Bloomsbury Press.
Pacific Legal Foundation. 2017. Environmental Regulations: Issues and Cases. https://www.pacificlegal.org/issuesandcases#tab3/. Accessed February 11, 2017.
Pavelko, M. T., D. B. Wood, and R. J. Laczniak. 1999. *Las Vegas, NV: Gambling with Water in the Desert*. US Geological Survey Circular 1182. https://pubs.usgs.gov/circ/circ1182/pdf/08LasVegas.pdf/. Accessed February 11, 2017.
Pew Research Center. 2015. *Public and Scientists' Views on Science and Society*. January 29, 2015. http://www.pewinternet.org/2015/01/29/public-and-scientists-views-on-science-and-society/. Accessed February 12, 2017.
Pew Research Center. 2016. *The Politics of Climate*. October 4, 2016. http://www.pewinternet.org/2016/10/04/the-politics-of-climate/. Accessed February 11, 2017.
Pister, E. P. 1991. The Desert Fishes Council: Catalysts for change. In W. L. Minckley and J. E. Deacon, eds., *Battle against Extinction: Native Fish Management in the American West*, 55–68. Tucson: University of Arizona Press.
Popovich, N., L. Albeck-Ripka, and K. Pierre-Louis. 2018. 66 environmental rules on the way out under Trump. *New York Times*, January 29, 2018. https://www.nytimes.com/

interactive/2017/10/05/climate/trump-environment-rules-reversed.html. Accessed January 31, 2018.
Popper, K. 1959. *The Logic of Scientific Discovery*. New York: Basic Books.
Powell, J. W. 1879. *Report on the Lands of the Arid Region of the United States*. 2nd ed. Washington, DC: Government Printing Office.
Reisner, M. 1986. *Cadillac Desert: The American West and Its Disappearing Water*. New York: Viking Penguin.
Riggs, A. C., and J. E. Deacon. 2002. Connectivity in desert aquatic ecosystems: The Devils Hole story. In D. W. Sada, and S. E. Sharpe, eds., *Conference Proceedings, Spring-Fed Wetlands: Important Scientific and Cultural Resources of the Intermountain Region, May 7–9,2002*. Paradise, NV: Desert Research Institute. http://www.dri.edu/images/stories/conferences_and_workshops/spring-fed-wetlands/spring-fed-wetlands-riggs-deacon.pdf.
Rusinek, W. 1990. A preview of coming attractions—Wyoming v. United States and the reserved rights doctrine. *Ecology Law Quarterly* 17: 355–412.
Rutenberg, J. 2017. "Alternative facts" and the costs of Trump-branded reality. *New York Times*, January 22, 2017. https://www.nytimes.com/2017/01/22/business/media/alternative-facts-trump-brand.html/.
Sabo, J. L., T. Sinha, L. C. Bowling, G. H. W. Schoups, W. W. Wallender, M. E. Campana, K. A. Cherkauer, P. L. Fuller, W. L. Graf, J. W. Hopmans, J. S. Kominoski, C. Taylor, S. W. Trimble, R. H. Webb, and E. E. Wohl. 2010. Reclaiming freshwater sustainability in the Cadillac Desert. *Proceedings of the National Academy of Sciences* 107: 21263–69.
Sada, D. W., and G. L. Vinyard. 2002. Anthropogenic changes in biogeography of Great Basin aquatic biota. In R. Hershler, D. B. Madsen, and D. Currey, eds., *Great Basin Aquatic Systems History*, 277–93. Smithsonian Contributions to the Earth Sciences, 33.
Seager, R., and G. A. Vecchi. 2010. Greenhouse warming and the 21st century hydroclimate of southwestern North America. *Proceedings of the National Academy of Sciences* 107: 21277–82.
Shogren, E. 2017. US House changes its rules to ease federal land transfers. *High Country News*, January 4, 2017. http://www.hcn.org/articles/u-s-house-changes-its-rules-to-ease-federal-land-transfers.
SNWA (Southern Nevada Water Authority). 2009. *Water Resource Plan 09*. Southern Nevada Water Authority, Las Vegas, Nevada. https://www.blm.gov/style/medialib/blm/nv/groundwater_development/snwa/feis0/feis_appendices.Par.80891.File.dat/Appendix%20A_SNWA%20Water%20Resources%20Plan,% 202009.pdf.
SNWA (Southern Nevada Water Authority). 2017. *Third Drinking Water Intake at Lake Mead*. Southern Nevada Water Authority, Las Vegas, Nevada. https://www.snwa.com/about/regional_intake3.html/. Accessed February 11, 2017.
Stein, B. A., C. Scott, and N. Benton. 2008. Federal lands and endangered species: The role of military and other federal lands in sustaining biodiversity. *BioScience* 58: 339–47.
Supin, J. 2014. The moral universe: Barbara Kingsolver on writing, politics, and human nature. *The Sun*, March 2014,. http://thesunmagazine.org/issues/459/the_moral_universe/.
Suskind, R. 2004. Faith, certainty and the presidency of George W. Bush. *New York Times Magazine*, October 17, 2004. http://www.nytimes.com/2004/10/17/magazine/faith-certainty-and-the-presidency-of-george-w-bush.html.
Taylor, B. 2017. *The Hue and Cry at Our House: A Year Remembered*. New York: Penguin Books.
Theobald, D. M., W. R. Travis, M. A. Drummond, and E. S. Gordon. 2013. The changing Southwest. In G. Garfin, A. Jardine, R. Merideth, M. Black, and S. LeRoy, eds., *Assessment of Climate Change in the Southwest United States: A Report Prepared for the National Climate Assessment*, 37–55. A report by the Southwest Climate Alliance. Washington, DC: Island Press. Available at http://www.swcarr.arizona.edu/sites/default/files/ACCSWUS_Ch3.pdf. Accessed February 12, 2017.
Tra, C. I. 2015. *Population Forecasts: Long-Term Projections for Clark County, Nevada, 2015–2050*. UNLV Center for Business and Economic Research. http://www.clarkcountynv.gov/comprehensive-planning/demographics/Documents/2015_Population_Forecasts.pdf/. Accessed February 11, 2017.
Trump, D. J. 2012. @realDonaldTrump. November 6, 2012. https://twitter.com/realdonaldtrump/status/265895292191248385?lang=en. Accessed February 11, 2017.
Udall, B., and J. Overpeck. 2017. The twenty-first century Colorado River hot drought and implications for the future. *Water Resources Research* 53: 2404–18.
USBR (US Bureau of Reclamation). 2017a. Lake Mead at Hoover Dam, elevation (feet). https://www.usbr.gov/lc/region/g4000/hourly/mead-elv.html. Accessed July 7, 2017.
USBR (US Bureau of Reclamation). 2017b. Lower Colorado River operations schedule. https://www.usbr.gov/lc/region/g4000/hourly/rivops.html. Accessed February 12, 2017.
US Bureau of the Census. 2012. Population of the 100 largest cities and other urban areas in the United States: 1790 to 1990. https://www.census.gov/population/www/documentation/twps0027/twps0027.html. Accessed February 11, 2017.
US Bureau of the Census. 2017. 2005 interim state population projections. https://www.census.gov/population/projections/data/state/projectionsagesex.html. Accessed February 20, 2017.
US Drought Monitor. 2017. January 3, 2017. http://droughtmonitor.unl.edu/MapsAndData/MapArchive.aspx. Accessed February 12, 2012.
USEPA. 2017. Climate change indicators: Atmospheric concentrations of greenhouse gasses. Environmental Protection Agency. https://www.epa.gov/climate-indicators/climate-change-indicators-atmospheric-concentrations-greenhouse-gases. Accessed February 11, 2017.
USFWS (US Fish and Wildlife Service). 2018. Endangered species. US Fish and Wildlife Service. https://www.fws.gov/endangered/. Accessed February 1, 2018.
USGS. 2011. *Effects of Climate Change and Land Use on Water Resources in the Upper Colorado River Basin*. Fact Sheet 2010-3123. https://pubs.usgs.gov/fs/2010/3123/pdf/FS10-3123.pdf.
USGS. 2017. California drought. USGS California Water Science Center. https://ca.water.usgs.gov/data/drought/. Accessed February 11, 2017.

Utah State Legislature. 2010. H.J.R. 12 Enrolled. Climate Change Joint Resolution. http://le.utah.gov/~2010/bills/static/HJR012.html/. Accessed February 11, 2017.

Vincent, C. H., L. A. Hanson, and J. P. Bjelopera. 2014. *Federal Landownership: Overview and Data.* Washington, DC: Congressional Research Service. https://fas.org/sgp/crs/misc/R42346.pdf.

Wemple, E. 2016. CNN commentator Scottie Nell Hughes: Facts no longer exist. *Washington Post,* December 1, 2016. https://www.washingtonpost.com/blogs/erik-wemple/wp/2016/12/01/cnn-commentator-scottie-nell-hughes-facts-no-longer-exist/?utm_term=.af2ba819e2c8/.

Woodhouse, C. A., S. T. Gray, and D. M. Meko. 2006. Updated streamflow reconstructions for the upper Colorado River basin. *Water Resources Research* 42: W05415. https://doi.org/10.1029/2005WR004455.

Woodhouse, C. A., D. M. Meko, G. M. MacDonald, D. W. Stahle, and E. R. Cook. 2010. A 1,200-year perspective of 21st century drought in southwestern North America. *Proceedings of the National Academy of Sciences* 107: 21283–88.

Wulf, A. 2015. *The Invention of Nature: Alexander von Humboldt's New World.* New York: Alfred A. Knopf.

Yardley, W. 2016. Ryan Zinke, Trump's pick as Interior secretary, is all over the map on some key issues. *Los Angeles Times,* December 15, 2016. http://www.latimes.com/nation/la-na-pol-interior-zinke-2016-story.html. Accessed January 8, 2017.

Zaffos, J. 2016. House Republicans want to "repeal and replace" the ESA. *High Country News,* December 28, 2016. http://www.hcn.org/articles/house-republicans-may-try-to-repeal-and-replace-the-endangered-species-act/. Accessed February 11, 2017.

Zinke, R. 2017. *Jobs and Economic Growth.* https://zinke.house.gov/issues/jobs-and-economic-growth/. Accessed February 11, 2017.

26

Searching for Common Ground between Life and Extinction

Christopher W. Hoagstrom,
Kevin R. Bestgen,
David L. Propst, and
Jack E. Williams

Increasing water withdrawals (Xenopoulos et al. 2005), coupled with habitat alterations, water pollution, and biological invasions (Dudgeon 2010) are rapidly depleting global freshwater biodiversity. As highlighted throughout this volume, this trend is especially strong in North American deserts, where unsustainable use of scarce water supplies (Garrett et al., chap. 8, this volume; Udall, chap. 7, this volume), increasing aridity (Overpeck and Bonar, chap. 9, this volume), and nonnative species invasions (Albrecht et al., chap. 11, this volume; Moyle, chap. 4, this volume) bode ill for native aquatic species and natural aquatic ecosystems. Grim examples of population loss and ecosystem degradation abound (e.g., Boersma and Lytle, chap. 10; Lozano-Vilano et al., chap. 5; Stone and Morrison, chap. 16; and Williams and Sada, chap. 6, all in this volume).

The biota of aquatic ecosystems throughout the North American desert region (henceforth, desert waters) has been discovered relatively slowly and even more slowly appreciated, but is now revealed as highly diverse (e.g., Bogan et al., chap. 18; Hoagstrom et al., chap. 3; and Varela-Romero et al., chap. 19, all in this volume), even though many putative taxa still need further study for accurate inventory of the total sum of native aquatic biodiversity. Rigorous archiving of aquatic desert species remains important for documenting and cataloging this biodiversity (Cohen et al., chap. 13, this volume), and even relatively pristine ecosystems require monitoring to ensure protection (e.g., Reinthal et al., chap. 15, this volume). Imperiled populations that depend on threatened or degraded habitats require intensive conservation efforts (e.g., Echelle and Echelle, chap. 23, this volume; Lema et al., chap. 22, this volume). Monitoring of genetic diversity is also critical for effective conservation of

[Box 26.1] **Important Concepts for Conservation and Restoration of Aquatic Ecosystems in North American Deserts, in the Order Introduced in the Text**

Wicked problem: A complex problem lacking a clear-cut solution (Ludwig 2001). Ecological complexity, uncertainty inherent within management actions, administrative complexity in decision making, and diverse stakeholder values make ecosystem management a wicked problem (DeFries and Nagendra 2017).

Climatic optimum: A prehistoric period of unusual warmth. A Holocene (postglacial) optimum about 8,000–3,000 years ago produced warmer, drier ecosystems than in the twentieth century (Kirby et al. 2015).

Megadrought: A prehistoric drought with aridity comparable to or exceeding that of historical droughts, but prolonged through consecutive decades, in contrast to historical droughts of less than 10 years in duration (Cook et al. 2016).

Utilitarian conservation: A view of natural resources as most valuable when developed for beneficial uses; championed by Theodore Roosevelt and Gifford Pinchot, among others; opposed by John Muir (Worster 2008) and Aldo Leopold (Callicott 2000), among others.

Shifting-baseline syndrome: Overly optimistic perceptions of ecosystem health resulting from ignorance or denial of historical (preindustrial) levels of natural biodiversity and ecosystem productivity (Soga and Gaston 2018). Degraded, sterile, and depauperate ecosystems are mistakenly viewed as healthy, productive, and diverse, often because observers are unaware of, or have not experienced, pristine systems. Extinction debts are overlooked.

Ecological ratchet: A persistent impact that shifts an ecosystem into a degraded, depauperate, and unproductive state that is sustained indefinitely and over the long term (Pitcher 2001, Pinnegar and Engelhard 2008). Ratcheting occurs as multiple such impacts accrue.

Extinction debt: A situation in which ecosystem degradation has virtually doomed a species to extinction, but the species has not yet become extinct (Kuussaari et al. 2009). A simple example: A long-lived species persists as an aging population without recruitment and ultimately disappears when the persisting individuals perish due to their senescence. A more complex example: A short-lived species that requires high streamflows for annual recruitment sustains its population while water development causes incremental reductions in high streamflows, but those incremental reductions, when combined with inevitable drought, eventually produce a period altogether lacking high streamflows over enough consecutive years to eliminate the population after consecutive years of failed recruitment.

Post-environmental era: The period, from about 1978 to the present, during which organized, partisan opposition to enforcement of environmental legislation arose in the United States; characterized by chronic polarization of viewpoints, in contrast to the preceding environmental era, when landmark legislative actions establishing modern environmental protections were fashioned in a nonpartisan milieu.

Crisis discipline: A discipline requiring action despite incomplete information, such as emergency medicine and conservation biology (Soulé 1985).

Anthropocene: The ongoing human-dominated epoch (~1950–present) characterized by a distinct trajectory of rapidly increasing atmospheric CO_2 and climate warming attributable to industrial activities (Steffen et al. 2016).

Novel ecosystems: Environments and biotic communities absent from preindustrial drainages, which sometimes develop pseudo-natural ecosystems that support native species (Johnson 2002).

Conservation ethic: Aldo Leopold's alternative to utilitarian conservation, focused on ecosystem health and summarized by the quotation, "A thing is right when it tends to preserve the integrity, stability, and beauty of the biotic community. It is wrong when it tends otherwise" (Leopold 1949, 188).

imperiled populations (Turner et al., chap. 14, this volume).

Water insecurity (Grafton 2017), nonpoint-source pollution (Patterson et al. 2013), and biological invasions (Woodford et al. 2016), are *wicked problems* (see box 26.1 for descriptions of this and other concepts highlighted in the chapter text). Because 30 different areas of endemism each harbor a unique assemblage of endemics (Hoagstrom et al., chap. 3, this volume), each of these wicked problems must be confronted separately in each area if comprehensive conservation of native biodiversity is to be achieved. Such efforts would be tantamount to ecosystem conservation across the North American desert region. The measures needed in each area generally require long-term commitments to adaptive management (e.g., Bestgen et al., chap. 21, this volume; Propst et al., chap. 20, this volume). Progress varies case by case, and prospects for success are universally tenuous (e.g., Echelle and Echelle, chap. 23; Lema et al., chap. 22; and Wilson et al., chap. 24, all in this volume).

Thus, the goal of conserving desert waters can seem unreachable or even quixotic. Legal complications can be a major constraint on conservation actions and must be navigated carefully to sustain forward progress (Andersen and Brooks, chap. 12, this volume). Widespread anti-conservation and anti-science sentiments can further endanger progress and undermine research findings, threatening to even devalue ecosystem health and biodiversity (Norment, chap. 25, this volume). The following discourse explores a way forward.

LOOKING BACK

> Technological humans and native freshwater fishes are locked in a mortal competition for water, and indigenous fishes will not prevail unless we plan and dictate a scenario that *ignores short-term economic concerns*.
> —MINCKLEY AND DOUGLAS (1991, 17, emphasis added)

Prehistory

Some early researchers believed that most desert fish lineages originated in the relatively recent and wet Pleistocene Epoch, implying a limited evolutionary tolerance for drought (Smith 1981). Most lineages, however, actually originated much earlier, between the late Middle Miocene and Early Pliocene (~16–4 million years ago) (Smith et al. 2002; Hoagstrom et al., chap. 3, this volume). The ages of these lineages thus equal or exceed the age of the modern Grand Canyon (for instance) and, importantly, verify a heritage of long-term drought tolerance.

The intense aridity of the Late Miocene and Early Pliocene (Chapin 2008) implies that desert fish lineages arose as isolated drought specialists that survived wherever adequate waters persisted. Even the youngest endemic lineages, which do have Pleistocene origins, survived a postglacial *climatic optimum* during which tree lines within the Great Basin ascended and surface waters such as the Humboldt River, Pyramid Lake, Walker Lake, and Owens Lake desiccated (Grayson 2011). Further, even within the last 2,000 years, these lineages endured repeated *megadroughts*. In this light, historical extinctions of desert fishes (Williams and Sada, chap. 6, this volume) are not attributable to low drought tolerance, because in natural habitats, now-extinct populations previously survived extreme, prolonged droughts much more severe than those associated with their historical extinctions and extirpations. What made those earlier, prehistoric droughts less threatening was the absence of modern anthropogenic hazards and industrial ecosystem alterations.

Utilitarian Conservation

Industrial-scale dewatering in the twentieth century transformed desert waters (Minckley and Douglas 1991). As president, famed conservationist Theodore Roosevelt supported diverting water from Owens Valley, California, for use by the people of Los Angeles (Reisner 1993), with dire consequences for the waters and resident fishes of Owens Valley (Pister 1974). Similar activities desiccated Pyramid Lake, Nevada (Ono et al. 1983). Elsewhere, dams and water withdrawals severely altered or destroyed portions of the Colorado River and Rio Grande (Collier et al. 1996; Mueller and Marsh 2002). An anthropogenic version of megadrought had begun.

Roosevelt followed the then-fashionable *utilitarian* approach to conservation, viewing natural resources as most valuable when developed for beneficial uses (Reisner 1993). For example, the Newlands Reclamation Act (1902) authorized construction of Theodore Roosevelt Dam on the Salt River, Arizona. Dedicated in 1911, with former president Roosevelt present, this dam initiated industrial transformation of the Gila River basin (Fradkin 1968), presaging its severe environmental degradation (Miller 1961; Mueller and Marsh 2002).

Although these ecosystem impacts do not negate Roosevelt's positive contributions to conservation writ large (Meine 2001), they illustrate important complexities. Varied viewpoints exist among conservationists, and disagreements over conservation approaches are common. A strong conservation effort in one arena does not ensure comparable efforts elsewhere. A champion for conservation in one scenario may withhold support from, or side with the opposition to, conservation in another.

Utilitarian views remained prominent long after Roosevelt (Soulé 1985) and persist within the fields of fisheries (Pister 1987; Clarkson et al. 2005) and hydrology (Trush et al. 2000; Graf 2001). Nevertheless, conservationists concerned with ecosystem health have long questioned utilitarian conservation (Callicott 2000; Worster 2008). We consider ecosystem viability central not only to conserving desert waters, but also to establishing sustainable use of ecosystem benefits.

Shifting Baseline

Because other human activities concurrent with industrialization imposed novel stressors on desert waters, the suite of current threats has no prehistoric precedent to which native species could have adapted. For example, excessive watershed fragmentation isolates remnant populations, elevating their extinction risk (Perkin et al. 2015). Extirpations of natives can fundamentally change aquatic ecosystems by altering trophic dynamics (East et al. 2017; Boersma and Lytle, chap. 10, this volume). Further, invasive species are ubiquitous (e.g., Propst et al. 2008; Moyle, chap. 4, this volume) and are favored in degraded habitats (Olden et al. 2006), which are widespread. Because of these insidious and synergistic legacies, perhaps no living person can fully conceive how different contemporary ecosystems are from preindustrial ones.

However, historical awareness is essential to avoid a *shifting-baseline syndrome* wherein each new generation of conservationists and resource managers accepts the conditions they first encounter themselves as "normal" (Soga and Gaston 2018). Unfortunately, only anecdotes document the preindustrial conditions of most desert waters, and those conditions can seem unimaginable. For example, in 1915, Snyder (1924) found Bonneville-Snake bluehead sucker *Pantosteus virescens* predominant in the Weber River, with thousands visible from shore. With his field assistant, Carl Hubbs, Snyder captured 708 suckers in one seine haul (230–395 mm TL), which did not visibly diminish the school. In comparison, the remnant population, now fragmented into subpopulations, was estimated—based on multi-pass electrofishing, including all individuals longer than 150 mm TL (Webber et al. 2012)—at about 880 individuals. The total population hardly exceeded the number captured in one 1915 seine haul, just 100 years before. This comparison reveals the dramatic results of the cumulative anthropogenic influences affecting the Weber River, including dams, water withdrawals, river-channel degradation, and abundant nonnative brown trout *Salmo trutta*.

Enduring disturbances like these act as an *ecological ratchet* that perennially depresses ecosystem integrity (Perkin et al. 2015). Desert waters had already suffered ratchet-like impacts before all but the earliest biological surveys (Miller et al. 1991). Many natives persist only in degraded ecosystems, hovering close to their extinction thresholds (Osborne et al. 2014). Importantly, activities such as flow-regime restoration, effective nonnative species management or removal, habitat restoration, reduced fragmentation, and dam removal are necessary to reverse ratchet-like pressures (Amoros et al. 1987) and reduce *extinction debt*.

COMPLICATIONS

> History has shown that public understanding *precedes* attempts to correct major environmental abuse. Conservation legislation in the United States grew out of moral outrage over environmental abuses so obvious that even a Congress dedicated to economic development and pork-barrel politics *could not ignore them*.
> —DEACON WILLIAMS AND DEACON (1991, 109, emphases added)

The Rapid Transition to Post-environmental Politics

Early victories for desert fish conservation were enshrined in the national environmental movement

(Pister 1991; Deacon Williams and Deacon 1991) exemplified by a unanimous Senate and 345–4 House vote for passage of the Endangered Species Act (ESA). President Richard M. Nixon enthusiastically signed the legislation in 1973 (Petersen 1999). Thereafter, zeal for environmental protection waned as the realities and dilemmas of ESA enforcement emerged.

Thus, despite an election campaign that pledged to halt industrial water projects, President Jimmy Carter, ironically, strengthened support for them (Reisner 1993). Succumbing to intense political pressure, Carter did not oppose legislation exempting the Tellico Dam project from the ESA, even though the dam threatened the endangered snail darter *Percina tanasi*, and even though the president initially promised a veto (Plater 2008, 2013). This history exemplifies the phrase, "Ideology is the first casualty of water development" (Reisner 1993, 309). The snail darter controversy fueled a growing desire to weaken the original ESA, leading to its amendment in 1978—with support from President Carter—to allow more discretionary enforcement, which effectively reduced environmental protection (Stromberg 1978).

By 1978, the issue of endangered species protection was politicized along party lines—Democrats for, Republicans against (López and Sutter 2004). Nonetheless, this stereotype had exceptions. The fiscal policies of Ronald Reagan were unfriendly to pork-barrel proposals, which made large water development projects less likely (Reisner 1993). Further, Reagan's decidedly anti-conservation secretary of the interior, James Watt, authorized emergency listing of Ash Meadows pupfish *Cyprinodon nevadensis mionectes* and Ash Meadows speckled dace *Rhinichthys osculus nevadensis* as endangered in 1982, providing an unexpected catalyst for their conservation (Deacon and Williams 1991).

These encouraging exceptions aside, increased legal discretion within the amended ESA launched a *post-environmental era* just five years after the law's enactment. Hence, J. E. Deacon and W. L. Minckley (1991; Minckley 1991) sensed discretionary weakening in endangered species protection. Strong opposition to conservation became common under both Republican (Petersen 1999; Norment, chap. 25, this volume) and Democratic (Pulliam 1998a; Wilkinson 1998) presidents, as did political reversals of environmentally favorable decisions (Pulliam 1998b; Wagner 1999). Reliance on politics to ensure conservation became inherently risky because, whereas conservation required long-term vision to maintain functional ecosystems, politicians prioritized short-term gains to maintain their political popularity (Blignaut et al. 2014).

The Limitations of Science

Science alone is an imperfect means of ensuring conservation. Even scrupulous science can be ignored (Likens 2010), but the uncertainty inherent within science can also cause hesitancy to act (Clark 2002; DeFries and Nagendra 2017) or fuel delaying tactics (Havas et al. 1984; Andersen and Brooks, chap. 12, this volume). Scientists and nonscientists alike can labor under the illusion that quantitative studies produce unequivocally superior solutions, although they seldom do (Tukey 1960; Cullen 1990). Irrational desires to forestall action due to uncertainty can devalue legitimate, qualitative understanding (Behnke 1987; Piccolo 2011).

Sharing specialized scientific knowledge with the public can be difficult (Aronson et al. 2010). Many citizens distrust science because they are aware that scientists, as human beings, can face personal and professional pressures that may create bias, as revealed in several highly publicized scandals (Savan 1988). In extreme cases, such as the nuclear arms race, scientific achievements can produce innocent victims (Gallagher 1993), and such instances have resulted in legitimate distrust of scientists and bureaucrats among many citizens. Further, some scientists elect to serve special interests that systematically undermine well-founded evidence by casting doubt or promoting misinformation (Oreskes and Conway 2010). These and associated factors can combine to marginalize conservation efforts (e.g., Plater 2013; Andersen and Brooks, chap. 12, this volume).

Hence, broad collaborations (Knight et al. 2008; Barmuta et al. 2011), focused research and management strategies that address public concerns (Cullen 1990; Dunn and Laing 2017), effective communication (Pace et al. 2010; McMullin 2017), and cultivation of stakeholder trust (Horton et al. 2016; Young et al. 2016) are all necessary to overcome the barriers confronting conservation initiatives. Because conservation biology is a *crisis discipline*, opportunities to influence conservation policy require decisiveness (Dunn and Laing 2017). Nevertheless, sustained success requires long-term commitment (Lawson et al. 2017). Learning-by-doing and adaptive management approaches are needed to jointly improve management and increase understanding (Clark 2002; Laub et al. 2015). Conservation efforts must emphasize resource limitations as a frame that envelops a shared vision that represents all stakeholders and promotes public good as integral to conservation efforts (Rogers 2006; Naiman 2013).

FISH-EYE PERSPECTIVES

> Constant attention to the details of maintaining biotic diversity—the earth's life-support system—is not an option; it is mandatory.
>
> —DEACON AND MINCKLEY (1991, 413)

Post-environmental Conservation

Twenty-first-century conservation requires collaborative, participatory, and adaptive approaches (Barmuta et al. 2011). For example, a conservation team of state, federal, and tribal signatories established new wild populations of, and strengthened legal protection for, least chub *Iotichthys phlegethontis* following its classification as a candidate for listing under the ESA (USFWS 2010). Thereafter, the US Fish and Wildlife Service deemed listing unnecessary (USFWS 2014; Thompson et al. 2015).

Likewise, Modoc sucker *Catostomus microps* became the first desert fish delisted under the ESA through improved protection and new evidence of minimal risk from nonnatives, hybridization, climate change, and drought (USFWS 2015). Scientists, government personnel, and private landowners developed trust and collaborated to seize this opportunity for conservation (Echelle and Echelle, chap. 23, this volume).

An international collaboration planned, enacted (in spring 2014), and studied the results of an experimental pulse flow through the desiccated Colorado River delta. Documentation of its ecological (Kendy et al. 2017; Pitt et al. 2017) and cultural (Bark et al. 2016) benefits strengthened the parties' resolve to continue their efforts, and an updated collaboration agreement was signed in September 2017 (Garrick et al. 2017). Strained US-México relations under US President Donald Trump reportedly did not slow progress (Cornwall 2017).

Ecosystem Preservation and Restoration

The emerging *Anthropocene* climate will accelerate and expand desiccation of desert waters (Jaeger et al. 2014), which will tighten ecological ratchets and elevate extinction debts. Although industrial impacts remain the paramount threat (Tedesco et al. 2013), conservationists must also anticipate emerging threats from climate change (Krabbenhoft et al. 2014; Warren et al. 2014). Improved management of riparian vegetation (Seavy et al. 2009), flow-regime restoration (Poff et al. 2017), reduced water consumption (Xenopoulos et al. 2005), improved water security (Vörösmarty et al. 2010), and protection and restoration of climate refuges (Morelli et al. 2016) can all help.

Remnant ecosystems sustaining high biodiversity and sensitive species are high conservation priorities (Pool et al. 2013; Howard et al. 2018b). Many of these ecosystems are in less controlled streams and rivers that sustain more natural hydrology and geomorphology (Sabo et al. 2012), support high food-web diversity and food-chain length (Sabo et al. 2018), and maintain critical habitat features (Yackulic et al. 2014; Fraser et al. 2017). Often, these systems are relatively remote from cities and irrigation projects (e.g., Reinthal et al., chap. 15, this volume). Although preservation of whole ecosystems can be an arduous task, these remnant areas not only maintain natural ecosystem processes, but provide natural laboratories critical for understanding preindustrial ecosystems. And knowledge gained from these areas can guide restoration of degraded ecosystems.

Ecosystem restoration does not necessarily mean a return to preindustrial conditions, but implies reversal of ecological ratchets to resuscitate an ecosystem's functions (Kendy et al. 2017; Poff 2018) and realize its remaining natural potential (Echelle and Echelle, chap. 23, this volume). Focused actions such as reconnecting fragmented habitats (Pennock et al. 2018) or eliminating nonnative species (Propst et al., chap. 20, this volume) can improve such potential. Further, multifaceted restoration strategies have high potential for success (Walsworth and Budy 2015), as demonstrated by the positive outcomes of efforts combining restored streamflow with either increased habitat connectivity (Scoppettone and Rissler 1995, 2007) or invasive species removal (Marks et al. 2010).

Nevertheless, tremendous challenges remain (Franssen et al. 2014; Archdeacon 2016), and ecosystem restoration is not necessarily straightforward. For instance, the invasive riparian salt cedar tree (*Tamarix* spp.) can exacerbate ecosystem degradation along regulated rivers (Merritt and Poff 2010; Kui et al. 2017). Salt cedar removal in conjunction with flow-regime restoration can restore riverine habitats (e.g., Keller et al. 2014), but the tree's importance for songbirds, including the endangered riparian-obligate southwestern willow flycatcher *Empidonax traillii extimus*, makes salt cedar removal controversial (Sogge et al. 2008), increasing emphasis on revegetation where salt cedar is removed (Paxton et al. 2011).

In particularly perplexing instances, a species imperiled within its native range may, unexpectedly, become invasive elsewhere. When introduced to suitable remnant ecosystems, imperiled nonnatives may establish populations that threaten imperiled natives (Hoagstrom et al. 2010). For severely imperiled species, these nonnative populations may have conservation significance (Osborne et al. 2013). This creates a surreal situation in which conservation of invasive nonnatives may conflict with conservation of imperiled natives. Conservation of co-occurring natives can create similar conflicts, requiring innovative solutions protecting both species (Lema et al., chap. 22, this volume).

Novel ecosystems that develop within regulated rivers can also benefit natives. For instance, some native Colorado River fishes use reservoir-inflow habitats for spawning and rearing (Albrecht et al. 2017). Such scenarios provide additional management options (Bestgen et al., chap. 21, this volume), but require new emphasis on habitats traditionally viewed as unimportant (Backstrom et al. 2018).

FIAT LUX

> One of the facts hewn by *science* is that every river needs more people, and all people need more inventions, and hence more science; the good life depends on the indefinite extension of this chain of logic. That the good life on any river may likewise depend on the *perception* of its music, and the preservation of some music to perceive, is a form of doubt not yet entertained by science.
>
> —ALDO LEOPOLD (1949, 134, emphases added)

Effective ecosystem management begins with the recognition of resource limitations (Daly and Farley 2011). Conservation science must help our industrial society confront its legacy of unsustainable resource use and realize that ecosystem preservation and restoration can safeguard natural beauty, ecological viability, and ecosystem services, all with implications for human health and economic sustainability (Aronson et al. 2016). Environmental protections can also stimulate development of more environmentally favorable technologies (Taylor et al. 2005; Lee et al. 2010).

In his *conservation ethic*, Aldo Leopold (1933, 1949) insisted that scientists, citizens, and government officials consider a "biotic view" of land that recognizes the significance of functioning ecosystems. More than 80 years later, it remains critical for conservation scientists to communicate the legitimate economic values of functioning ecosystems (de Groot et al. 2012) and of ecosystem restoration (de Groot et al. 2013). Cultivating public perception of natural ecosystems as critical community features could ease this difficult task (Lopez 1990; Piccolo 2017). Although scientific advocacy has risks, maintaining transparency and integrity (Carroll et al. 2017) can allow scientists to play useful roles in developing public policy (Pielke 2007; Snelder et al. 2014). Medical scientists advocating public health have modeled the manner in which conservation scientists could advocate ecosystem health (Aronson et al. 2016), and the field of ecological economics provides a foundation for integrating ecosystem services into economic planning (Daly and Farley 2011).

The politicized setting of the post-environmental era requires bridging cultural gaps to develop sustainable solutions (Aronson et al. 2010; DeFries and Nagendra 2017). In the spirit of the conservation ethic, which calls for collective focus on conservation, inclusivity within conservation efforts can broaden and deepen societal support (Tallis and Lubchenco 2014). Strategies to overcome disagreements include reducing antagonism (Bliuc et al. 2015), unpacking root causes of opposition (Akerlof et al. 2013; Walsh and Tsurusaki 2014), and addressing sources of confusion (Plutzer et al. 2016).

Long-term collaborations can meld resource management with ecosystem conservation and restoration. In the Colorado River basin, rapidly declining water levels in Lakes Powell and Mead, exacerbated by high seepage and evaporative losses, have spurred new discussions regarding the practicalities of maintaining water storage in both reservoirs. Given decreasing water availability, it is prudent to examine alternative water management options that can benefit aquatic ecosystems and maintain human water supplies.

There is a need to better harmonize ecosystem conservation and restoration with the priorities and values of local communities (Pringle 2017). To the degree that these efforts remain peripheral to public appreciation, they are vulnerable to opposition or summary dismissal. As in the aforementioned case of the Colorado River delta, conservation efforts elsewhere should benefit from engaged public and media support, as well as from adaptive management approaches that incorporate representative stakeholder input. Rationales for conservation are diverse, and consistent use of multiple arguments can broaden its appeal (Tinch et al. 2018). Because the most compelling or germane rationale differs from one situation to another, collaborations are critical for identifying stakeholder values that align with each conservation effort (Howard et al. 2018a).

Ultimately, "society must *decide* to change the direction and character of its economic-development pathway" (Blignaut et al. 2014, 55, emphasis in original). Perhaps collaborative conservation, as discussed above, infused with citizen science (e.g., Maceda-Veiga et al. 2016), can speed the realization that human behavior must change to reflect an ideology of sustainable resource use, especially water use. Sustaining desert waters would conserve the aquatic ecosystems we value along with the water supplies we require. As concerned citizens, we must discover roles we can play to avoid creating a waterless desert devoid of aquatic life. We must all stand between life and extinction (Pister 1993), together.

> Under U.S. law, as we have seen, especially during the past decade, it is possible to promote practices that cause or accelerate the major ecosystem disruptions that are increasingly evident in western deserts and throughout the world. Let us not forget that it is equally possible under the law to promote policies leading toward *sustainable coexistence of human and nonhuman communities*.
>
> —MINCKLEY AND DEACON (1991, 403, emphasis added)

REFERENCES

Akerlof, K., E. W. Maibach, D. Fitzgerald, A. Y. Cedeno, and A. Neuman. 2013. Do people "personally experience" global warming, and if so how, and does it matter? *Global Environmental Change* 23: 81–91.

Albrecht, B. A., H. E. Mohn, R. Kegerries, M. C. McKinstry, R. Rogers, T. Francis, B. Hines, J. Stolberg, D. Ryden, D. Elverud, B. Schleicher, K. Creighton, B. Healy, and B. Senger. 2017. Use of inflow areas in two Colorado River basin reservoirs by the endangered razorback sucker (*Xyrauchen texanus*). *Western North American Naturalist* 77: 500–514.

Amoros, C., J. C. Rostan, G. Pautou, and J. P. Bravard. 1987. The reversible process concept applied to the environmental management of large river systems. *Environmental Management* 11: 607–17.

Archdeacon, T. P. 2016. Reduction in spring flow threatens Rio Grande silvery minnow: Trends in abundance during river intermittency. *Transactions of the American Fisheries Society* 145: 754–65.

Aronson, J. C., C. M. Blatt, and T. B. Aronson. 2016. Restoring ecosystem health to improve human health and well-being: Physicians and restoration ecologists unite in a common cause. *Ecology and Society* 21(4): 39.

Aronson, J., J. N. Blignaut, R. S. de Groot, A. Clewell, P. P. Lowry II, P. Woodworth, R. M. Cowling, D. Renison, J. Farley, C. Fontaine, D. Tongway, S. Levy, S. J. Milton, O. Rangel, B. Debrincat, and C. Birkinshaw. 2010. The road to sustainability must bridge three great divides. *Annals of the New York Academy of Sciences* 1185: 225–36.

Backstrom, A. C., G. E. Garrard, R. J. Hobbs, and S. A. Bekessy. 2018. Grappling with the social dimensions of novel ecosystems. *Frontiers in Ecology and the Environment* 16: 109–17.

Bark, R. H., C. J. Robinson, and K. W. Flessa. 2016. Tracking cultural ecosystem services: Water chasing the Colorado River restoration flow pulse. *Ecological Economics* 127: 165–72.

Barmuta, L. A., S. Linke, and E. Turak. 2011. Bridging the gap between "planning" and "doing" for biodiversity conservation in freshwaters. *Freshwater Biology* 56: 180–95.

Behnke, R. J. 1987. The illusion of technique and fisheries management. *Proceedings of the Desert Fishes Council* 19: 47–49.

Blignaut, J., J. Aronson, and R. de Groot. 2014. Restoration of natural capital: A key strategy on the path to sustainability. *Ecological Engineering* 65: 54–61.

Bliuc, A. M., C. McGarty, E. F. Thomas, G. Lala, M. Berndsen, and R. Misajon. 2015. Public division about climate change rooted in conflicting socio-political identities. *Nature Climate Change* 5: 226–29.

Callicott, J. B. 2000. Aldo Leopold and the foundations of ecosystem management. *Journal of Forestry* 98(5): 4–13.

Carroll, C., B. Hartl, G. T. Goldman, D. J. Rohlf, A. Treves, J. T. Kerr, E. G. Ritchie, R. T. Kingsford, K. E. Gibbs, M. Maron, and J. E. M. Watson. 2017. Defending the scientific integrity of conservation-policy processes. *Conservation Biology* 31: 967–75.

Chapin, C. E. 2008. Interplay of oceanographic and paleoclimate events with tectonism during middle to late Miocene sedimentation across the southwestern USA. *Geosphere* 4: 976–91.

Clark, M. J. 2002. Dealing with uncertainty: Adaptive approaches to sustainable river management. *Aquatic Conservation: Marine and Freshwater Ecosystems* 12: 347–63.

Clarkson, R. W., P. C. Marsh, S. E. Stefferud, and J. A. Stefferud. 2005. Conflicts between native fish and nonnative sport fish management in the southwestern United States. *Fisheries* 30(9): 20–27.

Collier, M., R. H. Webb, and J. C. Schmidt. 1996. *Dams and Rivers: Primer on the Downstream Effects of Dams.* US Geological Survey Circular 1126. 94 pp.

Cook, B. I., E. R. Cook, J. E. Smerdon, R. Seager, A. P. Williams, S. Coats, D. W. Stahle, and J. V. Díaz. 2016. North American megadroughts in the common era: Reconstruction and simulations. *Wiley Interdisciplinary Reviews: Climate Change* 7: 411–32.

Cornwall, W. 2017. US-Mexico water pact aims for a greener Colorado delta. *Science* 357: 635.

Cullen, P. 1990. The turbulent boundary between water science and water management. *Freshwater Biology* 24: 201–9.

Daly, H. E., and J. Farley. 2011. *Ecological Economics, Principles and Applications.* 2nd ed. Washington, DC: Island Press.

Deacon, J. E., and C. Deacon Williams. 1991. Ash Meadows and the legacy of the Devils Hole pupfish. In W. L. Minckley and J. E. Deacon, eds., *Battle against Extinction: Native Fish Management in the American West*, 69–87. Tucson: University of Arizona Press.

Deacon, J. E., and W. L. Minckley. 1991. Western fishes and the real world: The enigma of "endangered species" revisited. In W. L. Minckley and J. E. Deacon, eds., *Battle against Extinction: Native Fish Management in the American West*, 405–13. Tucson: University of Arizona Press.

Deacon Williams, C., and J. E. Deacon. 1991. Ethics, federal legislation, and litigation in the battle against extinction. In W. L. Minckley and J. E. Deacon, eds., *Battle against Extinction: Native Fish Management in the American West*, 109–21. Tucson: University of Arizona Press.

DeFries, R., and H. Nagendra. 2017. Ecosystem management as a wicked problem. *Science* 356: 265–70.

De Groot, R. S., J. Blignaut, S. van der Ploeg, J. Aronson, T. Elmqvist, and J. Farley. 2013. Benefits of investing in ecosystem restoration. *Conservation Biology* 27: 1286–93.

De Groot, R., L. Brander, S. van der Ploeg, R. Costanza, F. Bernard, L. Braat, M. Christie, N. Crossman, A. Ghermandi, L. Hein, S. Hussain, P. Kumar, A. McVittie, R. Portela, L. C. Rodríguez, P. ten Brink, and P. van Beukering. 2012. Global estimates of the value of ecosystems and their services in monetary units. *Ecosystem Services* 1: 50–61.

Dudgeon, D. 2010. Prospects for sustaining freshwater biodiversity in the 21st century: Linking ecosystem structure and function. *Current Opinion in Environmental Sustainability* 2: 422–30.

Dunn, G., and M. Laing. 2017. Policy-makers perspectives on credibility, relevance and legitimacy (CRELE). *Environmental Science and Policy* 76: 146–52.

East, J. L., C. Wilcut, and A. A. Pease. 2017. Aquatic food-web structure along a salinized dryland river. *Freshwater Biology* 62: 681–94.

Fradkin, P. L. 1968. *A River No More, the Colorado River and the West.* Tucson: University of Arizona Press.

Franssen, N. B., J. E. Davis, D. W. Ryden, and K. B. Gido. 2014. Fish community responses to mechanical removal of nonnative fishes in a large southwestern river. *Fisheries* 39: 352–63.

Fraser, G. S., D. L. Winkelman, K. R. Bestgen, and K. G. Thompson. 2017. Tributary use by imperiled flannelmouth and bluehead suckers in the upper Colorado River basin. *Transactions of the American Fisheries Society* 146: 858–70.

Gallagher, C. 1993. *American Ground Zero, the Secret Nuclear War.* Cambridge, MA: MIT Press.

Garrick, D. E., J. W. Hall, A. Dobson, R. Damania, R. Q. Grafton, R. Hope, C. Hepburn, R. Bark, F. Boltz, L. De Stefano, E. O'Donnell, N. Matthews, and A. Money. 2017. Valuing water for sustainable development. *Science* 358: 1003–5.

Graf, W. L. 2001. Damage control: Restoring the physical integrity of America's rivers. *Annals of the Association of American Geographers* 91: 1–27.

Grafton, R. Q. 2017. Responding to the "wicked problem" of water insecurity. *Water Resources Management* 31: 3023–41.

Grayson, D. K. 2011. *The Great Basin, a Natural Prehistory.* Rev. ed. Berkeley: University of California Press.

Havas, M., T. C. Hutchinson, and G. E. Likens. 1984. Red herrings in acid rain research. *Environmental Science and Technology* 18: 176A–86A.

Hoagstrom, C. W., N. D. Zymonas, S. R. Davenport, D. L. Propst, and J. E. Brooks. 2010. Rapid species replacements between fishes of the North American plains: A case history from the Pecos River. *Aquatic Invasions* 5: 141–53.

Horton, C. C., T. R. Peterson, P. Banerjee, and M. J. Peterson. 2016. Credibility and advocacy in conservation science. *Conservation Biology* 30: 23–32.

Howard, B., L. C. Braat, R. J. F. Bugter, E. Carmen, R. S. Hails, A. D. Watt, and J. C. Young. 2018a. Taking stock of the spectrum of arguments for biodiversity. *Biodiversity and Conservation* 27: 1561–74.

Howard, J. K., K. A. Fesenmyer, T. E. Grantham, J. H. Viers, P. R. Ode, P. B. Moyle, S. J. Kupferburg, J. L. Furnish, A. Rehn, J. Slusark, R. D. Mazor, N. R. Santos, R. A. Peek, and A. N. Wright. 2018b. A freshwater conservation blueprint for California: Prioritizing watersheds for freshwater biodiversity. *Freshwater Science* 37: 417–31.

Jaeger, K. L., J. D. Olden, and N. A. Pelland. 2014. Climate change poised to threaten hydrologic connectivity and endemic fishes in dryland streams. *Proceedings of the National Academy of Sciences* 111: 13894–99.

Johnson, W. C. 2002. Riparian vegetation diversity along regulated rivers: Contribution of novel and relict habitats. *Freshwater Biology* 47: 749–59.

Keller, D. L., B. G. Laub, P. Birdsey, and D. J. Dean. 2014. Effects of flooding and tamarisk removal on habitat for sensitive fish species in the San Rafael River, Utah: Implication for fish habitat enhancement and future restoration efforts. *Environmental Management* 54: 465–78.

Kendy, E., K. W. Flessa, K. J. Schlatter, C. A. de la Parra, O. M. Hinojosa Huerta, Y. K. Carrillo-Guerrero, and E. Guillen. 2017. Leveraging environmental flows to reform water management policy: Lessons learned from the 2014 Colorado River delta pulse flow. *Ecological Engineering* 106: 683–94.

Kirby, M. E., E. J. Knell, W. T. Anderson, M. S. Lachniet, J. Palermo, H. Eeg, R. Lucero, R. Murrieta, A. Arevalo, E. Silveira, and C. A. Hiner. 2015. Evidence for insolation and Pacific forcing of late glacial through Holocene climate in the central

Mojave Desert (Silver Lake, CA). *Quaternary Research* 84: 174–86.
Knight, A. T., R. M. Cowling, M. Rouget, A. Balmford, A. T. Lombard, and B. M. Campbell. 2008. Knowing but not doing: Selecting priority conservation areas and the research-implementation gap. *Conservation Biology* 22: 610–17.
Krabbenhoft, T. J., S. P. Platania, and T. F. Turner. 2014. Interannual variation in reproductive phenology in a riverine fish assemblage: Implications for predicting the effects of climate change and altered flow regimes. *Freshwater Biology* 59: 1744–54.
Kui, L., J. C. Stella, P. B. Shafroth, P. K. House, and A. C. Wilcox. 2017. The long-term legacy of geomorphic and riparian vegetation feedbacks on the dammed Bill Williams River, Arizona, USA. *Ecohydrology* 10: e1839.
Kuussaari, M., R. Bommarco, R. K. Heikkinen, A. Helm, J. Krauss, R. Lindborg, E. Öckinger, M. Pärtel, J. Pino, F. Rodá, C. Stefanescu, T. Teder, M. Zobel, and I. Steffan-Dewenter. 2009. Extinction debt: A challenge for biodiversity conservation. *Trends in Ecology and Evolution* 24: 564–71.
Laub, B. G., J. Jimenez, and P. Budy. 2015. Application of science-based restoration planning to a desert river system. *Environmental Management* 55: 1246–61.
Lawson, D. M., K. R. Hall, L. Yung, and C. A. F. Enquist. 2017. Building translational ecology communities of practice: Insights from the field. *Frontiers in Ecology and the Environment* 15: 569–77.
Lee, J., F. M. Veloso, D. A. Hounshell, and E. S. Rubin. 2010. Forcing technological change: A case of automobile emissions control technology development in the US. *Technovation* 30: 249–64.
Leopold, A. 1933. The conservation ethic. *Journal of Forestry* 31: 634–43.
Leopold, A. 1949 (2013). A Sand County Almanac, and Sketches Here and There. In C. Meine, ed.,. *Aldo Leopold*, 1–189. New York: Library of America.
Likens, G. E. 2010. The role of science in decision making: Does evidence-based science drive environmental policy? *Frontiers in Ecology and the Environment* 8: e1–e9.
Lopez, B. 1990. *The Rediscovery of North America*. Lexington: University Press of Kentucky.
López, E. J., and D. Sutter. 2004. Ignorance in congressional voting? Evidence from policy reversal on the Endangered Species Act. *Social Science Quarterly* 85: 891–912.
Ludwig, D. 2001. The era of management is over. *Ecosystems* 4: 758–64.
Maceda-Veiga, A., O. Domínguez-Domínguez, J. Escribano-Alacid, and J. Lyons. 2016. The aquarium hobby: Can sinners become saints in freshwater fish conservation? *Fish and Fisheries* 17: 860–74.
Marks, J. C., G. A. Haden, M. O'Neill, and C. Pace. 2010. Effects of flow restoration and exotic species removal on recovery of native fish: Lessons from a dam decommissioning. *Restoration Ecology* 18: 934–43.
McMullin, S. 2017. Can fisheries scientists win the war on science? *Fisheries* 42: 495–96.
Meine, C. 2001. Roosevelt, conservation, and the revival of democracy. *Conservation Biology* 15: 829–31.
Merritt, D. M., and N. L. Poff. 2010. Shifting dominance of riparian *Populus* and *Tamarix* along gradients of flow alteration in western North American rivers. *Ecological Applications* 20: 135–52.
Miller, R. R. 1961. Man and the changing fish fauna of the American Southwest. *Papers of the Michigan Academy of Science, Arts, and Letters* 46: 365–404.
Miller, R. R., C. Hubbs, and F. H. Miller. 1991. Ichthyological exploration of the American West: The Hubbs-Miller era, 1915–1950. In W. L. Minckley and J. E. Deacon, eds., *Battle against Extinction: Native Fish Management in the American West*, 19–40. Tucson: University of Arizona Press.
Minckley, W. L. 1991. Native fishes of the Grand Canyon region: An obituary. In *Colorado River Ecology and Dam Management*, 124–77. Washington, DC: National Academy Press.
Minckley, W. L., and J. E. Deacon. 1991. Epilogue: Swords of our fathers, paying the piper, and other clichés. In W. L. Minckley and J. E. Deacon, eds., *Battle against Extinction: Native Fish Management in the American West*, 403–4. Tucson: University of Arizona Press.
Minckley, W. L., and M. E. Douglas. 1991. Discovery and extinction of western fishes: A blink of an eye in geologic time. In W. L. Minckley and J. E. Deacon, eds., *Battle against Extinction: Native Fish Management in the American West*, 7–17. Tucson: University of Arizona Press.
Morelli, T. L., C. Daly, S. Z. Dobrowski, D. M. Dulen, J. L. Ebersole, S. T. Jackson, J. D. Lundquist, C. I. Miller, S. P. Maher, W. B. Monahan, K. R. Nydick, K. T. Redmond, S. C. Sawyer, S. Stock, and S. R. Beissinger. 2016. Managing climate change refugia for climate adaptation. *PLoS ONE* 11: e0159909.
Mueller, G. A., and P. C. Marsh. 2002. *Lost: A Desert River and Its Native Fishes: A Historical Perspective of the Lower Colorado River*. US Geological Survey Information and Technology Report USGS/BRD/ITR-2002-0010.
Naiman, R. J. 2013. Socio-ecological complexity and the restoration of river ecosystems. *Inland Waters* 3: 391–410.
Olden, J. D., N. L. Poff, and K. L. Bestgen. 2006. Life-history strategies predict fish invasions and extirpations in the Colorado River basin. *Ecological Monographs* 76: 25–40.
Ono, R. D., J. D. Williams, and A. Wagner. 1983. *Vanishing Fishes of North America*. Washington, DC: Stone Wall Press.
Oreskes, N., and E. M. Conway. 2010. *Merchants of Doubt*. New York: Bloomsbury Press.
Osborne, M. J., T. A. Diver, and T. F. Turner. 2013. Introduced populations as genetic reservoirs for imperiled species: A case study of the Arkansas River shiner (*Notropis girardi*). *Conservation Genetics* 23: 637–47.
Osborne, M. J., J. S. Perkin, K. B. Gido, and T. F. Turner. 2014. Comparative riverscape genetics reveals reservoirs of genetic diversity for conservation and restoration of Great Plains fishes. *Molecular Ecology* 23: 5663–79.
Pace, M. L., S. E. Hampton, K. E. Limburg, E. M. Bennett, E. M. Cook, A. E. Davis, J. M. Grove, K. Y. Kaneshiro, S. L. LaDeau, G. E. Likens, D. M. McKnight, D. C. Richardson, and D. L. Strayer. 2010. Communicating with the public: Opportunities and rewards for individual ecologists. *Frontiers in Ecology and the Environment* 8: 292–98.
Patterson, J. J., C. Smith, and J. Bellamy. 2013. Understanding enabling capacities for managing the "wicked problem" of

nonpoint source water pollution in catchments: A conceptual framework. *Journal of Environmental Management* 128: 441–52.
Paxton, E. H., T. C. Theimer, and M. K. Sogge. 2011. Tamarisk biocontrol using tamarisk beetles: Potential consequences for riparian birds in the southwestern United States. *Condor* 113: 255–65.
Pennock, C. A., D. Bender, J. Hofmeier, J. A. Mounts, R. Waters, V. D. Weaver, and K. B. Gido. 2018. Can fishways mitigate fragmentation effects on Great Plains fish communities? *Canadian Journal of Fisheries and Aquatic Sciences* 75: 121–30.
Perkin, J. S., K. B. Gido, K. H. Costigan, M. D. Daniels, and E. R. Johnson. 2015. Fragmentation and drying ratchet down Great Plains stream fish diversity. *Aquatic Conservation: Marine and Freshwater Ecosystems* 25: 639–55.
Petersen, S. 1999. Congress and charismatic megafauna: A legislative history of the Endangered Species Act. *Environmental Law* 29: 463–91.
Piccolo, J. J. 2011. Challenges in the conservation, rehabilitation and recovery of native stream salmonid populations: Beyond the 2010 Luarca symposium. *Ecology of Freshwater Fish* 20: 346–51.
Piccolo, J. J. 2017. Intrinsic values in nature: Objective good or simply half of an unhelpful dichotomy? *Journal for Nature Conservation* 37: 8–11.
Pielke, R. A., Jr. 2007. *The Honest Broker; Making Sense of Science in Policy and Politics*. Cambridge: Cambridge University Press.
Pinnegar, J. K., and G. H. Engelhard. 2008. The "shifting baseline" phenomenon: A global perspective. *Reviews in Fish Biology and Fisheries* 18: 1–16.
Pister, E. P. 1974. Desert fishes and their habitats. *Transactions of the American Fisheries Society* 103: 531–40.
Pister, E. P. 1987. A pilgrim's progress from group A to group B. In J. B. Callicott, ed., *Companion to* A Sand County Almanac, 221–32. Madison: University of Wisconsin Press.
Pister, E. P. 1991. The Desert Fishes Council: Catalyst for change. In W. L. Minckley and J. E. Deacon, eds., *Battle against Extinction: Native Fish Management in the American West*, 55–68. Tucson: University of Arizona Press.
Pister, E. P. 1993. Species in a bucket. *Natural History* 102: 14–17.
Pitcher, T. J. 2001. Fisheries managed to rebuild ecosystems? Reconstructing the past to salvage the future. *Ecological Applications* 11: 601–17.
Pitt, J., E. Kendy, K. Schlatter, O. Hinojosa-Huerta, K. Flessa, P. B. Shafroth, J. Ramírez-Hernández, P. Nagler, and E. P. Glenn. 2017. It takes more than water: Restoring the Colorado River delta. *Ecological Engineering* 106: 629–32.
Plater, Z. J. B. 2008. Tiny fish, big battle. *Tennessee Bar Journal* 44(4): 14–22, 38, 42.
Plater, Z. J. B. 2013. *The Snail Darter and the Dam: How Pork-Barrel Politics Endangered a Little Fish and Killed a River*. New Haven: Yale University Press.
Plutzer, E., M. McCaffrey, A. L. Hannah, J. Rosenau, M. Berbeco, and A. H. Reid. 2016. Climate confusion among US teachers. *Science* 351: 664–65.
Poff, N. L. 2018. Beyond the natural flow regime? Broadening the hydro-ecological foundation to meet environmental flows challenges in a non-stationary world. *Freshwater Biology* 63:1011–1021.
Poff, N. L., R. E. Tharme, and A. H. Arthington. 2017. Evolution of environmental flows assessment science, principles, and methodologies. In A. C. Horne, J. A. Webb, M. J. Stewardson, B. Richter, and M. Acreman, eds., *Water for the Environment*, 203–36. London: Elsevier.
Pool, T. K., A. L. Strecker, and J. D. Olden. 2013. Identifying preservation and restoration priority areas for desert fishes in an increasingly invaded world. *Environmental Management* 51: 631–41.
Pringle, R. M. 2017. Upgrading protected areas to conserve wild biodiversity. *Nature* 546: 91–99.
Propst, D. L., K. B. Gido, and J. A. Stefferud. 2008. Natural flow regimes, nonnative fishes, and native fish persistence in arid-land river systems. *Ecological Applications* 18: 1236–52.
Pulliam, H. R. 1998a. The political education of a biologist. Part 1. *Wildlife Society Bulletin* 26: 199–202.
Pulliam, H. R. 1998b. The political education of a biologist. Part 2. *Wildlife Society Bulletin* 26: 499–503.
Reisner, M. 1993. *Cadillac Desert: The American West and Its Disappearing Water.* Rev. ed. New York: Penguin Books.
Rogers, K. H. 2006. The real river management challenge: Integrating scientists, stakeholders, and service agencies. *River Research and Applications* 22: 269–80.
Sabo, J. L., K. Bestgen, W. Graf, T. Sinha, and E. E. Wohl. 2012. Dams in the Cadillac Desert: Downstream effects in a geomorphic context. *Annals of the New York Academy of Sciences* 1249: 227–46.
Sabo, J. L., M. Caron, R. Doucett, K. L. Dibble, A. Ruhí, J. C. Marks, B. A. Hungate, and T. A. Kennedy. 2018. Pulsed flows, tributary inputs and food-web structure in a highly regulated river. *Journal of Applied Ecology* 55: 1884–95.
Savan, B. 1988. *Science under Siege: The Myth of Objectivity in Scientific Research*. Montréal: CBC Enterprises.
Scoppettone, G. G., and P. H. Rissler. 1995. Endangered cui-ui of Pyramid Lake, Nevada. In E. T. LaRoe, G. S. Farris, C. E. Puckett, P. D. Doran, and M. J. Mac, eds., *Our Living Resources*, 323–24. Washington, DC: US Department of the Interior, National Biological Service.
Scoppettone, G. G., and P. H. Rissler. 2007. Effects of population increase on cui-ui growth and maturation. *Transactions of the American Fisheries Society* 136: 331–40.
Seavy, N. E., T. Gardali, G. H. Golet, F. T. Griggs, C. A. Howell, R. Kelsey, S. L. Small, J. H. Viers, and J. F. Weigand. 2009. Why climate change makes riparian restoration more important than ever: Recommendations for practice and research. *Ecological Restoration* 27: 330–38.
Smith, G. R., T. E. Dowling, K. W. Gobalet, T. Lugaski, D. K. Shiozawa, and R. P. Evans. 2002. Biogeography and timing of evolutionary events among Great Basin fishes. In R. Hershler, D. B. Madsen, and D. R. Currey, eds., *Great Basin Aquatic Systems History*, 175–234. Smithsonian Contributions to the Earth Sciences, 33.
Smith, M. L. 1981. Late Cenozoic fishes in the warm deserts of North America: A reinterpretation of desert adaptations. In R. J. Naiman and D. L. Soltz, eds., *Fishes in North American Deserts*, 11–38. New York: John Wiley & Sons.
Snelder, T. H., H. L. Rouse, P. A. Franklin, D. J. Booker, N. Norton, and J. Diettrich. 2014. The role of science in setting water

resource use limits: Case studies from New Zealand. *Hydrological Sciences Journal* 59: 844–59.
Snyder, J. O. 1924. Notes on certain catostomids of the Bonneville system, including the type of *Pantosteus virescens* Cope. *Proceedings of the US National Museum* 64(18): 1–6.
Soga, M., and K. J. Gaston. 2018. Shifting baseline syndrome: Causes, consequences, and implications. *Frontiers in Ecology and the Environment* 16: 222–30.
Sogge, M. K., S. J. Sferra, and E. H. Paxton. 2008. *Tamarix* as habitat for birds: Implications for riparian restoration in the southwestern United States. *Restoration Ecology* 16: 146–54.
Soulé, M. E. 1985. What is conservation biology? *BioScience* 35: 727–34.
Steffen, W., R. Leinfelder, J. Zalasiewicz, C. N. Waters, M. Williams, C. Summerhayes, A. D. Barnosky, A. Cearreta, P. Crutzen, M. Edgeworth, E. C. Ellis, I. J. Fairchild, A. Galuszka, J. Grinevald, A. Haywood, J. Iver do Sul, C. Jeandel, J. R. McNeill, E. Odada, N. Oreskes, A. Revkin, D. deB. Richter, J. Syvitski, D. Vidas, M. Wagreich, S. L. Wing, A. P. Wolfe, and H. J. Schellnhuber. 2016. Stratigraphic and earth system approaches to defining the Anthropocene. *Earth's Future* 4: 324–45.
Stromberg, D. B. 1978. The Endangered Species Act amendments of 1978: A step backwards? *Environmental Affairs* 7: 33–42.
Tallis, H., and J. Lubchenco. 2014. A call for inclusive conservation. *Nature* 515: 27–28.
Taylor, M. R., E. S. Rubin, and D. A. Hounshell. 2005. Control of SO_2 emissions from power plants: A case of induced technological innovation in the U.S. *Technological Forecasting and Social Change* 72: 697–718.
Tedesco, P. A., T. Oberdorff, J. F. Cornu, O. Beauchard, S. Brosse, H. H. Dürr, G. Grenouillet, F. Leprieur, C. Tisseuil, R. Zaiss, and B. Hugueny. 2013. A scenario for impacts of water availability loss due to climate change on riverine fish extinction rates. *Journal of Applied Ecology* 50: 1105–15.
Thompson, P. D., P. A. Webber, and C. D. Mellon. 2015. The role of introduced populations in the management and conservation of least chub. *Fisheries* 40: 546–56.
Tinch, R., R. Bugter, M. Blicharska, P. Harrison, J. Haslett, P. Jokinen, L. Mathiew, and E. Primmer. 2018. Arguments for biodiversity conservation: Factors influencing their observed effectiveness in European case studies. *Biodiversity and Conservation* 27: 1763–88.
Trush, W. J., S. M. McBain, and L. B. Leopold. 2000. Attributes of an alluvial river and their relation to water policy and management. *Proceedings of the National Academy of Sciences* 97: 11858–63.
Tukey, J. W. 1960. Conclusions vs. decisions. *Technometrics* 2: 423–33.
USFWS (US Fish and Wildlife Service). 2010. Endangered and threatened wildlife and plants: 12-month finding on a petition to list the least chub as threatened or endangered. *Federal Register* 75: 35398–424.
USFWS (US Fish and Wildlife Service). 2014. Endangered and threatened wildlife and plants: 12-month finding on the petition to list least chub as an endangered or threatened species. *Federal Register* 79: 51042–66.
USFWS (US Fish and Wildlife Service). 2015. Endangered and threatened wildlife and plants: Removal of the Modoc sucker from the federal list of endangered and threatened wildlife. *Federal Register* 80: 76235–49.
Vörösmarty, C. J., P. B. McIntyre, M. O. Gessner, D. Dudgeon, A. Prusevich, P. Green, S. Glidden, S. E. Bunn, C. A. Sullivan, C. R. Liermann, and P. M. Davies. 2010. Global threats to human water security and river biodiversity. *Nature* 467: 555–61.
Wagner, F. H. 1999. Whatever happened to the National Biological Survey? *BioScience* 49: 219–22.
Walsh, E. M., and B. K. Tsurusaki. 2014. Social controversy belongs in the climate science classroom. *Nature Climate Change* 4: 259–63.
Walsworth, T. E., and P. Budy. 2015. Integrating nonnative species in niche models to prioritize native fish restoration activity locations along a desert river corridor. *Transactions of the American Fisheries Society* 144: 667–81.
Warren, D. R., J. B. Dunham, and D. Hockman-Wert. 2014. Geographic variability in elevation and topographic constraints on the distribution of native and nonnative trout in the Great Basin. *Transactions of the American Fisheries Society* 143: 205–18.
Webber, P. A., P. D. Thompson, and P. Budy. 2012. Status and structure of two populations of the bluehead sucker (*Catostomus discobolus*) in the Weber River, Utah. *Southwestern Naturalist* 57: 267–76.
Wilkinson, T. 1998. *Science under Siege; The Politicians' War on Nature and Truth*. Boulder, CO: Johnson Books.
Woodford, D. J., D. M. Richardson, H. J. MacIsaac, N. E. Mandrak, B. W. van Wilgen, J. R. U. Wilson, and O. L. F. Weyl. 2016. Confronting the wicked problem of managing biological invasions. *NeoBiota* 31: 63–86.
Worster, D. 2008. *A Passion for Nature, the Life of John Muir*. Oxford: Oxford University Press.
Xenopoulos, M. A., D. M. Lodge, J. Alcamo, M. Märker, K. Schulze, and D. P. V. Vuurens. 2005. Scenarios for freshwater fish extinctions from climate change and water withdrawal. *Global Change Biology* 11: 1557–64.
Yackulic, C. B., M. D. Yard, J. Korman, and D. R. Haverbeke. 2014. A quantitative life history of endangered humpback chub that spawn in the Little Colorado River: Variation in movement, growth, and survival. *Ecology and Evolution* 4: 1006–18.
Young, J. C., K. Searle, A. Butler, P. Simmons, A. D. Watt, and A. Jordan. 2018. The role of trust in the resolution of conservation conflicts. *Biological Conservation* 195: 196–202.

Acknowledgments

Throughout the course of this project, we have benefited greatly from the expertise and encouragement of staff at the University of Chicago Press. Christie Henry, Editorial Director, provided support and guidance to get the project off in good form. Miranda Martin ably stepped in once Christie moved on. Scott Gast, Editor; Michaela Luckey, Editorial Associate; Mary Corrado, Manuscript Editor; and Norma Sims Roche, Copy Editor, skillfully guided the book through reviews, revision, and publication. The indexing skills of Sharon Hughes are greatly appreciated. Their patience and insightful suggestions made our work easier and contributed immeasurably to it being published. Our grateful appreciation to all.

We thank our respective institutions for their support as this volume was conceived, developed, and delivered. Our appreciation to the University of New Mexico, Trout Unlimited, Colorado State University, and Weber State University.

More than 60 authors participated in this book and more than 40 reviewers critiqued their efforts. Both groups are listed separately below. We are indebted to all for your time and wisdom.

Matt Mayfield of Trout Unlimited prepared many of the maps used herein and did so in a timely and highly professional manner. We greatly appreciate his efforts. Joe Tomelleri kindly allowed use of several of his fine fish illustrations. The art of W. H. Brandenburg graces the book's cover as well as the section title pages.

Perhaps our strongest notes of appreciation go to the Executive Committee and members of the Desert Fishes Council, who have been strong supporters of this project, as they were with the earlier *Battle against Extinction: Native Fish Management in the American West.* edited by W. L. Minckley and James Deacon. Members of the Council participated as contributors, reviewers, and funders. In particular, Kathryn Boyer, Stewart Reid, and Michael Bogan were instrumental in making the book a reality. Neither this volume nor the many desert denizens described herein would be here today without the tireless dedication of these colleagues. Thank you.

In addition to funding provided by the Desert Fishes Council, support for this volume was provided by the US Bureau of Land Management, USDA Forest Service, Texas Parks and Wildlife Department, and the Dixon Water Foundation. Stephanie Carman, Yvette Paroz, Dan Shively, Megan Bean, and Gary Garrett generously assisted with funding efforts.

Minck and Deacon left big shoes to fill. We hope we were up to the challenge. As is evidenced in these pages, the Battle against Extinction carries on.

Contributors

EDITORS

David L. Propst
University of New Mexico
Albuquerque, New Mexico, USA
dpropst@unm.edu

Jack E. Williams
Trout Unlimited
Medford, Oregon, USA
jwilliams@tu.org

Kevin R. Bestgen
Colorado State University
Fort Collins, Colorado, USA
kevin.bestgen@colostate.edu

Christopher W. Hoagstrom
Weber State University
Ogden, Utah, USA
christopherhoagstrom@weber.edu

CHAPTER AUTHORS

Brandon Albrecht
BIO-WEST, Inc.
Logan, Utah, USA
balbrecht@bio-west.com

Matthew E. Andersen
US Geological Survey
Reston, Virginia, USA
mandersen@usgs.gov

Carlos Alonso Ballesteros-Córdova
Universidad de Sonora
Hermosillo, Sonora, México
caballesteros0411@gmail.com

Megan G. Bean
Texas Parks and Wildlife Department
Mountain Home, Texas, USA
megan.bean@tpwd.texas.gov

Scott E. K. Bennett
US Geological Survey
Seattle, Washington, USA
sekbennett@usgs.gov

Heidi Blasius
Bureau of Land Management
Safford, Arizona, USA
hblasius@blm.gov

Kate S. Boersma
University of San Diego
San Diego, California USA
kateboersma@gmail.com

Michael T. Bogan
University of Arizona
Tucson, Arizona, USA
mbogan@email.arizona.edu

Scott A. Bonar
US Geological Survey
Tucson, Arizona, USA
sbonar@ag.arizona.edu

Kathryn Boyer
Oregon State University
Corvallis, Oregon, USA
kathryn.boyer@oregonstate.edu

James E. Brooks
University of New Mexico
Albuquerque, New Mexico, USA
arroyodejaime@gmail.com

Kevin C. Brown
University of California, Santa Barbara
Santa Barbara, California, USA
kbrown@es.ucsb.edu

Adam E. Cohen
The University of Texas at Austin
Austin, Texas, USA
acohen@austin.texas.edu

Armando J. Contreras-Balderas
San Nicolás de los Garza, Nuevo León, México
ajcb1951@gmail.com

Michael H. Darin
Northern Arizona University
Flagstaff, Arizona, USA
michael.darin@nau.edu

Cindy Deacon Williams
Environmental Consultants
Medford, Oregon USA
cdwill656@gmail.com

Thomas E. Dowling
Wayne State University
Detroit, Michigan, USA
thomas.dowling@wayne.edu

Alice F. Echelle
Oklahoma State University
Stillwater, Oklahoma, USA
aechelle@juno.com

Anthony A. Echelle
Oklahoma State University
Stillwater, Oklahoma, USA
anthony.echelle@okstate.edu

Robert J. Edwards
The University of Texas at Austin
Austin, Texas, USA
robertedwards@utexas.edu

Kurt D. Fausch
Colorado State University
Fort Collins, Colorado, USA
kurt.fausch@colostate.edu

Olin G. Feuerbacher
US Fish and Wildlife Service
Amargosa Valley, Nevada, USA
olin_feuerbacher@fws.gov

Lloyd T. Findley
Carretera al Varadero Nacional
Guaymas, Sonora, México
ltfindley@gmail.com

María Elena García-Ramírez
San Nicolás de los Garza, Nuevo León, México
mgarciar29@hotmail.com

Gary P. Garrett
The University of Texas at Austin
Austin, Texas, USA
garygarrett@utexas.edu

Jennifer M. Gumm
US Fish and Wildlife Service
Amargosa Valley, Nevada, USA
jennifer_gumm@fws.gov

Mark Haberstich
The Nature Conservancy
Wilcox, Arizona, USA
mhaberstich@tnc.org

Mark B. Hausner
Desert Research Institute
Las Vegas, Nevada, USA
mark.hausner@dri.edu

Dean A. Hendrickson
The University of Texas at Austin
Austin, Texas, USA
deanhend@austin.utexas.edu

Paul Holden
BIO-WEST, Inc., retired
Providence, Utah, USA
pholden442@gmail.com

Derek D. Houston
Western State Colorado University
Gunnison, Colorado, USA
derek.d.houston@gmail.com

Ron Kegerries
BIO-WEST, Inc.
Logan, Utah, USA
rkegerries@bio-west.com

Trevor J. Krabbenhoft
University of Buffalo
Buffalo, New York, USA
krabben@buffalo.edu

Sean C. Lema
California Polytechnic State University
San Luis Obispo, California, USA
slema@calpoly.edu

María de Lourdes Lozano-Vilano
San Nicholás de los Garza, Nuevo León, México
marlozan2006@gmail.com

David A. Lytle
Oregon State University
Corvallis, Oregon, USA
lytleda@oregonstate.edu

Norman Mercado-Silva
Universidad Autónoma del Estado de Morelos
Cuernavaca, Morelos, México
norman.mercado@uaem.mx

Chuck O. Minckley
Native Fish Advocate
Pomerene, Arizona, USA
cminckley@outlook.com

Jerry A. Monzingo
Gila National Forest
Silver City, New Mexico, USA
jmonzingo@fs.fed.us

Ryan R. Morrison
Colorado State University
Fort Collins, Colorado, USA
ryan.morrison@colostate.edu

Peter B. Moyle
University of California, Davis
Davis, California, USA
pbmoyle@ucdavis.edu

Dustin J. Myers
Gila National Forest
Silver City, New Mexico, USA
dustinmyers@fs.fed.us

Christopher Norment
College at Brockport SUNY
Brockport, New York, USA
cnorment@brockport.edu

Megan J. Osborne
University of New Mexico
Albuquerque, New Mexico, USA
mosborne@unm.edu

Jonathan T. Overpeck
University of Michigan
Ann Arbor, Michigan, USA
overpeck@umich.edu

Tyler J. Pilger
FISHBIO Environmental
Chico, California, USA
tylerpilger@fishbio.com

Edwin P. (Phil) Pister
Desert Fishes Council
Bishop, California, USA
phildesfish123@gmail.com

Peter N. Reinthal
The University of Arizona
Tucson, Arizona, USA
pnr@email.arizona.edu

Ron Rogers
BIO-WEST, Inc.
Logan, Utah, USA
rrogers@bio-west.com

Gorgonio Ruiz-Campos
Universidad Autónoma de Baja California
Ensenada, Baja California, México
gruiz@uabc.edu.mx

Donald W. Sada
Desert Research Institute
Reno, Nevada, USA
don.sada@dri.edu

Sergio Sánchez-Gonzalez
Escuela de Biologia
Universidad Autónoma de Sinaloa
Culiacan, SI, México
ssanchez@uass.uasnet.mx

Michael R. Schwemm
US Fish and Wildlife Service
Las Vegas, Nevada, USA
michael_schwemm@fws.gov

Lawrence E. Stevens
Museum of Northern Arizona
Flagstaff, Arizona, USA
lstevens@musnaz.org

Mark C. Stone
University of New Mexico
Albuquerque, New Mexico, USA
stone@unm.edu

Thomas F. Turner
University of New Mexico
Albuquerque, New Mexico, USA
turnert@unm.edu

Bradley H. Udall
Colorado State University
Fort Collins, Colorado, USA
bradley.udall@colostate.edu

Tom Udall
US Senate
Washington, DC, USA

Alejandro Varela-Romero
Universidad de Sonora
Hermosillo, Sonora, México
alejandro.varela@unison.mx

Kevin P. Wilson
Death Valley National Park
Pahrump, Nevada, USA
kevin_wilson@nps.gov

Koreen A. Zelasko
Colorado State University
Fort Collins, Colorado, USA
koreen.zelasko@colostate.edu

REVIEWERS

The following graciously provided their time to review one or more chapters in this book. Thank you!

Mark Belk
Timothy Birdsong
Heidi Blasius
Kathryn Boyer
James E. Brooks
Evan Carson
Julie Meka Carter
Tom Chart
Dan Dauwalter
Cindy Deacon Williams
Alice F. Echelle
Anthony A. Echelle
Robert J. Edwards
Kurt D. Fausch
Jennifer Fowler-Propst
Nathan Franssen
Stanley V. Gregory
Amy L. Haak
John Hamill
Robert Hershler
Stanley Hillyard
Mary Jane Keleher
Gail C. Kobetich
Jeffrey Lantow
Paul Marsh
Mauricio de la Maza-Benignos
Norman Mercado-Silva
Allison Pease
Quinton Phelps
Steven Platania
David Rogowski
Steven Ross
Dennis Shiozawa
David Speas
Abraham E. Springer
Tracy Stephens
Dale Turner
Peter Unmack
David Ward
Seth Wenger
James Whitney
John Wiens
Ellen Wohl

Index

The letter *f* following a page number denotes a figure, the letter *b* denotes a box, and the letter *t* denotes a table.

Abedus herberti (giant water bug), 156*f*
 diet of, 279
 dispersal potential of, 158
 extirpation of, 157
 habitat of, 156–59
 habits of, 277–78
 recolonization potential of, 158
 Sierra El Aguaje, 278*f*
 spring rehabilitation and, 262
 vulnerability of, 155
Abiquiu reservoir, 118
adaptation, gene flow and, 212
adaptive markers, 212–13
Agnostomus monticola (mountain mullet), 279, 280*f*, 280*t*
Agosia chrysogaster (longfin dace)
 abundance of, 229, 229*f*
 Aravaipa Creek populations, 228, 228*t*
 areas of endemism, 44*t*, 56*t*
 biology of, 228
 Black Canyon, 303
 Colorado River basin native fish, 318*t*
 distribution, 44, 47*t*, 56*t*, 226*t*, 228, 471
 Fossil Creek, 361
 genetic diversity of, 212
 Habitat Conservation Plans for, 188
 niches, 56*t*
 study of, 228–29
 Yaqui form, 188
Agosia spp., 47*t*
agriculture
 climate change impacts on, 119, 120
 groundwater use, 399
 surface water use, 399
 water use and, 110
Alamo Dam, Arizona, 251
Albuquerque, New Mexico
 aquifers of, 118
 population of, 117, 241*t*
Alburnops, 48*t*
 evolutionary clade, 42*t*
Algansea-Agosia
 areas of endemism, 47*t*, 56*t*
 evolutionary clade, 42*t*
 niches, 47*t*, 56*t*
alien species, battle against, 70–77. *See also* invasive species; nonnative species
All-American Canal, 240
allozyme studies, 207, 305
 Colorado River basin, 325
 cost of, 11
 Gila trout, 300
Álvarez del Villar, José, 20
Alvord area, description of, 41*t*
Alvord Basin streams, trout from, 101
Alvord cutthroat trout (*Oncorhynchus clarkii alvordensis*), 101–2
Amargosa Desert, Nevada, 379–89
Amargosa toad (*Anaxyrus nelsoni*), 261–62
Amazon molly (*Poecilia formosa*), 85
Ameiurus spp. (bullhead catfish), 70
 melas, 172*t*, 226, 227*t*, 231, 324
 natalis, 172*t*, 226, 227*t*, 230
 nebulosus, 172*t*, 226
American bullfrog (*Lithobates [Rana] catesbeianus*), 81, 262, 281
American desert region, extinct species of, 89–105
American eel (*Anguilla rostrata*), 128
Amigos del Pandeño, A.C., 367
Amistad Dam, 96
Amistad gambusia (*Gambusia amistadensis*), 96–97, 126
amphipods, valley floor springs and, 259
Anaxyrus spp.
 nelsoni, 261–62
 punctatus, 279
Anguilla rostrata (American eel), 128
Anthropocene, definition of, 408*b*
Anthropocene climate, 412
anti-conservation movements, 409
antimycin, use of, 304, 362
anti-science sentiments, 409
Apache Corporation, 129
Apache trout (*Oncorhynchus apache*)
 area of endemism, 50*t*
 captive propagation of, 9
 Colorado River basin native fish, 318*t*
 conservation concerns, 6
 distribution of, 296, 298
 habitat of, 143
 historic distribution of, 296–98
 management of, 175
 thermal sensitivity of, 143–44

apex-predator removal, ecological vacancies and, 157
Aquarius remigis (common water strider), 154*f*
aquatic ecosystems
  extinct species, 89–105
  southwestern waters, 74–75 (*see also specific systems*)
aquatic habitats
  conservation of species in, 3–12
  degradation of, 125
  restoration of, 353–74
aquatic macroinvertebrates (AMIs)
  conservation efforts, 98, 255
  crenophilic, 98
  Devils Hole, 382
  responses to fires, 264
  spring ecosystems, 261
aquifer–spring type distribution, 257–61
  flow, 261
  water chemistry, 261
Aravaipa Canyon Primitive Area, 228
Aravaipa Canyon Wilderness Area, 228
Aravaipa Creek, Arizona
  assemblage structure patterns, 232
  fish communities, 233*f*
  fish of, 18
  flow and populations, 232–34
  flow–fish abundance regressions, 234–35, 234*f*, 235*f*
  long-term monitoring, 225–37
  management by BLM, 227–28
  monitoring sites in, 227*f*, 228–31
  native species, 226*t*, 230*f*, 230*t*
  nonnative fish in, 226, 227*t*, 230–31
  species abundances in, 229–30
  study area, 225–36
  temperature changes, 144, 145*f*
Archilestes grandis (great spreadwing damselfly), 154*f*, 278*f*
areas of endemism, 36*f*, 39*t*–41*t*
  biogeographic subdivisions, 44–45
  boundaries, 39, 44–45
  compared using *S*, 41*f*
  definition of, 38–39
  North American desert fish, 42*f*, 59*f*
arid-land aquifers
  structure of, 261
  water quality, 261
arid-land springs
  ecology of, 262
  rehabilitation of, 262–63
arid-land streams, novel drought regimes, 153–65
Arizona snowfly (winter stoneflies, *Mesocapnia arizonensis*), 154*f*, 155
Arizona State University, history of, 17
Arizona Water Settlements Act, 244
Army Corp of Engineers, US, 118
arroyo chub (*Gila orcuttii*), 339
Arroyo El Palmar, 277–78, 279*f*
Arroyo El Tigre basin, 276, 281
Arroyo Los Chorros, 83–84
artificial environments, phenotypic consequences, 338
Arundo donax (giant reed), 128
Ash Creek, giant-water-bug populations, 158–59
Ash Meadows, Nevada, 379
  preservation plan for, 23
  spring sources in, 260
Ash Meadows Amargosa pupfish (*Cyprinodon nevadensis mionectes*), 6, 242, 382
Ash Meadows Fish Conservation Facility, Nevada, 344*f*, 384
Ash Meadows National Wildlife Refuge, Nevada, 355
  Devils Hole pupfish, 342
  establishment of, 6, 392, 397
  federal landholdings, 393
Ash Meadows poolfish (*Empetrichthys merriami*), 99, 100*f*
  emergency listing of, 411
  extinction of, 6, 341
  status of, 394
Ash Meadows pupfish (*Cyprinodon nevadensis mionectes*), 6, 342, 382
Ash Meadows speckled dace (*Rhinichthys osculus nevadensis*), protection of, 6
Assiminea pecos (Pecos assiminea), 131, 367
Assiniboine River, desert-like habitats, 38
assisted colonization, definition of, 336*b*, 347
assisted gene flow, 336*b*, 339
assisted recovery
  definition of, 336*b*, 337
  efforts toward, 337
  strategies for, 336*b*, 337*f*
Astacidae, recent extinctions, 93*t*
Astyanax mexicanus (Mexican tetra), 85
Ataeniobius toweri, 51*t*
Atherinella spp., 42*t*, 50*t*
Atherinopsidae
  areas of endemism, 43*t*, 50*t*
  lineages, 43*t*
  niches, 50*t*
Atlantic and Pacific Railroad, 299
Audubon Society, 10
Awaous banana, 280*t*

Babbitt, Bruce, 186
Baja, California
  habitats and fish of, 80–86
  Mexican killifish of, 79
  oases, 277, 281
Baja California killifish (*Fundulus lima*), 51*t*, 81, 82
Baja California Sur, 81–82
bajada springs, 259
bald eagle (*Haliaeetus leucocephalus*), 188
Balmorhea Springs Complex
  case study, 367–69
  habitats and fish of, 128–30
Balsas catfish (*Ictalurus balsanus*), 291
Baños de San Ignacio, Linares, Nuevo León, 85, 85*f*
Barber, W. E., 343
Barker, Elliot S., 299
Barton Creek, Texas, species distribution, 204
bass (*Micropterus* spp.)
  black, 70
  largemouth
    Cibola High Levee Pond and, 173–74
    extinctions caused by, 91
    habitat of, 83
    illicit stocking of, 321
    impact of introduction, 339
    predation by, 361
    specimen collection, 201
  Rio Grande largemouth, 131
bass, striped (*Morone saxatilis*), 320–21
Battle against Extinction (Minckley and Deacon), xii, xiv, xv, 6, 9, 11, 18, 23, 33, 70, 183–84, 207, 329, 380, 382*f*, 387, 391, 392, 396, 401
Bear Fire, 305
bedrock, permeability of, 276
behavioral management in Diamond Y Draw, 345–46
Behnke, Robert J., 30–33, 30*f*, 31*f*, 101, 299
Bennett, Hugh H., 146
Bidahochi area, 40*t*
Big Bend Conservation Alliance, 133
Big Bend gambusia (*Gambusia gaigei*)

habitat of, 127
rescue efforts, 354
Big Bend National Park, springs in, 127–28
Big Bend reach of Rio Grande, 127–28
habitats and fish of, 127–28
water management, 128
Big Dry Creek
Gila trout lineage, 310*t*
renovation of, 300
bighead pupfish (*Cyprinodon pachycephalus*), 83
Bill Burrud's Animal World, 5
Bill Williams River, Alamo Dam on, 251
biodiversity
conservation and, 61
North American desert fish, 37–56
value of specimen collections in, 199–206
biogeography, North American desert fish, 37–56
biological invasions, problems of, 409
black-necked garter snake (*Thamnophis cyrtopsis*), 279, 280*f*
black redhorse (*Moxostoma duquesnei*), 202
BLM Spring, 339, 340*f*
blotched gambusia (*Gambusia senilis*), 127
bluegill (*Lepomis macrochirus*), 99
bluehead sucker (*Pantosteus discobolus*), 176, 190, 210, 212, 216, 318*t*, 412
Blue Springs complex, Arizona, 261
blue tilapia (*Oreochromis aureus*), 86
Boa constrictor imperator (Mexican boa), 280
Bolsón de Cuatro Ciénegas fish, 83
Bolsón de Sandia fish, 84–85
Bonita Creek, 358
Bonner Fire, 302–3
Bonneville area, 40*t*
Bonneville cutthroat trout (*Oncorhynchus clarkii utah*), 190
Bonneville-Snake bluehead sucker (*Pantosteus virescens*), 410
bonytail (*Gila elegans*), 8*f*
Colorado River basin, 8*f*
Habitat Conservation Plans for, 188
poisoning of, 4
recruitment of, 173–74
reproduction of, 324–25
Boquillas Hot Springs, Texas, 127
Borax Lake, Oregon, 363*f*
Borax Lake chub (*Gila boraxobius*), 363–64
Borax Lake Chub Recovery Plan, 363–64
Bouse area, 40*t*
brook trout (*Salvelinus fontinalis*), 30
brown trout (*Salmo trutta*)
Colorado River basin nonnative fish, 172*t*
predation by, 361
removal of, 300
Weber River, 410
Brycon guatemalensis (Macabí tetra), 86
bullfrogs (*Lithobates [Rana] catesbeianus*), 262
bullhead (*Ameiurus* spp.), 70, 226
black, 172*t*, 227*t*, 231, 324
brown, 172*t*
yellow, 172*t*, 226, 227*t*, 230
bulrush (*Scirpus americanus*) encroachment, 265, 339, 346
Bureau of Land Management (BLM), Candidate Conservation Agreements and, 186
Bureau of Reclamation (USBR) reservoirs, US, 112, 117
Bush, George W., 395

Cabeza de Vaca area, description of, 40*t*
caddisfly (*Hesperophylax* sp.), 154*f*
Cadillac Desert (Reisner), 397
California golden trout (*Oncorhynchus mykiss aguabonita*), 34
California State Water Project, 399
California WaterFix, 148
Callibaetis sp. (mayfly), 154*f*, 278*f*
Cambarellus chihuahuae (Chihuahuan dwarf crayfish), 83, 91, 365
Cambaridae, recent extinctions, 93*t*
Camp Cady Wildlife Area, 340, 341
Campostoma spp.
areas of endemism, 45*t*, 46*t*
evolutionary clade, 42*t*
niches, 45*t*
ornatum, 128
spadiceum, 202
Candidate Conservation Agreements (CCAs), 186–87, 187*t*, 354
Candidate Conservation Agreements with Assurances (CCAAs), 186, 187, 187*t*
canyon treefrog (*Hyla arenicolor*), 279, 280*f*
Capitan area, 40*t*
Cappaert v. United States, 5, 381, 386, 391, 393
captive breeding stock
genomic consequences of, 215
razorback sucker, 325–26
captive propagation programs
definition of, 336, 336*b*
genetic management plans, 214–15
captive rearing, 336, 340*f*
captive rearing and breeding programs
definition of, 336*b*
design of, 215
captive stocks, funding of, 214
captivity, adaptation to, 214
carbonate mound complexes, 257
Carbonera pupfish (*Cyprinodon fontinalis*), 365, 365*f*
habitat of, 83
status of, 369
Carson, Rachel, 146
Carter, Jimmy, 411
Catarina pupfish (*Megupsilon aporus*)
culture of, 91
extinct, 81*t*, 92*t*
habitat, 84
catfish (Ictaluridae)
Amazon sailfin, 86
Balsas, 291
bullhead, 70, 226
black, 172*t*, 227*t*, 231, 324
brown, 172*t*
yellow, 172*t*, 226, 227*t*, 230
channel
Colorado River basin, 172*t*
conservation of, 285, 286*f*
hybridization of, 285
mechanical removal of, 173
México, 286*t*
nonnative stocking, 291
problems caused by, 170, 173, 290
status of, 290
Chihuahua, 126
conservation of, 285–93
distribution of, 286*f*, 287*f*
flathead, 171*t*, 291
headwater, 128, 131
Lacandon sea, 86
pale, 86
Rio Grande blue, 128, 285, 286*f*
vermiculated sailfin, 86
Yaqui
conservation of, 285
descriptions of, 287–89
Habitat Conservation Plans for, 188

San Bernardino National Wildlife Refuge, 354
status of, 290
Catinella land snail, 262
Catlow area, description of, 41*t*
Catostomidae
Aravaipa Creek, 226*t*
areas of endemism, 43*t*, 48*t*–49*t*
Colorado River basin, 171*t*, 318*t*
lineages, 43*t*
niches, 48*t*–49*t*
recent extinctions, 92*t*
Catostomus spp., 49*t*, 56*t*, 70
clarkii, 143, 226*t*, 303, 318*t*, 361, 362
evolutionary clade, 42*t*
latipinnis
Colorado River basin, 318*t*
conservation concerns, 190
conservation genetics, 216
inflow habitats, 178
restoration of, 174
San Pedro River, 226
microps, 359–61
distribution of, 359
habitat, 359
listing of, 412
occidentalis, 359
tahoensis, 98
cattail (*Typha* spp.), vegetation encroachment, 265
Center for Biological Diversity, 10, 26
center of survival, 61
centers of origin, diversification and, 61
Central Arizona Project (CAP), 113, 240
Central Valley Project, California, 399
Centrarchidae
areas of endemism, 43*t*, 53*t*
lineages, 43*t*
niches, 53*t*
Upper Colorado River basin, 171*t*
Cerro Prieto Geothermal Station, 80–81
Chamberlain, F. M., 298
channel catfish (*Ictalurus punctatus*)
Colorado River basin, 172*t*
conservation of, 285, 286*f*
hybridization of, 285
mechanical removal of, 173
México, 286*t*
nonnative stocking, 291
problems caused by, 170, 173, 290
status of, 290
Chapalichthyini, 51*t*
evolutionary clade, 42*t*
Characodon spp., 51*t*
Characodontini, 51*t*
Charco Azul pupfish (*Cyprinodon veronicae*), 84, 85
Chasmistes spp.
areas of endemism, 57*t*
cujus, 98
liorus liorus, 338
niches, 57*t*
Chiapa de Corzo cichlid (*Chiapaheros grammodes*), 86
Chiapaheros grammodes (Chiapa de Corzo cichlid), 86
Chiapas, fish and habitat of, 86
Chiapas cichlid (*Torichthys socolofi*), 86
Chihuahua, habitats and fish of, 82–83
Chihuahua catfish (*Ictalurus* sp.), 126
Chihuahuan Desert, Texas
ecoregion of, 125–35
flow of top five springs, 126*t*
Miller and C. L. Hubbs work in, 15
Native Fish Conservation Areas, 133*f*
springs and marshes in, 98
watersheds, 125
Chihuahuan dwarf crayfish (*Cambarellus chihuahuae*), 83, 91, 365
Chihuahua shiner (*Notropis chihuahua*), 83, 128
China Lake Naval Air Weapons Station, 340
chironomids, springs and, 260
Chirostoma spp.
areas of endemism, 50*t*
contrerasi, 22
evolutionary clade, 42*t*
niches, 50*t*
chub
arroyo, 339
Borax Lake, 363–64
Creek chub-plagopterin, 55*t*
area of endemism, 44*t*
niches, 44*t*
desert, 279
humpback
Colorado River basin, 6, 8, 8*f*, 73, 318*t*, 353
distribution of, 176, 318*t*, 322
Habitat Conservation Plans for, 188, 189, 249
Lake Mead, 322
poisoning of, 4
protection of, 73
Independence Valley tui chub, 91
least
areas of endemism, 56*t*
Conservation Agreement for, 190
niches, 56*t*
protection of, 412
rescue efforts, 354
Mohave tui
case study, 339–41
habitat of, 216
translocation of, 341
Oregon, 355–56, 355*f*
Oregon Chub Recovery Plan, 356
Oregon Chub Working Group, 356
roundtail
Aravaipa Creek, 226*t*, 230*f*
Colorado River basin native fish, 318*t*
conservation concerns, 190
distribution of, 4, 226*t*, 298, 318*t*
Fossil Creek, 296, 361
Gila River, 296
proposed listing of, 190
rotenone killing of, 4
thermal sensitivity of, 143
Salinas, 83
speckled
areas of endemism, 42
risks to, 128
taxa splits, 203–4
tui, 98
Virgin River, 174
Yaqui, 146, 188
Cibola High Levee Pond, 173, 321
Cichlidae
areas of endemism, 44*t*, 54*t*
lineages, 44*t*
niches, 54*t*
Upper Colorado River basin, 171*t*
cichlids
Chiapa de Corzo, 86
Chiapas, 86
Jewel, 83
Minckley's, 83
redhead, 86
Texas, 85
"Ciénegas" (Hendrickson and Minckley), 18
ciénegas, communities in, 126
Ciudad Juarez, population of, 117
cladistic patterns, areas of endemism, 42, 42*t*–44*t*, 44
Clark, Frances, 14
Clark, Laura, 14
Clark County Multispecies Habitat Conservation Plan, 399
Clark Hubbs Ciénega fish, 130, 368
Clark Hubbs Refuge, Big Bend, 127
Clean Water Act
impacts of, 9

piscicide application and, 302–3
Clear Creek gambusia (*Gambusia heterochir*), 27*f*
clear-cutting, watershed disturbance by, 247
climate change
biodiversity and, 161
Devils Hole pupfish and, 385–86
fish, aquatic systems and, 137–52
global mean temperature records, 138*f*
politics and, 393–94
potential impacts of, 142–46
southwestern water supplies and, 109–24
water supplies and, 328
See also global warming
climate models, trust in, 140
climatic optimum, 408*b*, 409
Clinton, William, 364
Closed Basin Project, 117
Clupeidae, Upper Colorado River basin, 171*t*
Coahuila area
description of, 40*t*
fish of, 83–84
Coahuilan box turtle (*Terrapene coahuila*), 19
Coccyzus americanus (yellow-billed cuckoo), 118
Cochiti Dam, New Mexico
floodplain changes, 243*f*
removal of, 118
Rio Grande flow and, 248*f*
Cochiti Reservoir, 118
Cochliopidae, recent extinctions, 93*t*
Codoma-Cyprinella, 47*t*–48*t*
evolutionary clade, 42*t*
Codoma spp., 47*t*–48*t*
collecting events, definition of, 200*b*
Colorado Plateau springs, 259
Colorado River
anthropogenic changes, 109
areas drained by, 317–19
floodplain changes, 242*f*
geography of, 110–15
inflow areas, 322
inflow channels, 175*f*
predictions of flow, 141*f*
recovery programs, 112
Upper Basin curtailment, 114
Colorado River basin
big river fish, 8*f*
flow reductions, 113
habitat and fish of, 169–70
invasive species in, 167–79
large-river fish conservation in, 317–33
map of, 111*f*, 168*f*
molecular investigations in, 325–26
native fish conservation, 9
native fish of, 318*t*–19*t*
water runoff and drought, 393–94
Colorado River Basin Project Act, 113
Colorado River Compact, 398
Article III section (d), 113
Article III section (e), 113
Lower Basin, 110
Upper Basin, 110
Colorado River cutthroat trout (*Oncorhynchus clarkii pleuriticus*), 32, 190, 217
Colorado River delta, 359
Comanche Creek ciénega, 130
Comanche Lake Spring
failure of, 130
flow of, 126*t*
Comanche Springs pupfish (*Cyprinodon elegans*), 28*f*
extirpation of, 131
refuge for, 367*f*
risks to, 129
status of, 367, 369
Comcàac (Seri) people, 273–74
common carp (*Cyprinus carpio*)
Aravaipa Creek, 227*t*, 231
habitat of, 81, 86
import of, 69
México, 290
Raycraft Ranch, 100
removal of, 170
Upper Colorado River basin, 171*t*
common water strider (*Aquarius remigis*), 154*f*
compromise restoration, Gila River, 354
computer simulation programs, 217–18
Conchos area, 40*t*
Conchos pupfish (*Cyprinodon eximius*), 128, 131
Conchos roundnose minnow (*Dionda* sp.), 128
Confederated Tribes of the Grand Ronde, 356
conservation
biodiversity and, 61
desert-endemic faunas, 61, 146–48
ecosystem-wide, 162
history of, 3–6
native fish resiliency and, 176
North American desert fish, 37–56
salvage and reintroduction, 162
Conservation Agreements
comparisons of, 187*t*
provisions of, 187*t*, 190
conservation biology, objectives of, 199
conservation ethic, 408*b*, 413. *See also* Land Ethic of Aldo Leopold
conservation genetics, 207–23
comparative methods in, 216
computational advances, 217–18
in practice, 209–18
toolbox, 208–9
Contreras-Balderas, Salvador, 16*f*, 20–22, 21*f*, 90
"Contributions to the Knowledge of the Ichthyofauna of the Rio San Juan, Province of Bravo, México" (Contreras-Balderas), 20
Conway, Kellyanne, 395
Corbicula fluminea (Japanese clam), 75
Corn Creek Springs, 340*f*
Pahrump poolfish in, 101, 341–42
Cottidae
areas of endemism, 43*t*
Colorado River basin, 319*t*
lineages, 43*t*
recent extinctions, 92*t*
Cottonball Marsh pupfish (*Cyprinodon salinus milleri*), 143
Cottopsis spp., 58*t*
evolutionary clade, 42*t*
Cottus greenei clade, 42*t*
Cottus spp., 58*t*
Cranston, Alan, 6*f*
crayfish (*Procambarus clarkii*), 100
Creek chub–plagopterin, 55*t*
area of endemism, 44*t*
niches, 44*t*
Crenichthys-Empetrichthys, 50*t*–51*t*
Crenichthys spp. (springfish), 318*t*
areas of endemism, 58*t*, 60
Deacon's study of, 23
nevadae (Railroad Valley springfish), 262
niches, 58*t*
crenobiontic taxa, springs and, 255–66
crenophilic macroinvertebrates, 98
crisis discipline, 408*b*, 411–12
Cross, Frank, 17, 23
Ctenopharyngodon idella (grass carp), 86
Cualac sp., 51*t*
evolutionary clade, 42*t*
Cuatro Ciénegas Basin, Coahuila, México, 18, 19
Cuatro Ciénegas pupfish (*Cyprinodon bifasciatus*), 83

Cuatro Ciénegas shiner (*Cyprinella xanthicara*), 47*t*, 83
Cub Fire, 305
cui-ui (*Chasmistes cujus*), 98
Culver, George Bliss, 13
Curicta pronotata (water scorpion), 154*f*
cutthroat trout (*Oncorhynchus clarkii*), 6, 32, 60, 98, 102*f*, 126–27, 143, 190, 201, 217, 318*t*. *See also specific trout*
Cycleptinae, evolutionary clade, 42*t*
Cycleptus spp.
  cf. *elongatus*, 49*t*
  risks to, 128
Cyprinella spp., 47*t*–48*t*
  alvarezdelvillari, 86
  bocagrande, 47*t*, 83, 365, 369
  faunal assembly, 60
  lutrensis, 171*t*, 353
    Aravaipa Creek, 226, 227*t*, 230, 232*f*
    eradication efforts, 174
    habitat of, 83
    predator, 353
    taxa splits, 204
    Virgin River, 174
  lutrensis blairi, 92*t*, 126, 201
  proserpina, 60, 131
  xanthicara, 83
Cyprinidae
  Aravaipa Creek, 226*t*
  areas of endemism, 42*t*, 44*t*, 55*t*
  Colorado River basin, 318*t*
  extinct, 81*t*
  lineages, 42*t*
  niches, 44*t*, 55*t*
  recent extinctions, 92*t*
  Upper Colorado River basin, 171*t*
Cyprinodon spp. (pupfish)
  albivelis, 82
  alvarezi, 84
  areas of endemism, 51*t*–52*t*, 60
  bifasciatus, 83
  bobmilleri, 85
  bovinus, 51*t*, 131, 214–15, 343–46, 355
  ceciliae, 84
  diabolis, 344*f*
    areas of, 39
    artificial refuge for, 338
    case study, 342–43
    conservation of, 4–5, 5*f*
    crises, 379–89
    Deacon's study of, 23
    litigation to save, 18
    Miller and C. L. Hubbs work on, 15
    population counts, 381–85, 382*f*, 385*f*
    protection of, 34
    RADseq studies of, 210
    range of, 5
    relocation of, 384–86
    reproductive stages, 385–86
    spawning of, 343, 386
  elegans, 28, 28*f*
    extirpation of, 131
    refuge for, 367, 367*f*
    risks to, 129
    status of, 367, 369
  evolutionary clade, 42*t*
  eximius, 128, 131
  extinct, 81*t*
  fontinalis, 83, 365, 365*f*, 369
  inmemoriam, 61*t*, 84, 92*t*
  julimes, 52*t*, 366
  longidorsalis, 84, 85
  macrolepis, 82
  macularius, conservation of, 148, 226, 226*f*, 358–59, 369
  nazas, 85–86
  nevadensis calidae, 99, 335, 397
  nevadensis mionectes, 6, 342, 382
  niches, 51*t*–52*t*
  pachycephalus, habitat of, 83
  radiosus, 4, 34, 338–39
  salinus milleri, 143
  salvadori, 22
  translocation of, 71
  variegatus
    invasion of Diamond Y Draw, 345, 368
    in Lake Balmorhea, 368
    range extension, 202–3
  veronicae, 84, 85
Cyprinodont Fishes of the Death Valley System of Eastern California and Southwestern Nevada (Miller), 99
Cyprinodontidae
  Aravaipa Creek, 226*t*
  areas of endemism, 43*t*, 51*t*
  Colorado River basin, 318*t*
  extinct, 81*t*
  extinct in the wild, 94*t*
  lineages, 43*t*
  niches, 51*t*
  recent extinctions, 92*t*
  Upper Colorado River basin, 171*t*–72*t*
Cyprinus carpio (common carp)
  Aravaipa Creek, 227*t*, 231
  habitat of, 81, 86
  import of, 69
  México, 290
  Raycraft Ranch, 100
  removal, 170
  Upper Colorado River basin, 171*t*
cytb gene, divergence distances, 289*t*
cytb sequences, 288*f*
cytochrome oxidase subunit I (COI), 287, 288

dace
  Ash Meadows speckled, 6
  desert, 260*f*
    areas of endemism, 55*t*
    geothermal springs, 260*f*
    niches, 55*t*
  Foskett speckled, 229, 355
  Las Vegas, 90, 92, 96, 247, 318*t*
  longfin
    Aravaipa Creek populations, 228
    areas of endemism, 56*t*
    Habitat Conservation Plans for, 188
    niches, 56*t*
    study of, 228–29
  See also spikedace (*Meda fulgida*)
dams
  ecosystem changes and, 75
  fish conservation and, 327
  impacts of, 3, 96, 240–41
damselflies (*Archilestes grandis*), 154*f*, 278*f*
data collection, standardization and, 200
Davenport, Guy, 397
Dávila, Francisco Manrique, 19
Davis Mountains runoff, 128
Deacon, James E., 6, 9, 17, 22–26, 24*f*, 25*f*, 90, 341, 381, 392, 396, 411
Dean, Mike, 27*f*
Death Valley model (DVM)
  of gene flow, 60
  of isolation, 211–12
Death Valley National Monument, 379
Death Valley National Park, 264–65
Death Valley National Park Aquatics Program, 383
Defenders of Wildlife, 227–28
delta smelt (*Hypomesus transpacificus*), 395
  competition with, 72

hatchery breeding of, 215
water conservation and, 399
Democratic Party, 392, 411
demographic parameter estimation, 11
dendritic geometry, 217–18
desert chub (*Gila* cf. *eremica*), 279
desert-endemic evolution
colonization, 59–60
desert fish, 59–61
faunal assemblies, 60–61
residency, 60
desert fish
census results, 39–42
economic value of, 4
Endangered Species Act and, 184–86, 184*t*
endemic lineage distributions, 39
extinction of, 61
factors endangering, 3–4
geographic distribution of, 38–39
imperilment of, 61
North American, 37–56
repatriation of, 174
See also specific fish and habitats
Desert Fishes Council, 4–6, 15, 16*f*, 18, 22–23, 28, 32, 34, 37, 146, 184*f*, 381, 396, 401
foundation of, 9, 391–92
founding members of, 4
geographic area of interest, 7*f*
meetings of, 6
desert fish habitat ethic, 61
Desert Fish Habitat Partnership, 132
desert pupfish (*Cyprinodon macularius*)
case study, 358–59
conservation of, 148, 358, 359, 369
Lago Volcano, 80–81
repatriation efforts, 226
deserts
definition of, 37, 90
dust blowing from, 115
North American, 37
desert spring ecosystems, 255–69
conservation efforts, 255–69, 335–51
distribution of, 256–61
ecohydrology of, 256–61
geography of, 256–57
geomorphology, 257–62
maps of the southwest US, 256*f*
rehabilitation of, 255–69
desert spring fish refuges, 336–38
desert sucker (*Catostomus clarkii*), 143, 226*t*, 303, 318*t*, 361, 362
desert tortoise (*Gopherus agassizii*), 188, 399
detritivores, 156
development, limitations on, 185
developmental plasticity, 338
Devils area, description of, 40*t*
Devils Hole, Nevada, 244*f*, 378*f*
Desert Fishes Council at, 5*f*
federal landholdings, 393
geology of, 380*b*
habitat of, 342–43, 380*b*
protection of, 397
water levels in, 5
Devils Hole pupfish (*Cyprinodon diabolis*), 344*f*
artificial refuge for, 338
case study, 342–43
conservation of, 4–5, 5*f*
crises, 379–89
Deacon's study of, 23
habitat of, 39
litigation to save, 18
Miller and C. L. Hubbs work on, 15
monitoring of, 380–85
population counts, 381–85, 382*f*, 385*f*
protection of, 34
RADseq studies of, 210
range of, 5
relocation of, 384–86
reproductive stages, 385–86
spawning of, 343, 386
Devils Hole Pupfish Recovery Plan, 386
Devils River Conservancy, 133
Devils River minnow (*Dionda diaboli*), 131
Devils River watershed, habitats and fish of, 131–32
Dexter National Fish Hatchery, 9, 320. *See also* Southwestern Native Aquatic Resources and Recovery Center, Dexter, New Mexico
Diamond Y Draw, Texas, 343–46, 345*f*, 346*f*
behavioral management for endemic species, 345–46
Diamond-Y Springs, Texas, 131
Diane Rehm Show, The, 394
Dionda spp.
areas of endemism, 45*t*–46*t*
argentosa, 60
cf. *episcopa*, 83, 129, 203
Conchos roundnose minnow, 128
diaboli, 46*t*, 52*t*, 60, 131
evolutionary clade, 42*t*
dispersal, 37, 44, 60, 69, 148, 154, 158, 216, 276, 323, 365
drought-induced, 158
modes of, 158
overland, 158
See also specific ecosystems and fish
diversions, impacts of, 240–41
Divide Fire, 301–3
diving beetles, 278*f*
new species of, 278
oasis habitats of, 277
Stictotarsus corvinus, 154*f*
Thermonectus marmoratus, 154*f*, 278*f*
Thermonectus nigrofasciatus, 154*f*
Thermonectus succinctus, 278*f*
DNA
environmental, 11, 209, 216–17
mitochondrial, 211, 305, 325
NextGen sequencing, 208, 210, 215
RADseq, 208, 210
dobsonflies (fishfly, *Neohermes filicornis*), 154*f*, 155
domestication
definition of, 336*b*, 337
evolutionary selection and, 214, 337–38
Dormitator latifrons (Pacific fat sleeper), 280*f*, 280*t*
drought-escape behaviors (DEBs), 158
droughts
dispersal induced by, 158
functional diversity and, 161
Gila trout losses in, 305
hydrologic extremes and, 155–56
novel regimes, 153–55, 161
politics and, 393–94
quantifying effects of, 154
resistance and resilience, 155–56
seasonal, 153, 155
streamflow and, 393–94
supraseasonal regimes, 153
See also megadroughts
dry-season refuges, aquatic invertebrate, 155
Duckwater Little Warm Spring, 262
Dude Creek, Gila trout, 310*t*
Dude Fire, Arizona, 303
Durango, habitats and fish, 85–86
dust blowing from deserts, 115
Dust Bowl droughts, 146
Dytiscidae, recent extinctions in, 93*t*

Early Pliocene, 409
East Sandia Spring, Texas, 367
Echo Bay, Las Vegas, 321
ecological ratchet, 408*b*, 410

ecological segregation, 59
EcoRestore, 148
ecosystems
anthropogenic disturbances to, 71
degradation of, 61
See also aquatic ecosystems; desert spring ecosystems; novel ecosystems; river ecosystems, ecological processes in; spring ecosystems
ectodysplasin gene, 213
Edwards, Bob, 27*f*
Edwards Aquifer lawsuit, 28
Edwards–Trinity Plateau Aquifer, Texas, 127, 131
Ejido Rancho Nuevo, Texas, 364–65
ejidos, 274–75
Elaeagnus angustifolia (Russian olive), 241
electrofishing, 73, 74, 170, 177, 303, 410
Eleotridae, 319*t*
Eleotris picta, 280*t*, 319*t*
Elephant Butte, 117
Elephant Butte Dam, 109
Elephant Butte Irrigation District, 115, 117
Elephant Butte Reservoir flows, 119
Elmidae, recent extinctions, 93*t*
El Niño–Southern Oscillation, 138
Elopidae, Colorado River basin, 319*t*
El Pandeño, Texas, 364*f*, 365–67, 366*f*
El Paso, Texas, 117
El Paso County Water Improvement District 1, 117
El Vado Dam, 245*f*
Emory, William H., 201
Empetrichthyinae, 58*t*
evolutionary clade, 42*t*
Empetrichthys spp.
areas of endemism, 50*t*–51*t*
latos latos, 101*f*
case study, 341–42
Deacon's study of, 23, 24
rescue efforts, 354
status of, 99, 394
merriami, 99, 100*f*
extinction of, 6, 341
status of, 394
poolfish of, 99–101
Empidonax traillii (southwestern willow flycatcher), 250, 413
impacts on, 118
Safe Harbor Agreement for, 188
endangered species, trends in, 102–3
Endangered Species Act (ESA) of 1973
application to fish, 183–98
changes to, 393, 411
cultural values and, 70
funding for, 193, 194
Gila trout distribution at the time of, 300–303
history of, 184–86, 185*f*, 194–95
impacts of, 9
implementation of, 186–88, 192–93, 193*f*
motivation by, 10
passage of, 6, 411
piscicide application and, 302
purpose of, 183*b*
restrictions of, 354
Endangered Species Preservation Act of 1966, 6, 300
endemism
areas of, 36*f*, 38–39, 39*t*–41*t*, 42*f*, 59*f*
Colorado River basin, 169
definition of, 38
Entosphenus spp., 55*t*
evolutionary clade, 42*t*
environmental DNA (eDNA)
description of, 11, 209
species detection using, 217
vertebrate, 216–17
environmental enrichment, 336*b*, 338
environmental flows, 251
environmental protections, politics and, 392–93
"Environments of the Bolsón of Cuatro Ciénegas, Coahuila, México" (Minckley), 18
E. O. Wilson Award, 26*f*
epigenetic profiling, 209
Eremichthys acros (desert dace)
areas of endemism, 55*t*
geothermal springs, 260*f*
niches, 55*t*
Esocidae, Upper Colorado River basin, 172*t*
Esox lucius (northern pike), 70, 170
Etheostoma spp.
areas of endemism, 54*t*
fonticola, 28, 54*t*
grahami, 131
niches, 54*t*
Eucinostomus currani (flag mojarra), 280*f*, 280*t*
evolution
desert-endemic faunas, 61
landscape genetics and, 212
evolutionary significant units (ESUs)
functional genomics and, 211
identification of, 212–13
policy setting and, 211
Exner's equation, 246
Exoglossum, 45*t*, 56*t*
evolutionary clade, 42*t*
extinct, definition of, 91*b*
extinct in the wild
aquatic taxa, 94*t*
definition of, 91*b*
extinction debts
definition of, 408*b*
reduction of, 410
extinction risk
prediction of, 213
refuges and, 337
extinctions
aquatic biota, 91–96
climate change and, 142–46
nomenclature of, 91*b*
North American desert fish, 92*t*–94*t*
North American deserts, 95*f*
trends in, 102–3
extinct species, trends in, 102–3
extirpated, definition of, 91*b*

fathead minnow (*Pimephales promelas*), 72, 173*t*, 227*t*, 228, 324
faunal filtering, 59–60
Felger, Richard, 275
Fintrol (piscicide), 362
fish capture kettles, 324*f*
Fishes (Jordan), 100*f*
Fishes of Arizona (Minckley), 18
Fishes of Texas Project (FoTX), 200–201
database evaluation, 201–3
historical data, 203
Native Fish Conservation Areas, 204
occurrence data, 204
range changes over time, 202–3
sampling program of 2008, 202
use by agencies, 204
"Fishes of the Big Blue River in Kansas" (Minckley), 17
"Fishes of the Río Yaqui Basin, México and United States" (Hendrickson), 18
fishfly (dobsonfly, *Neohermes filicornis*), 154*f*
Fish Lake pyrg (*Pyrgulopsis ruinosa*), 94*t*, 394
FishNet2, 200
fish transplantation programs, 143
fitness
assisted gene flow and, 339

domestication and, 215, 307, 325, 337–38
environmental enrichment and, 336*b*
inbreeding depression and, 336*b*
neutral markers and, 212
novel conditions and, 337
outbreeding depression and, 336*b*
flag mojarra (*Eucinostomus currani*), 280*f*, 280*t*
Flaming Gorge Dam, 4, 324
Flaming Gorge Reservoir, 322
flannelmouth sucker (*Catostomus latipinnis*)
Colorado River basin, 318*t*
conservation concerns, 190
conservation genetics, 216
inflow habitats, 178
restoration of, 174
San Pedro River, 226
flash flooding, adaptations to, 159–61
flathead catfish (*Pylodictis olivaris*), 171*t*, 291
floodplains
case studies, 355–59
fish capture kettle in, 324*f*
human impacts, 240
inundation frequency, 248*f*
flood risk reduction projects, 241
flow regimes
alien fish and, 73, 176–77
components of, 244*t*
floodplain connection, 328
minimally modified, 322, 327
restoration of, 176, 218, 240, 249–50, 410, 412–13
species abundance and, 232, 235
Foley, Roger, 381
food web diversity, 412
Foskett speckled dace (*Rhinichthys osculus* ssp.), 229, 355
Fossil Creek, Arizona, 358, 362*f*
eradication of alien fish, 72–73
fish, 361–62
flow of, 361–62
map of, 361*f*
stocking of, 362
Fossil Springs Diversion Dam, 361, 362*f*
founder effects, genetic diversity and, 305, 337, 341
fountain darter (*Etheostoma fonticola*), 28, 54*t*
Four Forest Restoration Initiative, 148
Fowles, John, 400
French Joe Canyon, 154
habitats and invertebrates of, 156–61
species richness in, 161
trophic interactions, 160*f*
freshwater, sustainability of, 394
Freshwater Fishes of México (Miller, Minckley, and Norris), 15, 18
functional diversity, 154
drought and, 161
functional genomics, 212
functionally extinct, definition of, 91*b*
functional trait analyses, 154
functional traits, 154
Fundulidae
areas of endemism, 43*t*, 51*t*
lineages, 43*t*
niches, 51*t*
Fundulus spp.
areas of endemism, 51*t*
evolutionary clade, 42*t*
lima, 51*t*, 81, 82
niches, 51*t*
zebrinus, 131
Furnace Creek, Death Valley, 4

Gadidae, Upper Colorado River basin, 172*t*, 327
Gambusia spp. (mosquitofish), 70
affinis
Aravaipa Creek, 227*t*, 231
Colorado River basin, 172*t*
habitat of, 81, 97
hybridization of, 97
interaction with Gila topminnow, 357
Lake Harriet, 342
mosquito abatement, 173
predator, 353
risks to, 131
alvarezi, 83
amistadensis, 96–97, 126
areas of endemism, 52*t*–53*t*
evolutionary clade, 42*t*
gaigei, 127, 354
geiseri, 131
georgei, 28
heterochir, 27*f*
interaction with *Cyprinodon bovinus*, 345–46, 345*f*, 346*f*
krumholzi, 131
marshi, 83
niches, 52*t*–53*t*
nobilis, 28, 28*f*
extirpation of, 131
habitat of, 345
Phantom Cave, 368
refuge, 367*f*
risks to, 129
status of, 367
senilis, 127
sexradiata, 86
speciosa, 85
Gammarus hyalleloides (amphipod)
risk to, 129
status of, 367
Gammarus pecos, 131
Gap Creek, Arizona, 300
Garrett, Gary, 27*f*
gas development, risk due to, 129–30
Gasterosteidae, Upper Colorado River basin, 172*t*
Gasterosteus aculeatus (threespine stickleback), 80, 213
gene flow
assisted, 210, 336*b*, 339, 341, 347
conservation genetics and, 217–18
evolutionary relationships and, 306*f*
isolation and, 158, 203, 211–12, 217, 305
local adaptation and, 212
refuge populations and, 337
gene flow models, 60–61, 211–12
genetic differentiation, geography and, 212
genetic diversity, 212
founder effects and, 337
loss of, 337
monitoring of, 407–8
population size and, 211–13
refuge populations, 337–38
genetic drift
assisted gene flow and, 336*b*
Devils Hole pupfish and, 386
differential, 337
divergence and, 158
effective population size and, 337
genetic load and, 336*b*
isolation and, 211–12, 305
modeling, 212
razorback sucker and, 326
refuge populations, 96, 213–14, 337–38, 346–47
genetic load, definition of, 336*b*, 343
genetic material, exchanges of, 11
genetic monitoring, 213–14
genetics, desert fish, 207–23
genetic sampling, temporal, 213
gene trees, incomplete lineage sorting in, 210
genomics

functional variation (adaptive), 212–13
neutral markers of variation, 212–13
in practice, 209–18
toolbox of, 208–9
genotyping, costs of, 214
"Geography of Western North American Freshwater Fishes" (Minckley), 18
geomorphology, desert springs, 257–62
geo-referencing, species labeling and, 201
geothermal springs, feed for, 260
giant reed (*Arundo donax*), 128
giant water bug (*Abedus herberti*), 156*f*
diet of, 279
dispersal potential of, 158
extirpation of, 157
habitat of, 156–59
habits of, 277–78
recolonization potential of, 158
Sierra El Aguaje, 278*f*
spring rehabilitation and, 262
vulnerability of, 155
Giffin Springs, Texas, 128, 367–68, 367*f*
Gila area, description of, 40*t*
Gila Primitive Area, Jenks Cabin Hatchery, 299
Gila River, 357*f*
mining resources of, 299
steelhead trout in, 295
Gila River, New Mexico
floodplain inundation frequency, 248–49, 248*f*
flow alterations, 244
flow regime, 74
inundation thresholds, 249*f*
volume of, 247
Gila River basin, industrial transformation of, 410
Gila River Basin Native Fishes Conservation Program, 354
Gila spp.
areas of endemism, 44*t*–45*t*, 55*t*
boraxobius, 363–64
cf. *eremica*, 279
cypha
Colorado River basin, 6, 8, 8*f*, 73, 318*t*, 353
distribution of, 176, 189, 322
Habitat Conservation Plans for, 188, 189, 249
Lake Mead, 322
poisoning of, 4
protection of, 73
elegans
Colorado River basin, 8*f*
Habitat Conservation Plans for, 188
poisoning of, 4
recruitment of, 173–74
reproduction of, 324–25
modesta, 83
niches, 44*t*–45*t*, 55*t*
orcuttii, 339
purpurea
fry of, 146
Habitat Conservation Plans for, 188
robusta
Aravaipa Creek, 226*t*, 230*t*
Colorado River basin native fish, 318*t*
conservation concerns, 143, 190, 318
distribution of, 4, 226*t*, 296, 318*t*
Fossil Creek, 361
Gila River, 296
proposed listing of, 190
rotenone killing of, 4
thermal sensitivity of, 143
seminuda, Virgin River sub-basin, 174
Gila topminnow (*Poeciliopsis occidentalis*)
Aravaipa Creek, 226*t*
case study, 356–58
Colorado River basin, 319*t*
mosquitofish and, 357
reestablishment efforts, 358*f*
repatriation efforts, 226
status of, 369
thermal tolerance of, 143
Gila trout (*Oncorhynchus gilae*), 296*f*
allozyme studies, 300
brood stock management, 307
conservation concerns, 6
conservation of, 295–315
current distribution of, 311*f*
future of, 310–12
genetic variations, 305
historical distribution of, 296–310, 297*f*
history of lineage, 309*t*–10*t*
management of, 175
origins of, 295–96
pre-ESA conservation efforts, 298–303
propagation protocols for, 305
rainbow trout and, 306*f*
Raspberry population, 307
recent conservation efforts, 303–12
repatriation programs, 312
spawning of, 301
Spruce lineage, 304
thermal sensitivity of, 143
"threatened" classification for, 305
Whiskey lineage, 304
Gila trout, populations of, 312
Gila Trout Management Plan, 300
Gila Trout Permits, 305
Gila Trout Recovery Plan
1993 revision of, 303
objective of, 300
public opinion on, 302
revision of, 300
Gila Trout Recovery Team, 300
Gilbert, Charles Henry, 14
Girardinichthyini, 51*t*
evolutionary clade, 42*t*
Glen Canyon Dam, 240
flow releases, 74
water temperatures, 322
Glen Canyon Dam Adaptive Management Program, 190, 191
Glennon, Robert, 399
Glenn's Ferry area, 41*t*
Glenwood Hatchery, Whitewater Creek, 299
Global Biodiversity Information Facility (GBIF), 200
global positioning system (GPS), data collection using, 11
global warming
effects of, 11
periods of, 138
See also climate change
Gobiesocidae, 44*t*, 54*t*
evolutionary clade, 42*t*
Gobiesox juniperoserrai, 54*t*
golden alga (*Prymnesium parvum*), 130
Goodeidae
areas of endemism, 43*t*, 50*t*–51*t*, 58*t*
Colorado River basin, 319*t*
extinct, 81*t*
extinct in the wild, 94*t*
lineages, 43*t*
niches, 50*t*–51*t*, 58*t*
recent extinctions, 93*t*
Goodenough Spring, Val Verde County, Texas, 96, 97*f*, 126*t*
Google Analytics, tracking via, 204
Goose Lake, 359
Goose Lake Basin, 361
Gopherus agassizii (desert tortoise), 188, 399
Graham Ranch Springs, 127
Grand Canyon

flow regime, 74
temperatures, 322
Grand Canyon National Park, Arizona, 257*f*
Grant County Water Pollution Nuisance Ordinance, 302
great spreadwing damselfly (*Archilestes grandis*), 154*f*, 278*f*
greenback cutthroat trout (*Oncorhynchus clarkii stomias*), 32, 217
greenhouse gases, slowing of, 146
Green River, Utah, 73, 322
Green River, Wyoming and Utah, 4
Green River basin, 169, 320, 322, 323, 324, 324*f*, 392
green sunfish (*Lepomis cyanellus*)
Aravaipa Creek, 226, 227*t*, 231
Colorado River basin, 171*t*
nonnative, 131, 171*t*, 226, 227*t*, 230–31
green swordtail (*Xiphophorus hellerii*), 81, 86
Grijalva livebearer (*Poeciliopsis hnilickai*), 86
groundwater depletion
agricultural use and, 399
Comanche Springs and, 130
desert fish endangered by, 3
Devils Hole, 380–87
groundwater recharge, 247
Gunnison River recovery programs, 112
guppy (*Poecilia reticulata*), 81
Guzmán Basin, Texas, 364*f*

Habitat Conservation Plans (HCPs), 186–88, 187*t*, 354
habitats
community persistence and, 212
definition of, 155
development and destruction of, 185
ecosystem degradation and fragmentation of, 61
gene flow and fragmentation of, 60
water use and destruction of, 61
Haliaeetus leucocephalus (bald eagle), 188
hanging gardens, 259
Hannapah Spring, California, 259*f*
Harney area, description of, 41*t*
hatcheries, 9, 320
adaptation to, 214
breeding programs, 215, 325, 338, 342
Glenwood Hatchery, Whitewater Creek, 299
Jenks Cabin Hatchery, 299
Lake Mead Fish Hatchery, 340*f*
translocations and, 214–16
Willow Beach National Fish Hatchery, 320
headwater catfish (*Ictalurus lupus*), 128, 131
areas of endemism, 42, 49*t*
niches, 49*t*
range of, 289
status of, 290
headwater model of gene flow, 60
Heartland Institute, 398
heat shock proteins, 155
Helianthus paradoxus (Pecos sunflower), 129
Hemichromis guttatus (Jewel cichlid), 83
herbaceous plants, adaptation of, 264
Herichthys spp.
areas of endemism, 54*t*
cf. *cyanoguttatus*, 85
evolutionary clade, 42*t*
minckleyi, 83
niches, 54*t*
Hershler, Robert, 91
Hesperoleucas symmetricus mitrulus, 55*t*
Hesperophylax sp. (caddisfly), 154*f*
Hickel, Walter, 5
Highland stoneroller (*Campostoma spadiceum*), 202
Hollis, Gary, 397
Hoover Dam, 109, 240, 242*f*, 321
Hoover Dam refuge management failure, 342, 386
Hualapai area, 40*t*
Hubbs, Carl Leavitt, 3, 13–14, 16*f*, 20, 90, 101, 200, 410
Hubbs, Clark, 3, 20, 26–29, 27*f*, 28*f*, 90
Hughes, Scottie Nell, 394
human impacts
climate change and, 138–42
ecosystem changes, 71
ecosystem degradation and, 61
ecosystem responses, 264–65
floodplains, 75, 118, 173, 188, 239, 240, 245, 247–50, 258*t*, 324–25, 355–57, 366
introgressive hybridization and, 211
reconciliation ecology, 75
removal of, 265–66
river, 240
Sierra El Aguaje, 282*f*
southwestern rivers, 239–53
watersheds, 240
water withdrawals and, 153
human population growth, 3
humpback chub (*Gila cypha*)
Colorado River basin, 6, 8, 8*f*, 73, 318*t*, 353
distribution of, 176, 318*t*, 322
Habitat Conservation Plans for, 188, 189, 322
Lake Mead, 322
poisoning of, 4
protection of, 73
Hurricane Jimena, 275, 279
hurricanes, 279
Hybognathus spp.
amarus
habitat of, 117
listing as endangered, 118
plains minnow and, 71
population size, 213–14
Safe Harbor Agreement for, 188
status of, 251, 354
areas of endemism, 48*t*
evolutionary clade, 42*t*
niches, 48*t*
placitus, 71
hybridization
catfish, 285, 290–91
creative potential of, 210
Gila trout, 296–98, 302, 304–5
Modoc sucker, 412
Mojave tui chub, 339
mosquitofish, 97
nonnative fish and, 9, 42, 167, 174
platyfish, 84
prevention efforts, 216
products of, 210–11
pupfish, 99, 342–43, 355, 359–60, 368, 382–84
risk of, 186, 190
sheepshead minnow, 186, 368
suckers, 45
trout, 101–2
uncertainty of relatedness and, 296
hybridization, introgressive. *See* introgressive hybridization
hydraulic fracturing, risk due to, 129–30
Hydrobiidae, recent extinctions, 94*t*
hydrophobic soils, formation of, 243
Hyla arenicolor (canyon treefrog), 279, 280*f*
Hypomesus transpacificus (delta smelt), 72, 215, 399

Ictaluridae
areas of endemism, 43*t*

lineages, 43*t*
Upper Colorado River basin, 172*t*
Ictalurus spp. (catfish)
australis, 285
balsanus, 291
classification of, 286*t*
conservation of, 285–93
distribution of, 287*f*
dugesii, 285, 289, 290
evolutionary clade, 42*t*
extirpations of, 126
furcatus spp., 128, 285, 286*f*
lupus, 128
areas of endemism, 42, 49*t*
divergence distances, 289*t*
extirpation, 131, 203
niches, 49*t*
range of, 289
status of, 286*f*, 290
meridionalis, 289
mexicanus, 289, 290
ochoterenai, 285
pricei, 285
descriptions of, 287–89
Habitat Conservation Plans for, 188
San Bernardino National Wildlife Refuge, 354
status of, 290
protected status of, 289–91
punctatus
Colorado River basin, 171*t*
conservation of, 285, 286*f*
hybridization of, 285
mechanical removal of, 173
México, 286*t*
nonnative stocking, 291
problems caused by, 170, 173, 290
status of, 290
iDigBio, 200
Imperial National Wildlife Refuge, 321
Imperial Pond, nonnative fish in, 174
inbreeding depression
definition of, 336*b*, 337
Devils Hole pupfish, 382
vulnerability to, 337
Independence Creek, Texas, 130
Independence Valley tui chub (*Siphateles bicolor isolate*), 91
infrastructure, anthropogenic changes and, 110
Inhofe, James, 395
Inland Fishes of the Greater Southwest (Minckley and Marsh), 18
intermittent-habitat specialist, 155
intermittent surface water, definition, 155
International Union for Conservation of Nature (IUCN)
Fishes of Texas Project data and, 204
Freshwater Fish Group, 15
species identified by, 184, 184*t*
introgressive hybridization
barriers to, 216
definition of, 336*b*
Gila trout, 305
mitigation of, 208
phylogenetic analysis and, 210–11
sheepshead minnow, 345
invasive species
Colorado River basin, 167–79
ecosystem degradation and, 61
riparian, 241
spring ecosystems and, 265
types of threats from, 167
See also alien species, battle against; nonnative species
invertebrates, aquatic
arid-land stream, 153–65
conservation of communities, 161–62
functional traits, 154
life history adaptations, 155–56
resistance and resilience of, 155–56
Sierra El Aguaje, 277–78
valley floor springs and, 259
Iotichthys phlegethontis (least chub)
areas of endemism, 56*t*
Conservation Agreement for, 190
niches, 56*t*
protection of, 412
rescue efforts, 354
Iron Creek
extant trout populations, 310*t*
Gila trout lineage, 309*t*
irrigation
aquifer pumping for, 130
Colorado River as source of, 110
desert fish endangered by, 3
See also agriculture; groundwater depletion
isolation
gene flow and, 60
reproductive, 276
zoogeographic models of, 211–12
Jackson Creek, Nevada, 101
jaguar (*Panthera onca*), 281
James River, Texas, 204
Japanese clam (*Corbicula fluminea*), 75
Jenks Cabin Hatchery, Gila Primitive Area, 299
Jewel cichlid (*Hemichromis guttatus*), 83
Jordan, David Starr, 13–14, 32, 100*f*
Journal of Ichthyology (trans. Behnke), 31*f*
Julimes pupfish (*Cyprinodon julimes*), 52*t*, 366
Julimes tryonia (*Tryonia julimesensis*), 93*t*, 98, 366
June sucker (*Chasmistes liorus liorus*), 338
June Sucker Recovery Implementation Program, 190
karstic terrains, 257
killifish
Baja California, 51*t*, 81, 82
largelip, 86
Mexican, 79
plains, 131
rainwater, 51*t*, 131
Kingsolver, Barbara, 396
Kinosternon (mud turtle), Sierra el Aguaje, 279
Klamath area, description of, 41*t*
Lacandon sea catfish (*Potamarius nelsoni*), 86
Lago Volcano, Mexicali Valley, 80–81
Laguna Salada, dry lakebed, 359
Lagunas de Colón, 86
Lahontan area, description of, 41*t*
Lahontan Basin, 97
Lahontan cutthroat trout (*Oncorhynchus clarkii henshawi*), 98, 188
Lake Harriet, 340*f*, 342
Lake Havasu, 320–21
Lake Mead, 112–14, 240, 321, 413
Lake Mead Fish Hatchery, 324, 340*f*
Lake Mohave, 173–74, 241
floodplain changes, 242*f*
hatchery-reared fish, 325
monitoring of, 320
razorback sucker, 320–21
Lake Powell, 112, 240
inflows, 175–76
water levels, 322, 413
water temperatures, 322
Lake Tuendae, 339–41, 340*f*
Lake Winnipeg habitats, 38
La Navaja Canyon oasis, 278, 282*f*
Land Ethic of Aldo Leopold, 61. *See also* conservation ethic

landscape genetics, hypothesis testing and, 212
Lane's relationship, 246
La Paloma pupfish (*Cyprinodon longidorsalis*), 52*t*, 84, 85
La Presita pupfish (*Cyprinodon ceciliae*), 84
largelip killifish (*Profundulus labialis*), 86
largemouth bass (*Micropterus salmoides*)
  Cibola High Levee Pond and, 173–74
  extinctions caused by, 91
  habitat of, 83
  illicit stocking of, 321, 339
  impact of introduction, 339
  predation by, 361
  specimen collection, 201
largemouth shiner (*Cyprinella bocagrande*), 47*t*, 83, 365, 369
largespring gambusia (*Gambusia geiseri*), 131
Las Vegas, Nevada, population of, 241*t*
Las Vegas Bay, Lake Mead, 321
Las Vegas dace (*Rhinichthys deaconi*), 90, 92, 96, 247, 318*t*
Late Miocene, 42*t*–44*t*, 44, 60, 409
La Trinidad pupfish (*Cyprinodon inmemoriam*), 84, 85
"Law of the River, The," 110
least chub (*Iotichthys phlegethontis*)
  areas of endemism, 56*t*
  Conservation Agreement for, 190
  niches, 56*t*
  protection of, 412
  rescue efforts, 354
Leon Springs, irrigation using, 131
Leon Springs pupfish (*Cyprinodon bovinus*), 51*t*, 131, 214–15, 343–46, 355
Leopold, A. Starker, 33
Leopold, Aldo, 4, 408*b*, 413
Lepidomeda spp., 55*t*
  altivelis, 394
  mollispinis, 190
Lepomis spp. (sunfish), 53*t*, 70
  cf. *megalotis*, 83
  cyanellus, 131, 171*t*, 226, 227*t*, 231
  evolutionary clade, 42*t*
  macrochirus, 99
lineage identity, phylogenetics and, 209–11
Lithobates spp.
  fisheri, 96, 394
  magnaocularis
    diet of, 279
    dispersal, 276
    Sierra El Aguaje, 278, 278*f*, 280*f*
  (Rana) catesbeianus, 81, 262, 281
Llano area, description of, 40*t*
loach minnow (*Rhinichthys cobitis*), 143, 217, 239–40. See also *Tiaroga cobitis* (loach minnow)
local evolutionary adaptation, 337
logperch (*Percina macrolepida*), 71
longear sunfish (*Lepomis* cf. *megalotis*), 83
longfin dace (*Agosia chrysogaster*)
  abundance of, 229, 229*f*
  Aravaipa Creek populations, 226*t*, 228
  areas of endemism, 44*t*, 56*t*
  biology of, 228
  Black Canyon, 303
  Colorado River basin native fish, 318*t*
  distribution of, 47*t*
  Fossil Creek, 361
  genetic diversity, 212
  Habitat Conservation Plans for, 188
  niches, 56*t*
  study of, 228–29
  Yaqui form, 188
longlip jumprock (*Moxostoma albidum*), 131, 202
longnose dace (*Rhinichthys cataractae*), 128
Longstreet Spring, extinct species of, 90–91
Long Term Ecosystem Monitoring Plan (LTEMP), 384
Loricariidae, Upper Colorado River basin, 172*t*
Los Angeles, California, population growth, 241*t*
Los Latos Pools, Pahrump poolfish in, 341
Lower Colorado River, 357*f*
Lower Colorado River Basin (LCRB)
  definition of, 169
  habitat and fish of, 173–75
  native trout of, 174–75
Lower Colorado River Multi-species Conservation Program, 10, 75, 250–51
Lucania parva (rainwater killifish), 51*t*, 131
  evolutionary clade, 42*t*
Lymnaeidae, recent extinctions, 94*t*

Macabí tetra (*Brycon guatemalensis*), 86
Mackenzie River, desert-like habitats, 38
Macrhybopsis aestivalis (speckled chub), 42, 47*t*, 128, 203–4
Macrhybopsis spp., 47*t*
macroinvertebrates, aquatic
  conservation efforts, 98, 255
  crenophilic, 98
  Devils Hole, 382
  responses to fires, 264
  spring ecosystems, 261
Magdalena (Mag) area, description of, 39*t*
Main Diamond Creek
  extant trout populations, 310*t*
  Gila trout, 299
  Gila trout lineage, 309*t*
Main-South Diamond Creek, 310*t*
management units (MUs), policy setting and, 211
"Man and the Changing Fish Fauna of the American Southwest" (Miller), 3–4
Manse Ranch, spring pool of, 99–101, 341
  Deacon and, 24–25, 24*f*
Manse Spring, Pahrump poolfish, 99–101, 341
Maple Grove Springs, Utah, 259*f*
mapping, species labeling and, 201
Maravillas red shiner (*Cyprinella lutrensis blairi*), 92*t*, 126, 201
Marsh, Dave, 27*f*
Marsh, E. G., Jr., 18
Marsh-Mathews, Edie, 27*f*
Martis Creek, California, 74
mayfly (*Callibaetis* sp.), 154*f*, 287*f*
McKelvey, Sharon, 397
McKenna Creek
  Gila trout, 299–300
  Gila x rainbow trout, 302
MC Spring, 341
Meda fulgida (spikedace)
  Aravaipa Creek distribution, 232*f*
  Aravaipa Creek populations, 228
  area of endemism, 44*t*
  detection of, 217
  niches, 44*t*
  population declines, 240
  study of, 228–29
  thermal sensitivity of, 143
megadroughts, 398
  definition of, 408*b*
  prehistory, 409
  risks of, 142
  See also droughts
Megupsilon aporus (Catarina pupfish), 51*t*, 84, 91

evolutionary clade, 42*t*
Menidia audens (Mississippi silverside), 72
Mescalero National Fish Hatchery (NFH), 303
Mesocapnia arizonensis (Arizona snowfly, winter stonefly), 154*f*, 155
mesopredators
resistance and resilience of, 157
trophic cascade, 160–61, 160*t*
metapopulations, diversity among, 212
metropolitan areas, population growth, 241*t*
Mexican blindcat (*Prietella phreatophila*), 131, 132
Mexican boa (*Boa constrictor imperator*), 280
Mexican Ichthyology Society, 21
Mexican killifish, Baja California, 79
Mexican mud turtle (*Kinosternon integrum*), 279
Mexican redhorse (*Moxostoma austrinum*), 128
Mexican stoneroller (*Campostoma ornatum*), 128
Mexican tetra (*Astyanax mexicanus*), 85
México
conservation status in, 79–88
extinct native fish of, 81*t*
sampling sites in, 80*f*
Microperca, 53*t*–54*t*
evolutionary clade, 42*t*
Micropterus spp.
black, 70
evolutionary clade, 42*t*
salmoides
Aravaipa creek, 227*t*, 230
Cibola High Levee pond, 173–74
Colorado River basin, 171*t*
extinctions caused by, 91
extirpation of, 83
habitat of, 74, 83
illicit stocking of, 321
impact of introduction, 339
predation by, 361
Rio Grande, 131
specimen collection, 201, 230
salmoides nuecensis, 131
microsatellite loci
Colorado River basin study, 325
genetic variations, 208
Gila trout, 305
Microvelia sp. (water strider), 154*f*, 278*f*
Middle Miocene, 42*t*–43*t*, 44, 409
Middle Rio Grande Endangered Species Collaborative Program, 10, 191, 251
Middle Rio Grande Project
dams, 240–41
USBR, 117
midges, springs and, 260
Miller, Robert Rush, 3–4, 13–17, 16*f*, 20, 24*f*, 90, 91, 99, 299, 378*f*
Minckley, W. L., 6, 9, 17–20, 18*f*, 90, 225–26, 341, 343, 396, 411
Minckley's cichlid (*Herichthys minckleyi*), 83
Mineral Springs, 339
mining operations, 3, 299
minnows
areas of endemism, 44*t*–45*t*, 55*t*–56*t*, 318*t*
Conchos roundnose, 128
Devils River, 131
fathead, 72, 173*t*, 227*t*, 228, 324
loach, 143, 217, 239–40
niches, 44*t*–45*t*, 55*t*–56*t*
plains, 71
Rio Grande silvery
habitat of, 117
listing as endangered, 118
plains minnow and, 71
population size, 213–14
Safe Harbor Agreement for, 188
status of, 251, 354
roundnose, 83, 129, 203
sheepshead, 186, 202–3, 345, 368
Miocene epoch, 38, 409
Mississippi River, lower, 38
Mississippi silverside (*Menidia audens*), 72
mitochondrial DNA-based assays, 305, 325
Moapa coriacea (Moapa dace)
areas of endemism, 55*t*
niches, 55*t*
spawning of, 145
Modoc sucker (*Catostomus microps*)
case study, 359–61
distribution of, 359
habitat, 359
listing of, 412
Mogollon Creek
evacuation of Gila trout from, 302
Gila trout populations, 312
renovation of, 300
Mohave tui chub (*Siphateles bicolor mohavensis*)
case study, 339–41
habitat of, 216, 340*t*
translocation of, 341
moisture gradients, 155
Mojave area, description of, 40*t*
Mojave River
climate change, 341
endemic fish, 339
nonnative fish, 341
molecular investigations, Colorado River basin, 325–26
monsoon seasons, flash flooding and, 160
Monterrey platyfish (*Xiphophorus couchianus*), 84, 94*t*
Montezuma Well, Arizona, 260*f*
Morelos Dam, Colorado River, 114, 240
Morisita's index, Aravaipa Creek, 233*f*
Morone saxatilis (striped bass), 320–21
Moronidae, Upper Colorado River basin, 172*t*
morphological divergence, genomic changes, 212
mosquitofish. See *Gambusia* spp. (mosquitofish)
Moxostoma spp.
albidum, 131, 202
areas of endemism, 49*t*
austrinum, 128
duquesnei, 202
niches, 49*t*
ranges of, 202
Mozambique tilapia (*Oreochromis mossambicus*), 82
mud turtle (*Kinosternon*), Sierra El Aguaje, 27
Mugilidae, Colorado River basin, 319*t*
Muir, John, 408*b*
Mule Spring, 339, 340*f*
mullet, mountain (*Agnostomus monticola*), 279, 280*f*, 280*t*
Murray-Darling River system, 73
museum lots, definition of, 200*b*
museum records, definition of, 200*b*
Mylocheilus, 56*t*
evolutionary clade, 42*t*

Nacapule Canyon, 277–78, 281, 282*f*
National Biodiversity Commission, 21
National Climate Assessment, US, 194
National Environmental Policy Act (NEPA), piscicide application and, 302
National Inventory of Dams database, 240
Native American tribal water rights, 113

native fish, impact of nonnative fish, 231–32. *See also* nonnative fish; *and specific fish*
Native Fish Conservation Areas, 204
native species
extinctions of, 74–75
invaders and, 71
Natural Diversity Database (NDD), 204
natural flow regime
alteration of, 132, 240, 244
components of, 244*t*
habitat complexity and, 176, 246
mimicry, 10, 73–75
natural selection, genetic markers and, 212
Nature Conservancy
Aravaipa Creek and, 18, 226, 228
Borax Lake and, 364
Deacon and, 26
Devils Hole and, 6
Diamond-Y Springs and, 131
Dolan Falls Preserve and, 131
East and West Sandia Springs, 130
C. Hubbs and, 28
Pupfish National Monument purchase by, 6
Safe Harbor Agreement and, 358
Sustainable Rivers Project and, 250
work of, 10
work with private landowners by, 250
Navajo Reservoir, fish kill, 4
Nazas pupfish (*Cyprinodon nazas*), 85–86
Neohermes filicornis (fishfly), 154*f*, 155
Neotropical silverside (*Chirostoma contrerasi*), 22
net annual primary productivity (NAPP), spring ecosystems, 262
Nevada pyrg (*Pyrgulopsis nevadensis*), 97–98
New Forks River, Wyoming, poisoning of, 4
Newlands Reclamation Act of 1902, 409
New Mexico Department of Game and Fish, 299
New Mexico Surface Water Quality Bureau, 302
New Mexico Wildlife Conservation Act of 1975, 300
NextGen DNA sequencing
breeding program design and, 215
description of, 208
evolutionary history issues with, 210
Nixon, Richard M., 411
nonnative fish, 320, 354
Aravaipa Creek, 226, 227*t*
Colorado River, 112, 171*t*, 353
electrofishing to remove, 303
exotic dilemma, 167
floods and, 144, 145, 235, 245
Fossil Creek, 361
impact on native fish, 231–32
mitigate effects, 169, 170, 174, 320, 354, 356
Mojave River, 341
negative effects of, 320, 359
piscicide eradication, 362
predation and competition, 319–21
Stewart Lake floodplain, 324
See also nonnative species; *and specific fish*
nonnative species
acceptance of, 74–75
American Southwest, 69–77
battle against, 70–77
control of, 73–74
cost of removal, 173
defense against, 72
desert fish endangered by, 3
ecosystem-driven approach to, 176–77
eradication of, 72–73
escape-prevention devices, 170
harm caused by, 176–77
hybridization with, 9
reconciliation ecology, 75
symbiosis, 71–72
See also nonnative fish; *and specific species*
nonpoint-source pollution, 409
Norma Oficial Mexicana list of Mexican Species at Risk, 79
North American deserts, census results, 39–42
North American western deserts, 7*f*
Northern Great Basin–Colorado Plateau subdivision, 54–58
areas of endemism, 36*f*, 55*t*–58*t*
description of, 40*t*–41*t*
niches, 55*t*–58*t*
Northern Great Plains area, 40*t*
northern pike (*Esox lucius*)
introduction of, 70
removal of, 170
Northwest México leopard frogs (*Lithobates magnaocularis*)
diet of, 279
dispersal, 276
Sierra El Aguaje, 278, 278*f*, 280*f*
Nosferatu spp., 54*t*
evolutionary clade, 42*t*
Notonecta lobata (redbacked backswimmer), 154*f*
"Notropis" longirostris clade, 48*t*
Notropis spp.
areas of endemism, 47*t*, 48*t*
braytoni, 128
buccula, 203
chihuahua, 83, 128
evolutionary clade, 42*t*
extinct, 81*t*
jemezanus
areas of endemism, 42
captive stock, 215
population declines, 239–40
range contraction, 203
risks to, 128, 131
megalops, 131
niches, 47*t*, 48*t*
orca, 126
oxyrhynchus, 202–3
simus pecosensis, 191, 213–14, 239–40
simus simus, 126, 394
texanus, 202
novel ecosystems, 413
definition of, 408*b*
nuclear-gene variation, 325–26
Nuevo León, 84–85
Nunes, Devin, 399
Nye County, Nevada, 393

oases
biodiversity of, 272–84
definition of, 275
Obama, Barack, 362, 396
occurrence records, 200*b*
Ogallala area, 40*t*
oil development, 129–30
Ojo Caliente refuge, 365*f*
Ojo de Agua de San Ignacio, 81–82
Ojo de Apodaca, 84
Ojo de Arrey, 82
Ojo de Julimes, 82–83
Ojo de la Concha, 85–86
Ojo del Potosí, 84
Ojo de San Diego, 83
Ojo de San Gregorio, 83
Ojo de Villa Lopez, 83
Ojo en San Pedro de Ocuila, 86
Ojo Hacienda Dolores, 82
Ojo Solo, 83, 365*f*
Oligocephalus, 54*t*
evolutionary clade, 42*t*

Omnibus Public Land Management Act, 362
Oncorhynchus spp. (trout), 50*t*
apache
area of endemism, 50*t*
captive propagation of, 9
Colorado River basin native fish, 318*t*
conservation concerns, 6
distribution of, 296, 298, 318*t*
historic distribution of, 296–98
management of, 175
thermal sensitivity of, 143
areas of endemism, 50*t*
clarkii, 57*t*, 102*f*, 318*t*
areas of endemism, 60
Behnke's work on, 32
evolutionary clade, 42*t*
clarkii alvordensis, 101–2
clarkii henshawi
Safe Harbor Agreements for, 188
status of, 98
clarkii pleuriticus, 32, 190, 217
clarkii seleniris, 6
clarkii stomias, 32, 217
clarkii utah, 190
clarkii virginalis
collection of, 201
extirpations of, 126–27
thermal sensitivity of, 143
gilae, 296*f*
brood stock management, 307
conservation concerns, 6
conservation of, 295–315
current distribution of, 311*f*
genetic variations, 305
historical distribution of, 296–98, 297*f*
management of, 175
origins of, 295–96
pre-ESA conservation efforts, 298–303
propagation protocols for, 305
rainbow trout and, 306*f*
Raspberry population, 307
repatriation programs, 312
spawning of, 301
Spruce lineage, 304
thermal sensitivity of, 143
"threatened" classification for, 305
Whiskey lineage, 304
mykiss, 73–74, 86
areas of endemism, 57*t*
evolutionary clade, 42*t*
Gila trout and, 306*f*
niches, 57*t*
piscicide removal of, 301
Trout Creek, 101
mykiss aguabonita, 34
Orconectes virilis (virile crayfish), 143
Oregon chub (*Oregonichthys crameri*), 355–56, 355*f*
Oregon Chub Recovery Plan, 356
Oregon Chub Working Group, 356
Oregonichthys crameri (Oregon chub), 355–56, 355*f*
Oreochromis spp. (tilapias), 70
aureus, 86
mossambicus, 82
ostracods, valley floor springs and, 259
otolith micro-increment analyses, 324
outbreeding depression, 336*b*, 338–39, 341
overgrazing, watershed disturbance by, 247
Owens pupfish (*Cyprinodon radiosus*)
case study, 338–39, 340*t*
preservation of, 34
saving of, 4
translocation of, 339
Owens Valley, fish of, 409
oxygen levels, drops in, 247

Pacific fat sleeper (*Dormitator latifrons*), 280, 280*f*, 280*t*
Pacific flyway, 112
Pacific Legal Foundation, 398
Pahranagat spinedace (*Lepidomeda altivelis*), 394
Pahrump poolfish (*Empetrichthys latos latos*), 101*f*
case study, 341–42
Deacon's study of, 23, 24
rescue efforts, 354
status of, 99, 394
translocation of, 341–42
Pahrump Ranch, 99–101
Pahrump Valley, 341
Paiute cutthroat trout (*Oncorhynchus clarkii seleniris*), 6
pale catfish (*Rhamdia guatemalensis*), 86
pallid sturgeon (*Scaphirhynchus albus*), 328
Paltothemis lineatipes (red rock skimmer), 154*f*
Panthera onca (jaguar), 281
Pantosteus spp.
areas of endemism, 49*t*, 57*t*
discobolus, 176, 190, 210, 212, 216, 318*t*, 412
evolutionary clade, 42*t*
niches, 49*t*, 57*t*
plebeius, 300
stream capture of, 45
virescens, 410
Pánuco area, description of, 40*t*
Paraneetroplus synspilus (redhead cichlid), 86
parentage-based genetic tagging (PBT), 214
Parras area, description of, 40*t*
passive integrated transponder (PIT) tags, 11
Pecos assiminea (*Assiminea pecos*), 131, 367
Pecos bluntnose shiner (*Notropis simus pecosensis*)
population declines, 239–40
population size, 213–14
protection of, 191
Pecos gambusia (*Gambusia nobilis*)
Balmorhea Springs Complex, 129
Diamond Y Draw, 367
extirpation of, 131
genetic drift, 211
habitat of, 345
C. Hubbs and, 28, 28*f*
interaction with Leon Springs pupfish, 345, 345*f*–46*f*
Phantom Cave, 368
refuge, 367*f*
status of, 367
Pecos River, New Mexico
invasive species, 71
salinity of, 130
Pecos River, Texas, watershed habitat and fish of, 130–31
Pecos sunflower (*Helianthus paradoxus*), 129
Pelicanus erythrorhynchos (white pelican), 98
Percidae
areas of endemism, 44*t*, 53*t*
lineages, 44*t*
niches, 53*t*
Upper Colorado River Basin, 172*t*
Percina spp.
areas of endemism, 53*t*
evolutionary clade, 42*t*
macrolepida, 71
niches, 53*t*
tanasi, 411
perennial surface water, definition, 155
Pesticide Use Permits, USFS, piscicide application and, 302

Petromyzontidae
  areas of endemism, 42*t*, 55*t*
  lineages, 42*t*
  niches, 55*t*
Phantom Cave snail (*Pyrgulopsis texana*), 129, 367
Phantom Lake, 129*f*
Phantom Lake Spring, Texas, 126*t*, 130, 367
Phantom Ranch, 321
phantom shiner (*Notropis orca*), 126
Phantom springsnail (*Tryonia cheatumi*), 129, 367
phenotypic integrity, refuge habitats and, 338
phenotypic plasticity, 218, 337
Phoenix, Arizona, population growth, 241*t*
Phylloicus mexicanus (sycamore caddisfly), 155, 157*f*, 159–61, 159*f*
phylogenetics
  analysis of Gila and rainbow trout, 306*f*
  lineage identity and, 209–11
phylogeny as a faunal filter, 60
Physidae, recent extinctions, 94*t*
pikeminnow, Colorado (*Ptychocheilus lucius*), 4, 318*t*
  areas of endemism, 45*t*, 55*t*
  Colorado River basin, 8*f*
  niches, 45*t*, 55*t*
  population declines, 239–40
  San Pedro River, 226
  stocking of, 170
Pimephales promelas (fathead minnow), 72, 171*t*, 227, 228, 231, 324
piscicides
  controversies around, 304
  Fossil Creek, 362
  nonnative control using, 167
  use of, 300
  See also antimycin, use of; rotenone
Pister, Edwin (Phil), 4, 4*b*, 11, 18, 23, 31–35, 34*f*, 146, 339
Pit area, description of, 41*t*
Pit River, brown trout stocking of, 361
Plagopterin, evolutionary clade, 42*t*
Plagopterus argentissimus (woundfin)
  areas of endemism, 44*t*, 55*t*
  Colorado River basin, 318*t*
  depletion of, 215
  niches, 44*t*, 55*t*
  Virgin River sub-basin, 174
plains killifish (*Fundulus zebrinus*), 131
plains minnow (*Hybognathus placitus*), 71
Plancterus, evolutionary clade, 43*t*, 51*t*
Platygobio clade, 47*t*
  evolutionary clade, 42*f*
Platygobio gracilis
  distribution of, 38
  endemism, 47*t*
  habitat of, 39
  niches, 47*t*
Pleistocene epoch, 42, 60, 130, 409, 410
Poecilia spp.
  evolutionary clade, 42*t*
  formosa, 85
  latipunctata, 52*t*
  reticulata, 81, 82
Poeciliidae
  Aravaipa Creek, 226*t*
  areas of endemism, 43*t*, 52*t*
  Colorado River basin, 319*t*
  extinct, 81*t*
  extinct in the wild, 94*t*
  lineages, 43*t*
  niches, 52*t*
  recent extinctions, 93*t*
  Upper Colorado River basin, 172*t*
Poeciliopsis spp.
  areas of endemism, 53*t*
  evolutionary clade, 42*f*
  gracilis, 86
  hnilickai, 86
  niches, 53*t*
  occidentalis
    case study, 356–58
    Colorado River basin, 319*t*
    mosquitofish and, 357
    reestablishment efforts, 358*f*
    repatriation efforts, 226
    status of, 369
    thermal tolerance of, 143
Point of Rocks Refuge, 382–83, 386
polymerase chain reactions (PCRs), 207
poolfish
  Ash Meadows, 99, 100*f*
    emergency listing of, 411
    extinction of, 6, 341
    status of, 394
  Pahrump, 101, 101*f*
    case study, 341–42
    Deacon's study of, 23, 24
    rescue efforts, 354
    status of, 99, 394
Popper, Karl, 401
populations
  conservations efforts, 407
  effective sizes of, 337
  genetic diversity, 211–12
  politics and growth of, 393–94
  refuge, 337–38
  translocation of, 216
  variation among, 325
  variation within, 325–26
porthole livebearer (*Poeciliopsis gracilis*), 86
post-environmental era
  conservation in, 412–13
  definition of, 408*b*, 411
  ESA and, 411
  transition to, 411
post-environmental politics, 410–11
Potamarius nelsoni (Lacandon sea catfish), 86
Potosí pupfish (*Cyprinodon alvarezi*), 84
Powell, John Wesley, 397–98
predation pressures, 69, 79, 89, 160–61, 167, 317
Prietella spp.
  area of endemism, 50*t*
  niches, 50*t*
  phreatophila, 131, 132
primers for Sanger sequencing, 208
Procambarus clarkii (crayfish), 100
Profundulus labialis (largelip killifish), 86
Pronatura Noreste, 366
proserpine shiner (*Cyprinella proserpina*), 60, 131
Prosopium spp., 57*t*
  evolutionary clade, 42*t*
  williamsoni, 50*t*, 318*t*
protein electrophoresis, 11
Pruitt, Scott, 393
Prymnesium parvum (golden alga), 130
Pterygoplichthys spp.
  disjunctivus, 86, 172*t*, 202
  pardalis, 86
Ptychocheilus lucius (Colorado pikeminnow), 4
  areas of endemism, 45*t*, 55*t*
  Colorado River basin, 8*f*
  niches, 45*t*, 55*t*
  population declines, 239–40
  San Pedro River, 226
  stocking of, 170
Pueblo rights, 117
pulse flow events, impacts of, 247–49
pupfish
  Ash Meadows, 6, 342
  Ash Meadows Amargosa, 382
  bighead, 83
  Carbonera, 365, 365*f*
    habitat of, 83
    status of, 369

Catarina
culture of, 91
extinct, 81*t*
habitat, 84
Charco Azul, 84, 85
Comanche Springs, 28*f*
Balmorhea Springs Complex, 129, 367, 367*f*
captive propagation of, 214
extirpation of, 131
C. Hubbs and, 28, 28*f*
refuge for, 367*f*, 368
risks to, 129, 131, 368
status of, 367, 369
Conchos, 128, 131
Cottonball Marsh, 143
Cuatro Ciénegas, 83
desert pupfish
Aravaipa Creek, 226*f*
case study, 358–59
conservation of, 148, 358, 359, 369
extirpation of, 226
Lago Volcano, 80–81
repatriation efforts, 146, 226
Devils Hole pupfish, 344*f*
artificial refuge for, 338
case study, 342–43
conservation of, 4–5, 5*f*
crises, 379–89
Deacon's study of, 23
habitat of, 39
litigation to save, 18
Miller and C. L. Hubbs work on, 15
monitoring of, 380–85
population counts, 381–85, 382*f*, 385*f*
protection of, 34
RADseq studies of, 210
range of, 5
relocation of, 384–86
reproductive stages, 385–86
spawning of, 343, 386
Devils Hole Pupfish Recovery Plan, 386
Julimes pupfish, 52*t*, 366
La Palma, 84, 85
La Presita, 84
La Trinidad, 81*t*, 84, 85, 92*t*
Leon Springs pupfish, 51*t*, 131, 214–15, 343–46, 355
Nazas, 85–86
Owens
case study, 338–39
preservation of, 34
saving of, 4
Potosí, 84
San Ignacio, 85
Tecopa
extinction of, 397
listing of, 335
status of, 99
whitefin, 82
Pupfish National Monument, 6
Putah Creek, California, 73–74
Pylodictis olivaris (flathead catfish), 171*t*, 291
Pyramid Lake, Nevada, 409
Pyrgulopsis spp. (springsnail)
collection of, 90–91
geothermal springs, 260*f*
nevadensis, 97–98
ruinosa, 94*t*, 394
texana, 129, 367

Railroad Valley area
description of, 40*t*
endemic fish, 60
Railroad Valley springfish (*Crenichthys nevadae*), 262
rainbow trout (*Oncorhynchus mykiss*)
Gila trout and, 306*f*
habitat of, 86
piscicide removal of, 301
population control of, 73–74
Trout Creek, 101
rainfall response behaviors (RRBs), dispersal and, 158
rainwater killifish (*Lucania parva*), 51*t*, 131
Rancho Las Cuedas oasis, 82, 82*f*
Rancho Nuevo, Texas, 364*f*
Rancho Nuevo Springs, 364–65
Raspberry Creek, Arizona, 303
Raspberry Gila trout population, 307
Raycraft Ranch, 100
razorback sucker (*Xyrauchen texanus*), 170, 322
adaptable habitat use by, 319
areas of endemism, 49*t*
captive breeding stock, 325–26
Colorado River basin, 8*f*, 323*t*
conservation efforts, 319–25
first year survival, 322–24
genetic variations, 326
historical distribution of, 316*f*, 317, 318*f*
Lake Mohave, 320–21
larva collection, 215, 328
larvae, 215, 321–22
lifespan of, 326
niches, 49*t*
poisoning of, 4
predation on, 321, 327
recruitment of, 173–74, 316
replacement of, 71
research on, 18, 316
resiliency of, 175–76
restoration of, 174, 316
San Pedro River, 226
spawning of, 323
stocking of, 170, 322, 323*t*
thermal sensitivity of, 143
Reagan, Ronald, 411
reconciliation ecology, nonnative species and, 75
recovery programs, provisions of, 187*t*, 188–90
redbacked backswimmer (*Notonecta lobata*), 154*f*
redbelly tilapia (*Tilapia* cf. *zillii*), 81, 82, 171*t*
Redfield Canyon, giant water bugs in, 159
redhead cichlid (*Paraneetroplus synspilus*), 86
red rock skimmer (*Paltothemis lineatipes*), 154*f*
red shiner (*Cyprinella lutrensis*)
Aravaipa Creek, 226, 230, 232*f*
eradication efforts, 174
habitat of, 83
predator, 353
taxa splits, 204
Virgin River, 174
red-spotted toad (*Anaxyrus punctatus*), 279
refuge, definition of, 335, 336*b*
refuge habitats, 162, 340*f*
artificial, 214
conservation goals, 347
definition of, 336, 336*b*
desert spring fish, 336–38
future of, 346–48
genetic variation in, 347
phenotypic integrity and, 338
recovery strategies, 347
risk reduction, 347
single-species, 338
stocking of, 368
terminology, 336*b*
trade-offs, 347
wetland fish, 336–38

Reisner, Marc, 397
Relictus solitarius, 55*t*
Reno, Nevada, population growth, 241*t*
Report on the Lands of the Arid Region of the United States (Powell), 397–98
Republican Party, 392, 411
reservoirs, 4, 112, 117–19, 322
fish conservation and, 327
inflows, 175–76, 321–22
resilience, ecological definition of, 155
resistance, ecological definition of, 155
restoration, definition of, 353
restoration projects, flow patterns and, 240
restriction fragment length polymorphisms (RFLPs), 207
restriction site-associated DNA sequencing (RADseq), 208, 210
Reynolds, J. T., 382
Rhamdia guatemalensis (pale catfish), 86
Rhinichthys spp.
areas of endemism, 45*t*, 56*t*
cataractae, 128
cobitis
areas of endemism, 45*t*
detection of, 217
niches, 45*t*
population declines, 239–40
thermal sensitivity of, 143
deaconi
extinct in wild, 96, 247
extinction of, 90
niches, 45*t*, 56*t*
osculus nevadensis, 6
osculus ssp., 229, 355
Richardson, Bill, 304
Richardsonius spp., 56*t*
balteatus, 38*t*
riffles, definition of, 275
Rift area, description of, 40*t*
Río Aguanaval, 86
Rio Chama, New Mexico, 115, 244
Rio Chama Flows Project, 251
Río Conchos, drainage of, 98
Río Conchos floodplain, 366
Rio Grande
anthropogenic changes, 109
Big Bend reach of, 127–28
Cochiti Dam operations, 248*f*
divertible flows of, 119
future flows of, 119
geography of, 115–19
volume of, 247
Rio Grande basin map, 116*f*
Rio Grande blue catfish (*Ictalurus furcatus* ssp.), 128, 285, 286*f*
Rio Grande blue sucker (*Cycleptus* sp.), 128
Rio Grande bluntnose shiner (*Notropis simus simus*), 126, 394
Rio Grande Compact, 115
Rio Grande cutthroat trout (*Oncorhynchus clarkii virginalis*)
collection of, 201
extirpations of, 126–27
thermal sensitivity of, 143
Rio Grande darter (*Etheostoma grahami*), 131
Rio Grande largemouth bass (*Micropterus salmoides nuecensis*), 131
Rio Grande Project dams, 240–41
Rio Grande shiner (*Notropis jemezanus*)
areas of endemism, 42
captive stock, 215
population declines, 239–40
risks to, 128, 131
Rio Grande silvery minnow (*Hybognathus amarus*)
collection data, 201
effective population size, 213–14
extirpation of, 126
habitat of, 117, 239
listing as endangered, 118
plains minnow and, 71
population size, 213–14
Safe Harbor Agreement for, 188
status of, 251, 354
Rio Grande sucker (*Pantosteus plebeius*), 300
stream capture, 45
Río Lacantun, fish and habitat of, 86
Río Santo Domingo, habitats and fish of, 80
riparian species, invasive, 241
riparian vegetation, recruitment of, 247
river beds, desiccation of, 247
river ecosystems, ecological processes in, 244
river engineering, geomorphic implications, 245–47
rivers
diversions, 240–41
event-based discharges, 247
human impacts, 240
restoration efforts, 249–50
seasonal discharges, 247
RNA sequencing (RNA-seq), 208–9
robust gambusia (*Gambusia marshi*), 83
Rocks Spring, Ash Meadows, 6*f*
Rocky Mountain snowpack, 112
Roosevelt, Theodore, 408*b*, 409–10
rotenone
San Juan River treatment with, 4
use of, 300, 304
"rough fish," definition of, 70
roundnose minnow (*Dionda* cf. *episcopa*), 83, 129, 203
roundtail chub (*Gila robusta*)
Aravaipa Creek, 226*t*, 230*t*
Colorado River basin native fish, 318*t*
conservation concerns, 143, 190, 316, 318
distribution of, 4, 226*t*, 296, 318*t*
Fossil Creek, 361
Gila River, 296
proposed listing of, 190
rotenone killing of, 4
thermal sensitivity of, 143
Ruiz-Campos, Gorgonio, 21*f*
Russell, I. C., 97
Russian olive (*Elaeagnus angustifolia*), 241
Rustler Aquifer, 131

Sacramento sucker (*Catostomus occidentalis*), 359
Sada, Don, 402
Safe Harbor Agreements (SHAs), 186, 187*t*, 188, 354, 356, 357
Sagebrush Rebellion, 185, 301
Salinas chub (*Gila modesta*), 83
salinity
fish ranges and, 203
increases in, 118
lake volumes and, 98
Pecos River, 130
Salton Sea and, 112
water levels and, 146
Salix exigua (willow), 259
Salmonidae. *See* trout (Salmonidae)
Salmo trutta (brown trout), 172*t*, 300, 361, 410
salt cedar (*Tamarix* spp.), 413
river engineering and, 241
sediment retention and, 128
Salton Sea, 112, 359
Salton Sink, 112
Salvelinus fontinalis (brook trout), 30
sample collection
data-directed, 202
remote, 218
See also data collection, standardization and; specimen collections

San Bernardino National Wildlife Refuge, 354
San Felipe Springs, flow of, 126*t*
Sanger sequencing
  description of, 208
  mitochondrial DNA genes, 211
  nuclear DNA genes, 211
San Ignacio pupfish (*Cyprinodon bobmilleri*), 85
San Javier Mission oasis, 82
San Juan–Chama Project, 117
San Juan River
  control of nonnative species in, 170–73
  inflow areas, 322
  inflows, 175–76
  recovery programs, 112
  rotenone treatment of, 4
San Juan River Basin Recovery Implementation Program (SJRIP), 10, 189
San Luis Valley, Colorado, 117
San Marcos gambusia (*Gambusia georgei*), 28
San Pedro River, historical species in, 226
San Solomon Springs, Texas, 367
  flow of, 126*t*
  outflow, 367
  swimming pool, 367*f*
Sapillo Creek, 300
Saragosa Spring, Texas, 367
Satan eurystomus, 50*t*
Scaphirhynchus spp.
  albus, 328
  platorynchus, 126
School Spring refuge, 386
science, limitations of, 411–12
Scirpus americanus (bulrush) encroachment, 265, 339, 346
sediment connectivity, transfer of, 245–46
sediment loading, 118
sediments
  channel aggregation and, 247
  deposition of, 245–46
  watershed disturbances and, 247
seeps, definition of, 275
selection, domestication and, 337–38
sequence capture, 209
sex ratios, 337
sharpnose shiner (*Notropis oxyrhynchus*), 202
  range contraction, 203
sheepshead minnow (*Cyprinodon variegatus*), 186
  invasion of Diamond Y Draw, 345, 368
  in Lake Balmorhea, 368
  range extension, 202–3
shifting-baseline syndrome, 408*b*, 410
shiners
  Chihuahua, 83, 128
  Cuatro Ciénegas, 83
  largemouth, 83, 369
  Maravillas red, 126
  Pecos bluntnose
    population declines, 239–40
    population size, 213–14
    protection of, 191
  phantom, 126
  proserpine, 131
  red
    Aravaipa Creek, 226, 232*f*
    eradication efforts, 174
    habitat of, 83
    taxa splits, 204
    Virgin River, 174
  Rio Grande
    areas of endemism, 42
    captive stock, 215
    population declines, 239–40
    risks to, 128, 131
  Rio Grande bluntnose, 126, 394
  sharpnose, 202
  smalleye, 203
  Tamaulipas, 128
  Tepehuán, 86
  weed, 202
  West Texas, 131
Shoshone Ponds, 341–42
shovelnose sturgeon (*Scaphirhynchus platorynchus*), 126
Sierra Club, work of, 10
Sierra El Aguaje, 272*f*
  aquatic and wetland plants, 277
  aquatic invertebrates, 277–78
  aquatic vertebrates, 278–80
  Arroyo El Palmar, 279*f*
  biodiversity, 276
  botanical surveys, 277
  conservation issues, 281
  deep history of, 275–76
  faults, 274*f*
  fauna of spring-fed habitats, 276
  freshwater habitats of, 274*f*, 275
  geology of, 275–76
  human impacts, 282*f*
  hydrology of, 275–76
  oases defined, 275
  riffles defined, 275
  seeps defined, 275
  tinajas defined, 275
Sierra Santa Úrsula, 276
single-nucleotide polymorphisms (SNPs), 208, 305
Siphateles spp.
  areas of endemism, 45*t*, 55*t*–56*t*, 60
  bicolor, 98
  bicolor isolate, 91
  bicolor mohavensis
    case study, 339–41
    habitat of, 216, 340*f*
    translocation of, 341
  niches, 45*t*, 55*t*–56*t*
Skiffia francesae, 81*t*, 94*t*
smalleye shiner (*Notropis buccula*), 203
smelt, delta (*Hypomesus transpacificus*)
  competition with, 72
  hatchery breeding of, 215
  water conservation and, 399
Smith, Nick, 301
Smithson, Charles, 400
snail darter (*Percina tanasi*), 411
Snyder, John Otterbein, 14, 410
Soil Conservation Service, 146
Soldier Meadow, Nevada, 260*f*
Solitary area, description of, 41*t*
Sonoran cliff chipmunk (*Tamias dorsalis sonorensis*), 280
Sonoran Desert, 275–76
Sørensen's coefficient of similarity (*S*), 38, 41*f*
South Diamond Creek
  extant trout populations, 310*t*
  Gila trout lineage, 309*t*
South Diamond Gila trout, 301
Southern California Bight, 39*t*
Southern Desert–Eastern Steppe subdivision, 36*f*, 39*t*–40*f*, 44*t*–54*t*
Southern Nevada Water Authority (SNWA), 25, 102–3, 392
southern platyfish (*Xiphophorus maculatus*), 84
South Tecopa Hot Springs channels, 99
Southwest area, climate and, 138–42
"Southwestern Fishes and the Enigma of 'Endangered Species'" (Minckley), 18
Southwestern Native Aquatic Resources and Recovery Center, Dexter, New Mexico, 214, 320
Southwestern Naturalist (ed. C. Hubbs), 29
Southwestern waters, aquatic ecosystems, 74–75

southwestern willow flycatcher (*Empidonax traillii*), 118, 188, 413
Spanish acequia rights, 117
spawning
  C. diabolis, 343, 386
  Gila trout, 301
  Moapa dace, 145
  water temperature and, 144
species, issues with concepts of, 211
species distribution models (SDMs), 204
Species of Greatest Conservation Need, 204
species richness, human impacts and, 264–65
specimen collections
  digitization efforts, 200
  documentation of, 199
  terms used, 200*b*
  value of, 199–206
  See also sample collection
specimens, definition of, 200*b*
speckled chub (*Macrhybopsis aestivalis*)
  areas of endemism, 42
  risks to, 128
  taxa splits, 203–4
spikedace (*Meda fulgida*)
  Aravaipa Creek distribution, 232*f*
  Aravaipa Creek populations, 228
  detection of, 217
  population declines, 240
  study of, 228–29
  thermal sensitivity of, 143
sponges, freshwater, 278*f*
spotfin gambusia (*Gambusia krumholzi*), 131
spring ecosystems
  biota, 261–62
  characteristics of, 258*t*
  classification of, 258*t*
  ecology of, 261–62
  inventory of, 263
  rehabilitation of, 262–65
  restoration of, 263–65
  Sierra El Aguaje, 275–76
  stewardship programs, 263
  vegetation, 264
  See also springs
Springer, Curtis Howe, 339
springfish (*Crenichthys* spp.), 23
spring isolation model of gene flow, 60
Spring Meadows Inc., 5
Spring Mountain Ranch State Park, 101, 342
springs
  classification of, 257
  ecological function of, 257
  flow of, 261
  forms of, 257
  See also desert spring ecosystems; spring ecosystems
springsnail (*Pyrgulopsis* spp.)
  Chihuahuan Desert, 98
  collection of, 90–91
  geothermal springs, 260*f*
Springs Preserve, Las Vegas, 342
Spruce Creek
  extant trout populations, 310*t*
  Gila trout in, 298
  Gila trout lineage, 310*t*
Spruce Fire impacts, 302
Spruce Gila trout, 300
stewardship programs
  goals of, 265
  rehabilitation plan development, 265
  rehabilitation process, 265
  spring ecosystems, 263
Stewart Lake floodplain, 324
Stictotarsus corvinus (diving beetle), 154*f*
stippled gambusia (*Gambusia sexradiata*), 86
streamflow modifications
  drought and, 393–94
  morphological responses, 246*f*
  river modification and, 247
  water quality and, 247
stream hierarchy model (SHM), 211–12
stream-hierarchy model of gene flow, 60
streams, catastrophic drying of, 157*f*
stream temperature models, 147
stressors
  nonstructural, 240
  structural, 240
striped bass (*Morone saxatilis*), 320–21
stumptooth minnow (*Stypodon signifer*)
  area of endemism, 48*t*
  extinct, 81*t*
  lack of cladistic placement, 42
  niches, 48*t*
Stypodon signifer (stumptooth minnow)
  area of endemism, 48*t*
  extinct, 81*t*
  lack of cladistic placement, 42
  niches, 48*t*
suckers (Catostomidae)
  bluehead, 176, 190, 210, 212, 216, 318*t*, 412
  Bonneville-Snake bluehead, 410
  desert, 143, 226*t*, 303, 318*t*, 361, 362
  flannelmouth
    Colorado River basin, 318*t*
    conservation concerns, 190
    conservation genetics, 216
    inflow habitats, 178
    restoration of, 174
    San Pedro River, 226
  June, 338
  June Sucker Recovery Implementation Program, 190
  Modoc
    case study, 359–61
    distribution of, 359
    habitat, 359
    listing of, 412
  razorback
    adaptable habitat use by, 319
    areas of endemism, 49*t*
    captive breeding stock, 325–26
    Colorado River basin, 8*f*, 323*t*
    conservation efforts, 319–25
    first year survival, 322–24
    genetic variations, 326
    historical distribution of, 316*f*, 317
    Lake Mohave, 320–21
    larva collection, 215, 328
    larvae, 321–22
    lifespan of, 326
    niches, 49*t*
    poisoning of, 4
    predation on, 321, 327
    recruitment of, 173–74, 316
    replacement of, 71
    research on, 18, 316
    resiliency of, 175–76
    restoration of, 174
    San Pedro River, 226
    spawning of, 323
    stocking of, 170, 322, 323*t*
    thermal sensitivity of, 143
  Rio Grande, 300
  Rio Grande blue, 128
  Sacramento, 359
  Tahoe, 98
sunburst diving beetle (*Thermonectus marmoratus*), 154*f*, 278*f*
sunfish (*Lepomis* spp.), 70
  bluegill, 99
  green
    Aravaipa Creek, 226, 227*t*, 231
    Colorado River basin, 171*t*
    risks to, 131
  longear, 83
Sunfish Falls, 362
surface waters
  agricultural use, 399

alterations in, 247
evaporative potential of, 153
maintenance of, 335
Suskind, Ron, 395
Suttkus, Royal D., 20
sycamore caddisfly (*Phylloicus mexicanus*), 155, 159–61, 159*f*
symbiosis, alien species and, 71–72

Tahoe sucker (*Catostomus tahoensis*), 98
Tamarix spp. (salt cedar), 413
river engineering and, 241
sediment retention and, 128
Tamaulipas shiner (*Notropis braytoni*), 128
Tamias dorsalis sonorensis (Sonoran cliff chipmunk), 280
Tampichthys spp., 48*t*
Tantilla yaquia (Yaqui black-headed snake), 280
targeted SNP genotyping, 208
Tecopa Hot Springs, 397
Tecopa pupfish (*Cyprinodon nevadensis calidae*)
extinction of, 397
listing of, 335
status of, 99
Tellico Dam, 411
temperature changes
aquatic organisms and, 247
lethal thermal maxima for fish, 144*f*
neuroendocrine processes and, 143
stream cooling, 147
Tepehuán shiner (*Cyprinella alvarezdelvillari*), 47*t*, 86
Terrapene coahuila (Coahuilan box turtle), 19
Texan aquifers, 124*f*
Texas cichlid (*Herichthys* cf. *cyanoguttatus*), 85
Texas Organization for Endangered Species, 368
Texas Parks and Wildlife Department (TPWD), 132, 200–201
Texas Rolling Plains area, 40*t*
Tex-Mex gambusia (*Gambusia speciosa*), 85
Thamnophis cyrtopsis (black-necked garter snake), 279, 280*f*
Theodore Roosevelt Dam, Arizona, 409
Thermonectus spp.
marmoratus, 154*f*, 278*f*
nigrofasciatus, 154*f*
succinctus, 278*f*
Thomas Creek, 359, 361
threespine stickleback (*Gasterosteus aculeatus*), 80, 213
Threloff, Doug, 381
Thunder River, spring ecosystem, 257*f*
Tiaroga cobitis (loach minnow), 45*t*
Aravaipa Creek, 226*t*
eDNA, 217
evolutionary clade, 42*t*
genetic differentiation, 212
population variation, 229
See also *Rhinichthys* spp.: *cobitis*
Tilapia cf. *zillii* (redbelly tilapia), 81, 82, 171*t*
tilapias (*Oreochromis* spp.), 70
timber production in 1879, 299
Timetable for Disaster (documentary), 5
tinajas, definition of, 275
toads (*Anaxyrus* spp.)
Amargosa, 261–62
red-spotted, 279
topminnow, Gila (*Poeciliopsis occidentalis*)
Aravaipa Creek, 226*t*
case study, 356–58
Colorado River basin, 319*t*
mosquitofish and, 357
reestablishment efforts, 358*f*
repatriation efforts, 226
status of, 369
thermal tolerance of, 143
Torichthys socolofi (Chiapas cichlid), 86
Toyah Creek, Texas, 367
trammel net haul, nonnative fish, 169*f*
Transactions of the American Fisheries Society (ed. C. Hubbs), 29
translocation
definition of, 336*b*
desert spring fish, 337
hatcheries and, 214–16
individuals from populations, 216
Mohave tui chub, 341
Owens pupfish, 339
Pahrump poolfish, 341–42
wetland fish, 337
"trash fish," definition of, 70
travertine, 257, 260–61
tribal rights, 115, 117*f*
Troglobites, 50*t*
evolutionary clade, 42*t*
Trogloglanis pattersoni, 50*t*
trout (Salmonidae), 70
Apache
conservation concerns, 6
historic distribution of, 296–98
management of, 175
thermal sensitivity of, 143
areas of endemism, 43*t*, 50*t*, 57*t*
brook, 30
brown, 300
predation by, 361
stocking of, 361
California golden, 34
Colorado River basin, 318*t*
cutthroat, 102*f*
Alvord, 101–2
areas of endemism, 60
Behnke's work on, 32
Bonneville, 190
Colorado River, 190
greenback, 32, 217
Lahontan, 98, 188
Paiute, 6
Rio Grande, 126–27, 143, 201
Gila, 296*f*
analysis of, 306*f*
brood stock management, 307
conservation of, 6, 295–315
current distribution of, 311*f*
future of, 310–12
genetic variations, 305
historical distribution of, 296–98, 297*f*
management of, 175
origins of, 295–96
pre-ESA conservation efforts, 298–303
propagation protocols for, 305
rainbow trout and, 306*f*
Raspberry population, 307
repatriation programs, 312
spawning of, 301
Spruce Creek lineage, 304
thermal sensitivity of, 143
"threatened" classification for, 305
Whiskey Creek lineage, 304
lineages, 43*t*
niches, 50*t*, 57*t*
rainbow, 73–74, 86
analysis of, 306*f*
areas of endemism, 57*t*
Gila trout and, 306*f*
niches, 57*t*
piscicide removal of, 301
Trout Creek, 101
Raspberry population of, 307
recent extinctions, 93*t*
South Diamond Creek, 309*t*, 310*t*
Spruce Creek lineage, 298, 310*t*
Upper Colorado River basin, 172*t*

Trout and Salmon of North America (Behnke), 101
Trout Unlimited, 304
Truman, Harry, 15
Trump, Donald J., 392–93, 395
Tryonia spp., 93*t*
- *cheatumi*, 129, 367
- *circumstriata*, 131
- *contrerasi*, 365
- *julimesensis*, 98, 366

Tuchman, Barbara, 400
Tucson, Arizona, population growth, 241*t*
tui chub (*Siphateles bicolor*), 98
- High Rock Spring, 92*t*
- *isolate*, 91
- *mohavensis*, 216, 339–41, 340*t*

Tularosa area, 40*t*
Typha spp. (cattail) vegetation encroachment, 265

Unionidae, recent extinctions, 94*t*
United States, Cappaert v., 5, 381, 386, 391, 393
units of conservation, definition of, 211
Upper Colorado River basin (UCRB)
- control of nonnative species in, 170–73, 316
- definition of, 169
- nonnative fish of, 171*t*–72*t*
- razorback sucker stocking of, 323*t*
- streams, 322
- water temperatures, 322

Upper Colorado River Endangered Fish Recovery Program (UCREFRP), 10, 328
- establishment of, 189
- goals of, 170
- public outreach by, 170

Upper Royal Arch Alcove, Arizona, 259*f*
Uranidea spp., 58*t*
- evolutionary clade, 42*t*

urbanization impacts on watersheds, 241–42
utilitarian conservation
- definition of, 408*b*
- overview of, 409–10

valley floor springs, characteristics of, 259–60
Valvatidae, recent extinctions, 94*t*
variable platyfish (*Xiphophorus variatus*), 84
Vegas Valley leopard frog (*Lithobates fisheri*), 96, 394
vegetation, destruction of, 3
vermiculated sailfin catfish (*Pterygoplichthys disjunctivus*), 86, 172*t*, 202
vertebrates
- aquatic, Sierra El Aguaje, 278–80
- environmental DNA (eDNA), 216–17
- riparian, 280–81
- *See also specific vertebrates*

VertNet, 200
Virgin Creek, Nevada, 101
Virgin River, Nevada, 174
Virgin River chub (*Gila seminuda*), 174
Virgin River Fishes Recovery Program, 190
Virgin spinedace (*Lepidomeda mollispinis*), 190
virile crayfish (*Orconectes virilis*), 143

Walker Lake, salinity of, 98
Wallow Fire, 307
water
- allocation in southwestern states, 10
- city usage of, 120
- depletion of, 3, 130, 185, 380–87, 399
- extinctions and supplies of, 89–90
- hydrologic extremes and, 155–56
- scarcity and politics of, 393–94
- shortages, 120
- temperature changes, 143
- unsustainable use of, 61
- withdrawals, 153
- *See also* groundwater depletion

water insecurity, 409
water scorpion (*Curicta pronotata*), 154*f*
watersheds
- human impacts, 240
- hydrologic alterations, 243–45
- modifications, 241–43
- modifications and sediment connectivity, 246

water strider (*Microvelia* sp.), 154*f*, 157
water tables, lowered, 125
Watt, James, 411
weed shiner (*Notropis texanus*), 202
Western chub-pikeminnow, evolutionary clade, 42*t*
Western minnows
- areas of endemism, 44*t*–45*t*, 55*t*–56*t*
- niches, 44*t*–45*t*, 55*t*–56*t*

western mosquitofish (*Gambusia affinis*)
- Aravaipa Creek, 227*t*, 231
- biological control of, 173
- Colorado River basin, 172*t*
- Diamond Y Springs, 131
- *Gambusia amistadensis* and, 97
- habitat of, 81, 97
- interaction with Gila topminnow, 357
- Lake Harriet, 342
- mosquito abatement, 173
- nonnative, 81, 353, 357
- predator, 353
- risks to, 131

West Fork Black River, Arizona, 143
West Fork Gila River, 304
West Sandia Spring, Texas, 367
West Texas shiner (*Notropis megalops*), 131
wetland fish, refuges for, 336–38
Whetstone Mountains, French Joe Canyon, 156
Whiskey Creek, Gila Wilderness, 301*f*
- after the Whitewater-Baldy Fire, 308*f*
- extant trout populations, 310*t*
- Gila trout lineage, 309*t*
- trout rescued from, 304

whitefin pupfish (*Cyprinodon albivelis*), 82
white pelican (*Pelicanus erythrorhynchos*), 98
White River, Colorado, 322
Whitewater-Baldy Fire, 307, 308*f*
Whitewater Creek, Glenwood Hatchery, 299
whole-genome resequencing, 209
wicked problems
- biological invasions, 409
- definition of, 408*b*
- nonpoint-source pollution, 409
- water insecurity, 409

WildEarth Guardians, 10
wildfires, Gila trout populations and, 302–10
Wileyichthys, 51*t*
- evolutionary clade, 42*t*

Willamette River, 355–56, 355*f*
Williams, Ted, 304
Willow Beach National Fish Hatchery, Arizona, 320, 342
Winters Doctrine, 113, 117
winter stoneflies (Arizona snowfly, *Mesocapnia arizonensis*), 154*f*, 155
Wolfcamp Survey Shale formation, 132

woundfin (*Plagopterus argentissimus*), 174
- areas of endemism, 44*t*, 55*t*
- Colorado River basin, 318*t*
- depletion of, 215
- niches, 44*t*, 55*t*
- Virgin River sub-basin, 174

Wright's Island model, 217–18

Xiphophorus spp.
- areas of endemism, 53*t*
- *captivus*, 51*t*
- *couchianus*, 81*t*, 84
- evolutionary clade, 42*t*
- *hellerii*, 81, 86
- *maculatus*, 84
- niches, 53*t*
- *variatus*, 84

Xyrauchen texanus (razorback sucker)
- adaptable habitat use by, 319
- areas of endemism, 49*t*
- captive breeding stock, 325–26
- Colorado River basin, 8*f*, 323*t*
- conservation efforts, 319–25
- first year survival, 322–24
- genetic variations, 326
- historical distribution of, 316*f*, 317
- Lake Mohave, 320–21
- larva collection, 215, 328
- larvae, 321–22
- lifespan of, 326
- niches, 49*t*
- poisoning of, 4
- predation on, 321, 327
- recruitment of, 173–74
- replacement of, 71
- research on, 18
- resiliency of, 175–76
- restoration of, 174
- San Pedro River, 226
- spawning of, 323
- stocking of, 170, 322, 323*t*
- thermal sensitivity of, 143

Yampa River, Colorado, 4, 322

Yaqui black-headed snake (*Tantilla yaquia*), 280

Yaqui catfish (*Ictalurus pricei*)
- conservation of, 285
- descriptions of, 287–89
- Habitat Conservation Plans for, 188
- San Bernardino National Wildlife Refuge, 354
- status of, 290

Yaqui chub (*Gila purpurea*), 146, 188

Yaqui-Mayo-Fuerte-Sinaloa, 47*t*

yellow-billed cuckoo (*Coccyzus americanus*), 118

yellowfin gambusia (*Gambusia alvarezi*), 83

Yuma Cove, 174

Yuriria, 47*t*
- evolutionary clade, 42*t*

Zacatecas, fish and habitat of, 86

Zinke, Ryan, 393

zoogeographic models of isolation, 211–12

"Zoogeography and Evolution of *Notropis lutrensis*" (Contreras-Balderas), 20